THE PHENOMENA OF FLUID MOTIONS

Robert S. Brodkey
Professor Emeritus, Chemical Engineering
The Ohio State University, Columbus, Ohio

DOVER PUBLICATIONS, INC.
New York

Copyright

Copyright © 1967 by Robert S. Brodkey.
All rights reserved under Pan American and International Copyright Conventions.

Published in Canada by General Publishing Company, Ltd., 30 Lesmill Road, Don Mills, Toronto, Ontario.
Published in the United Kingdom by Constable and Company, Ltd., 3 The Lanchesters, 162–164 Fulham Palace Road, London W6 9ER.

Bibliographical Note

This Dover edition, first published in 1995, is an unabridged, corrected republication of the work first published by Addison-Wesley Publishing Company, Reading, Massachusetts, 1967, in the "Addison-Wesley Series in Chemical Engineering."

Library of Congress Cataloging-in-Publication Data

Brodkey, Robert S.
    The phenomena of fluid motions / Robert S. Brodkey.
        p.   cm.
    Originally published: Reading, Mass. : Addison Wesley Pub. Co., 1967, in series: Addison-Wesley series in chemical engineering.
    Includes bibliographical references and indexes.
    ISBN 0-486-68605-1 (pbk.)
    1. Fluid dynamics. I. Title.
QA911.B69 1995
532'.051—dc20   95-2346
                                                                              CIP

Manufactured in the United States of America
Dover Publications, Inc., 31 East 2nd Street, Mineola, N.Y. 11501

PREFACE

This text is an outgrowth of the lecture notes I have used in teaching at The Ohio State University. At the time of writing, no single text was available which covered the subject material. In fact, the material could not even be found in a number of books, one or more for each part of the school year. Besides recognizing the need for a suitable single book, I wanted to present the background of fluid flow from the viewpoint of the transport phenomena, and to relate this transport to other transport mechanisms.

Since beginning students do not generally have a command of vector and tensor notation, and since it has appeared worthwhile to introduce the simplifying method of tensor notation at this time, I have presented the material as an integral part of the first few chapters of the book, rather than as an appendix section. The full tensor treatment has not been used; for example, the concepts of contravariant and covariant tensors have been introduced only where they are needed, which has allowed the use of the simpler Gibbs dyadic notation and a simplified cartesian tensor notation without superscripts. The problem of tensor components and physical components has been avoided by using the latter directly, and, as a result, the details of the metric tensor and Christoffel symbols did not have to be introduced. I feel that such a treatment makes a good beginning, and that it will prepare the student for further work both in the texts cited and in the literature of the various fields considered.

Thus my desires to gather together enough material for more than a one-year course, to treat certain areas not considered elsewhere, and to use newer approaches to the subject provide the motivation and, I hope, the justification for the presentation of this volume.

I have found that, within the limits of four quarter-year courses, I can cover the basic theory of the subject, the direct applications of the basic equations, and most of the extensions of those equations to the complex problems of turbulence, non-Newtonian flow, etc. More specifically, during the first quarter, I cover the introductory material of Chapters 1 through 11 (without the details of vector transformations) plus the phenomenological theories of turbulence in Chapter 14. This course is taken by most students and for those who terminate at the Masters level it may well be the only advanced course in fluids that they will have. During the second quarter, for more specialized students, I concentrate on the remainder of Chapter 14, the statistical theory of turbulence and mixing. Two courses are presented in the final quarter; in one of them I treat the non-

Newtonian material of Chapter 15, and in the other all of two-phase flow. Throughout the entire effort, I have found the excellent film clips of the Fluids Mechanics Committee to be most helpful in my teaching. With a shorter time limit, one would have to select the topics carefully, spending much less time on detailed theoretical development of the equations and transformations and more time on their applications and extensions. Specifically, for a one-semester course, one might use Chapters 1 through 5, selected topics from Chapters 6 through 11, and as much of Chapters 14 and 15 as possible.

I do not pretend that this book covers the subject of fluid motion completely; the field is far too broad for such a claim to be made about any single work. However, it is my hope that the material presented herein is both representative and basic enough to be of benefit to the graduate student. It is recognized that, for even the best of students, much of the material will prove difficult. Nevertheless, one can safely assume that it will become clearer as the student progresses, and that it will help him to understand more readily much of the current literature in the field of fluid transport. The necessity for advanced mathematics in modern chemical engineering will no doubt become obvious to the student, but he should keep in mind that it is a tool, and not an end in itself. To emphasize this point, I have used most of the complicated mathematical solutions to the basic differential equations in the examples, rather than in the main flow of the chapters.

This book should be regarded as a stepping-stone to the many complex and comprehensive references cited herein. In this sense, it should be of use to those practicing engineers concerned with obsolescence, as well as those who regard its subject as a new area of endeavor.

While I cannot hope to have presented a complete study of the field of fluid transport, I have made every effort to be complete within the scope of each topic included; that is, to present all the student needs to begin to understand each topic. Since what appears unnecessary to one may be essential to another, the attempt doubtless falls short of complete success.

Finally, I hope that the reader will inform me of any omissions or errors discovered, so that they may be corrected for other users and for any future edition.

Columbus, Ohio   R.S.B.
June 1967

TO THE MANY FRIENDS AND PLACES
THAT HAVE BEEN IMPORTANT INFLUENCES
IN MY LIFE,
AND TO CAROLYN AND PHIL,
WHO ARE PART OF IT

ACKNOWLEDGMENTS

It is difficult to recall all those who have in some way contributed to this work. Certainly my teachers at Berkeley and Madison deserve special mention: LeRoy Bromley, Ted Vermeulen, Charles Wilke, Olaf Hougen, Chuck Watson, and Bob Marshall. My colleagues at Ohio State have been most helpful with their criticism. I have learned much from Charles Dryden, Christie Geankoplis, Joe Koffolt, and Aldrich Syverson. Others have contributed by their papers, books, and conversations at the meetings of the AIChE. Of particular help have been L. V. Baldwin, S. G. Bankoff, R. L. Bates, R. B. Bird, D. C. Bogue, J. G. Brodnyan, S. Corrsin, A. E. Dukler, H. K. Fauske, A. G. Fredrickson, S. K. Friedlander, J. Gavis, T. J. Hanratty, P. Harriott, G. F. Hewitt, H. E. Hoelscher, R. H. Kraichnan, I. M. Krieger, C. E. Lapple, J. Lee, D. M. Meter, A. B. Metzner, R. S. Miller, M. Petrick, R. E. Rosensweig, P. N. Rowe, J. G. Savins, W. R. Schowalter, W. H. Schwarz, L. E. Scriven, J. D. Seader, J. C. Slattery, C. A. Sleicher, Jr., R. E. Sparks, D. G. Thomas, R. H. Wilhelm, and the many students with whom I have been associated. Helping in various stages of the preparation of this book were Edward T. Mahr, Martha Mahr, Martha Brodkey, Sylvia Bowman, Marjean Trau, Suzanne Phillips and one known only as Margharita.

To all those mentioned, to the publishers who have kindly given permission to reproduce drawings as cited, to those I have no doubt forgotten due to the length of time I have been working on this project, and to those I don't even know who helped by reviews, I extend my sincere thanks.

CONTENTS

PART I: **A Theoretical Background to Fluid Flow**

Chapter 1 **Introduction** 3
            Notation 4

Chapter 2 **Vector and Tensor Notation**

2–1 Introduction 7
2–2 Scalar Quantities and Vectors 7
2–3 The Scalar Product of Two Vectors 8
2–4 The Vector Product of Two Vectors 9
2–5 The Operators ∇, ∇^2, D/Dt 11
2–6 Tensors 13
2–7 Integrals 16
    A. Line or Curve Integral 16
    B. Surface Integral 17
    C. Differentiation of an Integral 17
2–8 Stokes' Theorem 18
2–9 Gauss' Theorem 18
2–10 Green's Theorems 18
2–11 Transformations 18

Chapter 3 **The Equations of Change**

3–1 Introduction 25
3–2 The Equations of Change in Terms of Fluxes 25
    A. The Flux Vectors 26
    B. The General Property Balance 30
    C. Explicit Expressions for the Equations of Change 32
    The Continuity Equation 32 · The Momentum
    Equation 33 · The Energy Equation 34 · Summary of
    the Equations of Change 35
    D. Comments 36
3–3 The Fluxes in Terms of the Transport Coefficients 36

Chapter 4 **The Navier-Stokes Equation**

4–1 Derivation 41
4–2 Summary 47

PART II: Applications of the Basic Flow Equations

Chapter 5 Introduction 53

Notation 56

Chapter 6 Ideal Flow

6–1 The Euler Equation of Motion 59
6–2 The Velocity Potential 60
6–3 The Stream Function and Two-Dimensional Potential Flow . . 61
6–4 Sources, Sinks, and Circulation 73
6–5 Vortex Motion 77

Chapter 7 Laminar Viscous Flow: Exact Solutions

7–1 The Flow Along a Flat Plate 84
7–2 The Flow Between Two Flat Plates 87
7–3 Hagen-Poiseuille Flow, or Flow in a Circular Pipe 90
7–4 Flow Between Rotating Cylinders 95
7–5 Compressible Flow 98

Chapter 8 Laminar Viscous Flow: Very Slow Motion

8–1 Stokes' Law 104
8–2 Oseen's Analysis 110
8–3 Alternate Shapes of Particles 111
8–4 Wall and Other Effects 112

Chapter 9 Laminar Viscous Flow: The Boundary Layer

9–1 The Boundary Layer Equations for Flow over a Flat Plate . . 117
9–2 Analysis of the Boundary Layer Equations 119
9–3 The Boundary Layer on a Flat Plate 120
9–4 Comparison of Commencement of Flow and
     Flow Along a Flat Plate 127
9–5 Flow in a Circular Pipe and Between Parallel Plates . . . 129
9–6 Similar Solutions 134
9–7 Separation 138
9–8 Other Items and Summary 140

Chapter 10 Integral Methods of Analysis

10–1 The General Integral Equation of Change 143
      A. The Integral Mass Balance 144
      B. The Integral Momentum Balance 146
      C. The von Kármán Theorem of Integral Momentum . . 148
      D. The Integral Energy Balance 153
      E. The Integral Angular Momentum Balance 154
10–2 The Bernoulli Equation for Ideal Flow 155
10–3 The Mechanical Energy Balance or the Engineering
      Bernoulli Equation 157

Chapter 11   Methods of Analysis

- 11–1   Inspection Analysis 162
- 11–2   Dimensional Analysis 165
- 11–3   Modeling 166
- 11–4   Dimensionless Groups; Completeness of Sets 166

Chapter 12   Compressible Flow

- 12–1   Thermodynamics 170
  - A.   The First Law 170
  - B.   The Second Law 171
  - C.   Specific Heats 171
  - D.   The Perfect Gas 171
    Isothermal Expansion   172 · Isentropic Expansion   172
- 12–2   Compressible Fluid Flow 173
  - A.   Propagation of a Small Pressure Wave (Velocity of Sound) . 173
  - B.   Velocity of Sound for a Perfect Gas 174
  - C.   Subsonic and Supersonic Flow 174
- 12–3   One-Dimensional, Steady, Adiabatic Flow of a Perfect Gas . . 178
  - A.   Adiabatic Flow 178
  - B.   Isentropic Flow 179
    General Considerations   179 · Converging Nozzles and Choking   184 · Converging-Diverging Nozzle   185
  - C.   Nonisentropic Flow 185
    General Considerations of Constant-Area Flow   185 · Shock in a Perfect Gas in Constant-Area Flow   189   Converging-Diverging Nozzle   192 · Wall Friction in a Perfect Gas in Constant-Area Flow   193 · Flow in Long Ducts   199
- 12–4   One-Dimensional, Steady Flow of a Gas in a Constant-Area Duct with Heat Transfer 206
  - A.   Heat Transfer without Friction 206
  - B.   Heat Transfer and Friction 211
    Definitions   211 · The General Equation   212   Isothermal Flow   212 · Nonisothermal Heat Transfer   215
- 12–5   Summary and Comments 220

PART III: Extensions of the Basic Flow Equations 223

Chapter 13   Introduction 225

Chapter 14   Turbulence and Mixing

- 14–1   Stability
- 14–2   The Reynolds Equation for Turbulent Motion 233
  - A.   The Equation 235
  - B.   Reynolds Stress 235
  - C.   Integration of the Reynolds Equation 236

14-3 Phenomenological Theories 239
   A. Boussinesq's Theory 239
   B. Prandtl's Mixing-Length Theory 240
   C. Taylor's Vorticity-Transport Theory 242
   D. Von Kármán's Similarity Hypothesis 243
   E. Dimensional-Analysis Approach 244
   F. Velocity Distribution for Turbulent Flow 245
      Separate Equations for Each Area of Flow 245 · Alternate Expressions for the Sublayer and Buffer Zones 247 Universal Velocity Distributions 249
   G. Friction Factors in Pipes and Velocity Distributions in Rough Pipes 253
   H. Reynolds Stresses 256
14-4 The Statistical Theory of Turbulent Flow 260
   A. Introduction to Terms and Definitions 261
      Description of Turbulence 261 · Correlation 262 Intensity 269 · Scale 272 · Spectrum 273 Probability Distribution 281 · Summary of Terms 281
   B. Equations of Statistical Turbulence 283
   C. Isotropic Turbulence 288
      Experimental and Measurement Methods 288 Theoretical Analysis 290
   D. Local Isotropic Turbulence 304
      The Reynolds Number 305 · Some Experimental Results 309
   E. Turbulent Shear Flow 310
      The Energy Balance 310 · Intermittency 316 Velocity Distribution 317
   F. Turbulent Dispersion 319
   G. Mixing 327
      Criteria for Mixing 330 · Mixing in an Isotropic Field 333 Chemical Reaction and Reactors 343 · Experimental Results 349
   Notation 353

Chapter 15 Non-Newtonian Phenomena

15-1 Rheological Characteristics of Materials 365
   A. Solids and Newtonian Fluids 366
   B. Non-Newtonian Materials 367
      Shear-Thinning Materials 368 · Shear-Thickening Materials 371 · Time-Dependent Systems 372 Normal Stress Effects 375
   C. Rheological Measurements 380
   D. Rheological Equations of State 384
      Empirical and Semi-empirical Equations 385 · Theoretical Approaches 390 · Constitutive Equations 397

15–2	Non-Newtonian Fluid Flow	405
	A. Viscometric Flows	406
	Capillary Flow 406 · Rotational Flow 419	
	B. Pipe and Other Flows	431
	The Reynolds Number 432 · Correlations for Pressure Drop 435 · The Critical Reynolds Number 439 Velocity Profiles 440 · Other Problems 443	
	Notation	445

Chapter 16   Multiphase Phenomena I: Pipe Flow

16–1	Two-Component Isothermal Flow	457
	A. Flow Patterns	457
	B. Pressure-Drop and Void-Fraction Correlations	465
	Equations of Two-Phase Flow 465 · Overall Correlations 470 · Specific Two-Phase Flow Problems and Analysis 480	
16–2	Adiabatic, Evaporating, One-Component Flow	494
	A. Equations of Flow with Interphase Transfer	495
	B. Overall Methods	498
	Friction-Factor Models 498 · Martinelli-Nelson Overall Approach 502	
	C. Specific Problems and Analyses	510
	Homogeneous Models 510 · Annular-Flow Models 511 · Critical Flow 512	
16–3	One-Component Two-Phase Flow with Heat Transfer	519
	A. Heat Transfer During Forced-Convection Boiling	520
	B. Overall Models	522
	C. Specific Flow Problems	524
	Nucleate Boiling Region 524 · Slug-Flow Boiling Region 525 · Annular-Flow Boiling Region 525 Critical Heat Flux or Burnout Region 526	
	Notation	528

Chapter 17   Multiphase Phenomena II: Free Flow

17–1	Formation of Drops and Bubbles	539
	A. Detachment of Drops and Bubbles	541
	Formation of Drops from Tips 541 · Formation of Bubbles from Orifices 543	
	B. Breakup of Jets	547
	The Rayleigh Jet 547 · Areas of Flow 549 Drop Distributions 551 · Further Observations 552	
	C. Atomization	555
	Breakup of Liquid Sheet 555 · Breakup of Drops 561 Atomizing Systems 565	
	D. Drop and Particle Size Distribution	571
	Distribution Analysis 571 · Some Experimental Results 578	

xiv CONTENTS

17–2	Motion of Single Drops and Bubbles	582
	A. The Effect of Circulation	583
	B. Some Experimental Evidence	584
	C. Analytical Representations	589
	D. Large Bubbles	591
	E. The Interface	595
	F. Mass and Heat Transfer	602
17–3	Interaction Effects for Drops and Bubbles	604
	Notation	609

Chapter 18 Multiphase Phenomena III: Solids–Fluid Flow

18–1	Introduction	619
18–2	Particle Behavior in Dilute Systems	621
	A. Motion Which is, in Effect, Single-Particle Motion	621
	Laminar Flow Conditions 621 · Turbulent Flow Conditions 624	
	B. Elementary Particle-to-Particle Interactions	628
18–3	Multiparticle Systems in Homogeneous Flow	629
	A. The Integral Approach for Vertical Systems	630
	The Equations 630 · The Viscosity Function 632 Operational Diagrams 635	
	B. Specific Analyses for Vertical Systems	636
	Particulate Fluidization 636 · Hindered Settling 640	
	C. Solids Transport Systems	643
18–4	Transition Between Flows	646
	A. Transition from Particulate to Aggregate Fluidization	646
	B. Minimum Transport Velocity	648
18–5	Multiparticle Systems in Nonhomogeneous Flow	651
	A. General Considerations	652
	B. Bubble Dynamics	653
	Bubble Formation at a Submerged Orifice 654 · Bubble Shape 655 · Motion 655 · The Ideal Bubble Model 661 · Bed Characteristics at Higher Velocities 670	
	C. Solid Dynamics	674
	D. Fluidized Bed Models	679
	The Two-Phase Model 680 · Multizone Model 684	
	E. Modified Systems	685
	F. Mass and Heat Transfer and Kinetics	686
	Notation	687
	Author Index	703
	Subject Index	725

PART 1
A THEORETICAL BACKGROUND TO FLUID FLOW

CHAPTER 1
INTRODUCTION

By way of introduction it will be well for us to outline briefly the entire text, and more specifically, the problems and methods necessary to develop the Navier-Stokes equation of fluid flow.

Part 1. A Theoretical Background to Fluid Flow
Part 2. Applications of the Basic Flow Equations
    A. Ideal flow
    B. Laminar viscous flow
    C. Integral methods
    D. Analysis methods
    E. Compressible flow
Part 3. Extensions of the Basic Flow Equations
    A. Turbulence
    B. Non-Newtonian flow
    C. Two-phase flow

By "a theoretical background to fluid flow," we mean consideration of the fundamentals leading to the establishment of the principles previously learned in undergraduate courses in fluid mechanics. In this text we will arrive at such familiar concepts as Newton's law of viscosity, laminar fluid flow, turbulent flow, Bernoulli's equation, analysis methods, etc. In the theoretical development, we will consider the transfer of momentum, the transfer of energy (heat transfer), and the transfer of mass (diffusion). The shorthand method of vector and tensor notation will be used. Chapter 2 is a brief introduction to the understanding of this notation.

Chapter 3 contains the development of the equations of change which describe the transport phenomena. To arrive at these equations, we must determine the transport fluxes (for example, q of heat transfer), and then express the basic differential equations in terms of these fluxes. The fluxes must in turn be expressed in terms of the transport coefficients: viscosity, thermal conductivity, and diffusivity. The equations of change for the general system can be used to develop the Navier-Stokes equation (Chapter 4).

Figure 1-1 is a block flow diagram for Part 1. It would be well to keep this in mind as the development is explained in Chapters 3 and 4. In Part 2, applications of the Navier-Stokes equation are given.

INTRODUCTION

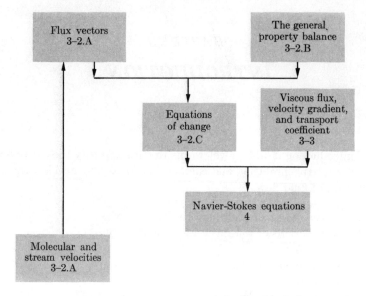

Fig. 1–1. Outline of the development of the Navier-Stokes equations.

GENERAL NOTATION FOR PART 1

a	scalar quantity
$\boldsymbol{A}, \boldsymbol{B}, A, B, A_x, A_y, A_z,$ B_x, B_y, B_z	vectors $\boldsymbol{A}$ and $\boldsymbol{B}$, their magnitudes, and components
$\boldsymbol{C}, \boldsymbol{D}$	Vectors $\boldsymbol{C}$ and $\boldsymbol{D}$
$\mathfrak{A}, \mathfrak{B}$	defined by Eq. (4–16)
c_p, c_v	heat capacities at constant pressure and volume
(D/Dt)	the substantial derivative
D_s	mass diffusivity
E	total energy
F	a function
$\boldsymbol{F}_s$	external force per unit mass acting on species s
h_i, h_j, h_k	coefficients in the general orthogonal coordinates
$\boldsymbol{i}, \boldsymbol{j}, \boldsymbol{k}$	unit vectors in cartesian coordinates
$\boldsymbol{J}$	mass flux vector
k	thermal conductivity
K_s	species s generated by chemical reaction
$\boldsymbol{L}, L, L_i$	a vector distance, its magnitude, and components
$\boldsymbol{N}$	the unit normal vector
$\boldsymbol{\mathsf{P}}, \mathsf{P}_{ij}$	the pressure tensor and its components
p	pressure
$\boldsymbol{q}$	the energy flux vector

r, θ, z	the values of L_i in cylindrical coordinates
r, θ, ϕ	the values of L_i in spherical coordinates
x, y, z	the values of L_i in cartesian coordinates
x_i	L_i for cartesian coordinates, general cartesian coordinate
$\mathbf{S}, S, S_x, S_y, S_z$	a directed area, its magnitude, and components
$\mathbf{T}, \mathbf{S}, \mathrm{T}_{ij}, \mathrm{S}_{ij}$	tensors and their compoments
$\mathbf{T}$	the tangent unit vector
T	temperature
t	time
$\mathbf{U}$	the unit tensor
$\mathbf{U}, U, U_x, U_y, U_z$	velocity vector, its magnitude, and components
u	internal energy
V	volume

Greek letters

γ	angle
δ_{ij}	Kronecker delta
θ	angle in cylindrical and spherical coordinates
κ	bulk viscosity
μ	viscosity
ν	μ/ρ, kinematic viscosity
ρ	density
$\bar{\bar{\tau}}, \bar{\tau}_{ij}$	shear stress and its components
ϕ	angle in spherical coordinates
$\mathbf{\Psi}$	the general flux vector
ψ	the general property per unit volume corresponding to the general flux vector
$\dot{\psi}_g$	generation of the general property per unit volume
$\mathbf{\Omega}$	vorticity vector
∇	del
∇^2	Laplacian operator. Lightface ∇ is used when a scalar quantity is indicated, that is, ∇^2 or one of the components of ∇.

Subscripts

s	pertaining to the species s
S	pertaining to the area S
V	pertaining to the volume V
i, j, k	indices
ds	diffusion velocity of species s

Overmarks

$\cdot$	per unit time
$\sim$	transposed tensor

The notation described below, of which only a few samples are shown, will be used throughout the entire text.

1. Scalar quantities: lightface italic and lightface Greek.
$$U \quad p \quad \tau \quad \psi$$
2. Vector quantities: boldface italic and boldface Greek.
$$\boldsymbol{U} \quad \boldsymbol{p} \quad \boldsymbol{\tau} \quad \boldsymbol{\Psi}$$
3. Components of a second-order tensor: lightface sans serif and lightface Greek with a double bar.
$$\mathsf{P} \quad \mathsf{d} \quad \bar{\bar{\tau}} \quad \bar{\bar{\psi}}$$
4. The full tensoral term: boldface sans serif and boldface Greek with a double bar.
$$\mathbf{\mathsf{P}} \quad \mathbf{\mathsf{d}} \quad \bar{\bar{\boldsymbol{\tau}}} \quad \bar{\bar{\boldsymbol{\epsilon}}}$$
5. Thermodynamic properties: lightface small capitals.
$$\text{H} \quad \text{S}$$

CHAPTER 2

VECTOR AND TENSOR NOTATION

2-1 INTRODUCTION

In the study of transport phenomena, the physical quantities to be considered, such as temperature, velocity, and shear stress, are scalar, vector, and tensor quantities respectively. In dealing with these quantities the convenient shorthand vector and cartesian tensor notation can be used to advantage. To ignore this tool would necessitate the use of a number of long, cumbersome equations in various coordinate systems where a single vector equation would suffice. For this reason, a summary of the necessary mathematics is provided in this chapter.

2-2 SCALAR QUANTITIES AND VECTORS

Scalar quantities are numbers which may be dimensional or dimensionless. They are physical quantities which do not require direction in space for their complete specification. Volume, density, viscosity, mass, and energy are examples of scalars.

Vector quantities need both magnitude and direction for their complete specification. Velocity and linear momentum are two good examples. *Speed* is the magnitude of the velocity vector and is a scalar quantity. Geometrically, a vector can be represented by a straight arrow in the direction of the vector, with its magnitude being shown by the length of the arrow compared to some chosen scale. Analytically, a vector can be represented by its projections on the coordinate axes (see Fig. 2–1). If $\boldsymbol{i}$, $\boldsymbol{j}$, and $\boldsymbol{k}$ are taken as unit vectors (magnitude unity) in the x-, y-, and z-directions, then

$$\boldsymbol{A} = \boldsymbol{i}A_x + \boldsymbol{j}A_y + \boldsymbol{k}A_z, \qquad (2\text{-}1)$$

$$A = |\boldsymbol{A}| = \sqrt{A_x^2 + A_y^2 + A_z^2}. \qquad (2\text{-}2)$$

The sum and difference of vectors can be obtained either geometrically or analytically. Geometrically, vectors are added by drawing the diagonal on a parallelogram constructed from the two vectors to be added. Analytically, the

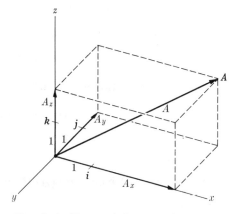

Fig. 2–1. Vector $\boldsymbol{A}$ in cartesian coordinate space.

components of the vectors are added and the result is the new vector. One can use the vector notation instead of a complicated coordinate notation for deriving theorems and obtaining relations between variables. If numerical calculations are to be made, it is usually necessary to introduce some coordinate system. For simplicity, this introduction is deferred as long as possible. In the text we shall introduce cartesian coordinates from time to time to illustrate the conversion from vectors to coordinate systems.

2-3 THE SCALAR PRODUCT OF TWO VECTORS

Let A and B be two vectors of magnitude A and B (see Fig. 2-2). The scalar or dot product of A and B is *defined* as the scalar

$$(A \cdot B) \equiv AB \cos \theta \qquad (2\text{-}3)$$
$$= A_x B_x + A_y B_y + A_z B_z$$
$$= \sum_i A_i B_i. \qquad (2\text{-}4)$$

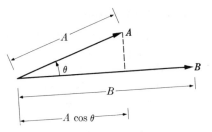

Fig. 2-2. Scalar product of two vectors.

This is the product obtained when the projection of A on B is multiplied by the magnitude of B. The notation $A_i B_i$ is commonly used and is called *cartesian tensor notation*. Whenever a double index appears, the summation may be dropped, and the double index implies a summation over the three values of i (x, y, and z).

From the definition,

$$(A \cdot B) = (B \cdot A) \qquad \text{(commutative)}, \qquad (2\text{-}5)$$
$$(A \cdot (B + C)) = (A \cdot B) + (A \cdot C) \qquad \text{(distributive)}, \qquad (2\text{-}6)$$
$$(A \cdot B)C \neq A(B \cdot C) \qquad (\textit{not} \text{ associative}). \qquad (2\text{-}7)$$

If the vectors are perpendicular,

$$(A \cdot B) = 0. \qquad (2\text{-}8)$$

From the definition of the unit vector,

$$(i \cdot i) = (j \cdot j) = (k \cdot k) = 1, \qquad (2\text{-}9)$$
$$(i \cdot j) = (j \cdot k) = (k \cdot i) = 0. \qquad (2\text{-}10)$$

EXAMPLE 2-1. Demonstrate $(A \cdot B)$ in cartesian coordinates (2-4).

Answer.

$$(A \cdot B) = ((iA_x + jA_y + kA_z) \cdot (iB_x + jB_y + kB_z)).$$

Using Eqs. (2-6), (2-9), and (2-10), we obtain

$$(A \cdot B) = (i \cdot i)A_x B_x + (j \cdot i)A_y B_x + (k \cdot i)A_z B_x + (i \cdot j)A_x B_y + (j \cdot j)A_y B_y$$
$$+ (k \cdot j)A_z B_y + (i \cdot k)A_x B_z + (j \cdot k)A_y B_z + (k \cdot k)A_z B_z$$
$$= A_x B_x + A_y B_y + A_z B_z.$$

2-4 THE VECTOR PRODUCT OF TWO VECTORS

The vector product is *defined* as a vector of magnitude

$$|(A \times B)| \equiv AB \sin \theta \qquad (2\text{-}11)$$

and of direction perpendicular to A and B, as determined by the right-hand rule.

From the definition, we obtain for the unit vectors

$$(i \times i) = (j \times j) = (k \times k) = 0, \qquad (2\text{-}12)$$

$$(j \times k) = i = -(k \times j),$$
$$(k \times i) = j = -(i \times k), \qquad (2\text{-}13)$$
$$(i \times j) = k = -(j \times i).$$

In cartesian coordinates,

$$(A \times B) = \begin{vmatrix} i & j & k \\ A_x & A_y & A_z \\ B_x & B_y & B_z \end{vmatrix}, \qquad (2\text{-}14)$$

where i, j, and k are unit vectors in the x-, y-, z-directions. Another means of expressing Eq. (2-14) is

$$(A \times B)_i = \sum_j \sum_k e_{ijk} A_j B_k.$$

Here the double indices could have been used to imply the summation, which gives

$$(A \times B)_i = e_{ijk} A_j B_k,$$

where e_{ijk} is called the permutation symbol and has the value of

    0 if any two indices are equal,
  +1 if $ijk = 123$, 231, or 312 (i.e, an *even* permutation of 123), or
  −1 if $ijk = 132$, 213, or 321 (i.e, an *odd* permutation of 123).

From the definition,

$$(A \times B) = -(B \times A) \qquad (not \text{ commutative}), \qquad (2\text{-}15)$$
$$(A \times (B + C)) = (A \times B) + (A \times C) \qquad (\text{distributive}), \qquad (2\text{-}16)$$
$$(A \times (B \times C)) \neq ((A \times B) \times C) \qquad (not \text{ associative}). \qquad (2\text{-}17)$$

If the vectors are parallel,

$$(A \times B) = 0. \qquad (2\text{-}18)$$

The triple vector product can be shown to be
$$(A \times (B \times C)) = (A \cdot C)B - (A \cdot B)C. \tag{2-19}$$
The product
$$(A \cdot (B \times C)) = (B \cdot (C \times A)) = (C \cdot (A \times B)) \tag{2-20}$$
is numerically equal to the volume of the parallelepiped formed with A, B, and C as the edges.

There are many other vector operations that can be proved [2, 3, 4, 7]*; however these are not needed in our work on vectors and so will be omitted.

EXAMPLE 2–2. Demonstrate $(A \times B)$ in cartesian coordinates, Eq. (2–14).

Answer. $(A \times B) = ((iA_x + jA_y + kA_z) \times (iB_x + jB_y + kB_z))$.

By the use of Eqs. (2–12), (2–13), and (2–16), we have

$$
\begin{aligned}
(A \times B) &= (i \times i)A_x B_x + (j \times i)A_y B_x + (k \times i)A_z B_x \\
&\quad + (i \times j)A_x B_y + (j \times j)A_y B_y + (k \times j)A_z B_y \\
&\quad + (i \times k)A_x B_z + (j \times k)A_y B_z + (k \times k)A_z B_z \\
&= -kA_y B_x + jA_z B_x + kA_x B_y - iA_z B_y - jA_x B_z + iA_y B_z \\
&= i(A_y B_z - A_z B_y) + j(A_z B_x - A_x B_z) + k(A_x B_y - A_y B_x).
\end{aligned}
$$

This is the same as

$$
\begin{vmatrix} i & j & k \\ A_x & A_y & A_z \\ B_x & B_y & B_z \end{vmatrix} = iA_y B_z + jA_z B_x + kA_x B_y - kA_y B_x - jA_x B_z - iA_z B_y
$$

$$
= i(A_y B_z - A_z B_y) + j(A_z B_x - A_x B_z) + k(A_x B_y - A_y B_x).
$$

EXAMPLE 2–3. Prove $(A \times B) = -(B \times A)$, Eq. (2–15).

Answer. $(A \times B)$ is above. Now,

$$
(B \times A) = \begin{vmatrix} i & j & k \\ B_x & B_y & B_z \\ A_x & A_y & A_z \end{vmatrix}
$$

$$
= i(A_z B_y - A_y B_z) + j(A_x B_z - A_z B_x) + k(A_y B_x - A_x B_y).
$$

This is the negative of the above determinant; therefore

$$(A \times B) = -(B \times A).$$

---

*Numbers in brackets are keyed to references at the end of each chapter.

2-5 THE OPERATORS ∇, ∇^2, AND D/Dt.

The derivative of a vector is the derivative of each part and is itself a vector:

$$\frac{\partial A}{\partial x} = \frac{\partial}{\partial x}(iA_x + jA_y + kA_z) = i\frac{\partial A_x}{\partial x} + j\frac{\partial A_y}{\partial x} + k\frac{\partial A_z}{\partial x}. \tag{2-21}$$

The operator known as *nabla* or *del* is *defined* as

$$\nabla \equiv i\frac{\partial}{\partial x} + j\frac{\partial}{\partial y} + k\frac{\partial}{\partial z}. \tag{2-22}$$

This operator can be used in scalar or vector multiplication, as follows.

Multiplication by a scalar:

$$\nabla a = \text{grad } a = i\frac{\partial a}{\partial x} + j\frac{\partial a}{\partial y} + k\frac{\partial a}{\partial z}. \tag{2-23}$$

Dot or scalar multiplication:

$$(\nabla \cdot A) = \text{div } A = \frac{\partial A_x}{\partial x} + \frac{\partial A_y}{\partial y} + \frac{\partial A_z}{\partial z} = \sum_i \frac{\partial A_i}{\partial x_i}. \tag{2-24}$$

Again the repeated index could be used to imply the summation over the three values of i; x_i is the general cartesian coordinate and takes on the values of x, y, and z.

Vector multiplication:

$$(\nabla \times A) = \text{curl } A = \text{rot } A = \begin{vmatrix} i & j & k \\ \frac{\partial}{\partial x} & \frac{\partial}{\partial y} & \frac{\partial}{\partial z} \\ A_x & A_y & A_z \end{vmatrix}, \tag{2-25}$$

or

$$(\nabla \times A)_i = \sum_j \sum_k e_{ijk}\left(\frac{\partial A_k}{\partial x_j}\right).$$

In the summation, x_j takes on the values x, y, and z as j is 1, 2, and 3. In each case of multiplication the product is distributive but not commutative or associative.

The divergence of the gradient of a scalar is the Laplacian operator on that scalar:

$$\nabla^2 = (\nabla \cdot \nabla) = \frac{\partial^2}{\partial x^2} + \frac{\partial^2}{\partial y^2} + \frac{\partial^2}{\partial z^2} = \sum_i \frac{\partial^2}{\partial x_i^2} = \frac{\partial^2}{\partial x_i \partial x_i}. \tag{2-26}$$

This operation may be applied to vectors as well as to scalars.

Another operator of interest is called the *substantial derivative*:

$$\frac{D}{Dt} = \frac{\partial}{\partial t} + (A \cdot \nabla). \tag{2-27}$$

Here A is a vector, but not necessarily the same as the quantity being operated on. The operator $(A \cdot \nabla)$ can be expanded to

$$(A \cdot \nabla) = A_x \frac{\partial}{\partial x} + A_y \frac{\partial}{\partial y} + A_z \frac{\partial}{\partial z} = \sum_i A_i \frac{\partial}{\partial x_i}; \qquad (2\text{-}28)$$

in the summation, x_i takes on the values of $x, y,$ and z. The following vector identity will be useful [7]:

$$(A \cdot \nabla) A = \tfrac{1}{2} \nabla (A \cdot A) - (A \times (\nabla \times A)). \qquad (2\text{-}29)$$

To provide some idea of the meaning of the substantial derivative, let us review other derivatives in terms of the motion of particles in a fluid stream. At any single reference point in the stream, if we were to follow the change in the number of particles at that point with time, we would have the partial derivative of the number with respect to time, a partial derivative because the number may also vary with position away from our reference point. The total derivative with respect to time could be obtained from a count over the entire field or from the chain rule. Here our view would have to be in a rapid random motion, and the components of this motion, dx/dt, etc., would have to be known:

$$\frac{dn}{dt} = \frac{\partial n}{\partial t} + \frac{\partial n}{\partial x} \frac{dx}{dt} + \frac{\partial n}{\partial y} \frac{dy}{dt} + \frac{\partial n}{\partial z} \frac{dz}{dt}.$$

In other words, the total derivative is the rate of change of the number of particles in the neighborhood of an observer moving with velocity components dx/dt, dy/dt, and dz/dt over the field.

In fluid flow theory, the vector A in Eq. (2–27) is specifically the stream velocity U. The substantial derivative is of the same form as the total derivative, but here the observer moves with the stream, and we substitute for dx/dt, etc., the components of the local fluid velocity. In the total derivative, the observer's motion is considered, and that derivative reflects his motion as well as that of the particles. The substantial derivative is the time change appearing to an observer who is moving with the stream; the partial derivative is the time change observed from a stationary point.

A number of other useful relations can be demonstrated:

$$(\nabla \cdot aA) = (\nabla a \cdot A) + a(\nabla \cdot A), \qquad (2\text{-}30)$$

$$(\nabla \cdot (AB)) = (A \cdot \nabla) B + B(\nabla \cdot A), \qquad (2\text{-}31)$$

$$(\nabla \times aA) = a(\nabla \times A) + (\nabla a \times A), \qquad (2\text{-}32)$$

$$(\nabla \cdot (\nabla \times A)) = 0, \qquad (2\text{-}33)$$

$$(\nabla \times (A \times B)) = (B \cdot (\nabla A)) - (A \cdot (\nabla B)) + A(\nabla \cdot B) - B(\nabla \cdot A). \qquad (2\text{-}34)$$

Terms such as (AB) and (∇A) will be explained in the next section.

EXAMPLE 2–4. Prove that $(\nabla \cdot a\mathbf{A}) = (\nabla a \cdot \mathbf{A}) + a(\nabla \cdot \mathbf{A})$, Eq. (2–30).

Answer. From Eq. (2–22),

$$(\nabla \cdot a\mathbf{A}) = \mathbf{i}\frac{\partial a\mathbf{A}}{\partial x} + \mathbf{j}\frac{\partial a\mathbf{A}}{\partial y} + \mathbf{k}\frac{\partial a\mathbf{A}}{\partial z} \qquad (a\mathbf{A} = \mathbf{i}aA_x + \mathbf{j}aA_y + \mathbf{k}aA_z)$$

$$= (\mathbf{i}\cdot\mathbf{i})\frac{\partial aA_x}{\partial x} + (\mathbf{j}\cdot\mathbf{j})\frac{\partial aA_y}{\partial y} + (\mathbf{k}\cdot\mathbf{k})\frac{\partial aA_z}{\partial z}$$

$$= a\frac{\partial A_x}{\partial x} + A_x\frac{\partial a}{\partial x} + a\frac{\partial A_y}{\partial y} + A_y\frac{\partial a}{\partial y} + a\frac{\partial A_z}{\partial z} + A_z\frac{\partial a}{\partial z}$$

$$= A_x\frac{\partial a}{\partial x} + A_y\frac{\partial a}{\partial y} + A_z\frac{\partial a}{\partial z} + a\left(\frac{\partial A_x}{\partial x} + \frac{\partial A_y}{\partial y} + \frac{\partial A_z}{\partial z}\right).$$

Since ∇a is the vector,

$$\nabla a = \mathbf{i}\frac{\partial a}{\partial x} + \mathbf{j}\frac{\partial a}{\partial y} + \mathbf{k}\frac{\partial a}{\partial z},$$

from Eq. (2–4), one obtains for the first part $(\mathbf{A}\cdot\nabla a)$, which equals $(\nabla a\cdot\mathbf{A})$ by Eq. (2–5). From Eq. (2–24) one obtains $a(\nabla\cdot\mathbf{A})$ for the second part; therefore

$$(\nabla\cdot a\mathbf{A}) = (\nabla a\cdot\mathbf{A}) + a(\nabla\cdot\mathbf{A}).$$

2-6 TENSORS

When there is a change in the coordinate system, tensors must follow certain transformation laws. If an array obeys these transformation laws, then we may manipulate it by means of tensor algebra, which is simply a method of performing a number of proved operations on the symbols used to represent the array. All we are doing is substituting a symbol for the array; any operation on the symbol must be performed exactly as though we were performing an analogous operation on the array.

Scalars are tensors of zero order, and vectors are tensors of the first order. The second-order tensor is an array of nine components expressed as

$$\mathbf{T} = \begin{pmatrix} T_{xx} & T_{xy} & T_{xz} \\ T_{yx} & T_{yy} & T_{yz} \\ T_{zx} & T_{zy} & T_{zz} \end{pmatrix} = T_{ij}. \tag{2–35}$$

The rows are associated with the i's and the columns with the j's. The tensor array is an ordered set of numbers (but is not a determinant, which is a certain sum and products of the numbers). The diagonal terms are those in which the two subscripts are the same; all others are the nondiagonal terms. If $T_{xy} = T_{yx}$, $T_{xz} = T_{zx}$, and $T_{yz} = T_{zy}$, then the tensor is symmetrical.

The transposed tensor of $\mathbf{T}$ is $\tilde{\mathbf{T}}$ and is formed by exchanging the rows and columns.

A unit tensor **U** is one in which each diagonal term is unity, and the nondiagonal terms are zero:

$$\mathbf{U} = \begin{pmatrix} 1 & 0 & 0 \\ 0 & 1 & 0 \\ 0 & 0 & 1 \end{pmatrix} = \delta_{ij}. \tag{2-36}$$

The latter form is called the Kronecker delta, that is,

$$\delta_{ij} = 1 \quad \text{when} \quad i = j$$

and

$$\delta_{ij} = 0 \quad \text{when} \quad i \neq j.$$

The sum of two tensors is defined as the set of terms obtained by the addition of corresponding components of the two tensors, and will be a tensor of the same order. In other words, the sum of two second-order tensors is a tensor, and will also have nine terms:

$$\mathbf{T} + \mathbf{S} = \begin{pmatrix} T_{xx} + S_{xx} & T_{xy} + S_{xy} & T_{xz} + S_{xz} \\ T_{yx} + S_{yx} & T_{yy} + S_{yy} & T_{yz} + S_{yz} \\ T_{zx} + S_{zx} & T_{zy} + S_{zy} & T_{zz} + S_{zz} \end{pmatrix}. \tag{2-37}$$

The two products of tensors are called the *outer* and *inner* products. The outer product is defined as the set of terms obtained by multiplying each component of the first tensor by the second tensor. The inner product is obtained from the outer product by an operation called *contraction*. As an example of the outer product, the product of a scalar multiplied by a second-order tensor is a tensor of order two, in which each term has been multiplied by the scalar. The product of a vector and a tensor will be of the third order and have 27 terms. The product of two tensors of the second order will be of the fourth order and have 81 terms. *The inner product will be two orders less than the outer product.* For example, the third-order tensor, when contracted, will be a vector; the inner product of the two second-order tensors will be a second-order tensor.

In cartesian coordinates, the terms of the contracted product are formed by multiplying, term by term, the rows of the first tensor by the columns of the second tensor; i.e., for a second-order tensor and a vector,

$$(\mathbf{T} \cdot \mathbf{A}) = \begin{pmatrix} T_{xx} & T_{xy} & T_{xz} \\ T_{yx} & T_{yy} & T_{yz} \\ T_{zx} & T_{zy} & T_{zz} \end{pmatrix} \begin{pmatrix} A_x \\ A_y \\ A_z \end{pmatrix} = \begin{pmatrix} T_{xx}A_x + T_{xy}A_y + T_{xz}A_z \\ T_{yx}A_x + T_{yy}A_y + T_{yz}A_z \\ T_{zx}A_x + T_{zy}A_y + T_{zz}A_z \end{pmatrix} = \mathbf{C}, \tag{2-38}$$

where the vector $\mathbf{A}$ has been denoted as

$$\mathbf{A} = \mathbf{i}A_x + \mathbf{j}A_y + \mathbf{k}A_z = \begin{pmatrix} A_x \\ A_y \\ A_z \end{pmatrix} \tag{2-39}$$

and is called a *column matrix* or *column vector*, since it has only one column. The first row is the ith component, the second is the jth, etc. It is important to note that even though the last group in Eq. (2–38) has nine terms, it is still a vector, since it has three components, each of which is the sum of three terms. The product $(\mathbf{T} \cdot A)$ is not commutative, since $(A \cdot \mathbf{T})$ is

$$(A \cdot \mathbf{T}) = (A_x \ A_y \ A_z) \begin{pmatrix} \mathsf{T}_{xx} & \mathsf{T}_{xy} & \mathsf{T}_{xz} \\ \mathsf{T}_{yx} & \mathsf{T}_{yy} & \mathsf{T}_{yz} \\ \mathsf{T}_{zx} & \mathsf{T}_{zy} & \mathsf{T}_{zz} \end{pmatrix}$$

$$= \begin{pmatrix} A_x\mathsf{T}_{xx} + A_y\mathsf{T}_{yx} + A_z\mathsf{T}_{zx} \\ A_x\mathsf{T}_{xy} + A_y\mathsf{T}_{yy} + A_z\mathsf{T}_{zy} \\ A_x\mathsf{T}_{xz} + A_y\mathsf{T}_{yz} + A_z\mathsf{T}_{zz} \end{pmatrix} = D, \qquad (2\text{–}40)$$

where the vector A is used as a *row matrix* or *row vector*. The two equations can be better expressed as

$$(\mathbf{T} \cdot A)_i = \sum_j \mathsf{T}_{ij} A_j \qquad (2\text{–}41)$$

and

$$(A \cdot \mathbf{T})_i = \sum_j A_j \mathsf{T}_{ji}. \qquad (2\text{–}42)$$

Furthermore,

$$(\mathbf{T} \cdot A) = (A \cdot \tilde{\mathbf{T}}). \qquad (2\text{–}43)$$

Of course, if $\mathbf{T}$ is symmetric, the dot product is commutative, since $\mathbf{T} = \tilde{\mathbf{T}}$. From this example, it follows that

$$(\mathbf{U} \cdot A) = (A \cdot \mathbf{U}) = A.$$

The inner product of a second-order tensor and a unit tensor is a second-order tensor, obtained by the procedure used in Eq. (2–38):

$$(\mathbf{U} \cdot \mathbf{T}) = \begin{pmatrix} 1 & 0 & 0 \\ 0 & 1 & 0 \\ 0 & 0 & 1 \end{pmatrix} \begin{pmatrix} \mathsf{T}_{xx} & \mathsf{T}_{xy} & \mathsf{T}_{xz} \\ \mathsf{T}_{yx} & \mathsf{T}_{yy} & \mathsf{T}_{yz} \\ \mathsf{T}_{zx} & \mathsf{T}_{zy} & \mathsf{T}_{zz} \end{pmatrix} = \begin{pmatrix} \mathsf{T}_{xx} & \mathsf{T}_{xy} & \mathsf{T}_{xz} \\ \mathsf{T}_{yx} & \mathsf{T}_{yy} & \mathsf{T}_{yz} \\ \mathsf{T}_{zx} & \mathsf{T}_{zy} & \mathsf{T}_{zz} \end{pmatrix} = \mathbf{T}. \qquad (2\text{–}44)$$

Similarly, the inner product of any two tensors can be obtained in cartesian coordinates. This inner product takes the form

$$(\mathbf{T} \cdot \mathbf{S})_{ij} = \sum_k \mathsf{T}_{ik} \mathsf{S}_{kj}, \qquad (2\text{–}45)$$

and is itself a tensor. For example, the first term would be

$$\mathsf{T}_{xx}\mathsf{S}_{xx} + \mathsf{T}_{xy}\mathsf{S}_{yx} + \mathsf{T}_{xz}\mathsf{S}_{zx}.$$

There are eight other terms (each a sum of three).

The cross product of a vector and second-order tensor can also be defined, and is a second-order tensor itself:

$$(A \times \mathsf{T})_{ij} = \sum_k \sum_l e_{ikl} A_k \mathsf{T}_{jl},$$
$$(\mathsf{T} \times A)_{ij} = \sum_k \sum_l e_{jkl} \mathsf{T}_{ik} A_l. \tag{2-46}$$

In the preceding summations, the double subscript could have been used to imply the summation, thus eliminating the summation sign.

The tensor of Eq. (2–45) is the result of the contraction of the outer product between two tensors. The resulting tensor could be contracted again, reducing its order again by two, and thus resulting in a scalar. This doubly contracted product is called the scalar product of two second-order tensors, or the double dot product, and is

$$(\mathsf{T} : \mathsf{S}) = \sum_i \sum_j \mathsf{T}_{ij} \mathsf{S}_{ji} = (\mathsf{S} : \mathsf{T}). \tag{2-47}$$

There are two sets of double subscripts, and summation must be performed over both sets; that is, i and j will each take on the values of x, y, and z. The various tensor products cited will appear in the next chapter when the equations of change (continuity, motion, energy) are derived.

In the physical problems to be considered in the following chapters, the tensor forms appear as the product of two vectors, and are called *dyads*. Such terms appeared in Eqs. (2–31) and (2–34), that is, (AB) and (∇A). These terms have nine components and involve all possible interactions of the two vectors. To illustrate the construction and tensor nature, we write

$$(AB) = \begin{pmatrix} A_x B_x & A_x B_y & A_x B_z \\ A_y B_x & A_y B_y & A_y B_z \\ A_z B_x & A_z B_y & A_z B_z \end{pmatrix} \text{ and } (\nabla A) = \begin{pmatrix} \dfrac{\partial A_x}{\partial x} & \dfrac{\partial A_y}{\partial x} & \dfrac{\partial A_z}{\partial x} \\ \dfrac{\partial A_x}{\partial y} & \dfrac{\partial A_y}{\partial y} & \dfrac{\partial A_z}{\partial y} \\ \dfrac{\partial A_x}{\partial z} & \dfrac{\partial A_y}{\partial z} & \dfrac{\partial A_z}{\partial z} \end{pmatrix}.$$

$$\tag{2-48}$$

The similarity between the above and Eq. (2–35) is apparent, and they can both be treated in exactly the same manner in addition and multiplication.

2-7 INTEGRALS

2-7.A Line or Curve Integral

The line or curve integral is given by

$$\int_C (X \, d\boldsymbol{L}) = \int_C (X\boldsymbol{T}) \, dL, \tag{2-49}$$

where dL is an element of arc on the curve, $d\mathbf{L}$ is a directed element of arc on the curve and is equal to $\mathbf{T}\,dL = \mathbf{i}\,dL_x + \mathbf{j}\,dL_y + \mathbf{k}\,dL_z$, or more simply $d\mathbf{L} = \mathbf{i}\,dx + \mathbf{j}\,dy + \mathbf{k}\,dz$. The tangent unit vector along the curve in the direction of integration is $\mathbf{T}$, and X is a scalar or vector $\mathbf{A}$. Multiplication may be scalar, vector, or tensor. For example,

$$\int_C (\mathbf{A}\cdot d\mathbf{L}) = \int_C (A_x\,dL_x + A_y\,dL_y + A_z\,dL_z)$$

$$= \int_C (A_x\,dx + A_y\,dy + A_z\,dz). \quad (2\text{–}50)$$

2-7.B Surface Integral

The surface integral is given by

$$\int_S (X\,d\mathbf{S}) = \int_S (X\mathbf{N})\,dS, \quad (2\text{–}51)$$

where dS is an element of surface, $d\mathbf{S}$ is an element of surface directed along the normal and is equal to $\mathbf{N}\,dS = \mathbf{i}\,dS_x + \mathbf{j}\,dS_y + \mathbf{k}\,dS_z$, and $\mathbf{N}$ is a normal unit vector directed outward from the surface. Since the projection of $\mathbf{N}$ on the surface is always zero,

$$\int_S \mathbf{N}\,dS = \int_S d\mathbf{S} = 0, \quad (2\text{–}52)$$

and X is a scalar or vector $\mathbf{A}$. Multiplication may be scalar, vector, or tensor. For example,

$$\int_S (\mathbf{A}\cdot d\mathbf{S}) = \int_S (A_x\,dS_x + A_y\,dS_y + A_z\,dS_z). \quad (2\text{–}53)$$

The surface integral of a vector is the flux of the vector throughout the surface.

2-7.C Differentiation of an Integral

The differentiation of a definite integral [4] is given by the Leibnitz formula

$$\frac{d}{du}\left(\int_{a(u)}^{b(u)} F(x,u)\,dx\right) = \int_a^b \frac{\partial F}{\partial u}\,dx + F(b,u)\left(\frac{db}{du}\right) - F(a,u)\left(\frac{da}{du}\right). \quad (2\text{–}54)$$

An extension of this is suggested by Bird, Stewart, and Lightfoot [6]:

$$\frac{d}{du}\left(\int_V \psi\,dV\right) = \int_V \left(\frac{\partial \psi}{\partial u}\right) dV + \oint_S \psi(\mathbf{U}_S\cdot d\mathbf{S}), \quad (2\text{–}55)$$

where ψ is a scalar, and $\mathbf{U}_S$ is the velocity of a surface element $d\mathbf{S}$. When a stationary element is being considered, the final term is zero.

2-8 STOKES' THEOREM

Stokes' theorem is expressed as

$$\oint_C (\boldsymbol{A} \cdot d\boldsymbol{L}) = \int_S ((\boldsymbol{\nabla} \times \boldsymbol{A}) \cdot d\boldsymbol{S}). \tag{2-56}$$

The line integral of a vector $\boldsymbol{A}$ around the periphery of a surface is equal to the surface integral of the curl of the vector $\boldsymbol{A}$ over the surface.

2-9 GAUSS' THEOREM

Gauss' theorem is given by

$$\int_V (\boldsymbol{\nabla} \cdot \boldsymbol{A}) \, dV = \oint (\boldsymbol{A} \cdot d\boldsymbol{S}) = \oint_S (\boldsymbol{A} \cdot \boldsymbol{N}) \, dS. \tag{2-57}$$

The volume integral of the divergence of a vector $\boldsymbol{A}$ over a volume is equal to the surface integral of the vector $\boldsymbol{A}$ over the closed surface surrounding the volume. In effect, the divergence of a vector $\boldsymbol{A}$ is the net flux $\boldsymbol{A}$ across a surface surrounding a unit volume.

2-10 GREEN'S THEOREMS

Green's first theorem is obtained from Gauss' by consideration of the divergence of $a\boldsymbol{\nabla} b$ in place of the vector $\boldsymbol{A}$:

$$\boldsymbol{A} = a\boldsymbol{\nabla} b.$$

Therefore, $(\boldsymbol{\nabla} \cdot \boldsymbol{A}) = a\nabla^2 b + (\boldsymbol{\nabla} a \cdot \boldsymbol{\nabla} b)$. Substitution in Eq. (2-57) gives

$$\int_V (a\nabla^2 b + (\boldsymbol{\nabla} a \cdot \boldsymbol{\nabla} b)) \, dV = \oint_S (a\boldsymbol{\nabla} b \cdot d\boldsymbol{S}). \tag{2-58}$$

Interchanging a and b and subtracting from Eq. (2-58) gives

$$\int_V (a\nabla^2 b - b\nabla^2 a) \, dV = \oint_S ((a\boldsymbol{\nabla} b - b\boldsymbol{\nabla} a) \cdot d\boldsymbol{S}), \tag{2-59}$$

which is the second form of Green's theorem.

2-11 TRANSFORMATIONS

One of the advantages of vector notation is that it allows ease of transformation between the various orthogonal coordinate systems. The cartesian coordinates have been given in the preceding sections of this chapter; in this section, the corresponding cylindrical and spherical systems are developed.

The line element was discussed in Section 2-7. A; in cartesian coordinates this is

$$d\boldsymbol{L} = \boldsymbol{T} \, dL = \boldsymbol{i} \, dL_x + \boldsymbol{j} \, dL_y + \boldsymbol{k} \, dL_z = \boldsymbol{i} \, dx + \boldsymbol{j} \, dy + \boldsymbol{k} \, dz.$$

From Eq. (2-2),

$$(dL)^2 = (dL_x)^2 + (dL_y)^2 + (dL_z)^2 = \sum_i dL_i \, dL_i. \tag{2-60}$$

In the transformed system, this becomes

$$(dL)^2 = \sum_i h_i^2 (dL_i)^2, \tag{2-61}$$

where the L_i's are the coordinates of the new system (that is, r, θ, z for cylindrical and r, θ, ϕ for spherical; the order is taken so as to maintain a right-hand system). The term x_i, previously used, will be reserved for cartesian coordinates; here L_i is used so as to imply any orthogonal coordinate system. Again the summation over the three values of the repeated index is implied. The h_i's may be functions of the L_i's. Equation (2–61) is restricted to orthogonal coordinate systems. The more general form is $(dL)^2 = \sum g_{ij}\, dL_i\, dL_j$, where g_{ij} is called the metric tensor. In the more restricted orthogonal system, the off-diagonal terms are zero, so that $i = j$ and one writes $g_{ii} = h_i^2$ for convenience. The h_i's are often called the scale factors, or metric coefficients.

In the new system, the vectors themselves still form a simple right-hand system; that is,

$$\boldsymbol{A} = \boldsymbol{i}_i A_i + \boldsymbol{i}_j A_j + \boldsymbol{i}_k A_k, \tag{2-62}$$

where the $\boldsymbol{i}$'s are the three base vectors ($\boldsymbol{i}$, $\boldsymbol{j}$, and $\boldsymbol{k}$ are unit vectors in cartesian coordinates, $\boldsymbol{i}_r$, $\boldsymbol{i}_\theta$, and $\boldsymbol{i}_z$ in cylindrical, and $\boldsymbol{i}_r$, $\boldsymbol{i}_\theta$, and $\boldsymbol{i}_\phi$ in spherical coordinates).

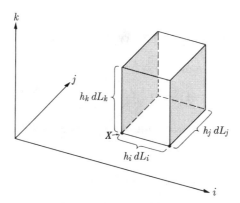

Fig. 2–3. *An element in a general orthogonal-coordinate system.*

By consideration of an infinitesimal element, whose diagonal is dL and whose faces are parallel to the L_i planes, we can derive the transformations for the various vector functions. The element is shown in Fig. 2–3, in which the length of the side in the transformed coordinate system is $h_i\, dL_i$ (no summation). The gradient in an arbitrary direction i is the limit of the change along that direction with decreasing length [4]; that is,

$$(\nabla a)_i = \lim_{h_i\, dL_i \to 0} \left[\frac{a(X + h_i\, dL_i) - a(X)}{h_i\, dL_i} \right] = \frac{1}{h_i} \frac{\partial a}{\partial L_i}. \tag{2-63}$$

In Eq. (2–63) the three values of i give the three components of the vector, the gradient of a.

By the use of the Gauss theorem (Eq. 2–57), the divergence can be obtained. The contribution to $\oint_S (\mathbf{A} \cdot d\mathbf{S})$ at X in the arbitrary direction i is, by Eq. (2–53), $-A_i h_j h_k \, dL_j \, dL_k$. At $X + h_i \, dL_i$, the contribution is

$$- A_i h_j h_k \, dL_j \, dL_k - \frac{\partial(A_i h_j h_k \, dL_j \, dL_k)}{\partial L_i} h_i \, dL_i,$$

and the net contribution for the arbitrary direction i is the difference, or

$$\frac{\partial(A_i h_i h_j h_k)}{\partial L_i} dL_i \, dL_j \, dL_k.$$

The arbitrary direction i will have three values, for example r, θ, and z in cylindrical coordinates. To maintain generality, we will continue to use i, j, and k with the understanding that the three directions can be generated from the expression $\sum_i^C (h_i h_j h_k A_i) dL_i \, dL_j \, dL_k$, where i will take on three values and where cyclic rotation of the indices is implied by the C, that is,

$$\begin{aligned}
&\text{if } i = r, &&\text{then } j = \theta &&\text{and } k = z, \\
&\text{if } i = \theta, &&\text{then } j = z &&\text{and } k = r, \\
&\text{if } i = z, &&\text{then } j = r &&\text{and } k = \theta.
\end{aligned}$$

Then

$$\lim_{V \to 0} \int_V (\mathbf{\nabla} \cdot \mathbf{A}) \, dV = (\mathbf{\nabla} \cdot \mathbf{A}) \, dV = (\mathbf{\nabla} \cdot \mathbf{A}) h_i h_j h_k \, dL_i \, dL_j \, dL_k.$$

Combining the last two expressions by use of Gauss' theorem, one obtains

$$(\mathbf{\nabla} \cdot \mathbf{A}) = \operatorname{div} \mathbf{A} = \frac{1}{h_i h_j h_k} \sum_i^C \frac{\partial h_j h_k A_i}{\partial L_i}, \tag{2–64}$$

where, of course, the cyclic rotation of the indices is still implied.

If $\mathbf{A} = \mathbf{\nabla}$, the Laplacian operator can be obtained:

$$\nabla^2 = \frac{1}{h_i h_j h_k} \sum_i^C \frac{\partial}{\partial L_i}\left(\frac{h_j h_k}{h_i} \frac{\partial}{\partial L_i}\right). \tag{2–65}$$

Once again the indices are rotated cyclically. The operator was obtained by considering the Laplacian of a scalar quantity. It is also valid if the quantity is a cartesian vector; that is, $\nabla^2 \mathbf{A} = \mathbf{i}\nabla^2 A_x + \mathbf{j}\nabla^2 A_y + \mathbf{k}\nabla^2 A_z$. However, it does not hold when a vector or tensor is transformed to curvilinear coordinates. Besides the terms given by Eq. (2–65), there are additional terms associated with the transformation; these will be considered further in Chapter 4.

From Stokes' theorem, $(\mathbf{\nabla} \times \mathbf{A})$ can be determined in a manner similar to the derivation of Eq. (2–64). The line integral about the area normal to the arbitrary direction i can be obtained as

$$\left[\frac{\partial}{\partial L_j}(A_k h_k) - \frac{\partial}{\partial L_k}(A_j h_j)\right] dL_j \, dL_k.$$

By Stokes' theorem, this is equal to $(\nabla \times A)_i h_j h_k dL_j dL_k$; thus

$$(\nabla \times A)_i = \frac{1}{h_j h_k}\left(\frac{\partial h_k A_k}{\partial L_j} - \frac{\partial h_j A_j}{\partial L_k}\right), \qquad (2\text{–}66)$$

where the indices are rotated cyclically.

The substantial derivative involves the term $(A \cdot \nabla)$. This can be obtained directly by using the vector and del in the transformed system [Eq. (2–62) for the vector and Eq. (2–63) for the components of del]. When we perform the required scalar multiplication and remember that the relations in Eqs. (2–9) and (2–10) are also valid in the transformed system (for the transformed unit vectors), we obtain

$$(A \cdot \nabla) = \sum_i \frac{A_i}{h_i} \frac{\partial}{\partial L_i} \qquad (2\text{–}67)$$

or

$$\frac{D}{Dt} = \frac{\partial}{\partial t} + \sum_i \frac{A_i}{h_i} \frac{\partial}{\partial L_i}. \qquad (2\text{–}68)$$

Like the Laplacian operator, the substantial derivative given above is valid only when operated on a scalar or cartesian vector (rectangular coordinates), and is not valid for a vector or tensor transformed to curvilinear coordinates. For this latter case, there are additional terms not given by Eq. (2–68) that appear in the transformation. These are associated with centrifugal and Coriolis forces and will be considered further in Chapter 4.

For *cartesian coordinates* there is no transformation made; thus from Eqs. (2–60) and (2–61) the h_i's are unity, and the equations already presented are obtained.

$$\nabla a = i\frac{\partial a}{\partial x} + j\frac{\partial a}{\partial y} + k\frac{\partial a}{\partial z}, \qquad (2\text{–}23)$$

$$(\nabla \cdot A) = \frac{\partial A_x}{\partial x} + \frac{\partial A_y}{\partial y} + \frac{\partial A_z}{\partial z}, \qquad (2\text{–}24)$$

$$(\nabla \times A) = i\left(\frac{\partial A_z}{\partial y} - \frac{\partial A_y}{\partial z}\right) + j\left(\frac{\partial A_x}{\partial z} - \frac{\partial A_z}{\partial x}\right) + k\left(\frac{\partial A_y}{\partial x} - \frac{\partial A_x}{\partial y}\right), \qquad (2\text{–}25)$$

$$\nabla^2 = \frac{\partial^2}{\partial x^2} + \frac{\partial^2}{\partial y^2} + \frac{\partial}{\partial z^2}, \qquad (2\text{–}26)$$

$$\frac{D}{Dt} = \frac{\partial}{\partial t} + A_x \frac{\partial}{\partial x} + A_y \frac{\partial}{\partial y} + A_z \frac{\partial}{\partial z}. \qquad (2\text{–}27)$$

For *cylindrical coordinates* (from Fig. 2–4),

$$x = r \cos \theta, \qquad y = r \sin \theta, \qquad z = z;$$

$$(dL)^2 = (dr)^2 + r^2 (d\theta)^2 + (dz)^2,$$

$$h_r = 1, \qquad h_\theta = r, \qquad h_z = 1. \qquad (2\text{–}69)$$

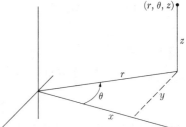

Fig. 2–4. *Cylindrical coordinate system.*

The derived quantities [from Eqs. (2–63) through (2–68)] corresponding to Eqs. (2–23) through (2–27) are:

$$\nabla a = i_r \frac{\partial a}{\partial r} + i_\theta \frac{1}{r} \frac{\partial a}{\partial \theta} + i_z \frac{\partial a}{\partial z}, \tag{2-70}$$

$$(\nabla \cdot A) = \frac{1}{r} \frac{\partial r A_r}{\partial r} + \frac{1}{r} \frac{\partial A_\theta}{\partial \theta} + \frac{\partial A_z}{\partial z}, \tag{2-71}$$

$$(\nabla \times A)_r = \frac{1}{r} \frac{\partial A_z}{\partial \theta} - \frac{\partial A_\theta}{\partial z},$$

$$(\nabla \times A)_\theta = \frac{\partial A_r}{\partial z} - \frac{\partial A_z}{\partial r}, \tag{2-72}$$

$$(\nabla \times A)_z = \frac{1}{r} \frac{\partial r A_\theta}{\partial r} - \frac{1}{r} \frac{\partial A_r}{\partial \theta},$$

$$\nabla^2 = \frac{\partial^2}{\partial r^2} + \frac{1}{r} \frac{\partial}{\partial r} + \frac{1}{r^2} \frac{\partial^2}{\partial \theta^2} + \frac{\partial^2}{\partial z^2}, \tag{2-73}$$

$$\frac{D}{Dt} = \frac{\partial}{\partial t} + A_r \frac{\partial}{\partial r} + \frac{A_\theta}{r} \frac{\partial}{\partial \theta} + A_z \frac{\partial}{\partial z}. \tag{2-74}$$

EXAMPLE 2–5. Demonstrate Eqs. (2–69).

Answer. The relations of x, y, and z in terms of r, θ, and z follow directly from the trigonometry of Fig. 2–4. From these we can obtain the derivatives:

$$dx = dr \cos \theta - r\, d\theta \sin \theta, \qquad dy = dr \sin \theta - r\, d\theta \cos \theta, \qquad dz = dz.$$

From Eq. (2–60), we obtain

$$(dL)^2 = (dr)^2(\cos^2 \theta + \sin^2 \theta) + r^2 (d\theta)^2(\cos^2 \theta + \sin^2 \theta) + (dz)^2$$
$$= (dr)^2 + r^2 (d\theta)^2 + (dz)^2.$$

Using Eq. (2–61) gives the values of the h_i's directly.

EXAMPLE 2–6. Evaluate $(\nabla a)_z$ in cylindrical coordinates (Eq. 2–70).

Answer. From Eq. (2–63), for $i = z$, we obtain $(\nabla a)_z = (1/h_z)(\partial a/\partial z)$. From Eq. (2–69), $h_z = 1$; thus $(\nabla a)_z = \partial a/\partial z$.

EXAMPLE 2–7. Evaluate $(\nabla \times A)_\theta$ in cylindrical coordinates (Eq. 2–72).

Answer. From Eq. (2–66), for $i = \theta, j = z$, and $k = r$, we obtain

$$(\nabla \times A)_\theta = \frac{1}{h_z h_r} \left(\frac{\partial h_r A_r}{\partial z} - \frac{\partial h_z A_z}{\partial r} \right).$$

From Eq. (2–69), $h_r = 1$ and $h_z = 1$; thus

$$(\nabla \times A)_\theta = \frac{\partial A_r}{\partial z} - \frac{\partial A_z}{\partial r}.$$

For *spherical coordinates* (from Fig. 2–5),

$$x = r \sin \theta \cos \phi,$$
$$y = r \sin \theta \sin \phi,$$
$$z = r \cos \theta;$$
$$(dL)^2 = (dr)^2 + r^2 (d\theta)^2 + r^2 \sin^2 \theta (d\phi)^2, \quad (2\text{--}75)$$

$h_r = 1, \quad h_\theta = r, \quad h_\phi = r \sin \theta.$

Fig. 2–5. Spherical coordinate system.

The derived values [From Eqs. (2–63) through (2–68)], which correspond to Eqs. (2–23) through (2–27) are:

$$\nabla a = \mathbf{i}_r \frac{\partial a}{\partial r} + \mathbf{i}_\theta \frac{1}{r} \frac{\partial a}{\partial \theta} + \mathbf{i}_\phi \frac{1}{r \sin \theta} \frac{\partial a}{\partial \phi}, \quad (2\text{--}76)$$

$$(\nabla \cdot A) = \frac{1}{r^2} \frac{\partial r^2 A_r}{\partial r} + \frac{1}{r \sin \theta} \frac{\partial A_\theta \sin \theta}{\partial \theta} + \frac{1}{r \sin \theta} \frac{\partial A_\phi}{\partial \phi}, \quad (2\text{--}77)$$

$$(\nabla \times A)_r = \frac{1}{r \sin \theta} \left(\frac{\partial A_\phi \sin \theta}{\partial \theta} - \frac{\partial A_\theta}{\partial \phi} \right),$$

$$(\nabla \times A)_\theta = \frac{1}{r} \left(\frac{1}{\sin \theta} \frac{\partial A_r}{\partial \phi} - \frac{\partial r A_\phi}{\partial r} \right), \quad (2\text{--}78)$$

$$(\nabla \times A)_\phi = \frac{1}{r} \left(\frac{\partial r A_\theta}{\partial r} - \frac{\partial A_r}{\partial \theta} \right),$$

$$\nabla^2 = \frac{1}{r^2} \left[\frac{\partial}{\partial r} \left(r^2 \frac{\partial}{\partial r} \right) \right] + \frac{1}{r^2 \sin \theta} \left[\frac{\partial}{\partial \theta} \left(\sin \theta \frac{\partial}{\partial \theta} \right) \right] + \frac{1}{r^2 \sin^2 \theta} \frac{\partial^2}{\partial \phi^2}, \quad (2\text{--}79)$$

$$\frac{D}{Dt} = \frac{\partial}{\partial t} + A_r \frac{\partial}{\partial r} + \frac{A_\theta}{r} \frac{\partial}{\partial \theta} + \frac{A_\phi}{r \sin \theta} \frac{\partial}{\partial \phi}. \quad (2\text{--}80)$$

EXAMPLE 2–8. Evaluate ∇^2 in spherical coordinates (Eq. 2–79).

Answer. Equation (2–65) can be directly combined with the values given in Eq. (2–75) to give

$$\nabla^2 = \frac{1}{r^2 \sin \theta} \left[\frac{\partial}{\partial r} \left(r^2 \sin \theta \frac{\partial}{\partial r} \right) + \frac{\partial}{\partial \theta} \left(\frac{r \sin \theta}{r} \frac{\partial}{\partial \theta} \right) + \frac{\partial}{\partial \phi} \left(\frac{r}{r \sin \theta} \frac{\partial}{\partial \phi} \right) \right].$$

Since $\theta \neq f(r \text{ or } \phi)$, and $r \neq f(\theta \text{ or } \phi)$, then

$$\nabla^2 = \frac{1}{r^2 \sin \theta} \left[\sin \theta \frac{\partial}{\partial r} \left(r^2 \frac{\partial}{\partial r} \right) + \frac{\partial}{\partial \theta} \left(\sin \theta \frac{\partial}{\partial \theta} \right) + \frac{1}{\sin \theta} \frac{\partial^2}{\partial \phi^2} \right]$$

$$= \frac{1}{r^2} \left[\frac{\partial}{\partial r} \left(r^2 \frac{\partial}{\partial r} \right) \right] + \frac{1}{r^2 \sin \theta} \left[\frac{\partial}{\partial \theta} \left(\sin \theta \frac{\partial}{\partial \theta} \right) \right] + \frac{1}{r^2 \sin^2 \theta} \frac{\partial^2}{\partial \phi^2}.$$

PROBLEMS

Given two vectors $\boldsymbol{A}$ and $\boldsymbol{B}$, and the scalar ϕ, determine in cartesian coordinates the following forms. In each case, show your work.

2–1. $(\boldsymbol{A}\cdot\boldsymbol{B})$ 2–2. $(\boldsymbol{A}\times\boldsymbol{B})$ 2–3. $(\boldsymbol{A}\cdot\nabla)\boldsymbol{B}$
2–4. $(\boldsymbol{A}\cdot\nabla)\phi$ 2–5. $(\nabla^2\boldsymbol{A})$ 2–6. $\nabla(\boldsymbol{A}\cdot\boldsymbol{B})$

Prove the following relationships.

2–7. $(\nabla\times\nabla\phi) = \text{curl grad }\phi = 0$
2–8. $(\nabla\cdot(\boldsymbol{A}\times\boldsymbol{B})) = (\boldsymbol{B}\cdot(\nabla\times\boldsymbol{A})) - (\boldsymbol{A}\cdot(\nabla\times\boldsymbol{B}))$
2–9. $(\nabla\times(\nabla\times\boldsymbol{A})) = \text{curl curl }\boldsymbol{A} = \nabla(\nabla\cdot\boldsymbol{A}) - (\nabla\cdot(\nabla\boldsymbol{A})) = \nabla(\nabla\cdot\boldsymbol{A}) - \nabla^2\boldsymbol{A}$

Write the following tensor forms in matrix form.

2–10. $(\boldsymbol{B}\boldsymbol{A})$ 2–11. $(\nabla\boldsymbol{A})$ 2–12. $\tilde{\boldsymbol{\mathsf{T}}}$ 2–13. $(\boldsymbol{A}\cdot\tilde{\boldsymbol{\mathsf{T}}})$ 2–14. $(\boldsymbol{\mathsf{S}}\cdot\boldsymbol{\mathsf{T}})_{xy}$

The following theorems can be found in references 4 and 7. Outline the proofs given there.

2–15. Stokes' theorem

2–16. Gauss' theorem

2–17. Green's two theorems

Obtain the following by using the transformations of Section 2–11. Express the results in both cylindrical and spherical coordinates.

2–18. $(\nabla a)_\theta$ 2–19. $(\nabla\cdot\boldsymbol{A})$ 2–20. $(\nabla\times A)_r$ 2–21. ∇^2 2–22. D/Dt

Obtain the metric coefficients for the following coordinate systems.

2–23. Parabolic coordinates

2–24. Prolate spheroidal coordinates

2–25. Oblate spheroidal coordinates

Note: See *Handbook of Mathematical Functions* (M. Abramowitz and I. A. Stegun, editors), p. 752, National Bureau of Standards, Washington, D.C. (1964).

REFERENCES

1. R. B. BIRD, *Advances in Chemical Engineering*, Vol. I, pp. 228–231, Academic Press, New York (1956).
2. G. E. HAY, *Vector and Tensor Analysis*, Dover Publications, New York (1953).
3. H. S. MICKLEY, T. K. SHERWOOD, and C. E. REED, *Applied Mathematics in Chemical Engineering*, McGraw-Hill, New York (1957).
4. L. A. PIPES, *Applied Mathematics for Engineers and Physicists*, McGraw-Hill, New York (1958).
5. L. PRANDTL and O.G. TIETJENS, *Fundamentals of Hydro- and Aeromechanics* (reprint of the 1934 edition), Dover Publications, New York (1957).
6. R. B. BIRD, W. E. STEWART, and E. N. LIGHTFOOT, *Transport Phenomena*, John Wiley & Sons, New York (1960); *see also* J. C. SLATTERY, *Chem. Eng. Sci.* 17, 895 (1962).
7. C. R. WYLIE, JR., *Advanced Engineering Mathematics*, 2nd ed., McGraw-Hill, New York (1960).

CHAPTER 3

THE EQUATIONS OF CHANGE

3-1 INTRODUCTION

The equations of change are statements of the conservation of mass, momentum, and energy. In this text we are specifically interested in fluid flow or momentum transfer. However, we will develop all three conservation equations, since the derivations for the mass and energy equations differ little from that of momentum transfer. In many of the problems of chemical engineering, one must consider not only fluid flow, but heat transfer, diffusion, and possibly chemical kinetics as well. For problems of this nature, an understanding of all the transport phenomena is important.

The overall equation of continuity (mass), the equation of continuity for each chemical species present, the equation of motion (momentum), and the energy balance equation comprise a series of coupled, nonlinear partial differential equations with no general solution. They provide a compact representation of the processes that occur in chemical engineering unit operations. Starting with these equations, and making judicious assumptions, one can arrive at analytical solutions such as Poiseuille's law, Stokes' law, and Bernoulli's equation. Where analytical methods are impossible, numerical methods can be applied to give solutions to such problems as flame propagation, shock waves, and flow around objects. Finally, where solutions cannot be obtained, empirical analysis is necessary; the form and number of the correlating groups can be obtained from an analysis of the differential equations of the system. These are the equations that govern the nonequilibrium processes in chemical engineering. (Thermodynamics is the study of equilibrium processes.)

The development of the equations of change from a molecular approach has been presented by Bird, Curtiss, and Hirschfelder [1, 4]. In some respects the macroscopic balance to be developed here parallels this, and from time to time, therefore, a comparison will be possible.

3-2 THE EQUATIONS OF CHANGE IN TERMS OF FLUXES

Let us consider a system of flowing fluid in which both mass and heat transfer are occurring, and in which there is a chemical reaction. To describe this system we need the equations of change, the thermal and caloric equations of state, and a knowledge of the chemical kinetics. The equations of change describe the changes in the local density, stream velocity, and temperature in terms of the fluxes of mass, momentum, and energy, and in terms of chemical kinetics. In the following analysis the equations will be derived from a macroscopic consideration.

3-2.A The Flux Vectors

Consider an element of volume V, moving at the mass average velocity U, of the system. This velocity is the weighted average of all the groups in the system; that is,

$$U = \frac{1}{\rho} \sum_s \rho_s U_s, \qquad (3\text{-}1)$$

where ρ_s is the density, and U_s is the average velocity of the sth group. To transfer any property associated with the groups out of V through its surface S, the groups must have a velocity different from U. If they were all traveling at U, none could pass through the surface, which is also traveling at U. This difference in velocity is called the *diffusion velocity* of the group, U_{ds}, and would be

$$U_{ds} = U_s - U. \qquad (3\text{-}2)$$

An important identity obtained from the above is

$$\sum_s \rho_s U_{ds} = \sum_s \rho_s (U_s - U) = \sum_s \rho_s U_s - \rho U,$$

since U is already an average over all s (Eq. 3-1). Combining this with Eq. (3-1), we get

$$\sum_s \rho_s U_{ds} = \rho U - \rho U = 0, \qquad (3\text{-}3)$$

which implies equimolar counterdiffusion.

In general, a property per unit volume ψ, can be associated with the volume V. The flux of this property through the volume's surface will be denoted by Ψ. The properties of interest are mass, momentum, and energy. The fluxes of these are vectors, which arise because of a concentration gradient in the property: mass transfer from a concentration gradient, momentum transfer or pressure drop from a velocity gradient, and heat transfer from a temperature gradient.

For mass, the property per unit volume is ρ_s. The flux J_s relative to U is $\rho_s U_{ds}$. For all groups in the volume, the property per unit volume is ρ, and the flux J is zero, since $\sum_s \rho_s U_{ds} = 0$ by Eq. (3-3). In other words, even though individual groups may diffuse in and out of the volume, there can be no net flux so long as a volume moving at U is considered. However, if a stationary element were considered, there would be a net flux J of ρU.

The diffusional transport of mass must occur by the relative motion of the molecular species, and thus the previous discussion in terms of the diffusion velocity is adequate. For momentum and energy, the mechanism giving rise to the property flux is more complicated in that additional mechanisms besides the molecular diffusion effect can give rise to property transfer. For example, the kinetic energy of a molecule is transferred when that molecule diffuses, but in addition, the kinetic energy can be transferred by successive collisions between molecules, even in the absence of a net diffusion.

For the molecular diffusion part, and again relative to U, the momentum per unit volume is $\rho_s U_{ds}$ for any one group. The momentum flux is $\rho_s U_{ds} U_{ds}$, which, when summed over all s, is

$$\mathbf{P} = \sum_s \rho_s U_{ds} U_{ds} = \begin{pmatrix} \mathsf{P}_{xx} & \mathsf{P}_{xy} & \mathsf{P}_{xz} \\ \mathsf{P}_{yx} & \mathsf{P}_{yy} & \mathsf{P}_{yz} \\ \mathsf{P}_{zx} & \mathsf{P}_{zy} & \mathsf{P}_{zz} \end{pmatrix}. \tag{3-4}$$

This is called the pressure tensor and contains nine terms. One example is

$$\mathsf{P}_{xy} = \sum_s \rho_s U_{dsx} U_{dsy} = \mathsf{P}_{yx}.$$

It is clearly symmetric, since

$$U_{dsx} U_{dsy} = U_{dsy} U_{dsx}.$$

Note that for the molecular effect part, the total momentum in the system is zero, since $\sum_s \rho_s U_{ds}$ is zero; however, there is a momentum flux $\mathbf{P}$, which is the pressure of the molecules on the walls of the system. The net zero momentum is a result of the random motion of the molecules. For the momentum flux, the off-diagonal or shear terms are zero, since there is no correlation between velocities in different directions. That is, a product such as $U_{dsx} U_{dsy}$ can have as many positive as negative values, and thus sums to zero. Conversely, the diagonal terms would always be positive, and it is each of these that equals the system pressure.

The pressure tensor is the representation of the normal and tangential stresses, and gives rise to the flux of momentum throughout the system. For the more general case, the pressure tensor is interpreted in terms of the source of momentum or surface forces. That is, flux of momentum (mass times velocity per unit time and unit area) is equivalent to a force interpretation (force per unit area times the gravitational conversion constant g_c). As will be seen later during the discussion of Eq. (3-26), this is equivalent to assuming that the surface forces causing acceleration by Newton's law of motion are due to pressure and shear. Each row of the tensor has three terms, one for the normal stress P_{xx}, and two for the tangential stresses P_{xy}, P_{xz}. The three normal stresses or diagonal elements P_{xx}, P_{yy}, and P_{zz} act in the x-, y-, and z-directions, respectively, and each is the force per unit area on a plane perpendicular to the direction in which it acts (see Fig. 3-1). These stresses contain the contributions of static pressure and a type of viscous term. The nature of these terms will become apparent later. The nondiagonal elements are shear stresses and will be related to the viscosity. Figure 3-1 shows three components acting at a point in space. These act at the center of a plane perpendicular to the x-direction (shaded). Figure 3-2 shows P_{xy} in more detail in a two-dimensional cross section. Interpreted as a shear stress, P_{xy} is the force per unit area exerted in the y-direction on a surface perpendicular to the x-direction. This stress gives rise to the flux of y-momentum through the surface perpendicular to the x-direction. The force is exerted in the y-direction and gives

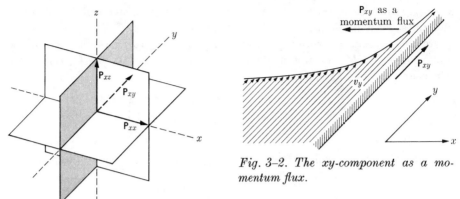

Fig. 3–2. The xy-component as a momentum flux.

Fig. 3–1. Pressure tensor on the yz-plane.

rise to a momentum flux in the x-direction, but it is the flux of y-momentum in the x-direction. Figures 3–3 and 3–4 show the same information for the y-direction plane; the center point is the same, and, for example, the terms P_{xx}, P_{yx}, and P_{zx} all act in the same x-direction and are coincident. These three terms would be required for an x-direction force balance. They are also the flux of the x-momentum in the three directions x, y, and z.

In our analysis, the point in the center of the three planes was considered to be moving with a stream velocity U; thus, the viewpoint is that of an observer moving with the stream.

The symmetry of the pressure tensor has already been noted. It can also be demonstrated by taking the moments of the forces per unit area on a small element $dx\,dy\,dz$, centered at the origin. Only shear stresses need be considered, since the normal and gravitational effects act through the center of mass and consequently have no moment. Another means of illustrating the symmetry is to divide the pressure tensor into symmetric and antisymmetric parts, and then prove that the antisymmetric part is identically zero. Because of the symmetry, the pressure tensor is defined by six terms. The necessary assumption of zero moment implies that the material is not magnetic, and that neither magnetic nor Coriolis forces are present. (These are odd functions of the velocity.) Further discussion on this point and on the general symmetry in viscous systems can be found in Cox's book on statistical mechanics [3]. Other effects can occur in non-Newtonian systems, such as unequal normal stresses; these effects will be considered in the chapter on rheology (Chapter 15).

If the system is at equilibrium or at rest, all the viscous stresses are zero, and the normal stresses are equal to each other and to the static pressure. In other words, the pressure in a system at rest is independent of direction and acts normal to any surface. The tensor terms are then

$$\mathsf{P}_{xx} = \mathsf{P}_{yy} = \mathsf{P}_{zz} = \text{static pressure} = p, \tag{3-5}$$

$$\mathsf{P}_{xy} = \mathsf{P}_{xz} = \mathsf{P}_{yx} = \mathsf{P}_{yz} = \mathsf{P}_{zx} = \mathsf{P}_{zy} = 0. \tag{3-6}$$

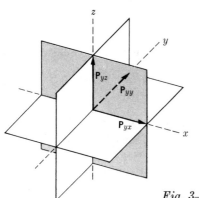
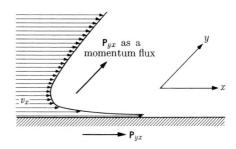

Fig. 3–4. *The yx-component as a momentum flux.*

Fig. 3–3. *Pressure tensor on the xz-plane.*

In this text we use P_{ij} to represent terms of the pressure tensor. Many authors use $\bar{\bar{\tau}}_{ij}$; We use $\bar{\bar{\tau}}_{ij}$ as the shear terms, so that $\mathsf{P}_{xy} = \bar{\bar{\tau}}_{xy}$, etc., for the nondiagonal terms, and $\mathsf{P}_{xx} = p + \bar{\bar{\tau}}_{xx}$ etc., for the diagonal terms. Many books use $-p$; however, if $\bar{\bar{\tau}}$ is taken as the flux of momentum, rather than as the deformation force, p is positive.

For the molecular diffusion part, the kinetic energy per unit volume is $\frac{1}{2}\rho_s U_{ds}^2$, where U_{ds} is the magnitude of $\boldsymbol{U}_{ds}$ [kinetic energy is a scalar tensor, not a second-order tensor, that is to say, $U_{ds}^2 = (\boldsymbol{U}_{ds} \cdot \boldsymbol{U}_{ds})$]. The flux of kinetic energy then becomes

$$q = \sum_s \tfrac{1}{2}\rho_s U_{ds}^2 \boldsymbol{U}_{ds}. \tag{3-7}$$

When the energy balance is made, the internal energy will also have to be considered, and in this case q will be taken as the total flux by diffusion and conduction.

The fluxes that have been considered are $\boldsymbol{J}_s$, $\boldsymbol{J}$, P, and $\boldsymbol{q}$. The overall mass flux $\boldsymbol{J}$ is zero. The other fluxes may be broken down into several parts; for the nine components of P, the diagonal terms are the normal stresses, and the other six are the shear stresses. The normal stresses are associated with both the static pressure p and a normal shear. The stresses from shear are, of course, associated with viscous stresses. In cartesian coordinates,

$$\mathsf{P} = \begin{pmatrix} \mathsf{P}_{xx} & \mathsf{P}_{xy} & \mathsf{P}_{xz} \\ \mathsf{P}_{yx} & \mathsf{P}_{yy} & \mathsf{P}_{yz} \\ \mathsf{P}_{zx} & \mathsf{P}_{zy} & \mathsf{P}_{zz} \end{pmatrix} = \begin{pmatrix} p + \bar{\bar{\tau}}_{xx} & \bar{\bar{\tau}}_{xy} & \bar{\bar{\tau}}_{xz} \\ \bar{\bar{\tau}}_{yx} & p + \bar{\bar{\tau}}_{yy} & \bar{\bar{\tau}}_{yz} \\ \bar{\bar{\tau}}_{zx} & \bar{\bar{\tau}}_{zy} & p + \bar{\bar{\tau}}_{zz} \end{pmatrix}$$

$$= \begin{pmatrix} p & 0 & 0 \\ 0 & p & 0 \\ 0 & 0 & p \end{pmatrix} + \begin{pmatrix} \bar{\bar{\tau}}_{xx} & \bar{\bar{\tau}}_{xy} & \bar{\bar{\tau}}_{xz} \\ \bar{\bar{\tau}}_{yx} & \bar{\bar{\tau}}_{yy} & \bar{\bar{\tau}}_{yz} \\ \bar{\bar{\tau}}_{zx} & \bar{\bar{\tau}}_{zy} & \bar{\bar{\tau}}_{zz} \end{pmatrix} = p\begin{pmatrix} 1 & 0 & 0 \\ 0 & 1 & 0 \\ 0 & 0 & 1 \end{pmatrix} + \begin{pmatrix} \bar{\bar{\tau}}_{xx} & \bar{\bar{\tau}}_{xy} & \bar{\bar{\tau}}_{xz} \\ \bar{\bar{\tau}}_{yx} & \bar{\bar{\tau}}_{yy} & \bar{\bar{\tau}}_{yz} \\ \bar{\bar{\tau}}_{zx} & \bar{\bar{\tau}}_{zy} & \bar{\bar{\tau}}_{zz} \end{pmatrix}.$$

If **U** is the unit tensor, and $\bar{\bar{\tau}}$ refers to viscous components, then

$$\mathbf{P} = p\mathbf{U} + \bar{\bar{\tau}}. \tag{3-8}$$

The mass flux $\boldsymbol{J}_s$ will include components due to gradients in concentration, temperature, pressure, and external forces [1], that is,

$$\boldsymbol{J}_s = \boldsymbol{J}_s^\rho + \boldsymbol{J}_s^p + \boldsymbol{J}_s^F + \boldsymbol{J}_s^T. \tag{3-9}$$

The first term is the flux due to a concentration gradient, and is molecular diffusion. The second term is a mass flux due to a pressure gradient, and can usually be considered negligible. If this effect exists, it will be a mass transfer or separation of materials due to a gradient in pressure only. The system may be completely homogeneous so far as concentration is concerned. Such a system can be found in separations made in an ultracentrifuge. The third term is mass flux due to diffusion by external forces; an example is ion diffusion. The last term is the flux due to thermal gradient or thermal diffusion; it is small, except under those conditions which are designed to make it the important contribution, such as thermal diffusion columns.

Likewise, the energy flux $\boldsymbol{q}$ will depend on gradients of temperature, concentration, diffusion, and radiation [1], that is,

$$\boldsymbol{q} = \boldsymbol{q}^T + \boldsymbol{q}^d + \boldsymbol{q}^\rho + \boldsymbol{q}^r. \tag{3-10}$$

The direct effect is $\boldsymbol{q}^T$, which depends on the thermal conductivity and the temperature gradient. The diffusion effect is $\boldsymbol{q}^d$, and is the energy associated with diffusion of matter. The concentration effect term $\boldsymbol{q}^\rho$ is associated with the energy flux from the Dufour effect. This is an energy transfer caused by a concentration gradient in the absence of actual mass transfer. The heat carried because of mass transfer itself is $\boldsymbol{q}^d$. The term $\boldsymbol{q}^r$, or radiation, is usually considered a separate, independent means of energy transfer.

3-2.B The General Property Balance

A general property balance can be made on an element of volume V moving with the stream velocity $\boldsymbol{U}$. The property (mass, momentum, or energy) per unit volume, is denoted by ψ, the flux of the property by $\boldsymbol{\Psi}$. The net flow across the surface will be given by the closed surface integral of the flux,

$$\oint_S (\boldsymbol{\Psi} \cdot d\boldsymbol{S}). \tag{3-11}$$

The integral is negative for net inflow and positive for net outflow.

The generation of the property per unit volume and time is denoted by $\dot{\psi}_g$, and the generation within the volume is

$$\int_V \dot{\psi}_g \, dV. \tag{3-12}$$

The time rate of change of the property in the volume is

$$\frac{d}{dt}\left(\int_V \psi \, dV\right). \tag{3-13}$$

Since *the time rate of change within the volume must equal the net flow across the surface plus the generation*, the balance becomes

$$\frac{d}{dt}\left(\int_V \psi \, dV\right) = -\oint_S (\boldsymbol{\Psi} \cdot d\boldsymbol{S}) + \int_V \dot{\psi}_g \, dV. \tag{3-14}$$

The term on the left is the total change of the property in V. This can be rewritten (Eq. 2–55) as

$$\frac{d}{dt}\left(\int_V \psi \, dV\right) = \int_V (\partial \psi/\partial t) \, dV + \oint_S \psi(\boldsymbol{U} \cdot d\boldsymbol{S}),$$

where the partial term considers the time rate of change in V, and the additional term accounts for the change of the property associated with a movement of the surface at the mean velocity $\boldsymbol{U}$ of the element. The result is obtained from the differentiation of an integral. Equation (3–14) then becomes

$$\int_V (\partial \psi/\partial t) \, dV + \oint_S \psi(\boldsymbol{U} \cdot d\boldsymbol{S}) = -\oint_S (\boldsymbol{\Psi} \cdot d\boldsymbol{S}) + \int_V \dot{\psi}_g \, dV. \tag{3-15}$$

Gauss' theorem (Eq. 2–57) states that

$$\int_V (\boldsymbol{\nabla} \cdot \boldsymbol{A}) \, dV = \oint_S (\boldsymbol{A} \cdot d\boldsymbol{S}). \tag{3-16}$$

Combining Eqs. (3–15) and (3–16), we obtain

$$\int_V [(\partial \psi/\partial t) + (\boldsymbol{\nabla} \cdot \psi \boldsymbol{U}) + (\boldsymbol{\nabla} \cdot \boldsymbol{\Psi}) - \dot{\psi}_g] \, dV = 0. \tag{3-17}$$

Since this expression is true for any region in space,

$$(\partial \psi/\partial t) + (\boldsymbol{\nabla} \cdot \psi \boldsymbol{U}) + (\boldsymbol{\nabla} \cdot \boldsymbol{\Psi}) - \dot{\psi}_g = 0, \tag{3-18}$$

which is the general equation of change for a property ψ.

The element of volume could have been taken as stationary in the flow stream; thus, the second term would be zero. However, the fluxes would now contain this term; i.e., they would have to be derived so as to include the fluxes caused by the mass average velocity $\boldsymbol{U}$ in and out of the volume. Of course, the resulting equations would be the same.

In the molecular approach, the Enskog "general equation of change" is derived from the statistical-mechanical theory of a dilute monatomic gas. Equation (3–18) is a macroscopic balance, while Enskog's equation is a microscopic balance for the sth molecules. This equation must be summed over all s to obtain the equivalent of Eq. (3–18).

3-2.C Explicit Expressions for the Equations of Change

To obtain explicit expressions for the equations of change, one must substitute the properties and fluxes into the general equations. In this section, Eq. (3–18) will be used. In the molecular approach, a modified form of Enskog's equation is used. Since the molecular approach will ultimately be summed over all s, the results will be the same.

(a) The continuity equation. The macroscopic property of mass per unit volume is the density ρ. The flux of mass in this case is zero, since the element under consideration is moving with the stream velocity U. The generation of mass is zero. The general balance equation (3–18) becomes

$$\partial \rho / \partial t + (\mathbf{\nabla} \cdot \rho \mathbf{U}) = 0. \tag{3-19}$$

In terms of the substantial derivative,

$$D/Dt = (\partial/\partial t) + (\mathbf{U} \cdot \mathbf{\nabla}), \tag{3-20}$$

Eq. (3–19) becomes

$$\partial \rho / \partial t + \rho (\mathbf{\nabla} \cdot \mathbf{U}) + (\mathbf{U} \cdot (\mathbf{\nabla} \rho)) = 0$$

or

$$D\rho/Dt = -\rho (\mathbf{\nabla} \cdot \mathbf{U}). \tag{3-21}$$

In general, the substantial derivative has the term $(\mathbf{A} \cdot \mathbf{\nabla})$; however, the vector $\mathbf{A}$ is specifically $\mathbf{U}$ for continuity, motion, and energy equations. For any one species, the property is ρ_s, and the flux is $\mathbf{J}_s$ (not necessarily zero). An individual species can be generated by chemical reaction K_s. Using Eqs. (3–18) and (3–20) [as was done to obtain Eq. (3–21)] we obtain

$$D\rho_s/Dt = -\rho_s (\mathbf{\nabla} \cdot \mathbf{U}) - (\mathbf{\nabla} \cdot \mathbf{J}_s) + K_s. \tag{3-22}$$

If the material is incompressible, then the density is constant and one obtains from Eq. (3–21)

$$(\mathbf{\nabla} \cdot \mathbf{U}) = 0.$$

Equations (3–19), (3–21), and (3–22) are all forms of the equation of continuity, and give the change in density with time and position. Specifically, Eq. (3–22) contains terms which will account for a change in the mass density of species s with an expansion of the fluid, with diffusion processes, and because of chemical reactions. The overall equation of continuity (3–21) shows that the change in density occurs only because of an expansion of the fluid. If the density in a given volume is decreasing with time, then there must be an excess of outward over inward flow; the term $(\mathbf{\nabla} \cdot \mathbf{U})$ from Eq. (3–21) must be a positive increasing function or the divergence of the velocity. The term *divergence*, meaning excess of outward over inward flow, originated in this interpretation.

(b) The momentum equation. The macroscopic property of momentum per unit volume is ρU; the flux of momentum is the pressure tensor $\mathbf{P}$. However, unlike mass, momentum can be generated by field forces such as gravity. The general equation becomes

$$\partial \rho U/\partial t + (\nabla \cdot \rho(UU)) + (\nabla \cdot \mathbf{P}) - \sum_s \rho_s F_s = 0, \qquad (3\text{--}23)$$

where the F_s's are the field forces per unit mass. Equation (3–23) is a momentum-balance form of the equation of motion, and indicates that a change in the momentum of a fluid element at a point is due to a divergence of the bulk flow of momentum out of the element, to a gradient in the pressure tensor, and to field forces.

Using the identity (2–31), we have $(\nabla \cdot \rho(UU)) = (\rho U \cdot \nabla)U + U(\nabla \cdot \rho U)$; also

$$\frac{\partial \rho U}{\partial t} = \rho \frac{\partial U}{\partial t} + U \frac{\partial \rho}{\partial t}.$$

Equation (3–23) becomes

$$\rho \frac{\partial U}{\partial t} + U \frac{\partial \rho}{\partial t} = -(\nabla \cdot \mathbf{P}) - (\rho U \cdot \nabla)U - U(\nabla \cdot \rho U) + \sum_s \rho_s F_s.$$

Combining this with the continuity equation (3–19), we obtain

$$\rho(\partial U/\partial t) = -(\nabla \cdot \mathbf{P}) - (\rho U \cdot \nabla)U + \sum_s \rho_s F_s. \qquad (3\text{--}24)$$

In terms of the substantial derivative, this becomes

$$\rho(DU/Dt) = -(\nabla \cdot \mathbf{P}) + \sum_s \rho_s F_s. \qquad (3\text{--}25)$$

The velocity of an element of fluid moving with the stream velocity (Eq. 3–25) undergoes change (acceleration) because of the gradient in the pressure tensor and because of external forces. In the previous discussion, the pressure tensor was shown to be composed of two terms; therefore, the change in velocity can be broken down into a term for the acceleration due to a gradient in the static pressure, $-(1/\rho)\nabla p$, and a term for the deceleration of the fluid element because of viscous stresses, $-(1/\rho)(\nabla \cdot \bar{\bar{\tau}})$.

Equation (3–25), the equation of motion, is a mathematical expression of Newton's second law of motion, which is

$$d(mU)/dt = m(dU/dt) = X. \qquad (3\text{--}26)$$

The mass multiplied by the acceleration is called the *inertial force*, and by this law, must be equal to the external forces. The inertial forces take the form of $\rho(DU/Dt)$; the external forces due to pressure and shear surface forces take the form of $(\nabla \cdot \mathbf{P})$; while the field forces, such as gravity, are of the form of F. Equating these forces on a cubic element of volume in the system will give Eq. (3–25). Such an analysis can be found in the book by Knudsen and Katz [7].

(c) The energy equation. The macroscopic property of internal energy per unit volume is $\rho\mathrm{u}$, where u is the internal energy per unit mass. Kinetic energy,

$$\tfrac{1}{2}\rho U^2 = \tfrac{1}{2}\rho(\boldsymbol{U}\cdot\boldsymbol{U}),$$

is also associated with the volume. The energy flux due to heat flow by conduction and diffusion is denoted by $\boldsymbol{q}$. Generation of energy by electrical, chemical, or nuclear means will not be considered; however, work (force × distance) is done on the fluid. This consists of two parts: first, the work done on the fluid which is associated with volume forces (all field forces $\rho\boldsymbol{F}$); and second, the work done by the fluid which is associated with overcoming surface forces [pressure and viscous forces as described by the pressure tensor $(\boldsymbol{\mathsf{P}}\cdot d\boldsymbol{S})$]. The work per unit time (force × velocity) is

$$\int_V \dot{\psi}_g\, dV = \int_V \sum_s (\boldsymbol{U}_s\cdot\rho_s\boldsymbol{F}_s)\, dV - \oint_S \boldsymbol{U}\cdot\boldsymbol{\mathsf{P}}\cdot d\boldsymbol{S}.$$

Introducing Eq. (3–2) for $\boldsymbol{U}_s$ in the first term and Gauss' theorem Eq. (2–57) for the second, we get

$$\int_V \dot{\psi}_g\, dV = \int_V [\sum_s (\rho_s\boldsymbol{U}_{ds}\cdot\boldsymbol{F}_s) + (\boldsymbol{U}\cdot\sum_s \rho_s\boldsymbol{F}_s)]\, dV - \int_V (\boldsymbol{\nabla}\cdot(\boldsymbol{U}\cdot\boldsymbol{\mathsf{P}}))\, dV.$$

Further rearrangement gives

$$\dot{\psi}_g = \sum_s (\boldsymbol{J}_s\cdot\boldsymbol{F}_s) + (\boldsymbol{U}\cdot\sum_s \rho_s\boldsymbol{F}_s) - (\boldsymbol{\nabla}\cdot(\boldsymbol{U}\cdot\boldsymbol{\mathsf{P}})).$$

Equation (3–18) can now be written in terms of the internal plus kinetic energies:

$$\partial[\rho(\mathrm{u} + \tfrac{1}{2}U^2)]/\partial t + (\boldsymbol{\nabla}\cdot\rho(\mathrm{u} + \tfrac{1}{2}U^2)\boldsymbol{U}) + (\boldsymbol{\nabla}\cdot\boldsymbol{q}) + (\boldsymbol{\nabla}\cdot(\boldsymbol{U}\cdot\boldsymbol{\mathsf{P}}))$$
$$- (\boldsymbol{U}\cdot\sum_s \rho_s\boldsymbol{F}_s) - \sum_s (\boldsymbol{J}_s\cdot\boldsymbol{F}_s) = 0. \qquad (3\text{–}27)$$

With the aid of the continuity equation, the first two terms can be rearranged into the substantial derivative form (as was done for the motion equation). The fourth term can be replaced by the vector identity

$$(\boldsymbol{\nabla}\cdot(\boldsymbol{U}\cdot\boldsymbol{\mathsf{P}})) = (\boldsymbol{U}\cdot(\boldsymbol{\nabla}\cdot\boldsymbol{\mathsf{P}})) + ((\boldsymbol{\mathsf{P}}\cdot\boldsymbol{\nabla})\cdot\boldsymbol{U}) = (\boldsymbol{U}\cdot(\boldsymbol{\nabla}\cdot\boldsymbol{\mathsf{P}})) + (\boldsymbol{\mathsf{P}}:(\boldsymbol{\nabla}\boldsymbol{U})).$$

With these, Eq. (3–27) now becomes

$$\rho[D(\mathrm{u} + \tfrac{1}{2}U^2)/Dt] = -(\boldsymbol{\nabla}\cdot\boldsymbol{q}) - (\boldsymbol{\mathsf{P}}:(\boldsymbol{\nabla}\boldsymbol{U})) - (\boldsymbol{U}\cdot(\boldsymbol{\nabla}\cdot\boldsymbol{\mathsf{P}}))$$
$$+ (\boldsymbol{U}\cdot\sum_s \rho_s\boldsymbol{F}_s) + \sum_s (\boldsymbol{J}_s\cdot\boldsymbol{F}_s).$$

The substantial derivative can be separated into two parts:

$$D(\mathrm{u} + \tfrac{1}{2}U^2)/Dt = (D\mathrm{u}/Dt) + (\boldsymbol{U}\cdot(D\boldsymbol{U}/Dt)),$$

which, when combined with the above, gives

$$\rho(D\mathrm{u}/Dt) = -(\boldsymbol{\nabla}\cdot\boldsymbol{q}) - (\boldsymbol{\mathsf{P}}:(\boldsymbol{\nabla}\boldsymbol{U})) + \sum_s (\boldsymbol{J}_s\cdot\boldsymbol{F}_s)$$
$$+ (\boldsymbol{U}\cdot[-\rho(D\boldsymbol{U}/Dt) - (\boldsymbol{\nabla}\cdot\boldsymbol{\mathsf{P}}) + \sum_s \rho_s\boldsymbol{F}_s]).$$

The term in the brackets is zero, by Eq. (3-25); thus the simplified form for the internal energy equation is

$$\rho(D\mathrm{u}/Dt) = -(\boldsymbol{\nabla}\cdot\boldsymbol{q}) - (\mathbf{P} : (\boldsymbol{\nabla}\boldsymbol{U})) + \sum_s (\boldsymbol{J}_s\cdot\boldsymbol{F}_s). \tag{3-28}$$

The pressure tensor term can be considered as two terms, a pressure-volume-work term associated with the compression effects $-p(\boldsymbol{\nabla}\cdot\boldsymbol{U})$, and a viscous dissipation term $-(\bar{\bar{\boldsymbol{\tau}}}:(\boldsymbol{\nabla}\boldsymbol{U}))$. With these, Eq. (3-28) becomes

$$\rho(D\mathrm{u}/Dt) = -(\boldsymbol{\nabla}\cdot\boldsymbol{q}) - p(\boldsymbol{\nabla}\cdot\boldsymbol{U}) - (\bar{\bar{\boldsymbol{\tau}}}:(\boldsymbol{\nabla}\boldsymbol{U})) + \sum_s (\boldsymbol{J}_s\cdot\boldsymbol{F}_s). \tag{3-29}$$

The internal energy changes because of heat flux, expansion effects, viscous heating, and the overcoming of field forces.

Additional forms for the energy equation can be derived in terms of total energy, enthalpy, kinetic energy, and temperature. The equations and derivations can be found in reference 2. An important form for heat transfer analysis is the energy equation in terms of temperature. This can be obtained by expressing the internal energy in thermodynamic terms of pressure, volume, and temperature, and using this in Eq. (3-29). The result is

$$\rho c_v(DT/Dt) = -(\boldsymbol{\nabla}\cdot\boldsymbol{q}) - (\bar{\bar{\boldsymbol{\tau}}}:(\boldsymbol{\nabla}\boldsymbol{U})) - T(\partial p/\partial T)_\rho(\boldsymbol{\nabla}\cdot\boldsymbol{U}) + \sum_s (\boldsymbol{J}_s\cdot\boldsymbol{F}_s). \tag{3-30}$$

The equations of change can be used in solving the problems of fluid, heat, and mass transfer. The equations of state are also often needed. The equations of change are in terms of the flux vectors, and for general usefulness must be related to the transport coefficients: viscosity, thermal conductivity, and diffusivity. For the most complex solution, one would need a knowledge of the dependence of these properties upon temperature and pressure.

(d) Summary of the equations of change. A very excellent summary table has been presented by Bird, Stewart, and Lightfoot [2]. Their Table 10.4–1 contains most of the forms summarized below plus a number of additional related forms. Those derived in this text are:

Continuity

$$(\partial\rho/\partial t) + (\boldsymbol{\nabla}\cdot\rho\boldsymbol{U}) = 0, \tag{3-19}$$

$$D\rho_s/Dt = -\rho_s(\boldsymbol{\nabla}\cdot\boldsymbol{U}) - (\boldsymbol{\nabla}\cdot\boldsymbol{J}_s) + K_s, \tag{3-22}$$

$$D\rho/Dt = -\rho(\boldsymbol{\nabla}\cdot\boldsymbol{U}). \tag{3-21}$$

Motion

$$\rho(\partial\boldsymbol{U}/\partial t) + \rho(\boldsymbol{U}\cdot\boldsymbol{\nabla})\boldsymbol{U} = -(\boldsymbol{\nabla}\cdot\mathbf{P}) + \sum_s \rho_s\boldsymbol{F}_s, \tag{3-24}$$

$$\rho(D\boldsymbol{U}/Dt) = -(\boldsymbol{\nabla}\cdot\mathbf{P}) + \sum_s \rho_s\boldsymbol{F}_s, \tag{3-25}$$

$$\partial\rho\boldsymbol{U}/\partial t = -(\boldsymbol{\nabla}\cdot\mathbf{P}) - (\boldsymbol{\nabla}\cdot\rho(\boldsymbol{U}\boldsymbol{U})) + \sum_s \rho_s\boldsymbol{F}_s. \tag{3-23}$$

Energy

$$\rho(D\upsilon/Dt) = -(\nabla \cdot \boldsymbol{q}) - (\mathbf{P} : (\nabla \boldsymbol{U})) + \sum_s (\boldsymbol{J}_s \cdot \boldsymbol{F}_s), \tag{3-28}$$

$$\rho(DU/Dt) = -(\nabla \cdot \boldsymbol{q}) - p(\nabla \cdot \boldsymbol{U}) - (\bar{\bar{\tau}} : (\nabla \boldsymbol{U})) + \sum_s (\boldsymbol{J}_s \cdot \boldsymbol{F}_s), \tag{3-29}$$

$$\rho c_v(DT/Dt) = -(\nabla \cdot \boldsymbol{q}) - (\bar{\bar{\tau}} : (\nabla \boldsymbol{U})) - T(\partial p/\partial T)_\rho (\nabla \cdot \boldsymbol{U}) + \sum_s (\boldsymbol{J}_s \cdot \boldsymbol{F}_s). \tag{3-30}$$

3-2.D Comments

At this point we should consider the general applicability of the equations of change. Bird et al. [1] have pointed out that these equations are useful only when point properties are involved. When properties such as density, velocity, and temperature cannot be regarded as continuous, local properties have little meaning. Two examples of this are: the Knudsen gas, for which the container size is of the same order of magnitude as the mean free path, and the ideal shock wave, in which changes are abrupt in very short distances. In these cases, the equations relating changes in macroscopic variables have little meaning, since the properties themselves have no meaning.

Bird et al. mentioned turbulence and multiphase flow. Turbulence is considered to be a somewhat random flow of eddies superimposed on the overall flow, with the size of the smallest eddy estimated as being of the order of 0.1 mm or larger [5]. Since these are much larger than the mean free path for dense systems of gases or liquids, the turbulent motion can be considered as macroscopic, rather than molecular. The fluid can be considered as a continuum, and the equations of change are valid for relating point changes. In a later section the equations will be modified to consider the average turbulent system, rather than point changes only. The problem of multiphase flow is one of matching up the equations at the interface, since the equations will apply in each of the phases. Multiphase flow is also considered in later chapters.

3-3 THE FLUXES IN TERMS OF THE TRANSPORT COEFFICIENTS

The equations of change given in the preceding section express the changes in density, velocity and energy in terms of the fluxes of mass, momentum, and energy. The fluxes can be related to the corresponding gradients and transport coefficients, so that the equations of change become more usable. In this section, the required relation between the pressure tensor and velocity gradient will be obtained. An alternative derivation, based on the concepts of irreversible thermodynamics, can be made, in which the form of the gradients (driving forces) and the relations between the fluxes and gradients are derived. Cross effects, such as thermal diffusion etc., are derived as a natural consequence of this method [3, 4, 9, 10, 11].

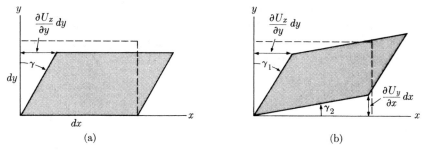

Fig. 3–5. One- and two-dimensional deformation of an element.

In laminar flow, momentum is transferred by a gradient type mechanism; for simple one-dimensional flow in the x-direction with a gradient in the y-direction, the flux is given by

$$\bar{\bar{\tau}}_{yx} = -\nu(\partial \rho U_x/\partial y), \tag{3-31}$$

where ρU_x is the concentration of momentum in the x-direction, ν is the momentum diffusivity or kinematic viscosity, and $\bar{\bar{\tau}}_{yx}$ is the momentum flux in the y-direction which is a result of the x-momentum concentration gradient. The negative sign is necessary, since any property is transferred down the gradient, i.e., from the high-velocity to the low-velocity area (see Fig. 3–4). For constant density, Eq. (3–31) reduces to Newton's familiar law:

$$\bar{\bar{\tau}}_{yx} = -\mu(\partial U_x/\partial y). \tag{3-32}$$

Equations (3–31) and (3–32) are valid only for the one-dimensional flow case (shown in Fig. 3–5a) where U_y is zero. The rate of deformation, or strain rate of the element (angular velocity of dy), is

$$\frac{\partial \gamma}{\partial t} = \frac{(\partial U_x/\partial y)dy}{dy} = \frac{\partial U_x}{\partial y},$$

or, combined with Eq. (3–32), is

$$\bar{\bar{\tau}}_{yx} = -\mu(\partial \gamma/\partial t). \tag{3-33}$$

For the two-dimensional case (Fig. 3–5b), both gradients contribute to the deformation; i.e., rotation of the axis to coincide with the lower side of the parallelogram will show that $\gamma = \gamma_1 + \gamma_2$; therefore

$$\frac{\partial \gamma}{\partial t} = \frac{\partial \gamma_1}{\partial t} + \frac{\partial \gamma_2}{\partial t}. \tag{3-34}$$

Thus

$$\bar{\bar{\tau}}_{yx} = -\mu\left(\frac{\partial \gamma_1}{\partial t} + \frac{\partial \gamma_2}{\partial t}\right) = -\mu\left(\frac{\partial U_x}{\partial y} + \frac{\partial U_y}{\partial x}\right). \tag{3-35}$$

In general,

$$\bar{\bar{\tau}}_{ji} = \bar{\bar{\tau}}_{ij} = -\mu\left(\frac{\partial U_i}{\partial x_j} + \frac{\partial U_j}{\partial x_i}\right) \quad i \neq j. \tag{3-36}$$

In the cartesian notation used above, i and j take the values x, y, and z, but $i \neq j$. The diagonal terms or normal stresses must contain, in addition to the shear terms, a bulk deformation. The additional deformation should involve the compressibility of a material and should disappear if the material is truly incompressible; i.e., if $(\nabla \cdot U) = 0$. Assuming the effect to be linear, we can write for the normal stresses $(i = j)$

$$\bar{\tau}_{ii} = -2\mu(\partial U_i/\partial x_i) - k(\nabla \cdot U), \tag{3-37}$$

where k is the proportionality constant to be determined.

A more general treatment [8] can be made if it is assumed that, as in Newton's law, all the $\bar{\tau}_{ij}$'s are linear functions of $\partial U_i/\partial x_j$. This assumption can in part be justified, since $\bar{\tau}$ must vanish for a constant U. Also, $\bar{\tau}$ must vanish if the fluid is in uniform rotation, which implies that the $\bar{\tau}_{ij}$'s must contain a symmetrical linear combination of the derivatives $\partial U_i/\partial x_j$; that is,

$$\frac{\partial U_i}{\partial x_j} + \frac{\partial U_j}{\partial x_i}.$$

The most general second-order tensor that meets these requirements is

$$\bar{\tau}_{ji} = a\left(\frac{\partial U_i}{\partial x_j} + \frac{\partial U_j}{\partial x_i}\right) + b\frac{\partial U_l}{\partial x_l}\delta_{ij},$$

which is simply a combination of Eqs. (3–36) and (3–37).

Adding all the $\bar{\tau}_{ii}$'s, we have

$$\bar{\tau}_{xx} + \bar{\tau}_{yy} + \bar{\tau}_{zz} = \sum_i \bar{\tau}_{ii} = -(2\mu + 3k)(\nabla \cdot U), \tag{3-38}$$

and combining this with the continuity equation, we obtain

$$\sum_i \bar{\tau}_{ii} = (2\mu + 3k)(1/\rho)(D\rho/Dt). \tag{3-39}$$

This equation would indicate that the normal viscous stresses are a function of the rate of density-change. In most cases this is not true, since the stresses are dependent only on the density and not on the rate at which it changes. For these cases the term on the right must be zero; therefore

$$k = -\tfrac{2}{3}\mu. \tag{3-40}$$

If this were not true, it would also imply that the pressure in an expanding or contracting fluid would deviate from the instantaneous static equilibrium value, since $\sum P_{ii} = 3p + \sum \bar{\tau}_{ii}$. In effect, Eq. (3–40) says that $\sum \bar{\tau}_{ii} = 0$. There are exceptions to Eq. (3–40), possibly in the case of sound propagation. For these cases, k is modified to be

$$k = -\tfrac{2}{3}\mu + \kappa, \tag{3-41}$$

where κ is called the bulk viscosity.

Equations (3–36) and (3–37) can be combined into one equation,

$$\bar{\tau}_{ji} = -\mu \left(\frac{\partial U_i}{\partial x_j} + \frac{\partial U_j}{\partial x_i}\right) + (\tfrac{2}{3}\mu - \kappa)\left(\frac{\partial U_l}{\partial x_l}\right)\delta_{ij}, \qquad (3\text{–}42)$$

or, in tensor notation, by using Eqs. (2–24) and (2–36):

$$\bar{\boldsymbol{\tau}} = -\mu[(\boldsymbol{\nabla}\boldsymbol{U}) + (\widetilde{\boldsymbol{\nabla}\boldsymbol{U}})] + (\tfrac{2}{3}\mu - \kappa)(\boldsymbol{\nabla}\cdot\boldsymbol{U})\boldsymbol{\mathsf{U}}, \qquad (3\text{–}43)$$

where $(\widetilde{\boldsymbol{\nabla}\boldsymbol{U}})$ is the transposed tensor of $(\boldsymbol{\nabla}\boldsymbol{U})$ (exchanged rows and columns).

Equation (3–43) can be written in matrix form, where it can be seen that all the expressions of (3–42) can be obtained by selecting the terms in corresponding locations in the tensors. It should be remembered that the tensor equation (3–43) is a shorthand notation for the nine components of the viscous tensor (3–42).

When the viscosity expression (3–43) is substituted into the equation of motion, the resulting equation is the most general form of the Navier-Stokes equation. Generally, the bulk viscosity is assumed to be zero. Bird et al. [1] have pointed out that in a dilute gas, the bulk viscosity is zero if the gas has no internal degree of freedom, but is finite for all other cases. The bulk viscosity is associated with the pure expansion of a fluid, as shown by Eqs. (3–39) through (3–41). Even though finite, the bulk viscosity is quite small, so that in most cases little error would be introduced by its neglect. For the special case of an incompressible fluid, $(\boldsymbol{\nabla}\cdot\boldsymbol{U}) = 0$, and the term including the bulk viscosity is zero. A discussion of bulk viscosity has been given by Karim and Rosenhead [6].

To complete our discussion we include an analysis of mass and heat transfer. The equation for heat transfer, analogous to (3–31), is

$$\boldsymbol{q} = -(k/\rho c_p)\boldsymbol{\nabla}\rho c_p T, \qquad (3\text{–}44)$$

where $k/\rho c_p$ is the thermal diffusivity and $\rho c_p T$ is the heat concentration. Expressed as the more common Fourier law, it is

$$\boldsymbol{q} = -k\,\boldsymbol{\nabla}T. \qquad (3\text{–}45)$$

For mass transfer, the analogous equation is Fick's first law,

$$\boldsymbol{J}_s = -D_s\boldsymbol{\nabla}\rho_s, \qquad (3\text{–}46)$$

where D_s is the mass diffusivity, and ρ_s is the mass concentration.

In the next chapter, the Navier-Stokes equation of motion will be obtained from the combination of Eq. (3–43) (relation between flux, gradient, and transport coefficient) with the equation of motion (in terms of flux).

PROBLEMS

3–1. Demonstrate that $\mathsf{P}_{xz} = \mathsf{P}_{zx}$.

3–2. Write the vector equation (3–24) in cartesian coordinates.

3–3. Write the tensor equation (3–43) in cartesian matrix form. Show that this is identical to Eq. (3–42) for both a diagonal and a nondiagonal term.

REFERENCES

1. R. B. Bird, C. F. Curtiss, and J. O. Hirschfelder, *Chem. Eng. Progr. Symposium Series* No. 16 **51**, 69–85 (1955).
2. R. B. Bird, W. E. Stewart, and E. N. Lightfoot, *Transport Phenomena*, John Wiley & Sons, New York (1960).
3. R. T. Cox, *Statistical Mechanics of Irreversible Changes*, pp. 10, 29–32, Johns Hopkins, Baltimore (1955).
4. J. O. Hirschfelder, C. F. Curtiss, and R. B. Bird, *Molecular Theory of Gases and Liquids*, John Wiley & Sons, New York (1954).
5. R. P. Hughes, *Ind. Eng. Chem.* **49**, 947 (1957).
6. S. M. Karim and L. Rosenhead, *Revs. Modern Phys.* **24**, 108 (1952).
7. J. G. Knudsen and D. L. Katz, *Fluid Dynamics and Heat Transfer*, McGraw-Hill, New York (1958).
8. L. D. Landau and E. M. Lifshitz, *Fluid Mechanics*, Addison-Wesley, Reading, Mass. (1959).
9. S. R. de Groot, *Thermodynamics of Irreversible Processes*, Interscience, Amsterdam (1951).
10. K. G. Denbigh, *Thermodynamics of Irreversible Processes*, C. C. Thomas, Springfield, Ill. (1955).
11. D. ter Haar, *Elements of Statistical Mechanics*, Rinehart, New York (1954).

CHAPTER 4

THE NAVIER-STOKES EQUATION

The Navier-Stokes equation [5] is used in the solution of many of the problems of fluid flow. This chapter will give the derivation of the equation from the equations of change, and the relations between the fluxes, gradients, and transport coefficients. Specifically, the equation comes from the combination of the viscous-flux equation and the equation of motion.

4-1 DERIVATION

The equation of motion (3–25) can be modified by splitting the pressure tensor into a static pressure term and a viscous tensor term according to Eq. (3–8), so that

$$\rho(DU/Dt) = -\nabla p - (\nabla \cdot \bar{\bar{\tau}}) + \sum_s \rho_s F_s. \tag{4-1}$$

When the bulk viscosity is assumed to be zero, the viscous tensor (3–43) reduces to

$$\bar{\bar{\tau}} = -\mu[(\nabla U) + (\widetilde{\nabla U})] + \tfrac{2}{3}\mu(\nabla \cdot U)\mathsf{U}. \tag{4-2}$$

Combining Eqs. (4–1) and (4–2), we obtain the Navier-Stokes equation for compressible fluids:

$$\rho(DU/Dt) = -\nabla p + \mu(\nabla \cdot (\nabla U)) + \mu(\nabla \cdot (\widetilde{\nabla U})) - \tfrac{2}{3}\mu\nabla(\nabla \cdot U) + \sum_s \rho_s F_s, \tag{4-3}$$

where U was eliminated by use of Eq. (2–44).

The Navier-Stokes equation (4–3) is not in the most familiar form; however, to reduce it, a vector identity is needed:

$$(\nabla \cdot (\widetilde{\nabla U})) = \nabla(\nabla \cdot U). \tag{4-4}$$

This identity can be proved by recourse to cartesian coordinates, and is shown in the next example.

EXAMPLE 4–1. Prove the identity $(\nabla \cdot (\widetilde{\nabla U})) = \nabla(\nabla \cdot U)$ (Eq. 4–4).

Answer. For the term on the left,

$$(\nabla U) = \begin{pmatrix} \partial U_x/\partial x & \partial U_y/\partial x & \partial U_z/\partial x \\ \partial U_x/\partial y & \partial U_y/\partial y & \partial U_z/\partial y \\ \partial U_x/\partial z & \partial U_y/\partial z & \partial U_z/\partial z \end{pmatrix}.$$

Therefore,

$$(\widetilde{\nabla U}) = \begin{pmatrix} \partial U_x/\partial x & \partial U_x/\partial y & \partial U_x/\partial z \\ \partial U_y/\partial x & \partial U_y/\partial y & \partial U_y/\partial z \\ \partial U_z/\partial x & \partial U_z/\partial y & \partial U_z/\partial z \end{pmatrix}.$$

According to Eq. (2–40) or (2–42),

$$(\nabla \cdot (\widetilde{\nabla U})) = \begin{pmatrix} \partial^2 U_x/\partial x^2 + \partial^2 U_y/\partial x\, \partial y + \partial^2 U_z/\partial x\, \partial z \\ \partial^2 U_x/\partial y\, \partial x + \partial^2 U_y/\partial y^2 + \partial^2 U_z/\partial y\, \partial z \\ \partial^2 U_x/\partial z\, \partial x + \partial^2 U_y/\partial z\, \partial y + \partial^2 U_z/\partial z^2 \end{pmatrix}.$$

For the term on the right,

$$(\nabla \cdot U) = \frac{\partial U_x}{\partial x} + \frac{\partial U_y}{\partial y} + \frac{\partial U_z}{\partial z},$$

and

$$\nabla(\nabla \cdot U) = i \frac{\partial(\nabla \cdot U)}{\partial x} + j \frac{\partial(\nabla \cdot U)}{\partial y} + k \frac{\partial(\nabla \cdot U)}{\partial z}.$$

Therefore,

$$\nabla(\nabla \cdot U) = \begin{pmatrix} \partial^2 U_x/\partial x^2 + \partial^2 U_y/\partial x\, \partial y + \partial^2 U_z/\partial x\, \partial z \\ \partial^2 U_x/\partial y\, \partial x + \partial^2 U_y/\partial y^2 + \partial^2 U_z/\partial y\, \partial z \\ \partial^2 U_x/\partial z\, \partial x + \partial^2 U_y/\partial z\, \partial y + \partial^2 U_z/\partial z^2 \end{pmatrix},$$

which is the same as the term on the left.

Another identity already noted is the Laplacian operator

$$(\nabla \cdot \nabla) = \nabla^2. \tag{4-5}$$

Combining the equalities of Eqs. (4–4) and (4–5) with Eq. (4–3) gives the more familiar form

$$\rho(DU/Dt) = -\nabla p + \tfrac{1}{3}\mu \nabla(\nabla \cdot U) + \mu \nabla^2 U + \sum_s \rho_s F_s. \tag{4-6}$$

EXAMPLE 4–2. Write the x-component equation for Eq. (4–6).

Answer.

$$\rho \frac{DU_x}{Dt} = -\frac{\partial p}{\partial x} + \tfrac{1}{3}\mu \frac{\partial(\nabla \cdot U)}{\partial x} + \mu \nabla^2 U_x + \sum_s \rho_s F_{sx},$$

where

$$(\nabla \cdot U) = \frac{\partial U_x}{\partial x} + \frac{\partial U_y}{\partial y} + \frac{\partial U_z}{\partial z},$$

and

$$\nabla^2 U_x = \frac{\partial^2 U_x}{\partial x^2} + \frac{\partial^2 U_x}{\partial y^2} + \frac{\partial^2 U_x}{\partial z^2}.$$

The above example illustrates the ease of conversion from the vector equation to cartesian coordinates. The same general procedure can be used to obtain other orthogonal coordinate systems; however, when curvilinear coordinates are considered, there are additional terms associated with the rotation of the new system when compared with rectangular coordinates. As noted in the section on transformations in Chapter 2, the additional terms arise from both the Laplacian operator and the substantial derivative when a vector is transformed to curvilinear coordinates. All the terms can be obtained by further rearrangement of Eq. (4–6) into a form that is easily reduced to the various coordinate systems. We can retain our definition of the Laplacian operator and the substantial derivative (Eqs. 2–65 and 2–68) and identify the additional terms.

A vector identity easily proved [1] is

$$\nabla(A \cdot B) = (A \times (\nabla \times B)) + (B \times (\nabla \times A)) + (A \cdot (\nabla B)) + (B \cdot (\nabla A)). \quad (4\text{--}7)$$

For $A \equiv B$, this gives

$$\nabla(A \cdot A) = 2(A \times (\nabla \times A)) + 2(A \cdot (\nabla A)). \quad (4\text{--}8)$$

Further,

$$(A \cdot (\nabla A)) = (A \cdot \nabla)A. \quad (4\text{--}9)$$

Thus, Eq. (4–8) becomes

$$\nabla(A \cdot A) = 2(A \times (\nabla \times A)) + 2(A \cdot \nabla)A, \quad (4\text{--}10)$$

which is identical to the form cited as Eq. (2–29). The last term of this equation is of the same form as the nonlinear term in the substantial derivative and will be substituted for that term.

Another vector identity [1] which can be easily proved in a manner similar to that used in Example 4–1 (Problem 2–9), is

$$(\nabla \times (\nabla \times A)) = \nabla(\nabla \cdot A) - (\nabla \cdot (\nabla A)) = \nabla(\nabla \cdot A) - \nabla^2 A. \quad (4\text{--}11)$$

The last term in this equation is the Laplacian operator on a vector, and can be used to replace that term in the equation of motion.

Substituting Eqs. (4–10) and (4–11) into Eq. (4–6), we get a completely general form:

$$(\partial U/\partial t) - (U \times (\nabla \times U)) + \tfrac{1}{2}\nabla(U \cdot U) = -(1/\rho)\nabla p + \tfrac{1}{3}\nu\nabla(\nabla \cdot U)$$
$$+ \nu[\nabla(\nabla \cdot U) - (\nabla \times (\nabla \times U))] + (1/\rho)\sum_s \rho_s F_s. \quad (4\text{--}12)$$

The term $(\nabla \times U)$ appears often and is called the vorticity vector:

$$\Omega = (\nabla \times U). \quad (4\text{--}13)$$

Further, the force term can often be expressed as a gradient of a scalar potential; that is,

$$(1/\rho)\sum_s \rho_s F_s = -\nabla \phi = -\text{grad } \phi. \quad (4\text{--}14)$$

Using Eqs. (4–13) and (4–14), we can write Eq. (4–12) as

$$(\partial U/\partial t) - (U \times \Omega) = -\nabla(p/\rho + \tfrac{1}{2}U^2 + \phi) + \tfrac{1}{3}\nu\nabla(\nabla \cdot U)$$
$$+ \nu[\nabla(\nabla \cdot U) - (\nabla \times \Omega)]. \quad (4\text{–}15)$$

EXAMPLE 4–3. Obtain the r-equation for cylindrical coordinates from Eq. (4–12).

Answer. Consider Eq. (4–12) term by term:

$$(\partial U/\partial t)_r = (\partial U_r/\partial t), \quad (\text{i})$$

$$(U \times (\nabla \times U))_r = U_\theta(\nabla \times U)_z - U_z(\nabla \times U)_\theta,$$

which is obtained from Eq. (2–14). The curl terms can be evaluated from Eq. (2–66) or (2–72), to give

$$(U \times (\nabla \times U))_r = U_\theta\left(\frac{1}{r}\frac{\partial r U_\theta}{\partial r} - \frac{1}{r}\frac{\partial U_r}{\partial \theta}\right) - U_z\left(\frac{\partial U_r}{\partial z} - \frac{\partial U_z}{\partial r}\right)$$

$$= \frac{U_\theta}{r}U_\theta + U_\theta\frac{\partial U_\theta}{\partial r} - \frac{U_\theta}{r}\frac{\partial U_r}{\partial \theta} - U_z\frac{\partial U_r}{\partial z} + U_z\frac{\partial U_z}{\partial r}. \quad (\text{ii})$$

Equation (2–4) gives

$$(U \cdot U) = U_r^2 + U_\theta^2 + U_z^2.$$

Equation (2–63) or (2–70) can be used to obtain the form for the third term:

$$\tfrac{1}{2}(\nabla(U \cdot U))_r = \tfrac{1}{2}\frac{\partial(U \cdot U)}{\partial r}$$

$$= \tfrac{1}{2}\frac{\partial(U_r^2 + U_\theta^2 + U_z^2)}{\partial r}$$

$$= U_r\frac{\partial U_r}{\partial r} + U_\theta\frac{\partial U_\theta}{\partial r} + U_z\frac{\partial U_z}{\partial r}. \quad (\text{iii})$$

This completes the terms on the left side of Eq. (4–12); these can be combined (i through iii) to give

$$\frac{\partial U_r}{\partial t} + U_r\frac{\partial U_r}{\partial r} + \frac{U_\theta}{r}\frac{\partial U_r}{\partial \theta} + U_z\frac{\partial U_r}{\partial z} - \frac{U_\theta^2}{r}. \quad (\text{iv})$$

This can be combined with Eq. (2–68) or (2–74) to give

$$\frac{DU_r}{Dt} - \frac{U_\theta^2}{r}. \quad (\text{v})$$

For the first term on the right, one obtains from Eq. (2–63) or (2–70),

$$\frac{1}{\rho}(\nabla p)_r = \frac{1}{\rho}\frac{\partial p}{\partial r}. \quad (\text{vi})$$

DERIVATION

Similarly, for the second term, one obtains

$$\tfrac{1}{3}\nu(\nabla(\nabla\cdot U))_r = \tfrac{1}{3}\nu(\partial(\nabla\cdot U)/\partial r), \qquad \text{(vii)}$$

where $(\nabla\cdot U)$ is given by Eq. (2–64) or (2–71); that is,

$$(\nabla\cdot U) = \frac{1}{r}\frac{\partial r U_r}{\partial r} + \frac{1}{r}\frac{\partial U_\theta}{\partial \theta} + \frac{\partial U_z}{\partial z}. \qquad \text{(viii)}$$

The third term is more complex and gives rise to the additional terms. The first part of this term is identical with Eqs. (vii) and (viii) except for the coefficient. Using these, one obtains

$$(\nabla(\nabla\cdot U))_r = \frac{\partial}{\partial r}\left(\frac{1}{r}\frac{\partial r U_r}{\partial r}\right) + \frac{\partial}{\partial r}\left(\frac{1}{r}\frac{\partial U_\theta}{\partial \theta}\right) + \frac{\partial^2 U_z}{\partial r\,\partial z}$$

$$= \frac{\partial^2 U_r}{\partial r^2} + \frac{1}{r}\frac{\partial U_r}{\partial r} - \frac{U_r}{r^2} + \frac{1}{r}\frac{\partial^2 U_\theta}{\partial r\,\partial \theta} - \frac{1}{r^2}\frac{\partial U_\theta}{\partial \theta} + \frac{\partial^2 U_z}{\partial r\,\partial z}. \qquad \text{(ix)}$$

The second part of this term can be evaluated from Eq. (2–66) or (2–72) as follows:

$$(\nabla\times(\nabla\times U))_r = \frac{1}{r}\frac{\partial(\nabla\times U)_z}{\partial \theta} - \frac{\partial(\nabla\times U)_\theta}{\partial z}$$

$$= \frac{1}{r}\frac{\partial}{\partial \theta}\left(\frac{1}{r}\frac{\partial r U_\theta}{\partial r} - \frac{1}{r}\frac{\partial U_r}{\partial \theta}\right) - \frac{\partial}{\partial z}\left(\frac{\partial U_r}{\partial z} - \frac{\partial U_z}{\partial r}\right)$$

$$= \frac{1}{r}\frac{\partial^2 U_\theta}{\partial r\,\partial \theta} + \frac{1}{r^2}\frac{\partial U_\theta}{\partial \theta} - \frac{1}{r^2}\frac{\partial^2 U_r}{\partial \theta^2} - \frac{\partial^2 U_r}{\partial z^2} + \frac{\partial^2 U_z}{\partial r\,\partial z}. \qquad \text{(x)}$$

Equations (ix) and (x) can be combined to give the expansion for the entire term:

$$\frac{\partial^2 U_r}{\partial r^2} + \frac{1}{r}\frac{\partial U_r}{\partial r} + \frac{1}{r^2}\frac{\partial^2 U_r}{\partial \theta^2} + \frac{\partial^2 U_r}{\partial z^2} - \frac{2}{r^2}\frac{\partial U_\theta}{\partial \theta} - \frac{U_r}{r^2}, \qquad \text{(xi)}$$

which, when combined with Eq. (2–65) or (2–73), gives

$$\nabla^2 U_r - \frac{2}{r^2}\frac{\partial U_\theta}{\partial \theta} - \frac{U_r}{r^2}. \qquad \text{(xii)}$$

The last term in Eq. (4–12) becomes

$$\frac{1}{\rho}\sum_s \rho_s F_{sr}. \qquad \text{(xiii)}$$

The combination of Eqs. (v), (vi), (vii), (xii), and (xiii) gives the final desired form:

$$\frac{DU_r}{Dt} - \frac{U_\theta^2}{r} = -\frac{1}{\rho}\frac{\partial p}{\partial r} + \tfrac{1}{3}\nu\frac{\partial(\nabla\cdot U)}{\partial r}$$

$$+ \nu\left(\nabla^2 U_r - \frac{2}{r^2}\frac{\partial U_\theta}{\partial \theta} - \frac{U_r}{r^2}\right) + \frac{1}{\rho}\sum_s \rho_s F_{sr}. \qquad \text{(xiv)}$$

In the foregoing example, the substantial derivative and the Laplacian operator, as defined by Eqs. (2–65) and (2–68), were recovered, and the final answer is expressed in terms of these. This can be done with the general orthogonal coordinates which involve the various h_i's; however, there is no simple general form for the resulting terms and so each additional term is listed. The general equation for the ith component is given below and an example of its use follows. To obtain the jth and kth equations, a cyclic rotation of the indexes is necessary; i.e., for the jth component, $i = j$, $j = k$, and $k = i$. The equation is

$$\frac{DU_i}{Dt} + \mathfrak{U}_i = -\frac{1}{\rho}\frac{1}{h_i}\frac{\partial p}{\partial L_i} + \tfrac{1}{3}\nu\frac{1}{h_i}\frac{\partial(\nabla \cdot U)}{\partial L_i} + \nu(\nabla^2 U_i + \mathfrak{B}_i) + \frac{1}{\rho}\sum_s \rho_s F_{si}, \tag{4-16}$$

where

$$\mathfrak{U}_i = \sum_j \frac{U_j}{h_j}\left(\frac{U_i}{h_i}\frac{\partial h_i}{\partial L_i} - \frac{U_j}{h_i}\frac{\partial h_j}{\partial L_i}\right)$$

$$= -\frac{U_j^2}{h_i h_j}\frac{\partial h_j}{\partial L_i} - \frac{U_k^2}{h_i h_k}\frac{\partial h_k}{\partial L_i} + \frac{U_i U_j}{h_i h_j}\frac{\partial h_i}{\partial L_j} + \frac{U_i U_k}{h_i h_k}\frac{\partial h_i}{\partial L_k}, \qquad i \neq j$$

and

$$\mathfrak{B}_i = -\frac{U_j}{h_i h_j^2 h_k}\frac{\partial h_j}{\partial L_i}\frac{\partial h_k}{\partial L_j} + \frac{U_i}{h_i h_j^2 h_k}\frac{\partial h_i}{\partial L_j}\frac{\partial h_k}{\partial L_j} + \frac{U_i}{h_i h_j h_k^2}\frac{\partial h_i}{\partial L_k}\frac{\partial h_j}{\partial L_k} - \frac{U_k}{h_i h_j h_k^2}\frac{\partial h_k}{\partial L_i}\frac{\partial h_j}{\partial L_k}$$

$$-\frac{U_j}{h_j}\frac{\partial}{\partial L_j}\left(\frac{1}{h_i h_j}\frac{\partial h_j}{\partial L_i}\right) + \frac{U_i}{h_j}\frac{\partial}{\partial L_j}\left(\frac{1}{h_i h_j}\frac{\partial h_i}{\partial L_j}\right) + \frac{U_i}{h_k}\frac{\partial}{\partial L_k}\left(\frac{1}{h_i h_k}\frac{\partial h_i}{\partial L_k}\right)$$

$$-\frac{U_k}{h_k}\frac{\partial}{\partial L_k}\left(\frac{1}{h_i h_k}\frac{\partial h_k}{\partial L_i}\right) + \frac{U_i}{h_i}\frac{\partial}{\partial L_i}\left(\frac{1}{h_i h_k}\frac{\partial h_k}{\partial L_i}\right) + \frac{U_i}{h_i}\frac{\partial}{\partial L_i}\left(\frac{1}{h_i h_j}\frac{\partial h_j}{\partial L_i}\right)$$

$$+\frac{U_j}{h_i}\frac{\partial}{\partial L_i}\left(\frac{1}{h_i h_j}\frac{\partial h_i}{\partial L_j}\right) + \frac{U_j}{h_i}\frac{\partial}{\partial L_i}\left(\frac{1}{h_j h_k}\frac{\partial h_k}{\partial L_j}\right) + \frac{U_k}{h_i}\frac{\partial}{\partial L_i}\left(\frac{1}{h_j h_k}\frac{\partial h_j}{\partial L_k}\right)$$

$$+\frac{U_k}{h_i}\frac{\partial}{\partial L_i}\left(\frac{1}{h_i h_k}\frac{\partial h_i}{\partial L_k}\right) - \frac{1}{h_i h_j^2}\frac{\partial h_j}{\partial L_i}\frac{\partial U_j}{\partial L_j} - \frac{1}{h_i h_k^2}\frac{\partial h_k}{\partial L_i}\frac{\partial U_k}{\partial L_k} + \frac{1}{h_i^2 h_j}\frac{\partial h_i}{\partial L_j}\frac{\partial U_j}{\partial L_i}$$

$$+\frac{1}{h_i^2 h_k}\frac{\partial h_i}{\partial L_k}\frac{\partial U_k}{\partial L_i} + \frac{1}{h_i}\frac{\partial U_j}{\partial L_j}\frac{\partial}{\partial L_i}\frac{1}{h_j} - \frac{1}{h_k}\frac{\partial U_k}{\partial L_i}\frac{\partial}{\partial L_k}\frac{1}{h_i} - \frac{1}{h_j}\frac{\partial U_j}{\partial L_i}\frac{\partial}{\partial L_j}\frac{1}{h_i}$$

$$+\frac{1}{h_i}\frac{\partial U_k}{\partial L_k}\frac{\partial}{\partial L_i}\frac{1}{h_k} + \frac{1}{h_i}\frac{\partial U_i}{\partial L_i}\frac{\partial}{\partial L_i}\frac{1}{h_i}. \qquad i \neq j$$

Although this is a rather impressive equation, its use is easy, as illustrated in the next example.

EXAMPLE 4–4. Obtain the θ-equation for cylindrical coordinates from Eq. (4–16).

Answer. The three values of the h_i's are

$$h_r = 1, \qquad h_\theta = r, \qquad h_z = 1.$$

The problem is to obtain the θ-equation, thus each subscript must be rotated one value, that is,

i	will become	j	or	θ,		
j	will become	k	or	z,		
k	will become	i	or	r,		
h_i	will become	h_j	or	h_θ	or	r,
h_j	will become	h_k	or	h_z	or	1,
h_k	will become	h_i	or	h_r	or	1.

From the above, *for the new subscripts,*

$$\frac{\partial h_j}{\partial L_i} = \frac{\partial r}{\partial r} = 1.$$

All other partials of h with respect to any variable are zero. In terms of Eq. (4–16) *as it is written,* this means that the only terms to be considered are those with

$$\frac{\partial h_i}{\partial L_k} \quad \text{or} \quad \frac{\partial}{\partial L_k}\frac{1}{h_i}.$$

Inspection of Eq. (4–16) shows that the only terms to be considered are the fourth in $\mathfrak{A}$, and the seventh, fourteenth, eighteenth and twentieth in $\mathfrak{B}$. In terms of the *new subscripts,* these are

$$\mathfrak{A} = \frac{U_j U_i}{h_j h_i}\frac{\partial h_j}{\partial L_i} = \frac{U_\theta U_r}{r},$$

$$\mathfrak{B} = \frac{U_j}{h_i}\frac{\partial}{\partial L_i}\left(\frac{1}{h_j h_i}\frac{\partial h_j}{\partial L_i}\right) + \frac{U_i}{h_j}\frac{\partial}{\partial L_j}\left(\frac{1}{h_j h_i}\frac{\partial h_j}{\partial L_i}\right) + \frac{1}{h_j^2 h_i}\frac{\partial h_j}{\partial L_i}\frac{\partial U_i}{\partial L_j} - \frac{1}{h_i}\frac{\partial U_i}{\partial L_j}\frac{\partial}{\partial L_i}\frac{1}{h_j}$$

$$= U_\theta\frac{\partial}{\partial r}\frac{1}{r} + \frac{U_r}{r}\frac{\partial}{\partial \theta}\frac{1}{r} + \frac{1}{r^2}\frac{\partial U_r}{\partial \theta} - \frac{\partial U_r}{\partial \theta}\frac{\partial}{\partial r}\frac{1}{r}$$

$$= -\frac{U_\theta}{r^2} + 0 + \frac{1}{r^2}\frac{\partial U_r}{\partial \theta} + \frac{1}{r^2}\frac{\partial U_r}{\partial \theta} = -\frac{U_\theta}{r^2} + \frac{2}{r^2}\frac{\partial U_r}{\partial \theta}.$$

The final equation can be written as

$$\frac{DU_\theta}{Dt} + \frac{U_r U_\theta}{r} = -\frac{1}{r\rho}\frac{\partial p}{\partial \theta} + \tfrac{1}{3}\nu\frac{1}{r}\frac{\partial(\nabla\cdot U)}{\partial \theta}$$

$$+ \nu\left(\nabla^2 U_\theta - \frac{U_\theta}{r^2} + \frac{2}{r^2}\frac{\partial U_r}{\partial \theta}\right) + \frac{1}{\rho}\sum_s \rho_s F_{s\theta}.$$

4-2 SUMMARY

For cartesian, cylindrical, and spherical coordinates, the Navier-Stokes equation can be summarized as follows:

$$\frac{DU_i}{Dt} + \mathfrak{A} = -\frac{1}{\rho}\frac{1}{h_i}\frac{\partial p}{\partial L_i}$$

$$+ \tfrac{1}{3}\nu\frac{1}{h_i}\frac{\partial(\nabla\cdot U)}{\partial L_i} + \nu(\nabla^2 U_i + \mathfrak{B}) + \frac{1}{\rho}\sum_s \rho_s F_{si}. \qquad (4\text{--}16)$$

Cartesian coordinates:

For $i = i = x$, $\quad h_i = 1$, $\quad L_i = x$, $\quad \mathfrak{A} = 0$, $\quad$ and $\quad \mathfrak{B} = 0$.
For $i = j = y$, $\quad h_j = 1$, $\quad L_j = y$, $\quad \mathfrak{A} = 0$, $\quad$ and $\quad \mathfrak{B} = 0$.
For $i = k = z$, $\quad h_k = 1$, $\quad L_k = z$, $\quad \mathfrak{A} = 0$, $\quad$ and $\quad \mathfrak{B} = 0$.

Cylindrical coordinates:

For $i = i = r$, $\quad h_i = 1$, $\quad L_i = r$,
$$\mathfrak{A} = -\frac{U_\theta^2}{r}, \quad \text{and} \quad \mathfrak{B} = -\frac{2}{r^2}\frac{\partial U_\theta}{\partial \theta} - \frac{U_r}{r^2}.$$

For $i = j = \theta$, $\quad h_j = r$, $\quad L_j = \theta$,
$$\mathfrak{A} = \frac{U_r U_\theta}{r}, \quad \text{and} \quad \mathfrak{B} = \frac{2}{r^2}\frac{\partial U_r}{\partial \theta} - \frac{U_\theta}{r^2}.$$

For $i = k = z$, $\quad h_k = 1$, $\quad L_k = z$, $\quad \mathfrak{A} = 0$, $\quad$ and $\quad \mathfrak{B} = 0$.

Spherical coordinates:

For $i = i = r$, $\quad h_i = 1$, $\quad L_i = r$,
$$\mathfrak{A} = -\frac{U_\theta^2}{r} - \frac{U_\phi^2}{r},$$

and
$$\mathfrak{B} = -\frac{2 U_r}{r^2} - \frac{2}{r^2}\frac{\partial U_\theta}{\partial \theta} - \frac{2 U_\theta}{r^2}\cot\theta - \frac{2}{r^2 \sin\theta}\frac{\partial U_\phi}{\partial \phi}.$$

For $i = j = \theta$, $\quad h_j = r$, $\quad L_j = \theta$,
$$\mathfrak{A} = \frac{U_r U_\theta}{r} - \frac{U_\phi^2}{r}\cot\theta$$

and
$$\mathfrak{B} = \frac{2}{r^2}\frac{\partial U_r}{\partial \theta} - \frac{U_\theta}{r^2 \sin^2\theta} - \frac{2\cos\theta}{r^2 \sin^2\theta}\frac{\partial U_\phi}{\partial \phi}.$$

For $i = k = \phi$, $\quad h_k = r\sin\theta$, $\quad L_k = \phi$, $\quad \mathfrak{A} = \frac{U_r U_\phi}{r} + \frac{U_\theta U_\phi}{r}\cot\theta$,

and
$$\mathfrak{B} = -\frac{U_\phi}{r^2 \sin^2\theta} + \frac{2}{r^2 \sin\theta}\frac{\partial U_r}{\partial \phi} + \frac{2\cos\theta}{r^2 \sin^2\theta}\frac{\partial U_\theta}{\partial \phi}.$$

The expressions for the substantial derivative, $(\boldsymbol{\nabla} \cdot \boldsymbol{U})$, and the Laplacian operator are given, for cartesian coordinates, by Eqs. (2–27), (2–24), and (2–26), respectively; for cylindrical coordinates, by Eqs. (2–74), (2–71), and (2–73), respectively; and for spherical coordinates, by Eqs. (2–80), (2–77), and (2–79), respectively.

In some cases, one is interested in the form of the viscous tensor given by Eq. (4–2). For cylindrical coordinates [6],

$$(\nabla U) = \begin{pmatrix} \dfrac{\partial U_r}{\partial r} & \dfrac{\partial U_\theta}{\partial r} & \dfrac{\partial U_z}{\partial r} \\ \dfrac{1}{r}\dfrac{\partial U_r}{\partial \theta} - \dfrac{U_\theta}{r} & \dfrac{1}{r}\dfrac{\partial U_\theta}{\partial \theta} + \dfrac{U_r}{r} & \dfrac{1}{r}\dfrac{\partial U_z}{\partial \theta} \\ \dfrac{\partial U_r}{\partial z} & \dfrac{\partial U_\theta}{\partial z} & \dfrac{\partial U_z}{\partial z} \end{pmatrix}, \qquad (4\text{--}17)$$

and for spherical coordinates [6],

$$(\nabla U) = \begin{pmatrix} \dfrac{\partial U_r}{\partial r} & \dfrac{\partial U_\theta}{\partial r} & \dfrac{\partial U_\phi}{\partial r} \\ \dfrac{1}{r}\dfrac{\partial U_r}{\partial \theta} - \dfrac{U_\theta}{r} & \dfrac{1}{r}\dfrac{\partial U_\theta}{\partial \theta} + \dfrac{U_r}{r} & \dfrac{1}{r}\dfrac{\partial U_\phi}{\partial \theta} \\ \dfrac{1}{r\sin\theta}\dfrac{\partial U_r}{\partial \phi} - \dfrac{U_\phi}{r} & \dfrac{1}{r\sin\theta}\dfrac{\partial U_\theta}{\partial \phi} - \dfrac{U_\phi}{r}\cot\theta & \dfrac{1}{r\sin\theta}\dfrac{\partial U_\phi}{\partial \phi} + \dfrac{U_r}{r} + \dfrac{U_\theta}{r}\cot\theta \end{pmatrix}.$$

$$(4\text{--}18)$$

In each case, the transposed tensor $(\widetilde{\nabla U})$ is easily found, and Eq. (4–2) evaluated in the two coordinate systems [2, 3, 4]. In addition, the dissipation term $(\bar{\bar{\tau}} : (\nabla U))$ can be obtained by using Eqs. (2–47), (4–17), and (4–18). A complete tabulation in terms of $\bar{\bar{\tau}}$ and combined with Eq. (4–2) can be found in Bird, Stewart, and Lightfoot [3].

The above transformations are restricted to Newtonian systems, since Eq. (4–2) was used. In some instances, one may want to use the equations with rheological laws descriptive of non-Newtonian systems. In such cases, he can repeat the analysis by using the new law for Eq. (4–2). An alternative procedure is to combine Eqs. (4–1) and (4–10) to give the general form of the Navier-Stokes equation in terms of $\bar{\bar{\tau}}$. The components of $(\nabla \cdot \bar{\bar{\tau}})$ for cartesian coordinates can be obtained directly from Eq. (2–42). For curvilinear coordinates, the problem is more complex. A convenient tabulation of the equations can be found in reference 3.

Some general observations about the Navier-Stokes equation will be presented in the next chapter, which will introduce the second part of this text on the application of this basic flow equation. Although the current chapter has primarily emphasized the Navier-Stokes equation, it should be pointed out that the mass (3–22) and energy (3–30) equations can be combined with the flux terms [(3–44) through (3–46)] to give convenient working equations. For temperature in an incompressible flow [$c_p = c_v$ and $(\nabla \cdot U) = 0$] with no external forces, Eqs. (3–30), (3–43), and (3–45) combine to give

$$\rho c_p (DT/Dt) = k\nabla^2 T + \mu[\{(\nabla U) + (\widetilde{\nabla U})\} : (\nabla U)]. \qquad (4\text{--}19)$$

The complex term on the right is the dissipation function for an incompressible flow. For mass transfer in an incompressible flow, in the absence of chemical reaction, Eqs. (3–22) and (3–46) combine to give

$$D\rho_s/Dt = D_s\nabla^2\rho_s, \qquad (4\text{–}20)$$

which is the flow form of Fick's second law of diffusion.

PROBLEMS

4–1. Write out the vector equation (4–3) in matrix form with cartesian coordinate terms. What is the x-component equation for this vector equation?

4–2. Write the z-component equation for Eq. (4–6).

4–3. What is the meaning of each term in Eq. (4–3)?

4–4. What is the meaning of each term in Eq. (4–6)?

4–5. Use Eq. (4–12) to obtain the x-component equation of the Navier-Stokes equation.

4–6. Use Eq. (4–12) to obtain the θ-component equation in cylindrical coordinates for the Navier-Stokes equation.

4–7. Use Eq. (4–15) to obtain the θ-component equation in spherical coordinates for the Navier-Stokes equation.

4–8. Use Eq. (4–16) to obtain the r-component equation in cylindrical coordinates for the Navier-Stokes equation.

4–9. Use Eq. (4–16) to obtain the θ-component equation in spherical coordinates for the Navier-Stokes equation.

REFERENCES

1. C. R. WYLIE, JR., *Advanced Engineering Mathematics*, McGraw-Hill, New York, (1960).
2. R. ARIS, *Vectors, Tensors, and the Basic Equations of Fluid Mechanics*, Prentice-Hall, Englewood Cliffs, N.J. (1962).
3. R. B. BIRD, W. E. STEWART, and E. N. LIGHTFOOT, *Transport Phenomena*, John Wiley & Sons, New York (1960).
4. L. D. LANDAU and E. M. LIFSHITZ, *Fluid Mechanics*, Addison-Wesley, Reading, Mass. (1959).
5. G. G. STOKES, *Trans. Cambridge Phil. Soc.* 8, 287–319 (1845).
6. M. C. WILLIAMS and R. B. BIRD, *Ind. Eng. Chem. Fund.* 3, 42–49 (1964).

PART 2
APPLICATION OF THE BASIC FLOW EQUATION

CHAPTER 5
INTRODUCTION

The Navier-Stokes equation, developed in the preceding chapter, is the starting point for solutions of many fluid flow problems. Before discussing the various possibilities, we will first look carefully at the equation. The first and last terms on the right-hand side of Eq. (4–6) give the rate of change of velocity due to pressure and external forces. These are the same for compressible and incompressible flow, and for frictionless flow. If the fluid is incompressible, the remaining term, $\mu \nabla^2 U$, is the viscosity effect and has the same form as Fick's law of diffusion or conduction. The equation is a nonlinear partial differential equation with no general solution, and may be of higher order, since the viscosity and thermal conductivity can be functions of temperature and thus of velocity. The assumption of an incompressible fluid is not enough to allow a general solution, because the equation is still nonlinear. Generally, in order to obtain a final solution, a number of assumptions will have to be made; these limitations on the validity of the final result must be recognized, so that misuse will not occur.

Some observations are possible about the general characteristics of the Navier-Stokes equation and the means of solving it. One class of solutions can be made by assuming ideal flow, i.e., the irrotational flow of an ideal fluid, which has constant density and zero viscosity. The continuity equation (3–21) becomes

$$(\nabla \cdot U) = 0, \tag{5-1}$$

and the irrotational assumption gives

$$(\nabla \times U) = 0. \tag{5-2}$$

The Navier-Stokes equation (4–12) for steady state and no external forces becomes

$$\tfrac{1}{2}\nabla(U \cdot U) = -(1/\rho)\nabla p. \tag{5-3}$$

It is not easy to solve Eqs. (5–1), (5–2), and (5–3) for the velocity and pressure; however, we can often assume that a velocity potential exists; that is,

$$U_x = \partial \phi/\partial x, \qquad U_y = \partial \phi/\partial y, \qquad U_z = \partial \phi/\partial z,$$

or

$$U = \nabla \phi. \tag{5-4}$$

Equation (5–1) gives

$$(\nabla \cdot U) = (\nabla \cdot \nabla \phi) = \nabla^2 \phi = 0. \tag{5-5}$$

Equation (5-2) is satisfied, since

$$(\nabla \times \nabla \phi) = 0 \tag{5-6}$$

can be proved to be a general vector identity. Furthermore, one can show that

$$\nabla^2 U = \nabla^2(\nabla \phi) = \nabla(\nabla^2 \phi) = 0. \tag{5-7}$$

Thus, for steady state and no external forces, the Navier-Stokes equation (4-12) reduces to (5-3) when Eqs. (5-5) and (5-6) are used. In other words, if a velocity potential exists, it is a solution of the ideal flow problem. Actually it is a solution of the Navier-Stokes equation even if the viscosity is finite; however, since the viscosity term disappears, the solution cannot satisfy the necessary "no slip" boundary condition at the wall. In the flow of viscous fluids, both the tangential and normal velocities at a surface are zero, whereas in the nonviscous case, only the normal velocity component is zero. This suggests the reason why the latter solution is extremely useful and approaches the more exact solutions when removed from a boundary. Birkhoff [1] has stated that the nonviscous solution is valid for low-viscosity fluids up to one-fourth the speed of sound, outside a rather wide belt behind a given obstacle, or outside a very thin "boundary layer" near the surface of the obstacle. In Chapter 6, ideal flow will be treated in detail, and a number of interesting problems will be considered, such as the ideal fluid flow around a corner and over a stationary or rotating cylinder.

If a simple geometry is assumed, exact solutions of the equations can be obtained for a number of steady- and unsteady-state problems. In these solutions, the viscosity is considered to be finite and constant (laminar viscous flow problems). Sample solutions, such as the flow in a pipe, are given in Chapter 7. The exact solutions have the least number of assumptions and offer the best systems for the measurement of viscosity.

Approximation to the Navier-Stokes equation can be made if the Reynolds number of the flow is very low; i.e., if the velocity is small enough to allow the inertial terms to be neglected. With this assumption, the Navier-Stokes equation becomes

$$\rho(\partial U/\partial t) = - \nabla p + \mu \nabla^2 U. \tag{5-8}$$

Equation (5-1) will apply also, since the velocity is so small that the liquid may be considered incompressible. If the divergence of Eq. (5-8) is taken, steady state assumed, and Eq. (5-1) used, we obtain

$$\nabla^2 p = 0. \tag{5-9}$$

Solution of the equations can be obtained for certain geometries, as in the flow over a sphere. These are the problems of very slow motion or Stokes' flow (settling, aerosols, etc.) and are considered in Chapter 8.

Viscous flow is rotational in nature, and this rotation or vorticity is distributed in the flow field. One problem is the determination of this distribution. The

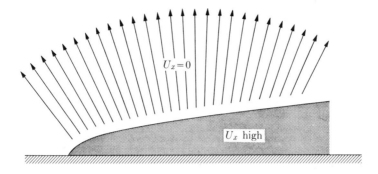

Fig. 5-1. Heat or vorticity transfer for two-dimensional incompressible flow.

equation of motion can be expressed in terms of vorticity transport. This equation is also nonlinear, and contains terms describing the transport of vorticity along the streamlines and deformation of the vortex tubes. For two-dimensional incompressible flow, the deformation terms vanish and the equation reduces to the same form as that for the temperature distribution in a heat-conducting medium. The effect of Reynolds number on vorticity can be visualized by exchanging the Reynolds number for the Peclet number of heat transfer. If the flow is slow, the temperature or vorticity will spread in all directions. If the flow is high, only a thin layer of fluid near the wall will be warmed, or the vorticity will be concentrated in the immediate neighborhood of the surface (see Fig. 5-1). Thus the Navier-Stokes equation gives us a "boundary layer" concept for large Reynolds numbers. Prandtl [4], in 1904, introduced this concept of the boundary layer in which the viscous effects are confined to a thin layer whose thickness tends to zero as the coefficient of viscosity approaches zero. The solution must be divided into two sections. The first, called the *interior limit problem*, is the solution for a nonviscous fluid. The other, the *boundary limit problem*, is the boundary layer solution. The boundary conditions of these two approximate solutions must be related. These important problems are introduced in Chapter 9 with the solution for the flow over a flat plate.

Equations of the nonlinear type, similar to the Navier-Stokes equation, have been studied; for example, Burgers [2] considered the equation

$$D\boldsymbol{U}/Dt = \mu\nabla^2\boldsymbol{U}. \tag{5-10}$$

This has no real flow meaning but, as indicated by Pai [3], it is a mathematical model of hydrodynamics. The solution of an energy equation of the form of (5-10) shows that there are four terms. The first is the total rate of change of kinetic energy; the second is the net flux of kinetic energy out of the system across its boundaries; the third is the rate of work done on the system at its boundaries; and the fourth is the total dissipation of energy by viscosity in the system. Comparison of the solution of (5-10) with the same equation linearized shows that the nonlinear terms account for energy brought to the system through the boundaries. This would be important in turbulence and in shock wave phenomena. Pai has summarized this work by noting that the nonlinear term contributes a wave of appreciable amplitude traveling in the positive x-direction. This wave eventually dissipates, and the nonlinear solution tends to the same form as the linearized solution, though with a smaller amplitude. The nonlinear term steepens the velocity gradient; however, because of the viscous damping effect, no discontinuity occurs.

Chapter 10 covers integral methods of analysis; these are the integrated or overall approaches. Included are the mass, momentum, and energy balances, the Bernoulli equation, and the mechanical energy balance. An important subsection covers the application of the integral momentum balance to the boundary layer, which allows approximate solutions to a number of difficult boundary layer problems.

When the nature of the problem does not allow any of the approaches suggested in Chapters 6 through 10, recourse is usually made to analysis methods, which may provide insight into the relations of the controlling variables. This inspection or dimensional approach is covered in Chapter 11. The final chapter of this part, Chapter 12, covers one-dimensional compressible flow problems, an exact solution of the Navier-Stokes equation, with the addition of an equation of state (the perfect gas law).

Laminar fluid flow theory is one of the most highly developed subject areas in fluid dynamics, and illustrates the usefulness of a fundamental theory as an aid in understanding these problems. However, in the process industries, laminar flow is not as commonly experienced as is turbulent flow; thus the number of practical applications is somewhat limited.

GENERAL NOTATION FOR PART 2

Where the notation of Part 2 is the same as that for Part 1, the symbols have not been repeated. The notation for Part 1 appears on p. 4.

A, A', B	constants
$\dot{A}$	time rate of change of angular momentum
a, b, c	points or constants
b	width of channel, perimeter of pipe
c	velocity of sound
$C_1, C_2, \ldots, C_n$	constants
c_p, c_v	specific heats at constant pressure and volume
C_D	drag coefficient
D	drag
D_s	total Stokes drag
d_0	diameter
f	friction factor (Fanning), fraction
f, F	functions

F	force
g	gravitational force
g_c	gravitational constant
G	Gibbs energy
H	enthalpy
I_n, J_n, Y_n	Bessel functions
k, K, K_c, K'	constants
L	length, function
l	mean free path
L_e	entrance length
M	Mach number
M_w	molecular weight
m	defined by Eq. (9–46)
n	constant, power
$N_{Re}, N_{Re'}, N_{Re''}, N_{Re,L}$	Reynolds numbers
N_{xx}	various dimensionless numbers defined in text
$\dot{P}$	time rate of change of momentum
Q	heat flow term, volumetric flow rate
R	universal gas constant
r_0	pipe radius
r	radius
s	entropy
T	torque, temperature
u, v	functions
v	specific volume
V_0	velocity at edge of boundary layer
V_x	relative velocity defined in Eq. (6–41)
w	mass flow rate
W_s	work term
X	net external force
Y_n	Bessel function

Greek letters

α	constant, kinetic energy correction term, angle
β	constant, momentum correction term, coefficient of sliding friction
γ	ratio of specific heats
Γ	circulation
δ	boundary layer thickness
δ^*	displacement thickness
ϵ	ratio of stagnation to wall temperature
ξ	defined by Eqs. (6–7) and (9–38)
η	defined by Eqs. (6–7), (7–8), and (7–41)
λ	defined by Eq. (9–40)

ζ	defined by Eq. (6–7)
π	3.1416 ...
σ	surface tension
ϕ	velocity potential
Φ	potential energy
ψ	stream function
θ	momentum thickness
ω	scalar potential, angular velocity

Subscripts

∞	infinity, far removed from the object or boundary layer
ave	average
ext	external
max	maximum
a	adiabatic
b	bounded system, back
c	corrected
d	discontinuous
e	exit
g	gravity
i	isothermal
l	liquid
o	outer, wall, reference, other, stagnation
s	solid, Stokes
t	terminal
vp	vapor pressure
w	wall
μ	viscous
ρ	inertial
σ	surface

Other symbols

$'$ $''$ $'''$	derivatives
$'$	transformed
$*$	Mach unity condition
$\cdot$	time rate of change (overmark)

REFERENCES

1. G. BIRKHOFF, *Hydrodynamics*, Dover Publications, New York (1955).
2. J. BURGERS, "A Mathematical Model Illustrating the Theory of Turbulence," *Advances in Applied Mechanics*, Vol. I, pp. 171–199, Academic Press, New York (1948).
3. SHIH-I PAI, *Viscous Flow Theory—Laminar Flow*, D. van Nostrand, Princeton, (1956).
4. L. PRANDTL, *Verhandlg. d. III Intern. Mathe. Kongr.*, Heidelberg (1904); *NACA TM* 452 (1928) translation.

CHAPTER 6
IDEAL FLOW

Ideal flow solutions are those obtained from the Navier-Stokes equation when the viscosity is assumed to be zero. As pointed out in the preceding chapter, such a concept will be valid where the viscosity is low, and where the field of interest is the main fluid stream, not the interactions of the stream with boundaries. In ideal flow the viscous effects are small when compared to the effects of inertial and external forces. One might inquire as to how close one can approach a boundary and still satisfy the basic assumptions of ideal flow. This question was considered by Prandtl, and as a result, he developed the boundary layer theory. So long as the boundary layer remains in contact with the surface, the flow pattern will be different from the ideal solution only inside a thin region near the surface; however, if the boundary layer breaks away, the ideal flow solution will not be valid. These problems of separation will be considered during the discussion of viscous flow; for now, we will consider ideal flow and remember its limitations near a boundary.

6-1 THE EULER EQUATION OF MOTION

The Euler equation of motion can be obtained from the Navier-Stokes equation by setting the viscosity equal to zero; that is,

$$\frac{D\boldsymbol{U}}{Dt} = \frac{\partial \boldsymbol{U}}{\partial t} + (\boldsymbol{U} \cdot \boldsymbol{\nabla})\boldsymbol{U} = -\frac{1}{\rho}\boldsymbol{\nabla}p + \frac{1}{\rho}\sum_s \rho_s \boldsymbol{F}_s. \tag{6-1}$$

The x-component of the velocity would be

$$\begin{aligned}\frac{DU_x}{Dt} &= \frac{\partial U_x}{\partial t} + U_x\frac{\partial U_x}{\partial x} + U_y\frac{\partial U_x}{\partial y} + U_z\frac{\partial U_x}{\partial z} \\ &= -\frac{1}{\rho}\frac{\partial p}{\partial x} + \frac{1}{\rho}\sum_s \rho_s F_{sx}.\end{aligned} \tag{6-2}$$

In most problems the only external force is gravity, which will be the same for all species:

$$\frac{1}{\rho}\sum_s \rho_s \boldsymbol{F}_s = \boldsymbol{g}. \tag{6-3}$$

There are five unknowns to be determined: the three components of the velocity, the pressure, and the density. There are five equations available: the three equations of (6–1), the equation of continuity, and an equation of state giving a relation between the pressure and the density. Consequently, with a given set of boundary conditions, the set of equations can be solved in principle. This is difficult, but under certain conditions simplifications are possible.

6-2 THE VELOCITY POTENTIAL

The velocity potential ϕ has already been considered in Eq. (5–4). It is defined as

$$U = \nabla \phi, \tag{6-4}$$

or

$$U_x = \partial\phi/\partial x, \qquad U_y = \partial\phi/\partial y, \qquad U_z = \partial\phi/\partial z.$$

The velocity potential can be pictured as analogous to an electric potential or voltage. It provides the driving force for the transport of fluid particles along a streamline, just as voltage provides the driving force for current.

As indicated in the preceding chapter (Eq. 5–6), the curl of the velocity is zero when Eq. (6–4) is used; that is,

$$\text{curl } U = (\nabla \times U) = (\nabla \times \nabla \phi) = 0. \tag{6-5}$$

This can be proved by taking the second partials of Eq. (6–4), and noting that each term of Eq. (6–5) is zero; that is,

$$\partial U_y/\partial z = \partial^2 \phi/\partial y\,\partial z, \qquad \partial U_z/\partial y = \partial^2 \phi/\partial z\,\partial y.$$

Therefore,

$$\partial U_y/\partial z = \partial U_z/\partial y \quad \text{or} \quad \partial U_y/\partial z - \partial U_z/\partial y = 0,$$

and the i-component of the curl U is zero. Likewise, it can be shown that the j- and k-components are also zero.

The curl U is not necessarily zero, and can be shown to be a function of the angular velocity ω. It is taken as (see Eq. 4–13)

$$(\nabla \times U) = \Omega, \tag{6-6}$$

where Ω is a vector of rotation of a fluid element or vorticity. The components of Ω are

$$\xi = \partial U_z/\partial y - \partial U_y/\partial z, \qquad \eta = \partial U_x/\partial z - \partial U_z/\partial x, \qquad \zeta = \partial U_y/\partial x - \partial U_x/\partial y. \tag{6-7}$$

Of course, these are also the components of the curl U, and are called the vorticity components. The angular velocity components would be $\frac{1}{2}\xi$, $\frac{1}{2}\eta$, and $\frac{1}{2}\zeta$. For cylindrical and spherical coordinates, the components can be written almost immediately since, as noted above, the components of Ω are components of $(\nabla \times U)$. These in turn can be obtained directly from Eq. (2–66) and are tabulated in Eqs. (2–72) and (2–78).

If the curl U is zero, there are no angular velocity terms, and there can be no rotation. This can be seen by considering Stokes' theorem (2–56),

$$\int_S (\nabla \times U) \cdot dS = \oint_C (U \cdot dL). \tag{6-8}$$

The term on the right-hand side is the integral of the tangential velocity along a curve, and can be pictured as the circulation Γ along this curve. If this circulation is zero, there is no rotation, and the curl U, which is $(\nabla \times U)$, must be zero. Lagrange has proved an important theorem [6, 9] which states that in an ideal fluid in which the density is constant or a function of the pressure only, rotation cannot be produced in the irrotational field when only irrotational forces exist. In other words, if a velocity potential exists, then the flow is irrotational and will remain so in the absence of rotational forces. Such forces, under the influence of which potential flow cannot occur, are viscous, convective, electrical, and magnetic forces. If the curl U is not zero, then circulation is important; problems of this kind are considered later in Section 6-5, on vortex motion.

Of particular interest is the class of incompressible flow problems that can be represented by a velocity potential

$$U = \nabla \phi, \tag{6-4}$$

which, as noted, implies that the flow is irrotational; that is,

$$(\nabla \times U) = 0. \tag{6-6}$$

The incompressible flow assumption, together with the fact that the flow is irrotational, reduces the Navier-Stokes equation (4–12) to the exact same form as the Euler equation (6–1), since for irrotational motion $\nabla(U \cdot U)$ is equal to $2((U \cdot \nabla)U)$ by Eq. (4–10). Thus, the assumption of incompressible potential flow is equivalent to assuming ideal flow. For this case,

$$(\nabla \cdot U) = (\nabla \cdot \nabla \phi) = \nabla^2 \phi = 0, \tag{6-9}$$

as already shown by Eq. (5–5). A solution of $\nabla^2 \phi = 0$, which is the Laplace equation, is a velocity potential for an ideal flow problem, and by Eq. (6–4) can be used to give the velocity field. The pressure field can be obtained from the motion equation (6–1) and the information about the velocity field provided by the solution of the Laplace equation. The Laplace equation has been treated extensively, and many solutions are known; furthermore, it is a linear equation, and the sums of solutions are also solutions. The problem with potential flow solutions is to select the proper ϕ, the one which will allow the boundary conditions to be satisfied.

6-3 THE STREAM FUNCTION AND TWO-DIMENSIONAL POTENTIAL FLOW

The Laplace equation (6–9) can be solved for a number of two- and three-dimensional problems. For the two-dimensional system, the introduction of the stream function ψ will allow a relatively simple mathematical solution. The mathematics are those of a complex variable [1, 8]. Consider a function $f(z)$ in a given region:

$$f(z) = f(x + iy) = u(x, y) + iv(x, y), \tag{6-10}$$

where i is $\sqrt{-1}$ (not to be confused with $\mathbf{i}$, the unit x-vector). For this function, the Cauchy-Riemann conditions,

$$\partial u/\partial x = \partial v/\partial y, \qquad \partial u/\partial y = -\partial v/\partial x, \tag{6-11}$$

are necessary and sufficient to ensure that the function is analytic in the region, if in that region u and v are real, continuous, and single-valued functions, and the four possible first partial derivatives are continuous. For functions of this nature, the second derivatives of the Cauchy-Riemann conditions are

$$\partial^2 u/\partial x^2 = \partial^2 v/\partial x\,\partial y, \qquad \partial^2 u/\partial y^2 = -\partial^2 v/\partial y\,\partial x.$$

This gives $\partial^2 u/\partial x^2 + \partial^2 u/\partial y^2 = 0$, which is the Laplace equation for u. Also v is harmonic; that is, it satisfies the Laplace equation. This can be shown by taking the other partial derivative of Eq. (6–11).

Let us now consider another analytic function:

$$F(z) = \phi + i\psi, \tag{6-12}$$

so that

$$\partial^2 \phi/\partial x^2 + \partial^2 \phi/\partial y^2 = 0, \tag{6-13}$$

and

$$\partial^2 \psi/\partial x^2 + \partial^2 \psi/\partial y^2 = 0. \tag{6-14}$$

The Cauchy-Riemann conditions give

$$\begin{aligned}\partial\phi/\partial x &= \partial\psi/\partial y, \\ \partial\phi/\partial y &= -\partial\psi/\partial x.\end{aligned} \tag{6-15}$$

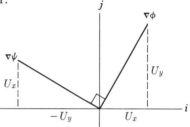

Fig. 6–1. Streamlines and velocity potentials.

If Eq. (6–13) is satisfied, the flow is potential, so that

$$U_x = \partial\phi/\partial x \qquad \text{and} \qquad U_y = \partial\phi/\partial y, \tag{6-16}$$

and by Eq. (6–15),

$$U_x = \partial\psi/\partial y \qquad \text{and} \qquad U_y = -\partial\psi/\partial x. \tag{6-17}$$

In addition,

$$\begin{aligned}\boldsymbol{\nabla}\phi &= \mathbf{i}(\partial\phi/\partial x) + \mathbf{j}(\partial\phi/\partial y) = \mathbf{i}U_x + \mathbf{j}U_y, \\ \boldsymbol{\nabla}\psi &= \mathbf{i}(\partial\psi/\partial x) + \mathbf{j}(\partial\psi/\partial y) = -\mathbf{i}U_y + \mathbf{j}U_x.\end{aligned} \tag{6-18}$$

The two terms of Eq. (6–18) are shown in Fig. 6–1; $\boldsymbol{\nabla}\phi$ and $\boldsymbol{\nabla}\psi$ are perpendicular to each other.

The family of curves of ϕ and ψ are orthogonal. The velocity vector is known to be perpendicular to the velocity potential, and must therefore be in the same direction as the stream function ψ. The family of curves of the stream function are the streamlines of flow, and are envelopes of points tangent to the velocity vectors. The stream function can be developed by an alternative method by assuming streamlines, and by showing that these are lines of constant stream function. Such a development can be found in references 5, 6, and 9. There are three types of lines to be considered. First, there are the streamlines of flow just mentioned, which map out the velocity direction at every point in the field. Second, there are path lines, which describe the paths of separate fluid particles (a small fluid element) with time. Finally, there are streak lines, which at a given instant are the loci of all fluid elements that had previously passed through some specific point in the flow field. These lines are of experimental interest in flow visualization studies; for example, one obtains a streak line when smoke or dye is injected into a moving stream.

For steady-state motion, all three lines are identical. The streamlines are fixed in the flow field with time. At each point, the fluid particle moves in the direction of the velocity, but since this is fixed, each fluid particle that originates on a given streamline must remain on that streamline. Thus, for steady-state motion, the fluid path is the same as the streamline. The streak line will coincide with the streamline and path line, since there is only one streamline at any specific point in the flow field, and each fluid element that had passed this point must have been on this streamline and must remain there. For steady state, streak lines can give useful information about the streamlines of flow and the paths of fluid elements.

For unsteady-state motions, the three lines do not coincide and are related in a complex manner. The path and streak lines can be obtained from a knowledge of the instantaneous streamlines. This can be done graphically. However, if analytical expressions are available, a mathematical solution is preferred. Some examples can be found in reference 11. For one specific case, Hama [4] has used a computer solution. He emphasized that caution must be exercised in interpreting unsteady flow phenomena when dye injection or marked particles are used in flow visualization techniques. The construction of a path and a streak line will be illustrated at a later point in this chapter.

The analogy of voltage to velocity potential has already been noted. Since in electrical conduction, the voltage distribution obeys the Laplace equation, an electrical analog can be used to experimentally solve potential flow problems. The method involves the use of a conducting plate or solution in the manner

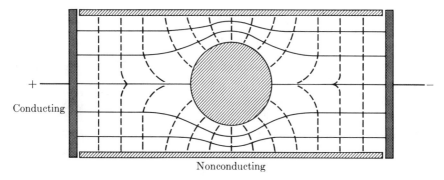

Fig. 6–2. Electrical analogy to ideal flow.

suggested by Fig. 6–2. The voltages, or potentials (dashed lines), are read, using a probe and voltmeter. The streamlines (solid lines) could be obtained directly by exchanging the conducting elements for the nonconducting, including the center disk in the figure.

The analytic approach to a specific flow geometry is not easy when the form of the potential is unknown. One simple method of studying potential flow problems is to assume various forms for the potential and then determine the nature of the flow field so described. Initially this method will be adopted here. If we have a function of the form of Eq. (6–12), and if the Cauchy-Riemann conditions (6–15) are satisfied, then the function is analytic and Eqs. (6–13) and (6–14) are valid. This function is a solution of some potential flow problem. Further, the equations are of the same form if ϕ and ψ are interchanged; thus, from any one solution, we can obtain another by considering the velocity potential as a stream function and vice-versa. This latter transformation can be illustrated from Eq. (6–12):

$$iF(z) = i(\phi + i\psi) = i\phi + i^2\psi = -\psi + i\phi.$$

An analytic function can be transformed from the z-plane to a z'-plane, where the new coordinates are x' and y'. If the transformation equation

$$z' = x' + iy' = f(z) = u + iv \qquad (6\text{–}19)$$

is analytic, and if $df(z)/dz \neq 0$, then the mapping or transformation will preserve angles between curves during transformation. Such a mapping is called conformal [1]. In conformal mapping, a function that is analytic is mapped into another analytic function, and will also satisfy the Cauchy-Riemann conditions. This implies that the new functions are solutions of potential flow problems. It is easy to see that, given a few simple potentials and the flows they describe, one can generate many more potentials and flow descriptions by interchanging the potential and stream functions and by transformations.

A simple example will allow clarification of the foregoing discussion on complex variables. Consider the analytic function

$$F(z) = Az \qquad (6\text{–}20)$$

or

$$F(z) = A(x + iy) = Ax + iAy.$$

Then $\phi = Ax$ and $\psi = Ay$. Equation (6–15) can be used to ascertain that the Cauchy-Riemann conditions are satisfied:

$$\partial\phi/\partial x = A = \partial\psi/\partial y, \qquad \partial\phi/\partial y = 0 = -\partial\psi/\partial x.$$

Since the conditions are satisfied, the flow is potential, and by Eq. (6–16),

$$U_x = A \quad \text{and} \quad U_y = 0.$$

Let us call this velocity U_∞, that is, $A \equiv U_\infty$. This flow is shown in Fig. 6–3

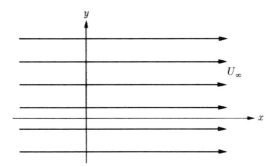

Fig. 6–3. Uniform ideal flow in the x-direction.

(U_∞ taken as positive) and is simply a flow in the x-direction with $U_x = U_\infty$; there is no y-component of velocity.

The flow of Fig. 6–3 is now known and can be mapped from the z-plane to some other z'-plane by the application of a conformal transformation or mapping. For example, consider the transformation

$$z = z'^2 \quad \text{or} \quad z = x'^2 - y'^2 + 2x'y'i. \tag{6-21}$$

Then $F(z)$ becomes

$$F(z) = U_\infty z = U_\infty (x'^2 - y'^2) + 2U_\infty x'y'i.$$

At this point, one can check to make certain that the transformation is analytic and the mapping conformal. The transformation will be analytic if the Cauchy-Riemann conditions (6–11) are satisfied; the mapping will be conformal if the function is analytic and if $df(z)/dz \neq 0$. Using Eq. (6–11), we get

$$\partial u/\partial x' = 2x'U_\infty = \partial v/\partial y',$$
$$\partial u/\partial y' = -2y'U_\infty = -\partial v/\partial x',$$
$$df(z)/dz = 2z' \neq 0.$$

Therefore, the transformation is analytic and the mapping conformal. The function $F(z)$ will also be analytic in the new z'-plane. However, this point could be checked separately.

Under this transformation the new ϕ and ψ become

$$\phi = U_\infty (x'^2 - y'^2) \quad \text{and} \quad \psi = 2U_\infty x'y'. \tag{6-22}$$

Since these expressions result from a conformal mapping of an analytic function, they should also be analytic and satisfy a potential flow problem. This point can be checked by using the Cauchy-Riemann conditions (6–15); that is,

$$\partial \phi/\partial x' = 2U_\infty x' = \partial \psi/\partial y', \quad \partial \phi/\partial y' = -2U_\infty y' = -\partial \psi/\partial x'.$$

Thus the functions given in Eq. (6–22) are a solution of some potential flow

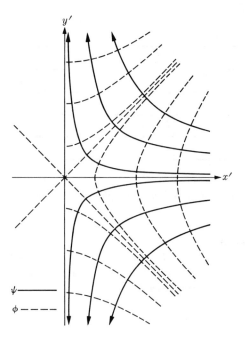

Fig. 6–4. *Potential flow around the inside of a corner or against a plate.*

problem. They are plotted in Fig. 6–4 for $+x'$ and $\pm y'$. The arrows for flow direction are for a negative U_∞; that is, to the left. The upper quarter could represent the potential flow around the inside of a sharp corner. The entire right half could be the potential flow against a flat plate (U_∞ is negative). From Eq. (6–16), the velocities in the z'-plane are

$$U_x = 2U_\infty x' \quad \text{and} \quad U_y = -2U_\infty y'$$

or

$$\boldsymbol{U} = 2U_\infty(x' - iy') = 2U_\infty(\boldsymbol{i}x' - \boldsymbol{j}y'). \tag{6-23}$$

The velocity at any point x' and y' is given by this equation. The absolute value of the velocity would be

$$U = \sqrt{U_x^2 + U_y^2} = \sqrt{4U_\infty^2 x'^2 + 4U_\infty^2 y'^2} = 2U_\infty \sqrt{x'^2 + y'^2}.$$

In other words, the speed along a streamline is proportional to the distance from the origin. Lines of constant velocity will be concentric circles about the origin.

Other transformations of the simple parallel flow function are possible by the use of other transformation equations instead of Eq. (6–21). For example, $z = z'^3$ would give the flow in one-third of the half-plane, just as z'^2 gave the flow in one-half of the half-plane, or a corner. If $z = z'$, then there is no change, and the flow is as in Fig. 6–3. For $z = z'^{2/3}$, the flow is outside a corner, and for $z'^{1/2}$, it is around the edge of a flat plate.

STREAM FUNCTION AND TWO-DIMENSIONAL POTENTIAL FLOW

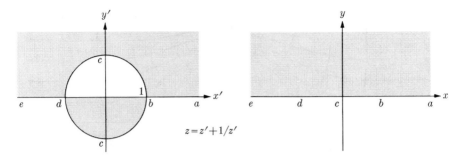

Fig. 6-5. *Transformation to outside a unit circle from the half plane.* [From R. V. Churchill, Complex Variables and Applications, *McGraw-Hill, New York (1948)*. *By permission.*]

Another interesting example which illustrates the means of developing specific transformations for specific problems is the two-dimensional flow around a long cylinder. If the radius is taken as unity, $x^2 + y^2 = 1$. Then, from a table of transformations [1], the upper half-plane around a unit circle can be mapped from a half-plane by the transformation

$$z = z' + 1/z'. \qquad (6\text{-}24)$$

This is shown as the shaded area of Fig. 6-5. For the plane,

$$F(z) = U_\infty z. \qquad (6\text{-}20)$$

Therefore,

$$F(z) = U_\infty(z' + 1/z'). \qquad (6\text{-}25)$$

The complex variable in polar coordinates can be expressed as

$$z = re^{i\theta} = r(\cos\theta + i\sin\theta) \quad \text{and} \quad 1/z = (1/r)e^{-i\theta} = (1/r)(\cos\theta - i\sin\theta). \qquad (6\text{-}26)$$

Equation (6-25) becomes

$$F(z) = U_\infty r' \cos\theta' + U_\infty r' i \sin\theta' + (U_\infty/r')\cos\theta' - (U_\infty/r')i\sin\theta'$$
$$= U_\infty(r' + 1/r')\cos\theta' + iU_\infty(r' - 1/r')\sin\theta'.$$

Thus

$$\phi = U_\infty(r' + 1/r')\cos\theta' \quad \text{and} \quad \psi = U_\infty(r' - 1/r')\sin\theta'. \qquad (6\text{-}27)$$

These are the potentials and streamlines for the flow outside a cylindrical shape. The upper half-plane gives the flow in the upper half-plane around a unit circle, while the lower half-plane maps into the lower half-plane around the circle. The same transformation will convert the half-plane into the inside of the opposite semicircle (see Fig. 6-5). The solution is the same, except that radii less than unity are considered. These mappings are illustrated in Fig. 6-6. The arrows for flow direction are for a positive U_∞ (to the right). The actual velocities can be

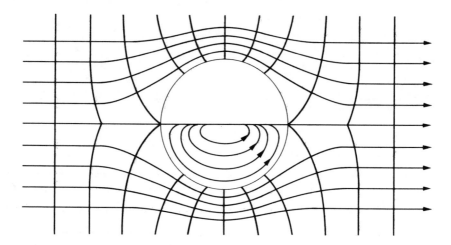

Fig. 6–6. Potential flow outside and inside a cylindrical shape. [*From L. Prandtl and O. G. Tietjens,* Fundamentals of Hydro- and Aeromechanics, *reprinted by permission of Dover Publications, Inc., New York 14, New York.*]

determined as follows. First, Eq. (6–18) can be transformed by the use of Eq. (2–70) to give

$$U_r = \frac{\partial \phi}{\partial r} = \frac{1}{r}\frac{\partial \psi}{\partial \theta} \quad \text{and} \quad U_\theta = \frac{1}{r}\frac{\partial \phi}{\partial \theta} = -\frac{\partial \psi}{\partial r}. \tag{6-28}$$

This can be combined with Eq. (6–27) to yield

$$U_r = U_\infty(1 - 1/r^2)\cos\theta \quad \text{and} \quad U_\theta = -U_\infty(1 + 1/r^2)\sin\theta,$$

where the primes on r and θ have been dropped for convenience. Far from the cylinder the velocity should be uniform and equal to U_∞. For example, at $r = \infty$ along $\theta = 0°$, the velocity equations above give $U_\theta = 0$ and $U_r = U_\infty$, as expected. The actual calculations for the streamlines are presented in the following example.

EXAMPLE 6–1. *For the potential flow problem described by Eq.* (6–27), *find several representative streamlines.*

Answer. The streamline $\psi = 0$, is given either by $\sin\theta$ or by $(r - 1/r)$ equal to zero; $\sin\theta = 0$ is the x-axis, and $(r - 1/r) = 0$ is the unit circle.

For convenience of calculation, let us set the streamline as $\psi = KU_\infty$. From Eq. (6–27), we obtain

$$r^2 - (Kr/\sin\theta) - 1 = 0$$

or

$$r = \frac{K \pm \sqrt{K^2 + 4\sin^2\theta}}{2\sin\theta}. \tag{i}$$

STREAM FUNCTION AND TWO-DIMENSIONAL POTENTIAL FLOW 69

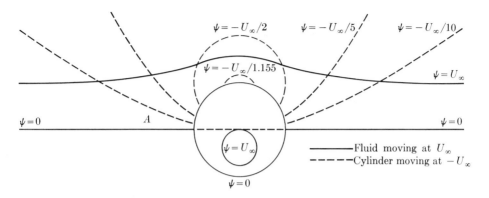

Fig. 6–7. The streamlines of flow about a cylindrical shape.

To illustrate, take $K = 1$ ($\psi = U_\infty$), assume values of θ, and calculate the corresponding values of r from (i). Some representative values are:

$\theta°$		$\sqrt{1 + 4\sin^2\theta}$	r	
0 or	180	1.000	0 or	∞
15	165	1.128	−0.247	4.11
30	150	1.414	−0.414	2.415
50	130	1.82	−0.535	1.84
70	110	2.13	−0.602	1.66
90		2.24	−0.620	1.62

These values are plotted in Fig. 6–7 (solid line). Other streamlines can be obtained by the use of different values for K.

In the preceding example, the streamlines for a steady motion were calculated. These lines would of course coincide with the fluid particle path lines and the streak lines formed by point dye injections. This same system can be used to illustrate the unsteady-state problem, if we assume that the cylinder is moving with some net velocity.

EXAMPLE 6–2. Calculate several of the instantaneous streamlines, assuming that a uniform velocity of $-U_\infty$ is superimposed on the entire field, including the cylinder. This will in effect change the problem to that of a cylinder moving with velocity $-U_\infty$ through a stationary fluid. As viewed from a position in the stationary fluid, the motion is unsteady as the cylinder approaches and passes the observer. As viewed by an observer moving with the cylinder, the motion is steady and is as shown in Fig. 6–7.

Answer. The uniform superimposed motion is given by $\psi = -U_\infty y$. Sums of analytic functions are analytic, and thus are solutions of the additive flow problem; that is,

$$\psi = U_\infty(r - 1/r)\sin\theta - U_\infty r \sin\theta,$$

since $y = r\sin\theta$. Hence $\psi = -(U_\infty/r)\sin\theta$. Values for $\psi = KU_\infty$, with K equal to $1/1.155$, $1/2$, $1/5$, and $1/10$, are shown as dashed lines in Fig. 6–7.

EXAMPLE 6–3. Illustrate the steps for the construction of a fluid particle path line for the element located at point A of Fig. 6–7. The flow is the same as in Example 6–2.

Answer. To know the direction that the fluid element will move at $t = 0$ at point A, we must know the instantaneous velocity through that point (streamline tangent). This can be obtained graphically by vector summation. The net velocity will be the sum of U_r, U_θ, and $-U_\infty$. To illustrate, for the point A, $\theta = 170°$ and $r = 2.0$; therefore

$$U_r = \tfrac{3}{4}U_\infty \cos 170° = -0.739 U_\infty,$$
$$U_\theta = -\tfrac{5}{4}U_\infty \sin 170° = -0.217 U_\infty.$$

The construction is shown in Fig. 6–8, where the resultant is noted as $\boldsymbol{U}_1$. The distance the fluid particle will move along this line during the time the cylinder moves a distance $-U_\infty t$ is $\boldsymbol{U}_1 t$. For convenience, assume that $t = 1$ for this first step; then the dashed line gives the new position of the cylinder. Point C relative

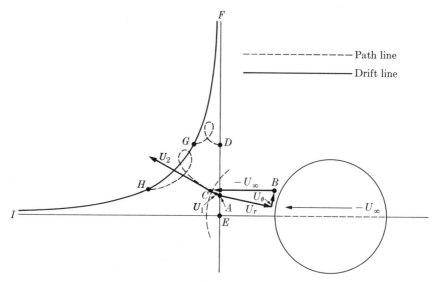

Fig. 6–8. Path lines and drift line of flow about a cylindrical shape.

to the cylinder's new position is exactly the same as though we had transposed point C to B, and then considered it relative to the original position. Its new position relative to the cylinder is then point B, or $\theta = 157°$ and $r = 1.15$. We can recalculate U_r and U_θ for this new position and then combine them with $-U_\infty$ to obtain U_2. As the curvature of the path line increases, steps considerably smaller than $t = 1$ must be taken in order to obtain the true shape of the path line shown. In the vicinity of the loop, the steps must be so small that it is more sensible and accurate to resort to numerical methods and a computer, or to attempt an analytical solution as did Darwin [2] and Lighthill [7].

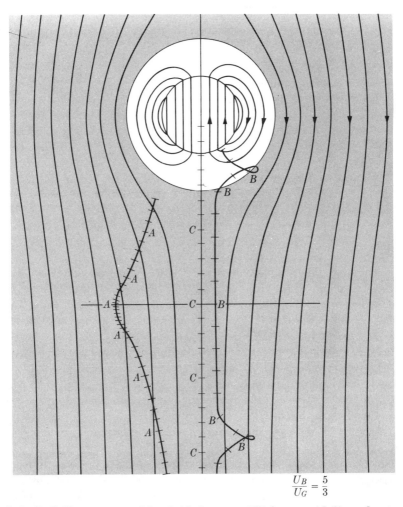

Fig. 6–9. *Path lines generated by fluid elements.* (*Ticks on path lines denote equal time intervals.*) [*From P. N. Rowe, Chem. Eng. Prog. Symposium Series No. 38, 58, 42 (1962). By permission.*]

Point D in Fig. 6-8 shows the start of a path line of a fluid element lying on the streamline calculated in Example 6-1. The locus of end points (e.g. G and H) of these paths traces the fluid drift ($FGHI$) of a marked line ($EADF$) as caused by the passage of the cylinder (see Fig. 18-13 for an experimental example).

Figure 6-9 shows two path lines obtained by a step-by-step computer solution for the case where a void moves through a moving fluid. In this specific case the void moves upward at a velocity of $\frac{5}{3} U_\infty$, while the stream moves downward at a velocity of U_∞. Some further examples can be found in Rowe and Partridge [10] and in Fig. 18-17.

EXAMPLE 6-4. Illustrate the construction of streak lines for the problem considered in Example 6-2.

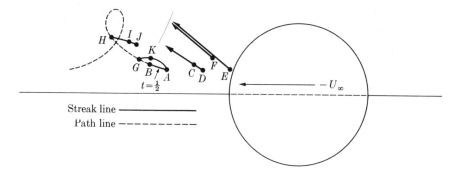

Fig. 6-10. Construction of a streak line.

Answer. The streak line is constructed (Fig. 6-10) in much the same manner as the path line. For point A, the dye can be injected only until the cylinder reaches this point; however, we can follow this streak line at subsequent times. The fluid element tagged by the dye at $t = 0$ will move along the previously calculated path line ($ABGH$). Let us take the first increment of time as $\frac{1}{2}$. The element which was dye-marked at point A at $t = 0$ will have moved to B. In the same time period, the cylinder will have moved $U_\infty t_1 = U_\infty/2$ closer to point A. Point A will now lie on a new path line closer to the cylinder (equivalent to point D). Point B will also lie closer to the cylinder (equivalent to point C). Since we wish to stop the dye injection when the cylinder arrives at point A, let us take $t_2 = \frac{1}{3}$. At the end of this period, point A will be to the moving cylinder as point E is to the cylinder's original position. In this time element, point B will have moved to G, and point A will have moved to K along the new path line vector now existing at point A, which, as previously noted, is shown at point D. The dye injection is now terminated, but we can continue to follow the streak line. Now let us take $t_3 = \frac{1}{2}$, which will move point G to H. Point K is now quite close to the cylinder (equivalent to point F) and moves along its path line to point I. Finally, in like manner, point A (now equivalent to point E) moves along its

6-4 SOURCES, SINKS, AND CIRCULATION

The analytic function $F(z) = Az$ (Eq. 6-20) represented the potential flow in the plane and was transformed to other, more complex flows. Another analytic function can be treated similarly:

$$F(z) = A' \ln z \qquad (6\text{-}29)$$

or

$$\begin{aligned} F(z) &= A' \ln (re^{i\theta}) \\ &= A' \ln r + A' \ln e^{i\theta} \\ &= A' \ln r + A'i\theta. \end{aligned}$$

This gives

$$\phi = A' \ln r, \qquad \psi = A'\theta. \qquad (6\text{-}30)$$

The streamlines are lines of constant angle, and the velocity potentials are lines of constant r, or circles. For a positive A', this flow (Fig. 6-11) is called a *source*. The negative of a source is called a *sink*.

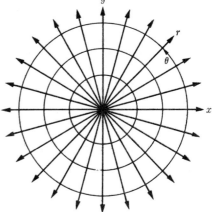

Fig. 6-11. Potential flow from a source.

Sources and sinks, in conjunction with other flows, can be used to develop many interesting new flow cases. The sums of analytic functions are analytic, so that any number of sources and sinks could be added to a potential flow and still result in potential flow. For the resulting system to be realistic (in balance), it is necessary for the flow into all of the sinks to balance that produced at the sources. As an example, we can consider the combination of a source, sink, and flow in the plane.

EXAMPLE 6-5. Obtain the streamlines for a source and sink of equal strength, located a distance $2a$ apart, with a uniform velocity superimposed (see Fig. 6-12).

Answer. The resulting stream function and potentials are

$$\begin{aligned} \phi &= A'(\ln r_{-a}) - A'(\ln r_{+a}) = A' \ln (r_{-a}/r_{+a}), \\ \psi &= A'\theta_{-a} - A'\theta_{+a} = A'(\theta_{-a} - \theta_{+a}), \end{aligned} \qquad (i)$$

which is simply the sum of the source $(A'\theta_{-a})$ at $-a$ and the sink $(A'\theta_{+a})$ at $+a$,

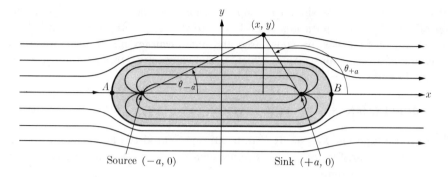

Fig. 6-12. *Potential flow about an elongated object.*

and of equal strength A'. Equation (i) can be transformed into cartesian notation, since

$$\tan \theta_{+a} = -\frac{y}{a-x} \quad \text{and} \quad \tan \theta_{-a} = \frac{y}{a+x}. \quad \text{(ii)}$$

Since

$$\tan(\theta_{-a} - \theta_{+a}) = \frac{\tan \theta_{-a} - \tan \theta_{+a}}{1 + \tan \theta_{-a} \times \tan \theta_{+a}} = \frac{2ya}{a^2 - x^2 - y^2}, \quad \text{(iii)}$$

it follows that

$$\psi = A' \tan^{-1}\left(\frac{2ya}{a^2 - x^2 - y^2}\right). \quad \text{(iv)}$$

If we now superimpose a uniform flow to the right, that is, $\psi = U_\infty y$, we obtain the final stream function

$$\psi = U_\infty y + A' \tan^{-1}\left(\frac{2ya}{a^2 - x^2 - y^2}\right). \quad \text{(v)}$$

The streamlines can be calculated for specific values of a, A', and U_∞ by letting ψ equal various values and calculating x from assumed values of y. The points A and B in the figure are stagnation points, where the velocity is zero. From Eqs. (v) and (6–17), we obtain

$$\begin{aligned} U_x &= \frac{\partial \psi}{\partial y} = U_\infty + A' \frac{2a(a^2 - x^2 + y^2)}{(a^2 - x^2 - y^2)^2 + 4y^2 a^2}, \\ U_y &= -\frac{\partial \psi}{\partial x} = A' \frac{4axy}{(a^2 - x^2 - y^2)^2 + 4y^2 a^2}. \end{aligned} \quad \text{(vi)}$$

At the stagnation points, $U_x = 0$ and $y = 0$, which gives

$$a^2 - x^2 = 2aA'/U_\infty \quad \text{(vii)}$$

or

$$x = \pm a\sqrt{1 - (2A'/aU_\infty)}.$$

SOURCES, SINKS, AND CIRCULATION

As already noted, ϕ and ψ can be interchanged, resulting in another solution of the potential flow problem. Therefore, Fig. 6–11 also represents a circular flow about a point, and is called a vortex. For this flow $iF(z) = -A'\theta + iA' \ln r$; that is,

$$\phi = -A'\theta \quad \text{and} \quad \psi = A' \ln r. \tag{6-31}$$

In polar coordinates, the conventional sign for a vortex is positive when it rotates counterclockwise.

Combining Eqs. (6–28) and (6–31), we have

$$U_r = 0 \quad \text{and} \quad U_\theta = -A'/r. \tag{6-32}$$

If A' is positive, then U_θ is negative and the rotation is clockwise; thus a positive A' describes a negative vortex.

The circulation about the vortex can be evaluated from the previously discussed circulation integral (6–8):

$$\Gamma = \oint_C \boldsymbol{U} \cdot d\boldsymbol{L}, \tag{6-33}$$

which, from Eq. (2–69), can be rewritten in polar coordinates as

$$\Gamma = \oint_C (\boldsymbol{U} \cdot \boldsymbol{T} \sqrt{dr^2 + r^2 d\theta^2 + dz^2}). \tag{6-34}$$

With the aid of Eq. (2–50), this equation reduces to

$$\Gamma = \oint U_\theta r \, d\theta, \tag{6-35}$$

since $U_r = U_z = 0$; U_θ is given by Eq. (6–32) and $d\theta$ varies from 0 to 2π. For any radius, Eq. (6–35) can be integrated to give the circulation, which is

$$\Gamma = -2\pi A'. \tag{6-36}$$

Combining Eqs. (6–31) and (6–36), we obtain

$$\phi = (\Gamma/2\pi)\theta \quad \text{and} \quad \psi = -(\Gamma/2\pi) \ln r. \tag{6-37}$$

A negative A' (positive vortex) implies a positive Γ, by Eq. (6–36); thus positive Γ means counterclockwise rotation, or a positive vortex.

In the previous section, the plane was mapped into the flow around a cylinder, and in this section, a vortex was developed. These two flows can be combined to give the flow around a cylinder, the surface of which has an imposed circulation. In this and similar cases, the velocity potentials and stream functions need only be added together; a complete new development is unnecessary. Combining Eqs. (6–27) and (6–37), we get

$$\phi = U_\infty(r + 1/r) \cos \theta + (\Gamma/2\pi)\theta$$

and

$$\psi = U_\infty(r - 1/r) \sin \theta - (\Gamma/2\pi) \ln r. \tag{6-38}$$

Some of these streamlines are shown in Fig. 6–13 for a negative $\Gamma < 4\pi U_\infty$. From Eqs. (6–28) and (6–38), we obtain

$$U_r = U_\infty(1 - 1/r^2) \cos \theta \quad \text{and} \quad U_\theta = -U_\infty(1 + 1/r^2) \sin \theta + \Gamma/2\pi r. \quad (6\text{--}39)$$

On the cylinder surface $r = 1$; thus

$$U_r = 0 \quad \text{and} \quad U_\theta = -2U_\infty \sin \theta + \Gamma/2\pi. \quad (6\text{--}40)$$

For a negative Γ, this implies that U_θ is increased when $\sin \theta$ is positive and decreased when $\sin \theta$ is negative. The velocity will be higher at the top of the cylinder. Because the kinetic energy is greater at the top, the pressure will be lower than in the slower velocity region at the bottom. The pressure difference, called the *Magnus effect*, provides a simple explanation for the lift found when circulation exists about an object in a fluid stream.

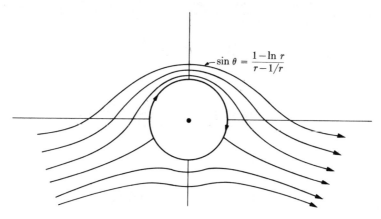

Fig. 6–13. Potential flow about a rotating cylinder (with circulation).

EXAMPLE 6–6. Calculate a typical streamline for the flow over a rotating cylinder (a cylinder with circulation).

Answer. The streamlines can be obtained by assuming $\psi = KU_\infty$ and $\Gamma = K'\pi U_\infty$. Equation (6–38) then becomes

$$K = (r - 1/r) \sin \theta - \tfrac{1}{2}K' \ln r. \quad (\text{i})$$

For various values of K (for one value of K'), θ can be calculated for assumed values of r. For example, for $K' = -2$ and $K = 0$ the equation is satisfied by all θ for $r = 1$; that is, the unit circle is a streamline. For $K = 1$ and $K' = -2$, we obtain from Eq. (i):

$$\sin \theta = \frac{1 - \ln r}{r - 1/r} \quad (\text{ii})$$

Thus for any selected r, the value of θ can be calculated. Equation (ii) is shown in Fig. 6–13.

In the motion just described, the flow is potential and therefore irrotational; however, there can be a constant circulation as given by Eq. (6–36). This paradox can be understood if it is realized that rotational motion refers to the rotation of the individual fluid particles, and not to the fluid as a whole. The vortex as described is really irrotational flow except at the infinitesimal singular point at $r = 0$. The circulation-line integral of Eq. (6–33) is zero everywhere except when it surrounds the singular point; it then has the value given by Eq. (6–36). For rotational motion, this integral is in general not zero at any time or place.

It should be realized that potential motion will not be accurate for all experimental conditions. The theory considers circulation but does not consider its cause. If circulation is induced by rotation of a cylinder, there is only a limited range of speeds at which the theory can be applied [3]. Nevertheless, it is interesting to note that the circulation effect about a cylinder has been used in ship propulsion. In 1927, the rotorship Barbara crossed the Atlantic twice. It was equipped with three cylinders 23 ft in diameter and 98 ft high, which rotated in the air at 150 rpm. A propulsion rotor is called a Flettner rotor after the man who first demonstrated the use of the Magnus effect for ship propulsion.

Flow about other shapes can be obtained by combinations of different potentials and additional transformations. In all these cases, there will be potential flow; however, there may be singular points so that a circulation exists. As an example, the flow of Fig. 6–13 can be transformed into a flat plate inclined at an angle to the horizontal. Another example would be to combine an eccentric circle with the unit circle and transform the result to obtain a shape that looks something like a wing profile (Joukowsky profile). For these and more complex cases one should refer to the classic books in the field [6, 9].

6-5 VORTEX MOTION

In the previous section a simple vortex motion was considered: the individual fluid particle in rotation at a singular point. The flow about this point was found to be potential and therefore irrotational. The singularity might well have been called a vortex point. However, vortex motion is usually associated with fluid particle rotations that occur along lines (vortex lines). In addition, the term *vortex motion* implies some form of circulation, and therefore one must include the induced potential flow about the rotating fluid particles. The following analysis should help to clarify the meaning of vortex motion.

For the general case of vortex motion, the curl U would be

$$(\nabla \times U) = \Omega. \tag{6-6}$$

The components of the vector of rotation (or vorticity vector) are given as Eq. (6–7). The divergence of the curl of a vector is zero (Eq. 2–33):

$$(\nabla \cdot (\nabla \times A)) = 0.$$

Therefore,

$$(\nabla \cdot \Omega) = 0. \tag{6-41}$$

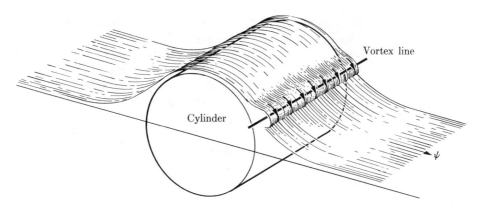

Fig. 6–14. Vortex line on a cylinder.

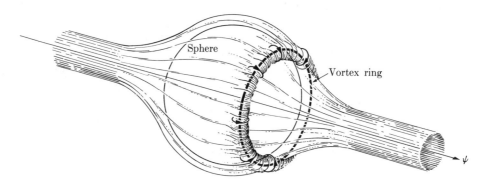

Fig. 6–15. Vortex ring on a sphere.

The divergence of the vorticity vector is zero, which is analogous to the zero divergence of the velocity for incompressible flow. To carry the analogy further, the streamlines of flow would be similar to lines of rotation. Just as the streamline is the envelope of points tangent to the velocity vectors, so the vortex line is the envelope to the axes of rotation of the fluid particles. A line drawn tangent to the axis of rotation at every point is the vortex line. The vortex line can be pictured, for example, as the line parallel to a cylinder, on which a vortex is forming (see Fig. 6–14). If a sphere is considered instead of a cylinder, a vortex ring is formed as shown in Fig. 6–15. If there are several vortex lines such that a series of lines can be drawn through some given closed curve, then a vortex tube is obtained, which is analogous to the stream tube. The fluid particles within this tube compose a vortex filament. Vortex lines, like streamlines, cannot end in the interior of a fluid field. The lines must be closed curves (as in Fig. 6–15) or end at a free surface or boundary (as in Fig. 6–14). The same, of course, would hold true for a vortex tube.

Lord Kelvin presented an important theorem on vortex motion, which states that in a field of irrotational forces an acceleration potential will exist, and the circulation will be a constant. The circulation integral has been given in Eq. (6–8) and again as Eq. (6–33). The circulation (of the integral around any closed curve in the system) is a constant; that is,

$$\Gamma = \int_S ((\nabla \times \boldsymbol{U}) \cdot d\boldsymbol{S}) = \int_S (\boldsymbol{\Omega} \cdot d\boldsymbol{S}) = \oint_C (\boldsymbol{U} \cdot d\boldsymbol{L}) = \text{const.} \quad (6\text{–}42)$$

If a fluid starts at rest with zero circulation, the circulation must then remain zero under the influence of irrotational forces. Another consequence of Kelvin's theorem is that a fluid particle within a vortex must remain within the tube. If we then consider an infinitesimal vortex tube (i.e., a vortex line), then a fluid particle on a vortex line will remain on the line during any motion and for all time.

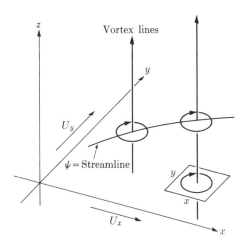

Fig. 6–16. *Circulation and vortex lines.*

The circulation integral as given by Eq. (6–42) can be used to determine the circulation or velocity field in the vicinity of a vortex filament. The flow outside the vortex will not be in rotation in the sense that the individual fluid particles are in rotation; therefore it will be an induced potential flow about the vortex and caused by the vortex. For a given vortex, the line integral over a closed curve that surrounds the vortex will have the value of the circulation and will remain a constant for that vortex. As an example, let us consider a rectilinear vortex line in two dimensions, that is, for a two-dimensional flow problem. Equation (6–7) reduces to

$$\xi = 0, \quad \eta = 0, \quad \zeta = \partial U_y/\partial x - \partial U_x/\partial y, \quad (6\text{–}43)$$

since $U_z = 0$ and U_y and U_x are functions only of x and y. The flow is in the xy-plane and the vortex lines are parallel to the z-axis. If we now consider a

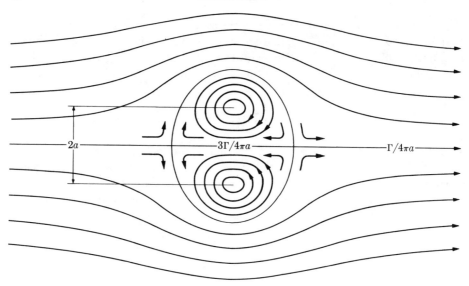

Fig. 6–17. Flow outside and inside a vortex pair. [From L. Prandtl and O. G. Tietjens, Fundamentals of Hydro- and Aeromechanics, reprinted by permission of Dover Publications, Inc., New York 14, New York.]

circle in the xy-plane that surrounds one of the vortex lines (see Fig. 6–16), then, from Eq. (6–42), the line integral about this circle becomes

$$\Gamma = \oint_C (\mathbf{U} \cdot d\mathbf{L}) = 2\pi U_\theta R. \tag{6-44}$$

The velocity U_θ is induced by the vortex line and is similar to the potential flow about the singular point. Just as in the singular point case, U_θ is inversely proportional to the distance from the vortex line.

Lamb [6] and Prandtl and Tietjens [9] have discussed in some detail the combination of several vortex filaments varying in thickness and distance from each other. Since the velocity about the vortex is the highest at the center and drops off toward the outside, they assume an equivalent vortex core of constant rotation and a potential motion exterior to the vortex core. The effect of increasing the size of the core, i.e., increasing the size of the vortex filament, is the same as moving the two vortex lines closer together. The flow for two vortex lines near each other is shown in Fig. 6–17.

EXAMPLE 6–7. Indicate the construction of the vortex pair in a uniform flow as shown in Fig. 6–17.

Answer. This problem is somewhat similar to Example 6–5, except that here a vortex pair of opposite rotation is located in the positions that were occupied by the source and sink. The streamlines and potentials of Example 6–5 become the potentials and streamlines of this problem, respectively. Thus

$$\phi = A'(\theta_{-a} - \theta_{+a}) \quad \text{and} \quad \psi = A' \ln (r_{-a}/r_{+a}), \tag{i}$$

where A' is given by Eq. (6–36), that is,

$$A' = -\Gamma/2\pi. \tag{ii}$$

In Fig. 6–17, Γ is negative, since in the upper half-plane the circulation is clockwise. The vortex at $-a$ induces a velocity on the vortex at $+a$ and vice versa. This velocity is given by Eq. (6–44) and is the equal and opposite velocity that must be added:

$$U = -\frac{\Gamma}{2\pi(2a)} = -\Gamma/4\pi a. \tag{iii}$$

This is a uniform flow in the plane, corresponding to a stream function of

$$\psi = -(\Gamma/4\pi a)y. \tag{iv}$$

The final stream function becomes the sum of Eqs. (i) and (iv):

$$\psi = -(\Gamma/4\pi a)y - (\Gamma/2\pi) \ln (r_{-a}/r_{+a}),$$
$$= -(\Gamma/2\pi)[(y/2a) + \ln (r_{-a}/r_{+a})]. \tag{v}$$

For $\psi = 0$, $\ln (r_{-a}/r_{+a}) = -y/2a$, which is the x-axis (that is, $r_{-a} = r_{+a}$ or $y = 0$) and the oval surrounding the two vortices. Figure 6–17 is not circular, and the streamlines inside do not bunch at the center line. Lamb gives the values of the semi-axes as approximately $2.09a$ and $1.73a$. The result is a closed oval streamline; i.e., the flow exterior to the oval is the same as it would be if the oval were a solid body, and the interior flow is a circulation about the vortex lines and moves with the vortex.

If the distance a is increased, the vortices will begin to separate. First a figure "8" form will develop, then two separate vortices will form. These will be similar to those observed when one runs his finger through a still pool of water. When the two vortices are formed, the curve for the free surface can be obtained if we include a gravity term in the z-direction. This curve would indicate a depression at the center of the vortex of $\omega b^2/g$, where b is the radius at the top.

PROBLEMS

6–1. Given the analytic transformation $z = z'^n$, find both the velocity potential and the stream function for the transformation of $F(z) = z$.

6–2. For Problem 6–1, where $n = 4$, draw several of the streamlines for the upper right quarter-plane.

6–3. Given an analytic function $F(z) = A \cosh z$, obtain the value of the velocity potential and streamlines, and suggest what type of boundary might give this sort of flow.

6–4. Repeat Problem 6–3, letting the analytic function be $F(z) = z + e^z$.

6–5. Given the stream function $\psi = Ar^2 \cos 2\theta = A(x^2 - y^2)$, obtain the value for the velocity at a point, and draw the streamline for values of ψ of 0, A, and $2A$.

6–6. Given the velocity potential and stream function in the form $x = A \cosh \phi \cos \psi$ and $y = A \sinh \phi \sin \psi$, draw this flow and suggest a boundary which will give this type of flow.

6–7. Transform the flow about a circular cylinder into the flow about a plate inclined to the stream. Use the transformation $z = a(z' + 1/z') + ib(z' + 1/z')$. Draw several of the streamlines.

6–8. Repeat Problem 7 for a circular cylinder with circulation. Choose the circulation so that one streamline corresponds to the plate on the downstream side.

6–9. The Joukowski airfoil can be obtained by the transformation $z = z' + 1/z'$ on a circular cross section eccentric to the origin. Taking a circle through $z = -1$ and enclosing $z = 1$, construct the equivalent airfoil. This may be done by adding the vector z to $1/z = (1/r)e^{-i\theta}$.

6–10. Derive Eq. 6–28.

6–11. What is the flow given by $\psi = (K/2\pi)\theta - U_\infty r \sin \theta$?

6–12. What is the flow given by $\phi = 3x + \ln r$?

6–13. Explain the difference between streamlines, streak lines, and particle paths for steady- and unsteady-state flows.

6–14. Establish the analytical solution for each line or path discussed in Problem 6–13, for the simple plane flow given by

$$U_x = 0, \qquad U_y = y, \qquad U_z = z/(1 + t).$$

6–15. For the flow around a circular cylinder with circulation, find the velocity at the surface of the cylinder. Use this result to obtain the pressure distribution equation at the surface under the assumption that the pressure is zero at infinity.

6–16. The drag on the cylinder is given by the integration of the x-component of the pressure force over the cylinder. Show that the drag is zero for the pressure distribution of Problem 6–15.

6–17. A line source of strength $K/2\pi$ is located on the x-axis at a distance $x = a$, and the y-axis is a rigid wall. Determine the position at which the velocity distribution along the wall reaches a maximum. The equation for the streamlines of this motion is $\psi = (K/2\pi) \tan^{-1} [2xy/(x^2 - y^2 - a^2)]$.

REFERENCES

1. R. V. CHURCHILL, *Complex Variables and Applications*, McGraw-Hill, New York (1948).
2. C. DARWIN, *Proc. Cambridge Phil. Soc.* **49**, 342 (1953).
3. H. L. DRYDEN, F. P. MURNAGHAM, and H. BATEMAN, *Hydrodynamics*, Dover Publications, New York (1956).
4. F. R. HAMA, *Phys. Fluids* **5**, 644–650 (1962).
5. J. G. KNUDSEN and D. L. KATZ, *Fluid Dynamics and Heat Transfer*, McGraw-Hill, New York (1958).

REFERENCES

6. H. LAMB, *Hydrodynamics* (6th ed., reprint of the 1932 edition), Dover Publications, New York (1945).
7. M. J. LIGHTHILL, *J. Fluid Mech.* **1**, 31 (1956).
8. H. S. MICKLEY, T. K. SHERWOOD, and C. E. REED, *Applied Mathematics in Chemical Engineering*, McGraw-Hill, New York (1957).
9. L. PRANDTL and O. G. TIETJENS, *Fundamentals of Hydro- and Aeromechanics* (reproduction of the 1934 edition), Dover Publications, New York.
10. P. N. ROWE and B. A. PARTRIDGE, *Chem. Eng. Sci.* **18**, 511–524 (1963).
11. C. A. TRUESDELL and R. TOUPIN, *Handbuch der Physik*, 2nd ed., Vol. III, Part 1 (S. Flugge, ed.), Springer, Berlin (1960).

CHAPTER 7

LAMINAR VISCOUS FLOW: EXACT SOLUTIONS

The ultimate solution to laminar viscous flow problems would be to have a general solution of the Navier-Stokes equation (4–12). Unfortunately, the equation is nonlinear, and there is no known method of obtaining an analytic solution. This chapter deals with exact solutions which are obtained after some simple geometric systems have been chosen, usually on the assumption of an incompressible fluid. In effect, the geometry is chosen so that

$$(U \times (\nabla \times U)) = \tfrac{1}{2} \nabla(U \cdot U), \tag{7-1}$$

and Eq. (4–12) becomes

$$\frac{\partial U}{\partial t} = -\frac{1}{\rho}(\nabla p) - \nu(\nabla \times (\nabla \times U)) + \frac{1}{\rho} \sum_s \rho_s F_s, \tag{7-2}$$

where incompressibility has been assumed. This equation can be solved for many cases, and as examples we will consider flow along and between flat plates, flow in a circular pipe, and flow between concentric rotating cylinders. The compressible flow in a shock wave will also be treated; here, besides the compressibility terms, the energy equation must also be considered, since the flow temperature will change according to some state equation. For further examples (such as the flow near an oscillating plate, convergent and divergent channels, stagnation point flow, and flow from a rotating disk), the reader should see the books by Pai [5] and Schlichting [8]. There is another group of solutions that can be classified as exact, at least in the sense that the total Navier-Stokes equations are used. These are the numerical computer calculations of the set of equations and boundary conditions by finite difference methods [10].

The exact solutions have the fewest assumptions, and thus offer the best systems for testing the equations of motion. Their validity has been well established, and these solutions have become the primary means for the measurement of viscosity.

7–1 THE FLOW ALONG A FLAT PLATE

The simplest unsteady flow involving a boundary is the movement of a flat plate through a semi-infinite incompressible fluid (Stokes' problem). In this problem (see Fig. 7–1), the conditions are

$$U_x = f_1(x,y,t), \qquad U_y = U_z = 0, \qquad p = f_2(x,y,t). \tag{7-3}$$

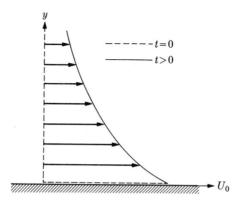

Fig. 7-1. Flow along a flat plate in motion.

Since the problem is two-dimensional, the variations in the z-direction are zero. From continuity, $(\nabla \cdot U) = 0$ or $\partial U_x/\partial x = 0$. Therefore, $U_x \neq f(x)$, and thus $U_x = f_1(y, t)$. The Navier-Stokes equation (4–6), (4–12), or (7–2) reduces to

$$\rho \frac{\partial U_x}{\partial t} = X_x - \frac{\partial p}{\partial x} + \mu \frac{\partial^2 U_x}{\partial y^2}, \qquad \frac{\partial p}{\partial y} = X_y, \qquad (7\text{–}4)$$

where X_x and X_y are used to represent the net external forces; that is,

$$X_x = \sum_s \rho_s F_{sx}, \qquad X_y = \sum_s \rho_s F_{sy}. \qquad (7\text{–}5)$$

If the force in the y-direction is gravity, then the second equation of (7–4) shows that the pressure varies linearly with the height. If the external force in the x-direction is zero, and if the pressure is constant along a parallel line very far from the plate, then the pressure must be constant along all other parallel lines. Thus, for this specific problem, dp/dx is zero. Equations (7–3) and (7–4) become

$$U_x = f_1(y, t), \qquad U_y = U_z = 0, \qquad p = f_2(y, t),$$
$$\rho \frac{\partial U_x}{\partial t} = \mu \frac{\partial^2 U_x}{\partial y^2} \quad \text{or} \quad \frac{\partial U_x}{\partial t} = \nu \frac{\partial^2 U_x}{\partial y^2}. \qquad (7\text{–}6)$$

If the plate starts from rest and is suddenly accelerated to the velocity U_0, then the boundary conditions are

$$\begin{array}{lll} t = 0, & U_x = 0, & \text{for all } y; \\ t > 0, & U_x = U_0, & y = 0; \\ t > 0, & U_x = 0, & y = \infty. \end{array} \qquad (7\text{–}7)$$

Equation (7–6) is of the same form as that for unsteady-state heat conduction, and the solution for the boundary conditions of Eq.(7–7) is

$$U_x/U_0 = 1 - (2/\sqrt{\pi}) \int_0^\eta e^{-\eta^2} d\eta, \qquad (7\text{–}8)$$

where $\eta = y/2\sqrt{vt}$. The solution is given in Example 7-1. The integral of Eq. (7-8) is the error function of statistics, and representative results are plotted in Fig. 7-2. For an η of 2.0, the error function is 0.99. This is sometimes used to define a boundary layer thickness (the distance into a fluid that viscous effects are apparent). For 99% of the decay of the error function or viscous effects, the boundary thickness would be

$$\delta = 4\sqrt{vt}. \qquad (7\text{-}9)$$

EXAMPLE 7-1. The solution of Eq. (7-6) with the boundary conditions (7-7) is given as Eq. (7-8). The solution is demonstrated in this example. Given

$$\frac{\partial U_x}{\partial t} = v \frac{\partial^2 U_x}{\partial y^2}, \qquad (7\text{-}6)$$

with boundary conditions

$$\begin{aligned} t &= 0, & U_x &= 0, & y &= y; \\ t &> 0, & U_x &= U_0, & y &= 0; \\ t &> 0, & U_x &= 0, & y &= \infty. \end{aligned} \qquad (7\text{-}7)$$

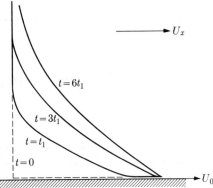

Fig. 7-2. Velocity distribution for flow along a flat plate in motion at various time intervals.

Answer. The solution can be obtained by applying the method of Laplace transforms [1, 3, 6] of linear differential equations. The problem is analogous to heating a semi-infinite solid with a constant wall temperature.

The Laplace transform is

$$f(s) = \int_0^\infty e^{-st} F(t)\, dt = L\{F(t)\}. \qquad (\text{i})$$

The symbol $L\{F(t)\}$ is used to represent the integral. An inverse transformation will have to be found (this is usually the most difficult part of the problem); it is denoted by

$$F(t) = L^{-1}\{f(s)\}. \qquad (\text{ii})$$

Tables are used to aid in its evaluation.

A theorem needed in this problem is

$$L\{\partial F/\partial t\} = s f(s) - F(0). \qquad (\text{iii})$$

Let us take $F(t) = U_x(y, t)$; thus

$$f(s) = L\{F(t)\} = L\{U_x(y, t)\} = Q(y, s). \qquad (\text{iv})$$

By theorem (iii) and the first boundary condition of Eq. (7-7),

$$L\{\partial U_x/\partial t\} = sQ - U_x(y, 0) = sQ. \qquad (\text{v})$$

By the use of (iv) and (v), Eq. (7–6) can be transformed to

$$sQ = \nu(\partial^2 Q/\partial y^2). \tag{vi}$$

When $y = 0$, $U_x = U_0 = sQ$ by the second boundary condition of Eq. (7–7) and Eq. (i); thus, at $y = 0$,

$$Q = U_0/s. \tag{vii}$$

Equation (vi) is a linear differential equation with constant coefficients, and has a general solution of the form

$$Q = ae^{-y\sqrt{s/\nu}} + be^{+y\sqrt{s/\nu}}. \tag{viii}$$

[See Eq. (ix) of Example 7–3.] The third boundary condition of Eq. (7–7) gives U_x as zero at $y = \infty$; therefore $b = 0$. Equation (vii) gives $a = U_0/s$; thus the final solution is

$$Q = (U_0/s)e^{-y\sqrt{s/\nu}} = L\{U_x\}. \tag{ix}$$

To find the inverse transform, let

$$A = y/\sqrt{\nu}. \tag{x}$$

Equation (ii) becomes

$$U_x = L^{-1}\{(U_0/s)e^{-A\sqrt{s}}\}. \tag{xi}$$

From a table of transforms, we find

$$U_x(y, t) = U_0[1 - \text{erf}(A/2\sqrt{t}\,)], \tag{xii}$$

where

$$\text{erf}(\eta) = (2/\sqrt{\pi})\int_0^\eta e^{-\eta^2} d\eta \tag{xiii}$$

The final result is

$$U_x/U_0 = 1 - (2/\sqrt{\pi})\int_0^\eta e^{-\eta^2} d\eta, \tag{7–8}$$

where $\eta = A/2\sqrt{t} = y/2\sqrt{\nu t}$.

7-2 THE FLOW BETWEEN TWO FLAT PLATES

For the flow between two flat plates (Fig. 7–3), Eqs. (7–3) and (7–4) are still valid. A pressure drop may have to exist in the x-direction to balance shear forces, depending on the relative movement of the fluid and walls. Let us assume that dp/dx is finite. If all the external forces are zero, $\partial p/\partial y$ is zero by Eq. (7–4), and the pressure becomes

$$p = f_2(x, t).$$

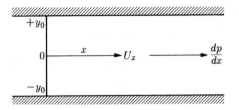

Fig. 7–3. Flow between two parallel flat plates (Couette flow).

In essence, we are making the assumption that hydrostatic pressure is negligible compared to dynamic pressure. However, the equation would be exact if the dynamic pressure were used instead of the total pressure, i.e., the difference between the total pressure and the hydrostatic pressure. In the following discussion it will be understood that p now refers to this difference.

Equations (7–3) and (7–4) now become

$$U_x = f_1(y, t), \qquad U_y = U_z = 0, \qquad p = f_2(x, t), \tag{7-10}$$

and

$$\rho(\partial U_x/\partial t) = -dp/dx + \mu(\partial^2 U_x/\partial y^2). \tag{7-11}$$

Further, dp/dx must be a constant or a function of time only, since by Eq. (7–10), $p \neq f(y)$ and $U_x \neq f(x)$.

Let us consider first the flow when one wall is at rest and the other is in motion ($U_1 = 0$, $U_2 = U_0$), called *Couette flow*. The boundary conditions are

$$\begin{aligned} U_x &= 0 & \text{at} \quad y &= -y_0, \\ U_x &= U_0 & \text{at} \quad y &= y_0, \end{aligned} \quad \text{for all } t. \tag{7-12}$$

For steady flow, Eq. (7–11) becomes

$$d^2 U_x/dy^2 = (1/\mu)(dp/dx). \tag{7-13}$$

The solution of Eqs. (7–12) and (7–13) can be obtained by direct integration, since dp/dx is constant. From the integration, we have

$$U_x = (1/2\mu)(dp/dx)y^2 + Ay + B. \tag{7-14}$$

The boundary conditions (7–12), when substituted into (7–14), give

$$(1/2\mu)(dp/dx)y_0^2 + Ay_0 + B = U_0, \qquad (1/2\mu)(dp/dx)y_0^2 - Ay_0 + B = 0.$$

Combining the above two equations, we find

$$A = \frac{U_0}{2y_0}, \qquad B = \frac{U_0}{2} - \frac{y_0^2}{2\mu}\frac{dp}{dx}.$$

From Eq. (7–14) and the values of the constants, the final result is

$$U_x = \frac{U_0}{2}\left(1 + \frac{y}{y_0}\right) - \frac{y_0^2}{2\mu}\frac{dp}{dx}\left(1 - \frac{y^2}{y_0^2}\right). \tag{7-15}$$

The solution is shown in Fig. 7–4. For $dp/dx = 0$, the velocity is linear across

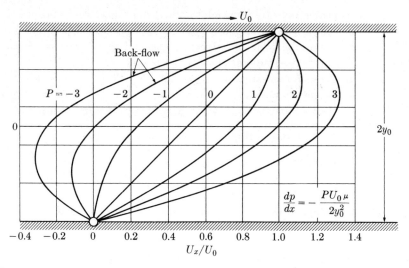

Fig. 7–4. Couette flow between two parallel flat plates. [From H. Schlichting, Boundary Layer Theory, McGraw-Hill, New York (1960). By permission.]

the fluid. This case is called *simple Couette flow*. For a pressure drop ($dp/dx < 0$), the velocity is positive everywhere, and for a pressure increase ($dp/dx > 0$), the velocity can become negative; i.e., a backflow can exist. The point of reversal would be that point at which $dU_x/dy = 0$ at $y = -y_0$. From Eq. (7–14), this occurs when

$$dp/dx = U_0\mu/2y_0^2. \tag{7-16}$$

For the commencement of this type of flow with $dp/dx = 0$, the equation is the same as that already given (7–6); however, the boundary conditions are different. Pai [5] has outlined the necessary steps for the solution which is given in Fig. 7–5.

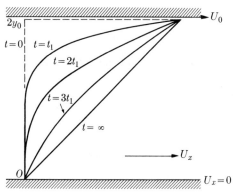

Fig. 7–5. The commencement of Couette flow between two parallel flat plates. [From Shih-I Pai, Viscous Flow Theory, Vol. I, Laminar Flow, D. Van Nostrand, Princeton, New Jersey (1956). By permission.]

Let us now consider the case in which the fluid moves between stationary walls ($U_1 = U_2 = 0$, called *plane Poiseuille flow*). In this case, the pressure gradient in the x-direction must be finite to balance the shear forces. However, both walls will affect the y-direction gradient to the same degree, so that $\partial p/\partial y$ still can be considered to be zero. For the steady-state system, Eq. (7–13) is valid, and the new boundary conditions are simply $U_x = 0$ at $y = \pm y_0$. Also at $y = 0$, we have $U_x = U_{\mathfrak{C}} = U_{x,\max}$, and dp/dx is constant everywhere. The constants of integration are

$$A = 0, \qquad B = U_{x,\max} = -(y_0^2/2\mu)(dp/dx)$$

and Eq. (7–14) becomes

$$U_x = U_{x,\max}(1 - y^2/y_0^2) = -(y_0^2/2\mu)(dp/dx)(1 - y^2/y_0^2) \qquad (7\text{–}17)$$

The velocity distribution in the liquid between the plates is parabolic at infinite time.

7–3 HAGEN-POISEUILLE FLOW, OR FLOW IN A CIRCULAR PIPE

The flow diagram is the same as that in Fig. 7–3; however, cylindrical coordinates should be used. The Navier-Stokes equation (4–12) for steady-state flow of an incompressible fluid becomes

$$dp/dz = \mu[d^2U_z/dr^2 + (1/r)(dU_z/dr)]. \qquad (7\text{–}18)$$

The boundary conditions are

$$dU_z/dr = 0 \quad \text{at} \quad r = 0 \quad \text{(symmetry)}$$
$$U_z = 0 \quad \text{at} \quad r = r_0, \quad dp/dz = \text{const.}$$

Equation (7–18) can be rearranged to

$$dp/dz = \mu(1/r)(d/dr)\,[r(dU_z/dr)],$$

which can be integrated:

$$(r^2/2)(dp/dz) = \mu r(dU_z/dr) + C_1, \qquad (7\text{–}19)$$

where $C_1 = 0$ from the boundary condition at the centerline. Integration again gives

$$(r^2/4)(dp/dz) = \mu U_z + C_2, \qquad (7\text{–}20)$$

where

$$C_2 = (r_0^2/4)(dp/dz)$$

from the boundary condition. Combined with the value of C_2, Eq. (7–20) becomes

$$U_z = -\frac{r_0^2}{4\mu}\frac{dp}{dz}\left(1 - \frac{r^2}{r_0^2}\right) = U_{z,\max}\left(1 - \frac{r^2}{r_0^2}\right). \qquad (7\text{–}21)$$

The flow rate Q can be obtained as follows:

$$Q = \int_0^{r_0} U_z 2\pi r \, dr = -\frac{dp}{dz} \frac{\pi r_0^2}{2\mu} \left(\frac{r^2}{2} - \frac{r^4}{4r_0^2} \right) \Big|_0^{r_0} = -\frac{\pi r_0^4}{8\mu} \frac{dp}{dz}. \quad (7\text{-}22)$$

This equation was obtained experimentally by Hagen and by Poiseuille. The amount of fluid passed in a capillary tube in unit time is proportional to the radius to the fourth power, to the pressure drop, and inversely to the length of the tube. The equation is the basis of the capillary tube viscometer. We know, or can measure directly, all the terms except the viscosity, and thus can calculate this property of the fluid.

The average velocity would be

$$U_{z,\text{ave}} = \frac{Q}{\pi r_0^2} = -\frac{r_0^2}{8\mu} \frac{dp}{dz} = \tfrac{1}{2} U_{z,\text{max}}, \quad (7\text{-}23)$$

where ave denotes the average across the cross section.

If a friction factor is defined as

$$f = -\tfrac{1}{2}(dp/dz)(d_0/\rho U_{z,\text{ave}}^2), \quad (7\text{-}24)$$

where d_0 is the pipe diameter, then a combination of Eqs. (7-23) and (7-24) gives

$$f = 16/N_{\text{Re}}. \quad (7\text{-}25)$$

If the effect of external forces is included and can be considered constant, so that it could be combined with dp/dx, then the flow rate would be

$$Q = (\pi r_0^4/8\mu)(-dp/dz + \rho F). \quad (7\text{-}26)$$

If slip at the boundary were allowed, the flow rate would become

$$Q = (\pi r_0^4/8\mu)(-dp/dz + \rho F)(1 + 4\lambda/r_0), \quad (7\text{-}27)$$

where $\lambda = \beta/\mu$, and β is the coefficient of sliding friction or the ratio of tangential forces to the relative velocity. Actually, there are no homogeneous materials for which this correction need be considered.

For unsteady flow, the differential equation in cyclindrical coordinates is

$$\frac{\partial U_z}{\partial t} = -\frac{1}{\rho} \frac{dp}{dz} + \nu \left(\frac{\partial^2 U_z}{\partial r^2} + \frac{1}{r} \frac{\partial U_z}{\partial r} \right), \quad (7\text{-}28)$$

dp/dz is still constant, and the boundary conditions are

$$t = 0, \quad U_z = 0, \quad \text{for all } r;$$
$$t > 0, \quad U_z = 0, \quad r = r_0;$$
$$t > 0, \quad \partial U_z/\partial r = 0, \quad r = 0.$$

Equation (7-28), with the above boundary conditions, has been solved by Szymanski [7]. In terms of Bessel functions, Szymanski's solution is

$$U_z = -\frac{r_0^2}{4\mu} \frac{dp}{dz} \left[1 - \frac{r^2}{r_0^2} - 8 \sum_{n=1}^{\infty} \frac{J_0(\alpha_n r/r_0)}{\alpha_n^3 J_0'(\alpha_n)} e^{-\alpha_n^2 \nu t/r_0^2} \right]. \quad (7\text{-}29)$$

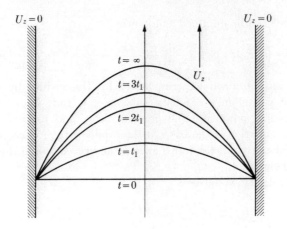

Fig. 7–6. The commencement of flow in a circular pipe. [From Shih-I Pai, Viscous Flow Theory, Vol. I, Laminar Flow, D. Van Nostrand, Princeton, New Jersey (1956). By permission.]

An outline of the solution is given as Example 7–2. A representation of the flow is shown in Fig. 7–6.

EXAMPLE 7–2. The solution to Eq. (7–28), with its accompanying boundary conditions, is given as Eq. (7–29). This example will illustrate that solution. We have

$$\frac{\partial U_z}{\partial t} = -\frac{1}{\rho}\frac{dp}{dz} + \nu\left(\frac{\partial^2 U_z}{\partial r^2} + \frac{1}{r}\frac{\partial U_z}{\partial r}\right). \tag{7–28}$$

The boundary conditions are

$$t = 0, \quad U_z = 0, \quad r = r;$$
$$t > 0, \quad U_z = 0, \quad r = r_0;$$
$$t > 0, \quad \partial U_z/\partial r = 0, \quad r = 0.$$

Answer. The problem reduces to that of solving the equation at the boundary of the flow. At the boundary, $r = r_0$, and the constant term $(1/\rho)(dp/dz)$ will vanish, since one of the boundary conditions is that both U_z and $\partial U_z/\partial r$ must be zero for $t = 0$. Thus Eq. (7–28), without the constant term, will be valid at the boundary for all time. This reduced equation can be solved by the method of separation of variables. Let

$$U_z = X(r)\,T(t). \tag{i}$$

Then

$$X\frac{\partial T}{\partial t} = \nu\left(T\frac{\partial^2 X}{\partial r^2} + \frac{T}{r}\frac{\partial X}{\partial r}\right), \tag{ii}$$

and

$$\frac{1}{T}\frac{\partial T}{\partial t} = \nu\left(\frac{1}{X}\frac{\partial^2 X}{\partial r^2} + \frac{1}{rX}\frac{\partial X}{\partial r}\right) = -a^2. \tag{iii}$$

This gives two ordinary differential equations:

$$(dT/dt) + a^2 T = 0, \qquad \text{(iv)}$$

$$(d^2 X/dr^2) + (1/r)(dX/dr) + (a^2/\nu)X = 0. \qquad \text{(v)}$$

The solution to (iv) is

$$T = C_1 e^{-a^2 t}. \qquad \text{(vi)}$$

The solution to (v) is obtained as follows. The transformation given below can be used to alter the equation:

$$r = e^u. \qquad \text{(vii)}$$

Then

$$dX/dr = e^{-u}(dX/du), \qquad \text{(viii)}$$

and

$$\frac{d^2 X}{dr^2} = e^{-2u}\left(\frac{d^2 X}{du^2} - \frac{dX}{du}\right). \qquad \text{(ix)}$$

[See Eqs. (ii) and (iii) of Example 7-3.] Substitution into Eq. (v) gives

$$d^2 X/du^2 + (a^2/\nu)e^{2u}X = 0, \qquad \text{(x)}$$

which is a reduced linear differential equation with variable coefficients. This reduced equation has a solution in terms of Bessel functions, and can be found in a table of equations. Its form is

$$d^2 y/dx^2 + (k^2 e^{2x} - n^2)y = 0, \qquad \text{(xi)}$$

and the solution is

$$y = C_1' J_n(ke^x) + C_2' Y_n(ke^x). \qquad \text{(xii)}$$

For Eq. (x), this becomes

$$X = C_1' J_0\left(\frac{a}{\sqrt{\nu}} e^u\right) + C_2' Y_0\left(\frac{a}{\sqrt{\nu}} e^u\right). \qquad \text{(xiii)}$$

Using the transformation of Eq. (vii), we get

$$X = C_1' J_0\left(\frac{ar}{\sqrt{\nu}}\right) + C_2' Y_0\left(\frac{ar}{\sqrt{\nu}}\right). \qquad \text{(xiv)}$$

Now, from (i), $U_z = f(X, T)$, and from the third boundary condition,

$$r = 0, \qquad U_z = U_{z,\max} \qquad \text{for} \quad t > 0.$$

From the tables of Bessel functions,

$$J_0(0) = 1 \quad \text{and} \quad Y_0(0) = \infty.$$

At $r = 0$, U_z must be finite, and X must be finite; consequently $C_2' = 0$. This gives

$$X = C_1' J_0(ar/\sqrt{\nu}). \qquad \text{(xv)}$$

The solution according to Eq. (i) can be obtained as the product of Eqs. (vi) and (xv); thus

$$U_z = C_1' J_0(ar/\sqrt{\nu}) e^{-a^2 t}. \quad \text{(xvi)}$$

This is the solution at the boundary. The second boundary condition at this point indicates that this velocity must be zero for all time. Since the exponential term is not zero, the Bessel function must be zero. However, there are an infinite number of zeros of this function. Szymanski [7] shows that if α_n represents the zeros of the Bessel function, then

$$a = \alpha_n \sqrt{\nu}/r_0 \quad \text{(xvii)}$$

and that for the infinite number of zeros, Eq. (xvi) becomes,

$$U_z = \sum_{n=1}^{\infty} C_n J_0(\alpha_n r/r_0) e^{-\alpha_n^2 \nu t/r_0^2}. \quad \text{(xviii)}$$

The solution of the complete differential equation (7–28) must have as one of its limits the solution at the boundary just considered. Inspection of Eq. (xviii) shows that although it is zero at the boundary, it will not be zero for any other r; that is, $\alpha_n r/r_0 \neq \alpha_n$. It should be noted that this is true for all r, even when $t = 0$. However, we know that under this condition the velocity must be zero everywhere. If the solution to the complete problem is *assumed* to be of the form

$$U_z = U_{z,\infty} + U_{z,\text{boundary problem}}, \quad \text{(xix)}$$

then at $t = 0$,

$$U_{z,\infty} = -U_{z,\text{boundary problem}} = -\sum_{n=1}^{\infty} C_n J_0(\alpha_n r/r_0), \quad \text{(xx)}$$

since $U_z = 0$. In addition, at $t = \infty$, the boundary problem is zero, since the exponential term goes to zero, and

$$U_z = U_{z,\infty}. \quad \text{(xxi)}$$

At $t = \infty$, the solution is the steady-state problem, and has been solved as Eq. (7–21). Using Eqs. (xviii), (xx), and (7–21), we obtain

$$U_{z,\infty} = -\frac{r_0^2}{4\mu} \frac{dp}{dz} \left(1 - \frac{r^2}{r_0^2}\right) = -\sum_{n=1}^{\infty} C_n J_0 \frac{\alpha_n r}{r_0}. \quad \text{(xxii)}$$

Szymanski [7] has used this equation to evaluate the constant C_n and found that it is

$$C_n = \frac{2(r_0^2/\mu)(dp/dz)}{\alpha_n^3 J_0'(\alpha_n)}. \quad \text{(xxiii)}$$

In essence, C_n has been selected so that Eq. (xviii) is zero everywhere when $t = \infty$, and equal to the steady-state solution when $t = 0$. Combining Eqs.

(xviii), (xix), (xxii), and (xxiii) gives the final desired solution:

$$U_z = -\frac{r_0^2}{4\mu}\frac{dp}{dz}\left[1 - \frac{r^2}{r_0^2} - 8\sum_{n=1}^{\infty}\frac{J_0(\alpha_n r/r_0)}{\alpha_n^3 J_0'(\alpha_n)}e^{-\alpha_n^2 \nu t/r_0^2}\right]. \quad (7\text{--}29)$$

Equation (xix) was an assumption only, and thus Eq. (7–29) should be substituted back into Eq. (7–28) to establish that it is an exact solution of the total general differential equation.

7-4 FLOW BETWEEN ROTATING CONCENTRIC CYLINDERS

To interpret the results from Couette rotational-type viscometers, we require solutions of the Navier-Stokes equation in cylindrical coordinates. The flow between two concentric cylinders is shown in Fig. 7–7.

In cylindrical coordinates, for steady state,

$$U_\theta = f_1(r), \qquad U_r = U_z = 0, \qquad p = f_2(r) \quad (p \text{ is dynamic pressure}).$$

The differential equations are (from the Navier-Stokes equation)

$$\frac{dp}{dr} = \frac{\rho}{r}U_\theta^2,$$

$$\frac{d^2 U_\theta}{dr^2} + \frac{1}{r}\frac{dU_\theta}{dr} - \frac{U_\theta}{r^2} = 0. \quad (7\text{--}30)$$

The boundary conditions will be

$$\begin{aligned} r &= r_i, & U_\theta &= r_i\omega_i, \\ r &= r_0, & U_\theta &= r_0\omega_0. \end{aligned} \quad (7\text{--}31)$$

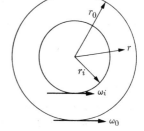

Fig. 7–7. *Flow between rotating concentric cylinders (Couette flow).*

The solution of Eq. (7–30), with the boundary conditions of (7–31), is given in Example 7–3; the results are

$$U_\theta = \left(\frac{1}{r_i^2 - r_0^2}\right)\left[r(r_i^2\omega_i - r_0^2\omega_0) + \frac{r_i^2 r_0^2}{r}(\omega_0 - \omega_i)\right], \quad (7\text{--}32)$$

$$p - p_i = \left[\frac{\rho}{(r_i^2 - r_0^2)^2}\right]\left[\frac{(r_i^2\omega_i - r_0^2\omega_0)^2(r^2 - r_i^2)}{2} + 2r_i^2 r_0^2(\omega_0 - \omega_i)(r_i^2\omega_i - r_0^2\omega_0)\ln\frac{r}{r_i}\right.$$

$$\left. + r_i^4 r_0^4(\omega_0 - \omega_i)^2\left(\frac{1}{r^2} - \frac{1}{r_i^2}\right)\right]. \quad (7\text{--}33)$$

Equation (7–32) is the basis for one method of viscosity measurement. For example, if the inner cylinder is rotated and the torque per unit height T is measured at the outer cylinder, then

$$T = \frac{Fr_0}{h} = 2\pi r_0^2 \tau_0 = -2\pi r_0^2 \mu\left(r\frac{\partial U_\theta/r}{\partial r}\right) = -4\pi\mu\left(\frac{r_i^2 r_0^2}{r_i^2 - r_0^2}\right)\omega_i. \quad (7\text{--}34)$$

For a cylinder rotating in a large fluid area, $\omega_0 = 0$ and $r_0 \to \infty$. From Eq. (7–32),

$$U_\theta = r_i^2 \omega_i / r, \qquad (7\text{–}35)$$

and the torque per unit height is from Eq. (7–34):

$$T = 4\pi \mu r_i^2 \omega_i. \qquad (7\text{–}36)$$

EXAMPLE 7–3. Find the solution for Eq. (7–30) with the boundary conditions of Eq. (7–31).

$$\frac{dp}{dr} = \frac{\rho}{r} U_\theta^2, \qquad \frac{d^2 U_\theta}{dr^2} + \frac{1}{r}\frac{dU_\theta}{dr} - \frac{U_\theta}{r^2} = 0, \qquad (7\text{–}30)$$

or, as an exact differential,

$$\frac{\partial}{\partial r}\left[\frac{1}{r}\frac{\partial}{\partial r}(rU_\theta)\right] = 0.$$

The boundary conditions are:

$$\begin{aligned} r &= r_i, & U_\theta &= r_i \omega_i, \\ r &= r_0, & U_\theta &= r_0 \omega_0. \end{aligned} \qquad (7\text{–}31)$$

Answer. There are two equations to be solved. The second equation must be solved first in order to obtain an expression for U_θ to be used in the first equation. The equation is a second-order linear differential equation with variable coefficients. However, since it is of the general form of Euler's equation, it can be reduced to constant coefficients by the transformation

$$r = e^u. \qquad (\text{i})$$

Thus

$$\frac{dU_\theta}{dr} = \frac{dU_\theta}{du}\frac{du}{dr} = e^{-u}\frac{dU_\theta}{du}, \qquad (\text{ii})$$

$$\frac{d^2 U_\theta}{dr^2} = \frac{du}{dr}\frac{d[(dU_\theta/du)e^{-u}]}{du} = e^{-2u}\left(\frac{d^2 U_\theta}{du^2} - \frac{dU_\theta}{du}\right). \qquad (\text{iii})$$

Combining these with Eq. (7–30), we obtain

$$(d^2 U_\theta/du^2)e^{-2u} - U_\theta e^{-2u} = 0, \qquad (\text{iv})$$

or

$$d^2 U_\theta/du^2 - U_\theta = 0. \qquad (\text{v})$$

Several methods of solving this equation are available, one of which is the operator D method.

$$L_2\{(D)y\} = F(x). \qquad (\text{vi})$$

For Eq. (v),
$$L_2\{(D)y\} = (D^2 - 1)y = 0. \qquad \text{(vii)}$$

The roots of $D^2 - 1$ are
$$m_1 = 1 \quad \text{and} \quad m_2 = -1. \qquad \text{(viii)}$$

Since the roots are real, the solution is of the form
$$y = C_1 e^{m_1 x} + C_2 e^{m_2 x}. \qquad \text{(ix)}$$

That is,
$$U_\theta = C_1 e^u + C_2 e^{-u}. \qquad \text{(x)}$$

Combining this with Eq. (i), we have
$$U_\theta = C_1 r + C_2/r. \qquad \text{(xi)}$$

Using the boundary conditions (7–31), we obtain
$$r_i \omega_i = C_1 r_i + C_2/r_i, \qquad r_o \omega_o = C_1 r_o + C_2/r_o, \qquad \text{(xii)}$$

and solving for C_1 and C_2, we have
$$C_1 = \frac{r_i^2 \omega_i - r_o^2 \omega_o}{r_i^2 - r_o^2}, \qquad C_2 = \frac{r_i^2 r_o^2 (\omega_o - \omega_i)}{r_i^2 - r_o^2}. \qquad \text{(xiii)}$$

Combining Eqs. (xi) and (xiii), we obtain
$$U_\theta = \frac{1}{r_i^2 - r_o^2}\left[r(r_i^2 \omega_i - r_o^2 \omega_o) + \frac{r_i^2 r_o^2}{r}(\omega_o - \omega_i)\right]. \qquad \text{(7–32)}$$

Combining the first equation of (7–30) with Eq. (xi) yields
$$dp = \frac{\rho}{r}\left(C_1 r + \frac{C_2}{r}\right)^2 dr = \rho\left(C_1^2 r + \frac{2C_1 C_2}{r} + \frac{C_2^2}{r^3}\right) dr. \qquad \text{(xiv)}$$

Integrating, we get
$$p \Big|_{p_i}^{p} = \rho\left(\frac{C_1^2 r^2}{2} + 2C_1 C_2 \ln r + \frac{C_2^2}{2r^2}\right)\Big|_{r_i}^{r},$$

and combining this with Eq. (xiii), we find
$$p - p_i = \frac{\rho}{(r_i^2 - r_o^2)^2}\Bigg[\frac{(r_i^2 \omega_i - r_o^2 \omega_o)^2 (r^2 - r_i^2)}{2} + 2r_i^2 r_o^2(\omega_o - \omega_i)(r_i^2 \omega_i - r_o^2 \omega_o)\ln \frac{r}{r_i}$$
$$+ r_i^4 r_o^4 (\omega_o - \omega_i)^2 \left(\frac{1}{r^2} - \frac{1}{r_i^2}\right)\Bigg]. \qquad \text{(7–33)}$$

Equation (xii) could be obtained by direct integration of the exact differential form of Eq. (7–30); however, the operator D method was used to illustrate the procedure.

7-5 COMPRESSIBLE FLOW

Another exact solution of special interest is the determination of the shock wave thickness in compressible flow. This is one of the few possible exact solutions involving the additional complexity that compressibility introduces. However, the solution can be obtained only for the steady one-dimensional case (one spatial coordinate).

For a constant viscosity, the motion equation (4–12) becomes

$$U_x \frac{dU_x}{dx} = -\frac{1}{\rho}\frac{dp}{dx} + \tfrac{4}{3}\nu \frac{d^2 U_x}{dx^2}. \tag{7-37}$$

If the viscosity were not constant, the equation would take the form

$$U_x \frac{dU_x}{dx} = -\frac{1}{\rho}\frac{dp}{dx} + \frac{4}{3\rho}\frac{d}{dx}\left(\mu \frac{dU_x}{dx}\right), \tag{7-38}$$

which can be deduced from Eqs. (7–1) and (7–2). The equation of continuity (3–19) becomes

$$\frac{d\rho U_x}{dx} = 0, \tag{7-39}$$

and the energy equations (3–30) and (3–43) for constant k reduce to

$$\rho c_v U_x \frac{dT}{dx} = k\frac{d^2 T}{dx^2} + \tfrac{4}{3}\mu\left(\frac{dU_x}{dx}\right)^2 - p\frac{dU_x}{dx}. \tag{7-40}$$

If k were not constant, the second term would have the form $d[k(dT/dx)]/dx$, which can be obtained from Eqs. (3–30) and (3–45). The third term is the one-dimensional dissipation function for compressible flow. It is obtained from $(\bar{\bar{\tau}}:(\nabla U))$ where $\bar{\bar{\tau}}$ is given by Eq. (3–43) (κ assumed zero). In this case, this is a simple operation since both (∇U) and $(\bar{\nabla} U)$ reduce to dU_x/dx:

$$-(\bar{\bar{\tau}}:(\nabla U)) = -\left(-\mu\frac{2dU_x}{dx} + \tfrac{2}{3}\mu\frac{dU_x}{dx}\right)\frac{dU_x}{dx} = \tfrac{4}{3}\mu\left(\frac{dU_x}{dx}\right)^2.$$

The final term requires the use of an equation of state, which in this case is assumed to be the ideal gas law ($p/\rho = RT/M_w$):

$$-T\left(\frac{\partial p}{\partial T}\right)_\rho (\nabla \cdot U) = -T\frac{\rho R}{M_w}\frac{dU_x}{dx} = -p\frac{dU_x}{dx}.$$

The system of equations just developed can be solved analytically for the restricted case of a Prandtl number of $\tfrac{3}{4}$ (see Example 7–4). The final solution for the velocity ratio is

$$(1 - V_x)/(V_x - \alpha)^\alpha = A \exp[\beta(1 - \alpha)M_{-\infty}\eta], \tag{7-41}$$

where $-\infty$ refers to a position far upstream from the shock. Here

$$V_x = U_x/U_{x,-\infty},$$

$$\alpha = [(\gamma-1)/(\gamma+1)] + [2/(\gamma+1)](1/M^2_{-\infty}),$$

$\gamma = c_p/c_v$, the ratio of specific heats,

$$\beta = \tfrac{9}{8}(\gamma+1)\sqrt{\pi/8\gamma},$$

$M_{-\infty} = U_{x,-\infty}/\sqrt{\gamma R T_{-\infty}/M_w}$, upstream Mach number [see (12–14) and (12–15)],

$$\eta = x/l_{-\infty}, \qquad \eta_{-\infty} = x_{-\infty}/l_{-\infty},$$

$l_{-\infty} = 3\nu_{-\infty}\sqrt{\pi M_w/8RT_{-\infty}}$, the mean free path at upstream conditions.

Figure 7–8 shows two distributions calculated for air, indicating that the major portion of the velocity change occurs in a small region at the center of the shock. In Chapter 12 we will make use of this result by considering the shock as a plane discontinuity with zero thickness. Morduchow and Libby [4] have introduced the temperature variation of viscosity, and the deviation of these modified results from those in Fig. 7–8 is quite small. Von Mises [9] extended the analysis to any Prandtl number, and Ludford [2] has discussed other one-dimensional steady flows by using the von Mises approach.

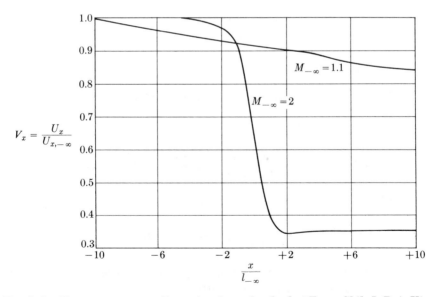

Fig. 7–8. Flow across a one-dimensional steady shock. [*From Shih-I Pai*, Viscous Flow Theory, Vol. I, Laminar Flow, D. Van Nostrand, Princeton, New Jersey (1956). By permission.]

EXAMPLE 7–4. Solve the system of equations (7–37), (7–39), and (7–40) for a Prandtl number of $\tfrac{3}{4}$.

Answer. Equation (7–39) can be integrated to

$$\rho U_x = \rho_{-\infty} U_{x,-\infty} = \text{const}, \tag{i}$$

where the $-\infty$ conditions are those far upstream. Combining Eq. (i) with the ideal gas law, we have

$$p = \frac{\rho R T}{M_w} = \frac{\rho_{-\infty} U_{x,-\infty} R T}{U_x M_w}. \tag{ii}$$

Differentiating Eq. (ii) we get

$$\frac{dp}{dx} = \frac{\rho_{-\infty} U_{x,-\infty} R}{M_w}\left(\frac{1}{U_x}\frac{dT}{dx} - \frac{T}{U_x^2}\frac{dU_x}{dx}\right) = \frac{\rho_{-\infty} U_{x,-\infty} R}{M_w U_x}\frac{dT}{dx} - \frac{p}{U_x}\frac{dU_x}{dx}. \tag{iii}$$

The energy equation (7–40) can be rewritten with the aid of Eq. (iii) to give

$$\rho U_x c_p \frac{dT}{dx} = k\frac{d^2 T}{dx^2} + \tfrac{4}{3}\mu\left(\frac{dU_x}{dx}\right)^2 + U_x\frac{dp}{dx}, \tag{iv}$$

where, from thermodynamics, $c_p = c_v + R/M_w$ for an ideal gas (see Eq. 12–5).

The equation of motion (7–37), Eq. (i), and the energy equation (iv) can be combined to give

$$c_p\frac{dT}{dx} = \frac{k}{\rho_{-\infty} U_{x,-\infty}}\frac{d^2 T}{dx^2} + \frac{4\mu}{3\rho_{-\infty} U_{x,-\infty}}\left(\frac{dU_x}{dx}\right)^2 - U_x\frac{dU_x}{dx} + \frac{4\mu}{3\rho_{-\infty} U_{x,-\infty}} U_x\frac{d^2 U_x}{dx^2},$$

or

$$c_p\frac{dT}{dx} = \frac{k}{\rho_{-\infty} U_{x,-\infty}}\frac{d^2 T}{dx^2} + \frac{4\mu}{3\rho_{-\infty} U_{x,-\infty}}\frac{d}{dx}\left(U_x\frac{dU_x}{dx}\right) - U_x\frac{dU_x}{dx}. \tag{v}$$

One integration results in

$$c_p T + \tfrac{1}{2}U_x^2 = \frac{k}{\rho_{-\infty} U_{x,-\infty}}\frac{dT}{dx} + \frac{4\mu U_x}{3\rho_{-\infty} U_{x,-\infty}}\frac{dU_x}{dx} + C_1,$$

which can be rearranged to

$$c_p T + \tfrac{1}{2}U_x^2 = \frac{k}{c_p \rho_{-\infty} U_{x,-\infty}}\left(\tfrac{4}{3}N_{\mathrm{Pr}}\frac{d\tfrac{1}{2}U_x^2}{dx} + c_p\frac{dT}{dx}\right) + C_1, \tag{vi}$$

where $N_{\mathrm{Pr}} = c_p\mu/k$. If N_{Pr} is assumed to be $\tfrac{3}{4}$ for convenience (which is a good approximation), then Eq. (vi) can be written as

$$c_p T + \tfrac{1}{2}U_x^2 = \frac{k}{c_p \rho_{-\infty} U_{x,-\infty}}\frac{d}{dx}(c_p T + \tfrac{1}{2}U_x^2) + C_1, \tag{vii}$$

and can be integrated to

$$c_p T + \tfrac{1}{2} U_x^2 = C_1 + C_2 \exp\left(\frac{c_p \rho_{-\infty} U_{x,-\infty} x}{k}\right). \tag{viii}$$

The constant C_2 must be zero, for the energy term on the left must remain finite even for infinite x; C_1 can be evaluated at $-\infty$ to give

$$c_p T + \tfrac{1}{2} U_x^2 = C_1 = c_p T_{-\infty} + \tfrac{1}{2} U_{x,-\infty}^2. \tag{ix}$$

This is an exact one-dimensional solution to the energy equation when the Prandtl number is $\tfrac{3}{4}$.

Equations (7–37) and (i) can be combined and integrated to give

$$\rho_{-\infty} U_{x,-\infty} U_x = -p + \tfrac{4}{3}\mu(dU_x/dx) + C_3. \tag{x}$$

At $-\infty$, $dU_x/dx = 0$, and

$$C_3 = p + \rho_{-\infty} U_{x,-\infty}^2 = \rho_{-\infty}[U_{x,-\infty}^2 + (RT_{-\infty}/M_w)]. \tag{xi}$$

Combining Eqs. (ii), (ix), and (x) and rearranging, we obtain

$$\frac{4\mu}{3\rho_{-\infty} U_{x,-\infty}} U_x \frac{dU_x}{dx} + \frac{C_3 U_x}{\rho_{-\infty} U_{x,-\infty}} - \frac{\gamma+1}{2\gamma} U_x^2 = \frac{\gamma-1}{\gamma} C_1, \tag{xii}$$

where $\gamma = c_p/c_v$, and the thermodynamic relation for an ideal gas (see Eq. 12–5),

$$c_p = \frac{R}{M_w} \frac{\gamma}{\gamma-1}, \tag{xiii}$$

has been used. With

$$\alpha = \frac{\gamma-1}{\gamma+1} + \frac{2}{\gamma+1}\frac{1}{M_{-\infty}^2}, \qquad \beta = \tfrac{9}{8}(\gamma+1)\sqrt{\pi/8\gamma},$$

$$M_{-\infty} = U_{x,-\infty}/\sqrt{\gamma RT_{-\infty}/M_w}, \qquad V_x = U_x/U_{x,-\infty},$$

$$\eta = x/l_{-\infty}, \qquad l_{-\infty} = 3\nu_{-\infty}\sqrt{\pi M_w/8RT_{-\infty}}.$$

Eq. (xii) becomes

$$V_x(dV_x/d\eta) = \beta M_{-\infty}(V_x - 1)(V_x - \alpha), \tag{xiv}$$

which can be integrated to

$$(1 - V_x)/(V_x - \alpha)^\alpha = C_4 \exp[\beta(1-\alpha)M_{-\infty}\eta]. \tag{xv}$$

Now, C_4 can be evaluated at $\eta = 0$, which gives

$$C_4 = (1 - V_{x0})/(V_{x0} - \alpha)^\alpha = A;$$

so Eq. (xv) becomes

$$(1 - V_x)/(V_x - \alpha)^\alpha = A \exp[\beta(1-\alpha)M_{-\infty}\eta]. \tag{7-41}$$

PROBLEMS

7-1. Several assumptions were made in the derivation of the Hagen-Poiseuille flow in a circular pipe. From a knowledge of these assumptions, what corrections would you make to extend this analysis to the most general case? Discuss any limitations that still exist after the corrections are made.

7-2. Solve the problem of the steady-state shape of a surface of a rotating liquid in a circular container.

7-3. Solve the problem of the steady-state axial flow of an incompressible fluid in an annulus. Consider the area away from the entrance.

7-4. Solve the problem of a laminar film flowing down a vertical flat plate. Consider the area away from the entrance.

7-5. Repeat Problem 7-4 for the flow inside a circular pipe.

7-6. The data in Table 7-1 were obtained from capillary tube measurements. Using these data, what can you say about the assumed boundary conditions in the derivation of the Hagen-Poiseuille pipe law? Show your reasoning. What practical use do the results of this study suggest?

7-7. The coefficient of friction, or the Fanning friction factor, is defined by
$$f = \tfrac{1}{2}(dp/dx)(d_0/U_{z,\text{ave}}^2).$$
What is the value of f in laminar flow in a circular pipe of diameter d_0? Express the result in terms of the Reynolds number.

7-8. Determine the laminar flow in a pipe of elliptical cross section.

7-9. A film of fluid of constant thickness flows down an inclined plane. Determine the flow due to gravity.

7-10. Determine the shear stress distribution and torque required to turn the outer shaft for the tangential laminar flow of an incompressible fluid between two vertical coaxial cylinders. The inner cylinder is at rest and the outer cylinder is rotating. End effects may be neglected. However, suggest how these might be accounted for.

TABLE 7-1

Time of efflux of a given volume of liquid	Diameter	Length	Pressure difference
Time, sec	cm	cm	gm/cm²
100,000.0	0.01	10	4
160.0	0.05	10	4
10.0	0.10	10	4
2.0	0.15	10	4
20.0	0.10	20	4
30.0	0.10	30	4
5.0	0.10	10	8
3.3	0.10	10	12

7-11. The rotation of a single cylinder in an infinite real fluid has been considered in the text [Eqs. (7–35) and (7–36)]. What is the frictionless flow that is equivalent to this real fluid flow? Show and explain the equivalence by comparing the solutions. What is the value of the constant in the ideal flow solution?

7-12. For two-dimensional flow of an incompressible fluid, show that the vorticity satisfies the diffusion equation.

REFERENCES

1. R. V. CHURCHILL, *Operational Mathematics*, McGraw-Hill, New York (1958).
2. G. S. S. LUDFORD, *J. Aeron. Sci.* **18**, 830–834 (1951).
3. H. S. MICKLEY, T. K. SHERWOOD, and C. E. REED, *Applied Mathematics in Chemical Engineering*, McGraw-Hill, New York (1957).
4. M. MORDUCHOW and P. A. LIBBY, *J. Aeron. Sci.* **16**, 674–684 (1949).
5. SHIH-I PAI, *Viscous Flow Theory—Laminar Flow*, Van Nostrand, Princeton, N.J. (1956).
6. L. A. PIPES, *Applied Mathematics for Engineers and Physicists*, McGraw-Hill, New York (1958).
7. P. SZYMANSKI, *J. math. pure et appliquée*, Series 9 **11**, 67–107 (1932).
8. H. SCHLICHTING, *Boundary Layer Theory*, McGraw-Hill, New York (1960).
9. R. VON MISES *J. Aeron. Sci.* **17**, 551–554 (1950).
10. J. E. FROMM, F. H. HARLOW, and J. E. WELCH, *Phys. Fluids* **6**, 975 (1963); **7**, 1147 (1964); **8**, 2182 (1965).

CHAPTER 8

LAMINAR VISCOUS FLOW: VERY SLOW MOTION

An approximate solution to the Navier-Stokes equation can be obtained for the case in which the Reynolds number, or the ratio of inertial to viscous forces, is very small. Under this condition, the inertial effects can be neglected, and the action of viscosity is considered to be controlling. This is, in essence, the assumption opposite to that made for an ideal flow. The main problem to be developed in this chapter is the very slow motion (or creeping motion) of a sphere in an infinite liquid medium, or its equivalent, the slow motion of a fluid over a standing sphere. These are the problems of settling, aerosols, fluidization, and other two-phase flows with very low relative motion. In addition to the sphere problem, which was originally solved by Stokes [24], other shapes, such as cylinders and ellipsoids, can be investigated. These will not be presented in detail here, but can be found in Lamb [14], and in the other references cited. Extensions of the theories to include wall and other effects will be noted.

8-1 STOKES' LAW

The problem of slow motion was briefly outlined in the discussion of the Navier-Stokes equation (Chapter 5). As noted, since the velocity is very small, the fluid can be considered incompressible, or $(\nabla \cdot U) = 0$. If no external forces exist, the Navier-Stokes equation (4–6) becomes

$$\rho(DU/Dt) = -(\nabla p) + \mu(\nabla^2 U). \tag{8–1}$$

For the Stokes approximation, the inertia term on the left can be neglected, since it is of the order of velocity squared as compared with the linear dependency of the other terms; that is,

$$\rho(DU/Dt) = \rho(\partial U/\partial t) + \rho(U \cdot \nabla)U = \rho(\partial U/\partial t), \tag{8–2}$$

where the second term on the right of the first equal sign approaches zero. For steady state, we obtain

$$\nabla p = \mu(\nabla^2 U), \tag{8–3}$$

which is of the same order as Eq. (8–1) and thus will require the same number of boundary conditions. This can be contrasted to ideal motion, in which the assumption of zero viscosity reduced the order of the equation and some of the boundary conditions could not be satisfied (i.e., the fluid slips at the wall). For a sphere located at the origin, in a fluid field moving in the x-direction at a velocity U_∞, the boundary conditions are

$$U_x = U_y = U_z = 0 \qquad \text{at} \qquad r = r_0,$$

and (8–4)

$$U_x = U_\infty, \qquad U_y = U_z = 0 \qquad \text{at} \qquad r = \infty.$$

The divergence of Eq. (8–3) gives

$$\nabla^2 p = 0, \qquad (8\text{–}5)$$

since $(\boldsymbol{\nabla} \cdot \nabla^2 \boldsymbol{U}) = \nabla^2 (\boldsymbol{\nabla} \cdot \boldsymbol{U}) = 0$ by continuity. Thus we see that the pressure in very slow motion is harmonic; i.e., it satisfies the Laplace equation. Many solutions to Eq. (8–5) are possible, but it is necessary to select the one that will satisfy the physical picture and the boundary conditions. One simple harmonic solution is

$$p = -Ax/r^3. \qquad (8\text{–}6)$$

This solution was selected so that the pressure on the upstream $(-x)$ side would be greater than that on the downstream $(+x)$ side (positive versus negative values), and so that the pressure would be zero at an infinite distance from the sphere. In this harmonic solution, A is a constant to be determined and

$$r^2 = x^2 + y^2 + z^2. \qquad (8\text{–}7)$$

Figure 8–1 provides an illustration of the system; the pressure curve is given for the distribution along the x-axis.

Combining Eqs. (8–3) and (8–6), we obtain

$$\nabla^2 \boldsymbol{U} = -\frac{A}{\mu} \boldsymbol{\nabla} \frac{x}{r^3}. \qquad (8\text{–}8)$$

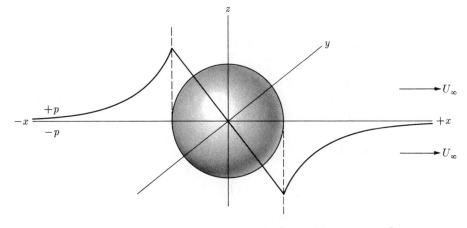

Fig. 8–1. Pressure distribution for very slow motion over a sphere.

This equation can be expressed in cartesian coordinates as

$$\nabla^2 U_x = \frac{A}{\mu}\left(\frac{3x^2}{r^5} - \frac{1}{r^3}\right), \qquad \nabla^2 U_y = \frac{A}{\mu}\frac{3xy}{r^5}, \qquad \nabla^2 U_z = \frac{A}{\mu}\frac{3xz}{r^5}. \qquad (8\text{--}9)$$

The integration of Eqs. (8–9) can be made by recognizing certain symmetry conditions [19]. Lamb [14] obtained a solution of Eq. (8–8) by an expansion into a series of solid harmonic functions. His solution for the velocity takes the form

$$U_x = U_\infty + \left(B + \frac{Ar^2}{6\mu}\right)\frac{\partial(x/r^3)}{\partial x} - \frac{2}{3}\frac{A}{\mu r},$$

$$U_y = \left(B + \frac{Ar^2}{6\mu}\right)\frac{\partial(x/r^3)}{\partial y}, \qquad (8\text{--}10)$$

$$U_z = \left(B + \frac{Ar^2}{6\mu}\right)\frac{\partial(x/r^3)}{\partial z}.$$

Using the boundary conditions of Eq. (8–4), we obtain

$$U_\infty - \frac{2}{3}\frac{A}{\mu r_0} = 0 \qquad \text{and} \qquad B + \frac{Ar_0^2}{6\mu} = 0,$$

or

$$A = \tfrac{3}{2}\mu r_0 U_\infty \qquad \text{and} \qquad B = -\tfrac{1}{4}r_0^3 U_\infty. \qquad (8\text{--}11)$$

Combining Eqs. (8–10) and (8–11) gives the velocities:

$$U_x = U_\infty\left(1 - \frac{r_0}{r}\right) + \tfrac{1}{4}U_\infty r_0(r^2 - r_0^2)\frac{\partial(x/r^3)}{\partial x},$$

$$U_y = \tfrac{1}{4}U_\infty r_0(r^2 - r_0^2)\frac{\partial(x/r^3)}{\partial y}, \qquad (8\text{--}12)$$

$$U_z = \tfrac{1}{4}U_\infty r_0(r^2 - r_0^2)\frac{\partial(x/r^3)}{\partial z}.$$

The negatives of the partials have been given as the terms in the parentheses of Eq. (8–9). These terms can be combined with Eq. (8–12) to give the velocities in terms of the radius of the sphere:

$$U_x = U_\infty\left[\frac{3}{4}\frac{r_0 x^2}{r^3}\left(\frac{r_0^2}{r^2} - 1\right) + 1 - \frac{1}{4}\frac{r_0}{r}\left(3 + \frac{r_0^2}{r^2}\right)\right],$$

$$U_y = U_\infty\left[\frac{3}{4}\frac{r_0 xy}{r^3}\left(\frac{r_0^2}{r^2} - 1\right)\right], \qquad (8\text{--}13)$$

$$U_z = U_\infty\left[\frac{3}{4}\frac{r_0 xz}{r^3}\left(\frac{r_0^2}{r^2} - 1\right)\right].$$

For the sphere, the motion along all planes through the x-axis is the same. For a system containing such an axis of symmetry, Stokes has shown that there exists a stream function somewhat analogous to the two-dimensional counterpart discussed in Chapter 6. The velocities in terms of a potential and the Stokes stream function ψ are

$$U_x = \frac{\partial \phi}{\partial x} = \frac{1}{y'} \frac{\partial \psi}{\partial y'},$$
$$U_{y'} = \frac{\partial \phi}{\partial y'} = -\frac{1}{y'} \frac{\partial \psi}{\partial x}, \quad (8\text{-}14)$$

where y' is used to denote the distance $\sqrt{y^2 + z^2}$, and is shown in Fig. 8–2. From the figure, it is clear that

$$x = r \cos \theta$$

and

$$y' = r \sin \theta,$$

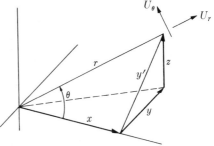

Fig. 8–2. Coordinate system relations.

where θ is in the plane under consideration. The radial velocity U_r and the velocity at right angles U_θ can be obtained as follows:

$$U_r = \frac{\partial \phi}{\partial r} = \frac{\partial \phi}{\partial y'} \frac{\partial y'}{\partial r} = \frac{-1}{r \sin \theta} \left(\frac{\partial \psi}{\partial x}\right) \sin \theta$$
$$= \frac{-1}{r} \frac{\partial \psi}{\partial \theta} \frac{\partial \theta}{\partial x} = \frac{-1}{r}\left(\frac{\partial \psi}{\partial \theta}\right)\left(\frac{-1}{r \sin \theta}\right)$$
$$= \frac{1}{r^2 \sin \theta} \frac{\partial \psi}{\partial \theta}.$$

and in a similar manner,

$$U_\theta = \frac{1}{r} \frac{\partial \phi}{\partial \theta} = \frac{-1}{r \sin \theta} \frac{\partial \psi}{\partial r}. \quad (8\text{-}15)$$

After some manipulation of Eq. (8–13), we can establish the radial velocity as

$$U_r = U_\infty \cos \theta \left[1 - \frac{3}{2}\left(\frac{r_0}{r}\right) + \frac{1}{2}\left(\frac{r_0}{r}\right)^3 \right],$$

since $U_r^2 = U_x^2 + U_y^2 + U_z^2$. This velocity, when combined with Eq. (8–15), gives the stream function

$$\psi = \tfrac{1}{2} U_\infty r^2 \sin^2 \theta \left[1 - \frac{3}{2}\left(\frac{r_0}{r}\right) + \frac{1}{2}\left(\frac{r_0}{r}\right)^3 \right]. \quad (8\text{-}16)$$

If the sphere is considered to move with the velocity $-U_\infty$, and the fluid remains at rest at infinity, then the stream function takes the form

$$\psi = \tfrac{3}{4} U_\infty r_0 r \sin^2 \theta \left[1 - \tfrac{1}{3}(r_0/r)^2 \right]. \quad (8\text{-}17)$$

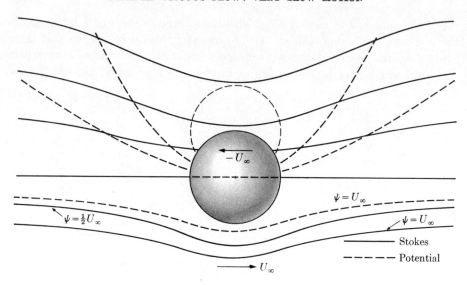

Fig. 8–3. Streamlines for flow over a sphere.

The latter instantaneous streamlines are shown in Fig. 8–3 (the upper half). For comparison, several streamlines calculated in Example 6–2 and plotted in Fig. 6–7 are shown. The difference is apparent, but is to be expected, since opposite assumptions were made in the two solutions. In the lower half of the figure, a streamline calculated from Eq. (8–16) is compared with the corresponding streamline from potential flow (from Example 6–1). These are the streamlines observed from a stationary sphere in a moving fluid and, as expected, are quite similar in both cases.

For the pressure, we take the value for A from Eq. (8–11), combine it with Eq. (8–6), and get

$$p = -\tfrac{3}{2}(\mu U_\infty r_0 x/r^3). \tag{8-18}$$

The drag on a sphere is the resultant of the pressure and frictional drag. The resultant is constant over all the sphere (otherwise the sphere would move) and is always in the x-direction. Since the resultant force is constant, one may evaluate it at $x = -r_0$, and this value will be valid over all the sphere. At $-r_0$, $x = -r_0$ and $r = r_0$, therefore the pressure is $\tfrac{3}{2}\mu U_\infty/r_0$. The total drag D_S is then equal to the force per unit area multiplied by the area ($x = r = r_0$), or

$$D_S = \tfrac{3}{2}(\mu U_\infty/r_0)4\pi r_0^2 = 6\pi \mu U_\infty r_0. \tag{8-19}$$

This is Stokes' formula for the drag of a sphere. The drag coefficient is

$$C_D = \frac{D_S}{\rho U_\infty^2 \pi r_0^2/2} = \frac{24}{2r_0 U_\infty/\nu} = \frac{24}{N_{\text{Re}}}, \tag{8-20}$$

where the Reynolds number is based on the diameter of the sphere.

The free settling velocity of a small particle can be obtained by a force balance, $6\pi\mu r_0 U_t + \frac{4}{3}\pi r_0^3 \rho_l g = \frac{4}{3}\pi r_0^3 \rho_s g$, thus

$$U_t = 2r_0^2(\rho_s - \rho_l)g/9\mu. \tag{8-21}$$

The movement of particles in fluid systems is very important, and much work has been done to determine, both by theory and experiment, the range of application of Stokes' law. For example, the law is found to be accurate up to a Reynolds number of 0.5 (which, for drops of water in air, would be diameters smaller than 0.1 mm). At a Reynolds number of 1, the law is about 7% low. At Reynolds numbers above 8, vortex rings form which are stable up to a number of 150, above which the rings become unstable and, from time to time, move off downstream.

EXAMPLE 8-1. Estimate the distance from a sphere at which its effect will be reduced by 90%.

Answer. The distance can be estimated from Eq. (8–13). The simplest case would be for $x = 0$, so that $U_y = U_z = 0$ and $U_x/U_\infty = 0.9$; that is, the velocity away from the sphere will have reached 90% of its final value. Equation (8–13) becomes

$$3\frac{r_0}{r} + \left(\frac{r_0}{r}\right)^3 = 4\left(1 - \frac{U_x}{U_\infty}\right).$$

Since r should be greater than r_0, the equation can be simplified to

$$\frac{r}{r_0} = \frac{\frac{3}{4}}{1 - U_x/U_\infty}.$$

For U_x/U_∞ of 0.9, r/r_0 is 7.5, and the simplification is justified to better than 1%. It is to be noted that the effect of the sphere is considerable. The effect is still 10% at 7.5 radii distant.

EXAMPLE 8-2. A 5-micron spherical water drop is dropped in air at 21°C. How long will it take the drop to reach 99% of its terminal velocity?

Answer. First, assume that the resistance to motion is given by Stokes' law. Second, use Newton's law of motion to obtain

$$\rho_s \tfrac{4}{3}\pi r_0^3 \frac{dU_z}{dt} = \tfrac{4}{3}\pi r_0^3(\rho_s - \rho_l)g - 6\pi\mu r_0 U_z.$$

Next, separate the variables and integrate between the limits $t = 0$, $U_z = 0$, and $t = t$, $U_z = fU_t$, where f is some fraction of the terminal velocity. After some simplification, this gives us

$$t = \frac{2r_0^2 \rho_s}{9\mu}\ln\left(\frac{1}{1-f}\right).$$

For the problem at hand,

$$r_0 = 2.5 \times 10^{-4} \text{ cm}, \qquad \rho_s = 1.0 \text{ gm/cm}^3,$$
$$\mu = 1.8 \times 10^{-4} \text{ poise}, \qquad f = 0.99,$$

which, when combined with the above equation, gives us a time of 3.6×10^{-4} sec. The Reynolds number should be obtained to check the assumption of Stokes' law. The value is slightly greater than 0.8, so that the assumption is reasonable.

8-2 OSEEN'S ANALYSIS

One way to improve on Stokes' equation would be to consider some of the higher-order terms of velocity, and thus remove part of the assumption that these terms are negligible. Such inertia forces would be small near the sphere, but at distances away from the sphere, they would be important. Thus one would expect these higher-order terms to cause considerable change in the flow pattern. The procedure used is to linearize Eq. (8–1) by allowing the inertia term to be $U_\infty \partial U/\partial x$ rather than the $(U \cdot \nabla)U$ of Eq. (8–2). This analysis was first presented by Oseen [17], and is good up to a Reynolds number of about 5. The drag becomes

$$C_D = (24/N_{\text{Re}'})(1 + \tfrac{3}{16} N_{\text{Re}'}). \tag{8–22}$$

At the sphere, the solutions of Oseen and Stokes are the same. Away from the sphere, the inertia term approximation becomes important and the flow pattern of Fig. 8–3 is completely changed; it is now like that in Fig. 8–4, as given by the stream function

$$\psi = \tfrac{3}{2} \nu r_0 (1 + \cos \theta)\left[1 - \exp\left(\frac{-U_\infty r(1 - \cos \theta)}{2\nu}\right)\right] - \frac{1}{4} \frac{U_\infty r_0^3}{r} \sin^2 \theta,$$

which, for small values of $U_\infty r/2\nu$, reduces to Eq. (8–17). The flow is not symmetric to $x = 0$, and thus appears something like a wake formation.

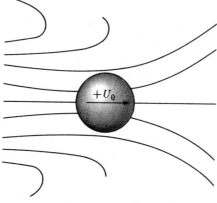

Fig. 8–4. Oseen's solution for the streamlines for flow over a sphere.

8-3 ALTERNATIVE SHAPES OF PARTICLES

The interest in extensions of slow-motion theory has been considerable because of the practical importance and the possibility of elegant mathematical solutions with relatively simple means for experimental confirmation. This section will summarize some of the more recent results of investigations in this field. Oberbeck [16] used the same approach as Stokes to solve for the first approximation for any ellipsoid with any orientation to the uniform flow stream. In the case of symmetry about an axis parallel to the stream, the Stokes stream function as previously mentioned can be used to describe the flow. Brenner [6] has given a generalized approach to the Stokes resistance of an arbitrary particle in an infinite medium.

Oseen [18] applied his approach to a number of ellipsoidal cases. The Stokes approach cannot be used for a cylindrical shape because the boundary conditions at the cylinder surface cannot be satisfied at the same time as the condition at infinity. However, the Oseen method can be used to solve this problem (Lamb [14], Art. 343). A combination of the Stokes and Oseen methods has been used to obtain higher approximations for the flow. Among these are analyses on a sphere, by Proudman and Pearson [22]; on a circular cylinder, by Kaplun and Lagerstrom [12]; and on ellipsoids of revolution, by Breach [1]. Payne and Pell [20, 21] have used a generalized axially symmetric potential flow theory to solve for the Stokes resistance for a class of axially symmetric bodies. They derived a relation between the drag and the stream function of the flow. Brenner [3] has suggested an equation to relate the Oseen drag to the Stokes drag for a particle of arbitrary shape. The result should be correct to the order of the first power of the Reynolds number. The only restriction on the equation is that the acting force on the particle must be parallel to its direction of motion. The equation is

$$\frac{D_O}{D_S} = 1 + \frac{D_S N_{Re''}}{16\pi\mu U_\infty r_0}, \tag{8-23}$$

where D_S is the total drag obtained from the Stokes method, D_O is the corresponding drag that would be obtained from the Oseen method, r_0 is any characteristic particle dimension, and the Reynolds number is of the form $r_0 U_\infty \rho/\mu$. In the same paper Brenner extended the analysis, which resulted in the above equation, to include motions in which the principal axis of the particle is not parallel to the stream velocity.

As an illustration, consider the flow past a circular disk moving broadside to the stream. The Stokes resistance was obtained by Oberbeck as a limiting case of the ellipsoid. From Lamb ([14], Art. 339), this is

$$D_S = (8/3\pi)(6\pi\mu U_\infty r_0) = 16\mu U_\infty r_0.$$

Using this drag in Eq. (8–23), we obtain

$$D_O = 16\mu U_\infty r_0 (1 + N_{Re''}/\pi), \tag{8-24}$$

in accord with Oseen's results. A similar result was obtained by Breach. However, in this case, the result was

$$D_O = 16\mu U_\infty r_0 \left[1 + \frac{N_{Re''}}{\pi} + \frac{8}{5}\left(\frac{N_{Re''}}{\pi}\right)^2 \log N_{Re''}\right].$$

In these equations, the Reynolds number is based on the disk radius rather than on the diameter.

The slow steady rotation of bodies in a viscous fluid is a closely allied field of study. Lamb ([14], Art. 337) considered the problem of a rotating sphere. Jeffery [10] considered two spheres and spheroids. Payne and Pell used their method to solve the same problems as Jeffery. Kanwal [11] extended the method to provide the solutions for a spindle, a torus, a lens, and certain special configurations of a lens, and discussed the limiting cases of a sphere and a circular disk.

8-4 WALL AND OTHER EFFECTS

The proximity of a wall can have a considerable effect on the results of the slow motion of particles, and thus has received considerable attention. Ladenburg [13] considered the motion of a settling sphere along the axis of a circular cylinder; he obtained the correction factor

$$\frac{D_b}{D_S} = 1 + 2.1044 \frac{r_0}{R},$$

where D_S is the Stokes drag in an infinite medium, D_b is the drag in the bounded system, r_0 is the radius of the sphere, R is the radius of the tube, and the numerical factor is from reference 7. Wakiya [27] solved the similar problem of the flow over a fixed sphere along the axis of a tube. He also solved [28] the problems of a falling spheroid along the axis of a circular tube and at one-fourth of the distance between two parallel walls. Haberman and Sayre [8] have solved a number of similar problems. Brenner and Happel [2] considered the problem of a moving particle at any position in the field within a circular cylinder. Brenner [4] has suggested a general theory, which provides a common form for most of the above flows. The only necessary information is a knowledge of the drag of the particle in an infinite medium, and the wall correction for a spherical particle. The theory is restricted to small values of r_0/R, since it is correct only to values of the second power of this ratio. In contrast, some of the solutions mentioned above have been worked out to much higher powers of the particle-boundary size ratio. Brenner's general correction to Stokes' law is

$$\frac{D_b}{D_S} = \frac{1 - U_0/U_\infty}{1 - kD_S/6\pi\mu U_\infty R}, \tag{8-25}$$

where U_0 is the fluid velocity relative to the particle velocity U_∞, and k is a constant which is independent of the particle shape and depends only on the bounding wall. Brenner obtained a number of values of k by considering the known solutions for the wall effect on a falling spherical particle. These were then shown to apply to other shapes, such as spheroids moving parallel to a single wall, two parallel walls, and along the axis of a circular cylinder (the solutions of Wakiya, previously referred to). Several of these values of k are: movement along the axis of a circular cylinder, 2.1044; falling perpendicular to a single infinite plane surface, with R being the distance from the particle center to the plane, $\frac{9}{8}$; falling parallel to the plane described above, $\frac{9}{16}$; and falling midway between and parallel to two of the planes, 1.004.

Other values of k and other checks of the theory are given in the references cited. In addition, consideration is given to the interaction between two particles in an infinite medium, and to the case in which the particle near a wall is in rotation. Brenner [5], in a more elaborate solution, considered the movement of a sphere toward a solid plane or free surface. As a practical application he considered the falling-ball viscometer and obtained improved estimates for the end corrections, which would be added to the correction for the cylindrical boundary obtained by Ladenburg, cited above. For a typical case, these might amount to 20% of the cylindrical boundary correction term, which would not be too great. Tanner [25] considered the falling-ball viscometer problem further, and concluded that the additional drag from the closed end is very small and usually negligible.

Brenner [6] also considered the relation between the Oseen and Stokes drag for an arbitrary particle falling along the axis of a circular cylinder. This is almost a combination of Eqs. (8–23) and (8–25), and is

$$\frac{D_{Ob}}{D_S} = 1 \bigg/ \bigg[\frac{1 - D_S N_{Re''}}{16\pi\mu U_\infty r_0} - \frac{D_S}{6\pi\mu U_\infty r_0 R}\frac{r_0}{R} L(x)\bigg], \qquad (8\text{–}26)$$

where $L(x)$ is a function of $x = \frac{1}{2}N_{Re''}(R/r_0)$, and has the following representative values:

$$L(0) = 2.1044, \quad L(0.5) = 1.76, \quad L(1) = 1.48, \quad L(2) = 1.04, \quad L(5) = 0.46.$$

Equation (8–26) reduces to (8–25) for the case in which the Reynolds number is very small, so that Stokes' type of analysis is valid. The value of k in Eq. (8–25) for the circular cylinder is the same as the value cited above for zero Reynolds number, $L(0)$.

When the Reynolds number is much higher than the applicable range of Stokes' law or Oseen's law, a numerical analysis of the Navier-Stokes equation is necessary. However, this is no longer a subject of slow motion but rather an exact numerical solution or a boundary layer problem. Plots of experimental drag coefficients versus Reynolds number can also be used. Such a plot for spheres and other shapes is shown in Fig. 8–5. This also is considered to be outside the realm of slow flow and is usually included as a part of two-phase flow (Chapter 17).

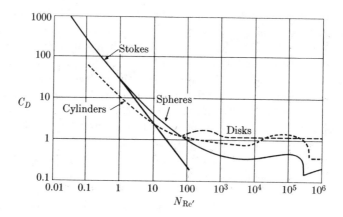

Fig. 8–5. The drag coefficient. [From V. L. Streeter, Fluid Mechanics, McGraw-Hill, New York (1958). By permission.]

When the particle is so small that its size approaches the mean free path of the fluid, the Cunningham correction must be made, so that

$$\frac{D_c}{D_s} = \frac{1}{1 + Al/r_0}, \qquad (8\text{-}27)$$

where D_c is the corrected drag, l is the mean free path as defined by $\mu = 0.350\rho l \bar{c}$ (in which $\bar{c}$ is the mean velocity of a gas particle), and A is a function of the gas, usually with a value between 0.7 and 1.0. Brownian motion can also be important and should be considered at diameters less than 3 microns; it is very important below 0.1 micron. When high concentrations are involved, hindered settling (or conversely fluidization) occurs. For these cases, multiparticle interactions must be considered. Some problems of this nature can be solved by the methods noted in this chapter, but we will not consider them until solid-gas and solid-liquid two-phase flows are treated (Chapter 18).

If the boundary is not a solid, one must consider the mobility of the interface. The general form of the correction [15] is

$$\frac{U_{tc}}{U_t} = \frac{3\mu_d + 3\mu + \gamma}{3\mu_d + 2\mu + \gamma}, \qquad (8\text{-}28)$$

where U_{tc} is the terminal velocity with the circulation considered, U_t is given by Eq. (8–21), μ_d is the discontinuous phase viscosity (either a drop or bubble), and γ arises from surface tension effects. The equation in which $\gamma = 0$ was derived by Rybcyński [23] and Hadamard [9]. For this case, if the discontinuous phase viscosity is small, as for air bubbles in a viscous fluid, then $3\mu/2\mu = \frac{3}{2}$. This implies that a bubble will rise one and one-half times faster than if it were a solid particle with the same density. The actual form of the term γ depends on the mechanism assumed. Levich [15] has discussed this at length and has provided detailed discussions and comparisons of the various ideas. Of particular importance is the experimental observation that for very small drops and bubbles, the term γ apparently predominates, since it is observed that these follow Stokes' law. For somewhat larger drops, the value of γ becomes unimportant and the Rybczyński-Hadamard correction is valid. These studies neglect the effect associated with inertia. Taylor and Acrivos [26] included these effects for low Reynolds numbers, so as to study the deformation and drag of the drop. The drop first deforms into an oblate spheroid and then approaches the shape of a spherical cap.

PROBLEMS

8-1. Prove that Eq. (8-6) is harmonic.

8-2. Calculate the velocity profile about a sphere in slow motion. First, consider the distribution when $x = 0$, then compare this with the case in which $x = 3r/r_0$.

8-3. What are the limitations of Stokes' formula for the drag of a sphere? Which of these limitations can be removed by correction factors?

8-4. A particle 5 microns in diameter ($\rho_s = 2.0$) is shot horizontally with a velocity of 25 ft/sec into still air. How far will it travel in the horizontal direction?

8-5. Oil droplets are to be separated from a water stream (50 gal/min) during the flow through a 10-ft-wide channel. An overflow for the oil is provided at the exit. The oil ($\rho_d = 0.8$) is in a uniform dispersion of 10-micron particles. The water temperature is 70°F. How long must the channel be to allow the separation to occur?

8-6. Two parallel, horizontal circular disks approach each other. The velocity of approach is constant and the contained fluid is displaced. Determine the flow and resistance to motion.

REFERENCES

1. D. S. Breach, *J. Fluid Mech.* **10**, 306–314 (1961).
2. H. Brenner and J. Happel, *J. Fluid Mech.* **4**, 195 (1958).
3. H. Brenner, *J. Fluid Mech.* **11**, 604–610 (1961).
4. H. Brenner, *J. Fluid Mech.* **12**, 35–48 (1962).
5. H. Brenner, *Chem. Eng. Sci.* **16**, 242–251 (1961).
6. H. Brenner, *Chem. Eng. Sci.* **18**, 1–25 (1963); **19**, 599–651, 703 (1964); **21**, 97–109 (1966).
7. O. Emersleban and H. Faxen, *Archiv. Mat. Astron. Fysik* **17**, No. 27, 1–28 (1923).
8. W. L. Haberman, and R. M. Sayre, *David Taylor Model Basin Report 1143* (Oct. 1958).
9. M. J. Hadamard, *Compt. Rend.* **152**, 1735 (1911).
10. G. B. Jeffery, *Proc. London Math. Soc.* **14**, 327 (1915).
11. R. P. Kanwal, *J. Fluid Mech.* **10**, 17–24 (1961).
12. S. Kaplun and P. Lagerstrom, *J. Math. Mech.* **6**, 585 (1957).
13. R. Ladenberg, *Ann. Physik* **23**, 447–458 (1907).

14. H. LAMB, *Hydrodynamics* (reprint of the 1932 edition), Dover Publications, New York (1945).
15. V. G. LEVICH, *Physicochemical Hydrodynamics*, Prentice-Hall, New York (1962).
16. A. OBERBECK, *Crelle* **81**, 62 (1876).
17. C. W. OSEEN, *Arkiv. Mat. Astron. Fysik* **6**, No. 29 (1911), **7**, No. 1 (1911); **9**, No. 16 (1913).
18. C. W. OSEEN, *Neuere Methoden und Ergebnisse in der Hydrodynamik*, Akademische Verlagsgesellschaft, Leipzig (1927).
19. SHIH-I PAI, *Viscous Flow Theory: Laminar Flow*, D. Van Nostrand, Princeton, N.J. (1956).
20. L. E. PAYNE and W. H. PELL, *J. Fluid Mech.* **7**, 529–549 (1960).
21. W. H. PELL and L. E. PAYNE, *Natl. Bur. Std. Report. No. 6474* (1959).
22. I. PROUDMAN, and J.R.A. PEARSON, *J. Fluid Mech.* **2**, 237 (1957).
23. D. P. RYBCZYŃSKI, *Bulletin intern. acad. sci. Cracovie* **A403**, 40 (1911).
24. G. G. STOKES, *Trans. Cambridge Phil. Soc.* **9**, No. 8 (1851).
25. R. I. TANNER, *J. Fluid Mech.* **17**, 161–170 (1963).
26. T. D. TAYLOR and A. ACRIVOS, *J. Fluid Mech.* **18**, 466–476 (1964).
27. S. WAKIYA, *J. Phys. Soc. Japan* **8**, 254 (1953).
28. S. WAKIYA, *J. Phys. Soc. Japan* **12**, 1130, 1318 (1957).

CHAPTER 9

LAMINAR VISCOUS FLOW: THE BOUNDARY LAYER

When inertial forces predominate over viscous forces, the approximation of very slow motion will not be valid. Neither is the approximation of ideal flow, since the boundary conditions for laminar viscous flow cannot be satisfied. Prandtl [21], in 1904, introduced the concept of the boundary layer as an approximation to the Navier-Stokes equation for cases of laminar flow at high Reynolds numbers. Since that time, this method has been the principal tool used to deal with these important cases in studies of fluid dynamics. There have been many solutions using the boundary layer approach, such as the flow in the entrance section of a channel; the flow over circular cylinders, wings, and rotating blades; and flow with suction or injection.

As suggested in Chapter 5 and implied in the preceding chapters, when the Reynolds number is large, the effect of the viscous forces is concentrated in a small region near the boundary. Prandtl therefore divided the problem into two parts: the flow of an ideal fluid exterior to the boundary layer, and the boundary layer problem. In ideal flow, inertial forces predominate and viscous forces are vanishingly small, while in boundary layer flow, both inertial and viscous forces are controlling factors.

In the next few sections, the boundary layer equations will be formulated, the means of solution considered, and the boundary layer problem for flow over a flat plate will be solved. Further sections will consider boundary layer thickness, drag, and flow in more complex systems.

9-1 THE BOUNDARY LAYER EQUATIONS FOR FLOW OVER A FLAT PLATE

Two-dimensional boundary layer flow over a flat plate is diagrammed in Fig. 9–1.

For incompressible flow in the absence of external forces, the Navier-Stokes equation and the continuity equation become

$$\frac{DU}{Dt} = -\frac{1}{\rho}\nabla p + \nu\nabla^2 U, \qquad (\nabla \cdot U) = 0. \tag{9-1}$$

The boundary conditions to be satisfied are

$$\begin{aligned} U_x = U_y = 0 &\quad \text{at} \quad y = 0 \quad \text{for all } x, \\ U_x = U_\infty &\quad \text{at} \quad y = \infty. \end{aligned} \tag{9-2}$$

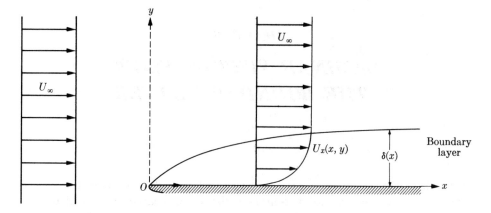

Fig. 9–1. Two-dimensional boundary layer over a flat plate.

The boundary layer thickness δ is assumed to be small compared with x, the length along the plate. At the front of the plate, there would be some question about this assumption; however, this point will be considered later.

To determine what terms might be eliminated, an order of magnitude estimation can be made on each quantity. Let us consider the free stream velocity U_∞ as the unit order of magnitude of velocity, the distance x, along the plate as the unit order of length, and t as the unit order of time. The velocity U_x would be of order 1, since it changes from 0 to U_∞ as y varies from 0 to δ. Further, since

$$U_x = \int_0^\delta (\partial U_x/\partial y)\, dy,$$

we can estimate the order of $\partial U_x/\partial y$ as $1/\delta$. That is, U_x is of order 1, and dy (integrated between 0 and δ) will be of order δ, thus $\partial U_x/\partial y$ must be of order $1/\delta$ to conserve the order on both sides of the equation:

$$1 = (1/\delta)\delta.$$

By a similar analysis, $\partial^2 U_x/\partial y^2$ would be of order $1/\delta^2$. The term $\partial U_x/\partial x$ must be of order 1, as would be $\partial^2 U_x/\partial x^2$; that is, $1/1 = 1$. In summary, the x-velocity terms are

$$U_x(1), \quad \frac{\partial U_x}{\partial y}\left(\frac{1}{\delta}\right), \quad \frac{\partial^2 U_x}{\partial y^2}\left(\frac{1}{\delta^2}\right), \quad \frac{\partial U_x}{\partial x}(1), \quad \frac{\partial^2 U_x}{\partial x^2}(1). \qquad (9\text{-}3)$$

From continuity (9-1), the term $\partial U_y/\partial y$ must be of order 1, since $\partial U_x/\partial x$ is of that order. Since

$$U_y = \int_0^\delta (\partial U_y/\partial y)\, dy,$$

U_y must be of order δ. An analysis similar to that used above will give the remaining terms, and in summary, the y-velocity terms are

$$U_y(\delta), \quad \frac{\partial U_y}{\partial y}(1), \quad \frac{\partial^2 U_y}{\partial y^2}\left(\frac{1}{\delta}\right), \quad \frac{\partial U_y}{\partial x}(\delta), \quad \frac{\partial^2 U_y}{\partial x^2}(\delta). \tag{9-4}$$

We combine Eqs. (9-3) and (9-4) with the Navier-Stokes equation (9-1), and obtain, for the x-direction,

$$\frac{\partial U_x}{\partial t} + U_x\frac{\partial U_x}{\partial x} + U_y\frac{\partial U_x}{\partial y} = -\frac{1}{\rho}\frac{\partial p}{\partial x} + \nu\left(\frac{\partial^2 U_x}{\partial x^2} + \frac{\partial^2 U_x}{\partial y^2}\right), \tag{9-5}$$

$$1/1 \qquad 1\cdot 1 \qquad \delta\cdot 1/\delta \qquad\qquad\qquad 1 \qquad 1/\delta^2$$

where the order of each term is given directly below the term. The term $\partial^2 U_x/\partial x^2$ can be neglected when compared with $\partial^2 U_x/\partial y^2$. For the y-direction, we have

$$\frac{\partial U_y}{\partial t} + U_x\frac{\partial U_y}{\partial x} + U_y\frac{\partial U_y}{\partial y} = -\frac{1}{\rho}\frac{\partial p}{\partial y} + \nu\left(\frac{\partial^2 U_y}{\partial x^2} + \frac{\partial^2 U_y}{\partial y^2}\right). \tag{9-6}$$

$$\delta/1 \qquad 1\cdot\delta \qquad \delta\cdot 1 \qquad\qquad\qquad \delta \qquad 1/\delta$$

The term $\partial^2 U_y/\partial x^2$ can be neglected when compared with $\partial^2 U_y/\partial y^2$.

If the inertial terms on the left of Eq. (9-5) are to be of the same order of magnitude as the viscous term on the right (i.e., both inertial and viscous forces are of importance in the boundary layer), then the kinematic viscosity must be of order δ^2. If this is so, all terms in Eq. (9-6) are of order δ, so that $(1/\rho)(dp/dy)$ must be no greater than of order δ. An integration across the boundary layer (as was done for the velocity) can be used to show that the change in pressure must be of order δ^2 and thus can be disregarded. Furthermore, Eq. (9-6) is of a smaller order than Eq. (9-5) and the continuity equation, and so need not be considered. The simplifications result in the equations for the boundary layer:

$$\frac{DU_x}{Dt} = -\frac{1}{\rho}\frac{dp}{dx} + \nu\frac{\partial^2 U_x}{\partial y^2}, \quad \frac{1}{\rho}\frac{dp}{dy} = 0, \quad \frac{\partial U_x}{\partial x} + \frac{\partial U_y}{\partial y} = 0. \tag{9-7}$$

The Reynolds number is $U_\infty x/\nu$, and for distances away from the leading edge is of the order of magnitude of $(1\cdot 1)/\delta^2 = 1/\delta^2$; that is, the Reynolds number must be large for this approximation to be valid.

9-2 ANALYSIS OF THE BOUNDARY LAYER EQUATIONS

The boundary layer equations are still not linear, but are simpler than the original ones. As noted, the pressure drop across the boundary layer is so small that it can be neglected, and therefore the pressure can be a function of x only. Furthermore, the pressure will be the same at any y and will equal that at the outer edge of the boundary layer. It is assumed to be determined by the flow exterior to the boundary layer and can be obtained from the ideal fluid flow solution of the Navier-Stokes equation (i.e., when $\nu = 0$). Thus we see that dp/dx is a given function and not an unknown of the boundary layer solution. This constitutes an additional and considerable simplification. For a uniform stream over a flat plate, the pressure is constant, or dp/dx is zero. For more complex systems, the pressure and the outer boundary condition for the boundary layer problem need not be constant, but can be functions of both x and t. In any event, they are determined from the ideal flow problem.

In the preceding discussion it has been assumed that the solution obtained for the outer edge of the boundary layer (from the boundary layer solution) will closely match that at the inner boundary of the ideal flow. Since we do not have the complete solution to the Navier-Stokes equation for the case in which the boundary layer is important, it is difficult to tell how adequate the approximation will be without some comparison with experimental results. Fortunately, past work has shown that the approximation is good so long as complex interactions between the main flow and the boundary layer do not occur.

9-3 THE BOUNDARY LAYER ON A FLAT PLATE

For steady-state conditions, Eq. (9–7) becomes

$$U_x \frac{\partial U_x}{\partial x} + U_y \frac{\partial U_x}{\partial y} = \nu \frac{\partial^2 U_x}{\partial y^2}, \qquad \frac{\partial U_x}{\partial x} + \frac{\partial U_y}{\partial y} = 0, \qquad (9\text{--}8)$$

since dp/dx is zero or at most of order δ [since (dp/dy) is at most of order δ]. The boundary conditions are

$$U_x = U_y = 0 \quad \text{at} \quad y = 0 \quad \text{for all } x,$$
$$U_x = U_\infty \quad \text{at} \quad y = \infty.$$

Prandtl suggested the problem in his original paper, and Blasius [3] later gave the solution to be outlined here.

The solution of the equations involves an amplification of the boundary layer area, normally of order δ, to unit order by using as the y-variable a transformed variable η, which will be of the form y/δ. The transformation can be established as follows: From a solution of the Navier-Stokes equation for a plane suddenly set into motion (Eq. 7–9), the boundary layer thickness was found to be of the form

$$\delta = K\sqrt{\nu t}. \qquad (9\text{--}9)$$

For steady state, t is replaced by x/U_∞, which is the time for a fluid element to travel a distance x in the outer flow. Since η is of the form y/δ, we can use Eq. (9–9) to write

$$\eta = y\sqrt{U_\infty/\nu x}, \qquad (9\text{--}10)$$

where, for simplicity, K is taken as unity.

A solution to the equation of continuity in the form of the stream function is assumed to be

$$U_x = \partial \psi/\partial y, \qquad U_y = -\partial \psi/\partial x. \qquad (9\text{--}11)$$

Further, the stream function can be expressed as

$$\psi = \sqrt{U_\infty \nu x}\, f(\eta). \qquad (9\text{--}12)$$

Equations (9-11) and (9-12) can be combined to give

$$U_x = U_\infty f', \qquad U_y = \tfrac{1}{2}\sqrt{U_\infty \nu/x}(\eta f' - f),$$

in which the prime denotes differentiation with respect to η. A combination of Eqs. (9-8) and (9-11) gives

$$\frac{\partial \psi}{\partial y}\frac{\partial^2 \psi}{\partial x \partial y} - \frac{\partial \psi}{\partial x}\frac{\partial^2 \psi}{\partial y^2} = \nu \frac{\partial^3 \psi}{\partial y^3}. \qquad (9\text{-}13)$$

This equation, when combined with Eqs. (9-10) and (9-12), gives an ordinary differential equation for the boundary layer:

$$ff'' + 2f''' = 0. \qquad (9\text{-}14)$$

[The factor of 2 could be eliminated by using η as one-half the value given by Eq. (9-10).] The boundary conditions become

$$\begin{aligned} f = f' = 0 &\quad \text{at} \quad \eta = 0, \\ f' = 1 &\quad \text{at} \quad \eta = \infty. \end{aligned} \qquad (9\text{-}15)$$

Equation (9-14) is still nonlinear and cannot be solved in closed form. However, by means of a power series expansion, a solution can be obtained. One of the constants in this solution must be obtained by means of an asymptotic series expansion or by an alternative numerical integration method. The solution is given in part in Example 9-1.

We can now continue our discussion of the matching of boundary conditions. Since the boundary layer solution is not valid for x very close to zero (the leading edge), we see that, from Eq. (9-10), an infinite η implies infinite y. In other words, the boundary condition (9-15) for $\eta = \infty$ or $y = \infty$ is specified at infinity, but with the desire that it match the inner edge of the ideal flow, which is at $y = 0$. In essence, we have assumed (and justified by comparison with experimental results) that the boundary layer solution at $y = \infty$ is close to the ideal flow problem at $y = 0$. We cannot rigorously prove that this is so.

The boundary layer solution of the flat plate problem can be summarized by plotting the velocity distribution as in Fig. 9-2.

Outside the boundary layer, as y approaches infinity, U_y approaches

$$0.865\, U_\infty \sqrt{\nu/U_\infty x} = 0.865\sqrt{U_\infty \nu/x}. \qquad (9\text{-}16)$$

Equation (9-16) shows that at the outer edge, the normal velocity component is not zero, although it is small. This small velocity is due to the necessity, because of continuity, of accounting for the deficiency of flow in the boundary layer which is caused by the retardation of the flow in the x-direction by the wall. It might be possible to eliminate this minor problem by a change in boundary conditions at $\eta = \infty$ from $U_x = U_\infty$ to $dU_x/d\eta = 0$. This would, in effect, set $U_y = 0$ and allow U_x to be greater than U_∞, as required by continuity. Even so, since this velocity is small (of order δ or less), the assumption of potential flow

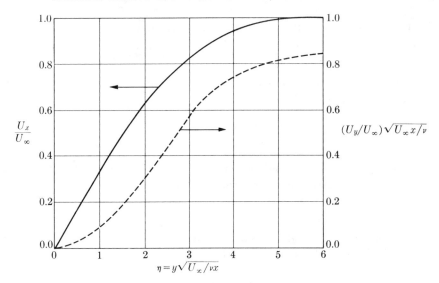

Fig. 9–2. Boundary-layer solution for flat-plate problem. [*From Shih-I Pai*, Viscous Flow Theory, Vol. I, Laminar Flow, *D. Van Nostrand, Princeton, New Jersey (1956). By permission.*]

along the flat plate with zero velocity normal to the wall is a good approximation of the flow exterior to the boundary layer.

EXAMPLE 9–1. Solve Eq. (9–14) with the following boundary conditions (9–15):

$$ff'' + 2f''' = 0, \qquad (9\text{--}14)$$

and

$$f = f' = 0 \quad \text{at} \quad \eta = 0, \quad f' = 1 \quad \text{at} \quad \eta = \infty.$$

Answer. We can expand $f(\eta)$ in a power series:

$$f(\eta) = A_0 + A_1\eta + \frac{A_2}{2!}\eta^2 + \frac{A_3}{3!}\eta^3 + \cdots + \frac{A_n}{n!}\eta^n. \qquad \text{(i)}$$

The first boundary condition gives $A_0 = A_1 = 0$. The derivatives of Eq. (i) are

$$f'(\eta) = A_2\eta + \frac{A_3}{2!}\eta^2 + \frac{A_4}{3!}\eta^3 + \cdots + \frac{A_n\eta^{n-1}}{(n-1)!}, \qquad \text{(ii)}$$

$$f''(\eta) = A_2 + A_3\eta + \frac{A_4}{2!}\eta^2 + \cdots + \frac{A_n\eta^{n-2}}{(n-2)!}, \qquad \text{(iii)}$$

$$f'''(\eta) = A_3 + A_4\eta + \frac{A_5}{2!}\eta^2 + \cdots + \frac{A_n\eta^{n-3}}{(n-3)!}. \qquad \text{(iv)}$$

Equations (i), (iii), and (iv) can be substituted into Eq. (9–14), and coefficients of like powers can be equated to show that all coefficients are zero except those

TABLE 9-1
Numerical Solution to Eq. (9-14) (after Howarth [11])

	$f(\eta)$	$f'(\eta)$	$f''(\eta)$
0.0	0.00000	0.00000	0.33206
0.4	0.02656	0.13277	0.33147
0.8	0.10611	0.26471	0.32739
1.2	0.23795	0.39378	0.31659
1.6	0.42032	0.51676	0.29667
2.0	0.65003	0.62977	0.26675
2.4	0.92230	0.72899	0.22809
2.8	1.23099	0.81152	0.18401
3.2	1.56911	0.87609	0.13913
3.6	1.92954	0.92333	0.09809
4.0	2.30576	0.95552	0.06424
4.4	2.69238	0.97587	0.03897
4.8	3.08534	0.98779	0.02187
5.2	3.48189	0.99425	0.01134
5.6	3.88031	0.99748	0.00543
6.0	4.27964	0.99898	0.00240

of the form A_{2+3n}. The A_2 is the lowest nonzero coefficient, and from Eq. (iii),

$$f''(0) = A_2 = \alpha, \tag{v}$$

where α is to be determined later from the boundary condition at infinity. The series becomes

$$f(\eta) = \sum_{n=0}^{\infty} \left(-\frac{1}{2}\right)^n \frac{C_n \alpha^{1+n} \eta^{2+3n}}{(2+3n)!}, \tag{vi}$$

where the first six values of C_n as given by Blasius are

$$C_0 = 1, \quad C_1 = 1, \quad C_2 = 11,$$
$$C_3 = 375, \quad C_4 = 27{,}897, \quad C_5 = 3{,}817{,}137.$$

The value for α can be obtained by an asymptotic series [3] or by a numerical integration method [27], and is $\alpha = 0.33206$. For this value of α, the terms f, f', and f'' can be obtained from Eq. (vi) and its derivatives [11]. The velocities U_x and U_y can also be evaluated from the tabulations of the values of f and its derivatives; that is,

$$U_x = \frac{\partial \psi}{\partial y} = \sqrt{U_\infty x \nu} \frac{\partial f}{\partial y} = \sqrt{U_\infty x \nu} \frac{\partial f}{\partial \eta} \frac{\partial \eta}{\partial y} = \sqrt{U_\infty x \nu} f' \sqrt{U_\infty/x\nu} = U_\infty f'. \tag{vii}$$

In a similar manner, U_y can be determined as

$$U_y = \tfrac{1}{2}\sqrt{U_\infty \nu/x}(\eta f' - f). \tag{viii}$$

Figure 9-2 is the result of the solution in terms of the velocities; some of the basic data are tabulated in Table 9-1.

The *drag* on the plate can be obtained from the wall shear stress:

$$\tau_w = -\mu\left(\frac{\partial U_x}{\partial y}\right)_w = -\mu\left(\frac{\partial U_\infty f'}{\partial y}\right)_w = -\mu U_\infty f''_w \sqrt{\frac{U_\infty}{\nu x}}.$$

At the wall, $\eta = 0$, so that $f''(0) = \alpha$ [by Eq. (v) of Example 9-1]; therefore

$$\tau_w = -\mu U_\infty \alpha \sqrt{U_\infty/\nu x}. \tag{9-17}$$

A *local drag coefficient* can be defined as

$$C_{D,x} = \frac{\tau_w}{\frac{1}{2}\rho U_\infty^2} = \frac{2\alpha}{\sqrt{xU_\infty/\nu}} = \frac{2\alpha}{\sqrt{N_{\text{Re},x}}}, \tag{9-18}$$

where 2α is 0.644, and the Reynolds number is based on the local position x. The *drag on one side* of the plate is the integrated shear stress over the area:

$$D = \mu U_\infty \alpha b \sqrt{U_\infty/\nu} \int_0^L (1/\sqrt{x})\, dx,$$

where b is the width and L is the length of the plate. Evaluation of the integral gives

$$D = 2\alpha U_\infty b \sqrt{\mu \rho L U_\infty}, \tag{9-19}$$

where 2α is 0.664. The drag coefficient will be

$$C_D = \frac{D}{\frac{1}{2}\rho U_\infty^2 A} = \frac{D}{\frac{1}{2}\rho U_\infty^2 Lb} = \frac{4\alpha}{\sqrt{U_\infty L/\nu}} = \frac{4\alpha}{\sqrt{N_{\text{Re},L}}}, \tag{9-20}$$

where 4α is 1.328, and the Reynolds number is now based on the plate length.

The *displacement thickness* of the boundary layer δ^* can be defined by the equation

$$U_\infty(y_1 - \delta^*) = \int_0^{y_1} U_x\, dy. \tag{9-21}$$

This thickness is a measure of the necessary displacement of the potential flow from the surface to offset the increasing size of the boundary layer with distance. The right-hand side of Eq. (9-21) is the volume flow per unit width in the boundary layer, and the left-hand side is the equivalent size necessary for the same volume flow at the free-stream velocity. The point y_1 is any point outside the boundary layer, which is pictured in Fig. 9-3. In the figure, the two areas A are equal, since the area B is common to both sides of Eq. (9-21). To obtain a value for δ^*, the following rearrangements can be made:

$$\delta^* = y_1 - \int_0^{y_1} \frac{U_x}{U_\infty}\, dy = \int_0^{y_1}\left(1 - \frac{U_x}{U_\infty}\right) dy.$$

Now $U_x/U_\infty = f'$ and $y = \eta/\sqrt{U_\infty/x\nu}$. Combining these equations gives

$$\delta^* = \sqrt{\nu x/U_\infty} \int_0^{\eta_1} (1 - f')\, d\eta. \tag{9-22}$$

Integration of this equation results in

$$\delta^* = \sqrt{\nu x/U_\infty}\,[\eta_1 - f(\eta_1)]. \tag{9-23}$$

Since the point η_1 lies outside the boundary layer, Eq. (9–23) can be simplified to

$$\delta^* = 1.73\sqrt{\nu x/U_\infty}. \tag{9-24}$$

For a high value of η_1 where f' is 1.0000, one can evaluate that term of Eq. (9–23) which is in brackets as 1.73. The streamlines for the potential flow outside the boundary layer are deflected by the amount δ^*.

The *boundary layer thickness* δ can only be approximated, since the solution of U_x/U_∞ approaches a value of unity asymptotically. If we use U_x/U_∞ as 0.99, we obtain a value for η of 5.0, or

$$\delta = 5.0\sqrt{\nu x/U_\infty}. \tag{9-25}$$

This boundary layer is about three times the displacement thickness.

The definition of *momentum thickness* θ is somewhat similar to that of displacement thickness. However, momentum thickness is so defined as to be a measure of the loss of momentum in the boundary layer as compared with a similar loss in the potential flow. It is defined as $\rho U_\infty^2 \theta = \rho \int_0^{y_1} U_x(U_\infty - U_x)\,dy$ or

$$\theta = \int_0^{y_1} \frac{U_x}{U_\infty}\left(1 - \frac{U_x}{U_\infty}\right) dy. \tag{9-26}$$

In a form similar to Eq. (9–22), this becomes

$$\theta = \sqrt{\nu x/U_\infty} \int_0^{\eta_1} f'(1 - f')\,d\eta, \tag{9-27}$$

where again η_1 is outside the boundary layer. Numerical evaluation gives

$$\theta = 0.664\sqrt{\nu x/U_\infty}. \tag{9-28}$$

A number of other definitions for the boundary layer can be made; two of these are shown as the straight lines in Fig. 9–4. The shaded areas are equal.

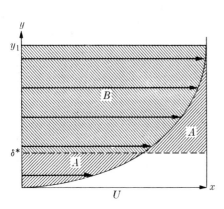

Fig. 9–3. Displacement-thickness definition.

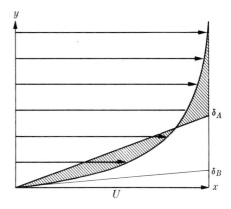

Fig. 9–4. Boundary-layer thickness definitions.

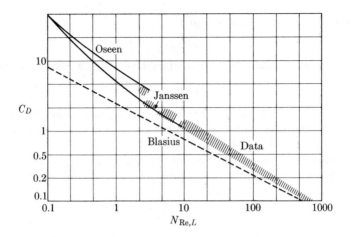

Fig. 9–5. Drag Coefficients. [From F. Janssen, Preprints of the 1956 Heat Transfer and Fluid Mechanics Institute, Stanford University Press, Stanford, California (1956). By permission.]

The preceding analysis of the boundary layer flow past a flat plate assumes that this layer is very small relative to the distance from the leading edge. Janssen [12], by means of an electric analog solution to the Navier-Stokes equation, obtained an estimate of the error introduced by the assumption. The new analysis indicated that disagreement becomes unimportant at a Reynolds number of about 1000. Figure 9–5 summarizes his results. The problem is also discussed by Van Dyke [28].

EXAMPLE 9–2. A one-foot-square section of a test wall in a wind tunnel is used as the test area for drag and other measurements. The forward edge of the test area is located one foot from the leading edge of the test wall. It is known that the test area lies within the laminar boundary layer. The free-stream air velocity is 10 ft/sec, and the temperature is 60°F. Estimate the various parameters that are descriptive of the boundary layer.

Answer. Consider first the various boundary layer thicknesses. For air at 60°F, $\nu = 1.6 \times 10^{-4}$ ft²/sec. The value of x varies between 1 and 2 ft over the test area. Thus

$$0.048 \text{ in.} < \sqrt{\nu x / U_\infty} < 0.067 \text{ in.}$$

Equations (9–24), (9–25), and (9–28) give

$$0.083 \text{ in.} < \delta^* < 0.116 \text{ in.},$$
$$0.24 \text{ in.} < \delta < 0.335 \text{ in.},$$
$$0.032 \text{ in.} < \theta < 0.045 \text{ in.}$$

The drag has to be calculated in two parts; i.e., the drag on the test area is the difference between the drag on the test wall from the leading edge to the end of the test area and the drag on the test wall up to the test area:

$$D_{1\text{-}2\text{ ft}} = D_{0\text{-}2\text{ ft}} - D_{0\text{-}1\text{ ft}}.$$

From Eq. (9–19), we have

$$D = 0.664 U_\infty b \sqrt{\mu \rho U_\infty} \, (\sqrt{2} - \sqrt{1}).$$

The air density at 60°F is 0.0765 lb/ft³. The plate width is one foot, so that the drag can be calculated as

$$D = 2.65 \times 10^{-4} \text{ lbf}.$$

9-4 COMPARISON OF COMMENCEMENT OF FLOW AND FLOW ALONG A FLAT PLATE

In Chapter 7 we considered the commencement of flow along a flat plate, in which the solution involved the error function. In the preceding section we discussed the boundary layer flow, and the velocity was given in Fig. 9–2. Both flows are plotted on the same coordinates in Fig. 9–6.

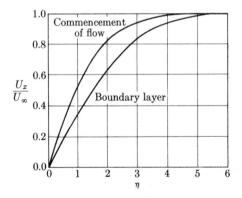

Fig. 9–6. Comparison of the commencement of flow along a flat plate with the boundary layer solution.

Because of the basic assumptions and the similarity of the flows, the two solutions should be somewhat similar. However, there is one difference that leads to a larger boundary layer and consequently a somewhat less steep velocity distribution for the boundary layer solution. In the exact solution for commencement of flow, U_y is zero for all t, which is not true in the boundary layer solution for any x/U_∞. In the boundary layer solution, the y-component of velocity tends to push out and enlarge the boundary layer. The data of Burgers [5] tend to fall between the two lines rather than on either one; however, the results of Nikuradse [18] check with the Blasius solution quite well.

TABLE 9-2

$\gamma(x)$	$(x/d_0)/N_{Re}$
100.00	0.0000065
20.00	0.000205
10.00	0.001045
6.00	0.003575
4.00	0.00838
2.00	0.02368
1.00	0.04488
0.80	0.05227
0.60	0.06198
0.40	0.0760
0.00	∞

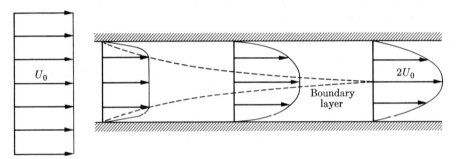

Fig. 9–7. *Boundary layer formation at the entrance to a pipe.*

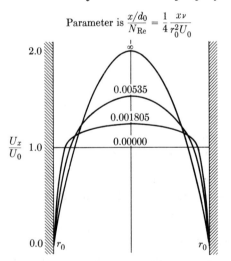

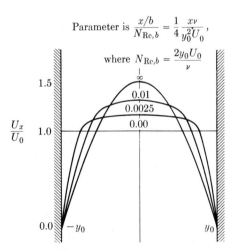

Fig. 9–8. *Langhaar's solution for the entrance to a pipe.* [*From the American Society of Mechanical Engineers,* Trans. ASME *(June 1942). By permission.*]

Fig. 9–9. *Schlichting's solution for the entrance between two parallel walls.*

9-5 FLOW IN A CIRCULAR PIPE AND BETWEEN PARALLEL PLATES

The flow in the inlet of a pipe is a subject of practical importance. The use of capillary viscometers for accurate viscosity measurements requires some means of estimation of the energy necessary to develop the velocity profile to its final steady-state form. If these energy losses can be reliably estimated from theory, a great deal of additional experimental effort can be avoided. Langhaar [15] and Sparrow et al. [26] treated the very difficult problem of flow in the inlet of a pipe. They linearized and solved the boundary layer equations by assuming that the acceleration is a function of the x-direction only. The linearization results in an approximate boundary layer solution rather than an exact one, as was the Blasius flat plate problem. The results of Langhaar's analysis are expressed in terms of Bessel functions and are reviewed in Example 9–3. The final equation is

$$\frac{U_x}{U_0} = \frac{I_0(\gamma) - I_0(\gamma r/r_0)}{I_2(\gamma)}, \qquad (9\text{--}29)$$

where I_0 and I_2 are modified Bessel functions of the first kind, of order zero and two, respectively. The values of γ, a function of x, are given in Table 9-2. The solution at any given distance down the pipe gives U_x/U_0 for all values of r. A boundary layer thickness could be defined as the value of r for U_x/U_0 of 0.99. Figure 9–7 gives the qualitative results with this boundary layer shown. Figure 9–8 shows the profiles (superimposed on one another) in more detail. The more recent analysis by Sparrow et al. gives quite similar results.

A velocity distribution similar to Fig. 9–8, for the flow between two walls of a channel, is shown in Fig. 9–9. This solution was given by Schlichting [24], who used a series expression for the boundary layer growth in the direction of flow and for the progressive deviation from the final parabolic profile in the direction opposite to the flow. At a point where both solutions are valid, the results are joined to give the final results for the entire inlet region. The flow becomes nearly fully developed at an $(x/b)/N_\text{Re}$ of 0.04, which corresponds to

$$L_e/b = 0.04\, N_\text{Re}, \qquad (9\text{--}30)$$

where b is the channel width, or $2y_0$.

The Langhaar analysis for the center-line velocity reaching 1.98 of the initial average velocity (99%) is

$$L_e/d_0 = 0.0567\, N_\text{Re}. \qquad (9\text{--}31)$$

For a flow with a Reynolds number of 2000, it would take a distance of 115 pipe diameters for the velocity distribution to be nearly parabolic. An earlier equivalent length determined by Boussinesq [4] was

$$L_e/d_0 = 0.065\, N_\text{Re}. \qquad (9\text{--}32)$$

The center-line velocities of the various developments for pipe flow can be compared; the results are reproduced as Fig. 9–10.

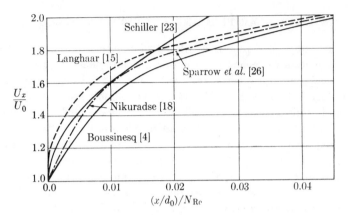

Fig. 9–10. *Comparison of pipe-flow entry-length solutions.*

For accurate viscosity measurements, one must be sure that the energy associated with the development of a parabolic velocity distribution is negligible, or that the results are corrected for the effect introduced. The pressure drop in the entrance region is summarized in Fig. 9–11 in terms of a friction factor averaged from the entrance ($x = 0$) to the point in question ($x = x$). Langhaar's theoretical curve (i), experimental data (ii), and a curve for the fully developed parabolic velocity distribution (iii) are given. Curve (ii) is given by

$$(x/d_0)f_{\mathrm{ave}} = 3.435\sqrt{(x/d_0)/N_{\mathrm{Re}}}, \tag{9-33}$$

which represents the data of Kline and Shapiro [13]. Curve (iii) is from the Hagen-Poiseuille law in terms of a friction factor:

$$(x/d_0)f_{\mathrm{ave}} = (16/N_{\mathrm{Re}})(x/d_0). \tag{9-34}$$

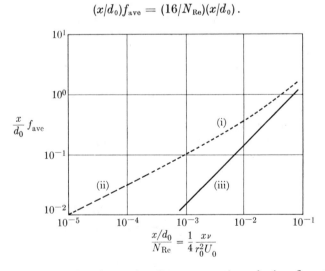

Fig. 9–11. *Average friction factor for the entry section of pipe flow.* [From J. G. Knudsen and D. L. Katz, Fluid Dynamics and Heat Transfer, McGraw-Hill, New York (1958). By permission.]

The pressure drop in the entrance section is greater than when the velocity distribution is fully established; furthermore, a correction is not needed if $(x/d_0)/N_{Re}$ is greater than 0.1. The analysis does not include the effects of the sudden contraction at the entrance and the sudden expansion at the exit; these will be discussed in the next chapter.

EXAMPLE 9–3. The flow in a circular pipe is a three-dimensional problem; the boundary layer equations (9–7) are

$$\frac{DU_x}{Dt} = -\frac{1}{\rho}\frac{dp}{dx} + \nu\left(\frac{\partial^2 U_x}{\partial y^2} + \frac{\partial^2 U_x}{\partial z^2}\right),$$

$$\frac{1}{\rho}\frac{dp}{dy} = \frac{1}{\rho}\frac{dp}{dz} = 0, \quad \text{(i)}$$

$$(\boldsymbol{\nabla}\cdot\boldsymbol{U}) = 0.$$

The solution to these equations for pipe flow was presented by Langhaar [15] with the following additional assumptions:

1. At $x = 0$, U_x is constant and equal to the average velocity U_0.
2. DU_x/Dt is of the form $\nu\beta^2 U_x$, which implies steady state and reduces Eq. (i) to a linear form.

Langhaar justified the form of the derivative given in assumption 2 for all points except those that lie within the boundary layer. In addition, when the boundary layer is fully developed, the assumptions are known to be valid. This example will illustrate the solution.

Answer. Combining assumption 2 with Eq. (i), we have

$$\frac{\partial^2 U_x}{\partial y^2} + \frac{\partial^2 U_x}{\partial z^2} - \beta^2 U_x = \alpha, \quad \text{(ii)}$$

where α and β are functions of x alone. A particular solution of (ii) is

$$U_x = -\alpha/\beta^2. \quad \text{(iii)}$$

The solution of the reduced equation

$$\frac{\partial^2 U_x}{\partial y^2} + \frac{\partial^2 U_x}{\partial z^2} - \beta^2 U_x = 0 \quad \text{(iv)}$$

can be obtained next. Equation (iv), in cylindrical coordinates, is

$$\frac{\partial^2 U_x}{\partial r^2} + \frac{1}{r}\frac{\partial U_x}{\partial r} - \beta^2 U_x = 0. \quad \text{(v)}$$

This equation is of the form of a differential equation whose solution can be expressed in terms of Bessel functions. The general equation and its solution are:

$$\frac{d^2u}{dr^2} + \frac{1-2a}{r}\frac{du}{dr} + \left[(bcr^{c-1})^2 + \frac{a^2-n^2c^2}{r^2}\right]u = 0, \quad \text{(vi)}$$

$$u = r^a[C_i J_n(br^c) + C_2 Y_n(br^c)]. \quad \text{(vii)}$$

For $a = 0$, $b = i\beta$, $c = 1$, and $n = 0$, Eq. (vi) reduces to Eq. (v). The solution (since Y_0 is infinity at $r = 0$ and thus $C_2 = 0$) is

$$U_x = C_1 J_0(i\beta r) = A I_0(\beta r), \tag{viii}$$

where I_0 is called the modified Bessel function of the first kind of order zero, and is defined as $I_n(r) = i^{-n} J_n(ir)$. The final solution for Eqs. (iii) and (viii) is

$$U_x = A I_0(\beta r) - \frac{\alpha}{\beta^2}. \tag{ix}$$

To evaluate A and α, note that at $r = r_0$, $U_x = 0$; thus

$$\alpha/\beta^2 = A I_0(\beta r_0) = A I_0(\gamma), \tag{x}$$

where $\gamma = \beta r_0$. Therefore

$$U_x = A[I_0(\gamma r/r_0) - I_0(\gamma)]. \tag{xi}$$

To evaluate A, from continuity,

$$\int_0^{r_0} r U_x \, dr = \tfrac{1}{2} r_0^2 U_0. \tag{xii}$$

Thus

$$A \int_0^{r_0} [r I_0(\gamma r/r_0) - r I_0(\gamma)] \, dr = \tfrac{1}{2} r_0^2 U_0.$$

Integration gives us

$$A \left[\frac{r_0}{\gamma} r I_1\left(\frac{\gamma r}{r_0}\right) - \frac{r^2}{2} I_0(\gamma) \right]_0^{r_0} = \tfrac{1}{2} r_0^2 U_0,$$

and substituting the limits results in

$$A \left[\frac{r_0^2}{\gamma} I_1(\gamma) - \frac{r_0^2}{2} I_0(\gamma) \right] = \tfrac{1}{2} r_0^2 U_0.$$

Rearranging gives us

$$A = U_0 / [(2/\gamma) I_1(\gamma) - I_0(\gamma)].$$

A general recurrence formula is

$$I_n(\gamma) = -(\gamma/2n)[I_{n+1}(\gamma) - I_{n-1}(\gamma)], \tag{xiii}$$

which, for $n = 1$, becomes

$$I_1(\gamma) = -(\gamma/2)[I_2(\gamma) - I_0(\gamma)]. \tag{xiv}$$

Combining this with the expression for A gives us

$$A = -\frac{U_0}{I_2(\gamma)}. \tag{xv}$$

FLOW IN A CIRCULAR PIPE AND BETWEEN PARALLEL PLATES 133

Using Eq. (xv) in (xi) gives the final result:

$$\frac{U_x}{U_0} = \frac{I_0(\gamma) - I_0(\gamma r/r_0)}{I_2(\gamma)}, \qquad (9\text{-}29)$$

where the term $\gamma = \beta r_0$ is a function of x only, since β is a function of x only. The problem that remains is to obtain a numerical solution for the value of γ as a function of x. Langhaar obtained the solution in terms of a parameter σ, which was equal to $(x/r_0)/N_{Re}$, where the Reynolds number was $r_0 U_0/\nu$. Table 9-2 gives his computed values in terms of the more common Reynolds number $d_0 U_0/\nu$.

EXAMPLE 9-4. Determine the viscosity of a material for which a pressure drop of 20 psi was measured across a test capillary for a flow rate of 1 liter/min. The length of the tube is 6 in. and it has an internal diameter of $\frac{1}{16}$ in. The contraction and expansion losses are assumed to be $1.2 U_0^2$. The specific gravity of the material is 1.0.

Answer. The necessity of correction cannot be directly established, since μ is unknown and the Reynolds number cannot be calculated. Therefore, to start, assume that no correction is needed:

$$U_0 = Q/A = 27.8 \text{ ft/sec}.$$

The contraction and expansion contributions will be

$$1.2 U_0^2/g_c = 1800 \text{ lbf/ft}^2$$

From Eq. (7-23), we have

$$\mu = (r_0^2/8 U_0)(dp/dz).$$

The net pressure drop dp/dz in the capillary will be

$$20 \times 144 - 1800 = 1080 \text{ lbf/ft}^2.$$

Thus the viscosity can be calculated as

$$0.0085 \text{ lb/ft-sec} = 12.6 \text{ cp}.$$

The Reynolds number can now be calculated:

$$N_{Re} = d_0 U_0/\nu = 1070 \quad \text{and} \quad (x/d_0)/N_{Re} = 0.09.$$

Therefore a correction from Fig. 9-11 is needed. This can be obtained from curves (i) and (iii), and gives a correction factor of $1.75/1.44 = 1.2$. Since the dimensions and velocity are constant, the correction can be applied directly to the viscosity and Reynolds number. Thus the second trial gives

$$0.00703 \text{ lb/ft-sec} = 10.4 \text{ cp},$$

$$N_{Re} = 1295, \quad (x/d_0)/N_{Re} = 0.075,$$

$$\text{correction factor} = 1.26.$$

9-6 SIMILAR SOLUTIONS

The boundary layer on a flat plate is one example of a class of problems called *similar solutions*. Even though the velocity distribution across the layer varies considerably as the flow progresses down the plate, it has only one form when expressed as the nondimensional velocity versus the nondimensional coordinate (see Fig. 9–2). Such systems are called similar, since the nondimensional plot or the differential equation (9–14) is not a function of x.

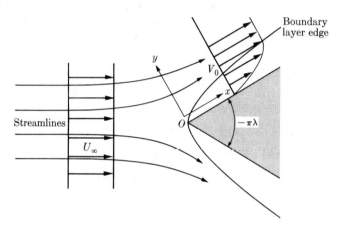

Fig. 9–12. Boundary-layer flow over a wedge.

There is a class of similar solutions for which we can obtain a general form for the differential equation. In general, the pressure drop in the x-direction will not vanish, but will continue to be determined from the flow exterior to the boundary layer; that is,

$$\frac{p}{\rho} + \frac{V_0^2}{2} = \text{const}$$

or

$$\frac{1}{\rho}\frac{dp}{dx} = -V_0 \frac{dV_0}{dx}, \qquad (9\text{–}35)$$

where V_0 denotes the velocity at the edge of the boundary layer. For steady state, the boundary layer equations (9–7) become

$$U_x \frac{\partial U_x}{\partial x} + U_y \frac{\partial U_x}{\partial y} = V_0 \frac{dV_0}{dx} + \nu \frac{\partial^2 U_x}{\partial y^2}, \qquad \frac{\partial U_x}{\partial x} + \frac{\partial U_y}{\partial y} = 0, \qquad (9\text{–}36)$$

with the boundary conditions of $U_x = U_y = 0$ at $y = 0$ and $U_x = V_0$ at $y = \infty$. The coordinates x and y are taken along the solid boundary (see Fig. 9–12).

The velocities in terms of the stream function (9–11) can be used to convert Eq. (9–36) to a form analogous to Eq. (9–13):

$$\frac{\partial \psi}{\partial y}\frac{\partial^2 \psi}{\partial x\,\partial y} - \frac{\partial \psi}{\partial x}\frac{\partial^2 \psi}{\partial y^2} = V_0\frac{dV_0}{dx} + \nu\frac{\partial^3 \psi}{\partial y^3}. \tag{9-37}$$

The reduction to an ordinary differential equation was first given by Falkner and Skan [7]. The modified approach used here was presented by Görtler [8] and has also been given by Meksyn [16]. A new variable can be defined as

$$\xi = \int_0^x (V_0/U_\infty)\,dx. \tag{9-38}$$

Analogous to Eqs. (9–10) and (9–12), the variables can be changed by

$$\eta = yV_0/\sqrt{2\nu U_\infty \xi} \tag{9-39}$$

and $\psi = \sqrt{2\nu U_\infty \xi}\,f(\xi, \eta)$. The velocities become

$$U_x = \partial \psi/\partial y = V_0 f',$$

$$-U_y = \frac{\partial \psi}{\partial x} = V_0\sqrt{\frac{\nu}{2\xi U_\infty}}\left[f + 2\xi\frac{\partial f}{\partial \xi} - (1+\lambda)\eta f'\right], \tag{9-40}$$

where $\lambda = -2\xi U_\infty(dV_0/dx)/V_0^2$. With these values, Eq. (9–37) can be transformed to

$$f''' + ff'' - \lambda(1 - f'^2) = 2\xi\left(f'\frac{\partial f'}{\partial \xi} - f''\frac{\partial f}{\partial \xi}\right), \tag{9-41}$$

where λ is now $-2\xi(d\ln V_0/d\xi)$. The boundary conditions are $f = f' = 0$ at $\eta = 0$ and $f' = 1$ at $\eta = \infty$, which are the same as in Eq. (9–15).

If f and f' are not functions of ξ, and λ is not a function of x, the resulting differential equation becomes

$$f''' + ff'' - \lambda(1 - f'^2) = 0, \tag{9-42}$$

$$\lambda = -2\xi(d\ln V_0/d\xi) = \text{const}, \tag{9-43}$$

and will not be a function of x; thus, a "similar solution" will exist.

For $\lambda \neq -2$, integration of Eq. (9–43) and evaluation of the integration constant yields

$$\ln V_0 = -(\lambda/2)\ln \xi + \ln U_\infty$$

or

$$V_0 = U_\infty \xi^{-\lambda/2}. \tag{9-44}$$

Using Eq. (9–38) for V_0 gives us $d\xi/dx = \xi^{-\lambda/2}$. Integrating and evaluating the constant from Eq. (9–38), we find that

$$\xi^{(\lambda+2)/2} = x. \tag{9-45}$$

Defining m as
$$m = -\frac{\lambda}{\lambda+2} \quad \text{or} \quad \lambda = -\frac{2m}{m+1}, \tag{9-46}$$

and combining with Eqs. (9-44) and (9-45), we obtain
$$V_0 = U_\infty x^m, \tag{9-47}$$

which is the form of the potential flow necessary to ensure that a similar solution exists. For $m=0$, $\lambda=0$, and the problem reduces to the previously studied case of the flat plate. For $\lambda \neq 0$ or $\neq -2$, Eq. (9-46) describes the potential flow past a wedge with an angle of $-\pi\lambda$ (see Fig. 9-12). For $m=1$, ($\lambda=-1$), the potential flow is that of a stagnation point, which is a good approximation for the front part of a circular or parabolic cylinder.

The quantities ξ, η, and ψ can be evaluated from Eqs. (9-38) and (9-39):
$$\xi = \int_0^x \frac{V_0}{U_\infty} dx = \int_0^x x^m dx = \frac{x^{m+1}}{m+1}. \tag{9-48}$$

Thus
$$\eta = \frac{yV_0}{\sqrt{2\nu U_\infty \xi}} = y\sqrt{\frac{(m+1)V_0}{2\nu x}} = y\sqrt{\frac{(m+1)U_\infty}{2\nu}} x^{(m-1)/2} \tag{9-49}$$

and
$$\psi = \sqrt{2\nu U_\infty \xi} f(\eta) = \sqrt{\frac{2\nu U_\infty}{m+1}} x^{(m+1)/2} f(\eta). \tag{9-50}$$

From the stream function (9-40), the velocities can be obtained:
$$U_x = \partial\psi/\partial y = V_0 f' = U_\infty x^m f', \tag{9-51}$$
$$-U_y = \frac{\partial\psi}{\partial x} = \sqrt{\frac{(m+1)\nu U_\infty}{2}} x^{(m-1)/2} \left(f + \frac{m-1}{m+1}\eta f'\right). \tag{9-52}$$

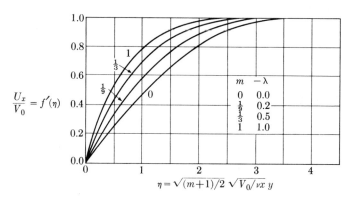

Fig. 9-13. Velocity distribution in the laminar boundary layer in the flow past a wedge given by $V_0 = U_\infty x^m$. The exponent m and the wedge angle $-\pi\lambda$ are connected through Eq. (9-46). [From H. Schlichting, Boundary Layer Theory, McGraw-Hill, New York (1960). By permission.]

Hartree [9] obtained a numerical solution to Eq. (9–42) for various values of m (or λ) via Eq. (9–46). The numerical values are available also in Knudsen and Katz [14], and are summarized in Fig. 9–13 in terms of the generalized coordinates of Eqs. (9–49) and (9–51) and the parameter m.

The *local drag coefficient* can be established:

$$\tau_w = -\mu \left(\frac{\partial U_x}{\partial y}\right)_w = -\mu V_0 \left(\frac{\partial f'}{\partial y}\right)_w = -\mu V_0 f''_w \sqrt{\frac{(m+1)U_\infty}{2\nu}} x^{(m-1)/2},$$

$$C_D = \frac{\tau_w}{\frac{1}{2}\rho U_\infty^2} = \sqrt{\frac{2\nu(m+1)}{U_\infty x}} f''_w = \frac{\sqrt{2(m+1)} f''_w}{\sqrt{N_{\text{Re},x}}}. \qquad (9\text{–}53)$$

For the flat plate limit, this equation reduces to (9–18), since f''_w of this problem is $\sqrt{2}\,\alpha$ when $m = 0$.

EXAMPLE 9–5. Plot the variation of the local drag coefficient as a function of wedge angle at a specific value of $N_{\text{Re},x}$.

Answer. Plot $C_D \sqrt{N_{\text{Re},x}} = \sqrt{2(m+1)}\, f''_w$ versus $-\pi\lambda$. As determined in the solution by Hartree [9], m and f''_w are unique functions of λ. Values of m (or $-\lambda$) between 0 and 1 are of interest, since these represent the extremes of the wedge, i.e., the flat plate and stagnation flow, respectively. This is tabulated as follows:

$-\lambda$	Angle	m	f''_w	$C_D \sqrt{N_{\text{Re},x}}$
0.0	0°	0	0.4696	0.664
0.2	36	$\frac{1}{9}$	0.686	1.020
0.4	72	$\frac{1}{4}$	0.854	1.351
0.6	108	$\frac{3}{7}$	0.996	1.686
0.8	144	$\frac{2}{3}$	1.120	2.042
1.0	180	1	1.2326	2.465

The results are plotted in Fig. 9–14. The rate of increase of the drag coefficient is linear with the angle of the wedge.

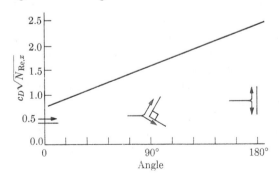

Fig. 9–14. Drag coefficient as a function of wedge angle.

9-7 SEPARATION

For the preceding problems, a mathematical solution was obtained for the boundary layer equations with given specific boundary conditions. When a pressure gradient exists, an additional complication arises if the gradient is adverse, i.e., if the pressure is increasing in the direction of flow (opposing the flow). This phenomenon is the *separation* of the boundary layer from its associated body. A flow over a cylinder, as shown in Fig. 9–15, is used to illustrate the problem.

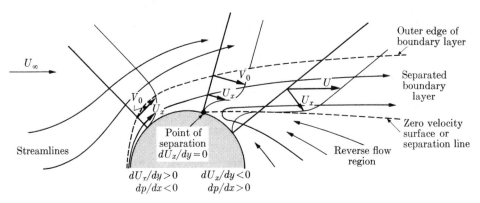

Fig. 9–15. The boundary layer over a cylindrical object.

The reason for separation can be analyzed as follows: As the flow passes the cylinder, it must accelerate over the forward portion and decelerate over the rear part. Over the forward part, the increase in velocity would be accompanied by a decrease in pressure ($dp/dx < 0$) or a favorable pressure gradient; over the rear part the converse would be true, and an adverse pressure gradient would exist. Since the pressure drop across the boundary layer is negligible, this adverse gradient would be felt at the wall. The gradient in pressure, as well as the shear force, causes a rapid deceleration of the fluid elements in the immediate vicinity of the wall. The deceleration continues until the fluid element comes to rest, at which point the viscous forces are zero, since the velocity is zero. However, the adverse pressure effect will continue to act, and the fluid will reverse and flow backward. The point at which the reversal first appears is that at which $(dU_x/dy)_w = 0$, and is the point of separation of the boundary layer from the surface. The separation must occur so that the flow can continue in the direction of the increasing pressure.

The actual point of separation cannot be accurately established from boundary layer theory. For example, by a series expansion method, Schlichting [25] found the separation point to be 108.8° from the forward stagnation point of a circular cylinder. Experimentally, this same point is found to be about 83°, but is known to be a function of the Reynolds number and the surface conditions. The main

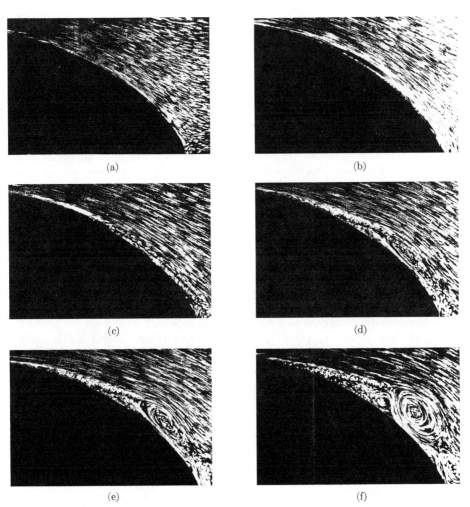

Fig. 9–16. Boundary-layer separation. [*From L. Prandtl and O. G. Tietjens, Applied Hydro- and Aeromechanics, reprinted by permission of United Engineering Trustees, Inc.*]

reason for the deviation of theory from experiments is believed to be the assumption of potential flow exterior to the boundary layer. The actual viscous effects apparently modify the pressure distribution enough so that potential flow is not even a good approximation in the area of separation.

Figure 9–16 shows several photographs of the development of separation under unsteady-state conditions. Figure 9–16(a) is the potential flow; in 9–16(b), the flow along the boundary has stopped; and in 9–16(c) the flow has started to reverse. Figure 9–16(d) shows the start of a vortex formation; 9–16(e) shows this flow further developed; and 9–16(f) shows a development in which the entire flow pattern has been changed. These rather drastic changes in flow patterns cause considerable alteration of the velocity distribution and the pressure drop.

As indicated in Chapter 6 on ideal flow, and as can be seen in Fig. 9–16, the flow resembles potential motion only for a short period after startup. Beyond this time, the flow is of a boundary layer nature and in many cases separation occurs. In the Magnus effect (Fig. 6–13) a starting vortex is shed from the lower rear side, and for the flow over airfoils, it is shed from the trailing edge. Kelvin's theorem on vortex motion (Eq. 6–42) states that the circulation must be constant, and so a fictitious bound vortex of rotation opposite to the starting vortex must be formed. This bound vortex will in turn be shed as a stopping vortex when all motion is stopped. The actual mechanism of lift is associated with this circulation, but not in exactly the manner described by potential flow (see discussion page 76). The lift depends primarily on the asymmetry of the flow about the object in question. This is caused by the point of separation being asymmetric with the axis of symmetry. In the Magnus effect this is caused by the rotation; for the airfoil it is caused by the nonsymmetrical geometry of the object.

9-8 OTHER ITEMS AND SUMMARY

With the preceding material as a basis, one can approach more complex problems involving such topics as axially symmetrical and three-dimensional boundary layers [17], higher approximations [28], unsteady boundary layer flow [20], flow with suction and injection, boundary layers in compressible flow [10], and thermal boundary layers. For these and other topics, the reader should see the references cited, especially the books by Schlichting [25], Pai [19], Dryden [6], Meksyn [16], and the review volume edited by Rosenhead [29].

Of particular interest are problems of mass and heat transfer and chemical reaction in boundary layer flows. The simplified boundary layer equations for steady state have already been given as Eq. (9–36), which must now be coupled with an additional equation for the scalar transfer obtained from a material or heat balance. The additional equation can be obtained from the equations of change (3–22) and (3–30), and the expressions for the fluxes in terms of the driving forces (3–44) through (3–46), that is, Eq. (4–19) or (4–20). The simplified form for steady state is

$$U_x \frac{\partial \theta}{\partial x} + U_y \frac{\partial \theta}{\partial y} = \delta \frac{\partial^2 \theta}{\partial y^2}, \tag{9-54}$$

where θ is either concentration or temperature, and correspondingly δ is the thermal or mass diffusivity. Although the equations for mass and heat transfer are the same, the boundary conditions are not necessarily so, and consequently there are cases where mass transfer is not the same as that for heat. This occurs because mass transfer involves the actual movement of material, and thus may contribute an appreciable interfacial velocity, that is, U_y will not be zero at the surface. As noted in the preceding paragraph, the heat-transfer problem has been extensively treated and references have been cited. The corresponding mass-transfer problem has been discussed by Acrivos [1, 2], who separated the analysis into two major cases. The first paper treats the problem of transfer from the main stream to the boundary and the second considers the reverse of this. In each case, asymptotic solutions were obtained and simple graphical interpolation forms suggested. The addition of chemical kinetics complicates the problem, especially if reactions greater than first order are considered.

An approximate solution for the boundary layer equations can be obtained by an overall analysis of the layer; that is, the boundary layer equations can be integrated to give an overall balance on momentum. This tells nothing about the velocity distribution in the fluid; in fact, a distribution of some form must be assumed in the solution. Von Kármán first applied the method, and it has since become known as the *von Kármán integral momentum theorem*. In the next chapter on integral methods, the theorem will be presented and the problem of flow along a flat plate will be solved, so that a comparison can be made with the more exact Blasius solution.

One subject which should be mentioned, but will not be discussed until later, is the transition to turbulence in the boundary layer. The material in this chapter has treated boundary layer theory as a large Reynolds number approximation to laminar viscous flow. However, laminar flow at high Reynolds numbers cannot be maintained for extensive distances in the direction of flow, and transition to a turbulent boundary layer occurs. This transition depends on such factors as the free-stream turbulence, surface conditions, and shape of the system. More details on this will be given in the chapter on turbulence.

PROBLEMS

9-1. Derive the expressions for the velocity components U_x and U_y for the flow along a flat plate.

9-2. From Table 9-1, compute the velocity distribution curve.

9-3. By numerical integration of this curve, compute the average-to-maximum velocity ratio.

9-4. Calculate the velocity profile by the method of Langhaar for flow in a circular pipe of 3 in. diameter, at a Reynolds number of 1200, and at a position 1 ft from the entrance. Compare this with the fully developed profile.

9-5. Consider the problem of flow over a flat plate with uniform suction. The amount of fluid removed by suction is small compared with the free stream velocity U_∞. This does not mean it can be totally neglected. Write the boundary layer equations for this flow. What are the boundary conditions? What is the difference between this problem and the flow over a flat plate without suction?

9-6. A simple exact solution can be obtained for Problem 9-5 for the case in which the velocity in the direction of flow is independent of length. The suction velocity must be of such a magnitude as to make this assumption correct. Assume steady state and solve for the normally defined boundary layer thickness.

9-7. Determine the displacement thickness for Problem 9-6.

9-8. Determine the shear stress at the wall for Problem 9-6.

REFERENCES

1. A. ACRIVOS, *AIChEJ* **6**, 410 (1960).
2. A. ACRIVOS, *J. Fluid Mech.* **12**, 337–357 (1962).
3. H. BLASIUS, *Z. Math. Physik* **56**, 1–37 (1908); *NACA TM 1256* translation.
4. J. BOUSSINESQ, *Compt. Rend.* **113**, 9, 49 (1891).
5. J. M. BURGERS, *Proc. Intern. Congr. Appl. Mech., First Congr.*, page 113, Delft, (1924).
6. H. L. DRYDEN, F. P. MURNAGHAN, and H. BATEMAN, *Hydrodynamics*, Dover Publications, New York (1956).
7. V. M. FALKNER, and S. W. SKAN, *Phil. Mag.* **12**, 865 (1931).
8. H. GÖRTLER, *J. Math. & Mech.* **6**, 1 (1957).
9. D. R. HARTREE, *Proc. Cambridge Phil. Soc.* **33**, 223 (1937).
10. W. D. HAYES, *Jet Propulsion* **26**, 270–274 (1956).
11. L. HOWARTH, *Proc. Roy. Soc. (London)* **A164**, 547 (1938).
12. F. JANSSEN, *1956 Heat Transfer and Fluid Mechanics Institute*, pp. 173–184, Stanford University, Stanford, Calif. (1956); *J. Fluid Mech.* **3**, 329–343 (1958).
13. S. J. KLINE, and A. H. SHAPIRO, *NACA TN 3048* (1953).
14. J. G. KNUDSEN, and D. L. KATZ, *Fluid Dynamics and Heat Transfer*, McGraw-Hill, New York (1958).
15. H. L. LANGHAAR, *Trans. ASME* **64**, A55 (1942).
16. D. MEKSYN, *New Methods in Laminar Boundary-Layer Theory*, Pergamon Press, London, (1961).
17. F. K. MOORE, *Advances in Applied Mechanics*, Vol. IV, pp. 160–228, Academic Press, New York (1956).
18. J. NIKURADSE, *Monographie, Zen. f. wiss. Berich.*, Berlin (1942).
19. SHIH-I PAI, *Viscous Flow Theory-Laminar Flow:* D. Van Nostrand, Princeton, N. J., (1956).
20. D. PIERCE, *J. Fluid Mech.* **11**, 460–464 (1961).
21. L. PRANDTL, *Verhandlg. d. III, Intern. Mathe. Kongr.*, Heidelberg (1904); *NACA TM* 452 (1928) translation.
22. L. PRANDTL, and O. G. TIETJENS, *Applied Hydro- and Aeromechanics* (reprint of the 1934 edition), Dover Publications, New York.
23. L. SCHILLER, *Physik. Z.* **23**, 14 (1922); *ZAMM* **2**, 96 (1922).
24. H. SCHLICHTING, *ZAMM* **14**, 368 (1934).
25. H. SCHLICHTING, *Boundary Layer Theory*, McGraw-Hill, New York (1960).
26. E. M. SPARROW, S. H. LIN, and T. S. LUNDGREN, *Phys. Fluids* **7**, 338–347 (1964); see also E. B. CHRISTIANSEN and H. E. LEMMON, *AIChEJ* **11**, 995–999 (1965).
27. C. TÖPFER, *Z. Math. Phys.* **60**, 397 (1912).
28. M. D. VAN DYKE, *J. Fluid Mech.* **14**, 161–177, 481–495 (1962); *see also* S. C. R. DENNIS and J. DUNWOODY, *ibid.* **24**, 577–595 (1966).
29. *Laminar Boundary Layers* (L. ROSENHEAD, editor), Oxford University Press, New York (1963).

CHAPTER 10

INTEGRAL METHODS OF ANALYSIS

In the preceding chapters, the equations of change were developed and applied to a variety of laminar flow problems. These equations may also be integrated over a finite volume of the system, thereby obtaining the overall balances of mass, momentum, and energy. Equations of this nature can give no information as to the gradients inside the system, but rather give only the relations between the properties at the inlet and outlet. Since the details of the flow inside the system are not needed in the analysis, the equations apply equally well to turbulent flows. There are five useful equations of this type: the overall balances of mass, momentum, angular momentum, energy, and mechanical energy (the Bernoulli equation). Each of these will be discussed, and their applications considered, in this chapter.

Two procedures can be used to obtain the overall balances of mass, momentum, and energy. The first is to integrate the detailed differential equations of change over a finite volume of the system; the second is to write a general overall balance and then consider the transfer of the properties. The first method is somewhat more complex than the second, and has been considered in detail by Bird [2]. The second procedure is more direct, since in the analysis only the overall effects are considered. This general method has been reviewed by Baron [1]. The analysis is quite similar to the development of the general property balance equation (3–18).

10-1 THE GENERAL INTEGRAL EQUATION OF CHANGE

The equation for the transfer of a property ψ from the volume V, which is moving at U, is similar to Eq. (3–14), that is,

$$\frac{d}{dt}\left(\int_V \psi \, dV\right) = -\oint_S (\mathbf{\Psi} \cdot d\mathbf{S}) + \int_V \dot{\psi}_g \, dV - \oint_S (\mathbf{\Psi}_o \cdot d\mathbf{S}), \qquad (10\text{--}1)$$

where $\mathbf{\Psi}$ is the flux of the property ψ, which is expressed per unit volume. The subscripts g and o represent the generation within the volume and the amount transported across the surface by means other than flow. The term on the left of the equation is the total rate of change of the property in the volume, and can be rewritten by using Eq. (2–55):

$$\int_V (\partial \psi / \partial t) \, dV + \oint_S \psi(\mathbf{U} \cdot d\mathbf{S}) = -\oint_S (\mathbf{\Psi} \cdot d\mathbf{S}) + \int_V \dot{\psi}_g \, dV - \oint_S (\mathbf{\Psi}_o \cdot d\mathbf{S}). \qquad (10\text{--}2)$$

The second term on the left accounts for the change of the property associated with the movement of the surface at U. This term can be re-expressed by using Eq. (2–51) and the definition of the scalar product (Eq. 2–3):

$$\oint_S \psi(U \cdot dS) = \oint_S \psi(U \cdot N)\, dS = \oint_S \psi U \cos\theta\, dS, \tag{10–3}$$

where N is the normal unit vector directed outward from the surface, and θ is the angle between U and N. As in the case of the general property balance, if a stationary system is considered, the total derivative will be identical to the partial derivative [the second term on the left of Eq. (10–2) will be zero]. However, the fluxes would then have to contain the transfer of property associated with the flow across the element's surface.

10-1.A The Integral Mass Balance

For mass, the first term of Eq. (10–2) is the time rate of change of mass, $\dot{M}$; the second term is the loss of mass per unit volume ρ from the volume; and the third, fourth, and fifth terms are zero, since the mass flux is zero (system moving at U), and mass is not generated or transferred by other methods. By use of Eq. (10–3), one can reduce Eq. (10–2) to

$$\dot{M} = -\oint_S \rho(U \cdot dS) = -\oint_S \rho U \cos\theta\, dS, \tag{10–4}$$

which is the *overall mass balance*.

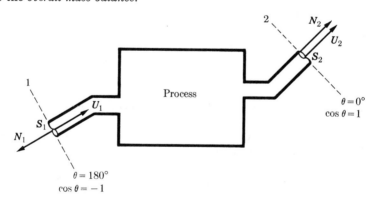

Fig. 10–1. System for integral mass balance.

A somewhat restricted system is shown in Fig. 10–1, in which the fluid enters at 1 and leaves at 2. The surfaces across the flow are selected so that they are perpendicular to the lines of flow. For this system, Eq. (10–4) becomes

$$\dot{M} = -\int_{S_1+S_2} \rho U \cos\theta\, dS = \int_{S_1} \rho U\, dS - \int_{S_2} \rho U\, dS = \rho_1 U_{1,\text{ave}} S_1 - \rho_2 U_{2,\text{ave}} S_2, \tag{10–5}$$

since at S_1, $\cos\theta = -1$, and at S_2, $\cos\theta = 1$. The average velocity U_{ave} is the average over the cross section at the point in question, and specifically for pipe flow is defined as

$$(U^n)_{\text{ave}} = \frac{\int_0^{2\pi}\int_0^{r_0} U^n r\, dr\, d\theta}{\int_0^{2\pi}\int_0^{r_0} r\, dr\, d\theta}. \qquad (10\text{–}6)$$

The denominator is simply the area, πr_0^2. The densities are assumed constant across the cross section. The mass flow rate is $w = \rho U_{\text{ave}} S$. Thus

$$\dot{M} = w_1 - w_2. \qquad (10\text{–}7)$$

For steady state, $\dot{M}$ is zero and $w_1 = w_2$, or

$$\rho_1 U_{1,\text{ave}} S_1 = \rho_2 U_{2,\text{ave}} S_2. \qquad (10\text{–}8)$$

For turbulent flows, U is generally replaced by $\bar{U}$, where $\bar{U}$ is the *time mean* of the instantaneous velocity vector U. This time mean must not be confused with the average across the pipe as given by Eq. (10–6). The replacement mentioned above in effect neglects the flow associated with the deviations from the mean. A more exact analysis for turbulent flows would be to use $U = \bar{U} + u$ and then time-average the equation; u is the deviation from the mean and by definition $\bar{u} = 0$. For the mass balance where $n = 1$, the approximation is without error, since $\overline{u_{\text{ave}}} = (\bar{u})_{\text{ave}} = 0$. For n other than one, this would not be true.

EXAMPLE 10–1. A tank of radius r_T is filled to a height H with a liquid of density ρ (Fig. 10–2). The fluid drains from the bottom of the tank through a hole with a radius of r_0. The flow can be considered to be given by Torricelli's law, $U_{\text{ave}}^2 = 2gh$, where h is the instantaneous height. What is the total time required to empty the tank?

Answer. From Eq. (10–5) or (10–7), we have

$$\dot{M} = d(\rho\pi r_T^2 h)/dt = w_1 - w_2 = 0 - \pi r_0^2 \rho\sqrt{2gh}.$$

Rearranging, we get

$$dh/\sqrt{h} = -(r_0^2/r_T^2)\sqrt{2g}\, dt.$$

Integrating gives us

$$2\sqrt{h} = -(r_0^2/r_T^2)\sqrt{2g}\, t + C.$$

At $t = 0$, $h = H$; therefore, $C = 2\sqrt{H}$, and

$$\sqrt{h} = \sqrt{H} - (r_0^2/r_T^2)\sqrt{g/2}\, t.$$

The total time of emptying is

$$t_{\text{total}} = (r_T^2/r_0^2)\sqrt{2H/g}.$$

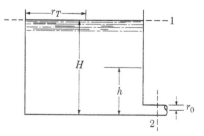

Fig. 10–2. Draining of fluid from a circular tank.

10-1.B The Integral Momentum Balance

The property of linear momentum per unit volume is $\rho \boldsymbol{U}$. In Eq. (10–2), the first term would be the time rate of change of momentum, $\dot{\boldsymbol{P}}$. Using Eq. (10–3) for the second term, we obtain

$$\oint_S \rho \boldsymbol{U}(\boldsymbol{U}\cdot d\boldsymbol{S}) = \oint_S \rho UU \cos\theta\, dS,$$

which reduces to $\int_{S_1+S_2} \rho UU \cos\theta\, dS$, since there is no flow through surfaces other than S_1 and S_2. The third term involves the momentum flux or the pressure tensor $\boldsymbol{\mathsf{P}} = p\boldsymbol{\mathsf{U}} + \bar{\bar{\boldsymbol{\tau}}}$. For the first part of this (Eqs. 2–44 and 2–51),

$$\oint_S (p\boldsymbol{\mathsf{U}}\cdot d\boldsymbol{S}) = \oint_S p\, d\boldsymbol{S} = \oint_S p\boldsymbol{N}\, dS,$$

and for the second,

$$\oint_S (\bar{\bar{\boldsymbol{\tau}}}\cdot d\boldsymbol{S}) = \oint_S (\bar{\bar{\boldsymbol{\tau}}}\cdot \boldsymbol{N})\, dS.$$

The actual breakdown used in practice is somewhat different. The pressure term contributes a momentum transfer due to pressure forces at the inlet and outlet of the system. That is,

$$\int_{S_1+S_2} p\boldsymbol{N}\, dS,$$

as above. The pressure term also contributes a pressure drag on the internal surfaces of the system; this is included in a term denoted as $\boldsymbol{F}_{\text{drag}}$. The viscous term can be broken up similarly. The momentum transfer associated with the viscous forces at the inlet and outlet is usually neglected. The viscous drag on the internal surfaces of the system is a major contributor, and is also included in $\boldsymbol{F}_{\text{drag}}$. In this analysis, $\boldsymbol{F}_{\text{drag}}$ is the force of the fluid on the solid. The fourth term of Eq. (10–2) is the generation of momentum as caused by the action of external forces on the fluid, such as the force of gravity. This is denoted as $\boldsymbol{F}_{\text{ext}}$. The final term is zero.

Combining the above into Eq. (10–2), we have the *overall momentum balance*

$$\dot{\boldsymbol{P}} = -\int_{S_1+S_2} \rho UU \cos\theta\, dS - \int_{S_1+S_2} p\boldsymbol{N}\, dS - \boldsymbol{F}_{\text{drag}} + \boldsymbol{F}_{\text{ext}}. \tag{10-9}$$

Again taking S_1 and S_2 perpendicular to the lines of flow, as in Fig. 10–1, we can rearrange Eq. (10–9) as

$$\dot{\boldsymbol{P}} = \int_{S_1} \rho UU\, dS - \int_{S_2} \rho UU\, dS - \int_{S_1} p\boldsymbol{N}\, dS - \int_{S_2} p\boldsymbol{N}\, dS - \boldsymbol{F}_{\text{drag}} + \boldsymbol{F}_{\text{ext}}.$$

Performing the integrations, we find that

$$\dot{\boldsymbol{P}} = [\rho_1(UU)_{1,\text{ave}}S_1 - \rho_2(UU)_{2,\text{ave}}S_2] + (-p_1\boldsymbol{N}_1 S_1 - p_2\boldsymbol{N}_2 S_2) - \boldsymbol{F}_{\text{drag}} + \boldsymbol{F}_{\text{ext}}. \tag{10-10}$$

The first difference in Eq. (10–10) can be expressed in terms of the mass flow rate w:

$$\dot{P} = \left[\frac{(UU)_{1,\text{ave}}}{U_{1,\text{ave}}}w_1 - \frac{(UU)_{2,\text{ave}}}{U_{2,\text{ave}}}w_2\right] + (-p_1 N_1 S_1 - p_2 N_2 S_2) - F_{\text{drag}} + F_{\text{ext}}. \quad (10\text{--}11)$$

Equation (10–11) is a vector equation and must be resolved into its component parts before being used. For example, let us reconsider the system of Fig. 10–1 in Fig. 10–3. For the x-direction, the following are true from the figure:

$$U_{1,x} = U_1 \cos \alpha_1, \quad U_{2,x} = U_2 \cos \alpha_2,$$
$$N_{1,x} = -\cos \alpha_1, \quad N_{2,x} = \cos \alpha_2.$$

Thus Eq. (10–11) becomes

$$\dot{P}_x = \left[\frac{(U_1^2)_{\text{ave}}}{U_{1,\text{ave}}}w_1 \cos \alpha_1 - \frac{(U_2^2)_{\text{ave}}}{U_{2,\text{ave}}}w_2 \cos \alpha_2\right] \\ + (p_1 S_1 \cos \alpha_1 - p_2 S_2 \cos \alpha_2) - F_{x,\text{drag}} + F_{x,\text{ext}}. \quad (10\text{--}12)$$

For the y-component, the cosines become sines; otherwise, the equation is the same.

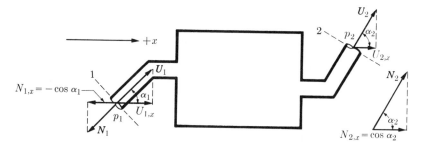

Fig. 10–3. System for integral momentum balance.

Quite often the term $(U^2)_{\text{ave}}/U_{\text{ave}}$ is replaced by U_{ave}/β, where β is $U_{\text{ave}}^2/(U^2)_{\text{ave}}$. Although β is usually assumed to be unity, it has been evaluated for a number of flows:

Turbulent flow: $\quad 0.99 > \beta > 0.95,$

One-seventh power law for
turbulent flow: $\quad \beta = 0.98,$

Laminar flow: $\quad \beta = \frac{3}{4}.$

EXAMPLE 10–2. What is the force acting on a horizontal 60° reducing bend, which has water flowing at a rate of 10 ft³/sec? The inlet pressure is 100 psig. The inlet diameter is 6 in., and the outlet 4 in. (Fig. 10–4).

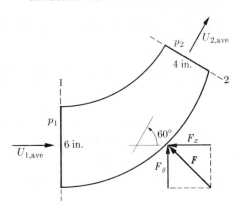

Fig. 10–4. Momentum balance on a reducing elbow.

Answer. The velocities $U_{1,\text{ave}}$ and $U_{2,\text{ave}}$ are determined from the volumetric flow rate, and are 51 and 114.5 ft/sec, respectively. The Reynolds number $d_0 U_{\text{ave}}/\nu$ of the flow varies from 2.4 to 3.7×10^6, and thus the assumption of $\beta = 1$ or $(U^2)_{\text{ave}}/U_{\text{ave}} = U_{\text{ave}}$ is valid. The inlet pressure is given as 100 psig; the outlet pressure p_2 can be calculated from this and the velocities by means of Bernoulli's equation, if it is assumed that the frictional loss is negligible. This result is 29.2 psig. For this problem,

$$\dot{P} = 0, \qquad w_1 = w_2 = 62.4\,(10) = 624 \text{ lb/sec},$$

and F_{ext} has a component in the z-direction only. Equation (10–12) becomes

$$F_{x,\text{drag}} = (624/32.174)(51 - 114.5 \cos 60°) + (100\pi 3^2 - 29.2\pi 2^2 \cos 60°)$$
$$= -120.8 + 2645 = 2524 \text{ lbf}.$$

The equation equivalent to (10–12) for the y-direction becomes

$$F_{y,\text{drag}} = (624/32.174)(0 - 114.5 \sin 60°) + (0 - 29.2\pi 2^2 \sin 60°)$$
$$= -1925 - 318 = -2243 \text{ lbf},$$

where $F_{x,\text{drag}}$ acts to the right and $F_{y,\text{drag}}$ acts downward. The forces exerted by the elbow (F_x and F_y in the figure) are opposite to those above, and are of equal magnitude.

10-1.C The Von Kármán Theorem of Integral Momentum

It was noted in the concluding remarks of the last chapter that an overall balance on momentum can be obtained for the boundary layer. As in the case of the other overall balances, there are two ways to obtain this balance: one method is to integrate the boundary layer equations, another is to use the overall balance approach. The first involves more detail, and reviews can be found in Pai [9] and Knudsen and Katz [8]. In the second method, which will be used here,

THE GENERAL INTEGRAL EQUATION OF CHANGE

the momentum balance as just developed is applied to the boundary layer shown in Fig. 10–5.

Steady state is assumed, external forces are neglected, and unit width is considered. The pressure across the layer is taken as constant by virtue of Eq. (9–7). With these limitations, the x-direction momentum equation becomes

$$0 = -\int_{ab+ad+cd} \rho U_x^2 \, dS + (p_{ab}S_{ab} - p_{cd}S_{cd}) - F_{x,\text{drag}}. \qquad (10\text{–}13)$$

The momentum associated with the mass flow (the first term) must be reconsidered, since flow occurs across all three sides. The mass flow into the element along ab is $\int_0^\delta \rho U_x \, dy$. The mass flow out through cd would be

$$\int_0^\delta \rho U_x \, dy + \frac{d}{dx}\left(\int_0^\delta \rho U_x \, dy\right) dx.$$

The difference must be equal to the inflow at the top ad:

$$\frac{d}{dx}\left(\int_0^\delta \rho U_x \, dy\right) dx.$$

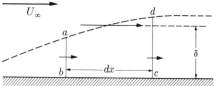

Fig. 10–5. Integral momentum balance on the boundary layer.

The momentum associated with the mass flow across ab is $\int_0^\delta \rho U_x^2 \, dy$. The loss of momentum through cd would be

$$\int_0^\delta \rho U_x^2 \, dy + \frac{d}{dx}\left(\int_0^\delta \rho U_x^2 \, dy\right) dx.$$

The momentum input at the top ad is the velocity U_∞ multiplied by the mass input:

$$U_\infty \frac{d}{dx}\left(\int_0^\delta \rho U_x \, dy\right) dx.$$

The net momentum inflow is the balance over the surfaces $ab + ad - cd$:

$$\int_0^\delta \rho U_x^2 \, dy + U_\infty \frac{d}{dx}\left(\int_0^\delta \rho U_x \, dy\right) dx - \int_0^\delta \rho U_x^2 \, dy - \frac{d}{dx}\left(\int_0^\delta \rho U_x^2 \, dy\right) dx.$$

This simplifies to

$$\frac{d}{dx}\left[\int_0^\delta \rho U_x(U_\infty - U_x) \, dy\right] dx. \qquad (10\text{–}14)$$

The second term is the pressure effect and is

$$p_{ab}\delta - p_{cd}\delta = p_{ab}\delta - [p_{ab} + (\partial p/\partial x)\, dx]\delta = -\delta(\partial p/\partial x)\, dx. \qquad (10\text{–}15)$$

The third term, or viscous drag force, is simply

$$-\tau_w \, dx. \qquad (10\text{–}16)$$

Substitution of Eqs. (10–14), (10–15), and (10–16) into (10–13) gives the *Von Kármán integral momentum equation*:

$$\frac{d}{dx}\left[\int_0^\delta \rho U_x(U_\infty - U_x)\,dy\right] = -\tau_w + \delta(\partial p/\partial x). \quad (10\text{–}17)$$

In order to solve this equation, an estimate must be made beforehand for the velocity U_x as a function of y, so that the integral can be determined.

For incompressible flow over a flat plate, dp/dx is zero and Eq. (10–17) becomes

$$\tau_w = -\rho \frac{d}{dx}\left[\int_0^{y_0} U_x(U_\infty - U_x)\,dy\right]. \quad (10\text{–}18)$$

The integration can be taken to any $y_0 \gg \delta$, since for all $y \gg \delta$, $U_x = U_\infty$, and the integral is zero. The velocity U_x can be assumed to be of the form

$$U_x/U_\infty = f(y/\delta) = f(\eta). \quad (10\text{–}19)$$

The boundary conditions become

$$\begin{aligned} y = 0, \quad \eta = 0, \quad U_x = 0, \quad f(\eta) = 0, \\ y = \delta, \quad \eta = 1, \quad U_x = U_\infty, \quad f(\eta) = 1. \end{aligned} \quad (10\text{–}20)$$

Combining Eqs. (10–18) and (10–19) and rearranging, we get

$$\tau_w = -\rho\, U_\infty^2 \frac{d}{dx}\left[\delta \int_0^1 f(1-f)\,d\eta\right].$$

Another expression for τ_w can be obtained from the relations at the boundary and Eq. (10–19):

$$\tau_w = -\mu\left(\frac{\partial U_x}{\partial y}\right)_w = -\mu\left(\frac{\partial U_x}{\partial f}\right)_w\left(\frac{\partial \eta}{\partial y}\right)_w\left(\frac{\partial f}{\partial \eta}\right)_w = -\mu\left(\frac{U_\infty}{\delta}\right)f'(0).$$

Equating these two expressions for τ_w, rearranging, and integrating, gives us

$$\frac{2\nu}{U_\infty}f'(0)x = \delta^2 \int_0^1 f(1-f)\,d\eta, \quad (10\text{–}21)$$

since $\delta = 0$ at $x = 0$, and since the integral is a numerical value for any given gradient and thus is not a function of either x or y.

The form of the velocity profile must now be assumed so as to satisfy the boundary conditions of (10–20) and other conditions that might be imposed. For example, one could use the simple linear relation $f(\eta) = \eta$. The integral term becomes

$$\int_0^1 \eta(1-\eta)\,d\eta,$$

which has a value of $\tfrac{1}{6}$. Furthermore, $f'(0) = 1$. Thus Eq. (10–21) becomes $2\nu x/U_\infty = \tfrac{1}{6}\delta^2$, or

$$\delta = 3.464\sqrt{\nu x/U_\infty}. \quad (10\text{–}22)$$

THE GENERAL INTEGRAL EQUATION OF CHANGE 151

TABLE 10-1

Comparison of the Integral Momentum Method and the Blasius Solution

Solution item*	Blasius	η	$\frac{3}{2}\eta - \frac{1}{2}\eta^3$	$2\eta - 2\eta^3 + \eta^4$
$\int_0^1 f(1-f)\,d\eta$	—	$\frac{1}{6}$ or 0.167	$\frac{39}{280}$ or 0.139	$\frac{37}{315}$ or 0.117
$f'(0)$	—	1	$\frac{3}{2}$	2
δ	~5	3.464	4.64	5.83
C_D	1.328	1.155	1.29	1.372
δ^*	1.729	1.732	1.740	1.752

*$\delta = K_1\sqrt{\nu x/U_\infty}$, $C_D = K_2/\sqrt{N_{\text{Re},L}}$, $\delta^* = K_3\sqrt{\nu x/U_\infty}$. The K values are the numbers in the table.

The result of using this simple approximation for the velocity gradient can be compared with the boundary layer thickness of $5\sqrt{\nu x/U_\infty}$ obtained from the Blasius solution. For a plate of width b and length L, the drag on one side is

$$D = b\int_0^L \tau_w\,dx = b\int_0^L \mu\left(\frac{U_\infty}{\delta}\right)f'(0)\,dx = \mu b U_\infty \left(\frac{1}{\sqrt{12}}\right)\sqrt{\frac{U_\infty}{\mu}}\int_0^L \left(\frac{1}{\sqrt{x}}\right)dx$$

$$= \left(\frac{2}{\sqrt{12}}\right)U_\infty b\sqrt{\mu\rho\,LU_\infty} = 0.578\,U_\infty b\sqrt{\mu\rho\,LU_\infty},$$

and the drag coefficient, similar to Eq. (9–20), becomes

$$C_D = 1.155/\sqrt{N_{\text{Re},L}}.$$

These various values and their comparison to the Blasius solution are given in Table 10-1; the agreement is quite good, considering the simple form selected for the velocity gradient.

Prandtl assumed a velocity distribution of the form

$$f(\eta) = \tfrac{3}{2}\eta - \tfrac{1}{2}\eta^3, \tag{10-23}$$

which satisfies the boundary conditions of Eq. (10–20). The analysis is the same as presented above, with the integral of Eq. (10–21) now being equal to $\frac{39}{280}$ and $f'(0) = \frac{3}{2}$. Pohlhausen assumed a distribution of the form

$$f(\eta) = 2\eta - 2\eta^3 + \eta^4. \tag{10-24}$$

The corresponding value for the integral is $\frac{37}{315}$, and $f'(0) = 2$. Table 10-1 includes the results of using these two velocity distributions.

The actual velocity distribution can be obtained by combining the expressions for the boundary layer thickness and for the assumed form of $f(\eta)$. For the linear distribution,

$$U_x/U_\infty = f(\eta) = \eta = y/\delta \quad\text{and}\quad \delta = 3.464\sqrt{\nu x/U_\infty},$$

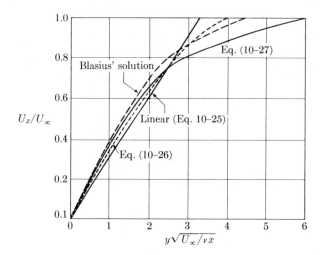

Fig. 10–6. Assumed velocity distributions and the result of the Blasius solution.

which combine to give

$$\frac{U_x}{U_\infty} = \left(\frac{1}{3.464}\right) y \sqrt{\frac{U_\infty}{\nu x}}. \tag{10-25}$$

For the velocity distribution of Eq. (10–23), we obtain

$$\frac{U_x}{U_\infty} = \left(\frac{3}{2 \times 4.64}\right) y \sqrt{\frac{U_\infty}{\nu x}} - \left(\frac{1}{2 \times 4.64^3}\right) y \left(\frac{U_\infty}{\nu x}\right)^{3/2}, \tag{10-26}$$

and for the distribution of Eq. (10–24), we have

$$\frac{U_x}{U_\infty} = \left(\frac{2}{5.83}\right) y \sqrt{\frac{U_\infty}{\nu x}} - \left(\frac{2}{5.83^3}\right) y^3 \left(\frac{U_\infty}{\nu x}\right)^{3/2} + \left(\frac{1}{5.83^4}\right) y^4 \left(\frac{U_\infty}{\nu x}\right)^2. \tag{10-27}$$

The three distributions above are compared with the Blasius solution in Fig. 10–6. This figure explains the results reported in Table 10-1. The boundary layer thickness depends on the approach of U_x/U_∞ to unity; the distributions vary just as the thickness does. Equation (10–22) shows that the drag coefficient depends on the slope of the distribution at the wall. The slopes do not vary appreciably; thus the coefficients in the table are all about the same value. Finally, the displacement thickness depends on the integral of the velocity over the boundary layer (Eq. 9–21); this should not change much, and hence these values are essentially constant.

To include the effect of a pressure gradient, Eq. (10–17) is generally put into an alternative form:

$$U_\infty^2 \left(\frac{d\theta}{dx}\right) + (2\theta + \delta^*) U_\infty \left(\frac{dU_\infty}{dx}\right) = -\frac{\tau_w}{\rho},$$

where δ^* and θ are the displacement and momentum thicknesses defined by Eqs. (9-21) and (9-26), respectively. For flow over a surface, a fourth-degree polynomial is assumed for the velocity distribution, and its form can be shown to be

$$\frac{U_x}{U_\infty} = f(\eta) = (2\eta - 2\eta^3 + \eta^4) - \left(\frac{\delta^2}{\mu U_\infty}\right)\left(\frac{dp}{dx}\right)\left(\frac{\eta}{6}\right)(1 - \eta)^3,$$

which reduces to Eq. (10-24) for zero pressure gradient. The results compare favorably with more exact solutions when the flow is accelerated (dp/dx negative); however, the results become less reliable as separation is approached in a retarded flow (dp/dx positive).

This brief introduction to approximate methods for the solution of the boundary layer equations is restricted to the problem of flow over a flat plate. More complex problems involving more complex geometries, pressure gradients, compressibility, non-Newtonian materials, higher approximations, and heat and mass transfer can be studied. The reader is referred to Schlichting [12], Pai [9], and other recent references [3, 4, 5, 7, 11, 15].

10-1.D The Integral Energy Balance

The third property to be considered is energy. The total energy per unit mass is denoted by E, and per unit volume by ρE. The first term of Eq. (10-2) would be the time rate of change of energy $\dot{\text{E}}$. Using Eq. (10-3) for the second term, we obtain

$$\int_S \rho\text{E}(\boldsymbol{U}\cdot d\boldsymbol{S}) = \int_S \rho\text{E}U \cos\theta\, dS = \int_{S_1+S_2} \rho\text{E}U \cos\theta\, dS,$$

since there is no flow through surfaces other than S_1 and S_2. The third term is associated with the energy flux $\boldsymbol{q}$. The fourth term, the generation of energy, can be neglected for most work, unless electrical or nuclear contributions exist. The energy associated with chemical reaction is included as a change in internal energy, which is a part of the total energy. The final term represents the transport of energy by nonflow methods, that is, by work on the system. This consists of shaft work and pressure-volume work over the inlet and outlet. The contribution of shear work is generally neglected. Combination of these considerations into Eq. (10-2) gives us

$$\dot{\text{E}} = -\int_{S_1+S_2} \rho\text{E}U \cos\theta\, dS + \dot{Q} + \dot{W}_s - \int_{S_1+S_2} pU\, dS. \tag{10-28}$$

If the surfaces S_1 and S_2 are taken perpendicular to the lines of flow, as in Fig. 10-1, Eq. (10-28) can be integrated for some known expression for the energy. The total energy consists of internal energy U, kinetic energy $\frac{1}{2}U^2$, and potential energy Φ, which, for a gravity system, is the height above some reference line; that is,

$$\text{E} = \text{U} + \tfrac{1}{2}U^2 + \Phi = \text{U} + \tfrac{1}{2}U^2 + gz. \tag{10-29}$$

Combining Eqs. (10–28) and (10–29) and integrating gives the *overall energy balance*:

$$\dot{E} = (\rho_1 \mathsf{u}_1 U_{1,\text{ave}} S_1 - \rho_2 \mathsf{u}_2 U_{2,\text{ave}} S_2) + \tfrac{1}{2}[\rho_1 (U_1^3)_{\text{ave}} S_1 - \rho_2 (U_2^3)_{\text{ave}} S_2]$$
$$+ g(\rho_1 z_1 U_{1,\text{ave}} S_1 - \rho_2 z_2 U_{2,\text{ave}} S_2) + \dot{Q} + \dot{W}_s \qquad (10\text{--}30)$$
$$+ (p_1 U_{1,\text{ave}} S_1 - p_2 U_{2,\text{ave}} S_2).$$

In terms of the mass flow rate this becomes

$$\dot{E} = (\mathsf{u}_1 w_1 - \mathsf{u}_2 w_2) + \frac{1}{2}\left[\frac{(U_1^3)_{\text{ave}}}{U_{1,\text{ave}}} w_1 - \frac{(U_2^3)_{\text{ave}}}{U_{2,\text{ave}}} w_2\right] + g(z_1 w_1 - z_2 w_2)$$
$$+ \dot{Q} + \dot{W}_s + \left(\frac{p_1}{\rho_1} w_1 - \frac{p_2}{\rho_2} w_2\right). \qquad (10\text{--}31)$$

For steady state, $\dot{E}$ is zero, and $w_1 = w_2$, so that Eq. (10–31) can be rewritten on a unit-mass basis as

$$(\mathsf{u}_1 - \mathsf{u}_2) + \frac{1}{2}\left[\frac{(U_1^3)_{\text{ave}}}{U_{1,\text{ave}}} - \frac{(U_2^3)_{\text{ave}}}{U_{2,\text{ave}}}\right] + g(z_1 - z_2) + \left(\frac{p_1}{\rho_1} - \frac{p_2}{\rho_2}\right) + Q + W_s = 0, \qquad (10\text{--}32)$$

where Q and W_s are the heat flow and work per unit mass of fluid flowing.

The term $\tfrac{1}{2}(U^3)_{\text{ave}}/U_{\text{ave}}$ can be replaced by $\tfrac{1}{2}U_{\text{ave}}^2/\alpha$, where α is $U_{\text{ave}}^3/(U^3)_{\text{ave}}$. While α is usually assumed to be unity, it has been evaluated for a number of flows:

Turbulent flow:	$0.88 < \alpha < 0.98$,
One-seventh power law for turbulent flow:	$\alpha = 0.94$,
Laminar flow:	$\alpha = \tfrac{1}{2}$.

10-1.E The Integral Angular Momentum Balance

The property, angular momentum per unit volume, is $\rho(\mathbf{R} \times \mathbf{U})$, where $\mathbf{R}$ is the position vector. The development of this balance parallels the overall linear momentum balance, and the equation analogous to (10–9) is

$$\dot{\mathbf{A}} = -\int_{S_1 + S_2} \rho(\mathbf{R} \times \mathbf{U}) U \cos\theta \, dS - \int_{S_1 + S_2} p(\mathbf{R} \times \mathbf{N}) \, dS - \mathbf{T}_{\text{drag}} + \mathbf{T}_{\text{ext}}, \qquad (10\text{--}33)$$

where $\dot{\mathbf{A}}$ is the time rate of change of angular momentum, $\mathbf{T}$ denotes a torque, and $\mathbf{T}_{\text{drag}}$ is the torque exerted on the boundary by the fluid. This equation can also be derived by integrating, over the system volume, the vector product of the position vector with the differential equation of change for momentum. This has recently been done by Slattery and Gaggioli [13].

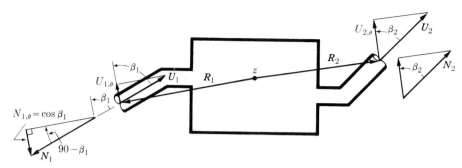

Fig. 10–7. System for integral angular momentum balance.

Performing the integrations in Eq. (10–33) and using the mass flow rate, we arrive at an equation analogous to (10–11):

$$\dot{A} = \left[\frac{(\boldsymbol{R} \times \boldsymbol{U}\,U)_{1,\text{ave}}}{U_{1,\text{ave}}} w_1 - \frac{(\boldsymbol{R} \times \boldsymbol{U}\,U)_{2,\text{ave}}}{U_{2,\text{ave}}} w_2 \right]$$
$$+ [-p_1 (\boldsymbol{R} \times \boldsymbol{N})_1 S_1 - p_2 (\boldsymbol{R} \times \boldsymbol{N})_2 S_2] - \boldsymbol{T}_{\text{drag}} - \boldsymbol{T}_{\text{ext}}, \quad (10\text{--}34)$$

which is a vector equation and must be resolved into its component parts, as was done for Eq. (10–11) to obtain (10–12) for the x-direction of momentum. However, in problems involving angular momentum, it is more common to use cylindrical coordinates. For example, let us consider again the system cited in the linear momentum balance (Fig. 10–7). For rotation about the z axis, the following are true from the figure and from Eq. (2–14):

$(\boldsymbol{R} \times \boldsymbol{U})_1 = R_1 U_{1,\theta} = R_1 U_1 \cos \beta_1,$

$(\boldsymbol{R} \times \boldsymbol{U})_2 = R_2 U_{2,\theta} = R_2 U_2 \cos \beta_2,$

$(\boldsymbol{R} \times \boldsymbol{N})_1 = R_1 N_{1,\theta} = R_1 \sin(90 - \beta_1) = R_1 \cos \beta_1,$

$(\boldsymbol{R} \times \boldsymbol{N})_2 = R_2 N_{2,\theta} = R_2 \cos \beta_2.$

Thus Eq. (10–34) becomes (for rotation about the z axis)

$$\dot{A}_z = \left[\frac{(R_1 U_1^2)_{\text{ave}}}{U_{1,\text{ave}}} w_1 \cos \beta_1 - \frac{(R_2 U_2^2)_{\text{ave}}}{U_{2,\text{ave}}} w_2 \cos \beta_2 \right]$$
$$- (p_1 R_1 S_1 \cos \beta_1 + p_2 R_2 S_2 \cos \beta_2) - T_{\theta,\text{drag}} + T_{\theta,\text{ext}}. \quad (10\text{--}35)$$

The application of the angular momentum balance to actual numerical problems is quite similar to the example already cited for the momentum balance.

10–2 THE BERNOULLI EQUATION FOR IDEAL FLOW

The Bernoulli equation is obtained by integration of the Euler equation of ideal flow along a streamline. The Euler equation (6–1) is

$$\frac{\partial \boldsymbol{U}}{\partial t} + (\boldsymbol{U} \cdot \boldsymbol{\nabla}) \boldsymbol{U} = -\frac{1}{\rho} \boldsymbol{\nabla} p + \frac{1}{\rho} \sum_s \rho_s \boldsymbol{F}_s. \quad (10\text{--}36)$$

If the force $\boldsymbol{F}$ can be expressed as a gradient of a scalar potential, $-\omega$, then the equation can be rewritten as

$$\frac{\partial \boldsymbol{U}}{\partial t} + (\boldsymbol{U}\cdot\boldsymbol{\nabla})\boldsymbol{U} = -\frac{1}{\rho}\boldsymbol{\nabla}p - \boldsymbol{\nabla}\omega. \tag{10-37}$$

This is to be integrated along a streamline with a differential distance of $d\boldsymbol{R}$:

$$\int_C \left(\frac{\partial \boldsymbol{U}}{\partial t}\cdot d\boldsymbol{R}\right) + \int_C ((\boldsymbol{U}\cdot\boldsymbol{\nabla})\boldsymbol{U}\cdot d\boldsymbol{R}) = -\int_C \left(\frac{\boldsymbol{\nabla}p}{\rho}\cdot d\boldsymbol{R}\right) - \int_C (\boldsymbol{\nabla}\omega\cdot d\boldsymbol{R}). \tag{10-38}$$

Terms of the form $(\boldsymbol{\nabla}a \cdot d\boldsymbol{R})$ can be transformed by use of the definitions of the gradient, position vector, and scalar product (2–1), (2–4), and (2–23) to give

$$(\boldsymbol{\nabla}a\cdot d\boldsymbol{R}) = da. \tag{10-39}$$

The first term of Eq. (10–38) is zero for steady state. For unsteady ideal irrotational flow, the velocity potential can be used, and the first term becomes

$$\int_C \left(\frac{\partial \boldsymbol{U}}{\partial t}\cdot d\boldsymbol{R}\right) = \int_C \left(\frac{\partial \boldsymbol{\nabla}\phi}{\partial t}\cdot d\boldsymbol{R}\right) = \int_C \left(\boldsymbol{\nabla}\frac{\partial \phi}{\partial t}\cdot d\boldsymbol{R}\right) = \int_C d\left(\frac{\partial \phi}{\partial t}\right) = \frac{\partial \phi}{\partial t}. \tag{10-40}$$

The strain tensor $(\boldsymbol{\nabla}\boldsymbol{U})$ is symmetric; that is to say, $(\boldsymbol{\nabla}\boldsymbol{U}) = (\widetilde{\boldsymbol{\nabla}\boldsymbol{U}})$, and the associative law is valid for the inner product of a vector (at either or both ends) with any number of tensors. Thus an identity analogous to Eq. (10–39) can be obtained for vectors; this identity is

$$((\widetilde{\boldsymbol{\nabla}\boldsymbol{U}})\cdot d\boldsymbol{R}) = (d\boldsymbol{R}\cdot(\widetilde{\boldsymbol{\nabla}\boldsymbol{U}})) = d\boldsymbol{U}.$$

Using this, one can write for the second term

$$\int_C ((\boldsymbol{U}\cdot\boldsymbol{\nabla})\boldsymbol{U}\cdot d\boldsymbol{R}) = \int_C (\boldsymbol{U}\cdot((\boldsymbol{\nabla}\boldsymbol{U})\cdot d\boldsymbol{R})) = \int_C (\boldsymbol{U}\cdot((\widetilde{\boldsymbol{\nabla}\boldsymbol{U}})\cdot d\boldsymbol{R}))$$
$$= \int_C (\boldsymbol{U}\cdot d\boldsymbol{U}) = \tfrac{1}{2}U^2. \tag{10-41}$$

The third and fourth terms follow directly from Eq. (10–39):

$$\int_C \left(\frac{\boldsymbol{\nabla}p}{\rho}\cdot d\boldsymbol{R}\right) = \int_C \frac{1}{\rho}dp, \tag{10-42}$$

$$\int_C (\boldsymbol{\nabla}\omega\cdot d\boldsymbol{R}) = \int_C d\omega = \omega. \tag{10-43}$$

Combining Eqs. (10–40) through (10–43) into (10–38) gives us Bernoulli's equation,

$$\partial\phi/\partial t + \tfrac{1}{2}U^2 + \omega + \int_C (1/\rho)\,dp = \text{const.} \tag{10-44}$$

THE MECHANICAL ENERGY BALANCE

If the potential ω is due to gravity only, then the equation becomes

$$\partial\phi/\partial t + \tfrac{1}{2}U^2 + gz + \int_C (1/\rho)\, dp = \text{const.} \tag{10-45}$$

The velocity term involves U^2 rather than $(U^3)_{\text{ave}}/U_{\text{ave}}$, because the equation is restricted to one streamline.

10-3 THE MECHANICAL ENERGY BALANCE OR THE ENGINEERING BERNOULLI EQUATION

The heat added to a system during a reversible process at steady state goes into an increase in the internal energy and into reversible work done during the expansion (as caused by the absorption of heat); that is,

$$Q = (\mathrm{U}_2 - \mathrm{U}_1) + \int_V p\, d(1/\rho). \tag{10-46}$$

For a real or irreversible process, the work done is less than the reversible work by the amount of the viscous dissipation within the system, or

$$Q = (\mathrm{U}_2 - \mathrm{U}_1) + \int_V p\, d(1/\rho) - F_{\text{losses}}. \tag{10-47}$$

The energy balance for steady state (10–32) can be combined with Eq. (10–47) to give one form of the mechanical energy balance.

$$\frac{1}{2}\left[\frac{(U_1^3)_{\text{ave}}}{U_{1,\text{ave}}} - \frac{(U_2^3)_{\text{ave}}}{U_{2,\text{ave}}}\right] + (gz_1 - gz_2) + \left(\frac{p_1}{\rho_1} - \frac{p_2}{\rho_2}\right) + \int_V p\, d\!\left(\frac{1}{\rho}\right) + W_s - F_{\text{losses}} = 0. \tag{10-48}$$

Another form is

$$\frac{1}{2}\left[\frac{(U_1^3)_{\text{ave}}}{U_{1,\text{ave}}} - \frac{(U_2^3)_{\text{ave}}}{U_{2,\text{ave}}}\right] + (gz_1 - gz_2) - \int_V \frac{dp}{\rho} + W_s - F_{\text{losses}} = 0, \tag{10-49}$$

since $p\, d(1/\rho) = d(p/\rho) - (1/\rho)\, dp$.

An alternate method of derivation, which is quite complex, involves integrating the scalar product of the velocity with the equation of motion. This method has been covered in detail in the works of Bird [2], Gaggioli [6], and Pings [10]. In spite of the complexity, this alternate method has an advantage in that it gives an explicit expression for the loss term for laminar flow; that is,

$$F_{\text{losses}} = \int_V (\bar{\bar{\tau}} : (\nabla U))\, dV, \tag{10-50}$$

where the double dot product or scalar product is given by Eq. (2–47).

For an isothermal constant composition process, the integral of Eq. (10–49) is the Gibbs free energy ΔG. For an isentropic constant composition process, the integral is the enthalpy change ΔH. For incompressible flow, the density can be considered constant and the integral is simply $\Delta p/\rho$. If the system is compressible, and if the pressure ratio is no greater than a factor of two, a good approximation is to use $v_{ave}(p_1 - p_2)$, where v_{ave} is the average specific volume between points 1 and 2; that is,

$$v_{ave} = \frac{1}{2}\left(\frac{1}{\rho_1} + \frac{1}{\rho_2}\right) = \frac{\rho_1 + \rho_2}{2\rho_1\rho_2}.$$

If the ideal gas approximation can be assumed, then the integral can be determined for both the isothermal and the isentropic systems. For isothermal flow,

$$\int \left(\frac{1}{\rho}\right) dp = \left(\frac{p_1}{\rho_1}\right) \int \left(\frac{1}{p}\right) dp = \left(\frac{p_1}{\rho_1}\right) \ln p \bigg|_1^2 = \left(\frac{p_1}{\rho_1}\right) \ln \left(\frac{p_2}{p_1}\right). \quad (10\text{-}51)$$

For isentropic flow,

$$\int \left(\frac{1}{\rho}\right) dp = -\Delta H = -\int c_p \, dT = -\int \left(\frac{\gamma R}{M_w(\gamma - 1)}\right) dT$$

$$= -\left(\frac{\gamma R}{M_w(\gamma - 1)}\right)(T_1 - T_2) = -\left(\frac{\gamma R T_1}{M_w(\gamma - 1)}\right) \quad (10\text{-}52)$$

$$\times \left[1 - \left(\frac{p_2}{p_1}\right)^{(\gamma-1)/\gamma}\right] = -\left(\frac{\gamma p_1}{\rho_1(\gamma - 1)}\right)\left[1 - \left(\frac{p_2}{p_1}\right)^{(\gamma-1)/\gamma}\right],$$

where $\gamma = c_p/c_v$.

The term for the losses can usually be determined by comparing the mechanical energy and momentum balances. In most cases, recourse to empirical information is necessary in order to evaluate the constants. The following examples will illustrate the estimations for these terms.

EXAMPLE 10-3. Determine the loss term for the sudden expansion of an incompressible fluid flowing turbulently in the pipe system shown in Fig. 10-8.

Answer. The control points 1 and 3 are taken so that the velocity distribution is uniform, but are not taken so far upstream and downstream that one would have to account for frictional forces. Since the flow is turbulent, it will be assumed

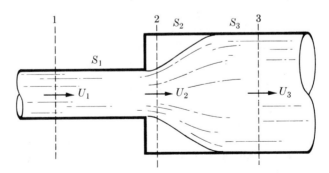

Fig. 10-8. Sudden expansion.

for simplicity that $\alpha = \beta = 1$. The mechanical energy balance between 1 and 2 gives us

$$U_1 = U_2, \quad p_1 = p_2. \tag{i}$$

Between 1 and 3,

$$\tfrac{1}{2}(U_1^2 - U_3^2) + (p_1 - p_3)/\rho = F_{\text{losses}}. \tag{ii}$$

The momentum balance equation (10–12) can be applied from 2 to 3. Since the drag and external forces are neglected, and since $S_2 = S_3$, this becomes

$$U_2 w_2 - U_3 w_3 + p_2 S_3 - p_3 S_3 = 0. \tag{iii}$$

Using the equivalent for the mass flow rate, we can modify this equation to

$$U_3(U_2 - U_3) + (p_2 - p_3)/\rho = 0. \tag{iv}$$

Comparison of Eqs. (i), (ii), and (iv) shows that

$$F_{\text{losses}} = \tfrac{1}{2} U_1^2 - \tfrac{1}{2} U_3^2 - U_1 U_3 + U_3^2 = \tfrac{1}{2}(U_1^2 - 2U_1 U_3 + U_3^2),$$

or

$$F_{\text{losses}} = \tfrac{1}{2}(U_1 - U_3)^2. \tag{v}$$

EXAMPLE 10–4. Determine the loss term for the sudden contraction shown in Fig. 10–9. Make the same general assumptions as in Example 10–3.

Answer. The loss between 1 and 2 is assumed zero, and all the loss occurs between 2 and 3; however, the restricted area at 2 is unknown (call it S_2'). Experimental data will be necessary to evaluate unknown constants that will have to be introduced. Equation (iii) of Example 10–3 becomes

$$\rho U_2^2 S_2' - \rho U_3^2 S_3 + p_2 S_3 - p_3 S_3 = 0. \tag{i}$$

Rearranging and using Eq. (10–8), we obtain

$$(p_2 - p_3)/\rho = U_3^2(1 - S_3/S_2'). \tag{ii}$$

Since S_3/S_2' is defined as $1/C_c$, and $(1 - S_3/S_2')^2$ is defined as K_c, then

$$(p_2 - p_3)/\rho = U_3^2(1 - 1/C_c) = U_3^2 \sqrt{K_c}. \tag{iii}$$

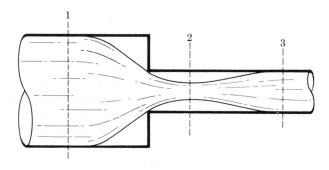

Fig. 10–9. Sudden reduction.

The mechanical energy balance is the same as in Eq. (ii) of Example 10–3; the velocity terms, however, must be rearranged. Proceeding as in that example, we find that the formula for the losses is

$$F_{\text{losses}} = (U_3^2/2)K_c. \qquad \text{(iv)}$$

The constants C_c and K_c are determined from experimental data. For example, for a drastic change in area, $C_c = 0.61$ and $K_c = 0.40$.

As a final example, the mechanical energy balance can be used to determine the velocity of an incompressible fluid stream from a Pitot tube measurement (see Fig. 10–10). Since the distances are short, friction can be neglected. Further, there will be no work or change in potential energy. Equation (10–48) reduces to

$$(U_1^3)_{\text{ave}}/2U_{1,\text{ave}} + (p_1 - p_2)/\rho = 0 \qquad \text{(10-53)}$$

or

$$\tfrac{1}{2}[(U_1^3)_{\text{ave}}/U_{1,\text{ave}}] = (p_2 - p_1)\rho. \qquad \text{(10-54)}$$

If the velocity term is taken as $U_{1,\text{ave}}^2$, Eq. (10–54) reduces to

$$U_{1,\text{ave}} = \sqrt{(2/\rho)(p_2 - p_1)}, \qquad \text{(10-55)}$$

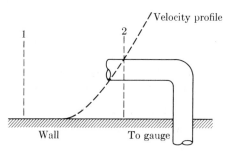

Fig. 10–10. Pitot tube.

which is the formula commonly used.

Because of the assumption made about the velocity term, an error is introduced by the use of Eq. (10–55) unless the velocity profile is perpendicular to the tube entrance. Specially designed multiport pitot tubes are used to eliminate this possibility. If the profile is not flat still another error can be introduced. If the shape of the profile and the geometry of the pitot tube are known, then a correction can be made. For laminar flow, the correction may be quite large. For turbulent flow, however, the profile is relatively flat and little or no correction is necessary. Knudsen and Katz [8] note that for laminar flow near a boundary, the average pressure difference measured may be 33% greater than that which exists at the centerline of the pitot tube.

In problems in which the friction losses are small and can be neglected, the energy balance, and often the momentum balance, can be used to obtain the same results as the mechanical energy balance. The mechanical energy balance will require a knowledge of a thermodynamic path to evaluate the integral term.

There are many other typical problems which require the use of the momentum and mechanical energy balances. For further examples, the reader can refer to the text in fluid mechanics by Streeter [14], and to other elementary books.

PROBLEMS

10-1. Compare the velocity distribution on a flat plate (Blasius' solution) with that obtained from the momentum balance method of Eqs. (10-26) and (10-27).

10-2. Derive the equation for a pitot tube, using a momentum balance. Consider a cylindrical space in front of the tube and assume that fluid is lost uniformly through the boundary, carrying its momentum with it.

10-3. What is the percentage of the pressure drop in a compressible flow system if one has assumed that the system is incompressible and if one wants to introduce not more than 10% error in the term representing the pressure difference? Assume the flow to be isothermal.

10-4. Repeat Problem 10-3 for adiabatic flow.

10-5. Evaluate α and β for laminar flow.

10-6. Derive the Von Kármán integral momentum theorem for the flow over a flat plate with uniform suction. If it is assumed that the suction velocity U_0 is small compared with the flow along the plate U_∞, the final equation is

$$\tau_w/\rho U_\infty^2 = U_0/U_\infty - \frac{d}{dx}\left(\delta \int_0^1 f(1-f)\, d\eta\right).$$

10-7. For the case of Problem 10-6, assume that the boundary layer has a finite thickness δ, and that the form of the velocity distribution is given by Eq. (10-19). Consider only the asymptotic solution when δ approaches a limiting value. Find this for the simple linear function and for Eqs. (10-23) and (10-24).

10-8. Compare the results of the exact solution without suction with the results of Problems 10-7 and 9-6.

REFERENCES

1. T. Baron and M. Souders, Jr. *Ency. Chem. Tech.* (R. E. Kirk and D. E. Othmer, eds.) **6**, 614–639 (1951).
2. R. B. Bird, *Chem. Eng. Sci.* **6**, 123–131 (1957).
3. R. B. Bird, W. E. Stewart, and E. N. Lightfoot, *Transport Phenomena*, John Wiley & Sons, New York (1960).
4. G. D. Bizzell and J. C. Slattery, *Chem. Eng. Sci.*, **17**, 777 (1962).
5. E.R.G. Eckert and R. M. Drake, *Heat and Mass Transfer*, 2nd ed., McGraw-Hill, New York (1959).
6. R. A., Gaggioli, *Chem. Eng. Sci.* **13**, 167–172 (1961).
7. O. T. Hanna, and J. E. Myers, *AIChEJ.* **7**, 437 (1961).
8. J. G. Knudsen and D. L. Katz, *Fluid Dynamics and Heat Transfer*, McGraw-Hill, New York (1958).
9. Shih-I Pai, *Viscous Flow Theory-Laminar Flow*, D. Van Nostrand, Princeton, N. J. (1956).
10. C. J. Pings, *Chem. Eng. Sci.* **17**, 947 (1962).
11. B. C. Sakiadis, *AIChEJ.* **7**, 26, 221, 467 (1961).
12. H. Schlichting, *Boundary Layer Theory*, McGraw-Hill, New York (1960).
13. J. C. Slattery and R. A. Gaggioli, *Chem. Eng. Sci.* **17**, 893 (1962).
14. V. L. Streeter, *Fluid Dynamics*, McGraw-Hill, New York (1958).
15. K. T. Yang, *J. Appl. Mech.* **28E**, 9, 470 (1961).

CHAPTER 11

METHODS OF ANALYSIS

For many problems, the Navier-Stokes equation cannot be solved exactly, or even approximately. Since theory cannot be used to obtain the desired answer, empirical methods must be employed. In most cases, it would be impractical to obtain the necessary information for an empirical correlation (or the desired design data) on the final system. Instead, models are tested in which the size, fluid conditions, and even the fluid may differ. The results of the model studies must be related to those expected for the final system. For those cases in which the Navier-Stokes equation cannot be solved, much of the necessary information can be obtained from the laws of similarity. One available method is called *inspection analysis* of the governing differential equations. Another is *dimensional analysis*. The latter is used when the variables are known, but the equations are not.

11-1 INSPECTION ANALYSIS

Inspection analysis is a method of determining what conditions are necessary for flows around geometrically similar bodies to be similar. Similarity exists when one case can be expressed in terms of the other case, with the differential equations and boundary conditions for the two cases remaining the same.

Consider the Navier-Stokes equation (4–6):

$$\frac{\partial U}{\partial t} + (U \cdot \nabla)U = -\frac{1}{\rho}\nabla p + \frac{1}{3}\frac{\mu}{\rho}\nabla(\nabla \cdot U) + \frac{\mu}{\rho}\nabla^2 U + \frac{1}{\rho}X, \qquad (11\text{--}1)$$

where the term $\sum_s \rho_s F_s$ has been replaced by an equivalent term X. Consider two flows with geometrically similar boundaries. If L is a length in the first flow, and L' the corresponding length in the second flow, then $L = c_1 L'$. This of course holds true for all points in the two systems; therefore $R = c_1 R'$. For two times in the two systems, $t = c_2 t'$. For the velocity terms, $U = R/t = c_1 R'/c_2 t' = (c_1/c_2) U'$. For other terms,

$$X = c_3 X', \qquad \rho = c_4 \rho', \qquad p = c_5 p', \qquad \mu = c_6 \mu'.$$

If we use Eq. (11–1) and the above, we can express the second flow field in terms of the first:

$$\frac{c_1}{c_2^2}\frac{\partial U'}{\partial t'} + \frac{c_1^2}{c_1 c_2^2}(U' \cdot \nabla')U' = \frac{c_3}{c_4}\frac{X'}{\rho'} - \frac{c_5}{c_1 c_4}\frac{1}{\rho'}(\nabla' p')$$

$$+ \frac{c_1 c_6}{c_1^2 c_2 c_4}\frac{\mu'}{\rho'}[\tfrac{1}{3}\nabla'(\nabla' \cdot U') + \nabla'^2 U'].$$

In order for the two equations to be the same, the constant factors must be equal:

$$c_1/c_2^2 = c_3/c_4 = c_5/c_1c_4 = c_6/c_1c_2c_4.$$
$$\quad\text{(a)}\qquad\text{(b)}\qquad\text{(c)}\qquad\text{(d)}$$

Consider each equality separately:
For equalities (a) and (d),

$$c_1^2 c_4 / c_2 c_6 = 1.$$

Substituting the values for the constants, we get

$$R^2 t' \rho \mu' / R'^2 t \rho' \mu = 1,$$

and rearranging to keep the real and model cases separate gives us

$$\frac{R^2 \rho / t \mu}{R'^2 \rho' / t' \mu'} = 1.$$

Noting that $R/t = U$, we obtain

$$\frac{RU\rho/\mu}{R'U'\rho'/\mu'} = 1 \quad \text{or} \quad \frac{RU\rho}{\mu} = \frac{R'U'\rho'}{\mu'}.$$

This means that $RU\rho/\mu$ must not change between the two systems, and thus it is an important dimensionless number:

$$RU\rho/\mu = N_{\text{Re}} = \text{Reynolds number.} \qquad (11\text{-}2)$$

The number involves the ratio of the (a) to (d) terms, which in effect is the ratio of the inertial to viscous forces.
For equalities (a) and (c),

$$c_1^2 c_4 / c_2^2 c_5 = 1$$

or

$$U^2 \rho / p = 1/N_{\text{Eu}} = 1/\text{Euler number} = \text{const.} \qquad (11\text{-}3)$$

The Euler number is the ratio of pressure to inertial forces, and is closely associated with the more common friction factor of Eq. (7–24). For compressible flow, the Euler number is split into two parts. If γ is the ratio of specific heats of the fluid and if the fluid obeys the ideal gas law, the local speed of sound is

$$c^2 = \gamma\, p/\rho = \gamma\, RT/M_w. \qquad (11\text{-}4)$$

Combining Eqs. (11-3) and (11-4), we obtain for the number, $(U^2/c^2)\gamma$. Thus, for the Euler number to remain constant, we must have

$$\gamma = \text{const} \qquad (11\text{-}5)$$

and

$$U/c = M = \text{Mach number} = \text{const.} \qquad (11\text{-}6)$$

For equalities (a) and (b),

$$c_1 c_4 / c_2^2 c_3 = 1 \quad \text{or} \quad U^2 \rho / RX = \text{const.}$$

If the body force X is caused by gravity only, then $X_z = \rho g$, and the number becomes

$$U^2/Rg = N_{\text{Fr}} = \text{Froude number,}$$

which is the ratio of inertial to gravitational forces.

The equations of motion for the two flows will be the same if their Reynolds, Mach, and Froude numbers, and the ratio of specific heats are equal. Inspection of the equation of continuity shows that no additional terms are introduced.

The energy equation (4–19) must be analyzed in a similar manner. Let

$$c_p = c_7 c_p', \quad T = c_8 T', \quad k = c_9 k',$$

so that for the equations of the two systems to be the same, we must have

$$\underset{(e)}{c_4 c_7 c_8 / c_2} = \underset{(f)}{c_9 c_8 / c_1^2} = \underset{(g)}{c_6 c_1^2 / c_1^2 c_2^2}. \tag{11-7}$$

For equalities (e) and (f),

$$c_1^2 c_4 c_7 / c_2 c_9 = 1$$

or

$$U R \rho c_p / k = N_{\text{Pe}} = \text{Peclet number} = \text{const.} \tag{11-8}$$

The Peclet number is the ratio of the heat transported by convection to that transported by conduction. The Peclet number divided by the Reynolds number is

$$c_p \mu / k = N_{\text{Pr}} = \text{Prandtl number,} \tag{11-9}$$

which depends only on the fluid properties. Since the Reynolds number must remain constant, as already discussed, this part of the analysis of the energy equation shows that for thermal similarity the Prandtl number must also stay constant. The Prandtl number is the ratio of the momentum diffusivity (kinematic viscosity) to the thermal diffusivity, or

$$N_{\text{Pr}} = \frac{c_p \mu}{k} = \frac{\nu}{k/\rho c_p}. \tag{11-10}$$

For equalities (f) and (g),

$$c_1^2 c_6 / c_2^2 c_8 c_9 = 1$$

or

$$\mu U^2 / kT = N_{\text{Br}} = \text{Brinkman number} = \text{const.} \tag{11-11}$$

This number is the ratio of the heat produced by viscous dissipation to that transported by molecular conduction.

TABLE 11-1

Summary of Dimensionless Numbers Presented in Text

Symbol	Number	Ratios
N_{Re}	$RU\rho/\mu$	Inertial to viscous forces
N_{Eu}	$p/\rho U^2$	Pressure to inertial forces
γ	c_p/c_v	Specific heats
M	U/c	Velocity to speed of sound
N_{Fr}	U^2/Rg	Inertial to gravitational forces
N_{Pe}	$UR\rho c_p/k$	Convection to molecular conduction heat transfer
N_{Pr}	$c_p\mu/k$	Momentum to thermal diffusivity
N_{Br}	$\mu U^2/kT$	Heat produced by viscous dissipation to molecular conduction heat transfer
$N_{\text{Pe},s}$	RU/D_s	Convective to molecular mass transfer
N_{Sc}	ν/D_s	Momentum to mass diffusivity
N_{We}	$U^2R\rho/\sigma$	Inertial to surface forces

The equation for mass transfer (4–20) can be used to introduce one more group. Let $D_s = c_{10} D'_s$, where D_s is the mass diffusivity. In order for the equations for the two systems to be the same, we must have $c_4/c_2 = c_{10}c_4/c_1^2$, which implies that

$$c_1^2/c_2 c_{10} = 1 \quad \text{or} \quad RU/D_s = N_{\text{Pe},s} = \text{mass transfer Peclet number} = \text{const.}$$

This number is the ratio of the convective to the molecular mass transfer, and if we divide it by the Reynolds number we obtain

$$\frac{\mu/\rho}{D_s} = \frac{\nu}{D_s} = N_{\text{Sc}} = \text{Schmidt number,} \qquad (11\text{–}12)$$

which is the ratio of momentum to mass diffusivities, and is analogous to the Prandtl number (11–10).

Finally, when a free surface is present, surface forces may be important. A good example is the surface force which must be overcome during the breakup of drops. The dimensionless ratio of inertial to surface forces is

$$U^2 R\rho/\sigma = N_{\text{We}} = \text{Weber number,} \qquad (11\text{–}13)$$

where σ is the surface tension.

The various numbers are summarized in Table 11-1.

11-2 DIMENSIONAL ANALYSIS

In a discussion of dimensional analysis, it is generally assumed that there are only five fundamental dimensions. These are usually taken to be the units of length, time, mass, temperature, and heat, but this selection is arbitrary. All other physical quantities can be expressed in terms of these fundamental units.

The main assumption of dimensional analysis is that the quantity under consideration is a function only of certain other known variables [1]. The validity of this assumption must be decided on the basis of considerable experimental knowledge. Since the procedure of dimensional analysis [2] is simple and can be found in most texts [4, 6, 8], it need not be given here. Dimensional analysis can be used without a knowledge of the equations of motion, and, if the assumption that all the variables are known is valid, it requires less information. There are cases, however, in which one may gain more information from inspectional than from dimensional analysis. As one example, let us consider the case of modeling of cavitation, as discussed by Birkhoff [1]. Until 1923, cavitation was generally stated to depend on the parameter $p/\frac{1}{2}\rho U^2$, which is directly suggested by dimensional analysis. In 1924, the effect of vapor pressure was observed and the modified parameter, $(p - p_{vp})/\frac{1}{2}\rho U^2$ was used. This parameter is suggested by inspectional analysis.

11-3 MODELING

In modeling, it is generally impossible to keep all the parameters of interest constant. Modeling methods are developed for each specific case, Reynolds models in some cases, Froude models in others. There are many other types, such as river and harbor, cavitation, and Mach models.

The various dimensionless groups suggest certain types of modeling. For example, the Reynolds number was obtained from the combination of the velocity change (inertial forces) and the viscosity contribution; external forces, such as gravity, and the compression term were not involved. Table 11-2 suggests

TABLE 11-2

Type of model	Of prime importance	To be neglected
N_{Re}	Viscosity	Gravity and compressibility
N_{Fr}	Gravity	Viscosity and compressibility
M	Compressibility	Gravity and viscosity

these uses. There are definite limitations in these suggestions. For example, pipe roughness and degree of turbulence have an effect on the use of the Reynolds number in pipe flow. There are many other limitations for other specific cases; for more detailed analyses of the problems involved, the books by Langhaar [4], Sedov [7], and Murphy [5] are suggested.

11-4 DIMENSIONLESS GROUPS: COMPLETENESS OF SETS

A complete set of dimensionless numbers is one in which a number outside the set can be expressed as a product of numbers in the set, but no number in the set can be identically expressed by means of other numbers in the set. A test for completeness [3, 4] can be made by representing forces or energies as points in space and the line connecting any two points as the dimensionless number, which is the ratio of the two forces in question. A complete set exists when the selected dimensionless numbers (lines connecting points) allow one to go from one point in the network to another over only one path.

EXAMPLE 11-1. Consider a system in which velocity, gravity, a length, and the system properties (density, viscosity, and surface tension) are important. Suggest a set of dimensionless numbers and test the set for completeness.

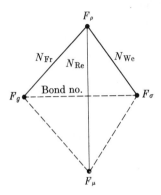

Fig. 11-1. *A complete dimensionless set.*

Answer. The forces associated with this system would be gravitational ($R^3\rho g$), inertial ($R^2U^2\rho$), viscous ($R\mu U$), and surface ($R\sigma$). The system could be expressed as a function of energies rather than of forces. In each case, the energy is the force multiplied by the characteristic distance. For example, the inertial force is $R^2U^2\rho$, and the inertial or kinetic energy would be $R^3U^2\rho$, which can be seen to be of the form mU^2. In Fig. 11-1, the forces are denoted as F_g, F_ρ, F_μ, and F_σ, for gravitational, inertial, viscous, and surface forces, respectively. The points form a tetrahedron in space, with six possible connecting lines. A complete set can be made up of three lines which do not form a triangle. (If a triangle were formed, one could not go from one point to another by one path only.) Other numbers can be constructed from the selected three. For example, the Bond number is

$$R^2\rho g/\sigma = F_g/F_\sigma = N_{\text{We}}/N_{\text{Fr}}.$$

If more than three groups are used, the set is overdetermined, and if less, the set is underdetermined.

An interesting representation is suggested by Fig. 11-1. Grassmann and Lemaire [3] have considered the system of numbers as analogous to a multicomponent phase diagram. That is, any surface of the tetrahedron of Fig. 11-1 would be similar to a triangular diagram if the numbers were expressed analogously to a mole fraction. A simple form, which varies from 0 to 1 while the number varies

from 0 to ∞, is
$$N/(N + 1). \tag{11-14}$$

Each leg of the tetrahedron would be expressed in the form of Eq. (11-14). For example,
$$N_{We}/(N_{We} + 1).$$

The point F_ρ would denote infinite values for all three dimensionless numbers, and the plane $F_g F_\sigma F_\mu$ would be the plane of zero value for all three numbers. The plane $F_g F_\rho F_\mu$ would be infinite for N_{We}, and values of this number between 0 and ∞ would lie on planes that pass through the points F_g and F_μ.

This method of representation is useful over only a small range of numbers (0.1 to 10), since low and high values are compressed at the extremes. A more general form that can be used to shift this range is
$$nN/(nN + 1), \tag{11-15}$$

where n is a numerical factor. This modified form is still useful over only a limited range of about two powers of ten. For large ranges of the system numbers, a right-angle system using logarithmic coordinates can be used. In this system, relations of type $N_{Re}^n N_{Fr}^m N_{We}^p$ are planes [3]. One plane of particular interest is $N_{Re}^4 N_{Fr} N_{We}^{-3}$, in which both the velocity and length disappear. This number depends only on properties and gravitation, and is
$$\rho\sigma^3/g\mu^4. \tag{11-16}$$

Thus, for a given liquid and for conditions that determine the density, viscosity, and surface tension, all points that describe this liquid must lie on this plane. There is an advantage in using this number as one of the complete set. The other two members of the set should be selected so that N_{Re}, N_{Fr}, and N_{We} can be extracted. This will ensure that the set is complete. The representation described here is useful in two-phase flow, and will be considered again in a later chapter.

PROBLEMS

11-1. Show that the Euler number can be obtained by inspection analysis of the Navier-Stokes equation for the limiting case of zero viscosity.

11-2. Use inspection analysis on the energy equations in terms of internal energy.

11-3. An infinite slab of thermal conductivity k is initially at a uniform temperature T_0. There is a uniform volume production of heat within the slab (G). The plate is cooled on each side by a fluid at T_0, and the heat transfer coefficient between the slab and the fluid is h. The equation and the boundary conditions are

$$\frac{\partial T}{\partial t} = \frac{k}{\rho c_p} \frac{\partial^2 T}{\partial x^2} + \frac{G}{\rho c_p},$$

$$\left(\frac{\partial T}{\partial x}\right)_{x=x_0} = \frac{h}{k}(T_{x=x_0} - T_0).$$

What are the dimensionless groups?

REFERENCES

1. G. Birkhoff, *Hydrodynamics*, Dover Publications, New York (1955).
2. E. Buckingham, *Phys. Rev.* **4**, 354–376 (1914).
3. P. Grassmann and L. H. Lemaire, *Chem.-Ing.-Tech.* **30**, 450–454 (1958).
4. H. L. Langhaar, *Dimensional Analysis and Theory of Models*, John Wiley & Sons, New York (1951).
5. G. Murphy, *Similitude in Engineering*, The Ronald Press, New York (1950).
6. R. Saint-Guilhem, *Les Principes de l'Analyse Dimensionelle*, Gauthier-Villars, Paris (1962).
7. L. I. Sedov, *Similarity and Dimensional Methods in Mechanics*, Academic Press, New York (1959).
8. I. H. Shames, *Mechanics of Fluids*, McGraw-Hill, New York (1962).

CHAPTER 12

COMPRESSIBLE FLOW

In the preceding chapters, applications of the basic equations were for the most part restricted to incompressible flow. Density could be assumed constant, and recourse to thermodynamics and an equation of state was not necessary. This chapter will be an introduction to some of the simple compressible-flow problems which involve these additional aspects. In many fields of engineering, much of the material in the present chapter would have been considered earlier because of its importance to certain specific areas (i.e., aeronautics and rocket research). However, in other fields of engineering, compressible flow has played a minor role and normally is discussed only briefly, if at all. The details of the developments are generally straightforward and in some cases are given here in outline form only. These derivations can be worked out or found in the various references already cited, or in Shapiro's treatises on *The Dynamics and Thermodynamics of Compressible Fluid Flow* [9]. As implied in the title of that book, thermodynamics is of prime importance in compressible flow, and therefore a short review of some thermodynamic principles will be given in the next section.

12-1 THERMODYNAMICS

12-1.A The First Law

The first law of thermodynamics, which is the statement of the law of conservation of energy, can be expressed in a number of ways. The most common form has already been given (Eq. 10–46):

$$Q = (\mathrm{u}_2 - \mathrm{u}_1) + \int p\, d(1/\rho) \tag{10-46}$$

or

$$dQ = d\mathrm{u} + p\, dv, \tag{12-1}$$

where $v = 1/\rho$ is the specific volume. Another statement is that the work done by a system in changing state is independent of the path and the manner of the work interchange, if the system and surroundings are at the same temperature during each step of the process. This of course implies that the temperature of the system can be defined. When applied to a flow system, the law becomes the macroscopic, or overall, energy balance (10–32).

12-1.B The Second Law

The second law of thermodynamics has been stated in many forms, each of which must imply that any system not worked upon will, on the average, approach a state of maximum probability [11]. Any actual process can be said to be irreversible, since it is generally impossible to restore each part of the system to its original state. An ideal, or reversible, process would be one in which all forms of dissipation of energy were eliminated. This would mean that no momentum transfer, heat conduction, or mass transfer would occur. For a reversible process, the increase in entropy for any system is defined as

$$\int \frac{dQ}{T}.$$

For any irreversible or real process, the increase in entropy is always greater than it is for an equivalent reversible system. Thus

$$\Delta s = s_2 - s_1 \geq \int_1^2 \frac{dQ}{T},$$

and is zero only for the adiabatic reversible or isentropic process.

12-1.C Specific Heats

From thermodynamics, the specific heats at constant pressure and constant volume are

$$c_p = (\partial H/\partial T)_p, \qquad c_v = (\partial U/\partial T)_v, \qquad c_p = c_v + [v - (\partial H/\partial p)_T](\partial p/\partial T)_v, \tag{12-2}$$

where the enthalpy is given by $H = U + pv$.

12-1.D The Perfect Gas

The perfect or ideal gas must fulfill two conditions: first, the equation of state that must be obeyed is $pV = nRT$ or

$$p/\rho = pv = RT/M_w, \tag{12-3}$$

where again v is the specific volume, R the universal gas constant, n the number of moles, and M_w the molecular weight. Second,

$$(\partial U/\partial v)_T = 0,$$

which follows directly from Eq. (12-3) and the second law of thermodynamics, and which implies that the internal energy is independent of the specific volume, and depends on temperature alone. The perfect gas is approximated by real gases at low pressures and high temperatures. Fortunately the behavior of a perfect gas is a reasonable approximation of the behavior of real gases under the normal conditions of interest here. The error introduced by this simplifying assumption can be less than that introduced by some of the necessary assumptions about hydrodynamics.

(a) Isothermal expansion. The internal energy remains constant, and from Eq. (12-1) we obtain, for a reversible system, $dQ = p\,dv$. For the perfect gas, integrating gives us

$$Q = \int_{v_1}^{v_2} p\,dv = \frac{RT}{M_w} \ln \frac{v_2}{v_1} = \frac{RT}{M_w} \ln \frac{p_1}{p_2}. \qquad (12\text{-}4)$$

(b) Isentropic expansion. In an adiabatic expansion $Q = 0$, and Eq. (12-1) becomes $d\mathrm{U} = -p\,dv$. Since Eq. (12-1) is valid only for a reversible system, this development is restricted to an isentropic process. From Eq. (12-2) and the knowledge that U is independent of the specific volume, we obtain $d\mathrm{U} = c_v\,dT$. Hence

$$p\,dv = -c_v\,dT \qquad \text{or} \qquad (RT/vM_w)\,dv = -c_v\,dT,$$

which can be rearranged to

$$(R/M_w)\,d(\ln v) = -c_v\,d(\ln T).$$

For the special case of constant c_v, this can be integrated to

$$Tv^{R/c_vM_w} = C_1,$$

where C_1 is a constant. For a perfect gas, Eq. (12-2) leads to

$$c_p - c_v = R/M_w$$

or the equivalent forms

$$c_p = \frac{R}{M_w}\frac{\gamma}{\gamma-1} \qquad \text{and} \qquad c_v = \frac{R}{M_w(\gamma-1)}. \qquad (12\text{-}5)$$

The first of these equations combined with the preceding equation gives

$$Tv^{(c_p/c_v)-1} = Tv^{\gamma-1} = C_1, \qquad (12\text{-}6)$$

which can be put into an equivalent form by combining it with Eq. (12-3)

$$pv^{c_p/c_v} = pv^{\gamma} = C_2. \qquad (12\text{-}7)$$

The specific volume can also be eliminated between Eqs. (12-6) and (12-7). In summary, the equations in terms of γ are

$$Tv^{\gamma-1} = C_1, \qquad pv^{\gamma} = C_2, \qquad T^{\gamma/(\gamma-1)}/p = C_3. \qquad (12\text{-}8)$$

In the free expansion of a perfect gas isolated from other systems, the entropy change can be obtained from the definition of entropy and Eq. (12-4). (Since the internal energy will not change, the temperature is constant.) The change is

$$\Delta s = (R/M_w) \ln (v_2/v_1). \qquad (12\text{-}9)$$

For the isentropic expansion, the entropy change is zero, as already noted. For an adiabatic or any ideal gas process, the entropy change is given by

$$\Delta s = c_v \ln[(T_2/T_1)(v_2/v_1)^{\gamma-1}] = c_v \ln[(p_2/p_1)(v_2/v_1)^{\gamma}] = c_v \ln[(T_2/T_1)^{\gamma}(p_2/p_1)^{-(\gamma-1)}], \quad (12\text{--}10)$$

all terms of which come from the combination of Eqs. (12–1) through (12–3).

12-2 COMPRESSIBLE FLUID FLOW

In many systems, the compressibility terms can be neglected; quite often, even with a change of 20% in the absolute pressure, the flow can be considered incompressible at the average pressure. When the flow is such that the constant density assumption is inadmissible, then the approach must be modified to include the rate of change of density with respect to pressure. In compressible flow the propagation of small pressure disturbances, such as might be expected for a sound wave, is of prime importance, and will be developed next.

12-2.A Propagation of a Small Pressure Wave (Velocity of Sound)

Consider a one-dimensional infinitesimal pressure wave moving in the x-direction. One can take the viewpoint of an observer moving with the wave so that the flow is relative to this moving position. In the absence of external and viscous forces, the Navier-Stokes equation (4–6) reduces to

$$\frac{\partial U_x}{\partial t} + U_x \frac{\partial U_x}{\partial x} = -\frac{1}{\rho}\frac{\partial p}{\partial x} = -\frac{1}{\rho}\frac{\partial p}{\partial \rho}\frac{\partial \rho}{\partial x}. \quad (12\text{--}11)$$

The equation of continuity (3–19) becomes

$$\frac{\partial \rho}{\partial t} + \frac{\partial \rho U_x}{\partial x} = 0.$$

For steady-state motion of the wave, the two equations become, respectively,

$$U_x \frac{\partial U_x}{\partial x} = -\frac{1}{\rho}\frac{\partial p}{\partial \rho}\frac{\partial \rho}{\partial x}, \qquad \rho \frac{\partial U_x}{\partial x} + U_x \frac{\partial \rho}{\partial x} = 0.$$

Combining these, we obtain

$$U_x^2 = \partial p/\partial \rho. \quad (12\text{--}12)$$

The velocity of flow relative to the wave (or the equivalent, velocity of propagation of the pressure wave) is

$$U_x = \sqrt{(\partial p/\partial \rho)_s}. \quad (12\text{--}13)$$

The restriction of constant entropy is applied, since the variations in pressure and temperature approach zero, and the process is so rapid that it is adiabatic (this has been confirmed experimentally). Since the wave propagation can be considered both reversible and adiabatic, it is isentropic, and will occur at a constant s.

12-2.B Velocity of Sound for a Perfect Gas

For a perfect gas, $p/\rho = RT/M_w$, from Eq. (12–3). From Eq. (12–7),

$$p/\rho^\gamma = C_2 \quad \text{or} \quad \ln p - \gamma \ln \rho = C_4.$$

Differentiating this, we get $dp/p = \gamma \, d\rho/\rho$, which is the same as $(\partial p/\partial \rho)_s = \gamma(p/\rho)$, since variations in temperature and pressure approach zero. Using this relation, Eq. (12–13), and the ideal gas law above, we can obtain

$$c = U_x = \sqrt{\gamma \, RT/M_w}, \tag{12–14}$$

where c is the velocity of sound. This equation is the same as Eq. (11–4) which was cited in the analysis in which the Mach number was derived. For air, the velocity in feet per second is given by the formula $c = 49.02\sqrt{T}$, where T is in °R.

Shapiro [9] points out that γ does not vary much, so that at a given temperature the speed of sound is a function of $\sqrt{1/M_w}$. At normal atmospheric temperatures, the speed of sound in air is 1100 ft/sec; in hydrogen, 4200 ft/sec; and in freon, about 300 ft/sec. As a consequence, in order to avoid excessive stresses in turbomachines, the rotor must not exceed a top speed of 1000–1500 ft/sec for air. In hydrogen compressors, the compressibility limitation is rarely a factor, while in freon compressors it is the principal limiting factor on the speed of rotation.

12-2.C Subsonic and Supersonic Flow

The Mach number was obtained in the last chapter by inspection analysis of the Navier-Stokes equation, and was

$$M = U/c, \tag{12–15}$$

where U is the stream velocity and c is the velocity of sound in the local medium. This ratio of stream speed to local velocity of sound profoundly affects the type of compressible flow which occurs.

Let us consider a point source of pressure disturbance moving with a velocity U_r relative to the stream velocity U. The pressure disturbance which moves at the speed of sound is propagated with a velocity c relative to U. In Fig. 12–1, the distance axis is given by the horizontal line. The numbers at the points denote starting points of the corresponding sound waves. The axis itself is the same in each part, but is shown in (a) only. In part (a) of the figure, $U_r = U$, or effectively *incompressible flow* ($M < 0.4$). The wave is shown for times 0, 1, 2 and 3 at time 3. Here the pressure wave travels uniformly in all directions. In (b), $U < U_r < c$ which is *subsonic flow* ($0.4 < M < 1$). The pressure wave is felt in all directions, but is not symmetrical. The wave "warns" the observer that the object is coming. In (c), $U_r = c > U$ or *transonic flow* ($M \sim 1$) at the speed of the wave. Here the object moves with the pressure wave and the zone of silence. The object does not "warn" of its coming. In (d), $U < U_r > c$ or *supersonic flow* ($1 < M \lesssim 3$); the pressure wave and zone of silence lag behind the object. The disturbance affects only that portion of the stream lying within a

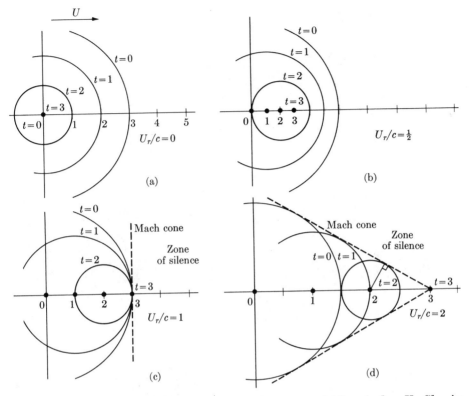

Fig. 12–1. Propagation of sound from a point source. [After Ascher H. Shapiro, The Dynamics and Thermodynamics of Compressible Fluid Flow, Vol. 1, The Ronald Press, New York (1953). By permission.]

cone, called the *Mach cone*. An observer does not hear the object moving overhead until the wave has passed his ear. By this time, the object has passed him. *Hypersonic flows* are cases in which $M \gtrsim 3$, and changes in the Mach number are due primarily to changes in c.

From Fig. 12–1, one can see that the angle of the Mach cone is related to the Mach number by

$$\sin \alpha = c/U_r = 1/M. \tag{12-16}$$

A measurement of the angle α on spark photographs permits the estimation of the Mach number of fast-moving objects, such as that shown in Fig. 12–2(a). Actually the flow over bodies is quite complex, and usually involves a two- or three-dimensional analysis, rather than the one-dimensional approach for duct flow which we shall present here. For flow over bodies, shocks can exist even at subsonic approach speeds, because of the necessary acceleration of the flow around the object. The complexity of the problem, which involves detached bow waves, boundary-layer separation, and oblique shocks, is indicated by the pictures in Fig. 12–2.

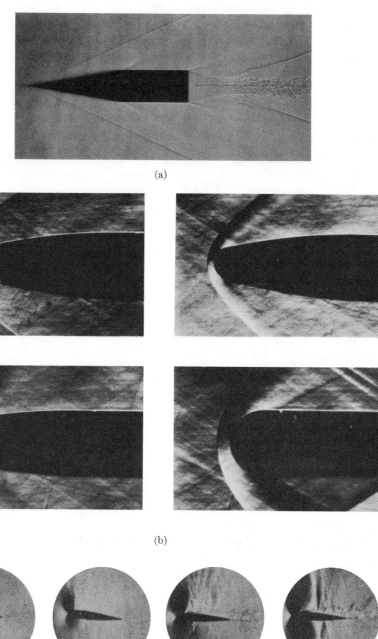

(a)

(b)

$M = 0.40$

$M = 0.60$

$M = 0.70$

$M = 0.80$

(c)

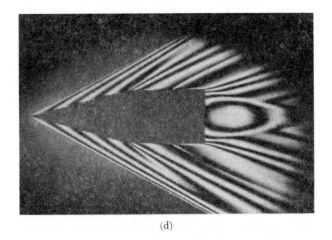

(d)

(e)

Fig. 12–2(a) Shadowgraph of a cone cylinder at supersonic speed in free flight at Mach 2.67. (b) Shock waves on axisymmetric bodies of revolution. Mach 1.86. (c) Complex shock formation at Mach 0.4, 0.6, 0.7, and 0.8, respectively. (d) Single-fringe interferogram of caliber .30, 30° cone cylinder at velocity of 2965 ft/sec. (e) Objects in Mach 2.0 air flow. In each case, $\rho U d/\mu = 1.2 \times 10^5$. Cone-cylinder model (left) has semi-apex angle of 15° and diameter of $\frac{3}{8}$ in. Sphere (right) has diameter of $\frac{3}{8}$ in. Parabolic wave aft of each object is caused by intersection of head wave with glass side walls. [Parts (a) and (d) from the Ballistic Research Laboratories, Aberdeen Proving Ground, Maryland. Part (c) from NACA Report 646, National Aeronautics and Space Administration. Parts (b) and (e) from the David Taylor Model Basin, official photographs of U.S. Navy. All parts by permission.]

12-3 ONE-DIMENSIONAL, STEADY, ADIABATIC FLOW OF A PERFECT GAS

The one-dimensional flow approximation is the application of relations that are valid along given streamlines to a flow geometry with a finite area. In making this approximation, one is assuming that the properties are uniform over any cross section. In any real case, this will not be true; thus one deals with some average property over the cross section. If the rate of change of properties *across* the flow is small in comparison with the rate of change of properties *with* the flow, then the errors introduced by the assumption will be slight. The errors can be minimized if the change in area of flow is small compared with length, if the radius of curvature is large relative to the diameter of the duct, and if the shape of the velocity and temperature profiles remains essentially constant over the length being considered. The assumption of a one-dimensional system is usually adequate for motion in ducts and stream tubes. However, as noted, it generally will not suffice for such problems as flows past objects, flow over turbine blades, or flow in ducts with very rapid changes in area. The one-dimensional approach can consider the variation of parameters with length of the system, but can give no information about variations normal to the flow, since these have been assumed constant. Clearly, a two- or three-dimensional analysis will be necessary to obtain this information.

12-3.A Adiabatic Flow

Certain essential relations can be developed by the assumption of adiabatic flow without the additional reversible restriction that isentropic flow implies. In thermodynamic calculations using the first law (Eq. 10–32 or 10–46), the term $u + p/\rho = u + pv'$ often appears and is defined as the enthalpy H. Equation (10–32), for no change in elevation or shaft work, becomes

$$u + p/\rho + \tfrac{1}{2}U^2 = H + \tfrac{1}{2}U^2 = C_5, \qquad (12\text{–}17)$$

since $Q = 0$ by the adiabatic assumption and $(U^3)_{\text{ave}}/U_{\text{ave}} = U^2$ or $\alpha = 1$ by the one-dimensional assumption. One state of particular interest is $U = 0$, or the *stagnation state*. For this case $H = C_5$ and is denoted as H_0. Other properties are noted similarly; i.e., the stagnation pressure (often called the total pressure) is p_0. However, this and the stagnation temperature T_0 depend on the path, and so these are taken as isentropic stagnation properties. It should be emphasized that the first law of thermodynamics states that for an adiabatic process, the enthalpy of a given state has the same value at all states which can be obtained from it adiabatically, and thus for the enthalpy the reversibility restriction is not necessary. The other stagnation values are those that would be obtained if the flow were to slow down to zero velocity by a reversible adiabatic process. This retardation process can be considered hypothetical, and as occurring at a specific point in the process. If the flow is isentropic, then the stagnation values will be the same everywhere; if not, the stagnation properties will vary from point to point but are always definable.

ONE-DIMENSIONAL, STEADY, ADIABATIC FLOW OF A PERFECT GAS

For an adiabatic process, $\Delta H = c_p \Delta T$. Thus, from Eq. (12–17) (with $C_5 = H_0$), the velocity becomes

$$U^2 = 2c_p(T_0 - T). \tag{12–18}$$

Combining Eqs. (12–5) and (12–18), we obtain

$$U^2 = \frac{2\gamma}{\gamma - 1} \frac{R}{M_w}(T_0 - T). \tag{12–19}$$

From this equation, it is clear that for adiabatic flow, the velocities at any two points are the same if the stagnation and actual temperatures are the same.

The temperature at a point can be expressed as a ratio to the stagnation temperature. From Eq. (12–18),

$$\frac{T_0}{T} = 1 + \frac{U^2}{2c_p T} = 1 + \frac{\gamma R}{2 M_w c_p}\left(\frac{U^2}{\gamma RT/M_w}\right).$$

Combining this with Eqs. (12–5) and (12–14), we have

$$\frac{T_0}{T} = 1 + \frac{\gamma - 1}{2}\frac{U^2}{c^2} = 1 + \frac{\gamma - 1}{2} M^2. \tag{12–20}$$

The *critical velocity* is defined as the velocity at the point where *the Mach number is unity*. This is

$$c^* = \sqrt{(2\gamma/(\gamma + 1))(RT_0/M_w)}, \tag{12–21}$$

which can be obtained from Eq. (12–14) by first obtaining, from Eq. (12–20), a relation between T_0 and T^* (the temperature at the point where the Mach number is unity). A modified Mach number will be useful in some later calculations; it is defined as

$$M^* = U/c^*. \tag{12–22}$$

The velocity of sound c, as given by Eq. (12–14), is a function of T, and thus will vary along the stream. In contrast, c^* is defined at one point and is constant for a given gas and stagnation temperature. Thus the use of M is complicated by variations in c, and M^* is more convenient to use.

For an adiabatic process, the flow per unit area or mass flux can be determined as follows:

$$J = \frac{w}{S} = \rho U = \frac{p M_w}{RT} U = \frac{pU}{\sqrt{\gamma RT/M_w}}\sqrt{\frac{\gamma M_w}{R}}\sqrt{\frac{T_0}{T}}\frac{1}{\sqrt{T_0}}$$

$$= \sqrt{\frac{\gamma M_w}{R}}\frac{p}{\sqrt{T_0}} M \sqrt{1 + \frac{\gamma - 1}{2} M^2}. \tag{12–23}$$

12-3.B Isentropic Flow

(a) General considerations. The analysis in the preceding section was for adiabatic flow, and thus will apply to the more restricted case of reversible adiabatic or isentropic flow. With the relations given in Eqs. (12-8) and (12-20), the following three equations can be obtained:

$$\frac{T_0}{T} = 1 + \frac{\gamma-1}{2}M^2, \qquad \frac{p_0}{p} = \left(1 + \frac{\gamma-1}{2}M^2\right)^{\gamma/(\gamma-1)},$$
$$\frac{\rho_0}{\rho} = \left(1 + \frac{\gamma-1}{2}M^2\right)^{1/(\gamma-1)}. \tag{12-24}$$

Further relations can be derived in terms of M^*, since it can be shown that

$$M^2 = \frac{2}{(1-\gamma) + (\gamma+1)/M^{*2}}.$$

Of course, these various equations are also valid at any specific point, even for nonisentropic flows. The stagnation properties would be those obtained by the hypothetical isentropic retardation process occurring at that point. If the entire flow is isentropic, these values are the same everywhere, if not, they vary from point to point.

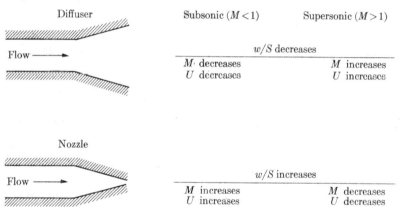

Fig. 12-3. *Compressible flow in diffusers and nozzles.*

The pressure may be eliminated between Eqs. (12-23) and (12-24) to give

$$J = \frac{w}{S} = \frac{\sqrt{\gamma M_w/R}(p_0/\sqrt{T_0})M}{\left(1 + \frac{\gamma-1}{2}M^2\right)^{(\gamma+1)/2(\gamma-1)}}. \tag{12-25}$$

Equation (12-25) can be shown to have a maximum value in J at the point at which the Mach number is unity. From Eq. (12-25), one can see immediately that J approaches zero if the Mach number approaches zero or infinity. An important implication of this can be illustrated by reference to Fig. 12-3. For a given set of upstream conditions, the mass flow w will be constant. Thus, the ratio w/S varies according to the variation in the cross-sectional area of the nozzle. Starting from a low velocity (some Mach number less than one), as in the lower left of the figure, the Mach number must increase as J increases (Eq. 12-25). However, the Mach number can increase only to the maximum value of $M = 1$,

ONE-DIMENSIONAL, STEADY, ADIABATIC FLOW OF A PEFECT GAS   181

since J has no way in which to decrease. To obtain a velocity corresponding to a Mach number greater than one, the value of w/S must in some manner be decreased (Eq. 12–25), and this can occur only in a diffuser, as shown in the upper right portion of Fig. 12–3. This result leads to the interesting conclusion that an accelerating stream starting from rest must first decrease in cross section and then increase in cross section as it passes the critical state. Another important consequence of this analysis is that in order for the flow to go from subsonic to supersonic, it must pass through a throat at $M = 1$. Since this is unity Mach, the conditions are those of the critical speed c^*, area S^*, etc.

In order to find the *maximum flow rate*, which will occur at $M = 1$, one need only let $M = 1$ in Eq. (12–25) to obtain

$$J_{max} = \frac{w}{S^*} = \sqrt{\left(\frac{\gamma M_w}{R} \frac{2}{\gamma+1}\right)^{(\gamma+1)/(\gamma-1)}} \frac{p_0}{\sqrt{T_0}}. \tag{12-26}$$

For a given gas, Eq. (12–26) shows that the maximum flow depends only on $p_0/\sqrt{T_0}$. For a given $p_0/\sqrt{T_0}$, the flow will be high for gases of high molecular weight and low for gases of low molecular weight. For air,

$$(w/S^*)(\sqrt{T_0}/p_0) = 0.532, \tag{12-27}$$

where w is in pounds per second, S^* is in square feet, T_0 is in degrees Rankine, and p_0 is in pounds force per square foot.

By use of the equations developed, we can express the mass flow or other variables in a number of different ways as functions of M, c/c_0, T/T_0, p/p_0, or ρ/ρ_0, Oswatitsch and Kuerti [6] have tabulated the possible interrelations. The formulas developed so far, and the other relations implied above, are tedious to use. For this reason, working charts and tables have been developed for isentropic flow. Shapiro [9] has provided these for $\gamma = 1.4$, and has given the source of his tables and an extensive list of other published tables and charts. A brief summary table for $\gamma = 1.4$ is given as Table 12–1. For values of γ other than 1.4, Fig. 12–4 has some of the same information; the data for this figure have come chiefly from computations made by Lapple [3]. Up to a Mach number of 0.3 (about 300 ft/sec for air at normal conditions), there is little change in density (about 4.4%). In many cases, this fact permits fluids such as air to be treated as though they are incompressible.

In the flow in real nozzles, geometrical effects sometimes introduce errors into the mass flow rate given by Eq. (12–26). In these cases, a multiplying factor, called the *discharge coefficient*, is used. For a well-designed nozzle, the coefficient is very close to unity at high Reynolds numbers.

EXAMPLE 12–1. A small supersonic demonstration nozzle (Fig. 12–5) is to be designed for a Mach number of two. The throat diameter is 0.25 in. Air is available in the laboratory at 100 psig and 70°F. Determine the mass flow needed, available test area, temperature, and pressure at this point. What is the exit pressure if the exit diameter is 0.35 in.?

TABLE 12-1

Insentropic Flow, Perfect Gas, $\gamma = 1.4$

M	M*	p/p_0	ρ/ρ_0	S/S^*	T/T_0	$(p/p_0)(S/S^*)$
0.00	0.00	1.000000	1.000000	∞	1.000000	∞
0.10	0.10943	0.99303	0.99502	5.8213	0.99800	5.7812
0.20	0.21822	0.97250	0.98027	2.9635	0.99206	2.8820
0.30	0.32572	0.93947	0.95638	2.0351	0.98232	1.9119
0.40	0.43133	0.89562	0.92428	1.5901	0.96899	1.4241
0.50	0.53452	0.84302	0.88517	1.3398	0.95238	1.12951
0.60	0.63480	0.78400	0.84045	1.1882	0.93284	0.93155
0.70	0.73179	0.72092	0.79158	1.09437	0.91075	0.78896
0.80	0.82514	0.65602	0.74000	1.03823	0.88652	0.68110
0.90	0.91460	0.59126	0.68704	1.00886	0.86058	0.59650
1.00	1.00000	0.52828	0.63394	1.00000	0.83333	0.52828
1.20	1.1582	0.41238	0.53114	1.03044	0.77640	0.42493
1.40	1.2999	0.31424	0.43742	1.1149	0.71839	0.35036
1.60	1.4254	0.23527	0.35573	1.2502	0.66138	0.29414
1.80	1.5360	0.17404	0.28682	1.4390	0.60680	0.25044
2.00	1.6330	0.12780	0.23005	1.6875	0.55556	0.21567
2.30	1.7563	0.07997	0.16458	2.1931	0.48591	0.17538
2.60	1.8572	0.05012	0.11787	2.8960	0.42517	0.14513
3.00	1.9640	0.02722	0.07623	4.2346	0.35714	0.11528
4.00	2.1381	0.00658	0.02766	10.719	0.23810	0.07059
7.00	2.3333	0.00024	0.00261	104.143	0.09259	0.02516
10.00	2.3904	0.00002	0.00050	535.938	0.04762	0.01263
∞	2.4495	0.00000	0.00000	∞	0.00000	0.00000

Answer. In order to use Table 12-1, we must establish the stagnation conditions. The stagnation pressure is defined as the pressure when the velocity is zero, or simply the inlet pressure when there is no flow. Therefore $p_0 = 100$ psig $= 114.7$ psia. The stagnation temperature is given as $T_0 = 70°F = 529.7°R$. We must determine S^*, which is the area where the Mach number is unity (the area of the throat): $S^* = 0.00034$ ft². From Table 12-1, for $M = 2$, we obtain

$$p/p_0 = 0.12780, \qquad T/T_0 = 0.55556, \qquad S/S^* = 1.6875.$$

With the known values of p_0, T_0, S^*, and the above, the desired values at the test section are

$$p = 14.7 \text{ psia, or atmospheric pressure,}$$

$$T = 294°R = -163.7°F,$$

$$S = 0.000574 \text{ ft}^2, \text{ or a diameter of } 0.324 \text{ in.}$$

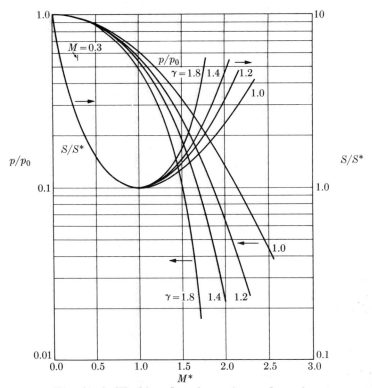

Fig. 12-4. Working chart for various values of γ.

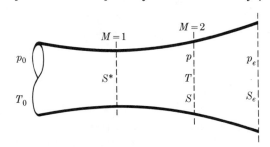

Fig. 12-5. Nozzle system for generation of supersonic flow.

Since the fluid is air, the mass flow rate can be determined from Eq. (12-27):

$$w = \frac{0.532 p_0 S^*}{\sqrt{T_0}} = 0.13 \text{ lb/sec}.$$

The exit area gives $S_e/S^* = 1.96$. From Table 12-1, the exit conditions are:

$$M_e = 2.17, \quad p_e/p_0 = 0.09715, \quad \text{or} \quad p_e = 11.15 \text{ psia}.$$

The nozzle exit will have to be under a vacuum of about $3\frac{1}{2}$ pounds to provide the proper design.

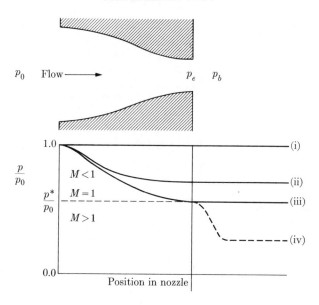

Fig. 12–6. Flow in a converging nozzle. [After Ascher H. Shapiro, The Dynamics and Thermodynamics of Compressible Fluid Flow, Vol. 1, The Ronald Press, New York (1953). By permission.]

(b) Converging nozzles and choking. An important effect called *choking* occurs because the mass flow rate is at a maximum when the Mach number is unity. Consider the flow, with constant stagnation conditions, in a converging nozzle (Fig. 12–6). The effect of changing the back pressure p_b can be described as follows. If $p_b = p_0$, there is no flow, since the pressure is constant throughout the nozzle (i). As p_b is reduced below p_0, the flow will commence, and will show a decreasing pressure with distance (ii). At condition (iii), the critical pressure ratio is reached in the throat; the velocity is sonic ($M = 1$) at this point, and the mass flow rate is at a maximum. Any further decrease in p_b would have to result in a decrease in the mass flow rate, since the flow is at a maximum. However, this cannot happen unless the stagnation conditions are changed. In other words, for constant stagnation conditions, any further decrease in p_b will not result in further change within the nozzle, and case (iv) is the same as case (iii). The dashed line part of (iv) cannot be predicted from the present analysis, since this part of the flow involves expansion waves and shocks. The simple convergent nozzle cannot give supersonic flows without a throat section. As the pressure p_b decreases below the critical level, the flow is *choked* to the critical flow rate in the nozzle.

This characteristic makes the simple convergent nozzle useful as a flow control regulator; when p_b/p_0 is less than the critical value, the flow is dependent only on p_0, T_0, and the area of the nozzle, and is independent of the back pressure p_b.

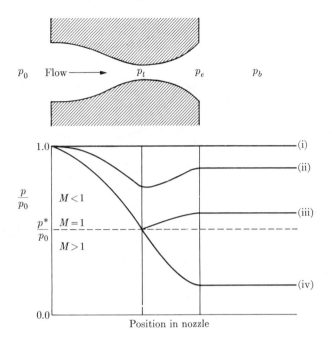

Fig. 12–7. Flow in a converging-diverging nozzle. [After Ascher H. Shapiro, The Dynamics and Thermodynamics of Compressible Fluid Flow, Vol. 1, The Ronald Press, New York (1953). By permission.]

(c) **Converging-diverging nozzles.** In the converging-diverging nozzle of Fig. 12–7, supersonic flows are possible. If $p_b = p_0$, there is no flow (i). When p_b is reduced below p_0, the flow is subsonic, and the pressure follows a line like (ii), similar to the converging-nozzle case. For condition (iii), the velocity in the throat becomes sonic, and no further change in p/p_0 in the convergent section will occur for a given set of conditions at p_0. Further reduction in p_b will result in supersonic flow being maintained beyond the throat (iv). The value of p/p_0 in the divergent section is exactly defined for isentropic flow, and corresponds to the area ratio of the nozzle, S/S^*. There is no isentropic, one-dimensional flow which will give values of p_b/p_0 between cases (iii) and (iv). As will be shown, these intermediate pressures result in shocks which, though adiabatic, are not isentropic, and thus cannot be predicted from the present analysis.

12-3.C Nonisentropic Flow

(a) **General considerations of constant area flow.** Under conditions for which a one-dimensional isentropic solution cannot be obtained, experimentation has led to the observation that irreversible (nonisentropic) discontinuities occur within the nozzle area. These are the normal shock waves cited above. Consider a compressible fluid flowing through a duct of constant cross section. Further, examine

186 COMPRESSIBLE FLOW

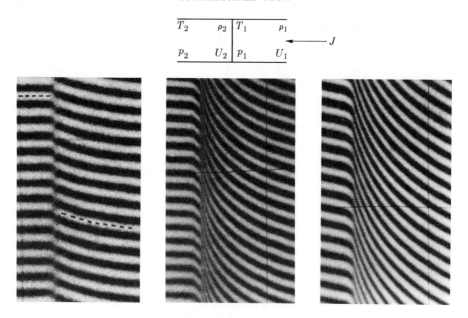

Fig. 12–8. Shock wave. [Photograph by permission of Dr. W. Bleakney, Palmer Physical Laboratory, Princeton University.]

only the relations on the two sides of the discontinuity or shock wave, since the change is experimentally and analytically (see Section 7-5) observed to occur over a very short distance. In addition, assume an irreversible adiabatic process with no change in elevation and no shaft work effects. The system under consideration is shown in Fig. 12–8, together with a series of photographs showing the formation of a shock wave. In these photographs the shock is moving from left to right, and would be held stationary by a flow to the left. Thus the sketch of the flow system has been drawn in reverse to correspond to the photographs. The displacement of the interference lines from a reference plane is a measure of the gas density, so that the density after the shock (on the left) is greater than before the shock (on the right).

The overall mass balance equation (10–8) for steady-state, one-dimensional flow in a duct of constant area gives us the continuity equation:

$$J = w/S = \rho U = \rho_1 U_1 = \rho_2 U_2. \tag{12-28}$$

The energy balance between sides 1 and 2 can be obtained from the integral energy balance equation (12–17):

$$\text{H} + \tfrac{1}{2}U^2 = \text{H}_1 + \tfrac{1}{2}U_1^2 = \text{H}_2 + \tfrac{1}{2}U_2^2 = \text{H}_0, \tag{12-29}$$

where $\text{H}_2 - \text{H}_1$ can be expressed as $c_p(T_2 - T_1)$. Combining Eqs. (12–28) and (12–29), we obtain

$$\text{H} = \text{H}_0 - (1/2\rho^2)J^2 = \text{H}_0 - \tfrac{1}{2}v^2 J^2. \tag{12-30}$$

For constant-area, one-dimensional flow without wall shear forces or external work, the momentum equation (10–12) for steady state reduces to

$$p + JU = p_1 + JU_1 = p_2 + JU_2.$$

Combining this with the equation of continuity (12–28), we have

$$p + J^2/\rho = p + J^2 v = p_1 + J^2/\rho_1 = p_2 + J^2/\rho_2. \tag{12–31}$$

Finally, the equation of state and thermodynamic relations give us

$$\text{H} = f(\text{S}, \rho) \quad \text{and} \quad \text{S} = \phi(p, \rho). \tag{12–32}$$

(i) The Fanno line. We can construct an H-S diagram (or an equivalent p-v diagram) from Eqs. (12–30) and (12–32). For example, for fixed stagnation conditions and mass flow rate, the value of H can be obtained for any ρ from Eq. (12–30). From Eq. (12–32), the value of S can be obtained from the value of H and ρ. These values can be plotted as in Fig. 12–9, and give what is called the *Fanno line*, which is the locus of points as determined by the energy, continuity, and state equations. Since the momentum equation was not used, the Fanno line represents only flows for the same J and H_0, but not for the same momentum. Changes along the Fanno line result from frictional effects.

Point a on the Fanno line is the state of maximum entropy of the system. For a change about a, the continuity equation $(\boldsymbol{\nabla} \cdot \rho \boldsymbol{U}) = 0$ gives, for one-dimensional flow,

$$d(\rho U) = 0$$

or

$$\rho \, dU + U \, d\rho = 0.$$

From the total energy equation (12–29),

$$d\text{H} + d(U^2/2) = 0$$

or

$$d\text{H} + U \, dU = 0.$$

The definition of enthalpy, the second law, and Eq. (12–1) can be combined to give the thermodynamic identity

$$d\text{H} = T \, ds + v \, dp.$$

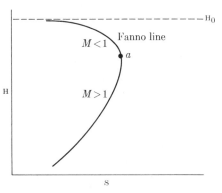

Fig. 12–9. The Fanno line.

Since $ds/d\text{H} = 0$ at point a, this reduces to $d\text{H} = v \, dp$. Combining the above relations, and solving for the velocity, we obtain

$$U = -\frac{d\text{H}}{dU} = -\frac{1}{\rho}\frac{dp}{dU} = \frac{dp/\rho}{U\,d\rho/\rho} = \frac{dp}{U\,d\rho} \quad \text{or} \quad U^2 = \left(\frac{dp}{d\rho}\right)_s.$$

The restriction of constant entropy comes from $ds/d\text{H} = 0$. The right-hand side is simply the square of the local sound velocity (Eqs. 12–13 and 12–14). Consequently, the Mach number at a is unity.

When the enthalpy is greater than that at point a, the fluid velocity is less than the local sound velocity (subsonic flow, $M < 1$). If the enthalpy is less than at point a, the fluid velocity is greater than the local sound velocity (supersonic, $M > 1$). In both regions, all possible spontaneous changes will go toward increasing entropy. The equilibrium state is reached when the entropy is at a maximum and, as just shown, where $M = 1$. In summary, if the flow is initially subsonic, frictional effects will cause the velocity and Mach number to increase, but if the flow is initially supersonic, the fluid velocity and Mach number will decrease. In both cases, the equilibrium Mach number is unity.

(ii) The Rayleigh line. The Fanno line is the locus of states defined by the energy, continuity, and state equations. In contrast, the Rayleigh line is the locus of states defined by the momentum, continuity, and state equations.

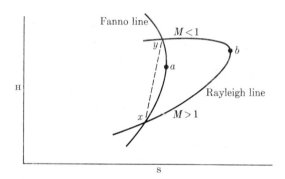

Fig. 12–10. The Rayleigh and Fanno lines.

To compute the locus, we can select a value of U_2, then obtain ρ_2 from the equation of continuity and p_2 from Eq. (12–31). We can then obtain H and S from the equation of state (12–32). Figure 12–10 is a plot of typical H and S values, and is called the *Rayleigh line*. The equivalent p-v values could also be plotted. The Rayleigh line does not restrict the stagnation enthalpies (energy equation not specified); consequently it represents heat transfer effects. It may be shown that point b on the Rayleigh line is at $M = 1$; the upper branch represents subsonic and the lower, supersonic flow.

The implications of the Fanno and Rayleigh lines are:
(1) Both lines imply one-dimensional, steady flow without area or elevation changes and without work effects.
(2) A given Fanno line allows changes in wall friction but does not allow heat transfer. It is restricted to a set mass flow rate per unit area (J) and to a set stagnation enthalpy (H_0). If J or H_0, or both, are varied, a family of Fanno lines is obtained.
(3) A given Rayleigh line allows heat transfer but does not allow wall friction. The stagnation enthalpy is not constant along this line. As in the case of the Fanno line, a family of curves can be obtained by varying J or p_0.

(4) The points of intersection of the two lines represent the locus of all changes in state carried out without area or elevation changes, heat or work transfer, and wall shear forces. In other words, the intersections satisfy energy, momentum, continuity, and state equations.

As previously pointed out, the normal shock occurs over a very short distance and is treated as a discontinuity. As such, there would be no area or elevation change, heat transfer, or wall friction. Thus the shock is subjected to the same restrictions as cited in (4) above and must satisfy all equations simultaneously. The conditions upstream and downstream of the shock would be specified by the points common to both the Fanno and Rayleigh lines (points x and y of Fig. 12–10). The direction of the process has not been restricted; however, for all fluids investigated, the point y lies to the right of x. From the second law of thermodynamics, the entropy must increase during an adiabatic change, and therefore a change can occur only from x to y. In the flow under consideration, subsonic flow will remain subsonic. Supersonic flow may change to a state of higher entropy by passing abruptly from point x to point y. This rapid change from supersonic to subsonic flow is the normal shock. The general shape of the Fanno and Rayleigh lines used to establish the shock direction was based on experimental observation. Further proofs and analyses of the direction of variation in shock waves can be made in a more analytical manner, especially if the analysis is restricted to a perfect gas. Discussion of this can be found in Shapiro [9] and Landau and Lifshitz [2].

In the present analysis of the shock, there are no intermediate states between points x and y. However, if a finite distance were to separate the beginning and end of the shock, intermediate states would be possible. In a real fluid, viscosity and thermal conductivity must be considered and a finite shock width does exist, as shown in Section 7-5. Furthermore, interactions with the boundary layer have not been considered. These cause rather marked deviations from the normal shock and can result in oblique shocks and separation. Our analysis will be restricted to the idealized normal shock case, since from the overall view this represents the actual end states quite satisfactorily.

(b) Shock in a perfect gas in constant area flow. The two sides of the normal shock lie on one Fanno line; thus the stagnation enthalpies, which in turn imply the stagnation temperatures, are the same on both sides of the shock. As a result of this, the equations of nonisentropic adiabatic flow apply.

The problem is to relate the Mach number and other properties on one side of the shock to the corresponding properties on the other side. For temperature, this is accomplished by the use of Eq. (12–20) to obtain the ratio of the two sides:

$$\frac{T_2}{T_1} = \frac{2 + (\gamma - 1)M_1^2}{2 + (\gamma - 1)M_2^2}. \tag{12-33}$$

Next, a relation between the pressure on the two sides can be obtained for the path of one Fanno line. To do this, we combine the continuity and state equations with the definition of the Mach number and Eq. (12–33). The result is an expression of the pressure ratio in terms of the Mach number:

$$\frac{p_2}{p_1} = \frac{M_1\sqrt{2 + (\gamma - 1)M_1^2}}{M_2\sqrt{2 + (\gamma - 1)M_2^2}}. \tag{12-34}$$

Another expression for the pressure ratio can be obtained by relating the two sides with the path of one Rayleigh line. This is done by combining the continuity and state equations and definitions with Eq. (12-31); the result is

$$\frac{p_2}{p_1} = \frac{1 + \gamma M_1^2}{1 + \gamma M_2^2}. \tag{12-35}$$

The shock is common to both the Fanno and Rayleigh lines, and Eqs. (12-34) and (12-35) must be equal. The resulting equation has two solutions; one is trivial ($M_1 = M_2$), and the other relates the Mach numbers on the two sides of the shock:

$$M_2^2 = \frac{2 + (\gamma - 1)M_1^2}{2\gamma M_1^2 - (\gamma - 1)}.$$

The density ratio can be obtained from the state equation, and the velocity ratio from the continuity equation. The ratio of stagnation pressures can be obtained from three equations, one for each side from Eq. (12-24), and either Eq. (12-34) or Eq. (12-35). All of these relations, either given or suggested, are easy to derive. They are hard to use, however, because of the tedious calculations necessary. To avoid this problem, a working table has been developed for $\gamma = 1.4$ which is similar to Table 12-1 in construction, and the same references apply. Table 12-2 is a brief summary of this table.

The values in Table 12-2 clearly indicate the direction of variation of the parameters in a shock wave:

$$M_1 > 1, \quad M_2 < 1, \quad U_1 > U_2,$$
$$\rho_2 > \rho_1 \quad \text{or} \quad v_1 > v_2,$$
$$T_2 > T_1, \quad p_2 > p_1, \quad p_{0,2} < p_{0,1}.$$

The actual pressure, temperature, and density increase while the stagnation pressure decreases. There is no limit on the pressure or temperature increase, but the density can increase only by a factor of six. The pressure ratio is often used as a measure of shock strength. As the pressure ratio decreases, the value of M_1 approaches unity. The value of M_2 also approaches unity, and the density change across the shock approaches zero. A shock of infinitesimal strength is identical with a sound wave. It follows that a very weak pressure disturbance travels with the speed of sound with respect to the fluid in which it is propagating. The normal shock (not weak) travels with a speed greater than the speed of sound, the ratio being M_1.

The moving shock wave can be treated by a simple transformation of a stationary coordinate system to a moving one. Such transformations are given by Shapiro [9].

TABLE 12-2

Normal Shock, Perfect Gas, $\gamma = 1.4$

M_1	M_2	p_2/p_1	U_1/U_2 and v_1/v_2	T_2/T_1	S_1^*/S_2^* and $p_{0,2}/p_{0,1}$	$p_{0,2}/p_1$
1.00	1.00000	1.00000	1.00000	1.00000	1.00000	1.8929
1.10	0.91177	1.2450	1.1691	1.06494	0.99892	2.1328
1.20	0.84217	1.5133	1.3416	1.1280	0.99280	2.4075
1.30	0.78596	1.8050	1.5157	1.1909	0.97935	2.7135
1.40	0.73971	2.1200	1.6896	1.2547	0.95819	3.0493
1.50	0.70109	2.4583	1.8621	1.3202	0.92978	3.4133
1.60	0.66844	2.8201	2.0317	1.3880	0.89520	3.8049
1.70	0.64055	3.2050	2.1977	1.4583	0.85573	4.2238
1.80	0.61650	3.6133	2.3592	1.5316	0.81268	4.6695
1.90	0.59562	4.0450	2.5157	1.6079	0.76735	5.1417
2.00	0.57735	4.5000	2.6666	1.6875	0.72088	5.6405
2.20	0.54706	5.4800	2.9512	1.8569	0.62812	6.7163
2.40	0.52312	6.5533	3.2119	2.0403	0.54015	7.8969
2.60	0.50387	7.7200	3.4489	2.2383	0.46012	9.1813
2.80	0.48817	8.9800	3.6635	2.4512	0.38946	10.596
3.00	0.47519	10.333	3.8571	2.6790	0.32834	12.061
3.50	0.45115	14.125	4.2608	3.3150	0.21295	16.242
4.00	0.43496	18.500	4.5714	4.0469	0.13876	21.068
4.50	0.42355	23.458	4.8119	4.8751	0.09170	26.539
5.00	0.41523	29.000	5.0000	5.8000	0.06172	32.654
6.00	0.40416	41.833	5.2683	7.9406	0.02965	46.815
7.00	0.39736	57.000	5.4444	10.469	0.01535	63.552
8.00	0.39289	74.500	5.5652	13.387	0.00849	82.865
9.00	0.38980	94.333	5.6512	16.693	0.00496	104.753
10.00	0.38757	116.500	5.7143	20.388	0.00304	129.217
∞	0.37796	∞	6.0000	∞	0.00000	∞

EXAMPLE 12–2. The air flow system of Example 12–1 is to be forced to go through a normal shock at the test area where the Mach number is 2 (Fig. 12–11). Calculate the velocity, pressure, and temperature after the shock, and determine the back pressure required.

Answer. The pressure, stagnation pressure, and temperature before the shock are known from Example 12–1:

$$p_1 = 14.7 \text{ psia}, \quad T_1 = 294°\text{R}, \quad p_{0,1} = 114.7 \text{ psia}.$$

The velocity of sound at the shock is

$$c_1 = 49.02\sqrt{T} = 857 \text{ ft/sec}.$$

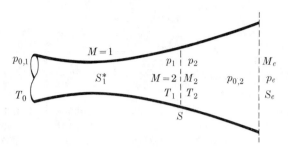

Fig. 12–11. *Nozzle system for supersonic flow with a shock.*

From Table 12-2,

$$M_2 = 0.57735,$$
$$U_1/U_2 = 2.6666 \quad \text{or} \quad U_2 = 643 \text{ ft/sec},$$
$$p_{0,2}/p_{0,1} = S_1^*/S_2^* = 0.72088 \quad \text{or} \quad p_{0,2} = 82.5 \text{ psia},$$
$$p_2/p_1 = 4.5000 \quad \text{or} \quad p_2 = 66.2 \text{ psia},$$
$$T_2/T_1 = 1.6875 \quad \text{or} \quad T_2 = 495°\text{R} = 35°\text{F}.$$

The subsonic flow after the shock involves an isentropic expansion to the exit. The pressure at this point must now be determined. Since the stagnation pressure changes across the shock, the starred reference conditions also change. The new value of S/S^* at the exit is

$$1.96 \times 0.72088 = 1.411,$$

which, from Table 12-1, corresponds to

$$M_e = 0.465,$$
$$p_e/p_{0,2} = 0.4991 \quad \text{or} \quad p_e = 41.2 \text{ psia} = 26.5 \text{ psig}.$$

(c) **Converging-diverging nozzle.** The converging-diverging nozzle in isentropic flow was discussed in Section 12-3.B(c). The effect of the normal shock can now be included with the information of that section. Figure 12–12 is the same as Fig. 12–7 for cases (i) through (iv). Reduction of p_b below that for case (iii) will result in the appearance of a normal shock downstream of the throat. The process immediately behind the shock comprises subsonic deceleration (v), as calculated in the last example. Beyond the shock, the value of the stagnation pressure changes from $p_{0,1}$ to $p_{0,2}$. The starred reference pressure p_2^* shows a corresponding change, since p^*/p_0 is fixed. The locus of points for $p_2^*/p_{0,1}$ is shown. For a given shock (nonisentropic flow), points above this line are subsonic, while those below are supersonic. If no shocks exist, the isentropic reference line p^*/p_0 applies. The locus of states just downstream of the shock is also shown; these are the points $p_2/p_{0,1}$, and can be computed from the normal shock table. As the back pressure is further lowered, the shock moves down the nozzle until, at (vi), it

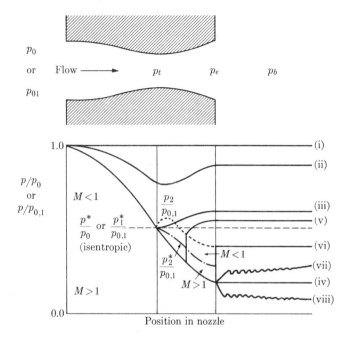

Fig. 12–12. Flow in converging-diverging nozzle with a shock. [After Ascher H. Shapiro, The Dynamics and Thermodynamics of Compressible Fluid Flow, Vol. 1, The Ronald Press, New York (1953). By permission.]

appears in the exit plane of the nozzle. Any further reduction of p_b results in the shock entering the area outside the nozzle and involves oblique shock waves, which cannot be treated on a one-dimensional basis (vii). At (iv), the design conditions, the exit plane pressure is identical to the back pressure. Reduction of p_b to (viii) results in oblique expansion waves which cannot be treated by one-dimensional theory. From condition (iii) onward, the nozzle is independent of back pressure, and corresponds to the flow pattern for the design condition. In all supersonic cases, the flow rate is independent of back pressure, and, since $M = 1$ at the throat, the rate may be calculated from Eq. (12–26). Of course, all these curves are ideal. Because of the neglect of the boundary layer effects, they only approximate the real case.

(d) **Wall friction in a perfect gas in constant-area flow.** The previous discussion of one-dimensional, constant-area flow did not rule out the possibility of influences of wall shear forces. If we again assume a one-dimensional steady flow in a constant-area duct with no external heat exchange, work, or change in elevation, the equations and analyses associated with the Fanno and Rayleigh lines will apply to the present case. Wall friction is associated with changes along a Fanno line. Since friction is an irreversible effect, all changes must cause an increase in entropy; otherwise the second law of thermodynamics would be

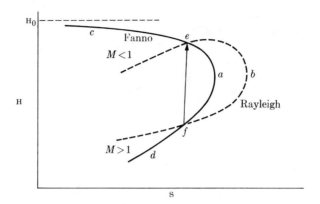

Fig. 12–13. Paths of change as a result of wall friction.

violated. Figure 12–13 illustrates the paths of the possible changes. If the flow is initially subsonic (point *c*), friction will cause an acceleration along the path *cea*. The flow cannot become supersonic without violating the second law. If the flow is initially supersonic (point *d*), two paths are possible. The first is a simple deceleration along the path *dfa* without a change to subsonic flow, and the second follows the path *dfea*, which includes deceleration of the flow, and a shock followed by a reacceleration of the resulting subsonic flow. The extent of travel along the various paths for a given set of conditions will be treated more fully later in the discussion.

The differential equation obtained by differentiating the definition of the Mach number for a perfect gas,

$$\frac{dM}{M} = \frac{dU}{U} - \frac{dT}{2T}, \tag{12-36}$$

can be combined with a like expression obtained from the energy equation (12–29),

$$c_p dT + U\, dU = 0, \tag{12-37}$$

to give two relations:

$$\frac{dU}{U} = \frac{2}{2 + (\gamma - 1)M^2} \frac{dM}{M}, \tag{12-38}$$

$$\frac{dT}{T} = -\frac{2(\gamma - 1)M^2}{2 + (\gamma - 1)M^2} \frac{dM}{M}, \tag{12-39}$$

in which Eqs. (12–5) and (12–14) have also been used. Equations (12–38) and (12–39) relate the Mach number to the stream velocity and the moving stream temperature, respectively. These two equations result from the perfect gas law and energy equations and are, therefore, not restricted to constant-area flow with friction, but only to adiabatic perfect gas conditions.

The differential equation from continuity,

$$d\rho/\rho + dU/U = 0, \tag{12-40}$$

which relates the density and velocity, can be combined with eq. (12-38) to give

$$\frac{d\rho}{\rho} = -\frac{2}{2 + (\gamma - 1)M^2}\frac{dM}{M}. \tag{12-41}$$

The gas law in differential form,

$$\frac{dp}{p} = \frac{d\rho}{\rho} + \frac{dT}{T}, \tag{12-42}$$

can be combined with Eqs. (12-39) and (12-41) to give

$$\frac{dp}{p} = -\frac{2[1 + (\gamma - 1)M^2]}{2 + (\gamma - 1)M^2}\frac{dM}{M}. \tag{12-43}$$

The stagnation pressure from Eq. (12-24) can be put into differential form and then combined with Eq. (12-43) to give

$$\frac{dp_0}{p_0} = \frac{2(M^2 - 1)}{2 + (\gamma - 1)M^2}\frac{dM}{M}. \tag{12-44}$$

Finally, from the momentum balance, the definition of the friction factor, and the hydraulic diameter, one can derive an expression for the mass flow in terms of the velocity and the Mach number. This relation can then be combined with the equation for velocity (12-38) and the expression for the pressure (12-42) to give the relation between the Mach number and the wall shear forces:

$$\frac{4f\,dx}{d_0} = \frac{4(1 - M^2)}{\gamma M^2[2 + (\gamma - 1)M^2]}\frac{dM}{M}. \tag{12-45}$$

Equation (12-45) can be used to substantiate the conclusions reached in the analysis of Fig. 12-13. The term on the left must always be positive since, according to the second law of thermodynamics, there can be no entropy decrease in an adiabatic process. For $\gamma > 1.0$, the equation indicates that for subsonic flow, friction must result in an increase in Mach number, and for supersonic flow, the change in Mach number must be negative. With this knowledge and Eqs. (12-38) through (12-45), we can list these effects:

	$M < 1$	$M > 1$
M	Increase	Decrease
p_0	Decrease	Decrease
p	Decrease	Increase
ρ	Decrease	Increase
T	Decrease	Increase
U	Increase	Decrease

Unexpectedly, friction accelerates a subsonic stream, and causes a pressure rise in a supersonic stream. However, one must remember that this is an adiabatic system, and that friction losses and temperature changes are not removed from the system by heat losses.

The series of equations which relate the variables to the Mach number must be integrated to give a series of working formulas. Since the longest pipe that can be used (without introducing a shock) is one in which $M = 1$ at the exit, one of the limits of the integration will be taken as unity, and be denoted by an asterisk (*). Equation (12–45) yields

$$\int_0^{L_{max}} \frac{4f\,dx}{d_0} = \frac{4\bar{f} L_{max}}{d_0} = \int_M^{M^*=1} \frac{2(1-M^2)}{\gamma M^4[2+(\gamma-1)M^2]} dM^2$$
$$= \frac{1-M^2}{\gamma M^2} + \frac{\gamma+1}{2\gamma} \ln \frac{(\gamma+1)M^2}{2+(\gamma-1)M^2}, \quad (12\text{--}46)$$

where $\bar{f}$ is the average friction coefficient with respect to length; that is,

$$\bar{f} = \frac{1}{L_{max}} \int_0^{L_{max}} f\,dx.$$

Since f is usually assumed constant, $f = \bar{f}$. The notation f will be used and, as usual, will imply an average. The other equations become

$$\frac{p_0}{p_0^*} = \frac{1}{M} \sqrt{\left[\frac{2+(\gamma-1)M^2}{\gamma+1}\right]^{(\gamma+1)/(\gamma-1)}}, \quad (12\text{--}47)$$

$$\frac{p}{p^*} = \frac{1}{M} \sqrt{\frac{\gamma+1}{2+(\gamma-1)M^2}}, \quad (12\text{--}48)$$

$$\frac{\rho}{\rho^*} = \frac{U^*}{U} = \frac{1}{M} \sqrt{\frac{2+(\gamma-1)M^2}{\gamma+1}}, \quad (12\text{--}49)$$

$$\frac{T}{T^*} = \frac{c^2}{c^{*2}} = \frac{\gamma+1}{2+(\gamma-1)M^2}. \quad (12\text{--}50)$$

The asterisk quantities ($M = 1$) are constant for any given constant-area adiabatic flow, and therefore may be used as reference values for relating the flow between any two points. For example, the ratio of the property to its critical value can be found at two Mach numbers being considered. The ratio of these two ratios gives the desired relation, since the critical value used would be the same for both Mach numbers and would cancel out:

$$\frac{T_2}{T_1} = \frac{(T/T^*)_2}{(T/T^*)_1}.$$

When we use Eq. (12–46), we need a knowledge of the friction factor. For subsonic flow, correlations of f as a function of the Reynolds number for a given degree of roughness of the pipe wall have been found to be satisfactory. Also in subsonic flow, the friction factor seems to be substantially independent of the

ONE-DIMENSIONAL, STEADY, ADIABATIC FLOW OF A PERFECT GAS

Mach number, if the distance from the duct inlet is greater than about 50 diameters. Near the duct inlet, entrance effects such as those already considered in this text contribute to the friction. For short pipes, the average apparent friction coefficient may be substantially in excess of that which would normally be predicted.

With supersonic flow, the available friction factor data are scanty. It is difficult to determine f because supersonic velocities cannot be maintained over a length of conduit exceeding about 100 pipe diameters. Consequently, the available friction factor measurements include entrance region effects. However, the values seem to be of the same order of magnitude as the values obtained from a conventional plot of friction factor versus Reynolds number. Even if this is not exactly true, it must be assumed so, since complete data are not available. The data which are known show scatter due mainly to effects outside the one-dimensional consideration. Some data of Keenan and Neumann [1] indicate that for tubes between 10 and 50 tube diameters in length, the average coefficient was from 0.002 to 0.004. The Mach range was 1.2 to 3, and the Reynolds number was from 25,000 to 700,000. The corresponding values for fully developed incompressible flow are 0.003 to 0.0065.

Equations (12–46) through (12–50) have been evaluated for $\gamma = 1.4$ (air) and are presented as a working table in Table 12-3. More detailed tables are available in the references already cited. To illustrate the importance of frictional effects, Table 12-4 has been evaluated from the data of Table 12-3. A value of 0.0025 has been taken as representative of the friction factor. The value for L_{max}/d_0 for $M = \infty$ can be obtained from the limit of Eq. (12–45) as $M \to +\infty$. The important point to note is that the L_{max}/d_0 is finite even for infinite Mach number. It takes only 30 pipe diameters to reduce the flow from $M = \infty$ to $M = 3$ and 20 more to reduce it to $M = 2$.

EXAMPLE 12–3. It is desired to attach a length of straight pipe to the nozzle discussed in Example 12–1 at the point where $M = 2$, in order to reduce the Mach number to 1.2. The pipe area is to be the same as the nozzle area at the point of attachment. How long a piece of pipe is needed? Assume that the friction factor is given by the standard correlation for a smooth pipe.

Answer. The Mach number at the straight pipe inlet will be 2.0; thus, from Table 12-3, $4fL_{max}/d_0 = 0.30499$ will reduce the Mach number to unity. At $M = 1.2$, $4fL_{max}/d_0 = 0.03364$, from the same table. The change or difference is the length necessary to reduce the Mach number from 2 to 1.2:

$$4fL/d_0 = 0.30499 - 0.03364 = 0.27135.$$

The temperature at the inlet to the pipe is 294°R, and, from Table 12-3, the temperature at the exit will be

$$\frac{T_2}{T_1} = \frac{0.93168}{0.66667} = 1.399 \quad \text{or} \quad T_2 = 411°R.$$

TABLE 12-3

Frictional, Adiabatic, Constant-Area Flow (Fanno Line)†

M	T/T^*	p/p^*	p_0/p_0^*	U/U^* and ρ^*/ρ	$4fL_{max}/d_0$
0.00	1.2000	∞	∞	0.00000	∞
0.10	1.1976	10.9435	5.8218	0.10943	66.922
0.20	1.1905	5.4555	2.9635	0.21822	14.533
0.30	1.1788	3.6190	2.0351	0.32572	5.2992
0.40	1.1628	2.6958	1.5901	0.43133	2.3085
0.50	1.1429	2.1381	1.3399	0.53453	1.06908
0.60	1.1194	1.7634	1.1882	0.63481	0.49081
0.70	1.09290	1.4934	1.09436	0.73179	0.20814
0.80	1.06383	1.2892	1.03823	0.82514	0.07229
0.90	1.03270	1.12913	1.00887	0.91459	0.014513
1.00	1.00000	1.00000	1.00000	1.00000	0.000000
1.20	0.93168	0.80436	1.03044	1.1583	0.03364
1.40	0.86207	0.66320	1.1149	1.2999	0.09974
1.60	0.79365	0.55679	1.2502	1.4254	0.17236
1.80	0.72816	0.47407	1.4390	1.5360	0.24189
2.00	0.66667	0.40825	1.6875	1.6330	0.30499
2.30	0.58309	0.33200	2.1931	1.7563	0.38623
2.60	0.51020	0.27473	2.8960	1.8571	0.45259
3.00	0.42857	0.21822	4.2346	1.9640	0.52216
4.00	0.28571	0.13363	10.719	2.1381	0.63306
7.00	0.11111	0.04762	104.14	2.3333	0.75281
10.00	0.05714	0.02390	535.94	2.3905	0.78683
∞	0.00000	0.00000	∞	2.4495	0.82153

†Perfect gas, $\gamma = 1.4$

TABLE 12-4*

M	0	0.3	0.5	0.8	1	2	3	∞
L_{max}/d_0	∞	530	107	7.2	0	30.5	52.2	82.2

*By permission from Shapiro [9].

The average temperature in the pipe is 353°R, or −107°F. The viscosity of air at this average temperature is 0.0134 cps, or 9.03×10^{-6} lb/ft-sec. The Reynolds number becomes

$$N_{Re} = d_0 w/\mu S = 6.77 \times 10^5,$$

where the diameter, area, and mass flow rate are obtained from Example 12–1. The corresponding Fanning friction factor for a smooth pipe is 0.0031. Thus

$$L/d_0 = 21.86 \quad \text{or} \quad L = 7.09 \text{ in.}$$

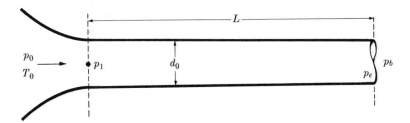

Fig. 12–14. Flow in a long duct.

(e) Flow in long ducts. Consider the flow in a duct fed by a convergent nozzle such as that shown in Fig. 12–14. Assume that the stagnation pressure and temperature are constant. As the exit pressure is reduced, the flow will commence and will be subsonic. At some pressure p^*, the flow will be such that the Mach number at the duct exit is unity. At any point upstream of the exit, the pressure will be greater and the flow will be subsonic. A further reduction in the back pressure will not cause a change in the flow in the system, since the exit Mach number cannot exceed unity. The flow at this point is choked. The flow in the convergent nozzle is isentropic, and that in the duct is adiabatic friction flow. Consequently, a combination of these two analyses will describe the system. Lapple [3] used this method to construct a series of charts for values of γ of 1.0, 1.4, and 1.8. These charts are given in Fig. 12–15, and the data in Table 12–5. Example 12–4 illustrates their construction and use.

The ratio of specific heats can vary from about 1.1 to 2.2, although the normal variation is from 1.3 to 1.6. Monatomic gases have a value of 1.67; gases such as air, oxygen, nitrogen, NO, CO, and hydrogen have values of about 1.4 (a value for air at $-79°C$ and 100 atm is reported as 2.2); the triatomic gases and methane have values of about 1.3; and finally, the organic vapors have values around 1.2, with propane being the lowest at 1.13.

EXAMPLE 12–4. (a) Illustrate the construction of Fig. 12–15 for a line of $4fL/d_0$ of 0.49081, and a γ of 1.4. Determine one point above and one point below the choked line. (b) If a convergent nozzle replaces the convergent-divergent nozzle of Example 12–3, what will be the discharge rate of air? Use the same length and diameter as in Example 12–3, and use the same upstream conditions.

Answer. (a) First, determine the limiting case (all points below the choked line) where the flow is critical ($M = 1$) at the pipe exit. For $4fL_{max}/d_0$ of 0.49081, Eq. (12–46) or Table 12–3 gives $M_1 = 0.6$. At point 1, the flow out of the nozzle is isentropic, so Eq. (12–25) can be used to obtain the flow-rate parameter:

$$J \frac{1}{p_0} \sqrt{\frac{RT_0}{M_w}} = \sqrt{1.4} \frac{M_1}{(1 + 0.2M_1^2)^3} = 0.574.$$

Thus a flow parameter of 0.574 corresponds to $4fL_{max}/d_0$ of 0.49081, and this will

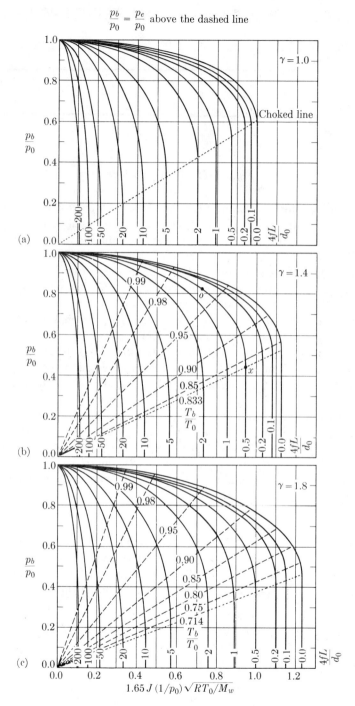

Fig. 12-15. Working figures for flow in long ducts. [From C. E. Lapple, Trans. AIChE **39**, 385–432 (1943). By permission.]

be true so long as the flow is critical at the exit of the duct. This is certainly always true at the limit of $p_b/p_0 = 0$.

The next step is to determine over what range of values of p_b/p_0 the flow will remain critical at the exit. For isentropic flow (Table 12-1), $p_1/p_0 = 0.78400$. From Table 12-3, $p_1/p^* = 1.7634$. If the flow is choked, $p^* = p_e = p_b$. Combining, we get

$$\frac{p_1}{p_0}\frac{p^*}{p_1} = \frac{p^*}{p_0} = \frac{p_e}{p_0} = \frac{p_b}{p_0} = 0.78400 \frac{1}{1.7634} = 0.445.$$

The flow will be choked for values of p_b/p_0 between 0 and 0.445; the flow parameter over this range is 0.574 or as plotted, $1.65(0.574) = 0.947$. The point x descriptive of this condition lies on the dotted line of Fig. 12-15. Its coordinates and parameters are

$$\frac{p_b}{p_0} = 0.445, \qquad 1.65 J \frac{1}{p_0}\sqrt{\frac{RT_0}{M_w}} = 0.947,$$

$$\frac{4fL}{d_0} = 0.49081, \qquad M_1 = 0.6, \qquad \frac{T_b}{T_0} = 0.833,$$

where the last ratio was obtained in a manner analogous to that of the pressure; that is,

$$\frac{T_1}{T_0}\frac{T^*}{T_1} = \frac{T^*}{T_0} = \frac{T_e}{T_0} = \frac{T_b}{T_0} = \frac{0.93284}{1.1194} = 0.833.$$

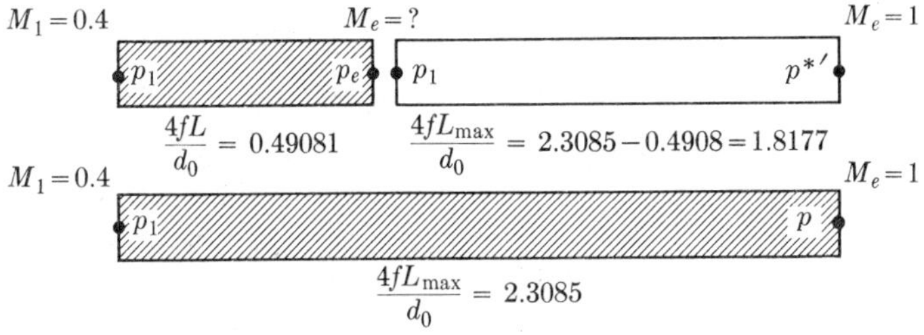

Fig. 12-16. Calculation procedure for Example 12-4.

To determine a point before choking occurs, we may proceed in several ways. One method is to set the Mach number at point 1. For example, take $M_1 = 0.4$ (see Fig. 12-16). The small shaded system is the problem of our current interest; it was just solved for $M_1 = 0.6$. We should like to solve the same problem for $M_1 = 0.4$. However, in this case, the flow at the exit will not be choked. The large shaded system shows a hypothetical case where $M_1 = 0.4$ and the exit is choked. This, of course, requires a longer system. The additional length required to cause the choking is just the difference in the values of $4fL/d_0$, and is shown as the unshaded system. This is the system we can treat, for its exit is choked and its inlet would correspond exactly to the outlet of the actual problem we wish to solve.

TABLE 12-5*

Values of $1.65 J\,(1/p_0)\sqrt{RT_0/M_w}$

γ	p_e/p_0	\multicolumn{11}{c	}{$4fL/d_0$}
		0	0.1	0.2	0.5	1.0	2	5	10	20	50	100	200
1.0	1.0	0.000	0.000	0.000	0.000	0.000	0.000	0.000	0.000	0.000	0.000	0.000	0.000
	0.9	0.681	0.636	0.602	0.553	0.490	0.404	0.289	0.213	0.155	0.099	0.072	0.050
	0.8	0.884	0.845	0.808	0.731	0.643	0.539	0.389	0.291	0.213	0.136	0.099	0.070
	0.7	0.975	0.935	0.900	0.819	0.728	0.618	0.456	0.344	0.252	0.161	0.116	0.082
	0.6	1.000	0.964	0.933	0.861	0.774	0.662	0.502	0.384	0.281	0.181	0.130	0.092
	0.5	1.000	0.965	0.936	0.870	0.792	0.689	0.527	0.409	0.302	0.196	0.140	0.100
	0.4	1.000	0.965	0.936	0.870	0.793	0.700	0.540	0.422	0.318	0.207	0.149	0.106
	0.3	1.000	0.965	0.936	0.870	0.793	0.700	0.546	0.430	0.327	0.214	0.154	0.110
	0.2	1.000	0.965	0.936	0.870	0.793	0.700	0.546	0.432	0.331	0.218	0.157	0.112
	0.1	1.000	0.965	0.936	0.870	0.793	0.700	0.546	0.432	0.331	0.219	0.158	0.113
	0.0	1.000	0.965	0.936	0.870	0.793	0.700	0.546	0.432	0.331	0.219	0.158	0.114
1.4	1.0	0.000	0.000	0.000	0.000	0.000	0.000	0.000	0.000	0.000	0.000	0.000	0.000
	0.9	0.693	0.640	0.614	0.563	0.491	0.406	0.290	0.214	0.156	0.100	0.072	0.050
	0.8	0.922	0.871	0.839	0.749	0.656	0.541	0.390	0.293	0.213	0.137	0.099	0.070
	0.7	1.052	0.998	0.953	0.856	0.752	0.626	0.459	0.345	0.253	0.163	0.117	0.082
	0.6	1.114	1.059	1.012	0.917	0.812	0.683	0.507	0.385	0.282	0.182	0.131	0.092

	1.129	1.078	1.034	0.943	0.842	0.716	0.538	0.411	0.303	0.198	0.141	0.101	0.5
	1.129	1.078	1.036	0.947	0.852	0.732	0.556	0.427	0.319	0.208	0.150	0.107	0.4
	1.129	1.078	1.036	0.947	0.852	0.737	0.563	0.437	0.328	0.215	0.155	0.111	0.3
	1.129	1.078	1.036	0.947	0.852	0.737	0.565	0.441	0.333	0.219	0.158	0.113	0.2
	1.129	1.078	1.036	0.947	0.852	0.737	0.565	0.441	0.334	0.220	0.159	0.115	0.1
	1.129	1.078	1.036	0.947	0.852	0.737	0.565	0.441	0.334	0.220	0.159	0.115	0.0
1.8	0.000	0.000	0.000	0.000	0.000	0.000	0.000	0.000	0.000	0.000	0.000	0.000	1.0
	0.699	0.659	0.620	0.569	0.493	0.407	0.291	0.215	0.157	0.101	0.072	0.050	0.9
	0.947	0.890	0.854	0.762	0.662	0.543	0.391	0.294	0.214	0.138	0.099	0.070	0.8
	1.096	1.032	0.983	0.879	0.768	0.637	0.462	0.346	0.254	0.164	0.117	0.082	0.7
	1.186	1.116	1.059	0.952	0.834	0.697	0.511	0.387	0.283	0.183	0.132	0.092	0.6
	1.226	1.152	1.098	0.988	0.868	0.733	0.543	0.413	0.304	0.199	0.142	0.101	0.5
	1.229	1.157	1.107	1.000	0.887	0.752	0.561	0.431	0.320	0.209	0.150	0.107	0.4
	1.229	1.157	1.107	1.001	0.891	0.761	0.572	0.442	0.329	0.216	0.156	0.111	0.3
	1.229	1.157	1.107	1.001	0.891	0.762	0.578	0.450	0.335	0.220	0.159	0.113	0.2
	1.229	1.157	1.107	1.001	0.891	0.762	0.578	0.451	0.339	0.221	0.160	0.115	0.1
	1.229	1.157	1.107	1.001	0.891	0.762	0.578	0.451	0.339	0.221	0.160	0.115	0.0

*From C. E. Lapple, *Trans. AIChE* **39**, 385–432 (1943). By permission.

To obtain the exit pressure, we proceed as follows: For $4fL_{max}/d_0$ of 2.3085 (from Table 12-3), $p_1/p^* = 2.6958$. For $4fL_{max}/d_0$ of 1.8177 (again from Table 12-3), $p_1/p^* = p_e/p^{*\prime} = p_e/p^* = 2.492$. For Mach number of 0.4, the isentropic table gives $p_1/p_0 = 0.89562$. The desired result is

$$\frac{p_1}{p_0}\frac{p^*}{p_1}\frac{p_e}{p^*} = \frac{p_e}{p_0} = 0.89562 \frac{2.492}{2.6958} = 0.823,$$

which is also p_b/p_0, since the flow is not choked. The corresponding Mach number M_e, from Table 12-3, is 0.433; the flow parameter can be calculated and is 0.43 or, as plotted, $1.65(0.43) = 0.71$. The coordinates and parameters of this specific point are

$$\frac{p_b}{p_0} = 0.823, \qquad 1.65 J \frac{1}{p_0}\sqrt{\frac{RT_0}{M_w}} = 0.71,$$

$$\frac{4fL}{d_0} = 0.49081, \qquad M_1 = 0.4, \qquad \frac{T_b}{T_0} = \frac{T_e}{T_0} = 0.964,$$

where the last ratio is determined from

$$\frac{T_1}{T_0}\frac{T_e}{T_1} = \frac{T_1}{T_0}\frac{(T/T^*)_e}{(T/T^*)_1} = \frac{T_e}{T_0} = \frac{T_b}{T_0} = \frac{(0.96899)(1.1565)}{1.1628} = 0.964.$$

The point o is plotted in Fig. 12-15.

(b) The value of $4fL/d_0$, from Example 12-3, is 0.30499. This value is based on the given length and diameter and a friction factor of 0.0031, which may not be valid for this problem. The stagnation conditions, from Example 12-1, are

$$p_0 = 114.7 \text{ psia}, \qquad T_0 = 529.7°R.$$

If the discharge is to atmospheric pressure ($p_b = 14.7$ psia), then $p_b/p_0 = 14.7/114.7 = 0.128$. In the figure, the point for this pressure ratio and the value of $4fL/d_0$ is in the choked-flow region. The value of the flow parameter as plotted is 1.025. The flow rate can be obtained, but care must be taken to check units:

$$w = \frac{(1.025/1.65)(114.7)(144\pi\, 0.324^2)}{4(144)\sqrt{10.73(144)(529.7)/32.17(29)}} = 0.198 \text{ lb/sec}.$$

The corresponding Mach number is 0.66. The maximum flow possible through the nozzle can be calculated from the isentropic flow equation (12-27), and is

$$w = 0.532 \frac{p_0 S^*}{\sqrt{T_0}} = 0.219 \text{ lb/sec}.$$

The answer is different from that in Example 12-1 because, in this case, S^* is based on 0.324 in. rather than $\frac{1}{4}$ in. The maximum flow can also be obtained from Fig. 12-15 by using the value of $4fL/d_0 = 0$.

The friction factor assumption can be checked as follows: From the figure, the ratio T_2/T_0 would be obtained from the choked line; this is 0.833. Therefore T_2 is 442°R. The average temperature in the pipe is 26°F, which for air corresponds to a viscosity of 0.0168 cps or 1.13×10^{-5} lb/ft-sec. The Reynolds number can be obtained from

$$N_{\text{Re}} = \frac{d_0 w}{\mu S} = 8.5 \times 10^5.$$

Since a smooth pipe was assumed in Example 12–3, this curve can be used to establish a value for f, which is 0.0030. Thus the assumed value of 0.0031 could be replaced and the problem recalculated. However, the values are close enough to make this unnecessary. For a steel pipe, a better value would have been about 0.004. If the exit pressure were not atmospheric, a different ratio would be used. However, the method is the same.

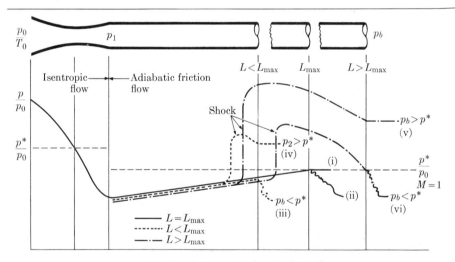

Fig. 12–17. Supersonic flow in long ducts.

Let us next consider a duct fed by a convergent-divergent nozzle (Fig. 12–17). For isentropic flow in the nozzle of a given area, there is only one Mach number greater than unity at the nozzle exit or duct inlet. Since the Mach number is fixed at the duct inlet, the value of $4fL_{\text{max}}/d_0$ is also fixed (Eq. 12–46). The flow can be divided into two cases: first, the one in which the length is less than or just equal to the maximum; and second, the one in which it is greater.

If the length is less than or just equal to L_{max}, the flow will depend on the exit pressure. For example, if $L = L_{\text{max}}$ and $p_b = p^*$, then the flow is just sonic at the exit, with no shock waves in or out of the duct. This is curve (i) in Fig. 12–17 and is the path dfa in Fig. 12–13. For a pressure p_b less than p^* (Fig. 12–17), the velocity will be sonic at the exit for $L = L_{\text{max}}$ (curve ii), and supersonic for

L less than $L_{\max}$ (curve iii). This is still on the path dfa in Fig. 12–13, but with a termination at a for curve (ii) and somewhere between f and a for curve (iii). The adjustment of the exit pressure will occur outside the duct. For a pressure greater than p^*, the velocity at the exit must be subsonic. Consequently, a shock must stand somewhere in the duct (curve iv). This is the path $dfea$ of Fig. 12–13, with a termination somewhere between e and a.

If the length is greater than $L_{\max}$, the flow cannot be supersonic in the entire duct. If the back pressure is greater than p^*, the exit pressure must correspond to the back pressure, and the flow must be subsonic at the exit (curve v). A shock will exist somewhere in the duct. For a pressure less than p^*, the exit velocity must be sonic. However, the flow must be subsonic up to this point, so that a shock must still exist in the tube between the super- and subsonic flows (curve vi). This final case is the complete path $dfea$ in Fig. 12–13.

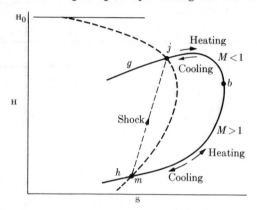

Fig. 12–18. *Paths of change as a result of heat transfer.*

12-4 ONE-DIMENSIONAL, STEADY FLOW OF A GAS IN A CONSTANT-AREA DUCT WITH HEAT TRANSFER

12-4.A Heat Transfer without Friction

Heat transfer in a constant-area duct is associated with changes along a Rayleigh line, and is measured by the change in the stagnation temperature. From the energy equation (12–29) we obtain

$$\frac{Q}{w} = c_p(T_2 - T_1) + \frac{U_2^2 - U_1^2}{2} = c_p(T_{0,2} - T_{0,1}). \qquad (12\text{–}51)$$

In contrast to friction, the entropy ($\Delta s \geq \int_1^2 dQ/T$) can be increased or decreased according to whether heat is transferred in or out of the system. Thus, as shown in Fig. 12–18, a variety of paths is possible. Before we discuss these, however, it should be pointed out that the case to be considered here is very limited in scope, since it is difficult to find a system in which friction does not also play an important part. One possible case might be a heat exchanger in which the heat flux is very high compared with the frictional loss. At any rate, area change and frictional effects have been considered previously and heat transfer will now be considered by itself. In the next section, the combined effects of heat transfer and friction will be treated.

If the flow is initially subsonic (point g in Fig. 12–18), heating will cause an acceleration along the path gjb. Additional heating cannot accelerate the flow further without violating the second law. However, if a cooling section were introduced at point b, two paths would be possible: (1) a deceleration back along path bjg or (2) an acceleration along path bmh. The actual path followed would depend on the back pressure. If the flow is initially supersonic (point h), several paths are possible: (1) heating over path hmb, (2) heating followed by a shock, and further heating over a path such as $hmjb$, and (3) various other combinations involving cooling.

A series of working formulas can be developed from the basic relations [9]. For example, the pressure ratio from the Rayleigh line analysis is given as Eq. (12–35). The pressure ratio in terms of the Mach number and temperature ratios can also be obtained from the perfect gas law, continuity, and the definition of the Mach number. Equation (12–3) can be expressed as

$$\frac{p_2}{p_1} = \frac{\rho_2}{\rho_1} \frac{T_2}{T_1}, \tag{12-52}$$

which, when combined with the continuity equation (12–28), gives

$$\frac{p_2}{p_1} = \frac{U_1}{U_2} \frac{T_2}{T_1}. \tag{12-53}$$

The definition of the Mach number can be rearranged to read

$$\frac{M_2}{M_1} = \frac{U_2/c_2}{U_1/c_1} = \frac{U_2\sqrt{\gamma\, RT_1/M_w}}{U_1\sqrt{\gamma\, RT_2/M_w}} = \frac{U_2}{U_1}\sqrt{\frac{T_1}{T_2}}, \tag{12-54}$$

then combined with the above result to give

$$\frac{p_2}{p_1} = \frac{M_1}{M_2}\sqrt{\frac{T_1}{T_2}\frac{T_2}{T_1}} = \frac{M_1}{M_2}\sqrt{\frac{T_2}{T_1}}. \tag{12-55}$$

The pressure ratio from Eq. (12–35), when combined with this, can be rearranged to

$$\frac{T_1}{T_2} = \left[\frac{(1+\gamma M_2^2)M_1^2}{(1+\gamma M_1^2)M_2^2}\right]^2. \tag{12-56}$$

If the asterisk notation is used to mark the conditions for which the Mach number is unity, this formula becomes

$$\frac{T}{T^*} = \left[\frac{(\gamma+1)M^2}{1+\gamma M^2}\right]^2. \tag{12-57}$$

TABLE 12-6

Frictionless Constant-Area Flow with Change in Stagnation Temperature (Rayleigh Line)†

M	T_0/T_0^*	T/T^*	p/p^*	p_0/p_0^*	ρ^*/ρ and U/U^*
0.00	0.00000	0.00000	2.4000	1.2679	0.00000
0.10	0.04678	0.05002	2.3669	1.2591	0.02367
0.20	0.17355	0.20661	2.2727	1.2346	0.09091
0.30	0.34686	0.40887	2.1314	1.1985	0.19183
0.40	0.52903	0.61515	1.9608	1.1566	0.31372
0.50	0.69136	0.79012	1.7778	1.1140	0.44445
0.60	0.81892	0.91670	1.5957	1.07525	0.57447
0.70	0.90850	0.99289	1.4235	1.04310	0.69751
0.80	0.96394	1.02548	1.2658	1.01934	0.81012
0.90	0.99207	1.02451	1.1246	1.00485	0.91097
1.00	1.00000	1.00000	1.00000	1.00000	1.00000
1.20	0.97872	0.91185	0.79576	1.01941	1.1459
1.40	0.93425	0.80540	0.64102	1.07765	1.2564
1.60	0.88419	0.70173	0.52356	1.1756	1.3403
1.80	0.83628	0.60894	0.43353	1.3159	1.4046
2.00	0.79339	0.52893	0.36364	1.5031	1.4545
2.30	0.73954	0.43122	0.28551	1.8860	1.5104
2.60	0.69699	0.35561	0.22936	2.4177	1.5505
3.00	0.65398	0.28028	0.17647	3.4244	1.5882
4.00	0.58909	0.16831	0.10256	8.2268	1.6410
7.00	0.52437	0.05826	0.03448	75.414	1.6896
10.00	0.50702	0.02897	0.01702	381.62	1.7021
∞	0.48980	0.00000	0.00000	∞	1.7143

†Perfect gas with $\gamma = 1.4$

In a similar manner, additional working formulas can be developed:

$$\frac{T_0}{T_0^*} = \frac{(\gamma + 1)M^2[2 + (\gamma - 1)M^2]}{(1 + \gamma M^2)^2}, \tag{12-58}$$

$$\frac{U}{U^*} = \frac{\rho^*}{\rho} = \frac{(\gamma + 1)M^2}{1 + \gamma M^2}, \tag{12-59}$$

$$\frac{p}{p^*} = \frac{\gamma + 1}{1 + \gamma M^2}, \tag{12-60}$$

$$\frac{p_0}{p_0^*} = \frac{\gamma + 1}{1 + \gamma M^2}\left[\frac{2 + (\gamma - 1)M^2}{\gamma + 1}\right]^{\gamma/(\gamma-1)} \tag{12-61}$$

From the formulas, calculations can be made to obtain working charts or tables; some selected values are presented in Table 12-6 for $\gamma = 1.4$. The same references apply as for the previous tables.

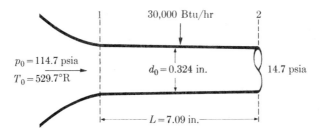

Fig. 12-19. Flow in a long duct with heat transfer.

EXAMPLE 12-5. In Example 12-4(b), a 7.09-in. duct, 0.324 in. in diameter, was attached to a convergent nozzle, and the system exhausted to the atmosphere. The flow rate was 0.198 lb/sec when friction was considered, and 0.219 lb/sec with no friction. The same system will now be solved for the case in which friction is neglected, and in which 30,000 Btu/hr are transfered to the system through the duct walls. No heat enters in the convergent nozzle area.

Answer. Figure 12-19 summarizes what is known about the system. The heat input causes a change in the stagnation temperature:

$$T_{0,2} - T_{0,1} = Q/wc_p = 30{,}000/3600w \cdot 0.24$$

or

$$T_{0,2} = T_{0,1} + (34.6/w) \quad \text{(with } w \text{ in lb/sec).}$$

Since the flow rate is unknown, it will have to be assumed. The upper limit to the flow is 0.219 lb/sec. As a first estimate, a value of 0.2 lb/sec will be satisfactory. For this value,

$$T_{0,2} = 529.7 + 173 = 702.7 \quad \text{and} \quad T_{0,2}/T_{0,1} = 1.326.$$

The assumed flow rate must be checked. The procedure is to determine the Mach number at 1 and, from the table, obtain the value of $(T_0/T_0^*)_1$. Then, from

$$\frac{T_{0,2}}{T_{0,1}} = \frac{(T_0/T_0^*)_2}{(T_0/T_0^*)_1}, \tag{i}$$

the ratio $(T_0/T_0^*)_2$ can be determined. With this ratio, a value for M_2 can be obtained from the table. If the value of $(T_0/T_0^*)_2$ should exceed unity, then a lower flow rate must be used. This is to be repeated until a value of w is found that makes $(T_0/T_0^*)_2$ equal to unity, since it is known that this term cannot exceed that value. The value of unity corresponds to choked flow, and must be checked by calculating the exit pressure and comparing it with the back pressure. If the exit pressure exceeds the back pressure, then the proper conditions have been found and the flow is choked. If the calculated value of $(T_0/T_0^*)_2$ is less than unity, the exit pressure can be calculated, and the trial and error continued until the calculated exit pressure equals the back pressure.

The Mach number at 1 can be obtained from the isentropic formula for the flow rate (Eq. 12–25). A plot or equation of w versus the Mach number will simplify the calculations. This is a unique relation, since the gas and stagnation conditions are fixed. For $w = 0.2$, the value of M_1 is 0.63. From Table 12-6, for a Mach number of 0.63, the value of $(T_0/T_0^*)_1$ is 0.8485. Using this and the previously obtained value in Eq. (i), we find that the value of $(T_0/T_0^*)_2$ is 1.125. For the exit to be choked, the value should be 1.000. Assume a flow of 0.175 and repeat the calculation:

$$M_1 = 0.54, \quad \left(\frac{T_0}{T_0^*}\right)_1 = 0.7302, \quad T_{0,2} - T_{0,1} = 198°,$$

$$\frac{T_{0,2}}{T_{0,1}} = 1.375, \quad \left(\frac{T_0}{T_0^*}\right)_2 = 1.006,$$

which is as close as can be estimated with the accuracy of the calculations.

In effect, it has been assumed that the flow is choked at the exit, and to check this assumption, the exit pressure must be determined. From Table 12-6 for $M_1 = 0.54$,

$$\left(\frac{T}{T^*}\right)_1 = 0.846, \quad \left(\frac{p}{p^*}\right)_1 = 1.70, \quad \left(\frac{p_0}{p_0^*}\right)_1 = 1.098.$$

From the isentropic Table 12-1,

$$\frac{p}{p_0} = 0.820, \quad \frac{T}{T_0} = 0.945.$$

For the pipe exit, all values of the ratios will be unity, since the flow was calculated for choking. The pressures and temperatures can now be calculated at point 1 and at the exit:

$$p_{0,1} = 114.7 \text{ psia},$$
$$p_1 = 0.820(114.7) = 94 \text{ psia},$$
$$T_1 = 0.945(529) = 500°\text{R} = 40°\text{F},$$
$$T_2 = T_1 \frac{(T/T^*)_2}{(T/T^*)_1} = 500 \frac{1}{0.846} = 591°\text{R} = 131°\text{F},$$
$$p_2 = p_1 \frac{(p/p^*)_2}{(p/p^*)_1} = 94 \frac{1}{1.70} = 55.3 \text{ psia},$$
$$p_{0,2} = p_{0,1} \frac{(p_0/p_0^*)_2}{(p_0/p_0^*)_1} = 114.7 \frac{1}{1.098} = 104 \text{ psia}.$$

The pressure at the exit for the assumed choked flow is 55.3 psia, which is greater than the 14.7 psia outside the duct; consequently the assumption of choked flow was correct. If the outside pressure were higher than 55.3 psia, then the flow would not be choked, and it would be necessary to recalculate the problem until the pressure outside and at the duct exit were matched.

12-4.B Heat Transfer and Friction

The analysis of combined friction and heat transfer in a constant-area duct requires that some of the basic relations be rederived, and that several new definitions be introduced.

(a) Definitions. The relation between the stream temperature T, its stagnation temperature T_0, and the adiabatic wall temperature T_{aw} (temperature of fluid adjacent to the wall) is defined as the *recovery factor* or r:

$$r = \frac{T_{aw} - T}{T_0 - T}. \tag{12-62}$$

At low stream velocities, T_{aw} is not significantly affected by the stream velocity. At high velocities, T_{aw} may be quite different from T. The wall drag retards the boundary layer, so that $T_{aw} \to T_0$. However, two effects can cause T_{aw} to differ from T_0. The first is heat transfer from the boundary layer to the mainstream, which would be proportional to the thermal diffusivity $k/\rho c_p$; and the second is dissipation of kinetic energy of the mainstream, which would be proportional to the momentum diffusivity μ/ρ. The ratio of these two is the Prandtl number $c_p\mu/k$, which will give some indication of the relative value of the boundary layer temperature and the stagnation temperature. For example, if the Prandtl number is less than one, the thermal term is greater than the momentum term; more heat would be transferred from the boundary layer than would be dissipated to it, and the temperature would be less than stagnation. If the two effects were equal, the Prandtl number would be one, and T_{aw} would be expected to approach T_0 rapidly. From Eq. (12-62), this would mean a recovery factor of unity. The recovery factor can be combined with Eq. (12-18), and expressed as

$$r = \frac{T_{aw} - T}{U^2/2c_p} = \frac{(T_{aw}/T) - 1}{[(\gamma - 1)/2]M^2}$$

or

$$\frac{T_{aw}}{T} = 1 + r\frac{\gamma - 1}{2}M^2. \tag{12-63}$$

For flat-plate flow or for a constant-area circular duct, r is closely approximated by

$$r = \sqrt{c_p\mu/k}. \tag{12-64}$$

The heat transfer coefficient is defined by using the temperature driving force between the wall T_w and the boundary layer T_{aw}:

$$dQ = h(T_w - T_{aw})\, dA. \tag{12-65}$$

Other definitions lead to heat transfer coefficients which might at times be negative and which usually depend to a great extent on the rate of heat transfer.

Finally, we shall use the very simple Reynolds analogy between heat and momentum transfer in turbulent flow:

$$h/c_p J = f/2. \tag{12-66}$$

(b) The general equation. Equation (12-20) is valid for relating the stream temperature to its local stagnation value. For a differential length of pipe of perimeter b, Eq. (12-65) becomes

$$dQ = hb(T_w - T_{aw})\,dx. \tag{12-67}$$

For the same differential length, Eq. (12-51) can be written as

$$dQ = c_p w\,dT_0. \tag{12-68}$$

Combining Eqs. (12-20) and (12-63) with Eqs. (12-67) and (12-68), we obtain

$$dQ = hb\,dx\left\{T_w - T_0\left[\frac{2 + r(\gamma - 1)M^2}{2 + (\gamma - 1)M^2}\right]\right\} = c_p w\,dT_0. \tag{12-69}$$

A simple force balance on the fluid flowing in the system, or application of either the momentum balance (10-10) or the mechanical energy balance (10-49), can be rearranged to give us

$$-\frac{1}{\rho}\,dp = \tfrac{1}{2}\,dU^2 + \frac{2fU^2\,dx}{d_0}, \tag{12-70}$$

where the losses have been expressed as a Fanning friction factor. Combining Eq. (12-70) with the perfect gas law (12-3) and (12-42), the definition of the Mach number (12-14), (12-15), and (12-36), and the differential form of the continuity equation (12-40), we have

$$\frac{4f\,dx}{d_0} = 2\left(\frac{1 - \gamma M^2}{\gamma M^2}\right)\frac{dM}{M} - \frac{1 + \gamma M^2}{\gamma M^2}\frac{dT}{T}. \tag{12-71}$$

Equations (12-20) and (12-71) can be combined to give

$$\frac{4f\,dx}{d_0} = \frac{4(1 - M^2)}{\gamma M^2[2 + (\gamma - 1)M^2]}\frac{dM}{M} - \frac{1 + \gamma M^2}{\gamma M^2}\frac{dT_0}{T_0}. \tag{12-72}$$

For adiabatic flow, $dT_0 = 0$, since there is no heat transfer. Equation (12-72) reduces to the adiabatic formula of Eq. (12-45).

Eliminating dL between Eqs. (12-69) and (12-72), we have the final desired differential equation for friction and heat transfer in a constant area duct:

$$\frac{fc_p J\,dT_0}{h\left\{T_w - T_0\left[\dfrac{2 + r(\gamma - 1)M^2}{2 + (\gamma - 1)M^2}\right]\right\}} = \frac{4(1 - M^2)}{\gamma M^2[2 + (\gamma - 1)M^2]}\frac{dM}{M} - \frac{1 + \gamma M^2}{\gamma M^2}\frac{dT_0}{T_0}. \tag{12-73}$$

(c) Isothermal flow. In very long ducts, such as those found in gas transportation lines, the case of isothermal flow with friction is a close approximation.

The Mach number is usually low. However, large pressure drops do exist because of the long lines.

For isothermal flow, $dT = 0$, and Eq. (12–71) reduces to

$$\frac{4f\,dx}{d_0} = 2\left(\frac{1 - \gamma M^2}{\gamma M^2}\right)\frac{dM}{M}. \tag{12–74}$$

This equation can be integrated directly between the entrance and exit of the line:

$$\int_0^L \frac{4f\,dx}{d_0} = \int_{M_1}^{M_2}\left(\frac{2}{\gamma M^3} - \frac{2}{M}\right)dM,$$

$$\frac{4fL}{d_0} = \left.\frac{1}{\gamma M^2} - 2\ln M\right]_{M_1}^{M_2},$$

$$\frac{4fL}{d_0} = \frac{1}{\gamma M_1^2}\left(1 - \frac{M_1^2}{M_2^2}\right) - \ln\left(\frac{M_2}{M_1}\right)^2. \tag{12–75}$$

From Eq. (12–55), for isothermal flow,

$$\frac{p_2}{p_1} = \frac{M_1}{M_2}. \tag{12–76}$$

Combining this with Eq. (12–75), we obtain a relation derived by Lobo, Friend, and Skaperdas [4]:

$$\frac{4fL}{d_0} = \frac{1 - (p_2/p_1)^2}{\gamma M_1^2} - \ln\left(\frac{p_1}{p_2}\right)^2. \tag{12–77}$$

This equation can be solved by trial and error, but Lobo, Friend, and Skaperdas proposed a method of avoiding this. Later Thomson [10] presented a nomograph which gave a direct solution to the equation, and which is reproduced here as Fig. 12–20. An example is given by the dashed line. If $\gamma = 1.4$, this corresponds to $M_1 = 0.313$ and $4fL/d_0 = 3.43$, and the pressure ratio $p_2/p_1 = 0.638$.

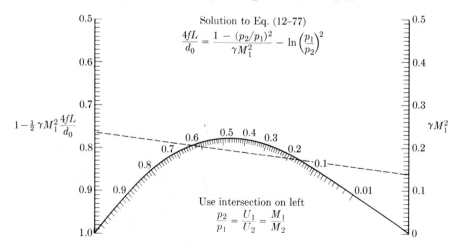

Fig. 12–20. Solution for Eq. (12–77). [*From* G. W. Thomson, Ind. Eng. Chem. **34**, 1485 (1942). *By permission.*]

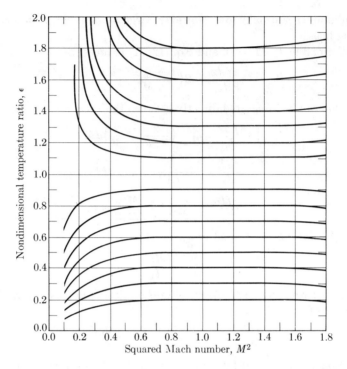

Fig. 12–21. *General isocline solution of Eq. (12–81) for nonadiabatic friction flow in tubes at uniform wall temperature.* [*From J. N. Nielson, NACA ARR No. 4 C16 (Oct. 1944). By permission.*]

Differentiating Eq. (12–20), and combining this with Eqs. (12–63), (12–67), and (12–68), we get

$$dx = \frac{c_p w}{hb} \frac{MT(\gamma - 1)\, dM}{T_w - T\{1 + r[(\gamma - 1)/2]M^2\}}. \quad (12\text{–}78)$$

Combining Eqs. (12–74) and (12–78), and using the Reynolds analogy (Eq. 12–66) and the hydraulic diameter, we obtain

$$\frac{T_w}{T} = \frac{\gamma(\gamma - 1)M^4}{1 - \gamma M^2} + r\frac{\gamma - 1}{2} M^2 + 1. \quad (12\text{–}79)$$

Equation (12–79) relates the wall temperature required for isothermal flow to the constant stream temperature and the Mach number. At low Mach numbers, the wall temperature must be greater than the stream temperature until the term $1 - \gamma M^2$ equals zero. In this region (M slightly less than $\sqrt{1/\gamma}$), the wall temperature must become extremely large, and impossibly large at $M = \sqrt{1/\gamma}$.

When $M > \sqrt{1/\gamma}$, the wall temperature will approach $-\infty$. Hence, in the region of M near $\sqrt{1/\gamma}$, it is impossible to maintain isothermal flow. Such flow is possible only over a narrow range where the Mach number is low. In the range of $M = \sqrt{1/\gamma}$, the flow will be more nearly adiabatic.

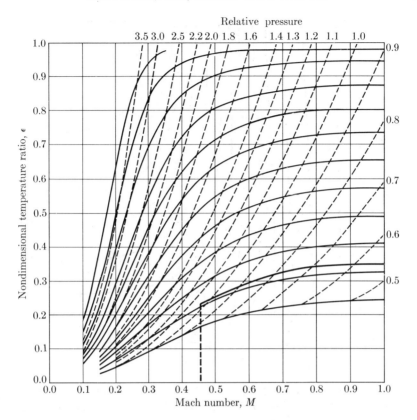

Fig. 12-22. *Isocline solution of Eq. (12-81) for air in a tube at a uniform wall temperature.* [*From J. N. Nielson*, NACA ARR No. 4 C16 (*Oct. 1944*). *By permission.*]

(d) Nonisothermal heat transfer. Equation (12-73) can be solved if the relation between T_w and one of the other variables is known. One case of prime importance is the *constant wall-temperature* heat exchanger. It will be assumed that the recovery factor is unity, and the Reynolds analogy is valid. The assumption that $r = 1$ implies that T_{aw} does not differ from T_0, which will be close if T_w is considerably in excess of T_{aw}. A new variable can be introduced:

$$\epsilon = \frac{T_0}{T_w}. \tag{12-80}$$

From Eqs. (12-73) and (12-80), we obtain

$$\frac{d\epsilon}{\epsilon(\epsilon - 1)} = \frac{(M^2 - 1)\,dM^2}{M^2\{1 + [(\gamma - 1)/2]M^2\}[\gamma M^2 + 1 + \epsilon(\gamma M^2 - 1)]}. \tag{12-81}$$

Equation (12-81) can be numerically integrated. Nielson's work [5] on this integration is presented in Figs. 12-21, 12-22, and 12-23 for $\gamma = 1.4$. Figure 12-21

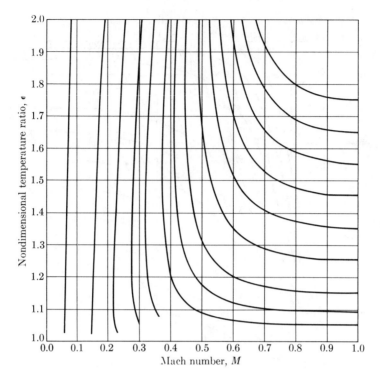

Fig. 12–23. Isocline solution of Eq. (12–81) for air in a tube at a uniform wall temperature. [From J. N. Nielson, NACA ARR No. 4 C16 (Oct. 1944). By permission.]

has two groups of curves; the lower, $\epsilon < 1$, is for heating and the upper, $\epsilon > 1$, is for cooling. Figure 12–22 is an amplification of the heating area of Fig. 12–21 plotted against Mach number rather than its square. Lines of constant

$$\sqrt{\epsilon}/M^2\{1 + [(\gamma - 1)/2]M^2\} \tag{12-82}$$

are also plotted (this term is the pressure). The value of p/p_0 is obtained from a ratio of final to initial values of Eq. (12–82). Figure 12–23 is similar to 12–22, but is for subsonic cooling.

Another useful relation can be obtained from Eqs. (12–67), (12–68), the Reynolds analogy (12–66), and $T_0 = T_{aw}$:

$$hb(T_w - T_{aw})\,dx = (fc_p J/2)b(T_w - T_{aw})\,dx = c_p w\,dT_0,$$

which reduces to $dT_0/(T_w - T_0) = (f/2)(b/S)\,dx$. Now $b/S = \pi d_0/(\pi d_0^2/4) = 4/d_0$, so that

$$\frac{dT_0}{T_w - T_0} = \frac{2f\,dx}{d_0}. \tag{12-83}$$

For constant T_w, Eq. (12–83) can be integrated to

$$\ln \frac{T_w - T_{0,1}}{T_w - T_{0,2}} = \frac{2f(L_2 - L_1)}{d_0}. \qquad (12\text{–}84)$$

This equation relates the pipe length to the conditions at any two points.

EXAMPLE 12–6. For the system of Example 12–5, determine the flow rate if friction and heat transfer are allowed. (a) Consider isothermal flow, and determine the heat load. (b) Consider a heat flow of 30,000 Btu/hr into the duct, and assume that the wall temperature can be maintained constant.

Answer. (a) From the analysis in the preceding problem, we would expect the flow to be choked at the exit. However, from the discussion on isothermal flow, it is clear that isothermal conditions cannot be maintained over the entire tube length. Therefore we shall assume that the flow is isothermal up to a Mach number given by $\sqrt{1/\gamma}$, and adiabatic from this point to the exit.

As a start, we can assume that f has the value estimated in Example 12–3, that is, 0.0031. However, we still cannot calculate the necessary $4fL/d_0$, since we do not know the length of the tube in isothermal flow.

We know the Mach number at the end of the isothermal section,

$$M_{2i} = \sqrt{1/\gamma} = 0.846, \qquad (i)$$

where the subscript i serves to remind us that this is the Mach number for the isothermal section. We assume that this value is the Mach number at the entrance to the adiabatic frictional flow section (M_{1a}), which will extend to the end of the tube (assumed for now to be choked). A $4fL_{\max}/d_0$ of 0.03920 is required to accelerate the flow from a M_{1a} of 0.846 to unity (from the Fanno analysis in Table 12-3). The $4fL/d_0$ corresponding to the assumed friction factor and the total length of pipe is 0.30499. The difference will be the isothermal section length:

$$\frac{4fL_i}{d_0} = 0.30499 - 0.03920 = 0.26579. \qquad (ii)$$

From Eq. (12–76), we have

$$\frac{p_{2i}}{p_{1i}} = \frac{M_{1i}}{\sqrt{1/\gamma}}. \qquad (iii)$$

Equation (12–77) becomes

$$\frac{4fL}{d_0} = \frac{1 - \gamma M_{1i}^2}{\gamma M_{1i}^2} + \ln \gamma M_{1i}^2. \qquad (iv)$$

Since we now know $4fL/d_0$, (Eq. ii), we can solve this equation numerically by trial and error or by the use of Fig. 12–20. By trial and error, we find that the value of M_{1i} is 0.611. From the chart constructed in Example 12–5 or from the equation, the flow rate is 0.192 lb/sec.

The Reynolds number must be determined in order to check the assumed friction factor of 0.0031. The viscosity can be obtained from the temperature. From Table 12-1, $(T/T_0)_{1i} = 0.93053$. Since

$$T_{0,1i} = 529.7°R, \quad T_{1i} = 494°R.$$

The section is isothermal, so the average temperature is 494°R or 34°F. The viscosity of air at this temperature is 0.0170 cps or 0.1143×10^{-5} lb/ft-sec. The Reynolds number is

$$N_{Re} = d_0 w/\mu S = 7.71 \times 10^5.$$

From the friction factor correlation for smooth pipe, we find that f is 0.0030, which is a reasonable check.

The heat load to maintain isothermal flow can be determined from

$$Q = c_p w \Delta T_0. \tag{v}$$

Since the flow is isothermal, T_{2i} is 494°R, and $T_{0,2i}$ can be calculated from Eq. (12–20) or Table 12-1, with M_2 given by (i): $T_{0,2i} = 494[1 + (0.2/1.4)] = 565°R$. Using Eq. (v), we get

$$Q = 0.24(0.192)(35.3)3600 = 5{,}851 \text{ Btu/hr}.$$

Finally, to check the assumption of choked flow, we must determine the pressure at the exit; we determine the pressure at the end of the isothermal section first. From the isentropic table, Table 12-1, we see that the pressure ratio at the entrance is $p_{1i}/p_{0,1i} = 0.77722$. Since $p_{0,1i}$ is 114.7 psia, $p_{1i} = 89.1$ psia. From (iii), we have

$$p_{2i} = \frac{p_{1i} M_{1i}}{\sqrt{1/\gamma}} = \frac{89.1(0.611)}{0.846} = 64.3 \text{ psia}.$$

This is the pressure at the entrance to the adiabatic friction section (p_{1a}) and can be used to obtain the exit pressure:

$$\frac{p_{1a}}{p_{2a}} = \frac{(p/p^*)_{1a}}{(p/p^*)_{2a}} = 1.2114, \quad p_{2a} = \frac{64.3}{1.2114} = 30.4 \text{ psia}.$$

Since the back pressure is 14.7 psia, the flow is choked as assumed, and the problem is solved. The flow from the end of the isothermal section to the exit is adiabatic, and thus requires no addition of heat.

(b) This part of the problem also involves making a number of assumptions and then checking. The assumption of $f = 0.0031$, $(4fL/d_0 = 0.30499)$ and choked flow appears to be reasonable. To use Fig. 12-22, we must have values of T_0 and of T_w. We know $T_{0,1}$, and if we assume a flow rate, we can estimate a value for $T_{0,2}$. We would expect the flow to be less than that with heat transfer alone. As a start, let us try 0.150 lb/sec. From the relation developed in Example 12-5, we have

$$T_{0,2} = T_{0,1} + (34.6/w) = 529.7 + 231 = 760.7°R.$$

We can estimate the value of T_w from Eq. (12–84):

$$T_w = 2210°R.$$

The ratio $(T_0/T_w)_2$ will enable us to use Fig. 12–22, since the flow has been assumed choked so that $M_2 = 1$:

$$\left(\frac{T_0}{T_w}\right)_2 = \frac{760.7}{2210} = 0.345.$$

This, together with the exit Mach number, can be used to establish the system flow line in Fig. 12–22. The ratio $(T_0/T_w)_1$ will enable us to establish M_1, from which we can determine the flow and check the assumed flow rate:

$$\left(\frac{T_0}{T_w}\right)_1 = \frac{529.7}{2210} = 0.24.$$

From the figure, the corresponding value of M_1 is 0.46. From the isentropic formula, the flow rate is 0.150, which is an excellent check for the assumed flow rate. If this were unsatisfactory, we would have to assume a new flow rate and repeat the calculation.

To check the assumed friction factor, we can determine the Reynolds number at the average temperature. From the isentropic table, the ratio of $(T/T_0)_1$ is 0.95940, and thus T_1 is 509°R. We can obtain T_2 in the same manner for $M_2 = 1$; its value is 634°R. The average temperature is 571°R or 111°F. For this temperature, the viscosity is 0.0187 cps or 0.126×10^{-5} lb/ft-sec. The Reynolds number is

$$N_{Re} = 5.5 \times 10^5.$$

The friction factor determined from the standard smooth tube correlation is 0.0033, which we shall assume to be a satisfactory check on the assumed value of 0.0031.

In order to check the assumption of choked flow, we must calculate the exit pressure. We can obtain this from Eq. (12–55), since we can determine p_1 to be 99.3 psia from the isentropic table, and since we know all the other terms. The value of p_2 is 52.9 psia, and since this is greater than 14.7 psia (the back pressure), the flow is choked.

We can compare the various problems solved as follows.

	w, lb/sec	M_1
Flow without friction, from Example 12–4	0.219	1
Flow with friction only, from Example 12–4	0.198	0.66
Flow with heat transfer only, from Example 12–5	0.175	0.54
Isothermal flow, from Example 12–6(a)	0.192	0.61
Flow with heat and friction, from Example 12–6(b)	0.150	0.46

The amount of heat necessary to maintain isothermal flow was less than that provided in the heat-transfer cases, so that we would expect an answer closer to the no-heat-transfer case. The lowest flow would be the flow under the combined effects of friction and heat transfer, and, as calculated, is very close to the sum of the individual effects; that is, $0.219 - (0.219 - 0.198) - (0.219 - 0.175) = 0.154$, which can be compared with the calculated value of 0.150 from Example 12–6(b).

12-5 SUMMARY AND COMMENTS

The approach used in our analysis of compressible flow was primarily the one-dimensional integral method discussed in Chapter 10. This is only an approximation to any real problem, but it is a reasonable one and is usually adequate for flow in ducts. We have avoided discussion of multidimensional problems and other complicating factors that would be required to analyze compressible flow over objects, interactions with and flow in boundary layers, oblique shocks, geometric variations in ducts that make a one-dimensional assumption inadequate, unsteady state problems, and the flow of real fluids. For the most part, these more complex problems are studied by the methods developed in the preceding chapters. In this text the only actual example we can cite is the one-dimensional steady flow of a viscous compressible fluid, in which we examined the shock-wave thickness of the normal shock (Section 7-5 and Example 7–4). Further examples are, or course, possible; i.e., if we assume that the Reynolds number is high enough for viscous forces to be restricted to a narrow boundary layer, then we can treat compressible flow over objects by an extension of the ideal-flow concepts presented in Chapter 6. Compressible boundary layer flow is a more complex matter, requiring treatment along the lines given in Chapter 9 for the exact boundary layer solutions and in Chapter 10 for the integral approximate solutions. Further references on the more complex aspects of compressible flow are Shapiro [9], Landau and Lifshitz [2], Pai [7], and Schlichting [8].

PROBLEMS

12–1. Figure 12–2(a) shows a projectile moving in still air. Assume the pressure to be atmospheric and the air temperature to be 70°F. (a) Calculate the Mach number of the projectile relative to the air and (b) calculate its velocity in ft/sec.

12–2. Calculate the velocity of sound in the following media: (a) air at 50°F, (b) hydrogen at 70°F, (c) water at atmospheric pressure and 70°F, (d) carbon dioxide at 40°F and 20 psia.

12–3. Show that the flow rate of Eq. (12–25) passes through a maximum.

12–4. Calculate the values found in Table 12-1 for Mach numbers of 0.50 and 2.00.

12–5. For a value of $p/p_0 > p^*/p_0$, compute curve (ii) of Fig. 12–6 and curve (ii) of Fig. 12–7.

12–6. Compute curve (iv) of Fig. 12–7.

PROBLEMS

12–7. Air flows through a pipe at a Mach number of 0.7 at a point at which the pressure is 20 psia and the temperature is 60°F. What would the temperature rise be on a small sphere immersed in the flow at this point?

12–8. Determine the flow of He, at 1000 psia and 60°F, through a convergent nozzle which has a throat whose diameter is $\frac{1}{4}$ in. The nozzle exhausts into the atmosphere.

12–9. An air stream flows at 0.1 lb/sec. in a duct whose diameter is 2 in. The stagnation temperature and pressure are measured and found to be 100°F and 25 psia, respectively. Calculate all the properties of the stream at a point at which the static pressure is 10 psia.

12–10. A diffuser has the following inlet conditions: $p_1 = 14.7$ psia, $U_1 = 300$ ft/sec, $T_1 = 500°$R, and $w = 100$ lb/sec. The area variation is such that the pressure varies linearly with axial distance. If U_2 at the exit is 100 ft/sec, find T_2, p_2, A_1, A_2. The fluid is air.

12–11. Calculate the values found in Table 12-2 for a Mach number of 2.0.

12–12. Show that point b in Fig. 12–10 occurs at the point where the Mach number is unity.

12–13. The conditions before a shock in air are: $p_1 = 10$ psia, $T_1 = 40°$F, $U_1 = 2000$ ft/sec. Find p_2, T_2, M_1, and M_2.

12–14. For a convergent-divergent nozzle designed for an isentropic Mach number of 2.0, calculate curves (iii), (iv), and (v) of Fig. 12–12. Assume reversible adiabatic flow before and after the shock.

12–15. Derive Eq. (12–45).

12–16. Compute $L_{\max}/D$ for $M = 0.5$ for the problem of Table 12-14.

12–17. Compute the values in Table 12-3 for a Mach number of 0.50 and 2.00.

12–18. Derive the Prandtl formula $M_1^* M_2^* = 1$ or $U_1 U_2 = c^{*2}$, which applies across a normal shock.

12–19. Derive Eq. (12–61).

12–20. Compute values for Table 12-6 for Mach numbers of 0.50 and 2.00.

12–21. A section of natural gas line, 24 in. in diameter, is 50 miles long. The gas enters the line from the first station at 100 psia and should feed the second station compressor at a pressure greater than 20 psia. Calculate the maximum permissible flow if the gas is at 70°F during its flow. Molecular weight is 18 and $\gamma = 1.3$.

12–22. Air enters a tube through a convergent-divergent nozzle at a Mach number of 1.5. Its stagnation temperature is 500°R and its stagnation pressure is 100 psia. The tube is being heated through the duct wall. Assume no friction and no shock. (a) What is the maximum amount of heat absorption? (b) What are the exit conditions?

12–23. On an H-S or T-S diagram, show the changes that are expected to occur for the following processes: (a) area change under adiabatic conditions, (b) friction without heat transfer, (c) both negative and positive heat transfer without friction, (d) heat transfer and friction. For each case note what happens for sonic, subsonic, and supersonic conditions. Indicate the type of line over which the change is taking place and note what happens to the stagnation terms.

12-24. Using the Bernoulli equation with a loss term,

$$dp/\rho + U\,dU + d(\text{loss}) = 0,$$

and other relations as necessary, show that for supersonic flow in a pipe, the velocity in the downstream direction must decrease.

12-25. For compressible flow with heat transfer and friction, derive the equation that relates the exit to inlet pressures of the duct to the exit and inlet Mach numbers and actual temperatures. What restrictions are there on this derived equation?

12-26. For a specific compressible flow with heat transfer and friction, it is found that the rate of heat transfer varies linearly from an initial maximum value at the inlet down to zero at the exit. How would you expect the term

$$\frac{T_{w,\max} - T_{0,1}}{T_{0,2} - T_{0,1}}$$

to vary with the friction parameter $4fL/D$?

REFERENCES

1. J. H. KEENAN, and E. P. NEUMANN, *J. Appl. Mech.* **13**, A91 (1946).
2. L. D. LANDAU, and E. M. LIFSHITZ, *Fluid Mechanics*, Addison-Wesley, Reading, Mass. (1959).
3. C. E. LAPPLE, *Trans. AIChE* **39**, 385–432 (1943).
4. W. E. LOBO, L. FRIEND, and G. T. SKAPERDAS, *Ind. Eng. Chem.* **34**, 821–823 (1942).
5. J. N. NIELSON, *NACA ARR L4C16* (Oct. 1944).
6. K. OSWATITSCH, and G. KUERTI, *Gas Dynamics*, Academic Press, New York (1956).
7. SHIH-I PAI, *Viscous Flow Theory-Laminar Flow*, D. Van Nostrand, Princeton, N. J. (1956).
8. H. SCHLICHTING, *Boundary Layer Theory*, McGraw-Hill, New York (1960).
9. A. H. SHAPIRO, *The Dynamics and Thermodynamics of Compressible Fluid Flow*, Vol. 1, The Ronald Press, New York (1953).
10. G. W. THOMSON, *Ind. Eng. Chem.* **34**, 1485 (1942).
11. K. S. PITZER, and L. BREWER, *Lewis and Randall-Thermodynamics*, McGraw-Hill, New York (1961).

PART 3

EXTENSIONS OF THE BASIC FLOW EQUATIONS

CHAPTER 13
INTRODUCTION

Part 2 of this text was concerned with exact and approximate solutions of the Navier-Stokes equation. Part 3 will emphasize the developments necessary to extend the analysis to problems beyond those of homogeneous Newtonian laminar flow. Solving these complex problems will usually be very difficult, and in a number of cases, solutions are impossible at the present time. Recourse will have to be made, therefore, to more approximate or empirical methods.

Turbulent flow (Chapter 14) is a major extension of the basic equations. To understand turbulence, it is important to understand the nature of the transition from laminar to turbulent flow, and so that is discussed first. The earlier attempts to formulate a simple mechanism for turbulence are treated, even though it has been established that these phenomenological models are generally inadequate. Because of a lack of a general solution to the turbulence problem, these methods are still essential for practical use in engineering analysis. The present hope for a more rational understanding of turbulence lies in the area of the statistical theory of turbulence. Because this theory holds promise of being a path to the solution of a number of perplexing problems, it is presented in detail. One of the problems to which it may hold the answer is the dependence of mixing on the turbulence within a mixing system. The principles of mixing on a macroscopic level have been known and used effectively for a number of years for the scale-up of mixing processes. However, due to the lack of information on mixing theory at a microscopic level, these scale-up methods have been somewhat empirical in nature. The ultimate goal is to estimate the degree of mixing from a knowledge of the properties of the components to be mixed and the turbulent characteristics of the mixing system. Some progress has been made on this particular problem, which is therefore discussed to illustrate the potential use of the statistical approach.

The development so far has been limited to the flow of Newtonian materials. However, many important substances are more complex in that the shear rate and shear stress are not simply related. The nature and classification of such non-Newtonian materials are discussed in the first part of Chapter 15. This is in the realm of *rheology*, which is the study of the deformation and flow of matter. A number of theoretical approaches are discussed in this same section (15-1), since one of the major problems has been the lack of adequate theoretical equations to describe flow characteristics. Equations with meaningful constants are needed because of the necessity of correlating the rheological properties with the fundamental structure of the material under consideration. Section 15-2 considers the flow characteristics of these materials and the means available for the treatment of data. The difference between Newtonian and non-Newtonian systems is emphasized for both laminar and turbulent flows. Among the areas of study covered are pipe and rotational flow, pressure drop correlations, the critical Reynolds number, and velocity profiles.

The remainder of the text is devoted to a discussion of multiphase flow. Chapter 16 treats the problems associated with liquid and gas systems under flow conditions, and discusses the types of flow experienced, the nature of the interface, the prediction of pressure loss, and the effects of added heat and mass transfer. For the most part, the descriptive equations and the associated boundary conditions in this area are so complex that overall or empirical methods are often used. Bubble and drop dynamics are considered in detail in Chapter 17, from the simple movement of a small single drop or bubble, to its breakup and ultimate atomization into a distribution of sizes. Such thorough treatment is necessary because of the extreme importance of this multiphase flow to the unit operations of chemical engineering (i.e., adsorption, distillation, spray drying, etc). Finally, in Chapter 18, the problems of solid–gas and solid–liquid flows are treated. This is the subject area of solids transport and fluidization. The nature and characteristics of particulate and aggregate motions are studied, as well as descriptive models necessary for any analytical approach.

Since the notation for the subsequent chapters is quite specific in some cases, a separate nomenclature list will be found at the end of each chapter.

CHAPTER 14

TURBULENCE AND MIXING

The Navier-Stokes equation of motion should be valid for turbulent flow, since the size of the smallest eddy is generally much greater than the mean free path of the molecules in the system. The turbulent motion can be considered macroscopic rather than molecular. The difficulty in applying the equations directly to the turbulence problem lies in the fact that the variables in the equations refer to instantaneous values at the point under consideration. These values vary to such a degree that little information can be gained by direct application of the basic theory, and thus some form of modification or extension is necessary. Reynolds modified the Navier-Stokes equation so that the variables would be time averages. When this is done, however, additional terms are introduced to account for the fluctuations in the flow. Two main areas have been studied: The first is the study of the equations as they were before modification. The purpose of this study is to gain some insight into the stability of laminar flow and to estimate the critical Reynolds number. The second area is concerned with an understanding of, and the obtaining of expressions for, the additional terms introduced in Reynolds' modification of the basic equations. This involves two main subdivisions: phenomenological theories and statistical turbulence. Mixing is an application of the latter.

14-1 STABILITY

The first experiments on the laminar-turbulent transition were made by Reynolds [197]. In these experiments, a fluid at rest in a tank was allowed to flow through a glass pipe. A thin stream of dye was injected at a point in the tank, and the motion of the dye was observed as it moved into the pipe and downstream with the fluid. At low velocities the dye moved in a straight line along the tube, indicating laminar flow. As the velocity was increased, the dye line became thinner and began a wave or sinuous motion. A further increase in velocity caused the dye line to break up into segments, or what is pictured as turbulent eddies.

The velocity at which laminar flow no longer exists is called the *upper critical Reynolds number*. When experimenters have taken elaborate precautions to still the fluid in the tank and to eliminate any disturbances at the entrance to the pipe (conditions made possible by the use of a bell-shaped entrance), an upper critical Reynolds number as high as 40,000 has been obtained. Taylor [241], using similar precautionary methods, observed a value of 32,000, even when there was a local bending of the pipe such that lateral velocities of one or two percent of the mean velocity existed. Reynolds, in his work, was able to observe a value of 12,000. A value of about 13,000 has been observed in rough tubes. The lowest value observed for this upper critical number is about 2300.

Consider now an experiment in which the flow is turbulent and the velocity is to be decreased. At some velocity the eddies will die out and the flow will become laminar; this point is called the lower critical velocity and corresponds to the *lower critical Reynolds number*. This number has been found to have a numerical value of about 2300, and does not depend on the roughness of the pipe [197].

Davies and White [57] have suggested a third critical Reynolds number, below which eddies in a fluid as it enters a pipe are not transmitted along the pipe. An approximate value for pipe flow is in the range of 200 to 400. Above this Reynolds number, the eddies will enter the pipe, but of course will die out.

More recent studies of the process of transition from laminar to turbulent flow abound in the literature. All of these attempt in some way to shed light on the mechanism of the transition. Out of this wealth of information has come the following qualitative picture of the process: The transition from laminar flow and the spread of turbulence to produce the final, fully developed turbulent flow are essentially continuous; i.e., the process is composed of a number of developing steps and is not a sudden single catastrophic change. In general, the transition process can be pictured as occurring in four steps: first, small two-dimensional waves are formed and are linearly amplified; second, the two-dimensional waves develop into finite three-dimensional waves and are amplified by nonlinear interactions (a three-dimensional disturbance usually starts with this step); third, a turbulent spot forms at some localized point in the flow; and finally, the turbulent spot propagates until it fills the entire flow field with turbulence.

The first step, the amplification of small two-dimensional waves, is one of the main areas of study described at the beginning of the chapter; i.e., the study of the Navier-Stokes equation before modification by time-averaging. For small disturbances or perturbations, the formulation of the physical problem in terms of differential equations is not difficult; however, the mathematical analysis necessary for the solution is usually formidable, and has been solved only for a few special flow cases. Theoretical efforts to solve the laminar instability problem have been extensively reviewed by Lin [151] and Schlichting [220]. Of particular interest is the case of flow between concentric rotating cylinders. The mathematical solution to the problem and the experimental confirmation of the nature of the predicted flow is a classic work in the field of stability. We shall not go into the mathematical details of Taylor's analysis [239]. Instead, we shall give a brief description of the problem and results: The steady flow between the cylinders is characterized by the solution already presented, which resulted in Eq. (7–32). The velocity terms are represented by an average and a fluctuating component. When we substitute these into the equations of motion, we obtain

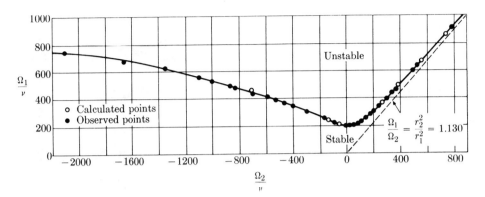

Fig. 14–1. Comparison between observed and calculated speeds at which instability first appears; $r_1 = 3.80$ cm and $r_2 = 4.035$ cm. [From G. I. Taylor, Phil. Trans. Roy. Soc. London 223A, 289–343 (1923). By permission.]

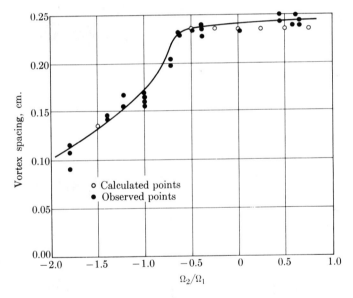

Fig. 14–2. Comparison of observed and predicted spacing of vortices for various relative velocities; $r_1 = 3.80$ cm and $r_2 = 4.035$ cm. [From G. I. Taylor, Phil. Trans. Roy. Soc. London 223A, 289–343 (1923). By permission.]

a series of equations for the small disturbances. We can then represent the disturbances as exponential functions and combine them with the small disturbance equations. With further mathematical manipulation, we can reduce the equations to two ordinary differential equations, which can be solved, although with difficulty. Figures 14–1 and 14–2 give plots of the calculated and experi-

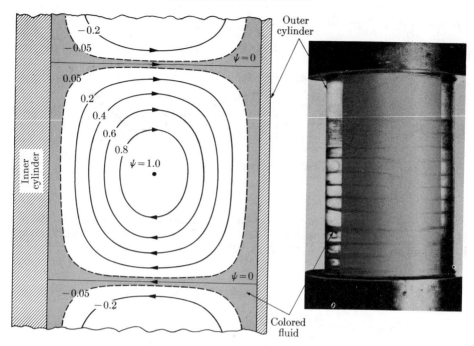

Fig. 14–3. Streamlines of motion after instability has set in: rotation in same direction. [From G. I. Taylor, Phil. Trans. Roy. Soc. London **223A**, 289–343 (1923). By permission.]

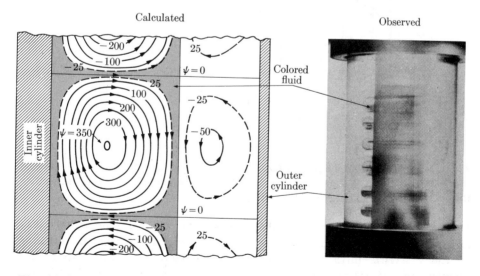

Fig. 14–4. Streamlines of motion after instability has set in: rotation in opposite directions. [From G. I. Taylor, Phil. Trans. Roy. Soc. London **223A**, 289–343 (1923). By permission.]

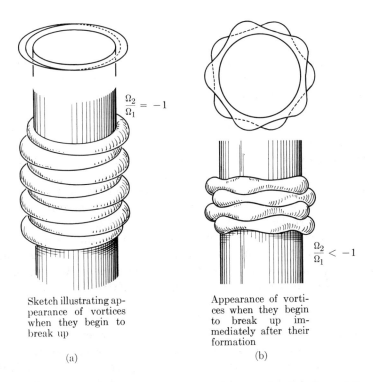

Fig. 14–5. *Inner and outer cylinders rotating in opposite directions.* [*From G. I. Taylor*, Phil Trans. Roy. Soc. London **223A**, *289–343 (1923). By permission.*]

mental results, and Figs. 14–3 and 14–4 show sketches and photographs from the same paper [239]. All these testify to the correctness of the basic instability theory for this problem. Figure 14–5 contains additional sketches describing the breakup of the observed vortices when the cylinders rotate in opposite directions. Of particular interest is the case of lowest stability; i.e., when the inner cylinder is rotated and the outer is at rest. For this limiting case,

$$\frac{\pi^4 \nu^2 (r_1 + r_2)}{2\Omega_1 r_1^2 (r_2 - r_1)^3} = 0.0571 \left(1 - 0.652 \frac{r_2 - r_1}{r_1}\right) + 0.00056 \left(1 - 0.652 \frac{r_2 - r_1}{r_1}\right)^{-1}.$$

It should be emphasized that Taylor used the growth of an infinitesimal disturbance to arrive at the instability just described. This instability is between two forms of laminar flow; turbulence results at higher values of relative velocity.

The possibility of instability in the boundary layer was proposed by Prandtl as early as 1912. The problem is of particular interest because of conflicting theories and experiments, which have now all been rationalized. The theoretical work of Tollmien and Schlichting for the solution of the resulting Orr-Sommerfeld equation showed that unstable waves could exist if the Reynolds number were greater than 420 (575 in Schlichting's analysis). The Reynolds number was

defined as
$$U_\infty \delta^*/\nu,$$
where U_∞ is the free-stream velocity and δ^* is the displacement thickness. Taylor [235] developed an alternate theory, which assumed that the transition is caused by a momentary separation due to an adverse pressure gradient associated with the free-stream turbulence.

In most of the early experimental work, the Tollmien-Schlichting instability was not observed; transition occurred quite abruptly without any preliminary oscillations. The experimental verification of the Tollmien-Schlichting waves by Schubauer and Skramstad [65, 217–219] pointed out the difficulty in the earlier experiments, which were conducted in wind tunnels with relatively high free-stream turbulence. In contrast, the later experiments were carried out in a tunnel with a turbulence level of less than 0.1% of the mean speed. It is now clear that when the free-stream turbulence is less than this value, the transition is preceded by the Tollmien-Schlichting instability; when greater, the transition follows the mechanism of the Taylor theory. The minimum Reynolds number for the occurrence of the Tollmien-Schlichting instability was found to be very close to 420.

Theoretical studies have been restricted for the most part to the two-dimensional aspects of boundary layer flows and flow between parallel boundaries. However, Squire [227] showed that one could transform the differential equation governing small three-dimensional disturbances into the corresponding two-dimensional problem. The net result was that the three-dimensional case, when transformed, was always more stable than it would have been if one had started with the two-dimensional case. The corresponding attempts to establish the two-dimensional instability in pipe flow have been unsuccessful (all have resulted in stability). It appears that pipe flow is stable to any two-dimensional perturbations. These solutions have been discussed and extended by Spielberg and Timan [225], who implied that the corresponding transformation by Squire for boundary layers does not exist for pipe flow and that therefore the stability to two-dimensional waves does not imply three-dimensional stability. Apparently some three-dimensional or finite disturbance is necessary to trigger the transition process. The experimental observation of the existence of laminar pipe flow at very high Reynolds numbers bears out these theoretical results.

The second stage of the transition process, which involves the formation of three-dimensional disturbances, has received much less attention, and investigations have been primarily directed toward an experimental understanding of the problem. The main studies on the boundary layer can be found in references [89–91, 115, 116, 126, 216]. Pipe flow studies have been much fewer in number [153–157, 255, 256]. The article by Morkovin [174] includes a general review of these problems. The exact nature of the three-dimensional picture is still unknown. Although a great deal has been learned from the above studies, a completely consistent picture is still lacking.

In the boundary layer the very small two-dimensional Tollmien-Schlichting waves apparently develop into three-dimensional arrangements in which, in the spanwise direction [116], there are valleys and peaks in the intensity of the waves. These waves are in some way related to the vortex loops described by Hama[90]. As the disturbance grows, the velocity distribution in the boundary layer is altered. At a certain time in the process, and at a certain location, the disturbance breaks down into the beginnings of a turbulent spot. There are strong indications [21] that the point of breakdown coincides with the point at which the rate of change in the kinetic energy of the mean velocity is at its maximum. Recently, however, there have been alternative explanations [268, 272] of the relationship between the mean velocity profile and the point at which the disturbances develop. Turbulent spots were first discussed by Emmons [67], and later in more detail by Schubauer and Klebanoff [216]. It might be that a similar situation exists in pipe flow, and, though there are few quantitative data to establish this, the information that is available [21] does provide additional support of the premise.

The final stage of the transition, which is the propagation and spread of the turbulent spot, has been considered by Schubauer and Klebanoff [216] for the boundary layer, and by Lindgren [153–157] and Rotta [209] for pipe flow. The results are discussed by Brodkey, Corrino, and Sanganhi [21], and in more detail by Coles [32].

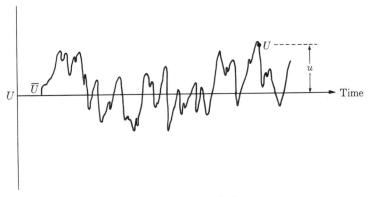

Fig. 14–6. Velocity in turbulent motion.

14-2 THE REYNOLDS EQUATION FOR TURBULENT MOTION

The Navier-Stokes equation should be valid for turbulent flow; however, as already noted, the equation would apply to the instantaneous velocity and would be most difficult to treat analytically. Thus it is necessary to use some statistical average and a measure of the deviation from that average. Figure 14–6 suggests the nature of the problem. If $U = iU_x + jU_y + kU_z$ is a point velocity in space, and if one can assume a time average to be the true average of U, then the instantaneous velocity can be written as a sum of its average and the instantaneous deviation from that average; that is,

$$\boldsymbol{U} = \bar{\boldsymbol{U}} + \boldsymbol{u}, \tag{14-1}$$

where $\bar{\boldsymbol{U}} = \boldsymbol{i}\bar{U}_x + \boldsymbol{j}\bar{U}_y + \boldsymbol{k}\bar{U}_z$ is the mean velocity at the point and $\boldsymbol{u} = \boldsymbol{i}u_x + \boldsymbol{j}u_y + \boldsymbol{k}u_z$ is the relative motion as a superimposed fluctuation on the mean.

Reynolds [198] formulated certain rules of approximation in calculation of mean values. They are not universally rigorous, but are good approximations when the fluctuations are sufficiently numerous and are distributed at random. These rules are: Quantities which have already been averaged may be considered as constants in subsequent averaging:

$$\bar{\bar{A}} = \bar{A}, \qquad \overline{\bar{A}B} = \bar{A}\bar{B}. \tag{14-2}$$

Averaging obeys the distributive law, $\overline{A + B} = \bar{A} + \bar{B}$, and

$$\overline{\partial A/\partial x} = \partial \bar{A}/\partial x. \tag{14-3}$$

From these rules, $\bar{A} = \overline{\bar{A} + a} = \bar{\bar{A}} + \bar{a} = \bar{A} + \bar{a}$; therefore

$$\bar{a} = 0. \tag{14-4}$$

Furthermore,

$$\overline{AB} = \overline{(\bar{A} + a)(\bar{B} + b)} = \overline{\bar{A}\bar{B}} + \overline{\bar{A}b} + \overline{\bar{B}a} + \overline{ab} = \bar{A}\bar{B} + \overline{ab}. \tag{14-5}$$

If A and B are vectors, such as velocity, then the term $\overline{\boldsymbol{U}\boldsymbol{U}}$ involves nine components, one of which is

$$\overline{U_x U_y} = \bar{U}_x \bar{U}_y + \overline{u_x u_y}.$$

The quantity $\overline{u_x u_y}$ is not necessarily zero, even though $\bar{u}_x$ and $\bar{u}_y$ are zero.

For a compressible fluid, the equation of continuity is

$$\partial \rho / \partial t = -(\boldsymbol{\nabla} \cdot \rho \boldsymbol{U}). \tag{3-19}$$

Combining this with Eq. (14-1) for the velocity and a similar equation for the density fluctuation ($\rho = \bar{\rho} + \tilde{\rho}$), we average both sides according to Eqs. (14-2), (14-3), and (14-4), and obtain

$$\partial \bar{\rho} / \partial t = -(\boldsymbol{\nabla} \cdot \bar{\rho} \bar{\boldsymbol{U}}) - (\boldsymbol{\nabla} \cdot \overline{\tilde{\rho} \boldsymbol{u}}). \tag{14-6}$$

The last term on the right is called the *turbulent impulse*. If the flow is incompressible, $\tilde{\rho} = 0$, $\bar{\rho}$ is constant, and

$$(\boldsymbol{\nabla} \cdot \bar{\boldsymbol{U}}) = 0. \tag{14-7}$$

This is the same as the point-continuity equation for incompressible flow, but it is now applied to the mean velocity at that point. The fluctuations in incompressible flow also obey the equation of continuity; this can be seen by

$$(\nabla \cdot U) = (\nabla \cdot \bar{U}) + (\nabla \cdot u) = 0,$$

from which Eq. (14-7) can be subtracted to give

$$(\nabla \cdot u) = 0. \tag{14-8}$$

14-2.A The Equation

The equation of motion for the flow of an incompressible fluid can be obtained from Eq. (4-6) or Eq. (4-12); however, a useful form is that analogous to Eq. (3-23), and is

$$\rho \frac{DU}{Dt} = \rho \frac{\partial U}{\partial t} + \rho (\nabla \cdot UU) = -\nabla p + \mu \nabla^2 U + \sum_s \rho_s F_s. \tag{14-9}$$

Each point property in this equation is replaced by its average and a fluctuating component, after which the entire equation is averaged. For the pressure, we use $p = \bar{p} + \tilde{p}$. After applying the Reynolds rules of averaging, we obtain the Reynolds equation of turbulent motion, which is

$$\rho \frac{D\bar{U}}{Dt} = \rho \frac{\partial \bar{U}}{\partial t} + \rho (\bar{U} \cdot \nabla) \bar{U} = -\nabla \bar{p} + \mu \nabla^2 \bar{U} + \sum_s \rho_s \bar{F}_s - (\nabla \cdot \rho \overline{uu}), \tag{14-10}$$

or, in cartesian tensor notation,

$$\rho \frac{D\bar{U}_i}{Dt} = \rho \frac{\partial \bar{U}_i}{\partial t} + \rho \bar{U}_j \frac{\partial \bar{U}_i}{\partial x_j} = -\frac{\partial \bar{p}}{\partial x_i} + \mu \frac{\partial^2 \bar{U}_i}{\partial x_j \partial x_j} + \sum_s \rho_s \bar{F}_{si} - \frac{\partial \overline{\rho u_i u_j}}{\partial x_j}. \tag{14-11}$$

These equations have the same form as the original equation, except that average properties now appear in place of point properties, and an additional term is added which is associated with the fluctuations. The Reynolds equation for incompressible turbulent motion cannot be solved, for there are more unknowns than there are available equations. The unknowns are the mean pressure, the three average velocity components, and the added fluctuation terms. It is interesting to note that the original equation was determinate, and that we are paying quite a penalty for the use of averages.

14-2.B Reynolds Stress

For incompressible flow, the additional term in Eq. (14-10) has nine components (six of which are different). The specific form which is called the *Reynolds, eddy,* or *turbulent stress* is

$$\rho \overline{uu}. \tag{14-12}$$

In cartesian coordinates, this is

$$\rho \begin{pmatrix} \overline{u_x^2} & \overline{u_x u_y} & \overline{u_x u_z} \\ \overline{u_y u_x} & \overline{u_y^2} & \overline{u_y u_z} \\ \overline{u_z u_x} & \overline{u_z u_y} & \overline{u_z^2} \end{pmatrix}. \tag{14-13}$$

The Reynolds stress can be considered a modification of the pressure tensor. For incompressible flow (from Eq. 3–43), this becomes

$$\bar{\bar{\tau}} = \bar{\bar{\tau}}_L + \bar{\bar{\tau}}_T = -\mu[(\nabla \bar{U}) + \widetilde{(\nabla \bar{U})}] + \rho \overline{uu}. \tag{14-14}$$

In cartesian tensor notation, the form follows that used in Eq. (3–42):

$$\bar{\tau}_{ij} = -\mu\left(\frac{\partial \bar{U}_i}{\partial x_j} + \frac{\partial \bar{U}_j}{\partial x_i}\right) + \rho \overline{u_i u_j}. \tag{14-15}$$

For compressible flow the problem is more complex, since the variation of density must be considered. The procedure is exactly the same, and the results can be expressed in the same manner (see Pai [188]). In the final equations, the turbulent impulse and third-order terms are introduced. For zero bulk viscosity, the Reynolds stress becomes

$$\bar{\bar{\tau}} = -\mu[(\nabla \bar{U}) + \widetilde{(\nabla \bar{U})}] + \tfrac{2}{3}\mu(\nabla \cdot \bar{U})\mathbf{U} + \bar{\rho}\overline{uu} + \bar{U}\overline{\rho u} + \overline{\tilde{\rho}uu}. \tag{14-16}$$

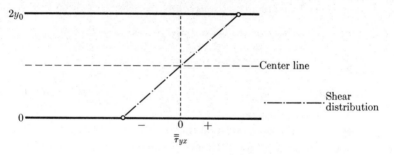

Fig. 14–7. Shear distribution for flow between two parallel walls.

14-2.C Integration of the Reynolds Equation

For flow between two walls or for pipe flow, the Reynolds equation can be partially integrated. Let us consider the simpler case of channel flow. The flow is in the x-direction and the distance across the flow, $2y_0$, is measured from one wall (see Fig. 14–7). The velocities of flow are

$$\bar{U}_x = f(x, y) \quad \text{and} \quad \bar{U}_y = \bar{U}_z = 0.$$

Here u_x, u_y, and u_z are functions of y only. The flow is assumed incompressible. Equation (14–7) thus becomes

$$\partial \bar{U}_x / \partial x = 0,$$

which shows that $\bar{U}_x$ is a function of y only, and not of x. From the equation of turbulent motion (14–10) for the z-direction, we obtain

$$\partial (\overline{u_y u_z}) / \partial y = 0.$$

Upon integrating, we find that

$$\overline{u_y u_z} = \text{const} = 0, \qquad (14\text{-}17)$$

since $\overline{u_y u_z} = 0$ at the wall.

The equations of motion for the x- and y-directions are

$$\frac{1}{\rho}\frac{\partial \bar{p}}{\partial x} - \nu \frac{d^2 \bar{U}_x}{dy^2} + \frac{d\overline{u_x u_y}}{dy} = 0, \qquad (14\text{-}18)$$

$$\frac{1}{\rho}\frac{\partial \bar{p}}{\partial y} + \frac{d\overline{u_y^2}}{dy} = 0. \qquad (14\text{-}19)$$

Differentiating Eq. (14–19) with respect to x yields

$$\frac{1}{\rho}\frac{d^2 \bar{p}}{dx\, dy} + \frac{d^2 \overline{u_y^2}}{dx\, dy} = 0.$$

Since the fluctuating velocities are not a function of x, the second term is zero. Thus

$$d^2\bar{p}/dx\, dy = 0,$$

and $d\bar{p}/dx$ is not a function of y. With this information, we can integrate Eqs. (14–18) and (14–19).

Equation (14–18) can be rearranged to give

$$\frac{1}{\rho}\frac{d\bar{p}}{dx} + \frac{d}{dy}\left(\overline{u_x u_y} - \nu \frac{d\bar{U}_x}{dy}\right) = 0. \qquad (14\text{-}20)$$

Integrating gives us

$$\frac{1}{\rho}\frac{d\bar{p}}{dx} y + \overline{u_x u_y} - \nu \frac{d\bar{U}_x}{dy} + C_1(x) = 0. \qquad (14\text{-}21)$$

The constant of integration is $-(1/\rho)(d\bar{p}/dx) y_0$, as determined from the boundary condition at the center line; that is, at $y = y_0$, by axial symmetry $\overline{u_x u_y} = 0$ and $(d\bar{U}_x/dy) = 0$. The boundary condition at $y = 2y_0$ is

$$\overline{u_x u_y} = 0, \qquad \overline{u_y^2} = 0,$$

and

$$\tau_w = -\mu \left(\frac{d\bar{U}_x}{dy}\right)_w \quad \text{or} \quad U^{*2} = -\nu \left(\frac{d\bar{U}_x}{dy}\right)_w,$$

where $U^{*2} = \tau_w/\rho$ by definition, and is called the *friction velocity*. When this boundary condition is combined with Eq. (14–21) [with $C_1(x)$ as evaluated], we obtain

$$\frac{1}{\rho}\frac{d\bar{p}}{dx} = -\frac{U^{*2}}{y_0}. \qquad (14\text{-}22)$$

Equations (14–21) and (14–22) can be combined to give the partially integrated equation for flow in the main stream direction:

$$-\left(\frac{y_0 - y}{y_0}\right)U^{*2} = -\nu\frac{d\bar{U}_x}{dy} + \overline{u_x u_y} \qquad (14\text{–}23)$$

or

$$-\rho\left(\frac{y_0 - y}{y_0}\right)U^{*2} = -\mu\frac{d\bar{U}_x}{dy} + \rho\overline{u_x u_y}.$$

This equation cannot be integrated further without some knowledge of the dependence of $\overline{u_x u_y}$ on y.

Equation (14–22) can be integrated to

$$\frac{\bar{p} - p_0}{\rho} = -\frac{U^{*2}x}{y_0} + C(y), \qquad (14\text{–}24)$$

where p_0 is the pressure at the wall at $x = 0$. At this point $\bar{p} = p_0$, and the constant of integration must be zero. Integration of Eq. (14–19) gives us

$$(\bar{p} - p_0)/\rho + \overline{u_y^2} + C_2(x) = 0. \qquad (14\text{–}25)$$

Combining Eq. (14–24) [with $C(y) = 0$] and Eq. (14–25) gives us

$$-U^{*2}x/y_0 + \overline{u_y^2} + C_2(x) = 0,$$

which, when combined with the boundary condition at $y = 0$, results in

$$C_2(x) = U^{*2}x/y_0.$$

This, together with Eq. (14–25), gives the y-direction equation as

$$-\frac{x}{y_0}U^{*2} = \frac{\bar{p} - p_0}{\rho} + \overline{u_y^2}. \qquad (14\text{–}26)$$

This equation shows that the mean pressure decreases with an increase in x. At a given cross section, the pressure in the fluid ($\bar{p}$) is less than at the wall (p_0), since $\overline{u_y^2}$ is positive.

Equation (14–23) is a simplified form of Eq. (14–15); that is, from Eq. (14–15), we have

$$\bar{\bar{\tau}}_{yx} = -\mu(\partial\bar{U}_x/\partial y) + \rho\overline{u_x u_y}. \qquad (14\text{–}27)$$

The linear variation of stress in the channel (see Fig. 14–7) is

$$\bar{\bar{\tau}}_{yx} = -[(y_0 - y)/y_0]\tau_w \qquad (14\text{–}28)$$

(that is, when $y = 2y_0$, $\bar{\bar{\tau}}_{yx} = \tau_w$). When this is combined with the definition of the friction velocity, we have

$$\bar{\bar{\tau}}_{yx} = -[(y_0 - y)/y_0]\rho U^{*2}. \qquad (14\text{–}29)$$

Combined with Eq. (14–27), this gives us Eq. (14–23) directly. Thus the derivation in this section supports the previously suggested representation of $\bar{\bar{\tau}}$ as a sum of the laminar and turbulent contributions.

14-3 PHENOMENOLOGICAL THEORIES

In order to obtain a solution to the partially integrated Reynolds equation for the mean flow direction (14–23), some functional dependency for the Reynolds stress $\rho \overline{u_x u_y}$ must be established. The aim of the phenomenological theories is to obtain such an expression directly. This is usually accomplished by assuming some simple mechanism for turbulence, and deriving from this the desired relation. A number of theories have been proposed and a few of these will be briefly discussed here; more extensive discussions can be found in references 96, 188, 220. The theories are all useful for the prediction of the mean velocity profile, and thus are important for the solution of a number of practical problems. But no matter what the mechanistic assumption, they all describe the velocity distribution equally well, and so they cannot be used to establish the mechanism of turbulent motion or provide insight into it.

14-3.A Boussinesq's Theory

In turbulence, two coefficients of viscosity can be defined: one real and the other apparent. The first is the molecular coefficient μ, and the second is a molar coefficient. The molecular coefficient of viscosity for Newtonian materials is independent of the Reynolds number, boundaries, and position in the fluid, whereas the other is dependent on all of these. This hypothesis was first suggested by Boussinesq [18], who introduced the exchange coefficient or eddy viscosity $\bar{\bar{\epsilon}}$, defined by

$$\overline{u_x u_y} = -\bar{\bar{\epsilon}}(d\bar{U}_x/dy), \tag{14-30}$$

which is to be used to represent the turbulence contribution. Thus, from Eqs. (14–27) and (14–29) (or 14–23), we obtain

$$\bar{\bar{\tau}}_{yx} = -\mu \frac{d\bar{U}_x}{dy} - \rho\bar{\bar{\epsilon}}\frac{d\bar{U}_x}{dy} = -\frac{y_0 - y}{y_0}\rho U^{*2},$$

or, in a form analogous to Eq. (3–31),

$$\bar{\bar{\tau}}_{yx} = -(\nu + \bar{\bar{\epsilon}})\frac{d\rho\bar{U}_x}{dy} = -\frac{y_0 - y}{y_0}\rho U^{*2}. \tag{14-31}$$

Similar assumptions can be made for the transfer of heat and mass:

$$q_y = -(\alpha + \epsilon_H)\left(\frac{d\rho c_p \bar{T}}{dy}\right), \quad J_{sy} = -(D_s + \epsilon_s)\left(\frac{d\bar{\rho}_s}{dy}\right), \tag{14-32}$$

which are the one-dimensional counterparts to Eqs. (3–44) and (3–46). From Eqs. (14–14) or (14–15) and (3–44) and (3–46), one can see that the eddy viscosity must be a second-order tensor (that is, $\bar{\bar{\epsilon}}$ as defined in Eq. (14–30) is really $\bar{\bar{\epsilon}}_{yx}$), and the eddy diffusivities of heat and mass are vectors (that is, ϵ_H and ϵ_s of Eq. (14–32) are really ϵ_{Hy} and ϵ_{sy}). In general, the forms would be $\bar{\bar{\epsilon}}$, ϵ_H, and ϵ_s for momentum, heat and mass, respectively.

The concept of eddy diffusivity has been applied to many engineering problems involving heat and mass transfer in turbulent flow. Generally the method is not the most satisfactory, and a great deal of caution must be exercised so that improper assumptions will not be made. For example, the $\bar{\epsilon}$'s are constant as an exception rather than as a rule, and it is questionable whether one can in any practical case assume that the various eddy diffusivities are equal to each other. A summary of the problems involved in the analysis, and further, more specific references are given in the review article by Sleicher [223].

14-3.B Prandtl's Mixing-Length Theory

Little further progress can be made unless the dependency of the eddy viscosity or eddy stresses on the variables of the system can be established. Further work based on some mechanistic picture of turbulence is necessary. The first of these theories was by Prandtl [191], who expressed the eddy stresses in terms of the mean velocity by means of a length l which was characteristic of the degree of turbulence and was somewhat analogous to the mean free path of the kinetic theory of gases. The length l is taken to be the length of path of a mass of fluid before it loses its individuality by mixing with its neighbors, and it is called the mixing or mixture length. The difference between laminar flow and turbulent flow is explained as the difference between the exchange of individual molecules between layers and the exchange of whole groups of molecules. In this theory, Prandtl assumed that the momentum of a group of particles or fluid mass in one layer is transferred to another layer.

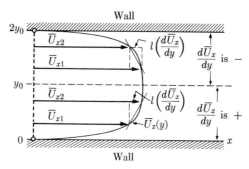

Fig. 14–8. Mixing length for flow between two parallel walls.

In Fig. 14–8, a mass of fluid is assumed to have a velocity $\bar{U}_{x1}$ in a stream. A velocity gradient exists in the direction perpendicular to the flow. Consider a mass of fluid to be displaced a distance l, in the $+y$ direction. The change in velocity is the difference between the velocity at the point of origin and that at its new position; to a first approximation, this is $l(d\bar{U}_x/dy)$. The fluctuation u_x experienced at point 2 is $-l(d\bar{U}_x/dy)$. The lateral motion u_y must be of the same order because of the assumed momentum conservation, and is taken to be proportional to u_x. The eddy stress becomes

$$\rho \overline{u_x u_y} = \rho l^2 (d\bar{U}_x/dy)^2. \tag{14-33}$$

A careful analysis of the signs of the fluctuations will indicate that the stress should be positive above the center line and negative below. Therefore it is more correct to write

$$\rho \overline{u_x u_y} = -\rho l^2 \left|\frac{d\bar{U}_x}{dy}\right|\frac{d\bar{U}_x}{dy}, \tag{14-34}$$

which is also in accord with the proper direction for momentum transfer to the walls. Equation (14-34) is Prandtl's equation for the turbulent stress in parallel flow. Equations (14-23) and (14-34) can be combined to give

$$l^2 \left|\frac{d\bar{U}_x}{dy}\right|\frac{d\bar{U}_x}{dy} + \nu \frac{d\bar{U}_x}{dy} = \frac{y_0 - y}{y_0} U^{*2}. \tag{14-35}$$

In order to solve this equation, one has to make some assumption about the mixing length, which is usually taken as directly proportional to the distance from the wall. In addition, Prandtl assumed that the shear stress is constant over the entire flow considered, and that the viscous contribution is negligible. Thus, for fully developed turbulent flow, the assumptions are $l = ky$ below the center line and $\bar{\tau}_{yx} = \tau_w$, which is equivalent to setting the right side of (14-35) to U^{*2}. Once these assumptions have been made, the equation may be written as $d\bar{U}_x/dy = U^*/ky$, which can be integrated to give

$$\bar{U}_x/U^* = (1/k) \ln y + C_1. \tag{14-36}$$

This is the logarithmic velocity distribution for turbulent flow, and, as will be shown later, it may be rearranged in the form

$$U^+ = (1/k) \ln y^+ + B, \tag{14-37}$$

where U^+ is $\bar{U}_x/U^*$ and $y^+ = yU^*/\nu$.

Equation (14-36) or (14-37) cannot satisfy the boundary condition at the wall ($\bar{U}_x = 0$) or at the center line ($d\bar{U}_x/dy = 0$). Another drawback is that from experimental work, k is about 0.4, making $l = 0.4y$ near the wall. The mixing length is not small when compared with the distance from the boundary, as it was originally assumed to be.

If viscous effects are not negligible and the assumption of constant shear stress is not made, Eq. (14-35) in nondimensional form becomes

$$l^{+2}\left|\frac{dU^+}{dy^+}\right|\frac{dU^+}{dy^+} + \frac{dU^+}{dy^+} + \frac{y^+}{y_0^+} - 1 = 0, \tag{14-38}$$

where

$$U^+ = \frac{\bar{U}_x}{U^*}, \quad y^+ = \frac{yU^*}{\nu}, \quad l^+ = \frac{lU^*}{\nu}, \quad \text{and} \quad y_0^+ = \frac{y_0 U^*}{\nu}.$$

Equation (14-38) can be solved for the velocity to give

$$U^+ = \int \left[\frac{-1 + \sqrt{1 + 4l^{+2}(1 - y^+/y_0^+)}}{2l^{+2}} \right] dy^+. \qquad (14\text{-}39)$$

This equation satisfies the boundary conditions at the wall and the center line for $l^+ = ky^+$. However, it is impossible to find a single value of k which will permit experimental velocity distributions to be matched both near the wall and in the area where Eq. (14-37) is valid. It is clear that the mixing length concept as discussed here cannot be used over the entire flow channel. Nevertheless, it is interesting to note that Eq. (14-39) can be put into asymptotic form at high Reynolds numbers ($y_0^+ \to \infty$):

$$U^+ = \int \frac{-1 + \sqrt{1 + 4l^{+2}}}{2l^{+2}} dy^+ = 2\int \frac{dy^+}{1 + \sqrt{1 + 4l^{+2}}}, \qquad (14\text{-}40)$$

and can be directly integrated for the case $l^+ = ky^+$ to give

$$U^+ = \frac{1 - \sqrt{1 + 4k^2 y^{+2}}}{2k^2 y^+} + \frac{1}{k} \ln[2ky^+ + \sqrt{1 + 4k^2 y^{+2}}], \qquad (14\text{-}41)$$

which is a form first suggested by Rotta [207].

14-3.C Taylor's Vorticity-Transport Theory

Taylor pointed out that because of pressure fluctuations there is no physical reason for assuming the conservation of momentum, and suggested that it is the moment of momentum or vorticity which is conserved. Assuming the viscous term of the Reynolds equation to be negligible, he wrote

$$d\bar{p}/dx = -\rho \overline{u_y \omega},$$

where ω denotes the fluctuation in vorticity obtained from $\Omega = \bar{\Omega} + \omega$. The expression for the mean value of vorticity is

$$\bar{\Omega} = -d\bar{U}_x/dy.$$

Assuming vorticity to be transferrable, he expressed it in terms of a mixing length and obtained

$$\frac{d\bar{p}}{dx} = -\rho |u_y| l_1 \frac{d\bar{\Omega}}{dy} = \rho |u_y| l_1 \frac{d^2 \bar{U}_x}{dy^2} = \rho l_\omega^2 \left| \frac{d\bar{U}_x}{dy} \right| \frac{d^2 \bar{U}_x}{dy^2},$$

where $l_\omega^2 = l l_1$. Since $d\bar{p}/dx = d\bar{\tau}_{yx}/dy$ for no viscous forces, a comparison with Prandtl's momentum transfer theory is possible. From Eq. (14-33) and the above, for $l \neq f(y)$,

$$\frac{d\bar{\tau}_{yx}}{dy} = \rho l^2 2 \left| \frac{d\bar{U}_x}{dy} \right| \frac{d^2 \bar{U}_x}{dy^2},$$

which differs from Eq. (14-41) by a factor of 2 only. The assumption of $l \neq f(y)$ is not valid in pipe flow, but should be reasonable in a free turbulent jet. Thus both theories will give identical velocity distributions for the turbulent jet. Predictions for heat or mass transfer will be different, however, with the experimental data tending to confirm Taylor's theory. In general, $l = f(y)$, and the results of the two theories are not the same. The particular vorticity transfer equation given above is correct only for two dimensions, but Taylor [240] developed a modified vorticity transport theory for three dimensions.

14-3.D Von Kármán's Similarity Hypothesis

Von Kármán's [110] approach was to consider the similarity of local flow patterns. He assumed, first, that viscosity is important only in the vicinity of the wall, and second, that the local flow pattern is statistically similar in the neighborhood of every point, with only the time and length scales variable. In parallel flow, the basic equations are identical to those of the Prandtl theory; however, the similarity hypothesis provides a different value for the mixing length because, by this hypothesis, one considers the ratio between like terms of a Taylor series expansion for the velocity at two points. The ratio between terms has the dimension of length and is taken to be proportional to the mixing length. Thus

$$\frac{(d\bar{U}_x/dy)_1}{(d^2\bar{U}_x/dy^2)_1} = \frac{(d\bar{U}_x/dy)_2}{(d^2\bar{U}_x/dy^2)_2} = \frac{l}{k}, \qquad (14\text{-}42)$$

where k is a universal constant. When Eqs. (14–34) and (14–42) are combined, the result is

$$\bar{\bar{\tau}}_{yx} = \rho k^2 \frac{(d\bar{U}_x/dy)^4}{(d^2\bar{U}_x/dy^2)^2},$$

or, in a form which accounts for the proper sign change,

$$\bar{\bar{\tau}}_{yx} = -\rho k^2 \left|\frac{(d\bar{U}_x/dy)^3}{(d^2\bar{U}_x/dy^2)^2}\right| \frac{d\bar{U}_x}{dy}. \qquad (14\text{-}43)$$

We can combine this equation with Eq. (14–29) and integrate it with the proper boundary conditions to obtain the velocity defect law, which takes the form

$$\frac{\bar{U}_{x,\max} - \bar{U}_x}{U^*} = f\left(\frac{y_0 - y}{y_0}\right), \qquad (14\text{-}44)$$

where $\bar{U}_{x,\max}$ is the maximum value of $\bar{U}_x$ (at y_0).

To illustrate, let us consider pipe flow in which r is the radius vector, y remains the distance from the wall (that is, y_0 is identically r_0), and z is the direction of flow (see Fig. 14–9). Equation (14–29) becomes

$$\bar{\bar{\tau}}_{rz} = \frac{r}{r_0}\tau_w = \left(1 - \frac{y}{r_0}\right)\tau_w, \qquad (14\text{-}45)$$

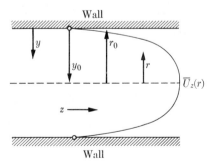

Fig. 14–9. Velocity distribution for pipe flow.

which, when combined with Eq. (14–43), gives us

$$\frac{d^2 \bar{U}_z/dy^2}{(d\bar{U}_z/dy)^2} = -\frac{k}{U^*}\frac{1}{\sqrt{1-(y/r_0)}}.$$

We can integrate this for $d\bar{U}_z/dy$, taking the boundary conditions as an infinite gradient at the wall:

$$d\bar{U}_z/dy = \frac{U^*}{2kr_0[1 - \sqrt{1 - (y/r_0)}]}. \qquad (14\text{--}46)$$

We can integrate the equation again for the velocity distribution by using a change in variable for $1 - (y/r_0)$, and by using the boundary condition of maximum velocity at the center line. The resulting equation is

$$\frac{\bar{U}_{z,\max} - \bar{U}_z}{U^*} = -\frac{1}{k}\left[\ln\left(1 - \sqrt{\frac{r}{r_0}}\right) + \sqrt{\frac{r}{r_0}}\right]. \qquad (14\text{--}47)$$

Equation (14–47) differs very little from the corresponding Prandtl equation, which is obtained by integrating Eq. (14–36) and using the center-line boundary condition. In terms of pipe radius, this result is

$$\frac{\bar{U}_{z,\max} - \bar{U}_z}{U^*} = -\frac{1}{k}\ln\left(1 - \frac{r}{r_0}\right). \qquad (14\text{--}48)$$

The best values for k are 0.36 and 0.4 for Eqs. (14–47) and (14–48), respectively. These values are based on the experimental work of Nikuradse [176].

14-3.E Dimensional-Analysis Approach

It is interesting to note that a combination of dimensional analysis and some experimental facts can lead to an adequate expression for the distribution of the mean velocity in turbulent flow. This is important when we realize that the previously presented theories were, to a large extent, developed for this same purpose. The mean velocity at a point should depend on the fluid properties ρ and μ, the distance between walls $2y_0$, the shear at the wall τ_w, and of course a distance representative of the point in question (y). By dimensional analysis, we obtain

$$\frac{\bar{U}_x}{U^*} = f\left(\frac{y_0 U^*}{\nu}, \frac{y}{y_0}\right). \qquad (14\text{--}49)$$

The main experimental fact necessary is Stanton's law,

$$\frac{\bar{U}_{x,\max} - \bar{U}_{x,\text{ave}}}{U^*} = \text{const} = A, \qquad (14\text{--}50)$$

where $\bar{U}_{x,\text{ave}}$ is the average of $\bar{U}_x$ across the flow area. We can verify the relation by integrating Eq. (14–48). By combining the above, changing variables, and finally differentiating, we can obtain $\bar{U}_x/U^* = A \ln (yU^*/\nu) + B$ or

$$U^+ = A \ln y^+ + B, \qquad (14\text{-}51)$$

which is comparable to the logarithmic velocity profile of Eq. (14–37). Refer to Pai [188] and Baron [5] for further details.

14-3.F Velocity Distribution for Turbulent Flow

(a) Separate equations for each area of flow. We can use the various theories presented in the previous sections to establish forms for the velocity distribution in turbulent flow, and we can obtain a number of limiting solutions for specific locations or areas.

(i) The viscous sublayer. If we assume that turbulence is negligible and that the shear stress is equal to that at the wall, then Eq. (14–23) becomes

$$\nu(d\bar{U}_x/dy) = U^{*2}. \qquad (14\text{-}52)$$

Integration gives us

$$\bar{U}_x/U^* = yU^*/\nu \qquad (14\text{-}53)$$

or

$$U^+ = y^+. \qquad (14\text{-}54)$$

(ii) The wall area. Beyond the area immediately adjacent to the wall, viscous forces become negligible. This is still close enough to the wall so that the local shear stress remains essentially equal to that at the wall. The above assumptions are identical to those used by Prandtl to obtain Eq. (14–36). At the edge of the laminar sublayer, a distance δ from the wall, the velocity is called $\bar{U}_\delta$. Using this as a boundary condition, we can evaluate the constant of integration C_1 in Eq. (14–36). The equation becomes

$$U^+ = \frac{\bar{U}_\delta}{U^*} + \frac{1}{k} \ln \frac{y}{\delta}. \qquad (14\text{-}55)$$

At the edge of the sublayer given by δ, Eq. (14–54) is valid, so that $U^{*2} = \nu \bar{U}_\delta/\delta$. Combining this with Eq. (14–55) gives us

$$U^+ = \frac{\bar{U}_\delta}{U^*} + \frac{1}{k} \ln \frac{U^{*2} y}{\nu \bar{U}_\delta} = \frac{\bar{U}_\delta}{U^*} + \frac{1}{k} \ln \frac{U^*}{\bar{U}_\delta} + \frac{1}{k} \ln y^+$$

or

$$U^+ = \frac{1}{k} \ln y^+ + B. \qquad (14\text{-}56)$$

Equation (14–41) can be simplified further for this same area of flow $(0 \ll y^+ \ll y_0^+)$:

$$U^+ = (1/k) \ln y^+ + (1/k)[\ln (4k) - 1], \qquad (14\text{-}57)$$

which in effect is another derivation of Eq. (14–56). However, the value of B is not independent of k and should be as given in Eq. (14–57).

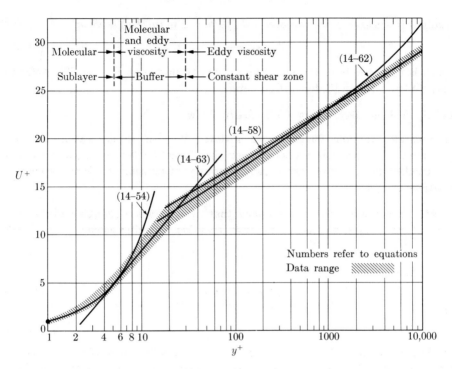

Fig. 14–10. *Universal turbulent velocity-distribution plot.*

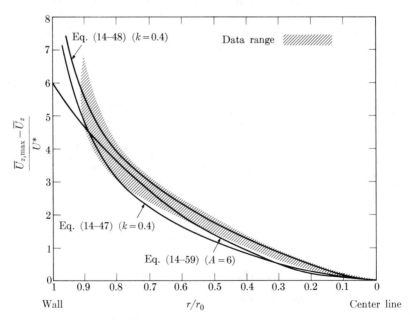

Fig. 14–11. *Velocity-defect plot.*

Nikuradse used

$$U^+ = 2.5 \ln y^+ + 5.5 \tag{14-58}$$

as the best fit for his data; but when we compare this with the value predicted from Eq. (14–57), it shows no agreement, and thus the assumed form of the mixing length ($l^+ = ky^+$) cannot be generally valid. It should be noted that the experimental equation (14–58) has been found to be reasonably valid even far beyond the point at which one would no longer assume that the shear stress was equal to that at the wall.

(iii) **The turbulent core.** Even though Eq. (14–58) has been extensively used in problems involving the turbulent core, there are equations that are more satisfactory. For example, the velocity-defect laws have already been cited (Eqs. 14–47 and 14–48). A simple parabolic form can often be used over limited ranges:

$$\frac{\bar{U}_{z,\max} - \bar{U}_z}{U^*} = A\left(\frac{r}{r_0}\right)^2. \tag{14-59}$$

Finally, Prandtl [192] presented what has become a well-known approximation for this region, based on Blasius' [17] friction-factor equation for $N_{\text{Re}} < 10^5$,

$$f = 0.0791(N_{\text{Re}})^{-1/4}. \tag{14-60}$$

The approximation is known as the one-seventh power law:

$$\frac{\bar{U}_z}{\bar{U}_{z,\max}} = \left(\frac{y}{r_0}\right)^{1/7}, \tag{14-61}$$

and, by the use of Eq. (14–60), it can also be expressed as

$$U^+ = 8.74(y^+)^{1/7}. \tag{14-62}$$

(iv) **The buffer zone.** In the area where viscous and turbulent stresses are both important, neither Eq. (14–54) nor Eq. (14–58) is valid. Reichart [195] has suggested an empirical logarithmic form for this area:

$$U^+ = 5.0 \ln y^+ - 3.05. \tag{14-63}$$

(v) **Summary.** Equations for the four areas are shown in Figs. 14–10 and 14–11. Although all the equations do a reasonable job of representing the velocity distribution, there is certainly a disadvantage in having so many equations represent what is actually a smooth continuous curve. Furthermore, in problems of mass and heat transfer, it is undesirable to have a system of equations for which discontinuities in the derivatives exist.

(b) **Alternate expressions for the sublayer and buffer zones.** Since it is undesirable to use a number of equations to represent a continuous velocity distribution, efforts have been made to reduce this number by combining the sublayer and buffer zones. Deissler [59] assumed an eddy viscosity of the form

$$\bar{\varepsilon} = -n^2 \bar{U}_z y [1 - \exp(-n^2 \bar{U}_z y/\nu)]. \tag{14-64}$$

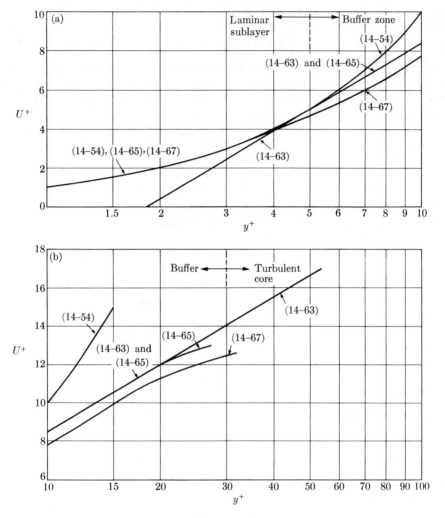

Fig. 14–12. Detail of universal velocity-distribution plot.

If we assume that the shear in the area near the wall is equal to that at the wall, we can combine Eq. (14–64) with Eq. (14–31), put them into nondimensional form, and integrate to obtain

$$U^+ = \int_0^{y^+} \frac{dy^+}{1 + n^2 U^+ y^+ [1 - \exp(-n^2 U^+ y^+)]}. \tag{14-65}$$

We cannot solve this equation directly, but we can evaluate it by numerical iteration methods. The curve is plotted in Fig. 14–12 for $n = 0.124$. The equation is valid only up to $y^+ = 26$. A simpler but less satisfactory form was given earlier by Deissler [58].

Lin et al. [152] have also derived an expression for the wall area that is not based on the laminar sublayer. When plotted, however, it is the same as Eq. (14–54), and does not apply for a y^+ greater than 5. The form equivalent to Eq. (14–64) is

$$\bar{\epsilon}/\nu = (y^+/14.5)^3, \tag{14-66}$$

which has a damping effect on turbulence as the wall is approached.

Hanratty [92] applied the concept of the penetration theory to replace the laminar sublayer. He assumed that a mass of fluid with a fixed velocity, penetrating to the wall, comes from the turbulent core, and has a fixed contact time with the wall θ_c. He further assumed that the transport of momentum with one such mass is represented by the diffusion equation

$$\partial \bar{U}_x/\partial \theta = \nu(\partial^2 \bar{U}_x/\partial y^2).$$

With suitable boundary conditions, this equation can be solved. The resulting equation has one constant, the ratio of the velocity at the edge of the turbulent core to the friction velocity. From the data of Deissler, this ratio can be estimated to be 13 (this is the value of U^+ at $y^+ = 26$). The final equation is

$$U^+ = 13 \int_0^1 \mathrm{erf}\,(y^+\sqrt{\pi})/(4.13\sqrt{\tau})\,d(\tau), \tag{14-67}$$

which is also plotted in Fig. 14–12. The value of U^+ falls off a bit above $y^+ = 20$, as shown in the figure. On the basis of the assumptions made, we would not expect the equation to be valid above $y^+ = 26$.

Clauser [30] wrote a journal article in which he reviewed some of the previous ideas, then extended the analysis, using the concept that a boundary layer system may be considered to be in equilibrium if the proper choice of dimensional parameters is made. For the outer region ($y^+ > 200$) he used a pseudolaminar system in which a very thin sublayer of a different fluid with lower viscosity was imagined to exist. With this picture in mind, Clauser was able to show that the turbulent distribution was quite similar to this pseudolaminar system. This approach is similar to that of using a parabolic velocity distribution for the turbulent core.

(c) **Universal velocity distributions.** Ideally, one single equation should represent the entire velocity distribution, and of course considerable effort has been directed toward this goal. Not only should it adequately represent the velocity distribution, but the universal equation should also be smooth (i.e., have a continuous first derivative), and should satisfy the boundary conditions at both the wall and the center line. If possible, it should also be applicable in the transition range between laminar and turbulent flow. Of course, a simple equation would be desirable. We do not yet have the final answer, although a number of worthwhile attempts have been made.

Equation (14–39) has already been suggested as one possible relation (or the asymptotic form given as Eq. 14–40). However, this equation with its *one* arbitrary

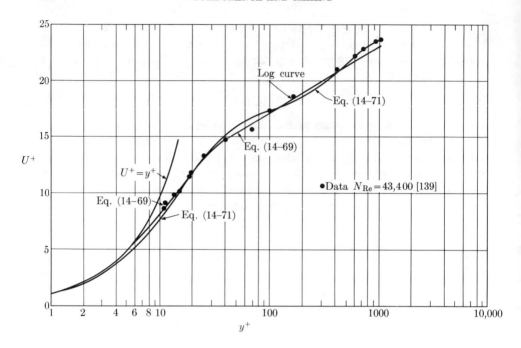

Fig. 14–13. Universal turbulent velocity-distribution plot.

constant was inadequate for representing the data. Van Driest [63] suggested a similar solution, using a further-modified form of the Prandtl mixing length. This modified form,

$$l^+ = ky^+(1 - e^{-y^+/A}), \tag{14-68}$$

which contains *two* arbitrary constants, was suggested by the theory of damping in a fluid when a simple harmonic oscillation is set up by a plate. For the asymptotic solution ($y_0^+ \to \infty$), Eq. (14–40) can be converted to

$$U^+ = 2\int \frac{dy^+}{1 + \sqrt{1 + 4k^2 y^{+2}[1 - \exp(-y^+/A)]^2}} \tag{14-69}$$

by using Eq. (14–68) for l^+. The equation fits the data quite well with $k = 0.4$ and $A = 36$ (see Fig. 14–13); however, it does require numerical integration for its solution and it cannot satisfy the boundary condition of

$$dU^+/dy^+ = 0 \quad \text{when} \quad y^+ = y_0^+ = \text{finite}.$$

This latter failure is a consequence of using the asymptotic solution rather than the complete solution [Eq. (14–40) rather than Eq. (14–39)].

Pai [186, 187] developed a solution to the complete Reynolds equations. He solved the steady-state problem both for flow between parallel plates and for flow in a circular pipe. Instead of assuming a form for the mixing length and then solving Eq. (14–35), Pai turned directly to Eq. (14–23), assumed a form for the final velocity distribution, and attempted to establish the parameters from the boundary conditions. The simplicity of Pai's final result, added to the fact that his equation can meet the boundary conditions, makes his method very attractive. However, it also has limitations.

For the velocity distribution for channel flow, Pai assumed that

$$\frac{\bar{U}_x}{\bar{U}_{x,\max}} = 1 + a_1\left(\frac{y}{y_0}\right)^2 + a_2\left(\frac{y}{y_0}\right)^{2m}, \tag{14-70}$$

where y is now measured from the center line. For pipe flow, r/r_0 replaces y/y_0 z replaces x, and the equation becomes

$$\frac{\bar{U}_z}{\bar{U}_{z,\max}} = 1 + a_1\left(\frac{r}{r_0}\right)^2 + a_2\left(\frac{r}{r_0}\right)^{2m}. \tag{14-71}$$

Only three terms are used because the number of known boundary conditions will allow the coefficients of only three terms to be completely determined. The power of the first term is set because of the known limiting laminar case. The parameter m is an integer greater than 2 and is set by the experimental data. One major requirement is symmetry of the velocity profile in cartesian coordinates. The symmetry of the first term is assured as y/y_0 changes sign because of the even power 2. In cylindrical coordinates the problem is not so apparent, since r/r_0 is always positive. The parameter m is taken as an integer greater than 2 in order to avoid the same nonsymmetry problem in cartesian coordinates.

The coefficients a_1 and a_2 can be obtained as follows: Since $\bar{U}_z = 0$ at $r = r_0$, Eq. (14-71) gives us

$$1 + a_1 + a_2 = 0. \tag{14-72}$$

For pipe flow, Eq. (14-23) takes the form

$$(r/r_0)U^{*2} = -\nu(d\bar{U}_z/dr) + \overline{u_r u_z}. \tag{14-73}$$

From this equation, when $r = r_0$, we have $\overline{u_r u_z} = 0$, and

$$(d\bar{U}_z/dr)_w = -U^{*2}/\nu. \tag{14-74}$$

After Eq. (14-71) is differentiated and then reduced to conditions at the wall, the result can be combined with Eq. (14-74) to give

$$-(U^{*2}/\nu)(r_0/\bar{U}_{z,\max}) = 2a_1 + 2ma_2. \tag{14-75}$$

Equations (14-72) and (14-75) can be solved for a_1 and a_2 to give

$$a_1 = \frac{s - m}{m - 1} \quad \text{and} \quad a_2 = \frac{1 - s}{m - 1}, \tag{14-76}$$

where

$$s = \frac{U^{*2}}{\nu}\frac{r_0}{2\bar{U}_{z,\max}} = \frac{y_0^+}{2U_0^+}, \tag{14-77}$$

in which

$$U_0^+ = \bar{U}_{z,\max}/U^*.$$

Equation (14–71) can be integrated to give the average velocity across the pipe. The volumetric flow rate is

$$Q = \bar{U}_{z,\text{ave}} \pi r_0^2 = 2 \int_0^{r_0} \pi \bar{U}_z r \, dr \quad \text{or} \quad \bar{U}_{z,\text{av}_3} = \frac{2}{r_0^2} \int_0^{r_0} \bar{U}_z r \, dr.$$

Combining this with Eq. (14–71) and integrating, we obtain

$$\frac{\bar{U}_{z,\text{ave}}}{\bar{U}_{z,\text{max}}} = 1 + \frac{a_1}{2} + \frac{a_2}{m+1}. \tag{14–78}$$

For the laminar flow limit, $s = 1$, which yields $a_1 = -1$ and $a_2 = 0$. Equation (14–78) reduces to $\bar{U}_{z,\text{ave}}/\bar{U}_{z,\text{max}} = \frac{1}{2}$, as expected.

Equations (14–71) and (14–78), with the constants given in (14–76), should be valid from the wall to the center line and from laminar through turbulent flow. The term m is a parameter which is expected to be a unique function of the Reynolds number. Brodkey et al. [19] have extended the analysis to apply to materials which are non-Newtonian and conform to the power law model.

The usefulness of the power-series representation for the velocity distribution depends on what we know about the functional dependence of the empirical integer parameter m on the Reynolds number. Correlations for this parameter for both Newtonian and non-Newtonian materials have been given by Brodkey [20]. However, actual velocity-distribution calculations disclosed certain limitations. The power series equation was found to represent the distribution over most of the turbulent core at all Reynolds numbers, but for Reynolds numbers greater than 10^5, deviations occurred at a y^+ of 75. At Reynolds numbers less than 10^5, the power-series equation provided a satisfactory representation over the entire flow field.

The parameter m was obtained from Eq. (14–78) and the experimental $\bar{U}_{z,\text{ave}}/\bar{U}_{z,\text{max}}$ data from the literature. An empirical correlation for the parameter can be expressed as the whole integer closest to the value given by

$$m = -0.617 + 8.211 \times 10^{-3} (N_{\text{Re}})^{0.786}. \tag{14–79}$$

It was also found that s [given by Eq. (14–77)] could be empirically correlated as a unique function of the Reynolds number [20]. In Eq. (14–79), the values of m are not based on a fit to a velocity profile, but rather are estimated on the basis of data concerning average conditions over the cross section. To be valuable, the equations, with the now-known value of m, must be able to predict satisfactorily the entire velocity profile. Thus, to establish their true worth, one must compare the equations with actual velocity profile data. Such a comparison has been made. As already noted, the representation is quite good below a Reynolds number of 10^5, and the result is a very convenient equation. It should be emphasized that this analysis is also satisfactory for profiles occurring during the transition between laminar and turbulent flow. The data for $N_{\text{Re}} = 43{,}400$ ($s = 23.45$, $m = 34$) are shown in Fig. 14–13. A value for y_0 of 1114 was calculated for these conditions from the useful relation

$$y_0^+ = \sqrt{f/8}\, N_{\text{Re}}, \tag{14-80}$$

which is in terms of the Fanning friction factor f, and is derived in the next section.

There have been several additional attempts for a universal velocity distribution for Newtonian materials. Spalding [224] suggested a totally empirical formula which is valid for turbulent flow only, and cannot be used in the transition region. Furthermore, the equation approaches the standard logarithmic form at high values of y^+, and thus cannot satisfy the boundary condition of symmetry at the center line. Another semiempirical formula was suggested by Gill and Scher [82] and applied to a heat transfer problem by Gill and Lee [83]. This equation, like that of van Driest, requires a numerical solution and a graphical representation. Einstein and Li [66] derived an equation based on the concept of an unsteady laminar sublayer. Their work can be used to explain fluctuations in pressure observed near the wall. Lee [141] combined Eq. (14–65) for the area near the wall with Eqs. (14–35) and (14–42) for the area away from the wall. He imposed, as a boundary condition, the requirement that both the velocity distribution and its derivative be continuous at the point of connection between the two equations. Since Lee used the entire equation (14–35) and not the asymptotic form, he was able to satisfy the symmetry condition at the center line. As proved to be the case with other complex problems, a computer solution was necessary.

There have been other solutions and more will be suggested in the future. It is hoped that a simple, completely smooth representation with a minimum of constants or parameters, one that meets all the boundary conditions, will someday be obtained.

14-3.G Friction Factors in Pipes and Velocity Distributions in Rough Pipes

The Fanning friction factor has been defined as

$$f = \frac{1}{2}\frac{d\bar{p}}{dz}\frac{d_0}{\rho \bar{U}_{z,\text{ave}}^2}, \tag{7-24}$$

which, for laminar flow, results in a simple expression in terms of the Reynolds number:

$$f = 16/N_{\text{Re}}. \tag{7-25}$$

For turbulent flow the results are not so simple. From a force or mechanical energy balance, pressure drop in the flow direction is $d\bar{p}/dz = 2\bar{\tau}_{rz}/r = 2\tau_w/r_0$. Combining this with the definition of the friction factor yields $\tau_w = \frac{1}{2}f\rho \bar{U}_{z,\text{ave}}^2$, which, when combined with the definition of U^*, results in

$$U^*/\bar{U}_{z,\text{ave}} = \sqrt{f/2}. \tag{14-81}$$

We can obtain the form of the friction factor for smooth pipes by starting with Stanton's law (Eq. 14–50) and Eq. (14–81):

$$\frac{\bar{U}_{z,\text{ave}}}{U^*} = \frac{\bar{U}_{z,\text{max}}}{U^*} - A = \sqrt{\frac{2}{f}}. \tag{14–82}$$

We can assume that, at the maximum velocity, the logarithmic equation is valid, so that

$$\bar{U}_{z,\text{max}}/U^* = (1/k) \ln y_0^+ + B. \tag{14–83}$$

Combining this with Eq. (14–82) gives us

$$\sqrt{2/f} = (1/k) \ln y_0^+ + B - A. \tag{14–84}$$

Now we have the following:

$$y_0^+ = \frac{U^* r_0}{\nu} = \frac{U^*}{\bar{U}_{z,\text{ave}}} \frac{\bar{U}_{z,\text{ave}} 2 r_0}{\nu} \frac{1}{2} = \sqrt{\frac{f}{2}} N_{\text{Re}} \frac{1}{2} = \sqrt{\frac{f}{8}} N_{\text{Re}},$$

which was previously presented as Eq. (14–80). The above, with Eq. (14–84), yields

$$\sqrt{2/f} = (1/k) \ln (N_{\text{Re}} \sqrt{f/8}) + B - A. \tag{14–85}$$

Von Kármán [110] determined the constant A by integrating the velocity equation (14–56) over the pipe radius to obtain an average velocity, and then combining this with Eqs. (14–82) and (14–83). The value he thus obtained was 3.75. With $A = 3.75$, $k = 0.4$, and $B = 5.5$, Eq. (14–85) becomes

$$1/\sqrt{f} = 4.06 \log (N_{\text{Re}} \sqrt{f}) - 0.60. \tag{14–86}$$

This equation has been verified by the experiments of Nikuradse [176]; however, a slight modification of the constants gave a better fit; that is,

$$1/\sqrt{f} = 4.0 \log (N_{\text{Re}} \sqrt{f}) - 0.40. \tag{14–87}$$

Equation (14–87) is compared with experimental data in Fig. (14–14); in addition, the correlation suggested by Blasius [17] at a much earlier date is included:

$$f = 0.0791 (N_{\text{Re}})^{-1/4}. \tag{14–88}$$

It was from this correlation of Blasius that Prandtl was able to derive the previously mentioned one-seventh power law. As the figure shows, this equation is satisfactory to a Reynolds number of 10^5.

The preceding analysis was restricted to smooth pipes. For rough pipe, as most commercial pipe is, some modification is necessary. If the Reynolds number were large, the laminar sublayer would be small and roughness would be controlling. On the other hand, if the Reynolds number were low and the sublayer relatively large, then the roughness would be buried in the sublayer, and the pipe would act as if it were smooth. In either case, the velocity distribution of Eq.

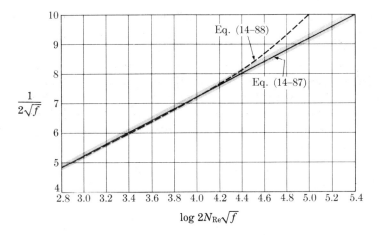

Fig. 14–14. Fanning friction factor in turbulent flow. [From H. Schlichting, Boundary Layer Theory, McGraw-Hill, New York (1960). By permission.]

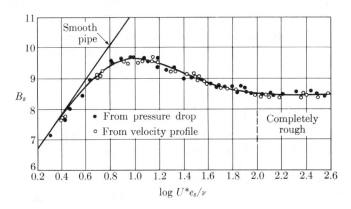

Fig. 14–15. Sand-roughness data. [From H. Schlichting, NACA TM 1218 (1949). By permission.]

(14–55) would be approximately valid, but the boundary conditions would be changed. For a smooth surface, a laminar sublayer was considered to exist, and the term δ was taken as $\bar{U}_\delta \nu / U^{*2}$. For a rough surface, δ is considered proportional to e, the average depth of the roughness. With this new assumption, Eq. (14–55) becomes

$$U^+ = (1/k) \ln(y/e) + B(eU^*/\nu). \tag{14–89}$$

The value of k should be the same; the new constant B will depend on the degree of roughness. For pipe made completely rough with sand particles, and in which roughness is the controlling factor, Nikuradse determined the value of B to be 8.5.

Figure 14–15 is a plot of B_s (the subscript s is used to denote data for flow in sand-roughened pipes) versus the roughness parameter. The surface is completely rough if U^*e_s/ν is greater than 100. If the parameter is less than 5, the roughness is buried, and the pipe acts as if it were smooth. Equation (14–89) can be expressed in a number of other ways which provide more generality. Further readings in this area include Clauser's review [30] and the article by Perry and Joubert [189]. The comments by Hinze [97] are also interesting.

Equations (14–83) through (14–85) are valid with the new value of B; however, the term inside the natural logarithm is now r_0/e rather than y_0^+. When we use the new value of B for the completely rough region, together with Eq. (14–85), the equation analogous to (14–86) is

$$1/\sqrt{f} = 4.00 \log (r_0/e) + 3.36. \tag{14-90}$$

A slightly better fit for the completely rough pipe flow data of Nikuradse is

$$1/\sqrt{f} = 4.0 \log (r_0/e) + 3.48. \tag{14-91}$$

The problem is far more complex for roughness elements that are not ideal. For example, one cannot predict the turbulent loss or velocity distribution for nonuniform, wavy, or widely different types or shapes of roughness. Generally, one attempts to express the more complex problem in terms of an equivalent sand roughness. This also applies to the universal friction-factor charts, which are well known and which will not be discussed further here. The reader is referred to the extensive discussion in Knudsen and Katz [117].

14-3.H Reynolds Stresses

From Eq. (14–23) for channel flow or (14–73) for pipe flow, one can obtain the Reynolds shear stress from a knowledge of the velocity gradient. The available data are shown in Fig. 14–16, where the shaded area illustrates the data of Laufer [139] and of Sandborn [211]. The data points close to the wall were obtained by Sandborn; however, he believed that the wall had a considerable effect on the heat transfer of his hot-wire anemometer, so he questioned these particular points. The velocity distribution given by Eq. (14–71) can be differentiated and combined with Eq. (14–73) to give

$$\frac{\overline{u_r u_z}}{U^{*2}} = \frac{r}{r_0}\left[1 - \left(\frac{r}{r_0}\right)^{2(m-1)}\right]\left(-\frac{ma_2}{s}\right). \tag{14-92}$$

Figure 14–16 shows the curve for $N_{\text{Re}} = 43{,}400$ ($s = 23.45$, $m = 34$); the agreement is good. Away from the wall, the Reynolds shear stress can be treated quite simply: $\bar{\tau}_{rz} = \rho \overline{u_r u_z} = (r/r_0)\tau_w$. Therefore we obtain

$$\overline{u_r u_z} = \frac{r}{r_0}\frac{\tau_w}{\rho} = \frac{r}{r_0}U^{*2}$$

or

$$\overline{u_r u_z}/U^{*2} = r/r_0. \tag{14-93}$$

We could also obtain this equation directly from Eq. (14–73) for $\nu \to 0$ or from Eq. (14–92), since only near the wall is the power term involving m important.

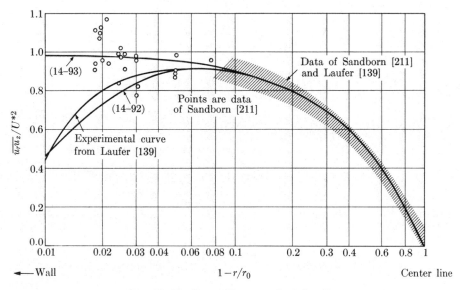

Fig. 14-16. Reynolds stress for pipe flow.

Laufer [139] modified Eq. (14–73) by multiplying by $(\nu/U^{*4})(d\bar{U}_z/dr)$ to give an equation which relates the total available energy due to pressure drop along the pipe,

$$\frac{r\nu}{r_0 U^{*2}} \frac{d\bar{U}_z}{dr}, \tag{14-94}$$

to the energy loss by direct viscous dissipation,

$$W_\mu = (\nu^2/U^{*4})(d\bar{U}_z/dr)^2, \tag{14-95}$$

and to that used in the production of turbulence,

$$\mathrm{Pr}_\tau = \overline{u_r u_z}\,(\nu/U^{*4})(d\bar{U}_z/dr). \tag{14-96}$$

Klebanoff [113] used the same method to modify the equivalent equation for boundary-layer flow. Schubauer [215] obtained a relation between these two representations, and thus permitted comparison. The representative curve is shown in Fig. 14–17. The dissipation term is directly related to the velocity profile, and the production term can be related to the profile by means of Eq. (14–73). Thus, if an equation for the velocity distribution is available, one can predict the curves of Fig. 14–17. The curves shown were obtained from Pai's equation for the velocity distribution; the data points are those of Klebanoff and Laufer. The deviation of the curve from the data points in the range of $y^+ = 15$ to $y^+ = 40$ is due to Pai's equation. If one uses the velocity distribution measured by Laufer, the agreement is much better. Equation (14–73), as modified by Laufer, can be written as

$$\mathrm{Pr}_\tau = (W_\mu)^{1/2}(r/r_0) - W_\mu. \tag{14-97}$$

258   TURBULENCE AND MIXING

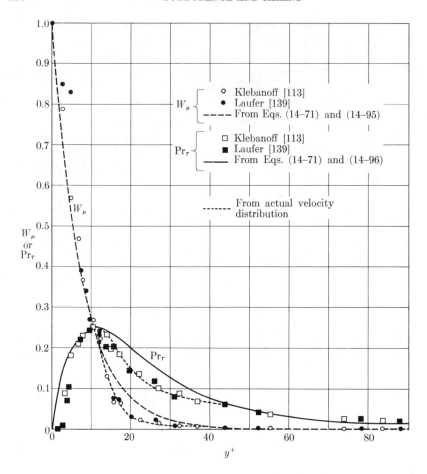

Fig. 14-17. *Direct viscous dissipation and turbulence production.*

In the area where $\Pr_\tau$ is a maximum, the approximation that r/r_0 is unity is valid. When we differentiate Eq. (14-97) (with $r = r_0$) with respect to W_μ and set the result equal to zero, we find that the maximum value occurs when $\Pr_\tau = W_\mu = \frac{1}{4}$. Thus, as a direct consequence of the equation, the shape near the maximum in $\Pr_\tau$ is fixed.

From the analysis we may conclude that viscous dissipation occurs mainly from the wall to a y^+ of approximately 30, and that at the maximum rate of production, the production and dissipation rates are equal ($y^+ \simeq 11$). This point is located at approximately the place where one would expect to find the edge of the laminar sublayer. Furthermore, the bulk of the energy converted into turbulence occurs in this same vicinity of the wall. We can now see why the wall and wall conditions are quite important in understanding turbulent shear flow. Schubauer pointed out that about 40% of the mean flow energy is dissipated directly as heat; the rest goes to production of turbulence, and then ultimately to heat. One-half of the production of turbulence occurs between the wall and a y^+ of approximately 60. Brodkey et al. [21] have shown that the maximum point for production of turbulence corresponds to the point at which the rate of change of kinetic energy is maximum.

The discussion of phenomenological theories has been restricted to pipe and boundary layer flow. But these are not the only geometries possible, and more theories—such as those dealing with jets, wakes, and annular flow—have been studied. The basic approaches to these other subjects are similar to those presented here; the reader should consult such references as Hinze [96], Schlichting [220], Pai [188], and Knudsen and Katz [117].

EXAMPLE 14-1. For a flow at a Reynolds number of 40,000, compare nondimensional forms of the mixing length as determined from the various theories. Use the Pai approach to establish the nature of the velocity profile and gradients. Make your calculations for the point at which $r/r_0 = 0.5$.

Answer. From Eq. (14–79) and reference [20], for $N_{Re} = 40,000$, $m = 33$ and $s = 22.2$. Thus $a_1 = -0.338$ and $a_2 = -0.662$, from Eq. (14–76). Equation (14–78) yields

$$\bar{U}_{z,\text{ave}}/\bar{U}_{z,\text{max}} = 0.811. \tag{i}$$

When one uses the above values, differentiation of Eq. (14–71) gives

$$(r_0/\bar{U}_{z,\text{max}})(d\bar{U}_z/dr) = -0.338, \tag{ii}$$

$$(r_0^2/\bar{U}_{z,\text{max}})(d^2\bar{U}_z/dr^2) = -0.676. \tag{iii}$$

Equation (14–71) itself yields

$$\bar{U}_z/\bar{U}_{z,\text{max}} = 0.9155. \tag{iv}$$

The Prandtl mixing length can be expressed as

$$l^+ = ky^+. \tag{v}$$

The term y_0^+ can be obtained from Eq. (14–80): $y_0^+ = 40{,}000\sqrt{0.0056/8} = 1060$, where the friction factor was obtained from Eq. (14–88). Then y^+ is $y^+ = (0.5)1060 = 530$. Using this in Eq. (v) yields $l^+ = 212$.

The mixing length in the Taylor vorticity theory is defined on page 242. The equation can be rearranged as follows.

$$\frac{d\bar{p}}{dz} = \frac{2\rho U^{*2}}{r_0} = \rho l_w^2 \frac{d\bar{U}_z}{dr}\frac{d^2\bar{U}_z}{dr^2}. \tag{vi}$$

Using (ii) and (iii) in (vi) results in

$$2U^{*2} = l_w^2(-0.338)(-0.676)(\bar{U}_{z,\text{max}}^2/r_0^2). \tag{vii}$$

One can obtain the ratio of the friction velocity to the maximum velocity by combining Eqs. (14–81) and (vi), which in turn can be combined with (vii) to give

$$l_w/r_0 = 0.127 \quad \text{or} \quad l_w^+ = (0.127)(1060) = 135.$$

The von Kármán similarity hypothesis gives the same result as the Prandtl mixing length theory. In all cases, one should note that the mixing length is not small when compared with the dimensionless radius y_0^+.

EXAMPLE 14–2. For the same conditions as given in the previous problem, calculate the ratio of eddy to molecular viscosity.

Answer. The ratio is obtained from Eq. (14–31). One nondimensional form is

$$y^+ = -(1 + \bar{\varepsilon}/\nu)(r_0/U^*)(d\bar{U}/dr),$$

which, for the data of Example 14–1, gives

$$\bar{\varepsilon}/\nu = -(530)(0.043)/(-0.338) - 1 = 66.5,$$

where $U^*/\bar{U}_{z,\max} = 0.043$.

14-4 THE STATISTICAL THEORY OF TURBULENT FLOW

In turbulent flow, momentum, heat, and mass transfer are to a large extent controlled by the degree of turbulence of the system. Consequently, an understanding of turbulence *per se* is necessary before we can analyze transport phenomena. In the previous section, phenomenological theories were developed in some detail. These mixing-length or semiempirical methods are very useful and, for the most part, they are the only methods of treating engineering problems. However, these theories were designed to evaluate the mean velocity distribution, and, as is now known, this distribution is relatively insensitive to the basic assumptions of the theories. For a more basic analysis of the problem, we should return to the Reynolds equation (14–10) or its partial integration (14–23), (14–26), (14–73). The major problem is to obtain additional relations between the unknowns, and for this purpose, we need theories and measurements which will enable us to evaluate the fluctuating velocity terms or Reynolds stresses. The present concept is that we can obtain the additional relationships, in part, if we remember that turbulent fluctuations are random in nature and can therefore be treated by means of statistics, as is done with molecules in statistical mechanics. To put it briefly, the mechanism of turbulent motion is of such a complex nature that at present we are unable to formulate a general model on which to base an analysis. Thus we approach the problem from a rigorous statistical theory into which we can introduce certain simplifying assumptions that will allow us to solve for some of the variables of interest. In the discussion to follow, we must view the material as simply a reasonable mathematical representation, with some limited mechanistic ideas injected. We simply do not know for certain the details of what physically occurs during the process of turbulence.

If we are to use statistical relations to define the fluctuating motion of turbulence, then we must restrict the meaning of turbulent motion to an irregular fluctuation about a mean value. Any regular motion, such as the von Kármán "vortex trail" in the wake of a cylinder, would not be considered turbulent motion. In other words, any motion which might have a regular periodicity is not considered to be turbulent.

It might be best, as we begin our discussion on turbulence, to propose a model—albeit a crude one—for the system. Let us assume that eddies range in size from the very smallest to the largest, which we might picture as being determined by the bounds of the system. In the most ideal case, the boundaries influence only these large eddies, and transfer energy to or from them. The larger eddies transfer their energy to the smaller eddies, etc., until the energy is transferred to the smallest of eddies. These smallest eddies lose their energy by viscous dissipation. As we shall see, a great deal of the theoretical work on turbulent flow is based on this somewhat intuitive model of turbulence.

14-4.A Introduction to Terms and Definitions

(a) Description of turbulence. In any theoretical approach to a difficult problem, we must make a number of simplifying assumptions. Frequently, we can devise experiments such that most of the assumptions can be satisfied either exactly or at least approximately. Turbulence, in this respect, is no different from any other subject.

The term *homogeneous turbulence* implies that the velocity fluctuations in the system are random, and that the average turbulent characteristics are independent of the position in the fluid, i.e., invariant to axis translation. We can further restrict the homogeneous system by assuming that in addition to its homogeneous nature, its velocity fluctuations are independent of the axis of reference, i.e., invariant to axis rotation and reflection. This restriction leads to *isotropic turbulence*, which by its definition is always homogeneous. To illustrate the difference between the two types of turbulence, consider the root-mean-square (rms) velocity fluctuations

$$u'_x = \sqrt{\overline{u_x^2}}, \quad u'_y = \sqrt{\overline{u_y^2}}, \quad \text{and} \quad u'_z = \sqrt{\overline{u_z^2}},$$

where u'_x, u'_y, and u'_z are used to simplify the notation. In homogeneous turbulence, the rms values can all be different, but each value must be constant over the entire turbulent field. In isotropic turbulence, spherical symmetry requires that the fluctuations be independent of the direction of reference, or that all the rms values are equal:

$$\sqrt{\overline{u_x^2}} = \sqrt{\overline{u_y^2}} = \sqrt{\overline{u_z^2}} \quad \text{or} \quad u'_x = u'_y = u'_z.$$

Isotropic homogeneous flow, by its very nature, has no cross velocity terms $(\overline{u_x u_y}, \overline{u_x u_z}, \overline{u_y u_z})$, because of random motion, which gives $u_x u_y$ just as many positive values as negative ones. Consequently, the average $\overline{u_x u_y}$ is zero.* In flows of this nature there are no shearing stresses and no gradients of the mean velocity.

---

*On the other hand, u_x^2 is always positive; thus $\overline{u_x^2}$ will have a finite value.

Isotropic turbulence is a constant space system, and thus the statistical quantities can vary only with time. Such a state of motion cannot easily be realized in experiments. Experimentally, it can be obtained approximately in the turbulence developed behind a properly designed grid, about which we shall say more when we reconsider the subject of isotropic turbulence.

The area of turbulent study which holds the greatest interest for engineers is *turbulent shear flow*. This flow is the modification of completely homogeneous flow to allow for shear stresses. Usually one or two of the Reynolds shearing stresses is zero. For example, in pipe flow, Eq. (14–17) shows that $\overline{u_r u_\theta} = 0$, and only $\overline{u_x u_r}$ is important. Turbulent shear flow in turn may be divided into flows that are nearly homogeneous in the direction of flow and those that are not homogeneous in the direction of flow. It has been found experimentally that the nearly homogeneous flows are those that are bounded, as in pipe flow, while the inhomogeneous shear flows are unrestricted systems, such as jets. Longitudinal homogeneity (or homogeneity in the direction of flow), arises from the fact that in pipe flow, turbulence is generated and $\overline{u_x u_r} \neq f(x)$; that is, there is no decay. One flow of importance, which has both characteristics depending on the location of study, is boundary-layer flow. The area near the wall is nearly homogeneous in the direction of flow, and that near the boundary, between the boundary layer and the ideal flow, is inhomogeneous. This aspect is associated with intermittency, which will be discussed later.

These basic definitions of the types of turbulence will prepare the way for a better understanding and further interpretation of the definitions necessary to describe the turbulent phenomena.

(b) Correlation. It was Taylor [237] who first suggested that a statistical correlation could be applied to fluctuating velocity terms in turbulence. He pointed out that no matter what may be meant by the diameter of an eddy, a high degree of correlation exists between the velocities at two points in space, if the separation between the points is small when compared with the eddy diameter. Conversely, if the points are taken so far apart that the space corresponds to many eddy diameters, little correlation can be expected (see Fig. 14–18).

Consider a statistical property of a random variable (in this case velocity) at two points separated by a distance vector r (that is, at x and $x + r$). An Eulerian correlation tensor (nine possible terms) at the two points can be defined by

$$\mathbf{Q}(r) = \overline{\mathbf{u}(x)\mathbf{u}(x + r)}. \qquad (14\text{–}98)$$

In the more commonly used cartesian general coordinates, Eq. (14–98) becomes

$$\mathbf{Q}_{ij}(r) = \overline{u_i(x)u_j(x + r)}. \qquad (14\text{–}99)$$

For zero separation ($r = 0$), Eq. (14–99) is the same as the velocity terms of the Reynolds stresses, and is sometimes called the *energy tensor*. A complete determination of this would permit us to evaluate the energy associated with the various velocity fluctuations.

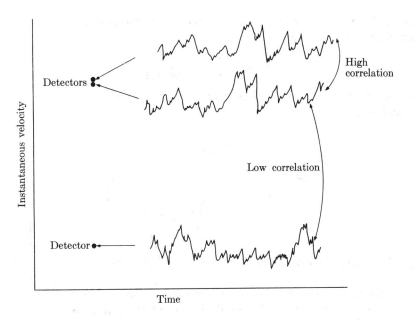

Fig. 14–18. *Velocity correlation.*

Ideally, averaging should be carried out for many equivalent points at one instant in time; i.e., there should be an ensemble average in which there are many equivalent systems. This is usually impossible, however, and we are generally forced to measure the fluctuations at a given point as the fluid moves relative to the measuring instrument. We assume that the Birkhoff theorem of statistical mechanics is valid for turbulence (though it is not proved). The equivalent theorem states that if the time considered is long enough, the average at one point is the same as an average over a large number of points at one time. Thus it is assumed that a time average is valid. Note that these correlations, and all subsequent terms, may be time dependent; however, this will not be shown except when confusion might otherwise result. For brevity, we shall drop the time argument; thus

$$\mathbf{Q}_{ij}(\mathbf{r}) = \mathbf{Q}_{ij}(\mathbf{r}, t).$$

Finally, to suggest time averaging implies that such an average can be obtained; that is, it implies that the system is at steady state or that the time the system takes to change is long compared with the time that is necessary to obtain the average.

A more useful correlation term is called the *Eulerian correlation function*, expressed by

$$\mathbf{R}(\mathbf{r}) = \frac{\overline{u(x)u(x+r)}}{u'(x)u'(x+r)}, \tag{14–100}$$

where $\boldsymbol{u}'$ is the rms vector velocity fluctuation at the point in question. In cartesian tensor notation, this is

$$\mathbf{R}_{ij}(\boldsymbol{r}) = \frac{\overline{u_i(\boldsymbol{x})u_j(\boldsymbol{x}+\boldsymbol{r})}}{u_i'(\boldsymbol{x})\,u_j'(\boldsymbol{x}+\boldsymbol{r})}, \tag{14-101}$$

where u_i' encompasses the three rms values u_x', u_y', and u_z', and can be different at $\boldsymbol{x}$ and at $\boldsymbol{x}+\boldsymbol{r}$.

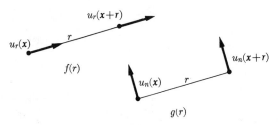

Fig. 14-19. *Velocity-fluctuation directions for double-velocity isotropic correlations.*

The situation would be considerably simpler if the turbulence were homogeneous, since a given rms value would be independent of separation. However, the term $\overline{u_i(\boldsymbol{x})u_j(\boldsymbol{x}+\boldsymbol{r})}$ would be dependent on $\boldsymbol{r}$. For the homogeneous case, Eq. (14-101) would become

$$\mathbf{R}_{ij}(\boldsymbol{r}) = \frac{\overline{u_i(\boldsymbol{x})u_j(\boldsymbol{x}+\boldsymbol{r})}}{u_i' u_j'}. \tag{14-102}$$

For isotropic turbulence, the rms fluctuating velocities are all equal, so that the equation becomes

$$\mathbf{R}_{ii}(\boldsymbol{r}) = \frac{\overline{u_i(\boldsymbol{x})u_i(\boldsymbol{x}+\boldsymbol{r})}}{u'^2}, \tag{14-103}$$

where u' is the rms fluctuating velocity and is independent of position and orientation of measurement. In isotropic turbulence all $i \neq j$ terms are zero, since these would change sign either under reflection or rotation and thus would not be isotropic. A dummy index pair ii has thus been used to show that the indices are identical. It can be shown [8] that only one scalar function is necessary to specify the velocity correlation in isotropic turbulence. This is usually taken as

$$f(r) = \frac{\overline{u_r(\boldsymbol{x})u_r(\boldsymbol{x}+\boldsymbol{r})}}{u'^2}, \tag{14-104}$$

where the subscript r means that the velocity is measured in the same direction (longitudinal) as the space vector $\boldsymbol{r}$ (see Fig. 14-19). Another correlation, which must be related to this one, is

$$g(r) = \frac{\overline{u_n(\boldsymbol{x})u_n(\boldsymbol{x}+\boldsymbol{r})}}{u'^2}, \tag{14-105}$$

where the subscript n means that the velocity is measured in a direction normal

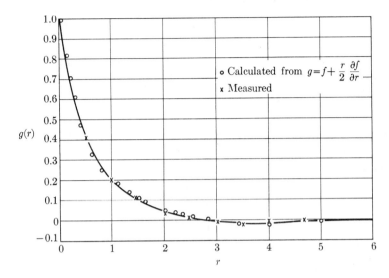

Fig. 14–20. Confirmation of the relation between $f(r)$ and $g(r)$. [*From J. O. Hinze, Turbulence, McGraw-Hill, New York (1959). By permission.*]

(lateral) to the space vector **r**. The term r is of course the magnitude of **r**. As just noted, these must be related, as only one scalar is needed to define the correlation in this totally symmetrical system. Batchelor [8] shows, by means of continuity of an incompressible fluid, that this relation is

$$g(r) = f(r) + (r/2)[\partial f(r)/\partial r] \qquad (14\text{--}106)$$

(see also Hinze [96] and MacPhail [169]). The experimental confirmation is shown in Fig. 14–20. The negative segment of $g(r)$ is associated with a negative correlation that must exist between velocities on opposite sides of an eddy. This can be seen also from Eq. (14–106). The term $f(r)$ is always positive and slowly approaches zero at large r's (about 5 in Fig. 14–20), while $\partial f(r)/\partial r$ is always negative and approaches zero at the same place. At some r less than 5, the second term becomes greater than the first (since r is increasing) and $g(r)$ becomes negative. Eventually both terms become zero, and $g(r)$ approaches zero from the negative side.

In Eqs. (14–104) and (14–105), the correlation functions would reduce to unity if the separation were reduced to zero. In a like manner, the functions would reduce to zero if the separation were large enough so that no correlation occurred; i.e., just as many positive as negative values would be possible.

EXAMPLE 14–3. In the final period of decay of turbulence behind a grid, the correlation function is given by Eq. (14–185):

$$f(r) = \exp(-r^2/8\nu t).$$

For this specific form, show that $g(r)$ becomes negative for large r.

Answer. The equation for $f(r)$ can be combined with Eq. (14–106) to give $g(r)$:

$$g(r) = \exp\left(\frac{-r^2}{8\nu t}\right) + \left(\frac{-r^2}{8\nu t}\right)\exp\left(\frac{-r^2}{8\nu t}\right) = \left[1 - \frac{r^2}{8\nu t}\right]\exp\left(\frac{-r^2}{8\nu t}\right).$$

In this relation, as r becomes large, the exponential term becomes zero and $g(r)$ becomes negative.

In the foregoing discussion, we have taken the Eulerian point of view; that is, the correlation function $\mathbf{R}_{ij}(\mathbf{r})$ is a correlation between the instantaneous velocity fluctuations separated by a distance $\mathbf{r}$. In some applications it is more convenient to consider the Lagrangian system of coordinates, by which we can correlate the velocity fluctuations of a fluid particle at two times along its path; thus

$$\mathbf{R}_{L_{ij}}(\tau) = \frac{\overline{v_i(t)\,v_j(t+\tau)}}{v'_i(t)\,v'_j(t+\tau)}, \qquad (14\text{–}107)$$

where τ is some increment of time. In isotropic turbulence, the form becomes much less complex, and we have

$$R_L(\tau) = \frac{\overline{v(t)\,v(t+\tau)}}{v'^2}. \qquad (14\text{–}108)$$

It is very important to note that v', in the Lagrangian system, is the rms velocity fluctuation of many particles averaged along their respective fluid particle paths, and not a time average at a point in space, as u' is in the Eulerian system. The reason why the velocity is denoted as v rather than u is to emphasize this difference. When the particle velocity is a stationary random function of time (i.e., steady state) in a homogeneous field, these two variances are equal. In fact, in an incompressible homogeneous field, averages of any statistical functions that depend on one point in space are equal in both the Eulerian and Langrangian frames of reference [162]. The Eulerian and Lagrangian correlations will not be the same. However, if we had additional information it is conceivable that we could relate them, although this would probably be empirical or approximate.

The desired measurement of the Eulerian correlation is not always possible because of probe interference; however, we might assume that the displacement is related to time by $x = t\bar{U}_x$, and that it replaces the space coordinate with an equivalent time. This would be an autocorrelation, in which we equated an Eulerian time correlation and an Eulerian space coorelation:

$$f(r) = \frac{\overline{u_r(\mathbf{x})\,u_r(\mathbf{x}+\mathbf{r})}}{u'^2} = \frac{\overline{u_r(t)\,u_r(t+\tau)}}{u'^2} = f(\tau). \qquad (14\text{–}109)$$

For isotropic turbulence, this equality has been shown to be valid by the experiments by Favre *et al.* [70]. For isotropic turbulence, the time correlation of Eq. (14–109) is of the same form as the Lagrangian coefficient of Eq. (14–108). However, Eq. (14–109) is an average of the product of the velocities u_r taken at

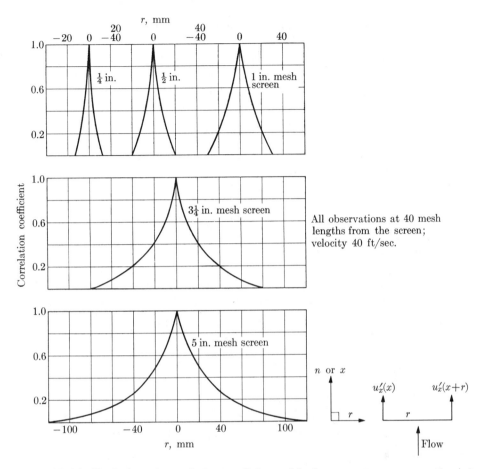

Fig. 14–21. Variation of correlation coefficient with the cross-stream separation of the hot wires. [From H. L. Dryden, Ind. Eng. Chem. **31**, *416–425 (1939). By permission.]*

two times for one point, while Eq. (14–108) is the average of the product of the velocities v taken at two times along particle paths, and of course involves velocities along those paths. Although the autocorrelation technique can be very convenient, it is limited to cases in which u'_x is much less than $\bar{U}_x$, and thus may not be valid in certain shear flows. Further discussion on this point is given by Hinze [96], Fisher and Davies [271], and Wills [282].

The correlation coefficient has been measured experimentally by hot-wire methods. The data in Fig. 14–20 have already been cited. Dryden et al. [64] obtained the results shown in Fig. 14–21 for the turbulence behind various grids in a wind tunnel. The specific correlation measured was

$$g(r) = \frac{\overline{u_x(x)\,u_x(x+r)}}{u'^2_x}, \qquad (14\text{–}110)$$

where $g(r)$ is in terms of u_x, since the direction x is normal to the separation distance r (shown in Fig. 14–21).

When the flow is not homogeneous, the shear stresses are finite, and a correlation exists between the cross-velocity terms, when the separation is zero or finite. For zero separation, Eq. (14–101) reduces to a point correlation, which is

$$R_{ij}(\boldsymbol{x}) = \frac{\overline{u_i(\boldsymbol{x})\,u_j(\boldsymbol{x})}}{u_i'(\boldsymbol{x})\,u_j'(\boldsymbol{x})}. \tag{14–111}$$

The functional dependence on $\boldsymbol{x}$ is usually dropped, since all terms are understood to be measured at a single point. However, this does not mean that the correlation does not vary from point to point. In the modified notation, Eq. (14–111) becomes

$$R_{ij} = \frac{\overline{u_i u_j}}{u_i'\,u_j'}. \tag{14–112}$$

In this point correlation function, if $i = j$, the function is trivial and always equal to unity. In turbulent shear flow, this correlation is of prime importance, and has been measured for a number of specific flows. It can be recognized as a nondimensional form for the cross-velocity part of the Reynolds stress. A specific example,

$$R_{zr} = \frac{\overline{u_z u_r}}{u_z'\,u_r'}, \tag{14–113}$$

is shown in Fig. 14–22 for pipe flow as a function of the radial position.

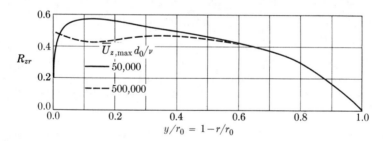

Fig. 14–22. The Reynolds stress for pipe flow. [From J. Laufer, NACA Report 1174 (1954). By permission.]

Other higher-order velocity correlations are possible. For example, the triple velocity correlation appears in conjunction with the double velocity correlation in the von Kármán-Howarth equation [111] (to be discussed later). The form of the triple velocity correlation is analogous to Eq. (14–99), and is

$$S_{ij,k}(\boldsymbol{r}) = \overline{u_i(\boldsymbol{x})\,u_j(\boldsymbol{x})\,u_k(\boldsymbol{x}+\boldsymbol{r})}. \tag{14–114}$$

This correlation tensor can have 27 terms (18 different) and is a third-order tensor. The comma in the subscript denotes the fact that two of the velocities, i and j, are measured at x, and one, k, is measured at $x + r$. For isotropic turbulence, the third-order correlation also depends on, or is determined by, a single scalar function. Batchelor [8] points out that in the triple velocity correlation, as in the double correlation, the important velocity fluctuations measured are either in the same direction as, or are normal to, the position vector r. There are seven nonzero components of Eq. (14–114). Of these, three are different, and their corresponding correlation functions are

$$k(r) = \overline{u_r(x)u_r(x)u_r(x + r)}/u'^3,$$
$$h(r) = \overline{u_n(x)u_n(x)u_r(x + r)}/u'^3,$$
$$q(r) = \overline{u_r(x)u_n(x)u_n(x + r)}/u'^3.$$

(14–115)

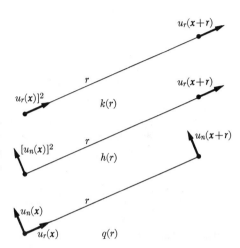

Fig. 14–23. Velocity-fluctuation directions for triple-velocity isotropic correlations.

They are shown in Fig. 14–23. Since only one scalar term is necessary in order to specify the triple velocity correlation in isotropic turbulence, the correlations of Eq. (14–115) must be related [96]:

$$h(r) = -\tfrac{1}{2}k(r),$$
$$q(r) = -\frac{1}{2r}\frac{\partial r^2 h(r)}{\partial r},$$
$$q(r) = \frac{1}{4r}\frac{\partial r^2 k(r)}{\partial r}.$$

(14–116)

In turbulent shear flows or inhomogeneous systems, triple velocity functions can exist at a single point, just as a double velocity function can exist at a single point (Eq. 14–112). The various functions have been measured for a few flows; however, the experimental errors are high and only qualitative results have been obtained.

(c) **Intensity.** The intensity of turbulence is defined as the rms value of the fluctuating velocities, or

$$u'_i = \sqrt{\overline{u_i^2}}.$$

The intensity, rms value, or variance of turbulence is sometimes expressed as a fraction or percentage of the mean flow velocity (for example, $u'_x/\bar{U}_x$) or as a fraction of the friction velocity U^*. Values for these velocities have been measured in pipes [139, 211], in a boundary layer over a flat plate [113], and in many other configurations, such as jets and wakes. The intensities for pipe flow are plotted as a fraction of the friction velocity in Fig. 14–24 (they are also often plotted as a fraction of the maximum or center-line velocity). In Fig. 14–24 there are some

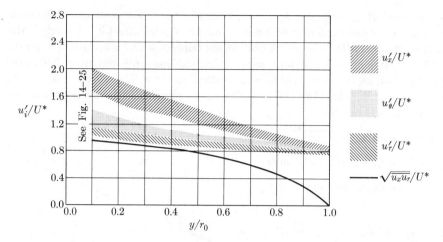

Fig. 14–24. *Turbulent velocities for pipe flow.* [*From J. Laufer*, NACA Report 1174 (*1954*). *By permission.*]

variations of the values with Reynolds number. These variations are not shown in the figure, except as a range of values reported in the literature. For reference, Fig. 14–24 also gives the shear stress over the pipe radius.

Near the center of the pipe the effect of Reynolds number is small, which indicates that viscosity in this area is negligible. At the center, all the rms values converge to nearly the same value, which indicates that the flow in this area may be isotropic. However, as we shall see later, this is apparently not true, because the spectra of the three rms fluctuating velocities are quite different. The longitudinal velocity fluctuation u'_x is the largest over the entire pipe. Note that this rms fluctuation is one that does not appear in the partially integrated Reynolds equation. However, it is evident that the energy from the mean flow goes to u_x motion, and this in turn transfers energy to the lateral motions. This conclusion is derived from the knowledge that the point of maximum production of turbulent energy (Fig. 14–17) is the point at which u'_x is a maximum, and at this point the other intensities are rather small (Fig. 14–25). For the area near the wall ($y/r_0 <$ 0.1), the effect of the Reynolds number is large; for y^+ less than 20, the rms fluctuating velocities are nearly independent of Reynolds number. Figure 14–25 shows the results.

The data obtained for boundary layer flow are plotted in Fig. 14–26. The boundary layer was 3 in. thick at the point of measurement, and the free stream velocity was 50 ft/sec. The dashed lines near the wall are extrapolations based on information obtained by Laufer for flow in a pipe. In the boundary layer, all the fluctuations of velocity went to zero as the free-stream boundary was approached. The free stream was a flow in the x-direction which was almost free of turbulence. This is contrasted with the pipe flow system in which the fluctuations in turbulence drop to a low value but not to zero. The data in Figs.

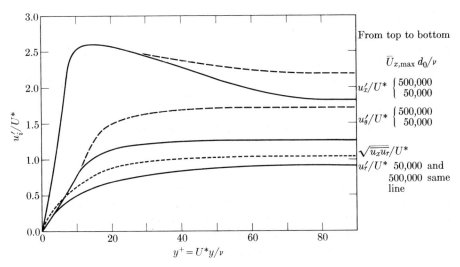

Fig. 14–25. Showing rms velocity distribution near wall of a pipe ($y/r_0 < 0.1$). [From J. Laufer, NACA Report 1174 (1954). By permission.]

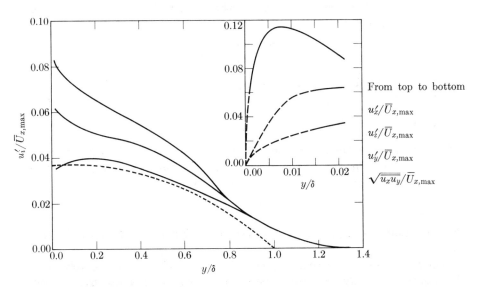

Fig. 14–26. Showing rms velocity distribution on a flat plate. [From P. S. Klebanoff, NACA Report 1247 (1955). By permission.]

14–24 and 14–26 can be compared, and, as shown by Schubauer [215], the comparison is excellent when we consider the phenomenon of intermittency. We shall discuss intermittency in the section on turbulent shear flow, and at that time we shall compare Figs. 14–24 and 14–26.

There have been many other measurements of the characteristic parameters of turbulence. Most of the air data are reviewed in Hinze [96]. A comprehensive set of measurements of the flow of air through pipes has been presented by Brookshire [23]. Pressure fluctuations, although more difficult to cope with, have recently been measured by several authors [112, 258–262]. Data on liquid systems are quite limited [4, 25, 31, 201]. The lack of extensive data here, like the lack of measurements of fluctuations of pressure, can be attributed to the relatively inadequate detection systems available. However, many investigations are in progress, and one can expect to see improved results appearing in the literature in the not-too-distant future.

(d) Scale. Scale, which we sometimes think of as being the average size of the eddies, has several possible definitions, depending on which correlation function we are using. If we first consider the Eulerian system of coordinates, we can define a scale as the area under the correlation-versus-separation curve, such as is shown in Fig. 14–21. In general,

$$\mathsf{L}_{ij} = \int \mathsf{R}_{ij}(r) \, dr, \tag{14-117}$$

where L_{ij} is the *Eulerian scale*. For isotropic turbulence [8], the longitudinal integral scale r is defined as

$$L_f = \int_0^\infty f(r) \, dr, \tag{14-118}$$

and the lateral integral scale n is defined as

$$L_g = \int_0^\infty g(r) \, dr. \tag{14-119}$$

This scale is also called the *transverse Eulerian scale*; that is, the separation vector is taken as being normal to the direction of flow. Since $f(r)$ and $g(r)$ are related, there must be a relation between the scales, which is

$$L_g = \tfrac{1}{2} L_f. \tag{14-120}$$

The notation l_2 or L_n is sometimes used for the Eulerian scale L_g, and l_0 is used for L_f.

The Lagrangian system follows the path of a particle, with the correlation given by Eq. (14–107). Although this system of coordinates is much easier to use in many cases, measurement of the corresponding correlation is complicated. However, the correlation has recently been estimated by means of measurement of turbulent diffusion in a turbulent shear flow [1–3] and in the decaying isotropic turbulence in the wake of a grid in a wind tunnel [251] (the method of measurement will be discussed later). A time scale based on the Lagrangian coordinate system would be

$$T_{\mathrm{L}} = \int_0^\infty R_{\mathrm{L}}(\tau) \, d\tau. \tag{14-121}$$

Several Lagrangian length scales can be defined. The transverse Lagrangian scale is defined as

$$L_\mathrm{L} = v'_y T_\mathrm{L}, \qquad (14\text{--}122)$$

where the velocity is Lagrangian, and another Lagrangian length scale is defined as

$$\Lambda_\mathrm{L} = \bar{U}_x T_\mathrm{L}. \qquad (14\text{--}123)$$

The term l_1 is sometimes used for L_L.

The relation between the Lagrangian and Eulerian scales has not been established by means of theory; the scales vary in many cases by some numerical value that depends on the nature of the motion. The value has to be determined by experiments. For now, it suffices to point out that the ratio of L_L/L_f has been reported as varying from 2 to 6.5 for various conditions. We shall make further comments on the relation between the Eulerian and the Lagrangian systems after we discuss turbulent diffusion, since this is essential to obtaining values in the Lagrangian system.

For very small separations, the correlation function approaches unity and is characteristic of the small eddies in the system. The shape of the correlation function near a separation distance of zero might in some way be used as a scale for the small eddies. The *microscale* of turbulence, which is related to this shape, is quite important; consequently, it will be further discussed and defined after we have introduced the mathematics necessary to the study of statistical turbulence.

(e) Spectrum. The equations to be presented shortly, which are descriptive of a turbulent field, involve double and triple correlations. We can considerably simplify the mathematics if we use Fourier transforms of the equations. The transform is simply a method of representing the complex random waveform of turbulent motion by what is equivalent to a sum of sine waves of various amplitudes and frequencies. (The sum of the sine waves must be equivalent to the original waveform.) Thus we can think of the fluctuation intensity as being distributed over a frequency range. In general, the spectrum is reported as a plot of the amplitude of the various sine waves against their respective frequencies. Mathematically, the analysis involves taking the transform of the various correlations already considered. However, one must be sure that a finite transform exists. If it does not, one can proceed in terms of *cumulants*, as discussed on page 21 of reference 8. For isotropic turbulence, double and triple correlation transformations exist without approximation. The transformed correlations have the form of an energy spectrum, and can provide insight into the distribution of the energy of turbulence over the frequencies of fluctuations. (The wave number, which is directly related to the frequency, can be pictured roughly as an inverse eddy size.) This important mathematical advance involving the application of the Fourier transform was made by Taylor [238]. He considered a one-dimensional spectrum, but Heisenberg [95] and others have extended the analysis to the three-dimensional spectrum.

274   TURBULENCE AND MIXING

The Fourier transform of the velocity correlation tensor (14–99) is called the *energy spectrum tensor*. In tensor notation this is given by

$$\bar{\Phi}_{ij}(\mathbf{k}) = (1/8\pi^3) \int_{-\infty}^{\infty} \mathbf{Q}_{ij}(\mathbf{r}) \exp(-i\mathbf{k}\cdot\mathbf{r})\, d\mathbf{r}, \qquad (14\text{–}124)$$

where the i in the exponent is $\sqrt{-1}$ and $\mathbf{k}$ is the wave number vector. The components of this vector are related to the frequency f and to the wavelength λ by

$$k_i = 2\pi/\lambda_i = 2\pi f_i/\bar{U}_i. \qquad (14\text{–}125)$$

Like the correlation on which it depends, the energy spectrum tensor is in general time dependent. For simplicity, this will not be shown, and we will use the abbreviated notation

$$\bar{\Phi}_{ij}(\mathbf{k}) = \bar{\Phi}_{ij}(\mathbf{k}, t).$$

A single Fourier element, $\mathbf{k}$, of the energy spectrum tensor is a shear wave covering all of physical space. The wavelength and frequency of this wave are given by Eq. (14–125). An illustration of a Fourier wave is given in Fig. 14–27. In effect, this figure shows one frequency of a complex turbulent waveform (in real physical space $\mathbf{x}$) which is transformed into wave number space ($\mathbf{k}$).

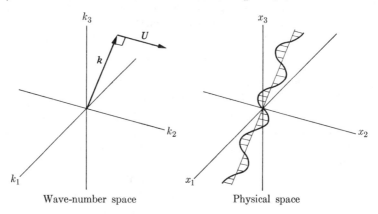

Fig. 14–27. *Representation of shear waves.* [*From S. Corrsin, J. Geophys. Res.* **64**, *2134–2150 (1959). By permission.*]

A Fourier series representation is used to define periodic functions; the accuracy of the representation is determined by the number of terms used in the series. Turbulence is a random function (nonperiodic) and an integral representation is used. The degree of approximation depends on the extent of the limits of integration; the representation is exact when the limits are infinite.

For the reverse transformation,

$$\mathbf{Q}_{ij}(\mathbf{r}) = \int_{-\infty}^{\infty} \bar{\Phi}_{ij}(\mathbf{k}) \exp(i\mathbf{k}\cdot\mathbf{r})\, d\mathbf{k}. \qquad (14\text{–}126)$$

The term $d\mathbf{k}$ is understood to mean $d\mathbf{k} = dk_i dk_j dk_k$. In other words, it is an element of volume in wave number space about the vector $\mathbf{k}$. The energy spectrum tensor is symmetrical, since its transform is symmetrical; thus, in the most general case, there are six possible different terms for the spectrum tensor. If $\mathbf{r} = 0$, Eq. (14-126) reduces to

$$\mathsf{Q}_{ij}(0) = \int_{-\infty}^{\infty} \bar{\bar{\Phi}}_{ij}(\mathbf{k})\, d\mathbf{k}. \tag{14-127}$$

The term $\mathsf{Q}_{ij}(0)$ has already been described as the energy tensor [see the discussion following Eq. (14-99)]. The energy spectrum tensor $\bar{\bar{\Phi}}_{ij}(\mathbf{k})$ can be pictured as being an energy density in wave number space. By analogy to mass, the amount of energy is equal to the total integral over the volume of the point densities times the differential volume. In this analogy, $\mathbf{k}$ is the volume, $d\mathbf{k}$ is the differential volume, $\bar{\bar{\Phi}}_{ij}(\mathbf{k})$ the density, and $\mathsf{Q}_{ij}(0)$ the mass.

In future sections, we shall encounter the triple velocity correlation. Its Fourier transformation is:

$$\bar{\bar{\omega}}_{ij,k}(\mathbf{k}) = (1/8\pi^3) \int_{-\infty}^{\infty} \mathsf{S}_{ij,k}(\mathbf{r}) \exp(-i\mathbf{k}\cdot\mathbf{r})\, d\mathbf{r}, \tag{14-128}$$

where $\mathsf{S}_{ij,k}(\mathbf{r})$ is given by Eq. (14-114). The reverse transformation is

$$\mathsf{S}_{ij,k}(\mathbf{r}) = \int_{-\infty}^{\infty} \bar{\bar{\omega}}_{ij,k}(\mathbf{k}) \exp(i\mathbf{k}\cdot\mathbf{r})\, d\mathbf{k}. \tag{14-129}$$

To summarize, the energy spectrum tensor is the transformation of the correlation tensor from Eulerian space to wave number space. The spectrum tensor indicates the way in which the energy associated with each velocity component is distributed over various wave numbers or frequencies. In essence, a Fourier analysis of the complex turbulent waveform shows how turbulent energy is distributed as a function of wave number or frequency. We have noted before that the wave number is often considered to be a measure of the reciprocal of an eddy size, but this should not be taken too literally; we should not infer from it that specific eddies of a given size actually rotate in the fluid. Rather, since any one Fourier element covers all of physical space, we should simply say "eddies of size k^{-1}" [41].

Since the energy spectrum tensor is so complicated, it has not been measured completely. One can measure the one-dimensional spectrum, proposed by Taylor, by means of an electronic harmonic analyzer operating on the output of a turbulence detector, such as a hot-wire anemometer. The one-dimensional spectrum is defined as the integration of the energy spectrum over all possible lateral values of $\mathbf{k}$; that is,

$$\bar{\bar{\phi}}_{ij}(k_l) = \int_{-\infty}^{\infty} \int_{-\infty}^{\infty} \bar{\bar{\Phi}}_{ij}(\mathbf{k})\, dk_m\, dk_n. \tag{14-130}$$

The subscripts lmn denote components, just as the subscripts ijk do, but the two

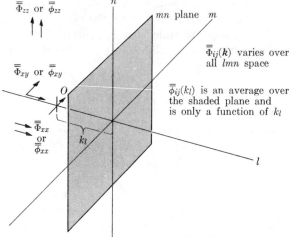

Fig. 14–28. *The one-dimensional spectrum.*

sets may operate separately. As shown in Fig. 14–28, the one-dimensional spectrum is a summation value of $\bar{\Phi}_{ij}(\boldsymbol{k})$ over a plane mn, which is located at a distance k_l from the origin O. Thus this spectrum $\bar{\phi}_{ij}(k_l)$, must depend on k_l. The one-dimensional spectrum is also the Fourier transformation of the corresponding correlation; that is,

$$\bar{\phi}_{ij}(k_l) = \left(\frac{1}{2\pi}\right) \int_{-\infty}^{\infty} \mathsf{Q}_{ij}(r_l) \exp(-ik_l r_l)\, dr_l. \tag{14-131}$$

In isotropic turbulence there would be a longitudinal one-dimensional spectrum ($i = j = r$) based on Eq. (14–104), and a lateral one-dimensional spectrum ($i = j = n$) based on Eq. (14–105).

The energy spectrum tensor is a function of the vector $\boldsymbol{k}$. An *integrated energy spectrum tensor*, which is a function of a scalar variable k, can be obtained by integrating the energy spectrum tensor over a spherical surface of radius $k = |\boldsymbol{k}|$. That is,

$$\mathsf{E}_{ij}(k) = \int_S \bar{\Phi}_{ij}(\boldsymbol{k})\, dS(k), \tag{14-132}$$

where $dS(k)$ is an element on the surface of the sphere of radius k. The term $\mathsf{E}_{ij}(k)$ can be pictured as the energy contribution per wave number, since it is a product of a density and an area term in wave number space. [Compare the forms of Eqs. (14–127) and (14–132).] The term $\mathsf{E}_{ij}(k)\, dk$ would be the contribution to the energy tensor $\mathsf{Q}_{ij}(0)$, from wave numbers in the range of k to $k + dk$, which would be the space between spheres of radii k and $k + dk$ in wave number space.

Another term associated with spectrum is the *three-dimensional energy spectrum function*, which is defined as

$$E(k) = \tfrac{1}{2}\mathsf{E}_{ii}(k), \tag{14-133}$$

and is equivalent to the kinetic energy density per wave number. The total kinetic energy is the integration of this result over all wave numbers:

$$\tfrac{1}{2}Q_{ii}(0) = \tfrac{1}{2}u_i'^2 = \tfrac{1}{2}(u_x'^2 + u_y'^2 + u_z'^2) = \int_0^\infty E(k)\,dk. \tag{14-134}$$

The foregoing spectrum equations are greatly simplified if we assume turbulence that is isotropic, for, as we have seen, the correlation tensor can be represented by a single scalar, and thus the same must hold true for the spectrum tensor. In addition, a number of interrelations can be derived for the terms so far defined; most of these can be found in books by Batchelor [8] and Hinze [96].

For isotropic turbulence and for a sphere of radius k in wave number space (the area is $4\pi k^2$), Eq. (14–132) can be integrated to give

$$E(k) = \tfrac{1}{2}\mathsf{E}_{ii}(k) = 4\pi k^2 \tfrac{1}{2}\bar{\bar{\Phi}}_{ii}(\boldsymbol{k}), \tag{14-135}$$

where all ij terms are zero by symmetry. The single scalar function $E(k)$ is generally used to describe the energy spectrum tensor. Equation (14–134), for the isotropic case, reduces to

$$\tfrac{1}{2}3u'^2 = \tfrac{3}{2}u'^2 = \int_0^\infty E(k)\,dk. \tag{14-136}$$

Direct experimental measurement of $E(k)$ has not been possible. The one-dimensional spectra can be measured experimentally and related to the three-dimensional spectrum function. For the longitudinal one-dimensional spectrum, the equation is

$$\bar{\bar{\phi}}_{rr}(k_l) = \int_{-\infty}^\infty \int_{-\infty}^\infty \bar{\bar{\Phi}}_{rr}(\boldsymbol{k})\,dk_m\,dk_n$$

$$= \frac{1}{2\pi}\int_{-\infty}^\infty u'^2 f(r_l) \exp(-ik_l r_l)\,dr_l$$

$$= \frac{1}{2}\int_{k_l}^\infty \left(1 - \frac{k_l^2}{k^2}\right)\frac{E(k)}{k}\,dk, \tag{14-137}$$

where k_l and r_l are associated with the separation distance $\boldsymbol{r}$. The inverse relation is

$$E(k) = k^3 \frac{d}{dk}\left[\frac{1}{k}\frac{d\bar{\bar{\phi}}_{rr}(k)}{dk}\right]. \tag{14-138}$$

Similar relations exist for the lateral (nn) one-dimensional spectrum; there the

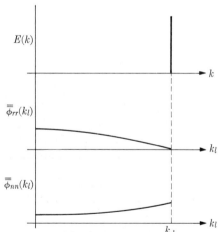

Fig. 14–29. Dirac-type three-dimensional isotropic spectrum and the two corresponding kinds of one-dimensional spectra. [From S. Corrsin, J. Geophys. Res. **64**, 2134–2150 (1959). By permission.]

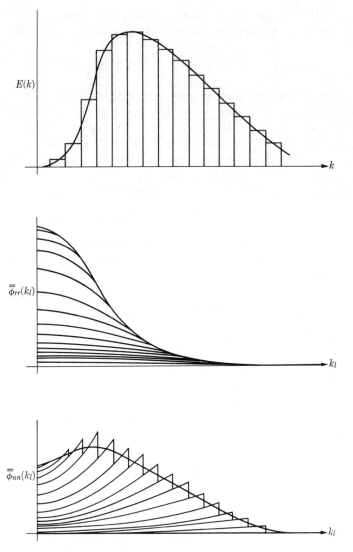

Fig. 14–30. An extension of Fig. 14–29 to a continuous $E(k)$. [From S. Corrsin, J. Geophys. Res. **64**, *2134–2150 (1959). By permission.*]

correlation is $g(r)$. The only additional difference is that in the last part of Eq. (14–137), a plus sign replaces the minus sign.

Corrsin [41] has illustrated the expected shape of the various spectra. The term $E(k)$ is represented as a Dirac function at some k_d (see Fig. 14–29). The last part of Eq. (14–137) (and its counterpart for the lateral spectrum) can be integrated to a parabolic form. Figure 14–30 shows the combination of many differential widths.

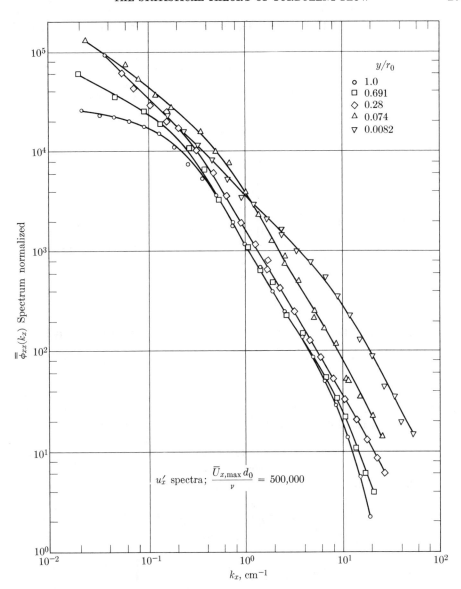

Fig. 14–31. Velocity spectrum during turbulent pipe flow. [From J. Laufer, NACA Report 1174 (1954). By permission.]

Favre *et al.* [70] have obtained excellent experimental confirmation of the validity of Eq. (14–137) in an isotropic field (see also Hinze [96]). In a later section we shall discuss the numerous spectra measurements that have been made. For now, we shall only give an example from the work of Laufer [139] on turbulent shear flow in pipes; see Fig. 14–31.

EXAMPLE 14–4. Extend Example 14–3 to obtain the spectrum terms $\bar{\phi}_{rr}(k)$ and $E(k)$.

Answer. Equation (14–137) can be rearranged as follows:

$$\bar{\phi}_{rr}(k_l) = \frac{1}{2\pi} \int_0^\infty u'^2 f(r_l) [\exp(ik_l r_l) + \exp(-ik_l r_l)] \, dr_l$$

$$= \frac{1}{\pi} \int_0^\infty u'^2 f(r_l) \cos(k_l r_l) \, dr_l. \tag{i}$$

Let $x^2 = r_l^2/8vt$ and $dx = dr_l/\sqrt{8vt}$, so that (for constant u')

$$\bar{\phi}_{rr}(k_l) = \frac{u'^2 \sqrt{8vt}}{\pi} \int_0^\infty e^{-x^2} \cos 2bx \, dx, \tag{ii}$$

where $b = \tfrac{1}{2} k_l \sqrt{8vt}$. The integral (ii) can be evaluated by reduction to known real integral forms by means of contour integration. The closed integral of $\exp(-z^2) = \exp(-x^2 - 2ixy + y^2)$ around the boundary shown in Fig. 14–32 is zero; that is,

$$\oint_C e^{-z^2} \, dz = 0 \tag{iii}$$

by Cauchy's integral theorem. Expressed as a sum of the various parts, it is

$$\int_{-a}^{a} e^{-x^2} dx + ie^{-a^2} \int_0^b e^{y^2 - 2iay} dy + e^{b^2} \int_a^{-a} e^{-x^2 - 2ibx} dx + ie^{-a^2} \int_b^0 e^{y^2 + 2iay} dy = 0. \tag{iv}$$

To obtain the desired limits, we let $a \to \infty$. The second and last terms approach zero by virtue of the fact that $\exp(-a^2)$ goes to zero. The integral

$$\int_0^\infty e^{-x^2} dx = \tfrac{1}{2}\sqrt{\pi}. \tag{v}$$

This, combined with (iv) (for the real part), gives

$$e^{b^2} \int_\infty^{-\infty} e^{-x^2} \cos 2bx \, dx = -\sqrt{\pi}, \qquad -2e^{b^2} \int_0^\infty e^{-x^2} \cos 2bx \, dx = -\sqrt{\pi},$$

or

$$\int_0^\infty e^{-x^2} \cos 2bx \, dx = \tfrac{1}{2}\sqrt{\pi} \, e^{-b^2}. \tag{vi}$$

When we use Eq. (vi), we can integrate Eq. (ii) to

$$\bar{\phi}_{rr}(k) = u'^2 \frac{\sqrt{8vt\pi}}{2\pi} \exp\left(\frac{-k^2 8vt}{4}\right)$$

$$= \frac{1}{2} u'^2 \sqrt{\frac{8vt}{\pi}} \exp(-2k^2 vt). \tag{vii}$$

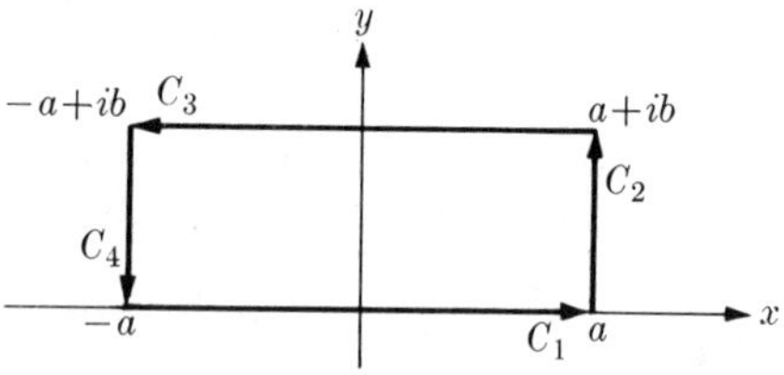

Fig. 14–32. Evaluation of integral by contour integration.

We can obtain $E(k)$ from Eq. (14–138).

$$\frac{d\bar{\phi}_{rr}(k)}{dk} = u'^2 \sqrt{\frac{8vt}{\pi}} (-2kvt) \exp(-2k^2 vt),$$

$$E(k) = k^3 u'^2 \sqrt{8vt/\pi}\, v^2 t^2 (8k) \exp(-2k^2 vt)$$
$$= 8u'^2 v^2 t^2 \sqrt{8vt/\pi}\, k^4 \exp(-2k^2 vt). \tag{viii}$$

This result shows that during the final period of decay, $E(k)$ behaves like k^4 as k approaches zero. It is possible that we might find the transform to (i) directly in one of the elaborate tables of Fourier transforms.

(f) **Probability distribution.** Turbulent motion cannot be fully described by just the correlation function or the spectral density of turbulence; we must also define the probability distribution of the turbulent fluctuations. Since the turbulence is random, we may use probability analysis. However, the fluctuations may not have a normal (i.e., Gaussian) distribution. Relatively few measurements have been made. Batchelor [8] has reviewed much of the work and noted that the experimental evidence points to a normal distribution for the velocity fluctuations at a single point. A normal distribution exists for the joint probability distribution $u_x(\boldsymbol{x} + \boldsymbol{r}) - u_x(\boldsymbol{x})$, if the separation distance is large, but the distribution is significantly non-normal for small values of $\boldsymbol{r}$.* We shall have occasion to return to this area of study, since we can make certain assumptions (associated with the probability distribution) to obtain a solution for the equations of turbulence.

(g) **Summary of terms.** The more important equations given in the preceding sections are:

(1) Eulerian correlation tensor,

$$\mathsf{Q}_{ij}(\boldsymbol{r}) = \overline{u_i(\boldsymbol{x})u_j(\boldsymbol{x}+\boldsymbol{r})}. \tag{14–99}$$

(2) Eulerian correlation function,

$$\mathsf{R}_{ij}(\boldsymbol{r}) = \frac{\overline{u_i(\boldsymbol{x})u_j(\boldsymbol{x}+\boldsymbol{r})}}{\overline{u'_i(\boldsymbol{x})u'_j(\boldsymbol{x}+\boldsymbol{r})}}. \tag{14–101}$$

(3) Eulerian correlation function for isotropic turbulence,

$$f(r) = \overline{u_r(\boldsymbol{x})u_r(\boldsymbol{x}+\boldsymbol{r})}/u'^2, \tag{14–104}$$

$$g(r) = \overline{u_n(\boldsymbol{x})u_n(\boldsymbol{x}+\boldsymbol{r})}/u'^2. \tag{14–105}$$

(4) Lagrangian correlation function,

$$\mathsf{R}_{L_{ij}}(\tau) = \frac{\overline{v_i(t)v_j(t+\tau)}}{\overline{v'_i(t)v'_j(t+\tau)}}. \tag{14–107}$$

---

*More recent measurements have been given by Frenkiel and Klebanoff [285].

(5) Lagrangian correlation function for isotropic turbulence,

$$R_L(\tau) = \frac{\overline{v(t)\,v(t+\tau)}}{v'^2}. \tag{14-108}$$

(6) Eulerian triple correlation function for isotropic turbulence,

$$k(r) = \overline{u_r(\boldsymbol{x})^2 u_r(\boldsymbol{x}+\boldsymbol{r})}/u'^3,$$
$$h(r) = \overline{u_n(\boldsymbol{x})^2 u_r(\boldsymbol{x}+\boldsymbol{r})}/u'^3, \tag{14-115}$$
$$q(r) = \overline{u_r(\boldsymbol{x})\,u_n(\boldsymbol{x})\,u_n(\boldsymbol{x}+\boldsymbol{r})}/u'^3.$$

(7) Intensity,

$$u'_x = \sqrt{\overline{u_x^2}}, \qquad u'_y = \sqrt{\overline{u_y^2}}, \qquad u'_z = \sqrt{\overline{u_z^2}}.$$

(8) Eulerian scales,

$$\mathsf{L}_{ij} = \int_0^\infty \mathsf{R}_{ij}(r)\,d\boldsymbol{r}, \tag{14-117}$$

$$L_f = \int_0^\infty f(r)\,dr, \tag{14-118}$$

$$L_g = \int_0^\infty g(r)\,dr, \tag{14-119}$$

$$L_g = \tfrac{1}{2} L_f. \tag{14-120}$$

(9) Lagrangian scales,

$$T_L = \int_0^\infty R_L(\tau)\,d\tau, \tag{14-121}$$

$$L_L = v'_y T_L. \tag{14-122}$$

(10) Energy spectrum tensor,

$$\bar{\Phi}_{ij}(\boldsymbol{k}) = \frac{1}{(2\pi)^3} \int_{-\infty}^\infty \mathsf{Q}_{ij}(\boldsymbol{r})\,e^{-i(\boldsymbol{k}\cdot\boldsymbol{r})}\,d\boldsymbol{r}. \tag{14-124}$$

(11) Energy tensor,

$$\mathsf{Q}_{ij}(0) = \int_{-\infty}^\infty \bar{\Phi}_{ij}(\boldsymbol{k})\,d\boldsymbol{k}. \tag{14-127}$$

(12) One-dimensional spectrum,

$$\bar{\phi}_{ij}(k_l) = \int_{-\infty}^\infty \int_{-\infty}^\infty \bar{\Phi}_{ij}(\boldsymbol{k})\,dk_m\,dk_n. \tag{14-130}$$

(13) Integrated energy spectrum tensor,

$$\mathsf{E}_{ij}(k) = \int_S \bar{\Phi}_{ij}(\boldsymbol{k})\,dS(k). \tag{14-132}$$

(14) Three-dimensional energy spectrum function,

$$E(k) = \tfrac{1}{2} \mathsf{E}_{ii}(k). \tag{14-133}$$

14-4.B Equations of Statistical Turbulence

An equation which relates the various correlation tensors can be derived from the Navier-Stokes equation written at two points. The equation of motion in terms of the instantaneous velocity has already been presented as Eq. (14–9). Let us now assume that the mean velocity $\bar{U}$ is constant in the region under consideration and is independent of time. When we derived the Reynolds equation, we did not take $\bar{U}$ as constant at the single arbitrary point considered. With the restriction of constant $\bar{U}$ and no external forces, Eq. (14–9) reduces to

$$\frac{D\mathbf{u}}{Dt} = \frac{\partial \mathbf{u}}{\partial t} + 2\bar{U}(\boldsymbol{\nabla}\cdot\mathbf{u}) + (\boldsymbol{\nabla}\cdot\mathbf{u}\mathbf{u}) = -\frac{1}{\rho}\boldsymbol{\nabla}p + \nu\nabla^2\mathbf{u}. \tag{14-139}$$

Let us first consider this equation as written at a point A, then multiply it by the fluctuating velocity at point B, $(\mathbf{u}_B)$, to obtain

$$\mathbf{u}_B\frac{\partial \mathbf{u}_A}{\partial t} + 2\bar{U}(\boldsymbol{\nabla}_A\cdot\mathbf{u}_A\mathbf{u}_B) + (\boldsymbol{\nabla}_A\cdot\mathbf{u}_A\mathbf{u}_A\mathbf{u}_B) = -\frac{1}{\rho}\boldsymbol{\nabla}_A p_A\mathbf{u}_B + \nu\nabla_A^2\mathbf{u}_A\mathbf{u}_B, \tag{14-140}$$

where we have considered $\mathbf{u}_B$ a constant in a differentation at point A. We can obtain a similar equation at point B by reversing the subscripts A and B:

$$\mathbf{u}_A\frac{\partial \mathbf{u}_B}{\partial t} + 2\bar{U}(\boldsymbol{\nabla}_B\cdot\mathbf{u}_A\mathbf{u}_B) + (\boldsymbol{\nabla}_B\cdot\mathbf{u}_A\mathbf{u}_B\mathbf{u}_B) = -\frac{1}{\rho}\boldsymbol{\nabla}_B p_B\mathbf{u}_A + \nu\nabla_B^2\mathbf{u}_A\mathbf{u}_B. \tag{14-141}$$

We can add Eqs. (14–140) and (14–141) and then average them to obtain

$$\frac{\partial \overline{\mathbf{u}_A\mathbf{u}_B}}{\partial t} + 2\bar{U}[(\boldsymbol{\nabla}_A\cdot\overline{\mathbf{u}_A\mathbf{u}_B}) + (\boldsymbol{\nabla}_B\cdot\overline{\mathbf{u}_A\mathbf{u}_B})] + (\boldsymbol{\nabla}_A\cdot\overline{\mathbf{u}_A\mathbf{u}_A\mathbf{u}_B}) + (\boldsymbol{\nabla}_B\cdot\overline{\mathbf{u}_A\mathbf{u}_B\mathbf{u}_B})$$
$$= -\frac{1}{\rho}[\boldsymbol{\nabla}_A \overline{p_A\mathbf{u}_B} + \boldsymbol{\nabla}_B \overline{p_B\mathbf{u}_A}] + \nu[\nabla_A^2\overline{\mathbf{u}_A\mathbf{u}_B} + \nabla_B^2\overline{\mathbf{u}_A\mathbf{u}_B}]. \tag{14-142}$$

We can write $\boldsymbol{\nabla}_A$ and $\boldsymbol{\nabla}_B$ in terms of the separation vector $\mathbf{r} = \mathbf{x}_B - \mathbf{x}_A$; that is,

$$\boldsymbol{\nabla}_A = -\boldsymbol{\nabla}_r, \qquad \boldsymbol{\nabla}_B = \boldsymbol{\nabla}_r, \qquad \nabla_A^2 = \nabla_B^2 = \nabla_r^2. \tag{14-143}$$

With these, Eq. (14–142) becomes

$$\frac{\partial \overline{\mathbf{u}_A\mathbf{u}_B}}{\partial t} + [\boldsymbol{\nabla}_r\cdot(\overline{\mathbf{u}_A\mathbf{u}_B\mathbf{u}_B} - \overline{\mathbf{u}_A\mathbf{u}_A\mathbf{u}_B})] = -\frac{1}{\rho}[\boldsymbol{\nabla}_r(\overline{p_B\mathbf{u}_A} - \overline{p_A\mathbf{u}_B})] + 2\nu\nabla_r^2\overline{\mathbf{u}_A\mathbf{u}_B}. \tag{14-144}$$

From Eqs. (14–98) and (14–99), we write $\overline{\mathbf{u}_A\mathbf{u}_B}$ as $\mathsf{Q}_{ij}(\mathbf{r})$. To convert to general cartesian notation, the subscript i has been associated with point A, and j with point B; thus, from Eq. (14–114), $\overline{\mathbf{u}_A\mathbf{u}_A\mathbf{u}_B}$ is $\mathsf{S}_{ik,j}(\mathbf{r})$ and $\overline{\mathbf{u}_A\mathbf{u}_B\mathbf{u}_B}$ is $\mathsf{S}_{i,jk}(-\mathbf{r})$ or $\mathsf{S}_{i,kj}(-\mathbf{r})$; we obtained the latter form because of the symmetry in the indices when both are at the same point ($\mathbf{u}_B$ is $\mathbf{u}_B$, no matter what the order). In general cartesian coordinate notation Eq. (14–144) becomes

$$\frac{\partial Q_{ij}(r)}{\partial t} - \frac{\partial [S_{ik,j}(r) - S_{i,kj}(-r)]}{\partial r_k} = -\frac{1}{\rho}\left(\frac{\partial \overline{p_B u_{iA}}}{\partial r_j} - \frac{\partial \overline{p_A u_{jB}}}{\partial r_i}\right) + 2\nu \frac{\partial^2 Q_{ij}(r)}{\partial r_k \partial r_k}. \quad (14\text{–}145)$$

In this equation, the summation over the repeated index is implied, and the minus sign on the argument of $S_{i,kj}(-r)$ is redundant when the comma is included, since it also implies that the correlation is from B to A. The unlike subscripts assume all possible combinations, as is apparent from the equation in tensor form (14–144).

The notation can be simplified by the following definitions and transforms to:

$$T_{ij}(r) = \frac{\partial}{\partial r_k}[S_{ik,j}(r) - S_{i,kj}(-r)] = \int_{-\infty}^{\infty} \bar{\omega}_{ij}(k)\, e^{i(k \cdot r)}\, dk \quad (14\text{–}146)$$

and

$$P_{ij}(r) = \frac{1}{\rho}\left(\frac{\partial \overline{p_B u_{iA}}}{\partial r_j} - \frac{\partial \overline{p_A u_{jB}}}{\partial r_i}\right) = \int_{-\infty}^{\infty} \bar{\pi}_{ij}(k)\, e^{i(k \cdot r)}\, dk. \quad (14\text{–}147)$$

These equations can be used to simplify Eq. (14–145) to

$$\frac{\partial Q_{ij}(r)}{\partial t} = T_{ij}(r) - P_{ij}(r) + 2\nu \frac{\partial^2 Q_{ij}(r)}{\partial r_k \partial r_k}, \quad (14\text{–}148)$$

and in turn transform it to

$$\partial \bar{\Phi}_{ij}(k)/\partial t = \bar{\omega}_{ij}(k) - \bar{\pi}_{ij}(k) - 2\nu k^2 \bar{\Phi}_{ij}(k). \quad (14\text{–}149)$$

Batchelor has considered the meaning of each term of Eq. (14–149), and points out that the nonlinear term $\bar{\omega}_{ij}(k)$ indicates a flow of energy between different wave numbers (for the same velocity component). It changes the wave number density of contributions to the energy tensor $\overline{u_i u_j}$, but does not affect the total contribution; i.e., its integral over all wave numbers is zero. The nonlinear pressure term $\bar{\pi}_{ij}(k)$ indicates a transfer of energy between different velocity components (for the same wave number), and gives the balance of total energy contributed by a small region dk in wave number space. These results are drawn in part from continuity, which shows that if the fluid is incompressible $P_{ii}(r)$, and thus $\bar{\pi}_{ii}(k)$, must be zero. Batchelor further comments that the mathematics do not define the direction of the transfer, but reasoning shows that the effect of inertia is directed toward the large wave numbers or small eddies, in which energy is lost due to viscous dissipation. The effect of pressure is to equalize the mean square velocity values. This effect must be normal or lateral, since continuity can be used to show that for incompressible fluids, the ii terms must be zero.

We can simplify Eqs. (14–148) and (14–149) considerably if we assume isotropic turbulence. For example, the velocity field is uniform, so that the lateral (ij) pressure terms must be zero. In isotropic turbulence, the only terms that concern us are the inertia terms, which provide the transfer of energy between the different wave numbers. Since for isotropic turbulence the various tensors that remain in

Eqs. (14–148) and (14–149) are known to reduce to scalar functions, the equations themselves will reduce to scalar equations. Equation (14–148) [for $j = i$ and $\mathsf{P}_{ij}(r) = 0$] becomes

$$\frac{\partial \mathsf{Q}_{ii}(r)}{\partial t} = \mathsf{T}_{ii}(r) + 2\nu \frac{\partial^2 \mathsf{Q}_{ii}(r)}{\partial r_k \partial r_k}, \qquad (14\text{–}150)$$

which can be summed over the repeated subscript i. For the $\mathsf{Q}_{ii}(r)$ term:

$$\mathsf{Q}_{ii}(r) = u'^2[f(r) + 2g(r)] = u'^2\left[3 + r\frac{\partial}{\partial r}\right]f(r), \qquad (14\text{–}151)$$

where we have used Eq. (14–106). The $\mathsf{T}_{ii}(r)$ term is more complicated:

$$\mathsf{T}_{ii}(r) = \frac{\partial}{\partial r_k}[\mathsf{S}_{ik,i}(r) - \mathsf{S}_{i,ki}(-r)] = \frac{\partial}{\partial r_k}[\mathsf{S}_{ik,i}(r) + \mathsf{S}_{ik,i}(r)] = 2\frac{\partial}{\partial r_k}\mathsf{S}_{ik,i}(r), \qquad (14\text{–}152)$$

where we have used invariance under reflection and symmetry of suffixes at a single point to change

$$\mathsf{S}_{i,kj}(-r) = -\mathsf{S}_{kj,i}(r) = -\mathsf{S}_{jk,i}(r).$$

The summation of $\mathsf{S}_{ik,i}(r)$ over the repeated index [8, 96] gives

$$\mathsf{S}_{ik,i}(r) = u'^3(r_k/r)[k(r) + 2q(r)],$$

which, when combined with Eq. (14–116) reduces to

$$\mathsf{S}_{ik,i}(r) = u'^3 \frac{r_k}{r}\left(2 + \frac{r}{2}\frac{\partial}{\partial r}\right)k(r). \qquad (14\text{–}153)$$

Equations (14–152) and (14–153) can be combined to give

$$\mathsf{T}_{ii}(r) = u'^3\left(3 + r\frac{\partial}{\partial r}\right)\left(\frac{4}{r} + \frac{\partial}{\partial r}\right)k(r), \qquad (14\text{–}154)$$

if we remember that, on summation over the repeated index,

$$\frac{\partial r_k \phi(r)}{\partial r_k} = \phi(r)\frac{\partial r_k}{\partial r_k} + r_k \frac{\partial \phi(r)}{\partial r_k} = 3\phi(r) + r\frac{\partial \phi(r)}{\partial r} = \left(3 + r\frac{\partial}{\partial r}\right)\phi(r),$$

where $\phi(r)$ is any function of r only. Substitution of Eqs. (14–151) and (14–154) into (14–150) gives

$$\left(3 + r\frac{\partial}{\partial r}\right)\frac{\partial u'^2 f(r)}{\partial t} - u'^3\left(3 + r\frac{\partial}{\partial r}\right)\left(\frac{4}{r} + \frac{\partial}{\partial r}\right)k(r)$$

$$- 2\nu u'^2\left(\frac{\partial^2}{\partial r^2} + \frac{2}{r}\frac{\partial}{\partial r}\right)\left(3 + r\frac{\partial}{\partial r}\right)f(r) = 0, \qquad (14\text{–}155)$$

where we have used the spherical coordinate equivalent of the Laplacian operator (2–79). Since

$$\left(\frac{\partial^2}{\partial r^2} + \frac{2}{r}\frac{\partial}{\partial r}\right)\left(3 + r\frac{\partial}{\partial r}\right) = \left(3 + r\frac{\partial}{\partial r}\right)\left(\frac{\partial^2}{\partial r^2} + \frac{4}{r}\frac{\partial}{\partial r}\right),$$

we can rewrite Eq. (14–155) as

$$\left(3 + r\frac{\partial}{\partial r}\right)\left[\frac{\partial u'^2 f(r)}{\partial t} - u'^2\left(\frac{4}{r} + \frac{\partial}{\partial r}\right)k(r) - 2\nu u'^2\left(\frac{\partial^2}{\partial r^2} + \frac{4}{r}\frac{\partial}{\partial r}\right)f(r)\right] = 0,$$
(14–156)

and we can integrate it by considering the entire large bracketed term as one variable. The result of the integration is

$$\frac{\partial u'^2 f(r)}{\partial t} = u'^3\left(\frac{4}{r} + \frac{\partial}{\partial r}\right)k(r) + 2\nu u'^2\left(\frac{\partial^2}{\partial r^2} + \frac{4}{r}\frac{\partial}{\partial r}\right)f(r), \quad (14\text{–}157)$$

where we have taken the constant of integration as zero, since the bracketed term must remain finite at $r = 0$. An equivalent form [96] is

$$\frac{\partial u'^2 f(r)}{\partial t} = u'^3 \frac{1}{r^4}\frac{\partial}{\partial r}[r^4 k(r)] + 2\nu u'^2 \frac{1}{r^4}\frac{\partial}{\partial r}\left\{r^4\left[\frac{\partial f(r)}{\partial r}\right]\right\}. \quad (14\text{–}158)$$

Equation (14–157) is the von Kármán-Howarth equation for isotropic turbulence, all the terms in which can be obtained experimentally; Stewart [230] has found that the equation is correct for grid-generated turbulence to within the experimental error of his measurements.

The spectrum equation (14–149) can be reduced by contracting subscripts and setting $\bar{\pi}_{ii}(\mathbf{k}) = 0$ [since $P_{ii}(\mathbf{r}) = 0$] to give

$$\partial \bar{\Phi}_{ii}(\mathbf{k})/\partial t = \bar{\omega}_{ii}(\mathbf{k}) - 2\nu k^2 \bar{\Phi}_{ii}(\mathbf{k}). \quad (14\text{–}159)$$

With Eq. (14–135) and the definition

$$T(k) = 4\pi k^2 \tfrac{1}{2}\bar{\omega}_{ii}(\mathbf{k}), \quad (14\text{–}160)$$

Eq. (14–159) becomes

$$\partial E(k)/\partial t = T(k) - 2\nu k^2 E(k). \quad (14\text{–}161)$$

The scalar $T(k)$ is associated with the transfer of energy between wave numbers or eddy sizes. Its integral over all wave numbers is zero. We can define a somewhat different transfer function,

$$S(k) = -\int_0^k T(k)\,dk, \quad (14\text{–}162)$$

which is the total energy transferred from eddies in the range from 0 to k to those in the range greater than k; that is, the flux of turbulent energy from a spherical shell of radius of wave number k. When we use this modification, Eq. (14–161) becomes

$$\frac{\partial E(k)}{\partial t} = -\frac{\partial S(k)}{\partial k} - 2\nu k^2 E(k). \quad (14\text{–}163)$$

We can establish the time rate of change of the kinetic energy of turbulent motion by integrating Eq. (14–161) over all wave numbers:

$$\partial \int_0^\infty E(k)\,dk \Big/ \partial t = \int_0^\infty T(k)\,dk - 2\nu \int_0^\infty k^2 E(k)\,dk. \qquad (14\text{–}164)$$

The first term is the rate of change of kinetic energy (Eq. 14–136) and the second term is zero, since, as just noted, $T(k)$ is the transfer between wave numbers, and there is no net transfer over all wave numbers. Simplifying gives us

$$-\frac{1}{2}\frac{\partial \overline{u_i^2}}{\partial t} = -\frac{3}{2}\frac{\partial \overline{u'^2}}{\partial t} = 2\nu \int_0^\infty k^2 E(k)\,dk = \epsilon, \qquad (14\text{–}165)$$

where ϵ is the rate of dissipation of turbulent energy.

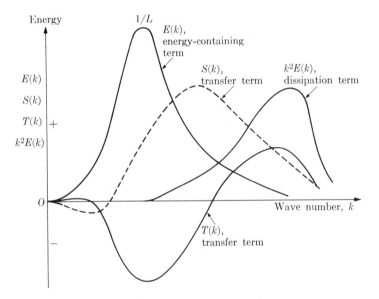

Fig. 14–33. *The scalar functions of turbulent motion.*

Equations (14–161) and (14–163) show that viscous dissipation reduces the energy from components of any one wave number, but has no effect on the transfer between wave numbers. Further, this energy dissipation is selective toward the high wave numbers, or small eddies, because of the strong effect of the k^2 factor. If we assume continuous decay, so that $\partial E(k)/\partial t$ is always negative, then $S(k)$ must always be positive and displaced toward the low wave numbers, or large eddies, since the energy must be transferred from these large eddies to the smaller ones where it will be dissipated by viscous forces. An approximate picture of these scalar functions is given in Fig. 14–33. The transfer function $T(k)$ is of the general form suggested by Proudman and Reid [194] and by Kraichnan [134]. We can obtain $S(k)$ by integration of this, by using Eq. (14–162).

Actually, Eqs. (14–157) and (14–161) cannot be solved, since in each case there are two unknowns and only one equation. We can construct a new equation for the triple velocity correlation in a manner analogous to that in which we constructed an equation for the double velocity correlation. In this new equation, however, because of nonlinearity, inertial terms of the fourth order appear, and if we were to write the fourth-order equation, the terms for the fifth order would appear. This constitutes one of the major problems of turbulence: Although any number of equations can be generated, there is always one more unknown than there are equations. This infinite hierarchy of dynamical equations gives rise to what is called *the closure problem of turbulence theory*. The problem is: By what means can we replace the indeterminate infinite set by a plausible determinate finite set, and thus gain some useful information about turbulence? The approaches of Proudman and Reid and of Kraichnan, which were mentioned above, constitute two such possible methods that we shall discuss.

14-4.C Isotropic Turbulence

The simplicity of the equations for isotropic turbulence allows extensive theoretical analysis. It is thus of great interest to know whether experiments are available to check the theoretical results.

(a) Experimental and measurement methods. The previously mentioned experiments with grids in wind tunnels are examples of stationary flows, in which decay is a function of distance from the grid rather than a function of time, as is homogeneous turbulence. In spite of this, the grid experiments can closely approximate homogeneous turbulence, if we consider a small segment of air moving away from the grid at the average stream velocity. The segment must be small enough to be considered homogeneous when compared with the inhomogeneity in the wind tunnel, but still large enough to be compared with the scale of turbulence. For such an experiment, the decay time would be $t_D = x/\bar{U}_x$, and the segment would be nearly homogeneous. In fact, these experiments are a good approximation of isotropic turbulence for distances of the order of 40 or more mesh lengths downstream. The strongest evidence (besides the near equality of the rms velocity values) was presented in Fig. 14–20, where the complete details of $g(r)$ can be obtained from $f(r)$ and the relation (14–106), which is valid for isotropic conditions only. In addition there are the experiments of Stewart [230] already cited, which show that one can predict within experimental error the triple velocity correlation $k(r)$ from $f(r)$ and Eq. (14–157), which is also valid only for isotropic conditions.

Isotropy is the most random state; thus any pressure forces would tend to cause an anisotropic system to approach isotropy. By passing the flow through a fine gauze or through a contracting duct to cause attenuation of the components in the direction of flow when compared with the lateral components, experimenters have shown that recovery to isotropic conditions is quite slow (of the same order as the rate of decay). Since it is known from experiments that the flow is nearly isotropic, it is clear that the observed isotropy is not due to the tendency of an anisotropic system to approach isotropy; rather, it must be associated with the coalescence of individual wakes and thorough mixing.

Much effort has gone into the design of experimental equipment to produce "simple" turbulent flows. There are limitations and complicating factors in any experimental setup. Fortunately, Corrsin [46], in an extensive review article, has considered the many aspects of experimental methods of producing and measuring turbulence. One should certainly consult his article before one attempts experimental work in the field.

Measurement of the rapidly varying quantities of turbulence has necessitated the development of unique experimental methods. Once one has defined the experimental system, one must consider the actual means of measurement to be used. A number of recent publications have covered this subject in some detail; consequently, it is not necessary to repeat the finer points here. The material that follows is a brief outline, listing references which contain detailed accounts.

A. General [46, 47, 96]
B. Hot-wire and hot-film anemometers as measuring devices
   1. Probes
      (a) Commercial [74, 100, 101, 158, 159, 287]
      (b) Special [74, 159]
   2. Electronics
      (a) Control: (i) constant current [73, 100, 124, 125], (ii) constant temperature [100, 101, 159, 253, 287]
      (b) Measurement: (i) intensity by rms meters [72, 101, 159, 287], (ii) correlation [70, 73, 222, 250, 287], (iii) microscale [138, 149, 243, 245], (iv) scale [243], (v) spectrum [73, 245, 270], (vi) other [245]
   3. Theory, analysis, and limitations [46, 73, 96, 253]
C. Other methods of measuring fluctuations in velocity
   1. Electric-discharge anemometer [76, 96, 245]
   2. Electromagnetic induction [87, 96, 119]
   3. Light scattering [231]
   4. Visual [69, 96, 153–156, 228, 252]
   5. Fiber anemometer [249]
D. Measuring fluctuations in temperature
   1. Hot wire [33, 73, 96, 171]
   2. Resistance thermometer [283]
E. Measuring fluctuations in concentration
   1. Hot wire [33, 96]
   2. Resistance thermometer [283]
   3. Conductivity [137, 168, 193]
   4. Light attenuation [140, 143, 221]
   5. Light scattering [204, 206, 257]
F. Measuring fluctuations in pressure
   1. Manometers in general [147]
   2. Ultrasonics [147, 258–262]

The list and the references are by no means complete; in many cases, only the most recent works have been noted, since these contain references to earlier work.

(b) Theoretical analysis. The solution to an ideal theoretical approach to turbulence is beyond our present capabilities. However, because the method of such an approach can be easily formulated, let us discuss it further. First, we assume that the initial mean velocity $\bar{U}$ (zero for isotropic conditions) and the initial infinite set of moments, which are the nonsimultaneous correlation tensors of all orders (also called space-time correlations), are known. The moments involve different times as well as different points; for example, the two-point, two-time moment is

$$\mathsf{Q}_{ij}(\boldsymbol{x}, t; \boldsymbol{x}', t') = \overline{u_i(\boldsymbol{x}, t) u_j(\boldsymbol{x}', t')}.$$

For isotropic turbulence,

$$\mathsf{Q}_{ij}(\boldsymbol{x}, t; \boldsymbol{x}', t') = \mathsf{Q}_{ij}(\boldsymbol{r}; t, t'), \qquad (14\text{--}166)$$

since $\boldsymbol{r} = \boldsymbol{x}' - \boldsymbol{x}$. Furthermore Eq. (14–166) reduces to Eq. (14–99) for simultaneous times. The initial moments specify the initial turbulent velocity $\boldsymbol{u}$. Our problem is to follow the subsequent development of these initial values with time. One procedure would be to use the unaveraged Navier-Stokes equation (14–139) in terms of U. From the initial $\bar{U}$ and the initial moments that define $\boldsymbol{u}$, we could construct the initial U and integrate forward in time. (Alternatively, we might start with a given initial random field for U). Of general interest are $\bar{U}$ and certain moments, which would have to be reconstructed from the solution. Furthermore, many different initial-value problems would have to be investigated. As Corrsin [43] pointed out, the numerical requirements of such a problem are well beyond the capabilities of existing machines.

Instead of using the ideal approach, we can construct the equations only for the moments of interest. (This is what von Kármán and Howarth did when they obtained their two-point equation.) Kraichnan [133] has given equations for the moments of interest, in their most complex nonsimultaneous form, and thus we need not repeat them here. For some analytic solutions, the simultaneous time moments are all that is required. These have also been presented [60, 61, 173, 194, 234], and they can easily be obtained from the nonsimultaneous forms by setting the time arguments equal. The infinite set of equations links together moments of all orders, and, as already pointed out, the equation for the nth-order moment will involve the $(n + 1)$th-order moments, a consequence of the nonlinearity of the Navier-Stokes equation. To close the set of equations so that a solution can be obtained is the closure problem already cited.

(i) Intuitive analysis. Before turning to specific closure approximations, let us consider a brief qualitative picture of the mechanism of isotropic turbulence. The inertial forces in the system give rise to the transfer term $S(k)$, which is a measure of the transfer of energy from larger eddies to smaller ones. In other words, since the motion of large eddies is unstable, it gives rise to smaller eddies. These smaller eddies are in turn unstable, and give rise to still smaller eddies; this process continues until eddies of the smallest size are formed. The mechanism of the transfer is not stated, and in fact we do not need to know it at this point. However, let us say that the mechanism of the transfer might be a breakup of the large eddies or a creation of new smaller eddies by some inertial stress–strain interaction. For all but the smallest eddies, the Reynolds number (based on the size of the eddy) is large, and dissipation by viscous forces is unimportant. As the size of the eddy decreases, the Reynolds number becomes smaller and smaller, and at some critical point viscous forces become important. The suggested picture or model is a cascade of energy from large to progressively smaller eddies, until the energy is lost to heat by the dissipation action associated with the smallest of eddies.

If the Reynolds number of the flow is high enough, we would expect a separation of the energy-containing eddies, characterized by the peak in the $E(k)$ curve, from the dissipation eddies, characterized by the peak in the $k^2 E(k)$ curve (see Fig. 14–33). Let k' be a representative wave number lying between these energy-containing and dissipation ranges. Integration of Eq. (14–163) between 0 and k' gives

$$\partial \int_0^{k'} E(k)\,dk / \partial t = -S(k') - 2\nu \int_0^{k'} k^2 E(k)\,dk. \tag{14–167}$$

Since for this case all the energy lies between 0 and k' and the dissipation between k' and ∞, this equation shows that the rate of change of the kinetic energy results only from a transfer from the range 0 to k' to the range k' to ∞. Within the range k' to ∞ there is clearly a balance given by (see Eq. 14–165)

$$\epsilon = S(k) = 2\nu \int_0^{\infty} k^2 E(k)\,dk, \tag{14–168}$$

which implies that $\partial E(k)/\partial t$ is negligible within the range. Thus there is a range of wave numbers associated with the small eddies which is statistically steady, responsible for the viscous dissipation, and is not directly dependent on the energy-containing eddies. This is generally known as the *universal equilibrium range*. The theory, first postulated by Kolmogoroff (120–122, 7), states that the eddies in this equilibrium state depend only on the rate of energy dissipation ϵ and on the kinematic viscosity ν. The kinematic viscosity determines the rate at which kinetic energy can be dissipated into heat.

The length and velocity parameters characteristic of this range can easily be found. The dimensions of ϵ and ν are $\epsilon(L^2 \theta^{-3})$ and $\nu(L^2 \theta^{-1})$. The only combinations for length and velocity are

$$\eta = (\nu^3/\epsilon)^{1/4}, \qquad v = (\nu\epsilon)^{1/4}. \tag{14–169}$$

The length η can be considered as an internal scale of local turbulence for the equilibrium range (usually called the *Kolmogoroff scale*) and can be contrasted with an external or integral scale L, given by Eq. (14–117), which would be descriptive of the overall turbulent motion. The scale L is a measure of the velocity fluctuations which cause a hot-wire unit to react, and thus is a measure of the eddies which contain the turbulent energy. To an order-of-magnitude approximation, L is associated with the point of the maximum in $E(k)$, the energy-containing group. In a like manner, η is associated with the viscous dissipation group $k^2 E(k)$. Very often the Kolmogoroff wave number is used; it is defined as

$$k_\eta = 1/\eta = (\epsilon/\nu^3)^{1/4}. \tag{14-170}$$

Townsend [246] has determined from his experimental work that, if the separation is large (that is, if the viscous dissipation eddies are entirely in the equilibrium area), the centroid of the equilibrium range (balance point) is near $0.5/\eta$. From Fig. 14–33, we can see that η is also a good measure of the smallest energy-containing eddy. Thus the smallest detectable energy-containing eddy or turbulent fluctuation is of the order of η.

From dimensional analysis, we can see that the form of the spectrum $E(k)$ in the equilibrium range is

$$E(k) = v^2 \eta \psi(k\eta), \tag{14-171}$$

where $\psi(k\eta)$ is a universal function. Combining Eqs. (14–169) and (14–171) gives us

$$E(k) = \nu^{5/4} \epsilon^{1/4} \psi(k\eta). \tag{14-172}$$

If the Reynolds number is very high, so that some of the large eddies in the universal equilibrium range are independent of viscous dissipation, then for these eddies the inertia term $S(k)$ or ϵ will be controlling. These eddies receive energy from the larger ones and then transmit this energy to the small dissipative eddies. If such a range is realized, then $\psi(k\eta)$ must be of such a form as to make Eq. (14–171) or (14–172) independent of the kinematic viscosity. The form is $(k\eta)^{-5/3}$ and Eq. (14–172) becomes

$$E(k) = \nu^{5/4} \epsilon^{1/4} A \left(\frac{\nu^{3/4}}{\epsilon^{1/4}}\right)^{-5/3} k^{-5/3} = A \epsilon^{2/3} k^{-5/3}. \tag{14-173}$$

This is called the *inertial subrange*. Batchelor has shown that the required Reynolds number is about twice that required for the equilibrium range to exist.

With this picture in mind, we can now turn to the various closure methods and in some cases make comparisons with Kolmogoroff's theory. The actual predictions of this theory have had excellent experimental confirmation [80–81, 84–85].

(ii) Phenomenological approximations. The most direct closure procedure is to find a nonzero relation between $E(k)$ and $T(k)$ or $S(k)$ in Eq. (14–161) or (14–163), and thus use only the first equation of the infinite set. The required extra relation is usually found by an analysis based on dimensional grounds or on an intuitive physical model. Such a dimensional form has been suggested by von Kármán [109]:

$$S(k) = 2\alpha \int_k^\infty k'^\gamma E(k')^\beta \, dk' \int_0^k (k'')^{1/2-\gamma} E(k'')^{3/2-\beta} \, dk'', \qquad (14\text{--}174)$$

where α is a constant greater than zero, and β and γ are also constants. Tanenbaum and Mintzer [233] have used the data of Stewart and Townsend [229] and Stewart [230] to establish empirically the best values of α, β, and γ. They suggest 0.60, $\frac{1}{4}$, and $-\frac{5}{2}$, respectively; however, good results were obtained for $\frac{1}{2} > \beta > 0$, and for $-\frac{3}{2} > \gamma > -\frac{7}{2}$.

Heisenberg [94] formulated a model analytically (see also von Weizsäcker [254]) which has the same form as Eq. (14–174). In this, $\alpha = \alpha_\mathrm{H}$, $\beta = \frac{1}{2}$, and $\gamma = -\frac{3}{2}$:

$$S(k) = 2\alpha_\mathrm{H} \int_k^\infty (k')^{-3/2} E(k')^{1/2} \, dk' \int_0^k k''^2 E(k'') \, dk''. \qquad (14\text{--}175)$$

The physical model is founded on the hypothesis that energy transfer between eddies occurs by means of a stress–strain mechanism, in which the strain is provided by the large eddies, and the resistance to strain or stress by the small eddies. In effect, Heisenberg assumed that

$$S(k) = 2\bar{\varepsilon}(k) \int_0^k k^2 E(k) \, dk \quad \text{and} \quad \bar{\varepsilon}(k) = \alpha_\mathrm{H} \int_k^\infty k^{-3/2} E(k)^{1/2} \, dk,$$

where $\bar{\varepsilon}(k)$ is a turbulent viscosity or eddy diffusivity at k and is only a function of smaller eddies characterized by a wave number larger than k. Although Heisenberg's theory of eddy diffusivity has been one of the most widely accepted approaches, it has been criticized [8, 96].

Kovasznay [123] has suggested a simple local transfer form, which dimensionally must be

$$S(k) = \alpha_\mathrm{K} k^{5/2} E(k)^{3/2}. \qquad (14\text{--}176)$$

In this case the transfer term at k depends on values at k only.

Obukhov [178] reasoned that the energy transfer across k was analogous to the energy transfer in a shear flow from the mean motion to the turbulence (via the Reynolds stresses); that is, the small-scale motion sees the large-scale motion as a mean flow. Obukhov thus establishes the mean squared strain rate of the large eddies ("mean flow") as

$$2 \int_0^k k^2 E(k) \, dk$$

and the kinetic energy of the small eddies ("Reynolds stress") as

$$\int_k^\infty E(k) \, dk,$$

from Eq. (14–136). The transfer across k is the stress times the rms rate of strain, or

$$S(k) = \alpha_0 \int_k^\infty E(k')\,dk' \left[\int_0^k k''^2 E(k'')\,dk'' \right]^{1/2}. \tag{14-177}$$

Equation (14–175), (14–176), or (14–177) can be used to close Eq. (14–163) to allow a solution for $E(k)$; however, this is generally done under the additional restriction of a universal equilibrium range. The assumption that implies this restriction is that a negligible amount of energy is contained in the eddies characterized by wave numbers greater than k. An exact result for Eqs. (14–175) and (14–163) under these conditions has been obtained by Bass [6].

$$E(k) = \left(\frac{8\epsilon}{9\alpha_\mathrm{H}} \right)^{2/3} k^{-5/3} \left(1 + \frac{8\nu^3}{3\alpha_\mathrm{H}^2 \epsilon} k^4 \right)^{-4/3}. \tag{14-178}$$

EXAMPLE 14–5. Solve the isotropic spectrum equation (14–163) by using Heisenberg's eddy diffusivity theory (14–175).

Answer. Equation (14–163) can be integrated between 0 and k [as was done to obtain Eq. (14–167)] to give

$$\partial \int_0^k E(k)\,dk/\partial t = -S(k) - 2\nu \int_0^k k^2 E(k)\,dk. \tag{i}$$

Now let us restrict the analysis to the universal equilibrium range so that a negligible amount of energy is contained in eddies characterized by wave numbers greater than k; that is, from Eqs. (14–136) and (14–165) we get

$$\epsilon = -\partial \int_0^\infty E(k)\,dk/\partial t \cong - \partial \int_0^k E(k)\,dk/\partial t. \tag{ii}$$

When we consider Eq. (ii), Eq. (i) reduces to

$$\epsilon = S(k) + 2\nu \int_0^k k^2 E(k)\,dk. \tag{iii}$$

We can combine Eqs. (14–175) and (iii) to give

$$\epsilon = 2\left[\nu + \alpha_\mathrm{H} \int_k^\infty k'^{-3/2} E(k')^{1/2}\,dk' \right] \int_0^k k''^2 E(k'')\,dk''. \tag{iv}$$

Now let

$$\phi(k) = \int_0^k k^2 E(k)\,dk \tag{v}$$

or

$$E(k) = (1/k^2)[d\phi(k)/dk], \tag{vi}$$

so that Eq. (iv) reduces to

$$\epsilon/2\phi(k) = \nu + \alpha_H \int_k^\infty k^{-5/2}[d\phi(k)/dk]^{1/2}\, dk. \qquad \text{(vii)}$$

Differentiation gives

$$\frac{\epsilon}{2\phi(k)^2}\frac{d\phi(k)}{dk} = \alpha_H k^{-5/2}\left[\frac{d\phi(k)}{dk}\right]^{1/2},$$

which can be simplified to

$$d\phi(k)/dk = 4\alpha_H^2 \phi(k)^4/\epsilon^2 k^5. \qquad \text{(viii)}$$

Integration gives us

$$\phi(k)^{-3} = (3\alpha_H^2/\epsilon^2)k^{-4} + (\epsilon/2\nu)^{-3},$$

since $\phi(k) \to \epsilon/2\nu$ as $k \to \infty$. Finally, from (vi),

$$E(k) = \left(\frac{8\epsilon}{9\alpha_H}\right)^{2/3} k^{-5/3}\left(1 + \frac{8\nu^3}{3\alpha_H^2 \epsilon}k^4\right)^{-4/3}. \qquad (14\text{-}178)$$

Reid [196] has summarized all the solutions for $E(k)$ and has in addition obtained the one-dimensional spectrum $\phi_{rr}(k)$ by numerical integration, using Eq. (14-137) (a closed solution can be obtained for the Kovasznay theory). Each approximation gives $E(k) \sim k^{-5/3}$ for small k [see Eq. (14-178) for the Heisenberg result]. For large k, Eq. (14-178) reduces to

$$E(k) \sim k^{-5/3}(k^4)^{-4/3} = k^{-7}.$$

For large k, Kovasznay's solution gives a complete cutoff at a finite value of k. The Obukhov approximation acts badly at high values of k, by actually going through an unreal minimum; thus, a sharp cutoff must be imposed. Because of the failure of this approximation at high values of k, it is generally not used. The results obtained by using the Heisenberg and the Kovasznay theories differ little, and the available data [for $\phi_{rr}(k)$] at very high values of k are inadequate to resolve the question of which is better [84]. All three theories are compared for $E(k)$ in Fig. 14-34, assuming a match for the $k^{-5/3}$ region at low k.

Pao [278] obtained a spectrum form for the same region by assuming that spectral elements are continuously transferred to larger wave numbers. He assumed that a form for the rate at which a spectral element is transferred, $\alpha(k)$, was a function of the energy dissipation and the wave number only, which by dimensional reasoning gives

$$\alpha(k) = \alpha_P \epsilon^{1/3} k^{5/3}.$$

The energy flux across k is given by

$$S(k) = \alpha(k)E(k) = \alpha_P \epsilon^{1/3} k^{5/3} E(k),$$

which can be combined with Eq. (14-163) for steady state, and integrated for $E(k)$. The constant of integration can be established by dimensional reasoning

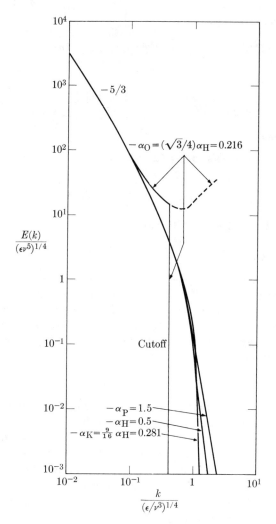

Fig. 14-34. *Phenomenological spectra.*

or by comparison with Eq. (14–173) for the inertial subrange, to give, for the final result,

$$E(k) = \alpha_P \epsilon^{2/3} k^{-5/3} \exp\left(-\tfrac{3}{2}\alpha_P \epsilon^{-1/3} k^{4/3}\right).$$

Pao has also obtained the one-dimensional spectrum by a numerical integration using Eq. (14–137) (different by a factor of two). The three-dimensional spectrum is compared with the others in Fig. 14–34. The one-dimensional spectrum, when compared with the experimental data, gives a satisfactory fit for α_P of 1.7.

In general the phenomenological approach does not have the mathematical elegance of some of the theories still to be considered. However, in spite of this, it is the only approach that has been capable of providing some of the working tools necessary for understanding turbulence and the closely related problem of turbulent mixing.

(iii) **Higher-order moment-discard approximation.** The simplest application of the higher-order moment-discard approximation is to use the first equation of the infinite set (14–157 or 14–161) and discard the third-order moment $k(r)$ or its transform $T(k)$. Batchelor has analyzed this problem, which turns out to be associated with the low wave numbers or the large eddies, and his argument can be seen from Fig. 14–33. As $k \to 0$, the inertia or transfer term $S(k)$ and the viscous term $k^2 E(k)$ fall off more rapidly than does the spectrum function $E(k)$. One can conclude that the big eddies have little interaction with the remainder of the turbulence, and thus change very little during the process of decay. In other words, in the vicinity of $k = 0$, the spectrum function changes little during decay. Near the end of the decay period, the energy in the smaller eddies will have been dissipated, and at some point in time, the controlling factor will be the supply of energy from the very large persistent eddies to viscous dissipation. When only the large eddies remain, the energy is concentrated in the direction of the low wave numbers and the inertia term is negligible. Thus, when we set the inertia term at zero, we should consider the final period of decay.

Kraichnan [133] has shown that this method of approximation is equivalent to an expansion in powers of the turbulent Reynolds number. The third-order moment discard is equivalent to the zeroth term in the Reynolds number expansion. The approximation is of course exact if carried out to the ∞th-order moment. The series truncated at several terms can only be expected to converge rapidly for very low turbulent Reynolds numbers, which, however, is reasonable for the final period of decay.

By taking the fourth moment of the von Kármán-Howarth equation (14–158) and assuming that

$$\lim_{r \to \infty} r^4 k(r) = 0, \qquad (14\text{--}179)$$

Loitsiansky [160] was able to show that

$$u'^2 \int_0^\infty r^4 f(r)\, dr = \text{const during decay}. \qquad (14\text{--}180)$$

This is equivalent [150] to

$$\lim_{k \to \infty} k^{-4} E(k) = \text{const during decay}, \qquad (14\text{--}181)$$

or to an $E(k)$ of the form

$$E(k) = C\, k^4 \qquad (14\text{--}182)$$

for $k \to 0$. Actually, Eq. (14–180) is not a general result, but is valid in the final stage of turbulent decay. This is associated with the assumption given as Eq.

(14–179). For a negligible inertia term (third-order moment discard), Eq. (14–161) reduces to the heat-conduction equation, which has the solution

$$E(k, t) = E(k, t_0) e^{-2\nu k^2 (t-t_0)}. \tag{14–183}$$

Here $E(k, t_0)$, and thus $E(k, t)$, would be of the form Ck^4. Substituting Eq. (14–183) into (14–134) and performing the integration gives the energy decay as

$$\tfrac{1}{2} u_i'^2 = A\,[\nu(t - t_0)]^{-5/2}, \tag{14–184}$$

where A is determined by the initial conditions. We can obtain the correlation function by taking the inverse transform of Eq. (14–183). The result is

$$f(r) = e^{-r^2/8\nu(t-t_0)}. \tag{14–185}$$

This correlation function was used in Example 14–3 to obtain $g(r)$ and in Example 14–4 to establish the form of $E(k)$ [given above as Eqs. (14–182) and (14–183)]. Combining Eqs. (14–184) and (viii) of Example 14–4 gives us

$$E(k) = C'\, k^4\, e^{-2\nu k^2(t-t_0)}.$$

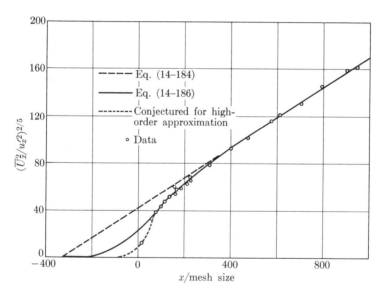

Fig. 14–35. *The final period of decay.* [*From R. G. Deissler*, Phys. Fluids **1**, *111–121 (1958). By permission.*]

The agreement with experimental data is quite good. For example, Fig. 14–35 shows the curve from Eq. (14–184) and the experimental data of Batchelor and Townsend [14]. In this figure there is also a curve which Deissler [60] obtained by deriving an equation for the triple velocity correlation and discarding the fourth-order moment. His final result is

$$\tfrac{1}{2} u_i'^2 = A\,[\nu(t - t_0)]^{-5/2} + (B/\nu)[\nu(t - t_0)]^{-7}. \tag{14–186}$$

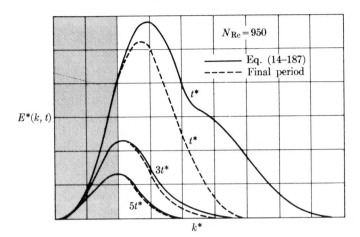

Fig. 14–36. The spectrum during the final period of decay. [From R. G. Deissler, Phys. Fluids 1, 111–121 (1958). By permission.]

The constants A and B are determined from the initial conditions. As one can see from Fig. 14–35, Eq. (14–186) represents the decay for times considerably before, as well as during, the final period of decay. One can obtain the corresponding spectrum equation, which Deissler gives as

$$E(k, t) = E(k, t_0) e^{-2\nu k^2(t-t_0)} - C_1 e^{-(3/2)\nu k^2(t-t_0)} P[\nu(t - t_0)], \qquad (14\text{--}187)$$

where C_1 is a constant dependent on the initial conditions, and $P[\nu(t - t_0)]$ is a series of terms in $\nu(t - t_0)$. A comparison of Eqs. (14–183) and (14–187) is given in Fig. 14–36. The stars denoting the parameters are in dimensionless form. Only the shadowed area is really valid, since no inertia term was used; i.e., only the large eddies are considered. Deissler [61] carried this approach on to the fourth-order equation, and neglected the inertial terms of the fifth order (fifth-order moment discard). He obtained improved agreement but pointed out that mathematical complexity would prevent this procedure from being carried out *ad infinitum*. This complexity apparently led Deissler to use Fourier transforms that do not exist as such. Ohji [180] has reconsidered the equations and has shown that the fifth-order moment discard is closely related to the cumulant-discard approximation (zero fourth-order cumulant) which will be considered in the next section.

Although the experimental confirmation appears to be excellent, several important questions and observations suggest that we use caution before we accept the conclusions. Batchelor and Stewart [13] have shown that the flow in the final decay period is anisotropic. Since the big eddies have little interaction with the remainder of the turbulence, they remain in essentially the same state throughout the decay, which implies that the large eddies were anisotropic when they formed at the grid. Thus, we would not expect the isotropic approximation used to be valid for the final period of decay. Batchelor [8] derived the same relations for homogeneous turbulence, thus eliminating the anisotropic objection. Batchelor and Proudman [12] reevaluated the results and determined that the only valid conclusion was that if the turbulence were homogeneous, the ultimate decay should be as $t^{-5/2}$ [10]. Corrsin [46] has estimated that the wall turbulence might account for as much as 25% of the measured turbulence at an x/mesh-size of 1000. If so, the true experimental decay would have been different from the $t^{-5/2}$ observed (it would be more positive), and thus even this confirmation of the theory would not be valid. More recently Tan and Ling [232] made decay measurements in a water flow in which the wall effect was very low. Their results strongly support a t^{-2} decay rather than a $t^{-5/2}$ one. Their theoretical justification for the results was based on a nonhomogeneity in the flow-direction model of noninteracting decaying vortex streaks (see also Rouse [210]). The exact cause of the difference between the predicted final decay period ($t^{-5/2}$) and these recent experiments is still not clear. It may well be associated with the actual nonhomogeneity in the flow direction.

(iv) **Cumulant-discard approximation.** From the mathematical standpoint the next logical step is to relate the highest-order moment in the system of equations to lower-order moments. To accomplish this, we postulate some statistical assumption for the joint-probability distribution of the turbulent field. Millionshtchikov [173] was the first to introduce the quasinormal approximation between the second- and fourth-order moments. In effect he assumed that the fourth-order moments were related to second-order moments as in a normal distribution. It is quasinormal, since for a true normal distribution, odd moments disappear because of symmetry; however, the problem would be trivial if the third-order moment were set equal to zero. Thus he assumed that the normal distribution was valid only between the fourth- and second-order moments. Specifically, for the four-point relation this is

$$\overline{u_i u_j' u_k'' u_l'''} = \overline{u_i u_j'}\,\overline{u_k'' u_l'''} + \overline{u_i u_k''}\,\overline{u_j' u_l'''} + \overline{u_i u_l'''}\,\overline{u_j' u_k''}, \qquad (14\text{--}188)$$

where the primes denote different positions; that is,

$$u_i(\boldsymbol{x}) = u_i, \qquad u_j(\boldsymbol{x}') = u_j', \qquad u_k(\boldsymbol{x}'') = u_k'', \qquad \text{etc.}$$

Various degenerate forms involving fewer points can easily be obtained by setting locations equal. A decided advantage to this closure approximation is that simultaneous time moments can be used. Kraichnan [133] has shown that the approximation is equivalent to higher-order-cumulant discard. In the Millionshtchikov relation, this would be the fourth-order-cumulant discard. The cumulants are a measure of the degree and complexity of the deviation of the actual probability distribution and its moments from that of a normal distribution.

The potential simplicity of the method led Proudman and Reid [194] and Tatsumi [234] to use the approximation for studying the dynamic behavior of turbulent motion. Their results are valid only for short periods of time after the initial specification of the turbulence spectrum. Ogura [183–185] has shown that for long times, the approximation leads to unreal results; i.e., a part of the spectrum decays so fast that it becomes negative, a physical impossibility. O'Brien and Francis [182] and Kraichnan [132] have shown that a similar phenomenon occurs in the case of a scalar field in turbulent motion (mixing). Unfortunately, most of our interest is in flows in which the Reynolds number is high and the approximation fails.

(v) Other closure methods. From a theoretical standpoint, the previously discussed closure methods leave much to be desired. More complex approximations have been suggested, but for the most part they are so complex that actual solution seems unlikely. One exception to this is the stochastic model approximation presented by Kraichnan [127–136]. The first approximation for isotropic turbulence is called the direct-interaction model and treats the third-order transfer term in an equation similar to (14–161). This term involves the distribution of energy across k, but with no net change. The major advantage of the approach is that the resulting equations describe a model which replaces the real problem. This model is described without approximation, and hence unreal effects, such as negative spectra, should be avoided. We must establish the degree of approximation to the real problem by asking how closely the model represents true turbulence as measured by experiments.

In the direct-interaction model we must use nonsimultaneous time arguments. To illustrate, in the more general form Eq. (14–161) is

$$\partial E(k; t, t')/\partial t = T(k; t, t') - \nu k^2 E(k; t, t'), \qquad (14\text{–}189)$$

where $E(k; t, t')$ is the nonsimultaneous three-dimensional energy spectrum function, and is related to the transform of the nonsimultaneous second-order moment $\mathbf{Q}_{ij}(\mathbf{r}; t, t')$, as given in Eq. (14–166). On the other hand, we can use the simultaneous moment

$$\mathbf{Q}_{ij}(\mathbf{r}; t, t) = \mathbf{Q}_{ij}(\mathbf{r}, t) = \mathbf{Q}_{ij}(\mathbf{r})$$

for the cumulant-discard approximation. In the direct-interaction approximation, we consider three modes or values of $\mathbf{k}$, and we retain the contributions which occur as a result of the direct elementary interaction among these three Fourier amplitudes. The elementary interactions are restricted to the modes that can form a triangle in wave number space; for example, if $\mathbf{k}$, $\mathbf{p}$, and $\mathbf{q}$ are three wave numbers, a direct-interaction contribution will occur so long as $\mathbf{k}$, $\mathbf{p}$, and $\mathbf{q}$ are related by the vector addition equation $\mathbf{k} = \mathbf{p} + \mathbf{q}$.

In the limit of a very large isotropic system all possible interactions are weak, but not negligible. In the specific model considered, more complex elementary interactions, involving longer paths in wave number space (i.e., including intermediate modes), do occur. They occur, however, in such a manner that in the limit of a large isotropic system their contributions cancel, making their total effect negligible when compared with the many weak direct interactions, which for the model do not cancel.

With the direct-interaction approximation, we can obtain an equation for the transfer term $T(k; t, t')$. It is found to contain the difference of two terms. The first term involves the flow of energy from p and q to k, that is, it is an absorption term. The second term is an emission term, and has just the opposite characteristics. The transfer term thus provides the paths for energy transfer between wave number modes, but once again there is no net change when all modes are summed. In the expression for the transfer term there appears a quantity called the *response function* $G(k; t, t')$. If the system of equations is to be closed, this term must also be evaluated. The response function maps out the relaxation process which occurs when a small perturbation occurs at k. The function is also controlled by the basic hydrodynamics described by the Navier-Stokes equation, and consequently obeys an equation quite similar to (14–189) [$G(k; t, t')$ replaces $E(k; t, t')$ and a new third-order transfer term, $H(k; t, t')$, is introduced in place of $T(k; t, t')$]. The term $H(k; t, t')$ is somewhat simpler than $T(k; t, t')$, but still complex. The final result of the approximation is a closed sequence of equations which involve $E(k; t, t')$ and $G(k; t, t')$, and, in terms of these, equations for the transfer functions $T(k; t, t')$ and $H(k; t, t')$. There is no point in reproducing the equations here, since they are available in the references cited; however, since this type of approximation appears to lead to consistent and reasonable results, further discussion is warranted.

There are two possible approaches to the solution of the complex integro-differential equations of the direct-interaction approximation: We can seek a direct numerical solution or make an investigation of asymptotic forms under further restricted conditions. In the inertial subrange, for which Kolmogoroff obtained Eq. (14–173), the direct interaction approximation is expected to be less accurate. Kraichnan found that the approximation took the form

$$E(k, t) \sim (\epsilon u')^{1/2} k^{-3/2}, \qquad (14\text{--}190)$$

which is small in terms of the difference in k, that is, $k^{1/6}$, but important in the additional term u', the rms velocity. The implication is that the inertial range is strongly dependent on the energy-containing range of eddies. An assumption which is basic to Kolmogoroff's analysis is that at higher wave numbers the energy-containing eddies do not affect the energy transfer and dissipation process. Kraichnan [135] found that the inconsistency was caused by the fact that the Eulerian formulation is inadequate to describe the effect of the convection of small-scale motion (high wave numbers) by that of a much larger scale (low wave numbers). If the convective effects are first removed from the basic flow equations, or if a Lagrangian approach is used [136], the direct-interaction approximation does yield Kolmogoroff's spectrum (14–173). The later approach has been called the Lagrangian-history direct-interaction approximation [LHDI]. At moderate Reynolds numbers, a reasonable description, free of inconsistencies, was obtained by direct numerical integration [134]. At high Reynolds numbers, the inertial subrange equation (14–190) was obtained; however, the general numerical accuracy was poor. A higher-order approximation is possible and improved results are expected. However, the higher approximation still suffers from an asymptotic inertial range of $k^{-3/2}$ [135]. Only when we use the Lagrangian approach [LHDI] does the direct-interaction approximation yield the Kolmogoroff form [136]. Kraichnan used numerical methods to find the value of 1.77 for the constant A of Eq. (14–173). The value compares favorably with the experimental values of 1.44 ± 0.19 from reference [84] and 1.57 from reference [81]. As an example of the reasonableness of the computation, Fig. 14–37 compares the experimental data for the one-dimensional dissipation spectrum $k^2\phi_{rr}(k, t)$, with comparable calculated results. Finally, it should be noted that the direct-interaction model can be generalized so that it is not restricted to isotropic turbulence [131]. However, even in its simplest form, it may find application to those shear flows for which the actual turbulence level is not too great [134].

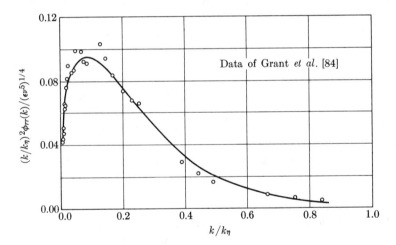

Fig. 14–37. Comparison of LHDI approximation and data. [From R. H. Kraichnan, Phys. Fluids **9**, 1738–1752 (1966). By permission.]

Several more generalized representations for turbulent motion have been formulated. These serve more as a consolidation scheme (usually a single dynamic equation). Even though there is little hope for their solution, they might be the basis by which new closure methods can be invented. The first of these formulations was by Hopf [98, 99], who formulated a dynamic equation for the time evolution of turbulence based on the characteristic functional* of a probability distribution of the velocities. His single equation is equivalent to the infinite set of moment equations. If a solution could be obtained for the characteristic func-

---

*A functional is a function whose value depends on *all* possible values of its argument. A function depends on *specific* values of its argument.

tional, all correlation functions could be obtained by differentiation of this. Lewis and Kraichnan [146] extended the analysis so that nonsimultaneous time arguments could be treated. Rosen [202, 203] has integrated the Hoff equation for the characteristic functional and expressed the resulting characteristic functional in terms of a functional integration. Rosen [203] has carried the analysis far enough to obtain (under certain assumptions of initial conditions) the limiting form during the *initial* period of decay in homogeneous turbulence. The initial period will be discussed later.

Wyld [264] has put forth a general formulation of the turbulence problem based on the perturbation technique used in quantum field theory. By an elaborate rearrangement of the perturbation series, he was able to express the problem in terms of three special functions which form a set of coupled integral equations, each involving an infinite series of increasingly complicated integral powers of the three functions (one of which is the velocity moment and another is associated with the response function). Again no attempt has been made to solve the three equations; however, by truncation, the set reduced to Chandrasekhar's equation [26, 27] for the lowest nontrivial order. Chandrasekhar's approximation is the cumulant-discard approximation equation (14-188) for nonsimultaneous time arguments. As a higher-order approximation, the set of equations reduces to the direct-interaction approximation. Kraichnan [133] has made some further comments on Wyld's approach.

Finally, let us mention the possibility of expressing the Navier-Stokes equation in Lagrangian form and using this as a starting point for a solution. Little progress has been made because of the continued nonlinearity of the system. However, there is some promise for this approach, since the nonlinearity arises in a different manner and the linearized form appears to be somewhat more realistic for describing turbulence than the corresponding linearized Eulerian equation. Geber [77], Pierson [190], Corrsin [43, 45], Lumley [162], and Kraichnan [136] have all contributed to this and associated problems.

14-4.D Local Isotropic Turbulence

The results of the preceding section on isotropic turbulence can, under certain restricted conditions, be applied to nonhomogeneous turbulence problems. For example, consider pipe flow at some relatively high Reynolds number. Throughout the pipe, the viscous forces along the wall provide the conditions necessary for turbulence formation; thus rotation, very large vortices, or large eddies arise from the interaction of the mean flow with the boundary. The scale of these large eddies would be comparable to the size of the system, or to the distance over which the mean flow velocity changes. In pipe flow, this size would be of the order of the pipe diameter. In the preceding section on isotropic turbulence, an intuitive picture was suggested that resulted in a model involving a cascade of energy from large to progressively smaller eddies. With this in mind, let us imagine that the boundary affects the largest eddies most strongly (in fact, creates them) and loses its effect as the process moves down the chain of eddies. At very high wave numbers or in small eddies, the effect of the boundaries is completely lost or is negligible. Thus the small eddies can be visualized as being independent of the boundaries or mean flow. In the nonhomogeneous system, pressure forces are expected to exist and, as already pointed out, would act in the proper direction to make the flow isotropic. Thus we can assume that, at a high enough Reynolds number, the motion of the small eddies is isotropic, because of the tendency created by the pressure forces and because the motion is independent of the boundaries and mean flow. Even though the system may be nonhomogeneous on the large scale, it may well be locally isotropic on the small scale, and thus an equilibrium range and inertial subrange might still be found. Batchelor [10] has cited a number of applications, which have already been made: relative dispersion in the atmosphere, radio-wave scattering, drop breakup by turbulent motion, mixing, and generation of magnetic fields by galactic clouds of ionized hydrogen.

Yet to be answered is the question of the nature and magnitude of the Reynolds number necessary for the existence of the various possible ranges. This question applies equally whether the flow is truly isotropic or only locally isotropic. It should be understood that any conclusions valid for local isotropic turbulence are immediately valid for fully isotropic turbulence over the same range of wave numbers.

(a) The Reynolds number. The Reynolds number is usually defined by using some characteristic velocity and length of the mean flow, for example, the average velocity and the pipe diameter. We emphasized above that, in the range of wave numbers in which local isotropic turbulence may exist, the turbulence is independent of the mean flow and the boundaries. For isotropic turbulence, there are no boundaries or mean motion. Thus the normal definition of the Reynolds number is inadequate to describe the problem and we must turn to more local parameters. For the velocity, it is natural to use the rms value

$$u'_i = \sqrt{3}\, u' \tag{14-191}$$

from Eqs. (14–134) and (14–136). For the length, the macroscale of Eq. (14–117) will not suffice, since it is associated with the large energy-containing eddies. Thus we introduce the dissipation length, or microscale of turbulence, which is defined as

$$\lambda^2 = -\frac{1}{[\partial^2 f(r)/\partial r^2]_{r=0}} = -\frac{1}{f''(0)}. \tag{14-192}$$

The corresponding Reynolds number would be

$$N_{\text{Re},\lambda} = \lambda u'/\nu = \lambda u'_i/\sqrt{3}\,\nu, \tag{14-193}$$

in which the latter form might find application in flows that are anisotropic locally, and in which the three components of the velocity are not equal. The problem that still exists is the measurement of λ, for which purpose we need several relations for λ.

The correlation function $f(r)$ can be expanded by a Maclaurin series,

$$f(r) = f(0) + rf'(0) + \tfrac{1}{2}r^2 f''(0) + \tfrac{1}{6}r^3 f'''(0) + \cdots . \qquad (14\text{-}194)$$

The odd power terms are zero since $f(r)$ is symmetric; that is, $f(r) = f(-r)$. Thus

$$f(r) = 1 - r^2/2\lambda^2 + \cdots . \qquad (14\text{-}195)$$

Equation (14-157), in the limit as $r \to 0$, gives us

$$\partial u'^2/\partial t = 2\nu u'^2 [f''(0) + 4f''(0)] = -10\nu u'^2/\lambda^2, \qquad (14\text{-}196)$$

where use has been made of Eq. (14-192) and the limits $f(0) = 1$ and $k'(0) = 0$. Equation (14-165) defines ϵ, and can be combined with Eq. (14-196) to give

$$\lambda^2 = 15\nu u'^2/\epsilon. \qquad (14\text{-}197)$$

Using Eqs. (14-136), (14-165), and (14-197), we can also express λ as

$$\lambda^2 = \frac{5 \int_0^\infty E(k)\, dk}{\int_0^\infty k^2 E(k)\, dk}. \qquad (14\text{-}198)$$

Two main procedures have been used to determine λ. The first method, based on Eq. (14-195), uses a parabola fitted to the curve of $f(r)$ versus r. Figure 14-38 illustrates this (actually λ is used here to bring together the curves at low r). The second method is based on a measurement of ϵ and Eq. (14-197). One means of obtaining ϵ [other than from spectrum measurements, that is, Eq. (14-165)]

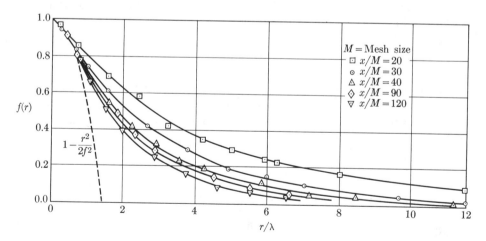

Fig. 14-38. Determination of the microscale. [From R. W. Stewart, and A. A. Townsend, Phil. Trans. Roy. Soc. London **243A**, 359-386 (1951). By permission.]

is from the relation

$$\epsilon = \nu\overline{[(\nabla u) + (\widetilde{\nabla u})]:(\nabla u)} = \nu\overline{(\partial u_i/\partial x_j + \partial u_j/\partial x_i)(\partial u_i/\partial x_j)}$$
$$= \tfrac{1}{2}\nu\overline{(\partial u_i/\partial x_j + \partial u_j/\partial x_i)^2}$$
$$= \nu\overline{(\partial u_i/\partial x_j)^2}$$
$$= 15\nu\overline{(\partial u_r/\partial r)^2} = \tfrac{15}{2}\nu\overline{(\partial u_n/\partial r)^2}, \qquad (14\text{-}199)$$

which is a measure of the dissipation of energy into heat. In other words, all the turbulent energy is eventually dissipated by the action of viscosity. The first and second lines come from averaging the viscous dissipation term in the energy equation already derived (4–17). We can easily prove the identity by summing the two tensors and performing the scalar multiplication, as indicated by Eq. (2–47). The third line of Eq. (14–199) is the form for homogeneous turbulence ($\overline{u_i u_j}$ terms are zero). The final line is the simplified form for isotropic turbulence, where the subscript r denotes the velocity in the direction of the radius vector and n is normal to this. For isotropic turbulence, Eqs. (14–197) and (14–199) can be combined to yield

$$\lambda^2 = u'^2/\overline{(\partial u_r/\partial r)^2}. \qquad (14\text{-}200)$$

Liepmann et al. [148] have measured the microscale by this method, by calculation from the spectrum, and by a third method based on the number of zeros in the velocity fluctuation. A comparison is given in Fig. 14–39, where λ^2 is plotted against the distance from the grid (initial period of decay). The zero-count method is based on the assumed statistical independence of u_x and du_x/dt. The theory gives

$$N^2(0) = u'^2/\pi^2\lambda^2. \qquad (14\text{-}201)$$

Liepmann and Robinson [149] have further investigated this method, and have concluded that statistical independence is valid to a few percent, and that the systematic difference shown in the figure cannot be due entirely to this assumption.

The slope of the line in Fig. 14–39 is $10\nu/\bar{U}_x$, which can be obtained from theory by assuming a power law decay for the velocity. That is,

$$u'^2 = A\, t^{-n}. \qquad (14\text{-}202)$$

Equations (14–165), (14–198), and (14–201) combine to give

$$\lambda^2 = 10\nu t/n, \qquad (14\text{-}203)$$

which, for $n = 1$ and $t = x/\bar{U}_x$, reduces to $\lambda^2 = (10\nu/\bar{U}_x)x$, as experimentally observed. The linear law ($n = 1$) holds for the initial period of decay to about 100 or more mesh diameters downstream [8]. We can also investigate the Reynolds number during the initial period of decay. By combining Eqs. (14–193),

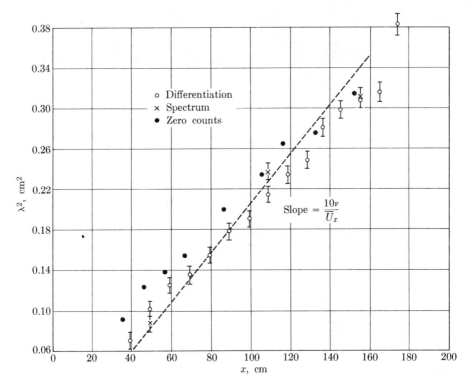

Fig. 14-39. Plot of λ^2 against x, with $\bar{U}_x = 630$ cm/sec and mesh size $= 1.68$ cm. [From H. W. Liepmann, J. Laufer, and K. Liepmann, NACA TN 2473 (1951). By permission.]

(14–202), and (14–203), we obtain

$$N_{\text{Re},\lambda} = \sqrt{10A/\nu n}\, t^{(1-n)/2}. \tag{14-204}$$

For $n = 1$, the number is constant at a value of $\sqrt{10A/\nu}$.

The microscale Reynolds number (14–193) can be empirically expressed in terms of macroscale parameters. Such a relation of dimensional validity is

$$\epsilon = -\tfrac{3}{2}(du'^2/dt) = \tfrac{3}{2}(A_1 u'^3/L_f). \tag{14-205}$$

The constant A_1 averages about 1.1, but varies from 0.8 to 1.4 with the Reynolds number; it is independent of the decay. Combining equations (14–197) and (14–205) gives

$$\lambda^2 = 10\nu L_f/A_1 u'. \tag{14-206}$$

From Eqs. (14–193) and (14–206), the macroscale Reynolds number is

$$N_{\text{Re},L_f} = L_f u'/\nu = (A_1/10)(N_{\text{Re},\lambda})^2. \tag{14-207}$$

The mesh Reynolds number for grid experiments in isotropic turbulence can be defined as

$$N_{\text{Re},M} = M\bar{U}_x/\nu, \qquad (14\text{--}208)$$

where M is the grid or mesh spacing. This Reynolds number cannot be related to N_{Re,L_f} exactly; however, Batchelor and Townsend give the empirical relation

$$N_{\text{Re},L_f} = (1/134)N_{\text{Re},M}. \qquad (14\text{--}209)$$

The critical value of the Reynolds number necessary for the existence of the equilibrium range or the inertial subrange is difficult to establish [38], simply because in the past it has not been easy to determine when a flow was isotropic. However, recent measurements at very high Reynolds numbers have shed some light on this problem.

(b) Some experimental results. An equilibrium range may exist in nearly any kind of flow if the Reynolds number is high enough and the very small eddies are not affected by the system boundaries. Because of the possibility of very high Reynolds numbers, experiments on turbulent shear and natural flows (i.e., the oceans and the atmosphere) are useful in local isotropic turbulence. It is necessary to show more than just the existence of a $-\frac{5}{3}$ spectrum range to establish that the flow is locally isotropic. One should also derive the longitudinal one-dimensional spectrum function from the corresponding lateral measurement and compare the calculated results with the actual measurements. This operation exactly parallels the correlation comparison which is obtained by using Eq. (14–106) and which is shown in Fig. 14–20. To further illustrate, the pipe flow measurements of Laufer presented in Fig. 14–31 clearly show a $-\frac{5}{3}$ region (one decade) for the spectrum at the center line in the direction of flow (longitudinal); however, his measurements show that the cross-stream or lateral spectrum does not have a corresponding range, and a calculation of $\phi_{rr}(k)$ from $\phi_{nn}(k)$ did not correspond to these experimental measurements. For these experiments, $N_{\text{Re},\lambda}$ was somewhat less than 300.

The best confirmation to date has been provided by the jet measurements of Gibson [80, 81], which showed about two decades of a $-\frac{5}{3}$ region, and an excellent confirmation was obtained by calculation of $\phi_{rr}(k)$ from $\phi_{nn}(k)$. The calculations did deviate at low wave numbers, showing that the flow was not completely isotropic, but rather locally isotropic at the higher wave numbers. For these experiments, $N_{\text{Re},\lambda}$ was 780. Grant et al. [84, 85] made measurements in a tidal channel at a $N_{\text{Re},\lambda}$ of about 3600, and observed three and a half decades of a $-\frac{5}{3}$ region. The lateral measurements were not available, but presumably the flow was locally isotropic. In both experiments the energy $\{\phi_{rr}(k)\}$ and the dissipation $\{k^2\phi_{rr}(k)\}$ spectrum were reasonably well separated, which is the condition necessary for the existence of an inertial subrange.

Many other measurements at lower values of $N_{\text{Re},\lambda}$ have been made. For example, Liepmann et al. [148] worked on grid turbulence at an estimated microscale Reynolds number of 142. In this set of experiments, there was a range where the $-\frac{5}{3}$ law was valid. Other experiments at even lower Reynolds number (30 or less) showed no $-\frac{5}{3}$ range. Klebanoff and Diehl [114] have reported wind-tunnel experiments and shear-flow studies in a turbulent boundary layer. For tunnel flow, the $-\frac{5}{3}$ range was missing at a microscale Reynolds number of about 40. For boundary layer flow, the Reynolds number was about 130. Figure 14–40 shows some of the results. The $-\frac{5}{3}$ range is clearly present. The data reported in this figure were recorded at points away from the wall; very near the wall no $-\frac{5}{3}$ region was found. A similar situation existed for Laufer's data, shown in Fig. 14–31. Klebanoff [113] has reported additional work on the boundary layer and obtained similar results. Sandborn and Braun [212] also studied the turbulent boundary layer for a microscale Reynolds number of about 75; there was no indication of a $-\frac{5}{3}$ range. Other results have been given by Dumas [270].

In general it appears that at a Reynolds number of about 130, a limited (about one decade) range of $-\frac{5}{3}$ spectra exist, but this by no means guarantees that the flow is isotropic. If it is a grid flow, it should be reasonably close to isotropic conditions; for shear flows however, the strong boundary effects apparently prevent local isotropic conditions from existing until quite high Reynolds numbers (somewhere between 300 and 780) are obtained.

14-4.E Turbulent Shear Flow

The entire section on phenomenological theories was concerned with turbulent shear flow. The Reynolds equation, the Reynolds stresses, and the mean flow energy balance were all a part of this analysis. These ideas will be extended here; however, the main concern will be with relations among the various statistical averages already cited, and only briefly with the mean velocity.

(a) The energy balance. The equation of motion can be converted to an expression for the kinetic energy of the fluctuation velocities. Each coordinate equation is multiplied by its own velocity, and the velocity and pressure fluctuations are introduced as

$$\boldsymbol{U} = \bar{\boldsymbol{U}} + \boldsymbol{u} \quad \text{and} \quad p = \bar{p} + \tilde{p}.$$

The equations are then summed and averaged according to the Reynolds rules of averaging. To obtain the final equation of the kinetic energy of the fluctuation, we subtract the mean-motion kinetic energy. To illustrate, let us consider the x-direction equation multiplied by U_x:

$$U_x \frac{\partial U_x}{\partial t} + U_x(\boldsymbol{U} \cdot \boldsymbol{\nabla}) U_x = -\frac{1}{\rho} U_x \frac{\partial p}{\partial x} + \nu U_x \nabla^2 U_x. \qquad (14\text{--}210)$$

This is a mixed notation; we could as well substitute the equalities

$$\boldsymbol{U} = U_j, \quad \boldsymbol{u} = u_j, \quad \boldsymbol{\nabla} = \frac{\partial}{\partial x_j}, \quad (\boldsymbol{U} \cdot \boldsymbol{\nabla}) = U_j \frac{\partial}{\partial x_j}, \quad \nabla^2 = \frac{\partial^2}{\partial x_j \partial x_j}.$$

However, we shall retain the less complex mixed notation.

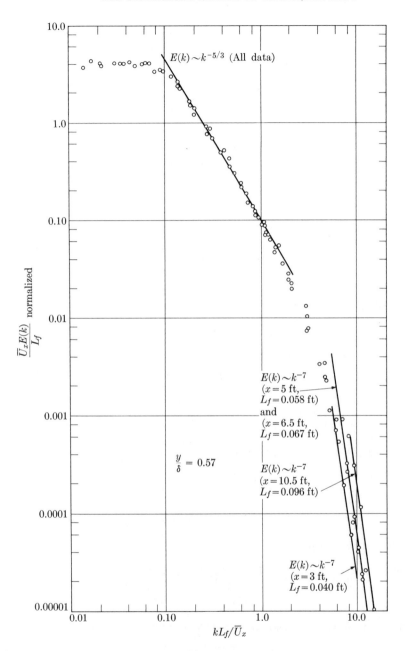

Fig. 14–40. Test of $-5/3$ law and -7 law for spectrum of u_x-fluctuations in turbulent boundary layer thickened by roughness for first two feet of surface. Here, $\bar{U}_x = 55$ ft/sec and $y/\delta = 0.57$. [From P. S. Klebanoff, NACA Report 1110 (1952). By permission.]

For an incompressible fluid

$$U_x(\boldsymbol{U}\cdot\boldsymbol{\nabla})U_x = U_x(\boldsymbol{\nabla}\cdot U_x\boldsymbol{U}) = \left(\boldsymbol{\nabla}\cdot\frac{U_x^2}{2}\boldsymbol{U}\right). \qquad (14\text{-}211)$$

Introducing this to help rearrange terms, and putting the equation into a kinetic energy form, we have

$$\frac{1}{2}\frac{\partial U_x^2}{\partial t} + \left(\boldsymbol{\nabla}\cdot\frac{U_x^2}{2}\boldsymbol{U}\right) = -\frac{1}{\rho}U_x\frac{\partial p}{\partial x} + \nu U_x\nabla^2 U_x. \qquad (14\text{-}212)$$

Representing the instantaneous velocities and pressure as their average and fluctuating parts, and summing the three coordinate equations and averaging, we get

$$\frac{1}{2}\frac{\partial}{\partial t}(\bar{U}_i^2 + \overline{u_i'^2}) + \left[\boldsymbol{\nabla}\cdot\left(\frac{\bar{U}_i^2}{2}\bar{\boldsymbol{U}} + \frac{\overline{u_i'^2}}{2}\bar{\boldsymbol{U}} + \bar{U}_i\overline{u_i\boldsymbol{u}} + \frac{\overline{u_i^2}}{2}\boldsymbol{u}\right)\right]$$
$$= -\frac{1}{\rho}(\boldsymbol{\nabla}\cdot\bar{p}\bar{\boldsymbol{U}}) - \frac{1}{\rho}(\boldsymbol{\nabla}\cdot\overline{\tilde{p}\boldsymbol{u}}) + \nu\bar{U}_i\nabla^2\bar{U}_i + \nu\overline{u_i\nabla^2 u_i}, \qquad (14\text{-}213)$$

where summation over the double index is implied and

$$\overline{u_i'^2} = \overline{u_i^2} = \overline{u_x^2} + \overline{u_y^2} + \overline{u_z^2}. \qquad (14\text{-}214)$$

We can treat the Reynolds equation [(14–10) or (14–11)] in a similar manner by multiplying each coordinate equation by its own average velocity, and summing to give

$$\frac{1}{2}\frac{\partial \bar{U}_i^2}{\partial t} + \left(\boldsymbol{\nabla}\cdot\frac{\bar{U}_i^2}{2}\bar{\boldsymbol{U}}\right) = -\frac{1}{\rho}(\boldsymbol{\nabla}\cdot\bar{p}\bar{\boldsymbol{U}}) + \nu\bar{U}_i\nabla^2\bar{U}_i - \bar{U}_i(\boldsymbol{\nabla}\cdot\overline{u_i\boldsymbol{u}}). \qquad (14\text{-}215)$$

Subtracting Eq. (14–215) from (14–213) gives

$$\frac{1}{2}\frac{\partial \overline{u_i'^2}}{\partial t} + \left[\boldsymbol{\nabla}\cdot\left(\frac{\overline{u_i'^2}}{2}\bar{\boldsymbol{U}} + \frac{\overline{u_i^2}}{2}\boldsymbol{u}\right)\right] + (\boldsymbol{\nabla}\cdot\bar{U}_i\overline{u_i\boldsymbol{u}}) - \bar{U}_i(\boldsymbol{\nabla}\cdot\overline{u_i\boldsymbol{u}})$$
$$+ \frac{1}{\rho}(\boldsymbol{\nabla}\cdot\overline{\tilde{p}\boldsymbol{u}}) - \nu\overline{u_i\nabla^2 u_i} = 0, \qquad (14\text{-}216)$$

which can be rearranged to

$$\underbrace{\frac{1}{2}\frac{\partial \overline{u_i'^2}}{\partial t}}_{(\text{I})} + \underbrace{(\overline{u_i\boldsymbol{u}}\cdot\boldsymbol{\nabla}\bar{U}_i)}_{(\text{II})} + \underbrace{\left[\boldsymbol{\nabla}\cdot\overline{\left(\frac{\tilde{p}}{\rho} + \frac{u_i^2}{2}\right)\boldsymbol{u}}\right]}_{(\text{III})} + \underbrace{\left(\boldsymbol{\nabla}\cdot\frac{\overline{u_i'^2}}{2}\bar{\boldsymbol{U}}\right)}_{(\text{IV})} - \underbrace{\nu\overline{u_i\nabla^2 u_i}}_{(\text{V})} = 0. \qquad (14\text{-}217)$$

The dissipation term can also be rearranged further [96, 113, 139]:

$$\underbrace{\nu\overline{u_i\nabla^2 u_i}}_{(\text{V})} = \underbrace{\nu\overline{\boldsymbol{\nabla} u_i[(\boldsymbol{\nabla}\boldsymbol{u}) + (\widetilde{\boldsymbol{\nabla}\boldsymbol{u}})]}}_{(\text{Va})} - \underbrace{\nu\overline{[(\boldsymbol{\nabla}\boldsymbol{u}) + (\widetilde{\boldsymbol{\nabla}\boldsymbol{u}})]:(\boldsymbol{\nabla}\boldsymbol{u})}}_{(\text{Vb})} = \underbrace{\nu\nabla^2(\overline{u_i'^2}/2)}_{(\text{Va})} - \underbrace{\nu(\overline{\boldsymbol{\nabla}\boldsymbol{u}})^2}_{(\text{Vb})}. \qquad (14\text{-}218)$$

The final term can be recognized as $\nu\overline{(\partial u_i/\partial x_j)^2}$ of Eq. (14–199). In complete cartesian tensor notation, Eqs. (14–217) and (14–218) combine to give

$$\frac{1}{2}\frac{\partial \overline{u_i'^2}}{\partial t} + \overline{u_i u_j}\frac{\partial \bar{U}_i}{\partial x_j} + \frac{\partial}{\partial x_j}\overline{\left[\left(\frac{\tilde{p}}{\rho} + \frac{u_i^2}{2}\right)u_j\right]} + \frac{\partial}{\partial x_j}\left(\frac{\overline{u_i'^2}}{2}\bar{U}_j\right) - \nu\frac{\partial^2}{\partial x_j \partial x_j}\frac{\overline{u_i'^2}}{2} + \nu\overline{\left(\frac{\partial u_i}{\partial x_j}\right)^2} = 0.$$

(I)　　　(II)　　　(III)　　　(IV)　　　(Vc)　　　(Vd)

(14–219)

Each term in Eqs. (14–217) to (14–219) represents a rate of change of turbulent energy per unit mass; specifically:

(I)   The time rate of change of the turbulent kinetic energy
(II) Production of turbulent kinetic energy by shearing stress (which is an energy transfer from the mean motion)
(III) Diffusion by kinetic and pressure effects
(IV) Transfer of energy by convection
(V)   The work done by viscous forces on the surface of a unit mass to change its kinetic energy as a whole [215]
(Va) The work done by viscous shear stresses of the turbulent motion
(Vb) Dissipation of turbulent kinetic energy
(Vc) Same as (Va) if the flow were homogeneous
(Vd) Same as (Vb) if the flow were homogeneous

For nonhomogeneous flow, the terms (Vc) and (Vd) are still valid, but their meaning must be altered; i.e., the net effect of terms (Vc) and (Vd) is the same as the net effect of (Va) and (Vb), which in turn are the same as term (V).

For isotropic turbulence, the dissipation term can be further simplified, as already cited in Eq. (14–199). The form might still be valid for high Reynolds-number flows, but this would have to be tested by experiment. Laufer [139] did test the assumption for his pipe flow work. He found that when the assumption was made, the dissipation across the pipe was less than the corresponding production by a factor of about 2.5. Thus each part of term (Vd) had to be measured separately and the simplified form was not valid. The measurements are not easy, and in a number of cases none have been made. Although not all dissipation terms can be measured, they can be approximated, and considerable information about the balance of energy can still be obtained. We can obtain the degree of approximation involved in estimating the unknown terms by again comparing the summation of the production with the dissipation across the pipe. Laufer was eventually able to obtain a balance to within 10%. If the simplified isotropic relation had been used, the balance would have differed by the factor of 2.5 just noted. The evaluation of the kinetic energy of term (III) necessitates the measurement of the triple velocity correlation. This measurement is subject to considerable experimental error. For example, Laufer found that the values away from the wall were considerably different for his high and low Reynolds number runs. He believed that these differences were due to experimental error, rather than to a real physical difference. Nevertheless, an estimate can be made for most terms and the results, as reported by Laufer, are shown in Fig. 14–41. Townsend [246] has made a small correction accounting for gradients near the wall. How-

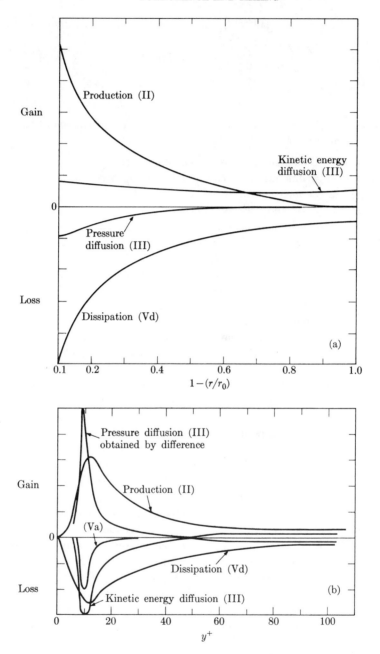

Fig. 14–41. Turbulent energy balance for pipe flow. [*From J. Laufer*, NACA Report 1174 (*1954*), *and G. B, Schubauer*, J. Appl. Phys. **25**, *188–196* (*1954*). *By permission.*]

ever, there is considerable question as to the exact magnitude near the wall of most of the terms shown in Fig. 14–41(b).

Figure 14–41(a) indicates that the energy production (II) is equal to the viscous dissipation (Vd) except at the pipe center, where the production is balanced by the kinetic energy diffusion term (III). Outside the viscous sublayer [Fig. 14–41(b)], the production (II) and dissipation (Vd) are nearly equal and opposite; however, in the sublayer region, all terms are of importance. The viscous transfer term (Va) becomes an important factor close to the wall. Transfer by convection (IV) is negligible everywhere.

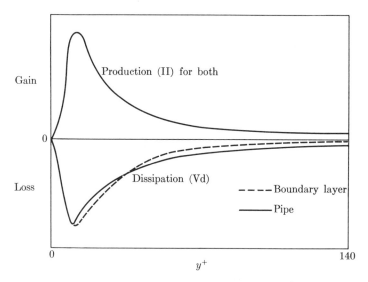

Fig. 14–42. Turbulent energy balance. [*From* G. B. Schubauer, J. Appl. Phys. **25**, 188 (1954). *By permission*.]

The production of turbulent energy (II) between pipe flow and the boundary layer has already been compared in Fig. 14–17. A similar comparison can be made for the dissipation term (Vd); this is done in Fig. 14–42.

Assuming that these figures are correct, one can draw the following conclusions for pipe and boundary layer flow.

1. There is a diffusion of kinetic energy from the wall toward the center. This results in a gain in kinetic energy beyond a y^+ of about 40, and a loss of kinetic energy from this point toward the wall. It is this diffusion that balances the imbalance between production (II) and dissipation (Vd).
2. There appears to be a diffusion of pressure energy which is in a direction opposite to that of the kinetic energy diffusion. However, it should be noted that this term has been obtained by difference and thus is subject to considerable error.

3. The diffusion of kinetic and the diffusion of pressure energies tend to balance and give a small net result; however, neither one separately can be considered small.

Considerable confusion still exists as to the exact meaning of the results of the turbulent energy balance, and its implications about a model of turbulent shear flow. Additional information would be desirable, as would more accurate methods of measurement near the wall. However, until more is known, the turbulent process as shown schematically in Fig. 14–43 will have to suffice.

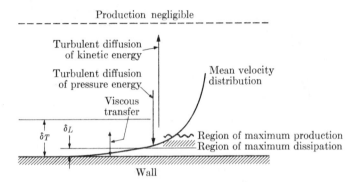

Fig. 14–43. Turbulent energy-transfer process. [*From J. O. Hinze*, Turbulence, *McGraw-Hill, New York (1959). By permission.*]

(b) Intermittency. As mentioned earlier, one can make a comparison of the intensities of turbulence for pipe flow (Fig. 14–24) and boundary layer flow (Fig. 14–26). For the latter, the intensity is felt outside the boundary layer; that is, at $y/\delta > 1$. This arises from the intermittent nature of a shear flow having a free boundary (see Fig. 14–44). The actual turbulent field at times extends beyond the average boundary layer thickness, and causes what appears to be an induced intensity beyond the outer edge.

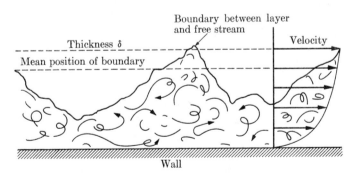

Fig. 14–44. The intermittent nature of the free boundary. [*From G. B. Schubauer, J. Appl. Phys.* **25***, 188 (1954). By permission.*]

This phenomenon was first observed by Corrsin and Kistler [51] and by Townsend [244], and has received much additional attention [32, 51, 114, 246]. Experimental methods have been developed to determine it accurately. The intermittency is usually represented as a factor γ which gives the fraction of time during which the flow is truly turbulent. For the boundary layer, it has been found that this factor follows a Gaussian integral curve with a mean value at $y/\delta = 0.78$, and a standard deviation of 0.14δ. This means that intermittency is usually found within the range of 0.4 to 1.2 y/δ.

Klebanoff [113] accurately measured the corresponding values of γ for his intensity data of Fig. 14–26. Schubauer [215] used these values and the intensity data to make a comparison with Laufer's pipe data (Fig. 14–24). The form plotted was $u_i'^2/\gamma U^{*2}$, and is shown in Fig. 14–45. The agreement is quite good; the deviation between Reynolds numbers is believed to be a Reynolds-number effect.

The basic implications of intermittency are not firmly established. Why it occurs and why it is Gaussian is not understood. Coles [32] has provided a review of what is known, and has attempted to define what is still unknown. One suggestion, made by Townsend [246], is that free turbulent flows have a double structure. The large eddy motions determine the intermittent nature, and the small-scale motions, within the large eddies, determine the turbulent properties. There is a degree of similarity between intermittency and certain stages of the laminar-turbulent transition. Intermittency, the turbulent spot, and intermittent slugs of turbulence in pipes in the transition region have the common characteristic of having a sharp boundary which separates the laminar and the turbulent parts. They also have in common the lack of a clear mechanistic picture of what is occurring in this stage or area of the flow. Clearly, further studies are needed to help us understand turbulence *per se*.

Even with the aid of energy balances and intermittency, our knowledge of turbulent shear flows is still rudimentary. One-dimensional spectra measurements (Figs. 14–31 and 14–40) have been made, and have been helpful. However, their interpretation is difficult and, at best, somewhat questionable. Measured correlations can contribute additional information but, again, much more (and more accurate) data are needed before any mechanism can be qualitatively established as correct. The experimental data available are much more extensive than cited here [96, 246]. Studies have been made on jets, wakes, channels, between rotating cylinders, pipes, and various boundary layer configurations. In most cases some turbulence parameters have been measured and energy balances attempted. Still lacking, however, are unifying concepts to tie together the information available on shear flows.

(c) **Velocity distribution.** The aim of the phenomenological theories was to establish the form of the mean velocity distribution. Each solution was generally restricted to a specific area of flow and involved several arbitrary constants. For the most part these theories involved rather simple assumptions or relatively

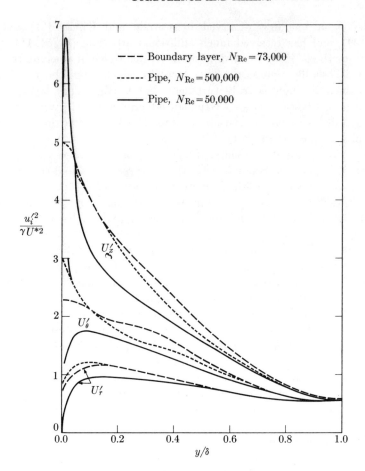

Fig. 14–45. *Turbulent velocities with intermittency considered.* [*From G. B. Schubauer, J. Appl. Phys.* **25**, *188 (1954). By permission.*]

simple intuitive analyses. In contrast, a more complex formulation of the problem can be made by using the Reynolds equation (14–23) in conjunction with an additional higher-order equation for $\overline{u_i u_j}$. This equation can be derived in much the same manner as the energy equation. As a result, additional terms appear and there are more unknowns than equations. The added terms, such as the pressure-velocity correlation, are very difficult to treat. Efforts along this line have been made by Chou [29], Rotta [208], Emmons [68], and Davidov [56]. The basic equations have been discussed by Hinze [96]. Approximations can be made for the added terms, and if enough estimates are made, a solution can be obtained; however, these added terms contain so many adjustable constants that conclusive proof of the theories cannot be made.

An alternative approach, of considerable merit, has been advanced by Malkus [165–167]. This theory treats mean motion only, but does so in a way that allows one to estimate the arbitrary constants; thus it goes beyond the phenomenological theories. In this approach, the Reynolds stress is expressed as a truncated (at n_0 terms) series of orthogonal functions (assumption 1). The derivatives of the mean velocity can be related to the series through the Reynolds equation (14–20 or 14–23). In the asymptotic form, the series can be replaced by a simpler function, which has the same properties of satisfying the boundary conditions (assumption 2, for example, zero velocities at boundaries, $d^2\bar{U}_x/dy^2 \geqslant 0$, etc.). When we integrate the function for the second derivative, we obtain a velocity defect law [such as Eq. (14–47)] and a logarithmic law (14–37). We obtain the constants (k and B) by means of a requirement of marginal stability (or linear stability theory, as discussed on page 228) of the last term of the orthogonal series (assumption 3). All this is accomplished under the condition that the energy dissipation is a maximum (assumption 4). The constant k is predicted as 3.0, or about 20% higher than the experimental value of 2.5. The constant B is close to 3, which can be compared with the experimental 5.5. Although the agreement is only fair, we must remember that the predictions are from theory and are in no way arbitrary. We can achieve variation of the constants only if we make a more elaborate analysis and eliminate assumptions. Some further discussion of the Malkus theory has been made by Townsend [248] and Spiegel [226].

14-4.F Turbulent Dispersion

The dispersion of a contaminant by turbulent motion is of fundamental importance in many problems. To illustrate: The oceanographer would like to be able to determine beforehand how radioactive wastes will disperse when they are discharged into the sea [103, 105]. The degree of air pollution and the proper design of smoke stacks depends on how well the turbulent wind can disperse smoke [43, 104, 118]. The time necessary for blending will depend on one's ability to disperse the material to be mixed. Such dispersions are often called *turbulent diffusion*. The analysis of turbulent dispersion will for the most part be concerned with cases in which there is no superimposed molecular diffusion, and in effect will consider the motion of marked fluid particles or elements.

A Lagrangian view can be used to help us gain some insight into the mechanism. Consider individual elements that leave a fixed point in space; first, in the case in which there is no mean motion, the various elements will be carried from the source by the turbulent eddies. We would expect that those caught in a large eddy (with generally large motion) would be carried further than those that were initially part of a small eddy. Thus at any instant in time there will be a distribution of elements about the point source. This may be easier to visualize if we superimpose a uniform mean velocity on the turbulent field. We expect that each element or particle leaving the point source will deviate from the linear path in a random manner, depending on the local nature of the turbulence.

We would observe the root-mean-squared deviation for the particles as a continued divergence, spread, or dispersion as the particles are carried downstream from the point source by the uniform mean velocity. This is an eddy motion, and can occur in the absence of molecular diffusion. In this discussion, the time mentioned in the first illustration has been replaced by the distance from the point source divided by the uniform mean velocity.

If we assume that the point source is located at the origin O, the distance a fluid element will travel (displacement) will be

$$\mathbf{X} = \int_0^t \boldsymbol{v}(t)\, dt, \tag{14-220}$$

where $\mathbf{X}$ is the random Lagrangian coordinate or random particle position along its path and $\boldsymbol{v}(t)$ is the turbulent velocity along that path. We use $\mathbf{X}$ and $\boldsymbol{v}(t)$ to distinguish the position and velocity along a path from $\boldsymbol{x}$ and $\boldsymbol{u}(x)$, the Eulerian position and velocity at a specific point in space. Both velocities may be a function of time when there is a decaying turbulence, but as we have noted, we shall not show a time dependence due to a nonstationary nature of the flow. In other words, $\mathbf{X}$ and $\boldsymbol{v}(t)$ are Lagrangian terms, while $\boldsymbol{x}$ and $\boldsymbol{u}(x) = \boldsymbol{u}(\boldsymbol{x},t)$ are Eulerian terms.

Restricting the analysis to the displacement in one direction (y) gives us

$$\mathrm{Y} = \int_0^t v_y(t)\, dt. \tag{14-221}$$

The mean-squared displacement, or variance, of Y can be obtained by integrating over many particles which start at different times from the point of origin [96]. Taylor [236] showed that for infinitesimal fluid particles the result could be expressed as

$$\tfrac{1}{2}(d\overline{\mathrm{Y}^2}/dt) = \int_0^t \overline{v_y(t)v_y(t+\tau)}\, d\tau, \tag{14-222}$$

or for homogeneous turbulence,

$$d\overline{\mathrm{Y}^2}/dt = 2\overline{v_y'(t)}^2 \int_0^t \mathrm{R}_{L_{yy}}(\tau)\, d\tau, \tag{14-223}$$

where $\mathrm{R}_{L_{yy}}(\tau)$ is the Lagrangian coefficient, as given by Eq. (14–107). It should be emphasized that the analysis is based on the particle path, and thus $v_y'(t)$ is the average of a large number of particle velocities measured along their respective paths. For an incompressible homogeneous stationary field, the Lagrangian variance is the same as the Eulerian, which is a time average at a point in space, or $v_y'(t) = u_y'(\boldsymbol{x})$. In a decaying field, such an assumption would not be valid [15, 162], and thus the equality will not be assumed in the analysis to follow. However, if the flow is isotropic, $\mathrm{R}_{L_{yy}}(\tau)$ can be simply written as $R_L(\tau)$, as given by Eq. (14–108).

For small values of τ, $R_L(\tau)$ approaches unity, and Eq. (14–223) can be integrated to give

$$\overline{\mathrm{Y}^2} = \overline{v_y'(t)}^2 t^2, \tag{14-224}$$

which shows that the spread of fluid particles is proportional to time, for small time. A modification of Eq. (14–223) by Kampé de Fériet [108] involved a partial integration to

$$\overline{Y^2} = 2v'_y(t)^2 \left[t \int_0^t R_L(\tau)\, d\tau - \int_0^t \tau R_L(\tau)\, d\tau \right] = 2v'_y(t)^2 \int_0^t (t - \tau) R_L(\tau)\, d\tau. \tag{14–225}$$

For large values of τ, $R_L(\tau)$ approaches zero exponentially and

$$\overline{Y^2} = 2v'_y(t)^2 \, (tT_L - \text{const}), \tag{14–226}$$

where T_L is the Lagrangian time scale given by Eq. (14–121). The spread is proportional to the square root of the time. For intermediate values of time, the dependency of $R_L(\tau)$ must be known as a function of time. If we can assume that the form of the correlation coefficient is

$$R_L(\tau) = e^{-\tau/T_L}, \tag{14–227}$$

then Eq. (14–226) becomes

$$\overline{Y^2} = 2v'_y(t)^2 \, (tT_L - T_L^2). \tag{14–228}$$

We can obtain the general form by combining Eq. (14–227), the definition of T_L, and Eq. (14–223). As given by Hanratty et al. [93], it is

$$\overline{Y^2} = 2v'_y(t)^2 \, T_L^2 [(t/T_L) - (1 - e^{-t/T_L})]. \tag{14–229}$$

In the same way as we arrived at a definition of the coefficient of molecular diffusion, we can define an eddy diffusion coefficient as

$$\epsilon_s = \overline{Y^2}/2t, \tag{14–230}$$

and for very long periods of time $(t > 10T_L)$, we can combine this with Eq. (14–226) and (14–122) to give us

$$\epsilon_s = v'_y(t)^2 \, T_L = v'_y(t) \, L_L. \tag{14–231}$$

If we allow ϵ_s to be a function of time, then we can combine Eq. (14–229) with the above to give

$$\epsilon_s = v'_y(t)^2 \, T_L[1 + (T_L/t)(e^{-t/T_L} - 1)], \tag{14–232}$$

which becomes approximately constant for long times of the order of $10T_L$.

There is excellent agreement between the available experimental data and the theory for turbulent dispersion [96]. In addition, agreement has even been found between experiment and theory for fluidization [93], which we shall discuss further when we consider motion in fluidized beds. At this point, let us again emphasize that diffusion or dispersion is caused by turbulent motion only, and is not due to molecular diffusion.

The single-particle dispersion has also been extended to the relative dispersion between two particles or elements. The form for the mean-squared displacement is nearly the same as Eq. (14–223), but here the Lagrangian correlation depends on a velocity difference and is not stationary in time:

$$d\overline{Y^2}/dt = 2[v_y(t) - v_{y'}(t)]^2 \int_0^t R_{\Delta L}(t, \tau)\, d\tau, \qquad (14\text{--}233)$$

where $R_{\Delta L}(t, \tau)$ is the Lagrangian correlation based on the velocity difference, and the y' denotes the second particle. Richardson [200] experimentally observed that at large Reynolds numbers the relative dispersion is given by

$$d\overline{Y^2}/dt \sim (\overline{Y^2})^{2/3}. \qquad (14\text{--}234)$$

Obukhov [177] suggested that if the Reynolds number is high enough and the dispersion lies within the inertial subrange, $d\overline{Y^2}/dt$ depends only on $\overline{Y^2}$ and ϵ, the total energy dissipation rate. Dimensional reasoning gives us

$$d\overline{Y^2}/dt \sim \epsilon^{1/3}(\overline{Y^2})^{2/3}, \qquad (14\text{--}235)$$

which corresponds to the preceding equation. Further comments and reading can be found in the review article by Corrsin [45] and the recent contributions by Kraichnan [136].

Dispersion in turbulent flow provides the experimental means for the measurement of Lagrangian parameters. Relations between the Lagrangian and Eulerian systems would be most helpful, because of the difficulty of Lagrangian measurements and the relative ease of obtaining Eulerian results. Unfortunately, there is no unique relation between the Eulerian and Lagrangian correlations, except at the functional level previously described in conjunction with the formulation of the turbulence problem by Hopf [43, 161, 162]. Thus, when we relate the two systems, we must make approximations, and only experimental verification can provide the degree of this [40, 163].

Mickelsen [170], Baldwin and Walsh [2], and Baldwin and Mickelsen [3] approached the problem empirically by using the hypothesis that the Eulerian and Lagrangian correlation coefficients are of the same form. They obtained data on the spreading of mass and heat (partially corrected for molecular diffusion). They also, by an incremental trial-and-error procedure, obtained the Lagrangian correlation coefficient, whose double integral [by Eq. (14–223)] gives the observed spread. The procedure used was convenient since, as a first estimate, they used the Eulerian correlation, and thus also empirically related the two systems. Figure 14–16 is the empirical relation for the transformation of an Eulerian into a Lagrangian correlation. The value of $f(r)$ becomes $R_L(\tau)$ at a transformed time given by

$$\tau = Kr/v_y'(t), \qquad (14\text{--}236)$$

where r is in the direction of flow. The velocity variance is that of the particle; it was established from the diffusion experiments, and is essentially equivalent to

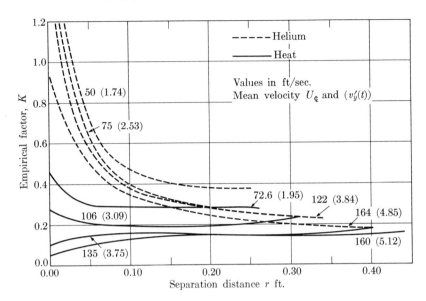

Fig. 14–46. Empirical factor for Eulerian and Lagrangian correlation coefficients. [From L. V. Baldwin and W. R. Mickelsen, J. Eng. Mech. 88, 37–69, 151–153 (1962). By permission.]

the lateral Eulerian variance. These values are noted in Fig. 14–46. The difference between the results obtained by using helium and those obtained by using change of temperature is somewhat disturbing. If corrections for molecular diffusion and other effects have all been accounted for, then under a given set of conditions the results should be the same. The differences for small separations are quite apparent. A constant value of K for any one velocity would imply that the correlations are of the same general form but this is clearly not so. To obtain the turbulent spread, it is necessary to perform a double integration (14–223) of the transformed Eulerian anemometer data. One could also obtain the turbulent spread by calculating a double integral of the Eulerian correlation with respect to (Kr)-space, and transforming the axis by the constant $1/v'_y(t)^2$.

Less empirical approaches to an Eulerian-Lagrangian relation are limited in number. An interesting method, suggested by Burgers [24], tested by Baldwin and Walsh [2], and extended by Deissler [62], involves the use of an Eulerian space-time correlation, defined for isotropic turbulence as

$$f(r, \tau) = \overline{u_r(\boldsymbol{x}, t) u_r(\boldsymbol{x} + \boldsymbol{r}, t + \tau)}/u'^2. \tag{14–237}$$

This might approximate the Lagrangian correlation when r is taken as $\bar{U}_x \tau$. The suggestion is that the Eulerian correlation convected with the flow can be used for the Lagrangian correlation. By this method, we neglect any effect of particle displacement, which might be critical, but which should depend on the spread

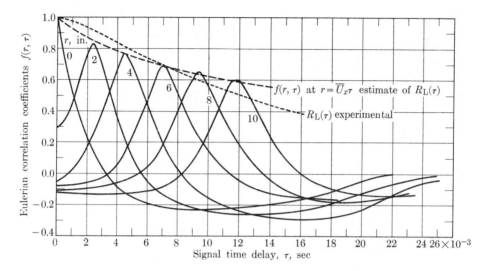

*Fig. 14–47. General Eulerian correlation coefficient as a function of space and time. [From L. V. Baldwin, and T. J. Walsh, AIChEJ **6**, 53–61 (1961). By permission.]*

relative to the convected distance. In other words, the approximation might be good for short times when the spread is small, and become increasingly poorer for longer times. Figure 14–47 shows an experimental comparison, and gives values of $f(r, \tau)$ versus τ for various values of r. The loci of the maxima are within 10% of $r/\bar{U}_x$ and the dashed line estimates $R_\mathrm{L}(\tau)$. The experimental result at this velocity is shown by the broken line. The two curves show a definite resemblance. Calculation of the spread from the estimated correlation was in fair agreement for short times and became increasingly poorer for longer times, as we would expect from the foregoing discussion. The predicted eddy diffusivities were as much as 2.3 times those observed. Considering the lack of alternative methods, this is still a helpful engineering result, since it is the only one that deals directly with the correlations.

Somewhat less ambitious, but still useful, would be estimates from Eulerian data for the various scales associated with turbulent dispersion. Such estimates might be useful in Eqs. (14–228) through (14–232). For example, Corrsin [49] has derived a number of relations for extremely high Reynolds numbers by assuming that the entire spectrum is in the inertial subrange. He found that the integral Eulerian scale L_f was approximately related to the Lagrangian length scale L_L (14–122) by $L_\mathrm{L}/L_f = A^{3/2}/B$, where A is the same as in Eq. (14–173), and, from the data of Grant et al. [84], is about 1.5. Corrsin [45] has estimated B to be about unity. Thus the ratio would be approximately 2, which can be compared with reported experimental values ranging from 2 to 6.5, and Corrsin's result is certainly reasonable.

Heisenberg [95] made an estimate for the microscale at extremely high Reynolds numbers which can be expressed [95, 251] in the form

$$(\lambda/\tau_L u'_x)^2 = (12.6/\alpha + 4.46\alpha)(1/N_{\text{Re},\lambda}), \tag{14-238}$$

where τ_L is the Lagrangian time microscale defined in a manner similar to that used to define other microscales:

$$\tau_L^2 = -2/[R_L(0)]'', \tag{14-239}$$

where the double prime denotes the second derivative. The constant α has been set by experiments. Heisenberg has suggested a value of 0.8; Proudman has found that 0.45 gives good agreement, and Reid has found that the constant varies between 0.20 and 0.62. Beek and Miller [16] have obtained a value of 0.7 for pipe flow from the data of Laufer [139]. Using the 0.45 value gives an equation also obtained by Corrsin (for $B = 1$) [49], which is

$$(\lambda/\tau_L u'_x)^2 = 30/N_{\text{Re},\lambda}. \tag{14-240}$$

Another useful relation obtained by Corrsin [45] by the same type of analysis is a ratio of the Lagrangian microscale to the integral scale:

$$\tau_L/T_L = 3/\sqrt{N_{\text{Re},\lambda}}. \tag{14-241}$$

If we use α as 0.7, τ_L from Eq. (14-238) varies from 5.1 to 6.3 times the experimental values obtained by Baldwin [1]. In this estimate, the Reynolds number was 84% of the isotropic value; the remaining required values are tabulated in reference [1]. We cannot expect the accuracy to be good, since λ was calculated from the Eulerian spectral density data and τ_L was obtained from the diffusion data by a method that requires extrapolation of a poorly defined curve to zero time. When we remember the assumptions in the theory and the difficulty in estimating τ_L from diffusion data, we must consider this approximation a good check.

The theory presented so far in this section is actually restricted to dispersion by eddy motion from a source in a static field, or from a source moving within a uniform velocity field. The steady-state dispersion from a fixed source in a uniform velocity field is simply a problem of coordinate transformation. The work of Fleishman and Frenkiel [71], which has been discussed by Hinze [96], illustrates the occurrence of back diffusion in a flowing system. The limit of this effect is the case of zero mean velocity, in which the back diffusion is exactly the same in magnitude as the forward diffusion. The equations for short or long time are the same as Eqs. (14-224) and (14-226), except for the transformation $t = x/\bar{U}_x$. It is clear that a plot of $\overline{Y^2}$ versus x would first increase with x^2 (Eq. 14-224), and then, finally, would increase with x (Eq. 14-226). For intermediate distances, a transformation of Eq. (14-229) would give an approximation to the spread. The data of Kalinske and Pien [107] verify this.

In most systems, the effects of molecular diffusion are masked by turbulent dispersion. This is not always true, however, especially in gas systems using heat as the contaminant or a light gas molecule such as H_2 or He. In these cases some correction for the effect of molecular diffusion is desirable. Unfortunately, it appears that molecular diffusion cannot always simply be added to eddy diffusion. Saffman [213, 214] has modified the work of Townsend [247] and Batchelor and Townsend [15] to illustrate the effect for short times:

$$\overline{Y^2} = \overline{Y_T^2} + 2Dt - \tfrac{1}{9}D\omega'^4 t^3, \qquad (14\text{-}242)$$

where $\overline{Y^2}$ is the total mean squared spread, $\overline{Y_T^2}$ is the Taylor or turbulent contribution, D is the diffusivity, and ω' is the rms vorticity of the turbulent motion. If there is no molecular diffusion, then the dispersion is identical to the infinitesimal fluid particle spread as given by either Eq. (14–223) or (14–225). The interaction term [third term on the right of (14–242)] decreases the effect that molecular diffusion would have if the molecular and turbulent actions were independent and additive. The decrease in total dispersion occurs because the scalar quantity is transported away from the source at a lower speed than the infinitesimal fluid particles are. If we consider a small spot of scalar quantity, the interaction term causes an increase in the dispersion *relative to its centroid*, but the centroid lags behind the infinitesimal fluid particles just enough to give a net negative effect, as shown in Eq. (14–242). This effect of molecular diffusion is very much akin to the turbulent mixing problem to be discussed in the next section. The dispersive action of the eddy motion increases the region in which the contaminant will be found by distorting and drawing out the original contaminant volume. In the absence of molecular diffusion, the original volume will remain the same, although it will cover a greater area. Molecular diffusion, on the other hand, will cause the volume to increase relative to its position, and turbulence enhances this increase by providing increased concentration gradients through the distortion and pulling action. Thus there is an enhancement or increase in the dispersion which is caused by the interaction of the turbulence and molecular diffusion, but this is relative to the centroid of the region as it is convected downstream. This is the effect derived by Townsend, and Saffman [213] shows that it is

$$\overline{Y_c^2} = 2Dt + \tfrac{2}{9}D\omega'^4 t^3 + \cdots, \qquad (14\text{-}243)$$

where the subscript c means "relative to the centroid." As noted above, the negative correction in Eq. (14–242) is due to the fact that the material being dispersed moves more slowly than the fluid particles. Relative to its origin, the dispersion of the centroid is [213]

$$\overline{Y^2} = \overline{Y_c^2} - \tfrac{1}{3}D\omega'^4 t^3, \qquad (14\text{-}244)$$

which, when combined with Eq. (14–243), gives (14–242). We see that $\overline{Y^2}$ is related to the correlation along a molecule path, while $\overline{Y_c^2}$ is related to the correlation along a fluid particle path. In both cases the equation is of the form of (14–223) or (14–225).

An additional aspect of the problem which has caused some disagreement in experimental work is the means of injection of the scalar contaminant. Any finite source will in some way disrupt the velocity field. This problem has been discussed by Flint et al. [75], who investigated turbulent diffusion in the central core of a pipe. Kada and Hanratty [106] have shown that the presence of solids in the flow system does not appreciably affect the turbulent diffusion if the particles move with the stream; that is, if the slip velocity is small. For large slip velocities, the effect is large.

In shear flow, when there is a gradient in the mean velocity, the radial distribution of a scalar contaminant, such as temperature or mass, may no longer be symmetrical, but rather may be skewed in some manner. A discussion of this problem is given by Hinze [96, 97] and by Batchelor and Townsend [15].

14-4.G Mixing

A visualization or model of mixing depends, to a great extent, on one's definition of the term "mixture." We shall use *mixing* to mean any blending into one mass, and *mixture* to mean "a complex of two or more ingredients which do not bear a fixed proportion to one another and which, however thoroughly commingled, are conceived as retaining a separate existence." The terminology used in discussing mixing can cause confusion, since there is generally a free interchange of words (e.g., blending, mixing, dispersion) and their meanings. Fortunately, certain well-defined mixing criteria have been established that go far to remove ambiguity and confusion and allow a more quantitative evaluation than has previously been possible. However, before turning to these, let us discuss the mixing process a little further.

Molecular diffusion is a product of relative molecular motion. In any system in which there are two kinds of molecules, if we wait long enough, the molecules will intermingle and form a uniform mixture on a submicroscopic scale (by submicroscopic, we mean larger than molecular, but still not visible even with the best microscope). This view is consistent with the definition of a mixture, for we know that if we were to use a molecular scale we would still observe individual molecules of the two kinds, and these would always retain their separate identities. The ultimate in any mixing process would be this submicroscopic homogeneity, in which molecules would be uniformly distributed over the field. However, the molecular diffusion process alone is generally not fast enough for present-day processing needs. In some systems, molecular diffusion is so slow as to be completely negligible in any reasonably finite time; the processing of polymers which have a high molecular weight is a good example of this state. If turbulence can be generated, then eddy diffusion effects can be used to aid the mixing process. However, although one can uniformly disperse a material that has no molecular diffusivity (e.g., two immiscible fluids), this does not necessarily produce submicroscopic mixing in the sense previously defined. [If the diffusivity is not zero, the dispersion is enhanced by molecular diffusion, but not additively, as already shown by Eq. (14–242).] In the mixing process, dispersion enhances submicroscopic mixing by providing an increased area for diffusion. In turbulent mixing, the dispersive action is always present and aids the molecular diffusion. In contrast, dispersion can occur in the absence of molecular diffusion effects.

Before we attempt to establish a more detailed model for the mixing process, let us consider the criteria previously mentioned. Danckwerts [54] formulated these so as to provide a measure of the level of mixing, and called this measurement the *goodness* of mixing. These criteria can be treated mathematically, but for now, we need only give qualitative definition to the terms. The *scale of segregation* is a measure of the size of the unmixed clumps of the pure components. As these clumps are spread out, the scale of mixing is reduced; this is represented by the drawings on the top line in Fig. 14–48. The second criterion is the *intensity of segregation*, which describes the effect of molecular diffusion on the mixing process. It is a measure of the difference in concentration between neighboring clumps of fluid. The intensity, for each value of the scale, is illustrated by the columns of Fig. 14–48. With the aid of these terms, we can now discuss a model of turbulent mixing.

The turbulent process can be used to spread out the fluid elements to some limiting point; however, because of the macroscopic nature of turbulence, one would not expect the ultimate level to be anywhere near molecular size. Since energy is required for this reduction in scale, the limiting scale should be associated with the smallest of the energy-containing eddies. This might be considered as the eddy size η (Eq. 14–169), which characterizes the dissipation range, but which is also an approximate measure of the tail of the energy curve $E(k)$, shown in Fig. 14–33. One might also use as a measure the microscale, which is defined by Eq. (14–192). In any case, this size will be large when compared with molecular dimensions. This reduction of scale without consideration of molecular diffusion is shown as the top row of Fig. 14–48, and can also be interpreted in terms of wave numbers. Clearly, as one progresses to the right in the figure he sees that there has been a transfer from low wave numbers to higher wave numbers. No matter how far the scale is reduced (or the wave number increased), the components are still pure. Any one of these levels in scale might be considered mixed, depending on the size observed. From the viewpoint of submicroscopic homogeneity (when molecules are uniformly distributed over the field) none are mixed. Without molecular diffusion, this ultimate mixing cannot be obtained.

Molecular diffusion permits molecules to move across the boundaries of the liquid elements, thus reducing the difference between elements. This reduction in intensity occurs with or without turbulence, but turbulence can help speed the process by increasing the area covered by the clumps, thus allowing more area for molecular diffusion. When diffusion has reduced the intensity of segregation to zero, the system is mixed, and the molecules are distributed uniformly over the field. Various degrees of this combined process are shown in Fig. 14–48. In systems in which reaction is to occur, the need for submicroscopic mixing is apparent, for without it, the only chemical reaction that could occur would be

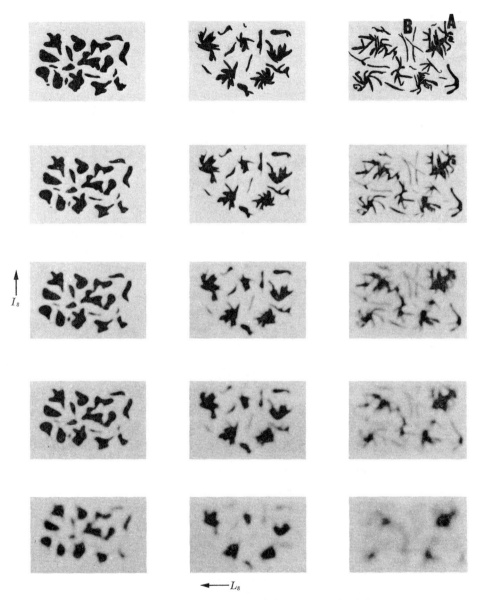

Fig. 14–48. *A representation of the criteria of mixing.*

on the surface of the fluid clumps. Danckwerts [55] has discussed the importance of the degree of mixing of two reactants. The intensity of segregation must be reduced rapidly in order to avoid local spots of concentrated reactant and the undesirable side reactions usually associated with such spots. In jet mixing, the scale of segregation is reduced by eddy motion, while molecular diffusion reduces the intensity. If a solid product occurs, its particle size will be a function of the rate of reduction of segregation. In the quenching of a hot gas reaction mixture, the freezing of the reaction products depends on the reduction of segregation. In the jet flame, oxygen is obtained from the surrounding air, and the nature of the flame depends on the segregation of the gases. In laminar flames, the mixing is poor because the scale of segregation is high. The flame occurs along a surface and is controlled by the molecular diffusion across that surface. In a turbulent flame, eddy diffusion reduces the scale, providing more area for molecular diffusion and thus more contacts for burning.

(a) Criteria for mixing. Two aspects of mixing have been mentioned: first, the degree to which the material to be mixed by the turbulent action has been spread out, as measured by the scale of segregation; second, the approach to uniformity by the action of molecular diffusion, as measured by the intensity of segregation. Danckwerts [54] has defined these measures so that they describe the mixing process and still can be estimated from measurable statistical values. The only major restriction on the parameters is that they cannot be applied to cases in which gross segregation occurs, i.e., in which the liquids are in two nearly equal parts. This restriction implies that the parameters are approximately uniform over the mixing field, thus eliminating consideration of the first part of the mixing process, in which the two liquids are initially brought together.

(i) Scale of segregation. The scale of segregation is analogous to the scale of turbulence (Eq. 14–117); however, since concentration is a scalar, there is only one term instead of nine in the correlation:

$$g_s(r) = \frac{\overline{a(\mathbf{x})a(\mathbf{x}+\mathbf{r})}}{a'^2}, \quad (14\text{--}245)$$

where $g_s(r)$ is the Eulerian concentration correlation, a is the fluctuation $A - \bar{A}$ (A is the concentration fraction of the liquid A, $\bar{A}$ is the average), and a' is the rms fluctuation $\sqrt{\overline{a^2}}$. A linear scale L_s is defined as

$$L_s = \int g_s(r)\,d\mathbf{r}. \quad (14\text{--}246)$$

A volume scale V_s is defined as

$$V_s = \int 2\pi\,r^2 g_s(r)\,d\mathbf{r}. \quad (14\text{--}247)$$

Either scale can be used, depending on the exact application. Danckwerts has considered various methods of measurement of the two scales, as follows
  (1) Obtaining $g_s(r)$ versus r and integrating.
  (2) Measuring the statistical variation of component A along a line of distance r (possibly by photoelectric measurement), and using the derived relation

$$L_s = a_r'^2/2ra'^2, \quad (14\text{--}248)$$

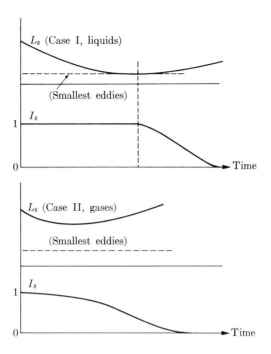

Fig. 14–49. *The intensity and scale of segregation during mixing.*

where $a_r'^2$ is the mean square of the concentration fraction over the distance r. One necessary restriction in this method is that $r \gg L_s$; however, if r is too large, it will be difficult to obtain $a_r'^2$.

(3) A procedure analogous to (2), using the variation in a volume V, and

$$V_s = a_V'^2/2Va'^2, \qquad (14\text{–}249)$$

where $a_V'^2$ is the mean square fluctuation (fraction) in the volume V.

(4) Several other methods, and their limitations, are given by Danckwerts [54].

Because the scale of segregation is an average over relatively wide values of r, it is a good measure of the large-scale process (breakup of the eddies), but not of the small (diffusional) process. In the liquid system with slow molecular diffusion, the scale would decrease rapidly to some small value (smallest eddy size), and then increase slowly as molecular diffusion completed the mixing (see Fig. 14–49, Case I). The increase in scale is due to an apparent increase in eddy size because of outward diffusion. The value would increase indefinitely with r, since $g_s(r)$ is unity everywhere in the uniform medium. In a gas, in which the molecular diffusion is very rapid, the scale may not be reduced appreciably before diffusional effects become controlling (Case II). For the rapid-diffusion gas system, and in the latter part of the liquid mixing process, when the scale is small, the intensity of segregation gives a better description of the degree of mixing.

(ii) Intensity of segregation. The intensity of segregation (also called the degree of segregation) is defined as

$$I_s = \overline{a'^2}/\bar{A}\bar{B} = \overline{a'^2}/a_0'^2, \tag{14–250}$$

and is measured at a point for enough time to obtain a true average. The subscript 0 refers to the initial value. This term is unity for complete segregation; that is,

$$a_0'^2 = \overline{(A_0 - \bar{A})^2} = \overline{\bar{B}(0 - \bar{A})^2 + \bar{A}(1 - \bar{A})^2} = \bar{B}\bar{A}^2 + \bar{A}\bar{B}^2 = \bar{A}\bar{B}(\bar{A} + \bar{B}) = \bar{A}\bar{B},*$$

and drops to zero when the mixture is uniform ($\overline{a'^2} = 0$). If there were no diffusion, and only the smallest possible eddies were present, the value of I_s would still be unity; thus the intensity of segregation as defined is a good measure of the diffusional process (see Fig. 14–49). Equation (14–250) gives the simplest form of the intensity of segregation and is defined as a function of time-averaged variables at a point. This of course assumes that such an average can be obtained; i.e., that the system is at steady state or is changing slowly when compared with the time necessary to obtain the average. For a complete definition of a given system, one would have to specify the variation of I_s over the entire volume. As a simple example, consider plug flow, in which two fluids are to be mixed. It will be assumed that each fluid is initially distributed uniformly across the pipe cross section on a macroscopic scale under the condition of complete segregation ($I_s = 1$). As the fluid moves down the tube in plug flow, mixing will occur as a result of the turbulent field and diffusion, and the value of I_s will decrease to zero in the limit of molecular uniformity. Actually I_s must be measured over some small but finite volume. If this volume is too small, submicroscopic variations are detected (statistical fluctuations in the number of molecules present), and if the volume is too large, the measurement becomes insensitive and approaches the average value of the system. For many problems (such as nonideal mixers used for reactors), a detailed study of the variation of I_s over the entire reactor is not desirable, and some space average of the entire system is used. We shall consider this briefly in a later section.

Although measuring the intensity of segregation is difficult, a number of attempts have been made. These have been discussed by Brodkey [22].

Because of the nature of the turbulent mixing problem, little is known about the actual mechanism. Consequently, we use an approach similar to that used for statistical turbulence. In other words, we formulate the problem in terms of statistical averages, without reference to any specific model. There are two important aspects of the problem. First, experimental information interpreted in terms of the theory may provide insight into the actual mechanistic contribution of turbulence and of molecular diffusion to mixing. Second, with reasonable approximations for the terms and boundary conditions, the theory can be used

---

*This is true since $\bar{A} + \bar{B} = 1$ and $A_0 = 1$ for an $\bar{A}$ fraction of the total volume and 0 for a $\bar{B}$ fraction ($\bar{B} = 1 - \bar{A}$):

to predict the time of mixing under specific mixing conditions. Admittedly, the work has not progressed in either direction as far as we would like; however, as will be seen, some progress has been made along both lines.

The turbulent mixing problem was developed by Obukhov [179] and Corrsin [35, 36] in much the same manner as was done for turbulent motion. As in the motion problem, the formulation of the turbulent mixing problem in terms of various statistical moments gives rise to a set of infinite equations, the closure of which cannot be deduced from the theory itself. Turbulent mixing is much simpler than turbulent motion, since the latter is described by a nonlinear equation that is of one tensor degree higher order than turbulent mixing. However, the mixing problem still involves the same nonlinearity difficulties when expressed in terms of moments. Moreover, turbulent mixing is strongly dependent on the nature of the velocity field. The closure of the infinite set of moment equations in mixing is analogous to that of the motion problem, but for a different reason. In turbulent motion, the closure problem is a direct consequence of the nonlinearity of the original equation of motion. In mixing, however, it is caused by the nonlinearity in the stochastic variables, even though the scalar conservation equation is essentially linear. The problem is a result of choosing a statistical approach and using averages, and for this a substantial penalty must be paid.

(b) Mixing in an isotropic field. Consider mixing in an isotropic turbulent field (for both the scalar and velocity fluctuation). By definition, neither mean velocity or mean concentration is present (i.e., fluctuations of velocity and concentration are about zero). The usual method of obtaining the desired relation is like the derivation of the von Kármán-Howarth equation; that is, two equations of mass conservation at two different points are multiplied by the scalar quantity fluctuation at the other point, then the two are added and averaged. If we wish to obtain the spectrum form, we transform the equation into wave number space. In physical space it is

$$\frac{\partial a'^2 g_s(r)}{\partial t} - 2a'^2 u' \left[\frac{\partial k_s(r)}{\partial r} + \frac{2k_s(r)}{r} \right] = 2Da'^2 \left[\frac{\partial^2 g_s(r)}{\partial r^2} + \frac{2}{r} \frac{\partial g_s(r)}{\partial r} \right], \quad (14\text{--}251)$$

which we can see is like the von Kármán-Howarth equation (14–157). The subscript s refers to scalar terms as contrasted to velocity. In terms of spectrum, the equation becomes

$$\partial E_s(k)/\partial t = T_s(k) - 2Dk^2 E_s(k). \quad (14\text{--}252)$$

This is similar to the velocity spectrum equation (14–161). A scalar transfer function $S_s(k)$ can be defined, as was done for velocity (14–162):

$$S_s(k) = -\int_0^k T_s(k)\, dk, \quad (14\text{--}253)$$

and Eq. (14–252) becomes

$$\partial E_s(k)/\partial t = -\partial S_s(k)/\partial k - 2Dk^2 E_s(k). \quad (14\text{--}254)$$

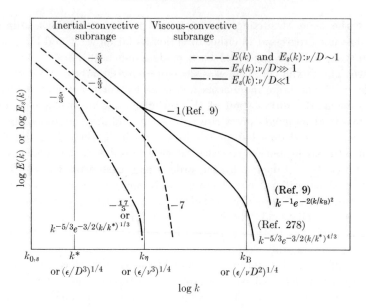

Fig. 14–50. Intuitive and phenomenological spectrum results.

Methods like those used for Eq. (14–161) can be used to solve this equation. We shall briefly discuss the phenomenological approximation which involves relating $S_s(k)$ to $E_s(k)$ by an equation analogous to that suggested by Heisenberg (14–175). We shall also mention some of the other closure procedures which parallel those already discussed in conjunction with the velocity closure problem. Before this, however, it will be well to discuss the mixing mechanism further.

Figure 14–50 is a composite of the intuitive and phenomenological results obtained previously on the velocity spectrum and from somewhat similar analyses for the scalar field [9, 11, 35, 36, 44, 179, 278, 279]. For example, in the viscous-convective subrange, owing to the great difference in ν and D, $(\nu/D \gg 1)$, the spectrum of concentration fluctuations extends much further into the high wave number range than that of the velocity fluctuations. One would expect a range of $E_s(k)$ (conc^2-cm), which would depend on ϵ_s (conc^2/sec), the effect of the large velocity eddies, stretching out the scalar blob [a strain rate parameter γ (sec^{-1})], and on k (cm^{-1}). But $E_s(k)$ would not depend on the diffusivity, since the diffusivity would be important beyond the high wave number end of this subrange. Dimensionally, one must have

$$E_s(k) \sim (\epsilon_s/\gamma) k^{-1}, \qquad (14\text{–}255)$$

which is the form derived by Batchelor [9] by both dimensional and analytical arguments. Pao [278] has suggested that the form for the very low Schmidt number is also valid for the high Schmidt-number range, with of course an extensive shift to the right because of the low diffusivity. However, recent experimental results by Nye and Brodley [286] confirm Eq. (14–255). The high Schmidt-number range is of major importance in liquid mixing systems, since the mass diffusivity is low. The qualitative effects of the velocity eddies on the scalar blobs have been shown in Fig. 14–48. In the top right drawing, the blob marked A is of the general shape one would expect from the action of eddies smaller than the blob (i.e., a situation that might exist in the inertial-convective subrange). The blob marked B has the shape one would expect as a result of a pulling action exerted by eddies of large velocity on a smaller blob (i.e., a situation that might exist in the viscous-convective subrange).

(i) **Consideration in physical space.** Some useful relations can be obtained directly from Eq. (14–251) without recourse to the spectrum form (14–254). Equation (14–251), in the limit as $r \to 0$, gives us

$$\frac{da'^2}{dt} = 2Da'^2[g_s''(0) + 2g_s''(0)] = 2Da'^2\left(\frac{2}{\lambda_s^2} + \frac{4}{\lambda_s^2}\right) = -\frac{12Da'^2}{\lambda_s^2}, \quad (14\text{–}256)$$

where, as in Eq. (14–192),

$$\lambda_s^2 = -2/g_s''(0), \quad (14\text{–}257)$$

and the limits $g_s(0) = 1$ and $k_s'(0) = 0$ have been used. Equation (14–256) can be compared to the velocity counterpart given as Eq. (14–196). Equation (14–256) can be rearranged to

$$\frac{1}{a'^2}\frac{da'^2}{dt} = \frac{1}{I_s}\frac{dI_s}{dt} = -\frac{12D}{\lambda_s^2}, \quad (14\text{–}258)$$

which has a very simple form. The solution of such first-order equations can be obtained readily by quadrature, provided that the time dependence of λ_s is specified. Corrsin [37] has suggested a roundabout path in order to impose some restrictions on the turbulent field which would justify the assumption that λ_s has a very weak time dependence. He has further suggested that for a low Schmidt number,

$$\lambda_s^2/\lambda^2 = 2D/\nu. \quad (14\text{–}259)$$

With the use of this, we can integrate Eq. (14–258) to give an exponential decay law,

$$I_s = e^{-t/\tau}, \quad (14\text{–}260)$$

where

$$\tau = \lambda_s^2/12D = \lambda^2/6\nu. \quad (14\text{–}261)$$

In a different approach, Corrsin [48] has used the spectrum analysis of Batchelor [9] to extend these results to the high Schmidt-number range. He has used an equation analogous to (14–198), specifically,

$$\lambda_s^2 = 6 \frac{\int_0^\infty E_s(k)\,dk}{\int_0^\infty k^2 E_s(k)\,dk}, \qquad (14\text{–}262)$$

to establish λ_s and thus the time constant from Eq. (14–261). First, for the low-Schmidt-number case ($N_{\text{Sc}} \leqslant 1$), Corrsin integrated the spectrum curve from $k_{0,s}$ to k_η (14–170), assessed the order of magnitude of the various terms after integration, and obtained the final result, which is

$$\tau = \frac{\lambda_s^2}{12D} = \frac{2}{3 - N_{\text{Sc}}^2} \frac{1}{(\epsilon k_{0,s}^2)^{1/3}} = \left(\frac{5}{\pi}\right)^{2/3} \frac{2}{3 - N_{\text{Sc}}^2} \left(\frac{L_s^2}{\epsilon}\right)^{1/3}, \qquad (14\text{–}263)$$

where L_s is the scalar macroscale defined by Eq. (14–246), ϵ is the velocity energy dissipation per unit mass given by Eq. (14–165), and $k_{0,s}$ is the wave number representative of the large scalar blobs, which would be analogous to k_0, associated with the energy-containing eddies. The term $k_{0,s}$ has been eliminated from the last part of Eq. (14–263) by the approximate relation

$$k_{0,s} = (\pi/5) L_s^{-1}. \qquad (14\text{–}264)$$

Finally, Corrsin has shown that Eq. (14–263) can be expressed in a form analogous to (and more general than) Eq. (14–261); that is,

$$\tau = \frac{\lambda_s^2}{12D} = \frac{\lambda^2}{10\nu} \left(\frac{k_0}{k_{0,s}}\right)^{2/3} \frac{2}{3 - N_{\text{Sc}}^2}. \qquad (14\text{–}265)$$

For this same case, Rosensweig [205] has used for the spectrum the asymptotic form of the von Kármán interpolation formula at high Reynolds numbers. He has obtained a result which is equivalent to

$$\tau \cong \frac{1}{1.44(\epsilon k_{0,s}^2)^{1/3}} = \frac{1}{1.19} \left(\frac{L_s^2}{\epsilon}\right)^{2/3}. \qquad (14\text{–}266)$$

For a Schmidt number of unity, the difference between this and Eq. (14–263) is a factor of 1.44. For a zero Schmidt number, the difference is reduced to a factor of 1.44/1.5.

For the high-Schmidt-number case, the time constant can be obtained by integration of the spectrum curve (a composite of the $-\frac{5}{3}$ and -1 subranges) from $k_{0,s}$ to k_B, to give

$$\tau = \frac{\lambda_s^2}{12D} = \frac{1}{2}\left[\frac{3}{(\epsilon k_{0,s}^2)^{1/3}} + \left(\frac{\nu}{\epsilon}\right)^{1/2} \ln N_{\text{Sc}}\right] = \frac{1}{2}\left[3\left(\frac{5}{\pi}\right)^{2/3}\left(\frac{L_s^2}{\epsilon}\right)^{1/3} + \left(\frac{\nu}{\epsilon}\right)^{1/2} \ln N_{\text{Sc}}\right]. \qquad (14\text{–}267)$$

This latter form removes a criticism [140] of the earlier work: that it failed to depend on the Schmidt number. As can be seen from Eq. (14–267), the requirement of no mixing in the limit of infinite Schmidt number is now satisfied ($\tau = \infty$, which gives $I_s = 1$ for all time).

Corrsin [37, 48] has related the results of mixing without reaction to the more commonly measured variables of turbulence and mixing. Equation (14–263) suggests that the time constant should scale with L_s^2/ϵ. This is also the most important term in Eq. (14–267). Thus one would want

$$L_s'^2/\epsilon' = L_s^2/\epsilon, \qquad (14\text{–}268)$$

where the prime denotes the scaled-up system. The velocity energy dissipation per unit mass is related to the power by

$$\epsilon = \mathcal{E}P/M, \qquad (14\text{–}269)$$

where $\mathcal{E}$ is the efficiency of turbulent production and M is the fluid mass. The combination of Eqs. (14–268) and (14–269) gives us

$$\mathcal{E}'P'/M'L_s'^2 = \mathcal{E}P/ML_s^2. \qquad (14\text{–}270)$$

The mass M scales with the geometry or length cubed, K^3, and L_s scales with some length K_s. If the efficiency varies in the scale-up, then the result for the power is

$$P' = K_\epsilon K^3 K_s^2 P. \qquad (14\text{–}271)$$

If the efficiency is constant,

$$P' = K^3 K_s^2 P. \qquad (14\text{–}272)$$

This specific result was presented by Corrsin [48] and is a little different from that presented in reference [37], which was

$$P' = K^5 P. \qquad (14\text{–}273)$$

While K depends on geometry, K_s would be expected to depend more on the nature of the injection of the material to be mixed. The fifth power can be derived also from pure blending relations (i.e., the number of tank turnovers held constant on scale-up); however, the empirical value is nearer to the fourth than the fifth power. From Eq. (14–272), one would expect between a third- and fifth-power dependency.

Corrsin [48] has considered the scaling for the high-Schmidt-number case in more detail. As a second approximation, he obtained from Eq. (14–267)

$$P' = K^3 K_s^2 \left[1 - \left(\frac{\pi}{5}\right)^{2/3} \frac{\nu^{1/2} \ln N_{\text{Sc}}}{\epsilon^{1/6} L_s^{2/3}} \left(1 - \frac{1}{K_s}\right)\right] P. \qquad (14\text{–}274)$$

Here the scaling depends on actual physical properties. If the scalar injection were somehow maintained at the same level, the power would scale as K^3, just as before. However, if $K_s \cong K$, the scaling would be less than K^5, which is in good accord with experimental observations. Equation (14–274) was derived for the important case of scale-up of the same liquid or better for a constant value of $\nu(\ln N_{\text{Sc}})^2$.

The equation cannot be used for scale-up of two different fluids in the same mixing vessel. In the unlikely event that the second term of Eq. (14–267) would be controlling, which it normally would not, the scale-up would be given by

$$P/P' = K^3(\nu'/\nu)(\ln N'_{\text{Sc}}/\ln N_{\text{Sc}})^2,$$

and would only apply where the diffusion was so low and controlling that the turbulence in the system would have adequate time to wipe out any effect of the nature of the scalar injection. For the case of different liquids, in which all factors are of importance, a crude scale-up might be based on Eq. (14–274), with K_s modified appropriately by the property ratio shown above.

Without reference to the moment equations, Hughes [102] approached the problem by assuming local isotropic turbulence to be valid. The time necessary for the final diffusion was assumed to be similar to that given by Einstein's diffusion equation:

$$t = \eta^2/2D, \qquad (14\text{--}275)$$

where η has been used to replace the mean displacement of a particle and is the characteristic length of a small-scale velocity energy eddy as defined by Eq. (14–169). The energy dissipation can be obtained from Eq. (14–205). Equations (14–169), (14–205), and (14–275) combine to give

$$\eta = (L_f/1.65)^{1/4}(\nu/u')^{3/4}. \qquad (14\text{--}276)$$

Hughes has used the equation to compute the minimum diffusion time for CO_2 in several gases and liquids. The results are reasonable if the Schmidt number is not too high. For a high N_{Sc}, Eq. (14–275) has too high a dependency on the molecular diffusion, and does not account for the interaction between the turbulence and diffusion. However, it should give an upper limit to the mixing time.

(ii) **Consideration in wave-number space.** Corrsin [34] has suggested using the spectrum equation (14–254) with the Heisenberg eddy-diffusivity type of transfer function for the scalar field [analogous to (14–175)]. Thus the rate of change of the scalar spectrum may be expressed as [16]

$$\frac{\partial}{\partial t}\int_0^k E_s(k)\,dk = -2(D + \bar{\bar{\epsilon}})\int_0^k k^2 E_s(k)\,dk, \qquad (14\text{--}277)$$

where

$$\bar{\bar{\epsilon}} = \alpha_s \int_k^\infty \left(\frac{E(k)}{k^3}\right)^{1/2} dk,$$

where α_s is a numerical factor. Since Eq. (14–277) is analogous to Heisenberg's eddy-viscosity type of analysis, one can expect not only that it will contain the good points of Heisenberg's transfer function but that it will also be subject to criticisms such as have been put forward by Hinze [96]. Beek and Miller [16] integrated the equations numerically for $E_s(k)$, and subsequently obtained from this the scalar intensity from

$$I_s = \frac{a'^2}{a_0'^2} = \frac{1}{a_0'^2} \int_0^\infty E_s(k)\, dk. \qquad (14\text{-}278)$$

The value of I_s is a function only of the Reynolds and Schmidt numbers. The former defines the turbulent field, and the latter defines the relative rate of molecular diffusion. Beek and Miller had to introduce a number of assumptions in addition to Eq. (14-277). For example, they adopted a physical model of isotropic turbulence which is fictitiously enclosed in a pipe and conveyed by a uniform mean velocity. Turbulent shear flow does provide such a stationary turbulent velocity field, but it is certainly not isotropic. Nevertheless this is one of the few available solutions to the problem of turbulent mixing and so it should be considered further.

In the analysis, the largest velocity eddy was taken as one-quarter of the pipe diameter;

$$1/k_0 = d_0/4. \qquad (14\text{-}279)$$

Multi-injection of the material to be mixed was introduced by using, as one parameter, a reduced wave number of the largest concentration eddy:

$$\chi_0 = \sqrt{N}, \qquad (14\text{-}280)$$

where N is the number of nozzles. Another parameter is

$$\alpha' = D k_0 / u'_x,$$

which can be combined with Eq. (14-279) and definitions to give

$$\alpha' = \left(\frac{1}{N_{\text{Sc}}}\right)\left(\frac{1}{N_{\text{Re}}}\right)\frac{4}{u'_x/\bar{U}_x} \qquad (14\text{-}281)$$

In order to obtain a unique solution of Eq. (14-277), we have to specify at least two boundary conditions. One is the initial scalar spectrum, which would depend on the injection system for the scalar contaminant. A Dirac (impulse) initial spectrum seems reasonable, which would mean that blobs of exactly uniform size are initially released. We can visualize this uniformity as being achieved by uniform injectors in a pipeline system or injectors at the wall of a mixing vessel. If the initial spectrum is defined by a Gaussian distribution, it can be pictured as a wide initial distribution with some average scale of segregation, but still an intensity of segregation of unity. We might visualize this distribution as being caused by the mixing occurring in a pipe beyond a mixing tee or by injection directly into the impeller area in a mixing vessel. There is, of course, a degree of arbitrariness involved in deciding on the relation between the initial eddy distribution and the physical dimension which characterizes the injection system. Beek and Miller have proposed to fix the initial wave number as $\sqrt{N}\, k_0$ by using Eq. (14-280). There is little information available by which one can evaluate this particular assumption.

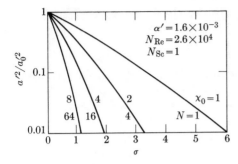

*Fig. 14–51. The decay of intensity for low Schmidt number. [From J. Beek, Jr., and R. S. Miller, Chem. Eng. Prog. Symposium, Series No. 25 **55**, 23–28 (1959). By permission.]*

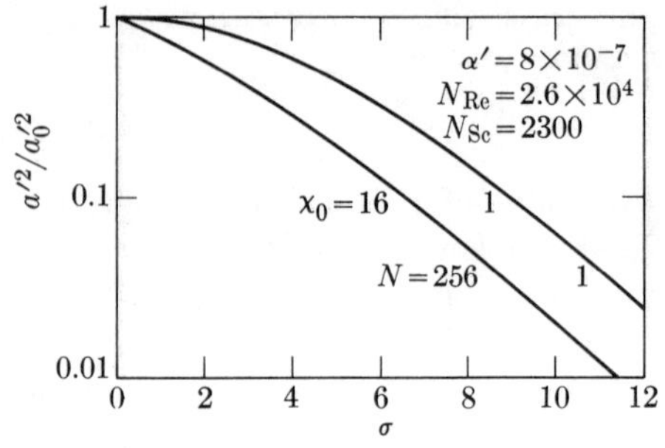

*Fig. 14–52. The decay of intensity for high Schmidt number. [From J. Beek, Jr., and R. S. Miller, Chem. Eng. Prog. Symposium, Series No. 25 **55**, 23–28 (1959). By permission.]*

Figures 14–51 and 14–52 give the final results of the work for $\alpha' = 1.6 \times 10^{-3}$ and 8×10^{-7}, respectively. The Schmidt numbers were 1 and 2300, respectively, and the Reynolds number is 2.6×10^4 for both figures. The symbol σ denotes a dimensionless time of mixing, given by

$$\sigma = k_0 u'_x t. \tag{14–282}$$

The difference between gases and liquids for an increasing initial wave number, which can be obtained by multiple injection, is striking. The effect on gases is quite large; for liquids, it is quite small.

Figure 14–53 gives the 99% decay time of the concentration intensity in liquids and in gases as a function of the Reynolds number. For the specific case of one nozzle, mixing of liquids appears to be greatly influenced by the Reynolds number, while for gases there is little effect. This is misleading, however, since σ is dependent on the Reynolds number through u'_x. If one computes an actual mixing time for a representative case by using Fig. 14–53 and estimated values of k_0 and u'_x, one finds that the time of mixing for both the liquid and gas systems is highly dependent on the Reynolds number. Actually, σ can be better visualized as a measure of the pipe length needed to obtain the mixing. The decrease in mixing time for the gas just compensates for the increase in velocity, and thus σ, the mixing length, remains relatively constant. For liquids, the

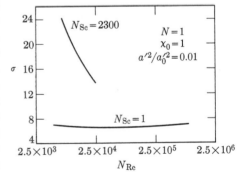

*Fig. 14–53. Decay of 99% as a function of Reynolds number. [From J. Beek, Jr., and R. S. Miller, Chem. Eng. Prog. Symposium, Series No. 25 **55**, 23–28 (1959). By permission.]*

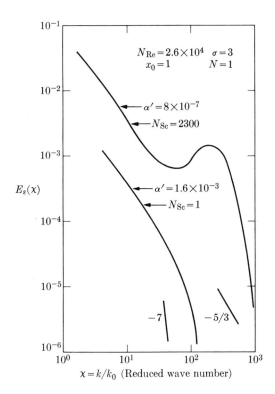

Fig. 14–54. The scalar spectrum in dimensionless form. [From J. Beek, Jr., and R. S. Miller, Shell Development Company, Emeryville, California (1960). By permission.]

effect of the increase in turbulence more than offsets the increase in velocity, thus resulting in a net decrease in the length of pipe required.

In the solution of Eq. (14–277), we obtain values of $E_s(k)$ as functions of k and t. In Fig. 14–54, which shows typical spectrum curves in dimensionless form, $E_s(\chi)$ is plotted against $\chi = k/k_0$, with σ of Eq. (14–282) as a parameter. The general shapes of the $E_s(\chi)$-curves are considerably different for low and high Schmidt numbers. For the low-Schmidt-number case, the various equilibrium subranges for $E_s(k)$ essentially overlap those for $E(k)$. Thus one might expect that $E_s(k)$ and $E(k)$ are similar in shape. On the other hand, for the high-Schmidt-number range, $E_s(k)$ extends far beyond the cutoff wave number for $E(k)$. The convective subrange of the scalar field overlaps the dissipation subrange of the velocity field, thus apparently creating a state of imbalance between the convective and dissipative subranges for the scalar field. This is indicated by a hump in the higher wave-number range for the scalar field, which seems to mark the separation between the two subranges. However, there is no reliable experimental verification of the suggested shape.

It has often been suggested that the Reynolds number be used for scale-up of pipeline mixers. We may question the validity of this and base our analysis on the foregoing material. For constant Reynolds number, $u'_x/\bar{U}_x$ is constant. Since $\bar{U}_x$ must decrease in order to offset the increase in d_0 for the same Reynolds number, u'_x must also decrease. Furthermore, by Eq. (14–279), if d_0 increases, k_0 must decrease. From Eq. (14–282), one sees that t must increase considerably for a given σ with an increase in d_0 (actually with d_0^2). However, since the time can be taken as $L/\bar{U}_x$, t increases only with d_0. Therefore, to obtain the same degree of mixing, one must also increase the length in proportion to the increase in d_0. This is, in effect, a requirement of geometric similarity. Even scale-up on the same velocity will not be sufficient, since the increase in Reynolds number will not change $u'_x/\bar{U}_x$ appreciably, which, by Eq. (14–279), must offset the decrease in k_0. Thus, in all cases, to obtain the same degree of mixing one must also design for an even greater length than in the smaller test section.

A more refined approach to the problem of homogeneous isotropic turbulent mixing can be made by closing the first two equations of the infinite set by making some reasonable assumption to relate the second- and fourth-order moments. This is the cumulant-discard approximation already used for the velocity field (page 300). The solution is mathematically complex, and as already indicated for long times, the approximation leads to unreal results; i.e., a part of the spectrum decays so fast that it becomes negative, which is a physical impossibility [132, 182]. Consequently, this derivation and the specific results [140, 181, 182] will not be discussed further here.

By the direct-interaction approximation, Lee [276] has numerically calculated the decay of the scalar field for a stationary isotropic turbulence. This parallels the calculation made by Kraichnan for the decay of the velocity field, shown in Fig. 14–37. The solution was well behaved and there were no negative spectrum segments; however, the solution was limited by computational time to relatively low values of the mixing parameters (i.e., Schmidt number and mixing time). Nevertheless, these were still considerably larger than would be expected for gas systems, and demonstrated the sensitivity of the decay to inlet conditions. In order to circumvent the problem of long computational times, Lee [277] combined the above-mentioned cumulant-discard approximation with a dampening function suggested from the direct-interaction approximation, and again obtained a well-behaved solution. The amount of computer time required for this latter calculation is quite small. The results compare well with the direct-interaction approximation, although they are less accurate. Lee further compared the results of the various closure approximations for the same set of mixing conditions, and was able to demonstrate the inadequacy of all but the above two methods.

Although not as much progress has been made as one would like to see, the calculations based on the idealized homogeneous isotropic turbulence are valuable because they might represent an upper limit to the time of mixing required. From data on the velocity spectrum in pipes (Fig. 14–31), the eddy size appears smaller and the turbulent intensity greater near the wall than at the center line, at which approximate isotropic conditions have the best chance of existing. Thus the mixing at the wall should be more rapid than at the center line. As a result of this, the time of decay along the center line may be a good measure of the total time required.

The problem of mixing in a shear field has not been considered in detail because of the complexity of the simpler isotropic problem. However, equations can be formulated to describe the variation of mean concentration and the decay of the concentration fluctuations. Brodkey [22] and Lee [140] have presented one possible set of equations, but no attempt at solving them has been made. A more simplified approach was suggested, but the basic hypothesis of the analysis is still unproved.

(c) **Chemical reaction and reactors.** An important subject area of mixing is the prediction of the effect of turbulence on chemical reactions. There are two main classes into which reactor problems can be placed. First, the mixing may be *between two or more streams* entering the reactor. This class of problems may be further subdivided into systems in which mixing is between components of a reaction which are *initially separated* (for example, $A + B$), or into systems in which mixing is between a diluent and components of a reaction which are *initially together* (for example $A + A$). In the limit of zero diffusivity for the first case (mixing between reactants), the only place in which a reaction can occur is on the hypothetical surface between the reactants, and the only effect of mixing is to increase the contact area. Clearly, this is close to the case in which a reaction takes place *between* components in two immiscible streams. In the second case the mixing is a simple blending of a reaction mixture with a diluent. The reaction mixture itself, which is introduced as a uniform stream, may consist of one or more components. Here the reaction occurs no matter what the diffusivity; however, in some cases the diffusion can have a marked effect. The second main class of reactor problems is the mixing of a homogeneous fluid entering the reactor. A better description of this is *self-mixing* [145]; however, the use of this term is not widespread in the literature. In slowly reacting systems that are reasonably well mixed, the mixing is essentially complete before an appreciable amount of reaction has occurred, and such systems can be treated as a problem of self-mixing.

Very little work has been done on the problem of reaction between *initially separated* components. On the theoretical side, Toor [242] has considered the enhancement of mixing caused by a rapid irreversible reaction between dilute materials with equal diffusivities. He showed that if the mixture were stoichiometric, the fractional conversion would be equal to the degree of mixing, that is, $1 - \sqrt{I_s}$. We can easily see the equivalence if we remember that the final mixing to a completely homogeneous system involves diffusion. For simple mixing, each molecule that diffuses from the high- to the low-concentration area has a double reduction effect on the driving force by simultaneously reducing the high concentration and increasing the low. Of course, the diffusion is equal molal, and the solvent is also transferred. Now consider the rapid irreversible reaction in stoi-

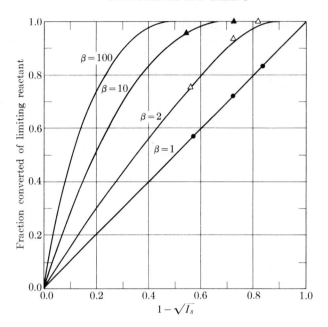

Fig. 14-55. Fractional conversion as a function of accomplished mixing. [From H. L. Toor, AIChEJ **8**, 70 (1962). By permission.]

chiometric balance; the scalar material diffuses into the other material and reacts, and by equal molal counter-diffusion, the reverse occurs and also reacts. The net double effect is still present; however, the double reduction now occurs in the scalar field. On the average, one quantity of scalar material has moved out and reacted, and one has been removed by reaction within the field. Thus, for this special case, the reduction in diffusional driving force is the same, and the conversion is equivalent to the degree of mixing. For nonstoichiometric mixtures, Toor obtained an estimate for the increase in conversion as a function of the square root of the intensity of segregation, by assuming a normal distribution of the concentration fluctuations about its mean. The results are shown in Fig. 14-55. The symbol β represents the ratio aC_{a0}/bC_{b0} from the reaction $aC_a + bC_b$ = products, and the subscript 0 refers to initial values. Toor concluded that "increasing the concentration of one species ... is most effective when a high conversion is required and the mixture is initially close to stoichiometric." These results can be used as an extension to any theory of the mixing of nonreacting materials. Some verification of this theory is available. Keller et al. [273] measured the turbulent field behind a grid and were able to predict the conversion for an ammonium hydroxide–acetic acid reaction. Vassilatos and Toor [281] obtained similar results for four different neutralization reactions (ionic) in a specially designed jet reactor. In their case the actual turbulent field was unknown, so that $1 - \sqrt{I_s}$ was established from the kinetic experiments at $\beta = 1$ and the 45°

line on Fig. 14–55. Some of these results are shown in the figure. Similar experiments have been reported by Saidel and Hoelscher [280] for a rapid irreversible second-order reaction in the wake of a cylinder. A direct comparison was not possible, however, because of the different geometries.

Isotropic theory as previously presented could be applied to mixing-vessel problems with some degree of approximation. However, in this case the turbulent field will decay in certain regions of the mixer. Probably the only area which would approximate a constant turbulent field is that in the immediate vicinity of the stirrer.

The local mixing problem is simplified if one can assume a well-stirred system. Rosensweig [205] has used this assumption to obtain a conservation statement for the concentration fluctuations in the vessel. This is

$$I_s = 1 - \epsilon_s \tau'/\bar{A}\bar{B}, \tag{14-283}$$

where τ' is the mean residence time and ϵ_s is the scalar energy dissipation, similar to ϵ for the velocity energy. An alternate form is possible, since

$$\epsilon_s = -da'^2/dt, \tag{14-284}$$

which is given by Eq. (14–256). These two equations can be combined to give

$$I_s = \frac{1}{1 + (12D/\lambda_s^2)\tau'} = \frac{1}{1 + \tau'/\tau}. \tag{14-285}$$

The term τ can be estimated from Eqs. (14–261) through (14–267), depending on the system. The conservation statement is as yet unproved, since no data are available for comparison.

Experimental work has also been limited. Some turbulence measurements have been reported by Cutter [53] and by Kim and Manning [274]; Manning and Wilhelm [168] have surveyed the concentration field with a conductivity probe. Rice et al. [199] considered the scale of mixing in a stirred vessel by following an acid–base reaction with an indicator. The system is pictured as the breakdown of the injected stream, by the turbulence, into small fixed-geometry fluid elements, and the subsequent reaction of these as controlled by the rate of molecular diffusion. The data of Manning and Wilhelm were found to be consistent with the model. This is a simple phenomenological view of the process; it bypasses the details of the turbulent motion of the fluid, and in this respect is limited.

Corrsin [39] considered the simplest case of the blending of a single component undergoing reaction (the *initially together* problem). For a first-order system, the equations are linear and it is easy to determine the effects of diffusion and reaction. In the derivation of Eq. (14–251), the added reaction term is $-2k_1 a'^2$, which carries through the analysis, and Eq. (14–258) becomes

$$(1/I_s)(dI_s/dt) = -12D/\lambda_s^2 - 2k_1, \tag{14-286}$$

where k_1 is the first-order reaction velocity constant. Under the assumptions made previously, the equations through (14-261) follow directly; i.e., for Eq. (14-261), the modification gives

$$\tau = \frac{1}{6\nu/\lambda^2 + 2k_1}, \qquad (14\text{-}287)$$

which can be used in Eq. (14-260). As before, the diffusivity contributes to the decay of the fluctuation intensity but has no effect on the mean concentration. The reaction, even in the absence of diffusion, contributes to the decay of the fluctuation intensity [Eqs. (14-260) and (14-287)], but does so at exactly the same rate as it causes the decay in the mean concentration; that is,

$$I_s = \overline{a'^2}/\overline{a_0'^2} = e^{-2k_1 t}, \qquad (14\text{-}288)$$

which can be compared with

$$(\bar{A}/\bar{A}_0)^2 = e^{-2k_1 t}, \qquad (14\text{-}289)$$

as obtained directly from the rate equation $d\bar{A}/dt = -k_1\bar{A}$. An important conclusion from Eq. (14-289) is that mixing, and consequently molecular diffusion, has no effect on the conversion of a first-order reaction. The system can best be pictured as a reacting mixture being mixed with a solvent stream. As before, the diffusivity reduces the concentration fluctuations; superimposed on this is the reduction by the reaction, at a fixed rate, of the rms intensity and mean concentration. A second-order reaction is far more complicated, and has been considered only for certain limiting conditions, such as extremely low turbulence and very slow or very fast reactions [39]. Even if more than one component is involved, the reaction can occur in the absence of molecular diffusion if the components are initially together. For reactions other than first-order ones, however, diffusion affects the degree of conversion. This can be illustrated by the second-order reaction rate equation:

$$-d\bar{A}/dt = k_2(\bar{A}^2 + \overline{a'^2}),$$

where the diffusivity influences $\overline{a'^2}$ and thus the rate.

Corrsin, in a series of articles, has considered the effect of a first-order reaction on the shape of the spectra shown in Fig. 14-50. The spectra of the reactants under various conditions [44], of the products [42], and of a slightly exothermic reaction [50] have all been considered. Further extensions have been given by Pao [279].

For many systems the definition of I_s as a time average at a point is adequate for following the mixing problem, and is the ideal definition to use if its variation over the system being studied is meaningful and is not too complicated. This is the case for the pipe mixer already cited, for which the variation of I_s in both the radial and axial directions can be measured. Another good example is the misnamed *ideal mixer* (to be called the *well-stirred mixer* here), in which the contents are the same at every point, and are equal to the exit conditions. It is easy to imagine that this system might not be mixed at all, or might be partially mixed in terms of the true local mixing I_s. For example, if the molecular diffusion is zero, complete segregation exists locally regardless of the homogeneity on the average; thus I_s is unity. For finite diffusion, the value of I_s depends on the rate of mixing by turbulence. Perfect mixing can be approached if the mixing time to local homogeneity is much less than the average residence time ($\tau' = V/Q$, where V is the mixer volume and Q is the rate of volumetric flow through the system). In any case, I_s has one value which is constant over the entire well-stirred mixer.

Under many conditions a time average of terms at a point is not adequate. To illustrate, let us consider a *nonideal mixer*, in which there is a region of long holdup of one of the materials to be mixed. If we measure I_s locally, its value may be nearly zero within and outside the holdup region; yet, since there are gross variations in concentration, we cannot assume that the system is mixed. Thus the use of I_s as defined is not meaningful, and must be restricted to systems that are initially (and continually) uniformly dispersed on a macroscopic scale. For systems in which gross variations occur, a space average at one time would be better [55]. This would be of the same form as Eq. (14–250), and would give a single value of the intensity for the entire system. The new intensity of segregation, which we shall call $I_{s'}$, would have the same properties as I_s. For the continuous nonideal mixer, the contents are not the same at every point system, and thus $I_{s'}$ cannot be zero.

The second main class of mixing problem (*self-mixing*) has been more extensively studied, and a good discussion of it can be found in the text by Levenspiel [145]. The use of the same general terminology for the very different physical process of mixing and self-mixing has led to some confusion. Mixing has already been considered in some detail, and to provide the proper perspective, a short discussion of self-mixing is in order. Self-mixing is also called back-mixing, and is associated with the mixing of fluid elements that have been in the reactor or mixer for different lengths of time. By the very fact that a homogeneous fluid is involved, not the mixing of two different streams, molecular diffusion plays no part in the problem. Thus an alternate definition of the intensity of segregation is necessary.

The term I_α is an intensity of segregation which is based on the use of the "age of a fluid at a point" rather than the previously used "concentration at a point" [55]. If $\bar{\alpha}$ represents the mean age of the molecules in the system at some given time, an rms deviation of all the molecules in the system from this average can be determined:

$$\alpha' = \sqrt{\overline{(\alpha - \bar{\alpha})^2}}. \tag{14–290}$$

A similar rms deviation can be determined for all of the points within the system; that is,

$$\alpha'_P = \sqrt{\overline{(\alpha_P - \bar{\alpha})^2}}, \tag{14–291}$$

where α_P is the mean age at a point. The intensity of segregation is defined as

$$I_\alpha = \overline{\alpha_P'^2}/\overline{\alpha'^2}. \qquad (14\text{-}292)$$

If the system is segregated, each molecule is surrounded by other molecules of the same age. Thus the mean age at a point (α_P) must be the same as the age of the molecules (α) in that point. Since this is true everywhere, the deviation of all points and molecules must be the same, and I_α is unity. If the system is uniform, the mean age of the molecules at any point is the same and is equal to the mean age of all molecules in the system, so that $\alpha_P = \bar{\alpha}$ everywhere $(\overline{\alpha_P'^2} = 0)$, and I_α is zero. In a nonideal system there is a lower limit for I_α, a lower limit that can be estimated, by the method proposed by Zwietering [265], from data concerning the distribution of residence time.

The classical limiting cases of a plug-flow pipeline (no radial variation) and the well-stirred mixer are of interest. Along the axial direction of the pipe, I_α must be unity, since, within any point, molecules of the same age as the point are contained (plug-flow assumption). The mean-square point deviation, then, is the same as the mean-square molecule deviation. For the well-stirred mixer, I_α is zero, because there is complete self-mixing of the various elements and the mean age of the molecules at any point is the same and is equal to the mean age of all molecules in the system. The plug-flow pipe or reactor and the well-stirred mixer clearly represent the extremes of I_α.

Depending on the mixing by the turbulence and the geometry of the system, the nonideal mixer has a value of I_α that ranges from unity down to some minimum value. The minimum I_α corresponds to the state of *maximum mixedness*, and can be determined for any given distribution of residence time [265]. The assumptions necessary in the evaluation are covered elsewhere [55, 145, 265] and need not be repeated here.

Possibly a clearer picture can be obtained from the work of Curl [52] and Miller et al. [172], who have shown that a similar situation can exist in dispersed-phase systems in which the reaction takes place in the dispersed phase only. The effect, a back-mixing, is caused by coalescence and redispersion of the drops. When this effect is zero, the reaction occurs under completely segregated conditions. At very high rates of coalescence and redispersion, the reactor approaches well-mixed conditions.

The effect of I_α on conversion of a chemical reaction depends on the order of the reaction [55]. When there is complete segregation, a gradient in concentration exists because of the reaction; the average reaction rate in the entire reactor depends on the average of the local rates, that is, on $\overline{A^n}$. For complete mixing $(I_\alpha = 0)$ the rate depends on $\bar{A}^n$. For a first-order reaction, the two averages are equal $(\overline{A^1} = \bar{A}^1)$, and there is thus no effect on a first-order reaction. For $n > 1$, $\overline{A^n} > \bar{A}^n$, and the reaction is faster in the completely segregated case. Just the opposite is true for $n < 1$.

From the foregoing discussion, it is clear that the distribution of residence time is also an important characteristic of mixing systems. The distributions are known for most of the limiting cases, but for the nonideal mixer some model for the fluid motion and back-mixing in the vessel is necessary. The degree of nonideality depends on the particular geometry and operating characteristics selected, and thus one would expect a great number of possible models, depending on such variables as type of stirrer, number and placement of baffles, location of injectors, etc. To consider this in detail is beyond the scope of this section, and the reader is referred to the references [28, 86, 88, 144, 145, 164, 175, 263, 266, 267, 269, 275, 284], which treat models designed to represent the distribution of the fluid residence time, the effect of the distribution on conversion for segregation and maximum mixedness, and some comparison with experimental data. These models generally involve the use of short-circuiting (bypassing or channeling), stagnant zones (dead space), or a transfer of fluid between a piston flow and the bulk of the fluid.

(d) Experimental results. In the past, most experiments have dealt with wind tunnels or pipe flow, in which temperature was the scalar quantity in a relatively isotropic turbulent field of air. Lee and Brodkey [22, 140, 142, 143] and Nye and Brodkey [286] studied the mixing of a secondary dye stream in a nondecaying turbulent field of water (using a three-inch pipe system), while Gibson and Schwarz [78, 79] studied the mixing of salt solution in a decaying turbulent field behind a grid (using a six-inch water-tunnel system). In all these experiments, the authors made extensive measurements of the concentration and velocity fields (mean and fluctuating values and spectra). The important difference between these and earlier experiments was the use of mass as the scalar quantity in a water system, which meant that the Schmidt number was high and thus the molecular mixing process was an important factor in the length of the mixing time.

Lee and Brodkey showed that for their injection system, the concentration fluctuation intensity was always higher at the center of the pipe. In other words, the concentration fluctuation intensity persisted longer at the center line than anywhere else in the pipe. They then compared their results for the decay at the center line with the predictions from isotropic theory. They did not imply by this that the flow was isotropic, but rather that isotropic theory could provide a crude estimate of the mixing at the center line, even though the flow was of a shear type. If the prediction was reasonable, then it would also serve as an estimate of the total mixing time required. This would constitute a use of isotropic theory, not a proof.

They obtained the time constant for the decay as follows: First, they calculated λ from the one-dimensional spectrum results (velocity). They combined Eqs. (14–263) and (14–265) to obtain

$$(5/\pi)^{2/3}(L_s^2/\epsilon)^{1/3} = (\lambda^2/10\nu)(k_0/k_{0,s})^{2/3}, \qquad (14\text{–}293)$$

which, together with the value of the microscale and the assumption that $k_0 = k_{0,s}$, gives a value for the term

$$S = (5/\pi)^{2/3}(L_s^2/\epsilon)^{1/3}. \qquad (14\text{–}294)$$

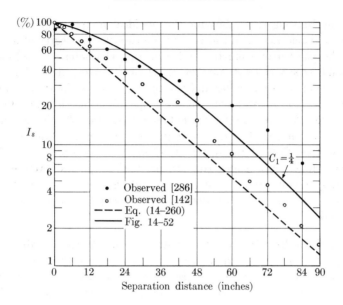

Fig. 14–56. Intensity of segregation. (Decay at the center line.) The curve for Eq. (14–260) is based on the experimental value of λ and ϵ, or on a value of $C_1 = \frac{1}{2}$.

They used this in turn in Eq. (14–267), with ϵ from velocity spectrum measurements, to give the time constant for the decay. Two main assumptions were made in this analysis. First, $k_0 = k_{0,s}$ was assumed valid, since the measurements were made after the dye was dispersed across the pipe cross-section. Second, the term S of Eq. (14–294) was estimated from the equation for the low-Schmidt-number case, since the scalar macroscale should depend on the manner of injection and not on the Schmidt-number level. The final result is compared with the data in Fig. 14–56. Since the analysis is not an exact solution to the isotropic mixing problem and is even more approximate for the shear flow, the coefficients are not expected to be exactly correct. The test of the equations is very stringent, in that ratios of low to high Schmidt-number experiments are not used so that the absolute values of the constants fall out. Thus one must consider the curve to be a good estimate of mixing. A more complete analysis of the decay as a function of radial position is given by Gegner and Brodkey [142], and an analysis in terms of the time constant is given by Nye and Brodkey [286].

For general use in mixing calculations the estimation of the microscale from the velocity spectrum is not desirable, since this measurement may not always be available. Thus the following semi-empirical method is a possible alternate. The microscale can be obtained from the empirical relation (14–206). The macroscale can be approximated [96] by

$$L_f = \tfrac{3}{4}(1/k_0). \tag{14–295}$$

The average size of the energy-containing eddies, unlike that of the small eddies, is influenced by the linear dimension of the pipe from which the eddies receive energy directly. The following relation can then be written on a dimensional basis:

$$1/k_0 = C_1 r_0, \qquad (14\text{-}296)$$

where C_1 is dimensionless; Beek and Miller [16] suggested for C_1 a value of $\frac{1}{2}$. Combining Eqs. (14-206), (14-295), and (14-296) into (14-294), one can obtain an estimate for the term S for the low-Schmidt-number case,

$$S = (5/\pi)^{2/3} (L_s^2/\epsilon)^{1/3} = 0.341\, r_0/u_x', \qquad (14\text{-}297)$$

where k_0 has been assumed to be equal to $k_{0,s}$. This estimate of S would be used in Eq. (14-267) to estimate the time constant. For the one case available, the value of the term S from the estimate was less than 3% higher than the value of the term S obtained from using the experimental value of λ.

The result obtained by Beek and Miller can also be compared with the experimental points, if one expresses the time σ in terms of the pipe length. Their curve, estimated for a value of C_1 of $\frac{1}{4}$, is also shown in Fig. 14-56. The value of C_1 is lower than expected, but the actual shape of the decay curve is closer to that observed than the straight line of Eq. (14-260).

Spectra results are reported in the references already cited. In agreement with the theory of Batchelor, Gibson and Schwarz [79] found an indication of a -1 spectral range. Lee and Brodkey [142] showed that isotropic conditions did not exist in their work. Nye and Brodkey [286], with a much-improved probe system [143], clearly established the existence of the -1 spectral region.

EXAMPLE 14-6. Estimate the time constant for the mixing at the center line in a pipe. Use the data obtained in Example 14-1. The pipe diameter is three inches, the Schmidt number is 1000, and the solvent fluid is water.

Answer. With the value of y_0^+ and the pipe diameter, we can obtain

$$U^*/\nu = 8480 \text{ ft.}^{-1}$$

From this, and from the ratio of the friction to maximum velocities, we can determine the maximum velocity as 2.12 ft/sec. From Fig. 14-24, $u_x'/U^* = 0.8$, and is not a function of the Reynolds number at the center line. The parameter S of Eq. (14-297) becomes 0.585. Using Eqs. (14-279) and (14-295) in (14-205) gives an estimate for the dissipation energy of $0.0136 \text{ ft}^2/\text{sec}^3$ or $12.6 \text{ cm}^2/\text{sec}^3$, which can be compared with the value of $14.9 \text{ cm}^2/\text{sec}^3$ reported in reference [142] for nearly the same conditions. With the above information, we see that Eq. (14-267) becomes

$$\tau = \tfrac{1}{2}[3(0.585) + (0.01/12.6)^{1/2} \ln(1000)] = \tfrac{1}{2}(1.95) = 0.95 \text{ sec.}$$

Under conditions that were almost identical, a time constant of 0.932 sec was reported in reference [142].

PROBLEMS

14-1. Draw the velocity distribution and compute the thickness of the laminar sublayer for the flow of water at room temperature in a pipe one foot in diameter. The average velocity is 6 ft/sec.

14-2. For the same average velocity given in Problem 1, obtain the laminar profile and compare this with the turbulent profile from Problem 1.

14-3. Repeat the solution of Problem 1 for a rough pipe ($e/r_0 = 0.01$). Assume the pressure gradient is the same as in Problem 1.

14-4. Plot a curve of $\bar{\epsilon}/U^* r_0$ as a function of y/r_0.

14-5. Derive the Reynolds shearing stress, using the assumption of a mixing length. What are the boundary conditions? Repeat the derivation, but start with the equations obtained from the integration of the Reynolds equations. What are the boundary conditions? Compare the two sets of conditions and indicate which satisfies the real physical problem.

14-6. Describe the physical model of turbulence as suggested by the theory of local isotropic turbulence. What would be the effect of Reynolds number?

14-7. Explain and compare the Eulerian and Lagrangian correlations, scales. Show the steps necessary to derive:

14-8. Eq. (14-10) from (14-9).

14-9. Eq. (14-141) from (14-139).

14-10. Eq. (14-142) from (14-140) and (14-141).

14-11. Eq. (14-163) from (14-145).

14-12. Eq. (14-164) from (14-149).

14-13. For Eq. (14-114), what are the seven nonzero components, and to which scalar correlation function of Eq. (14-115) is each equal?

14-14. Discuss and compare (using illustrations) Eulerian correlation and the corresponding spectrum terms.

14-15. What is the meaning of the statement that the Prandtl mixing length is independent of the shear stress? Is the statement valid? What are the implications?

14-16. What are the basic assumptions behind the von Kármán-Howarth equation (14-157) and behind the mixing equation (14-251)?

14-17. Discuss the basic premise and validity of using autocorrelation.

14-18. What is the value of the transfer function when integrated over all of wave number space?

14-19. In the inertial subrange of the universal equilibrium range, two particles are at a small distance x apart, where $x \gg \eta$. Determine the order of magnitude of the time τ required for the particles to move apart to a distance y, where $x \ll y \ll L_f$.

14-20. What is the problem in estimating the eddy viscosity at the center line in pipe flow?

14-21. What shape of correlation curve would you expect to obtain by the measurement on a field of uniform-sized spherical eddies?

14-22. Find an approximate expression for the mixing-length distribution for the one-seventh power law (14-61). Express the final result in terms of the friction velocity, maximum velocity, and geometry variables of the system.

NOTATION FOR CHAPTER 14

A	constant
$\bar{A}, \bar{B}$	average concentration fraction of A or B in a mixture
A, B	instantaneous concentration fraction of a scalar quantity
a, b	concentration-fraction fluctuation of a scalar quantity
a', b'	rms value of the concentration-fraction fluctuation
a'_r	rms value of the concentration-fraction fluctuation over a distance r
a'_V	rms value of the concentration-fraction fluctuation over a volume V
c, C	constants
D	mass diffusivity
d_0	diameter
$\mathsf{E}_{ij}(k)$	integrated energy spectrum tensor (14–132)
$E(k)$	three-dimensional energy spectrum function (14–133)
e	depth of roughness, natural log base
$\boldsymbol{F}_s$	external force
f	Fanning friction factor
$f(r), g(r)$	isotropic correlation functions (14–104), (14–105)
$g_s(r)$	scalar correlation function (14–245)
$h(r), k(r), q(r)$	isotropic triple velocity correlation function (14–115)
I_s, I_α	intensity of segregation (14–250), (14–292)
i	$\sqrt{-1}$
$\boldsymbol{i}, \boldsymbol{j}, \boldsymbol{k}$	unit vectors in the x, y, z directions
$\boldsymbol{k}, \boldsymbol{p}, \boldsymbol{q}, \boldsymbol{k}_i$	wave-number vectors
k	$|\boldsymbol{k}|$, constant
k_0	wave number of largest velocity eddy
k_η	Kolmogoroff wave number (14–170)
k_1, k_2	first- and second-order reaction velocity constants
K	constant
L_{ij}	Eulerian scale (14–117)
L_f, L_g	isotropic Eulerian scales (14–118), (14–119)
L_L	Lagrangian length scale (14–122)
L_s	linear scalar scale (14–248)
l	mixing length
m	parameter
M	mass
n	a power
N	number of nozzles
N_Re	Reynolds number, $d_0 \bar{U}_{x,\mathrm{ave}}/\nu$
$N_{\mathrm{Re},\lambda}$	Reynolds number, $\lambda u'/\nu$
N_Sc	Schmidt number, ν/D
P	power
$\mathsf{P}_{ij}(\boldsymbol{r})$	pressure term (14–147)

p	pressure
$\Pr_\tau$	defined by (14–96)
$\mathbf{Q}(r)$, $\mathbf{Q}_{ij}(r)$	velocity correlation tensor (14–98), (14–99)
$\mathbf{Q}_{ij}(0)$	energy tensor (14–127)
r	vector distance
$R_{ij}(r)$	velocity correlation function (14–101)
$R_{L_{ij}}(\tau)$	Lagrangian velocity correlation function (14–107)
$R_\mathrm{L}(\tau)$	Lagrangian isotropic velocity correlation function (14–108)
r	$\|r\| = \sqrt{x^2 + y^2 + z^2}$
$\mathsf{S}_{ij,k}(r)$	triple velocity correlation tensor (14–114)

Greek letters

δ^*	displacement thickness
δ	thickness
ϵ	energy dissipation
$\bar{\bar{\epsilon}}$	eddy viscosity coefficient (momentum diffusivity)
ϵ_H	eddy diffusivity for heat
ϵ_M	eddy diffusivity for mass
η	Kolmogoroff length (14–169)
$\mathcal{E}$	efficiency
Λ_L	Lagragian length scale
λ	microscale of turbulence
μ	viscosity
ν	kinematic viscosity
σ	time parameter
ρ	density
τ	time constant (14–261)
τ'	mean residence time
$\bar{\bar{\tau}}$	total shear stress (also $\bar{\bar{\tau}}_{ij}$)
τ_L	Lagrangian time microscale (14–239)
Ω	angular velocity, vorticity
ω	vorticity fluctuation
$\bar{\bar{\omega}}_{ij,k}(k)$	Fourier transform of $\mathsf{S}_{ij,k}(r)$
$\bar{\Phi}_{ij}(k)$	energy spectrum tensor (14–124)
$\bar{\phi}_{ij}(k_l)$	one-dimensional spectrum function (14–130)
χ_0	$\sqrt{N}$
ψ	streamline
$\psi(k\eta)$	a function (14–171)

Subscripts

A, B	pertaining to points A and B
0	initial value, pertaining to the wall
L	laminar
s	pertaining to species s, pertaining to the scalar contaminant

T	turbulent
w	pertaining to the wall
1	inner
2	outer
ave	average across the cross section
max	maximum value
H, K, O, B, P	Heisenberg, Kovasznay, Obukhov, Batchelor, and Pao, respectively

Overmarks

—	average at a point
$\sim$	fluctuating part, transposed tensor

Other symbols

∞	infinity
$\vert\ \vert$	the absolute value of
$'$	rms value of
$', '', '''$	different points in space; first, second, and third derivatives

REFERENCES

1. L. V. BALDWIN, Ph. D. Thesis, Dept. Chem. Engr., Case Institute of Technology, Cleveland, Ohio (1958).
2. L. V. BALDWIN and T. J. WALSH, *AIChEJ* **7**, 53–61 (1961).
3. L. V. BALDWIN and W. R. MICKELSEN, *J. Eng. Mech.* **88**, 37–69, 151–153 (1962).
4. S. G. BANKOFF and R. S. ROSLER, *Rev. Sci. Inst.* **33**, 1209 (1962).
5. T. BARON and M. SOUDERS, Jr., *Ency. Chem. Tech.* (R. E. Kirk and D. F. Othmer, editors) **6**, 614–639 (1951).
6. J. BASS, *Compt. rend. Paris* **228**, 228 (1949).
7. G. K. BATCHELOR, *Proc. Cambridge Phil. Soc.* **43**, 533–559 (1947).
8. G. K. BATCHELOR, *Theory of Homogeneous Turbulence*, Cambridge University Press, Cambridge (1953).
9. G. K. BATCHELOR, *J. Fluid Mech.* **5**, 113–133 (1959).
10. G. K. BATCHELOR, in *Mécanique de la Turbulence*, p. 85, Editions du Centre National de la Recherche Scientifique, Paris (1962).
11. G. K. BATCHELOR, I. D. HOWELLS, and A. A. TOWNSEND, *J. Fluid Mech.*, **5**, 134–139 (1959).
12. G. K. BATCHELOR and I. PROUDMAN, *Phil. Trans. Roy. Soc. London*, **248A**, 369 (1956).
13. G. K. BATCHELOR and R. W. STEWART, *Quart. J. Mech. Appl. Math.* **3**, 1 (1950).
14. G. K. BATCHELOR and A. A. TOWNSEND, *Proc. Roy. Soc. (London)* **194A**, 527 (1948).
15. G. K. BATCHELOR and A. A. TOWNSEND, in *Surveys in Mechanics*, (G. K. Batchelor and R. M. Davies, editors), 352–399, Cambridge University Press, Cambridge (1956).
16. J. BEEK, JR. and R. S. MILLER, *Chem. Eng. Progr. Symposium Series* No. 25 **55**, 23 (1959); private communication.

17. H. BLASIUS, *Mitt. Forschungsarbeit, VDI* **131**, 1–34 (1913).
18. T. V. BOUSSINESQ, *Mém. prés. par. div. sav., Paris* **23**, 46 (1877).
19. R. S. BRODKEY, J. LEE, and R. C. CHASE, *AIChEJ* **7**, 392–393 (1961).
20. R. S. BRODKEY, *AIChEJ* **9**, 448–451 (1963).
21. R. S., BRODKEY, E. R. CORINO, and P. K. SANGHANI, paper presented at AIChE meeting, Chicago (Dec. 1962).
22. R. S. BRODKEY, in *Mixing, Theory and Practice*, Vol. I (J. Gray and V. Uhl, editors), Chapter 2; Academic Press, New York (1966).
23. A. BROOKSHIRE, PH. D. THESIS, Louisiana State University, Baton Rouge, La. (1961).
24. J. M. BURGERS, *Rept. E-34.1*, Hydrodynamics Lab., California Institute of Technology, Pasadena, Calif. (July 1951).
25. J. E. CERMAK and L. V. BALDWIN, paper presented at AIChE meeting, Houston, Texas (Dec. 1963).
26. S. CHANDRASEKHAR, *Proc. Roy. Soc. (London)* **229A**, 1 (1955).
27. S. CHANDRASEKHAR, *Phys. Rev.* **102**, 941 (1956)
28. A. CHOLETTE, and L. CLOUTIER, *Can J. Chem. Eng.* **37**, 105, 112 (1959).
29. P. Y. CHOU, *Quart. Appl. Math.* **3**, 38–54 (1945).
30. F. H. CLAUSER, in *Advances in Applied Mechanics* **4**, 1–52, Academic Press, New York (1956).
31. M. F. COHEN, M. S. thesis, The Ohio State University, Columbus, Ohio (1962); J. KNOX, *ibid.* (1966); G. W. MCKEE, *ibid.* (1966).
32. D. COLES, in *Mécanique de la Turbulence*, p. 229, Editions du Centre National de la Recherche Scientifique, Paris (1962).
33. S. CORRSIN, *NACA TN 1864* (1949).
34. S. CORRSIN, *J. Appl. Phys.* **22**, 469–473 (1951).
35. S. CORRSIN, *J. Aeron. Sci.*, **18**, 417 (1951).
36. S. CORRSIN, *Proc. First Iowa Thermodynamics Symposium*, Iowa State University, Ames, Iowa (1953).
37. S. CORRSIN, *AIChEJ* **3**, 329–330 (1957).
38. S. CORRSIN, *NACA RM 58B 11* (May, 1958).
39. S. CORRSIN, *Phys. Fluids* **1**, 42–47 (1958).
40. S. CORRSIN, in "Atmospheric Diffusion and Air Pollution," *Adv. in Geophys.* **6**, 161–164, 441–446, Academic Press, New York (1959).
41. S. CORRSIN, *J. Geophys. Res.* **64**, 2134–2150 (1959).
42. S. CORRSIN, in *Proceedings Symposium on Fluid Dynamics and Applied Mathematics*, pp. 105–124, University of Maryland, Gordon and Breach, New York (1961).
43. S. CORRSIN, *American Scientist* **49**, 300–325 (1961).
44. S. CORRSIN, *J. Fluid Mech.* **11**, 407–416 (1961); *Phys. Fluids* **7**, 1156–1162 (1964).
45. S. CORRSIN, in *Mécanique de la Turbulence*, p. 27, Editions du Centre National de la Recherche Scientifique, Paris (1962).
46. S. CORRSIN, in *Handbuch der Physik*, 2nd ed., Vol. VIII, Part 2 (S. Flügge and C. Truesdell, editors), Springer, Berlin (1963).

REFERENCES

47. S. CORRSIN, in *Encyclopaedic Dictionary of Physics*, Macmillan (Pergamon Press), New York (in press).
48. S. CORRSIN, *AIChEJ* **10**, 870–877 (1964).
49. S. CORRSIN, *J. Atmospheric Sci.* **20**, 115–119 (1963).
50. S. CORRSIN, *Chemical Reaction in Homogeneous Turbulent Fields*, presented at AIAA meeting, New York (Jan. 1964).
51. S. CORRSIN and A. L. KISTLER, *NACA Report 1244* (1955); supersedes *NACA TN 3133* and *Wartime Report W-94* (1946).
52. R. L. CURL, *AIChEJ* **9**, 175–181 (1963).
53. L. A. CUTTER, *AIChEJ* **12**, 35–45 (1966).
54. P. V. DANCKWERTS, *Appl. Sci. Res.* **A3**, 279–296 (1953).
55. P. V. DANCKWERTS, *Chem. Eng. Sci.* **8**, 93 (1958).
56. B. I. DAVIDOV, *Compt. Rend. (Doklady) Acad. Sci., URSS* **127**, 768 (1959).
57. S. J. DAVIES and C. M. WHITE, *Proc. Roy. Soc. (London)* **119A**, 92–107, (1928).
58. R. G. DEISSLER, *NACA TN 2138* (1950).
59. R. G. DEISSLER, *NACA Rept. 1210* (1955); supersedes *NACA TN 3145*.
60. R. G. DEISSLER, *Phys. Fluids* **1**, 111–121 (1958).
61. R. G. DEISSLER, *ibid.* **3**, 176–187 (1960).
62. R. G. DEISSLER, *NASA TR R-96* (1961).
63. E. R. VAN DRIEST, *Paper XII, Heat Transfer and Fluid Mechanics Institute*, University of California, Los Angeles, (June 1955); *J. Aeron Sci.* **23**, 1007 (1956).
64. H. L. DRYDEN, G. B. SCHUBAUER, W. C. MOCK, JR. and H. K. SKRAMSTAD, *NACA-TR 581* (1937).
65. H. L. DRYDEN, *Advances Appl. Mech.* **1**, 1–40 (1948).
66. H. A. EINSTEIN and H. LI, *Paper XIII, Heat Transfer and Fluid Mechanics Institute*, University of California, Los Angeles (June 1955).
67. H. W. EMMONS, *J. Aeron. Sci.* **18**, 490–498 (1951).
68. H. W. EMMONS, in *Proc. Second U.S. Congress Appl. Mech.* p. 1 (1954).
69. A. FAGE, *Phil. Mag.* [7] **21**, 80 (1936).
70. A. FAVRE, J. GAVIGLIO, and R. DUMAS, *Recherche aéron., Paris* **32**, 21 (1953).
71. B. A. FLEISHMAN, and F. N. FRENKIEL, *J. Meteorol.* **12**, 141 (1955).
72. Flow Corporation, *Bulletin 50C*, Watertown, Mass.
73. Flow Corporation, "Selected Topics in Hot-Wire Anemometer Theory," *Bulletin 25*, Watertown, Mass.
74. Flow Corporation, "Probes," *Bulletin 15B*, Watertown, Mass.
75. D. L. FLINT, H. KADA, and T. J. HANRATTY, *AIChEJ* **6**, 325 (1960).
76. W. FUCKS, *Z. Physik* **137**, 49 (1954).
77. R. GERBER, *Proceedings Foursquare Institute*, p. 157, University of Grenoble (1949).
78. C. H. GIBSON, Ph. D. thesis, Stanford University, Stanford, Calif. (1962).
79. C. H. GIBSON and W. H. SCHWARZ, *J. Fluid Mech.* **16**, 357, 365 (1963).
80. M. M. GIBSON, *Nature* **195**, 1281 (1962).
81. M. M. GIBSON, *J. Fluid Mech.* **15**, 161–173 (1963).
82. W. N. GILL and M. SCHER, *AIChEJ* **7**, 61 (1961).

83. W. N. Gill and S. M. Lee, *AIChEJ* **8**, 303 (1962).
84. H. L. Grant, R. W. Stewart, and A. Moilliet, *J. Fluid Mech.* **12**, 241–268 (1962).
85. H. L. Grant and A. Moilliet, *J. Fluid Mech.* **13**, 237–240 (1962).
86. R. E. Greenhalgh, R. L. Johnson, and H. D. Nott, *Chem. Eng. Progr.* **55**, No. 2, 44, 48 (1959).
87. L. M. Grossman and A. F. Charwatt, *Rev. Sci. Instr.* **23**, 741 (1952).
88. E. B. Gutoff, *AIChEJ* **6**, 347 (1960).
89. F. R. Hama, *Phys. Fluids* **2**, 664 (1959).
90. F. R. Hama, *Tech. Note No. BN-195*, University of Maryland, College Park, Md. (1960).
91. F. R. Hama, J. D. Long, and J. C. Hegarty, *J. Appl. Phys.* **28**, 388 (1957).
92. T. J. Hanratty, *AIChEJ* **2**, 359–362 (1956).
93. T. J. Hanratty, G. Latimen, and R. H. Wilhelm, *AIChEJ* **2**, 372–380 (1956).
94. W. Heisenberg, *Proc. Roy. Soc. (London)* **195A**, 402–406 (1948).
95. W. Heisenberg, *Z. Physik* **124**, 628 (1948); translation *NACA TM 1431*.
96. J. O. Hinze, *Turbulence*, McGraw-Hill, New York (1959).
97. J. O. Hinze, in *Mécanique de la Turbulence*, pp. 63, 129, Editions du Centre National de la Recherche Scientifique, Paris (1962).
98. E. Hopf. *J. Rat. Mech. Anal.*, **1**, 87 (1952).
99. E. Hopf and E. W. Titt, *J. Rat. Mech. Anal.*, **2**, 587 (1953).
100. P. G. Hubbard, Bulletin 37, *Studies in Engineering*, State University of Iowa, Iowa City, Iowa (1957).
101. Hubbard Instrument Co., *Model IIHR Catalog*, Iowa City Iowa.
102. R. R. Hughes, *Ind. Eng. Chem.* **49**, 947–955 (1957).
103. Eiichi Inoue, *Proc. Tenth Japan Natl. Cong. Appl. Mech.*, p. 217 (1960).
104. Eiichi Inoue, *Metol. Research Notes* **11**, 332–339 (1960).
105. J. Joseph and H. Sendner, *J. Geophys. Res.*, **67**, 3201–3205 (1962).
106. H. Kada and T. J. Hanratty, *AIChEJ* **6**, 624 (1960).
107. A. A. Kalinske, and C. L. Pien, *Ind. Eng. Chem.* **36**, 220 (1944).
108. J. Kampé de Fériet, *Ann. Soc. Sci. Bruxelles, Ser. I* **59**, 145 (1939).
109. T. von Kármán, *Compt. Rend. Paris* **226**, 2108 (1948).
110. T. von Kármán, *J. Aeron. Sci.* **1**, 1–20 (1934).
111. T. von Kármán and L. Howarth, *Proc. Roy. Soc. (London)* **164A**, 192–215 (1938).
112. A. L. Kistler and W. S. Chen, *J. Fluid Mech.* **16**, 41–64 (1963).
113. P. S. Klebanoff, *NACA Report 1247* (1955); supersedes *NACA TN 3178*.
114. P. S. Klebanoff, and Z. W. Diehl, *NACA Report 1110* (1952); supersedes *NACA TN 2475*.
115. P. S. Klebanoff and K. D. Tidstrom, *NASA Tech. Note D-195* (1959); *see also* P. Bradshaw, *J. Fluid Mech.* **22**, 679 (1965); S. C. Crow, *J. Fluid Mech.* **24**, 153–164 (1966).
116. P. S. Klebanoff K. D. Tidstrom, and L. M. Sargent, *J. Fluid Mech.* **12**, 1 (1962).

117. J. G. KNUDSEN and D. KATZ, *Fluid Dynamics and Heat Transfer*, McGraw-Hill, New York (1958).
118. O. KOFOED-HANSEN, *J. Geophys. Res.* **67**, 3217–3221 (1962).
119. A. KOLIN, *Science* **130**, 1088 (1959).
120. A. N. KOLMOGOROFF, *Compt. Rend. (Doklady) Acad. Sci. URSS* **30**, 301 (1941).
121. A. N. KOLMOGOROFF, *ibid.* **31**, 538 (1941).
122. A. N. KOLMOGOROFF, *ibid.* **32**, 16 (1941).
123. L. S. G. KOVASZNAY, *J. Aeron. Sci.* **15**, 745 (1948).
124. L. S. G. KOVASZNAY, *NACA Report 1209* (1954).
125. L. S. G. KOVASZNAY, *Appl. Mech. Rev.*, **12**, 375 (1959).
126. L. S. G. KOVASZNAY, in *Aeronautics and Astronautics*, p. 161, Pergamon Press, New York (1960).
127. R. H. KRAICHNAN, *Phys. Rev.* **107**, 1485–1490 (1957); **109**, 1407–1422 (1958).
128. R. H. KRAICHNAN. *Phys. Fluids* **1**, 358–359 (1958); **6**, 1603 (1963); **7**, 1048–1062, 1163–1177 (1964); **8**, 552 (1965).
129. R. H. KRAICHNAN, *J. Fluid Mechs.* **5**, 497–543 (1959).
130. R. H. KRAICHNAN, *Res. Report No. HSN-1*, New York University, New York (March 1959).
131. R. H. KRAICHNAN, *J. Math. Phys.* **2**, 124 (1961); **3**, 205 (1962).
132. R. H. KRAICHNAN, *Proc. Thirteenth Symp. Appl. Math.* p. 199, Amer. Math. Soc., Providence, R. I. (1962).
133. R. H. KRAICHNAN, in *Mécanique de la Turbulence*, p. 99, Editions du Centre National de la Recherche Scientifique, Paris (1962).
134. R. H. KRAICHNAN, *Phys. Fluids* **7**, 1030–1048 (1964); **8**, 210 (1965).
135. R. H. KRAICHNAN, *ibid.* **7**, 1723–1734 (1964).
136. R. H. KRAICHNAN, *ibid.* **8**, 575–598, 995 (1965); **9**, 1728–1752, 1884, 1937–1943 (1966).
137. D. E. LAMB, F. S. MANNING, and R. H. WILHELM, *AIChEJ* **6**, 682–685 (1960).
138. J. LAUFER, *NACA TN 2123* (1950).
139. J. LAUFER, *NACA Report 1174* (1954); supersedes *NACA TN 2954*.
140. J. LEE, Ph. D. thesis, The Ohio State University, Columbus, Ohio (1962).
141. J. LEE, *Appl. Sci. Research* **14A**, 250–252, (1965).
142. J. LEE and R. S. BRODKEY, *AIChEJ* **10**, 187–193 (1964); R. S. BRODKEY, *ibid.* **12**, 403–404 (1966); J. P. GEGNER and R. S. BRODKEY, *ibid.* **12**, 817–819 (1966).
143. J. LEE and R. S. BRODKEY, *Rev. Sci. Instr.* **34**, 1086–1090 (1963); J. O. NYE and R. S. BRODKEY, *ibid.* **38**, 26–28 (1967).
144. O. LEVENSPIEL, *Can J. Chem. Eng.* **40**, 135 (1962).
145. O. LEVENSPIEL, *Chemical Reaction Engineering*, John Wiley & Sons, New York (1962).
146. R. M. LEWIS, and R. H. KRAICHNAN, *Res. Report HSN-5 (Nonr-285(33))*, New York University, New York (1962).
147. P. LIENARD, *Groupe Consultatif Pour la Recherche et la Réalisation Aéronautiques, Report 170* (1958).
148. H. W. LIEPMANN, J. LAUFER, and K. LIEPMANN, *NACA TN 2473* (1951).

149. H. W. LIEPMANN and M. S. ROBINSON, *NACA TN 3037* (1953).
150. C. C. LIN, *Proc. First Symp. Appl. Math.* p. 81, Amer. Math. Soc., Providence, R. I. (1947).
151. C. C. LIN, *Theory of Hydrodynamic Stability*, Cambridge University Press, Cambridge (1955).
152. C. S. LIN, R. W. MOULTON, and G. L. PUTNAM, *Ind. Eng. Chem.* **45**, 636–646 (1953).
153. E. R. LINDGREN, *Arkiv Fysik* **7**, 293 (1953).
154. E. R. LINDGREN, *ibid.* **15**, 97–119, 503–519 (1959).
155. E. R. LINDGREN, *ibid.* **16**, 101–112 (1959).
156. E. R. LINDGREN, *ibid.* **18**, 449–464, 533–541 (1960).
157. E. R. LINDGREN, *Appl. Sci. Res.* **4A**, 313 (1954).
158. S. C. LING and P. G. HUBBARD, *J. Aeron. Sci.* **23**, 890 (1956).
159. Lintronic Laboratories, *Model 40-W Catalog*, Silver Springs, Md.
160. L. G. LOITSIANSKY, *Report 440*, Central Aero-Hydrodynamic Inst., Moscow (1939); translation *NACA TM 1079*.
161. J. L. LUMLEY, *J. Math. Phys.* **3**, 309–312 (1962).
162. J. L. LUMLEY, in *Mécanique de la Turbulence*, p. 17, Editions du Centre National de la Recherche Scientifique, Paris (1962).
163. J. L. LUMLEY and S. CORRSIN, in "Atmospheric Diffusion and Air Pollution," *Adv. in Geophys.* **6**, 179–183, Academic Press, New York (1959).
164. G. R. MARR and E. F. JOHNSON, *Chem. Eng. Progr. Symposium Series* No. 36 **55**, 109 (1961).
165. W. V. R. MALKUS, *Proc. Roy. Soc. (London)*, **225A**, 196 (1954).
166. W. V. R. MALKUS, *J. Fluid Mech.* **1**, 521–539 (1956).
167. W. C. REYNOLDS and W. G. TIEDERMAN, *J. Fluid Mech.* **27**, 253–272 (1967).
168. F. S. MANNING and R. H. WILHELM, *AIChEJ* **9**, 12 (1963).
169. A. C. MACPHAIL, *J. Aeron. Sci.* **8**, 73 (1940).
170. W. R. MICKELSEN, *NACA Tech. Note 3570* (1955).
171. R. R. MILLS, A. L. KISTLER, V. O'BRIEN, and S. CORRSIN, *NACA TN 4288* (1958).
172. R. S. MILLER, J. L. RALPH, R. L. CURL, and G. D. TOWELL, *AIChEJ* **9**, 196–202 (1963).
173. M. MILLIONSHTCHIKOV, *Compt. Rend. (Doklady) Acad. Sci. URSS* **32**, 615 (1941).
174. M. V. MORKOVIN, *Trans. ASME* **80**, 1121 (1958).
175. P. NAOR and R. SHINNAR, *Ind. Eng. Chem., Fund.* **2**, 278–286 (1963).
176. J. NIKURADSE, *VDI-Forschungshelf*, No. 356 (1932).
177. A. M. OBUKHOV, *Izvest. Akad. Nauk. SSSR, Ser. Geogr. i. Geofiz.*, **5**, 453–463 (1941); translation *TIL/T4822* (61-13005).
178. A. M. OBUKHOV, *Compt. Rend. (Doklady) Acad. Sci. URSS* **32**, 19 (1941).
179. A. M. OBUKHOV, *Izvest. Akad. Nauk. SSSR, Ser. Geogr. i. Geofiz.* **13**, 58 (1949).
180. M. OHJI, *Reports of Res. Inst. for Appl. Mech.* **10**, No. 38, 33–43, Kyushu University, Japan (1962); *J. Phys. Soc. Japan* **17**, 132 (1962); see also R. G. DEISSLER, *Phys. Fluids* **8**, 2106–2107 (1965).

REFERENCES

181. E. E. O'BRIEN, *Phys. Fluids* **6**, 1016–1020 (1963).
182. E. E. O'BRIEN and G. C. FRANCIS, *J. Fluid Mech.* **13**, 369 (1962).
183. Y. OGURA, *Phys. Fluids* **5**, 395–401 (1962).
184. Y. OGURA, *J. Geophys. Res.* **67**, 3143–3149 (1962).
185. Y. OGURA, *J. Fluid Mech.* **16**, 33–40 (1963).
186. S.-I. PAI, *J. Appl. Mech.* **20**, 109–114 (1953).
187. S.-I. PAI, *J. Franklin Inst.* **256**, 337–352 (1953).
188. S.-I. PAI, *Viscous Flow Theory, Vol. II-Turbulent Flow*, D. Van Nostrand, Princeton, (1957).
189. A. E. PERRY and P. N. JOUBERT, *J. Fluid Mech.* **17**, 193 (1963).
190. W. J. PIERSON, JR., *J. Geophys. Res.* **67**, 3151–60 (1962).
191. L. PRANDTL, *VDI* **77**, 105–114 (1933); *NACA TM 720* (1933).
192. L. PRANDTL, *Ergeb. Aerodyn. Versuchanstalt. Göttingen*, **3**, 1 (1927).
193. J. M. PRAUSNITZ and R. H. WILHELM, *Rev. Sci. Instr.* **27**, 941 (1956).
194. I. PROUDMAN and W. H. REID, *Phil. Trans. Roy. Soc. London* **247A**, 163 (1954).
195. J. REICHART, *NACA TM 1047* (1943).
196. W. H. REID, *Phys. Fluids* **3**, 72 (1960).
197. O. REYNOLDS, *Trans. Roy. Soc. (London)* **174A**, 935–982 (1883).
198. O. REYNOLDS, *ibid* **186A**, 123–164 (1895).
199. A. W. RICE, H. L. TOOR, and F. S. MANNING, *AIChEJ* **10**, 125–129 (1964).
200. L. F. RICHARDSON, *Proc. Roy. Soc. (London)* **110A**, 709 (1926).
201. R. S. ROSLER and S. G. BANKOFF, *AIChEJ* **9**, 672–676 (1963).
202. GERALD ROSEN, *Phys. Fluids* **3**, 519–524 (1960).
203. GERALD ROSEN, *ibid.* **3**, 525–528.
204. R. E. ROSENSWEIG, Sc. D. thesis, Massachusetts Institute of Technology, Cambridge, Mass. (1960).
205. R. E. ROSENSWEIG, *AIChEJ* **10**, 91–97 (1964).
206. R. E. ROSENSWEIG, H. C. HOTTEL, and G. C. WILLIAMS, *Chem. Eng. Sci.* **15**, 111 (1961).
207. J. ROTTA, *Ingen.-Arch.* **18**, 277 (1950).
208. J. ROTTA, *Z. Physik.* **129**, 547–572; **131**, 51 (1951).
209. J. ROTTA, *Ingen.-Arch.* **24**, 258–281 (1956).
210. H. ROUSE, *American Scientist* **51**, 285–314 (1963).
211. V. A. SANDBORN, *NACA TN 3266* (1955).
212. V. A. SANDBORN and W. H. BRAUN, *NACA TN 3761* (1956).
213. P. G. SAFFMAN, *J. Fluid Mech.* **8**, 273–283 (1960).
214. P. G. SAFFMAN, in *Mécanique de la Turbulence*, p. 53, Editions du Centre National de la Recherche Scientifique, Paris (1962).
215. G. B. SCHUBAUER, *J. Appl. Phys.* **25**, 188–196 (1954).
216. G. B. SCHUBAUER and P. S. KLEBANOFF, *NACA Report 1289* (1956).
217. G. B. SCHUBAUER and H. K. SKRAMSTAD, *NACA Wartime Report W-8* (1943).
218. G. B. SCHUBAUER and H. K. SKRAMSTAD, *J. Res. NBS* **38**, 251 (1947).
219. G. B. SCHUBAUER, and H. K. SKRAMSTAD, *NACA Report 909* (1949).

220. H. Schlichting, *Boundary Layer Theory* 4th ed., McGraw-Hill, New York (1960).
221. L. M. Schwartz, *Chem. Eng. Sci.* **18**, 223–226 (1963).
222. T. Skinner, *NACA TN 3682* (1956).
223. C. A. Sleicher, Jr., in *Modern Chemical Engineering* (A. Acrivos, editor), p. 45, Reinhold, New York (1963).
224. D. B. Spalding, *J. Appl. Mech.* **28E**, 455 (1961).
225. K. Spielberg and H. Timan, *J. Appl. Mech.* **82**, 381 (1960).
226. E. A. Spiegel, in *Mécanique de la Turbulence*, p. 181, Editions du Centre National de la Recherche Scientifique, Paris (1962).
227. H. B. Squire, *Proc. Roy. Soc. (London)* **142A**, 612 (1933).
228. J. R. Stalder, and E. G. Slack, *NACA TN 2263* (1951).
229. R. W. Stewart and A. A. Townsend, *Phil. Trans. Roy. Soc. London* **243A**, 359 (1951).
230. R. W. Stewart, *Proc. Cambridge Phil. Soc.* **47**, 146–157 (1951).
231. H. A. Stine and W. Winovich, *NACA TN 3719* (1956).
232. H. S. Tan and S. C. Ling, *Phys. Fluids* **6**, 1693–1699 (1963); see also D. A. Lee, *ibid.* **8**, 1911–1913 (1965), and R. G. Deissler, *ibid.* **8**, 2106–2107 (1965).
233. B. S. Tanenbaum and D. Mintzer, *Phys. Fluids* **3**, 529 (1960).
234. T. Tatsumi, *Proc. Roy. Soc. (London)*, **239A**, 16–45 (1957).
235. G. I. Taylor, *Proc. Roy. Soc. (London)* **156A**, 307–317 (1936).
236. G. I. Taylor, *Proc. London Math. Soc.* **20**, 196–212 (1921).
237. G. I. Taylor, *Proc. Roy. Soc. (London)* **151A**, 421 (1935).
238. G. I. Taylor, *ibid.* **164A**, 15, 476–490 (1938).
239. G. I. Taylor, *Phil. Trans. Roy. Soc. London* **223A**, 289–343 (1923); see also such references as S. Chandrasekhar, *Hydrodynamic and Hydromagnetic Stability*, Oxford University Press, London (1961); S. Chandrasekhar and D. D. Elbert, *Proc. Roy. Soc. (London)* **268A**, 145 (1962); E. M. Sparrow, W. D. Munro, and V. K. Jonsson, *J. Fluid Mech.*, **20**, 35–46 (1964); R. L. Duty, and W. H. Reid, *ibid.* 81–94; D. L. Harris, and W. H. Reid, *ibid.*, 95–101; D. Coles, *ibid.* **21**, 385–425 (1965).
240. G. I. Taylor, *Proc. Roy. Soc. (London)* **135A**, 685 (1932).
241. G. I. Taylor, in *Mécanique de la Turbulence*, p. 249, Editions du Centre National de la Recherche Scientifique, Paris (1962).
242. H. L. Toor, *AIChEJ* **8**, 70 (1962).
243. A. A. Townsend, *Proc. Cambridge Phil. Soc.* **43**, 560 (1947).
244. A. A. Townsend, *Australian J. Sci. Res.* **1A**, 161–174 (1948).
245. A. A. Townsend, *Rept. Progr. Phys.* **15**, 135 (1952).
246. A. A. Townsend, *The Structure of Turbulent Shear Flow*, Cambridge University Press, Cambridge (1956).
247. A. A. Townsend, *Proc. Roy. Soc. (London)* **224A**, 487 (1954).
248. A. A. Townsend, in *Mécanique de la Turbulence*, p. 167, Editions du Centre National de la Recherche Scientifique, Paris (1962).
249. D. J. Tritton, *J. Fluid Mech.* **16**, 269–312 (1963).
250. M. Tucker, *J. Sci. Instr.* **29**, 327 (1952).

251. M. S. UBEROI and S. CORRSIN, *NACA Rept. 1142* (1953); supersedes *NACA TN 2710*.
252. M. S. UBEROI, and L. S. G. KOVASZNAY, *J. Appl. Phys.* **26**, 19 (1955).
253. M. S. UBEROI, and L. S. G. KOVASZNAY, *Quart. Appl. Math.* **10**, 375 (1953).
254. C. F. VON WEIZSÄCKER, *Z. Physik.* **124**, 614 (1948).
255. J. R. WESKE, *Tech. Note BN-47*, University of Maryland, College Park, Md. (1955).
256. J. R. WESKE, *Tech. Note BN-91*, University of Maryland, College Park, Md. (1957).
257. R. H. WILHELM, Y. G. KIM, and D. B. CLEMONS, paper presented at AIChE meeting, Houston (Dec. 1963).
258. W. W. WILLMARTH, *J. Aeron. Sci.* **25**, 335 (1958).
259. W. W. WILLMARTH, *NACA TN 4139* (1958).
260. W. W. WILLMARTH, *Rev. Sci. Instr.* **29**, 218–222 (1958).
261. W. W. WILLMARTH, *NASA Mem. 3-17-59W* (1959).
262. W. W. WILLMARTH and C. E. WOOLDRIDGE, *J. Fluid Mech.* **14**, 187–210 (1962); W. W. WILLMARTH, *ibid.* **21**, 107–109 (1965); and F. W. Roos, *ibid.* **21**, 81–94 (1965).
263. D. WOLF, and W. RESNICK, *Ind. Eng. Chem., Fund.* **2**, 287–293 (1963).
264. H. W. WYLD, JR., *Ann. Phys.* **14**, 143–165 (1961); see also the newer developments by S. F. EDWARDS, *J. Fluid Mech.* **18**, 239–273 (1964) and by J. R. HERRING, *Phys. Fluids* **8**, 2219 (1965); **9** 2106 (1966), and comments on these methods by R. H. KRAICHNAN, *Symposium on the Dynamics of Fluids and Plasmas*, University of Maryland, College Park, Md., Academic Press, New York (in press).
265. TH. N. ZWIETERING, *Chem. Eng. Sci.* **11**, 1–15 (1959).
266. R. ARIS, *Can. J. Chem. Eng.* **40**, 87 (1962).
267. R. ARIS, *The Optimal Design of Chemical Reactors*, Academic Press, New York (1961).
268. R. BETCHOV and W. O. CRIMINALE, *Phys. Fluids*, **7**, 1920–1926 (1964).
269. J. M. DOUGLAS, *Chem. Eng. Progr. Symposium Series No. 48* **60**, 1 (1964).
270. R. DUMAS, *A L'étude des Spectres de Turbulence*, No. 404, Publications Scientiques et Techniques, du Ministére, Paris (1964).
271. M. J. FISHER and P.O.A.L. DAVIES, *J. Fluid Mech.* **18**, 97–116 (1964).
272. A. E. GILL, *J. Fluid Mech.* **21**, 145–172, 503–511 (1965).
273. R. N. KEELER, E. E. PETERSEN, and J. M. PRAUSNITZ, *AIChEJ* **11**, 221–227 (1965).
274. W. J. KIM and F. S. MANNING, *AIChEJ* **10**, 747 (1964).
275. P. LA ROSA, and F. S. MANNING, *Can. J. Chem. Eng.* **42**, 65, 282 (1964).
276. J. LEE, *Phys. Fluids* **8**, 1647–1658 (1965).
277. J. LEE, *ibid.* **9**, 363–372, 1753–1763 (1966).
278. Y. H. PAO, *Phys. Fluids* **8**, 1063–1075 (1965); *Di-82-0369, Report. 90*, Flight Sci. Lab., Boeing Sci. Res. Labs. (1964).
279. Y. H. PAO, *AIAAJ* **2**, 1550–1559 (1964).

280. G. M. Saidel and H. E. Hoelscher, *AIChEJ* **11**, 1058–1063 (1965).
281. G. Vassilatos and H. L. Toor, *AIChEJ* **11**, 666–673 (1965).
282. J. A. B. Wills, *J. Fluid Mech.* **20**, 417–432 (1964).
283. R. A. B. Willson, and P. V. Danckwerts, *Chem. Eng. Sci.* **19**, 885 (1964)
284. G. R. Worrell, and L. C. Eagleton, *Can. J. Chem. Eng.* **42**, 254 (1964).
285. F. N. Frenkiel and P. S. Klebanoff, *Phys. Fluids* **8**, 2291–2293 (1965).
286. J. O. Nye and R. S. Brodkey, *J. Fluid Mech.* (1967).
287. Zitzewitz Engineering Associates, *DISA Catalog*, Wyckoff, N. J.

CHAPTER 15

NON-NEWTONIAN PHENOMENA

In most of our preceding discussion we have made the basic assumption that we were dealing with Newtonian fluids. Unfortunately, however, nature is not so simple, and many materials such as drilling muds, clay coatings and other suspensions, certain oils and greases, polymer melts, elastomers, and many emulsions are non-Newtonian. The investigation of their properties by consideration of their molecular characteristics has been carried forward by the physical chemist and is an important part of *rheology*, the study of the deformation and flow of matter. Another part of the problem is the formulation of constitutive equations (also called rheological equations of state), which are descriptive of particular materials [e.g., Eq. (3–43) is the constitutive equation, or the rheological equation of state, of the Newtonian fluid]. Still another part is the use of these equations in conjunction with the various *conservation relations* (i.e., the equations of change, the so-called field equations) to solve specific problems. From a rigorous mathematical viewpoint, these latter aspects of rheology can be considered as a part of continuum mechanics. More recently, the practical engineering problems of industry have necessitated analytical and empirical solutions of problems based on simplified descriptions of the actual flow characteristics of real materials. This has led to an applied or engineering approach which clearly should draw as much as possible from the more basic physico-chemical work and continuum mechanics.

Because the subject of rheology has many facets, the formal definitions and terminology used in the field are not very well standardized. They will be introduced as needed, and an attempt will be made to use those accepted by the majority of investigators. A complete notation is provided at the end of the chapter.

15-1 RHEOLOGICAL CHARACTERISTICS OF MATERIALS

In the first sections, we shall review the relations between shear stress and shear rate (or strain) for a number of classes of materials. For this purpose we shall use the basic shear diagram which relates the variables in any given small element in the material. Although such a diagram is not always easy to obtain, we shall nevertheless assume that it is available. The various means of measuring rheological data will be briefly discussed, but the exact nature of the results and their relation to the basic shear diagram will be reserved for the second part of the chapter. Finally we shall treat equations, both empirical and theoretical, that can be used to describe the characteristics of non-Newtonian materials. It should be clearly understood that many of the classes of materials and descriptive equations to be presented are idealizations, and that these ideal systems only approximate real cases.

15-1.A Solids and Newtonian Fluids

Many solids, such as steel, have a stress-strain relation which is linear, although nonlinear relations for solids are also possible. Furthermore, if the stress is maintained below certain limits, the same curve is retraced as the strain is gradually diminished so that when the stress is completely released, there is no permanent deformation (see Fig. 15-1). The stresses are kept small enough so that the complications associated with large deformations are not introduced. The analysis of small deformations is a part of the study of linear elasticity.

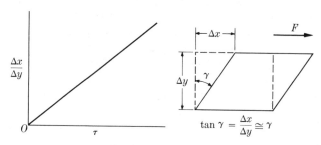

Fig. 15-1. Hooke solid.

There are two types of deformations to be considered: the *dilatational or extensional deformation*, in which a change in size (but not shape) causes a change in density; and the *shear deformation*, in which there is a change in shape (but not size). The fractional deformations are called strains, and are respectively: dilatational strain ($e_v = \Delta\rho/\rho_{\text{mean}}$), extensional strain ($e_l = \Delta l/l_{\text{mean}}$, where l is the length), and shear strain ($e_\tau = \gamma \cong \Delta x/\Delta y$, where Δx is the difference in the displacement of two parallel planes separated by a distance Δy). See Fig. 3-5 also. *Hooke's law* is the classical definition of linear elasticity, and can be expressed as the linear relations between the above-mentioned strains and the corresponding stresses:

$$p = k e_v, \tag{15-1}$$

where p is the pressure and k is the bulk modulus or modulus of volume elasticity;

$$T = Y e_l, \tag{15-2}$$

where T is the tensile stress and Y is Young's modulus; and

$$\tau = -G e_\tau = -\gamma G, \tag{15-3}$$

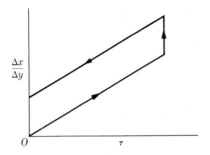 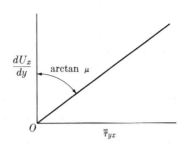

Fig. 15–2. St. Venant body.   Fig. 15–3. Newtonian fluid.

where τ is the shearing stress and G is the shear modulus or rigidity. In addition, Poisson's ratio σ is defined as the reciprocal of the ratio of the linear extension to the accompanying contraction e_c; that is,

$$\sigma = e_c/e_l. \tag{15-4}$$

There are a number of relations among the various moduli, and if we know any two of them we can determine all four.

Another ideal material is the *St. Venant body* (Fig. 15–2), which deforms elastically up to some yield stress. When the stress is released, the elastic fraction of the total deformation is regained.

A Newtonian fluid is defined by Eq. (3–31):

$$\bar{\bar{\tau}}_{yx} = -\mu(dU_x/dy), \tag{15-5}$$

where $\bar{\bar{\tau}}_{yx}$ is the shearing stress and dU_x/dy is the shear rate. Figure 15–3 illustrates the shear rate versus the stress curve. The viscosity is μ, and its reciprocal is known as the *fluidity*.

15-1.B Non-Newtonian Materials

It appears that, under extreme conditions (such as very high shear stress), all fluids, even those normally considered Newtonian, deviate from Newton's law (15–5). Non-Newtonian materials can be classified as shear thinning or shear thickening. In addition, they can be classified according to their time dependency, viscoelasticity, and the extent to which they exhibit the effects of normal stress. There has been much confusion over the classification of non-Newtonian materials, partly because some materials behave in more than one way, depending on conditions of experimentation, degree of concentration, etc. Adjectives used in earlier attempts at general classification, such as plastic, dilatant, thixotropic, rheopectic, etc., have been found to be inadequate because of their restrictive meaning, since they sometimes imply a specific mechanism for an observed non-Newtonian characteristic. Nevertheless, they are still in common use and we shall use them in our discussion when we wish to make specific subclassifications. Throughout the discussion, the various differences in usage will be pointed out.

An apparent viscosity, which will not be a constant, is like a Newtonian fluid, in that it can be defined by Eq. (15-5):

$$\bar{\bar{\tau}}_{yx} = -\mu_a(dU_x/dy). \tag{15-6}$$

Here the shear stress and shear rate are derived from the basic shear diagram, and μ_a is called the *true apparent viscosity*.

(a) Shear-thinning materials. Figure 15-4 presents curves of the shear rate versus the shear stress and the apparent viscosity versus the shear stress for the ideal Bingham plastic material [9]. The true apparent viscosity, as defined by Eq. (15-6), decreases with increasing shear; thus this material is a subclass of the more general shear-thinning group. The flow curve is characterized by a rate of shear that is linearly dependent on the shear stress above some given yield value τ_0. If the ordinate is shifted to τ_0, the viscosity defined by Newton's law is called the *plastic viscosity*, μ'. Its reciprocal is called the *mobility*. The rheological equation of state for this material is

$$\bar{\bar{\tau}}_{yx} - \tau_0 = -\mu'(dU_x/dy). \tag{15-7}$$

Most materials are not ideal in nature, but some can be approximated by this equation; for example, certain suspensions [192] (such as thorium dioxide, clay, and talc) can be approximated by it, as can paints, printing inks, toothpaste, and drilling muds.

After Bingham and Green [10] introduced the ideal plastic material, Bingham found that some shear-thinning materials clearly did not have a yield-stress value and were true liquids, even though their viscosity, as defined by Eq. (15-6), was not constant. Examples of such materials are: high-polymer solutions, such as rubber in toluene [158] and cellulose esters in organic solvents [149]; high polymers such as polyethylene [115, 152]; and emulsions such as GR-S

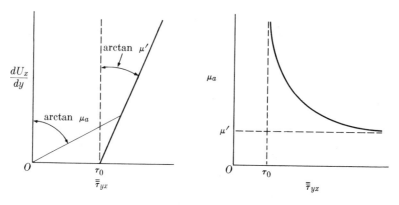

Fig. 15-4. Ideal Bingham plastic material.

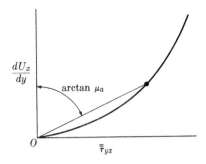

Fig. 15–5. Shear-thinning material.

rubber latex [103]. The observed flow curve or basic shear diagram for such materials is shown in Fig. 15–5. It is primarily in the case of these materials that confusion in terminology exists. Williamson [221] introduced the term *pseudoplastic* to describe these materials. Ostwald [141] used the expression "a material with structural viscosity." And more recently Reiner [158] has suggested that materials which possess the rheological property of viscosity, regardless of whether or not it is a constant, should be called *Newtonian liquids*. He feels that a simple Newtonian liquid is one that has a constant viscosity, and that a system which has variable viscosity should be called a generalized Newtonian liquid with structural viscosity.

Ostwald [141] pointed out that the curve of Fig. 15–5 was incomplete, and that if laminar flow could be maintained, the curve (now known as the Ostwald curve) should look like that in Fig. 15–6. (A log-log plot is included for comparison.)

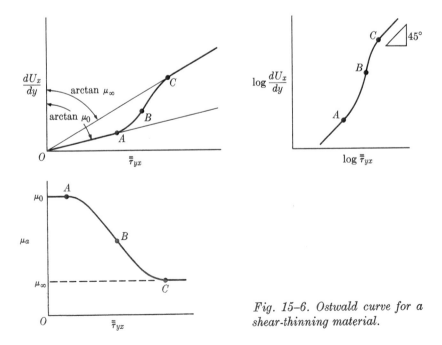

Fig. 15–6. Ostwald curve for a shear-thinning material.

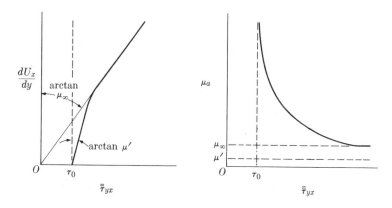

Fig. 15–7. *Ideal Bingham plastic material with upper Newtonian range.*

In the literature the normally reported data on pseudoplastics terminate somewhere between A and C, because of the limitations of the instruments used. There are, however, data to verify the existence of an upper Newtonian range [25, 26, 142, 152]. Furthermore, one would expect that a material such as the ideal Bingham plastic would also show an upper Newtonian range if one could obtain a high-enough shear rate [116]. This is suggested by Fig. 15–7 (compare with Fig. 15–4). There would be no lower Newtonian range, and the viscosity would eventually become constant (μ_∞) at high shear rates instead of approaching μ'. One must be careful not to mistake the onset of turbulence for the upper Newtonian range, since turbulence causes an increase in viscosity (eddy viscosity is always greater than molecular viscosity) [169].

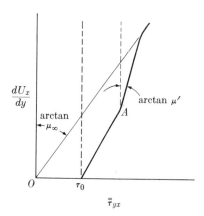

Fig. 15–8. *False-body material.*

The ideal Bingham plastic and the fluid with structural viscosity are the two main subclasses of shear-thinning materials. There are others that are different in various small degrees. For example, there is the false-body material (see Fig. 15–8), which is a variation of the ideal Bingham plastic (compare with Fig. 15–7).

The section from τ_0 to A of the curve which describes the false body is linear; the rest of the curve is similar to that of the ideal Bingham plastic. This section depiciting the false body can be interpreted [169] as showing that there is no flow until the material passes its point of yield stress, and that there is then a range of relaxation of the overall continuous structure, and finally a relatively rapid breakdown of the structure itself. The relaxation is linear and occurs until the material reaches the point A. The structural breakdown starts at this same point. The Bingham plastic material is similar; however, there is no relaxation (τ_0 to A), but rather an immediate breakdown of structure. This immediate breakdown implies a collapse of the continuous structure into smaller groups, which in turn break down into even smaller groups, until the system is completely dispersed at μ_∞.

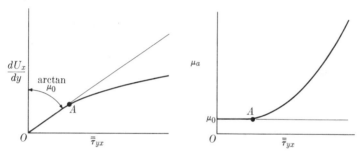

Fig. 15–9. Shear-thickening material.

(b) **Shear-thickening materials.** A shear-thickening material is characterized by the basic shear diagram shown in Fig. 15–9. Materials that exhibit *dilatancy* are considered to be one specific subclass. Dilatancy is a term introduced by Reynolds [162], and is used to describe an increase in rigidity which takes place in materials when they are closely packed. The increase is associated with an increase in volume, or a dilatant effect. Thus dilatancy is associated only with high-concentration suspensions. Reynolds discovered dilatancy during an investigation whose purpose was to arrive at a mechanical theory of the ether. In his experiments he used flexible bags filled with shot, glass balls, or quartz sand, and discovered that if a bag was filled with water (and if all air bubbles were excluded), it was rigid. The volume change can readily be seen if the bag or balloon is connected to a glass tube. On distortion the liquid level falls rather than rises as would be expected if the balloon were empty. Reynolds concluded that any distortion necessitates a change in volume. He gave, as a common example, the dilatancy of sand. When the water-to-sand ratio is such that there is just enough water to fill all the voids, and when the volume of the voids is at a minimum, any shear applied to force the material to flow disturbs the position of the particles and causes a dilation of the voids. This leads to a situation in which the total volume of the voids is greater than the volume of the water present. This results in an apparent partial dryness which increases the resistance of the material to shearing stress. The dryness is a result of the time necessary for capillary forces to provide the additional water required for complete saturation. When the pressure is removed, the sand becomes wet because the voids contract, and the water which has become excess escapes at the surface.

Some authors have misused the term dilatancy to mean any increase in viscosity for any reason, and as a result some materials that show an increase in viscosity with shear have been included with the dilatant group even though they do not dilate (for example, a $0.005N$ suspension of ammonium oleate). These materials should be referred to simply as shear-thickening materials.

True dilatancy can probably exist in any suspension; the only requirement is that a concentration be obtainable which is high enough so that the material can exist in closely packed form. Dilatancy is usually interpreted successfully in terms of varying degrees of packing. For example, the densest packing of spheres is about 74%. One of the less-dense packings is 37%, and thus it is not surprising to find that starch particles which are very nearly spherical begin to show dilatant effects at a concentration of about 40%. For highly nonspherical particles, the concentration might be considerably less, as in the case of iron oxide (12%).

(c) **Time-dependent systems.** Any confusion of terminology which we have encountered so far is mild compared with what we shall encounter when we begin to discuss time-dependent effects. The reason is, in part, that any material undergoing change is time dependent if the time of observation is decreased to the point at which one is following a rate of change to an equilibrium state (that is, if the time necessary to make a measurement is short when compared to the response time of the material). In the discussion to follow, let us assume that an adequate measuring device is available, so that any changes in the variables with time can be easily and accurately followed.

(i) **Thinning with time.** When a material is tested at a constant shear rate and the measured stress (or apparent viscosity) decreases as the experiment progresses, we say that the material undergoes thinning with time. If the broken structure, after it stands for awhile at zero stress, can reform and recover its original viscosity, then we call the material *thixotropic*. Originally the term thixotropic was associated with the reversible isothermal gel-sol-gel transformation; i.e., on mixing, the gel breaks down into a less viscous liquid, and then, when it has rested, returns to a gel. More generally, the term has become descriptive of any *reversible time-dependent breakdown–reformation mechanism*. Thinning can also occur because of an irreversible shear-induced breakdown of the molecular structure of the system; this is not what happens during a reversible thixotropic process. An example is the breakdown of carbon-to-carbon bonds in a polymer of high molecular weight. In such a case, the original viscosity is not recovered on standing.

A thixotropic material can be pictured as a material that undergoes a slow reaction from some unbroken structure to its equilibrium state. For now, let us restrict our discussion to shear-thinning materials which have structural viscosity (pseudoplastic materials), as shown in Fig. 15–6. The characteristics to be expected in such a material are illustrated in Fig. 15–10.

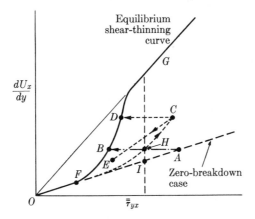

Fig. 15–10. Ostwald curve, including thinning with time.

At extremely low rates of shear (OF), the original structure is not modified and remains Newtonian at μ_0. If high-shear measurements could somehow be made instantaneously, so that there was no time for breakdown, we would expect the material to continue to be Newtonian (for example, at point A). Now, if the material were to be maintained under shear at this level, the stress would decrease until an equilibrium level (point B) was reached. Many such measurements would define the complete equilibrium curve $OFBDG$. The only difference between this curve and that of Fig. 15–6 is that in the present case we have assumed that the measurements could be made fast enough to follow the time-rate-of-change effect. Thus it is clear that the degree of thixotropy depends on one's ability to make rapid measurements. For this reason some authors include in the thixotropic group all materials mentioned previously as being shear-thinning materials. We shall not follow their example; rather, we shall associate thixotropy with a measurable time dependency. We shall consider the previous classifications (shear thinning or thickening) as equilibrium curves, because equilibrium conditions were obtained by choice or before a measurement could be completed.

Some actual experiments are made with instruments that provide a rate of shear that continuously and uniformly increases and decreases. These experiments have yielded curves such as $OFHCE$, a curve which is most difficult to interpret, since the degree of breakdown along it is not constant. Curves which are more meaningful and easier to understand are those which show decreasing stress with time at constant shear, or vice versa. The curve $OFHCE$ corresponds to the classical definition of a thixotropic material that we have used here; i.e., the curve that is obtained during the increasing–decreasing shear experiment (when there is a measurable time effect) has a hysteresis loop [213].

Although we have used a shear-thinning material in our example, it is also possible to have a shear-thickening material that exhibits thinning with time. Several dilatant materials, such as carbon black, zinc oxide, and iron oxide in water, show a loop in the curves of their increasing–decreasing shear experiments. There are probably two effects occurring: a nonattractive dilatancy (which is the most important), and a smaller attractive force which breaks down on shearing. This latter attractive structure is slow in forming, and gives the thixotropic effect. Figure 15–11 illustrates this, and parallels Fig. 15–10.

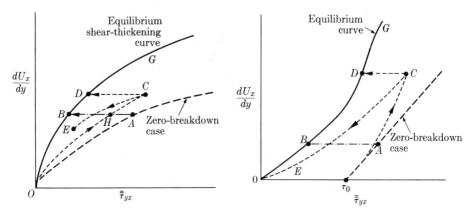

Fig. 15–11. Shear-thickening material, thinning with time.

Fig. 15–12. Thinning with time for a material with a yield.

The description of a low-shear region for a material with a yield value is somewhat complicated. Figure 15–12 shows the observed time-dependency curve, which also parallels that of Fig. 15–10. A typical example of this type of material was measured by Roller and Stoddard [168]. They used a bentonite clay at 8% in water and showed that it had a definite yield value on the up-curve ($\tau_0 C$). However, when the material reached the equilibrium curve ($OBDG$), there was no question that the curve passed through the origin. Furthermore, the down-curve (CE) bent in the same direction. For such a material, the curves must ultimately pass through the origin, which can be explained as follows [168]: in order to make a measurement at a given rate of shear, the yield stress, τ_0, must be exceeded, and the continuous structure must be broken down to somewhat smaller units. If the shear starts out at point C of Fig. 15–12, and enough time is allowed, the stress will decrease until equilibrium is reached at a point D. No matter how small the rate of shear selected, in order to obtain this rate the yield stress must be exceeded, and the continuous structure must be broken down. Under the conditions of no continuous structure, the resistance to shear is viscous in nature (not solid, since that would give rise to τ_0), and at an infinitesimal rate of shear, the shear stress is also infinitesimal. The only way to return to the point τ_0 is to reform the continuous structure; but, as just noted, this is prevented by the shear rate. Thus the curve passes through the origin.

(ii) **Thickening with time.** A material can show characteristics which are exactly the opposite of those previously described as being evidence of thixotropic behavior. Such a material might be called *antithixotropic* [57, 123]. An example of an antithixotropic material is a $0.005N$ suspension of ammonium oleate which shows an increase in viscosity under certain shear conditions, and a breakdown to the original structure when at rest. Unfortunately, however, most authors have used the misnomer *rheopexy* to describe such materials.

Rheopexy also leads to a structural buildup, but it can occur only in materials that are thixotropic. Quite often after breakdown, when slow buildup is occurring, the buildup can be accelerated by gentle shear. Freundlich and Juliusburger [59] gave as an example an aqueous paste of 42% gypsum powder, which was quite thixotropic. After shaking, the material resolidified in about 40 minutes; however, if gentle shear was applied, the material solidified in about 20 seconds. Their explanation was that gentle shear brings particles into contact at a much more accelerated pace than is possible when the material stands, and contact is provided by Brownian movement. Rheopexy is thus a time buildup of structure induced by mild shear. While antithixotropy (i.e., buildup followed by breakdown) is the opposite of thixotropy, a rheopectic material is also thixotropic.

(d) **Normal stress effects.** Non-Newtonian materials are either shear thinning or shear thickening. Furthermore, these materials can exhibit an apparent viscosity that changes with time, depending on the rapidity of measurement. To ascertain these characteristics, one need deal only with the basic shear diagram, which can be obtained from flow experiments that involve simple shear only. One such example would be a steady pipe flow away from the entrance (cylindrical coordinates r, θ, z) for which dU_z/dr is the only component of the shear-rate tensor. This can in turn be related to $\bar{\tau}_{rz}$ to give the basic shear diagram. We shall see later that even if normal stresses ($\bar{\tau}_{ii}$ terms) exist, simple shear experiments lead to valid results which can be used to construct the basic shear diagram. However, experiments can also be designed to measure other components of the stress tensor.

Those materials which show finite normal stresses have elastic as well as fluid characteristics, and are normally called *viscoelastic*. One can hypothesize a rheological equation of state for a viscous nonelastic substance (the Reiner-Rivlin fluid) which would be expected to show normal stress effects. However, such a material has not been found to exist in reality. Most of the detailed representations will be presented in the section on rheological equations of state. In this section we shall discuss general characteristics, simple limiting equations, and anomalous effects to which these materials give rise.

Experiments can be designed to emphasize various characteristics of viscoelastic substances. As already noted, simple shear tests bring out the effects of viscosity.

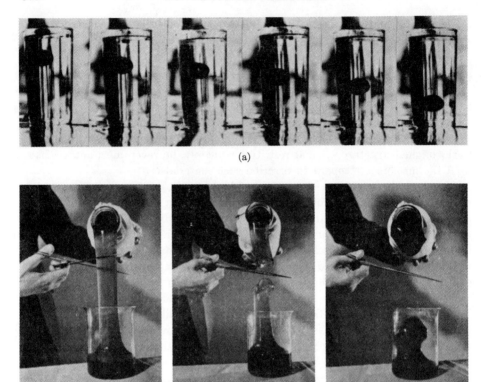

Fig. 15–13. (a) *Purely viscous.* (b) *Viscoelastic.* [*From S. A. Lodge,* Elastic Liquids, *Academic Press, New York (1964). By permission.*] (c) *Viscoelastic.* [*From ESSO Research and Engineering Co. By permission.*]

Figure 15–13 presents a qualitative experiment designed to show both elastic and viscous properties. The first sequence (a) shows a purely viscous material, while the second (b) shows a material that is viscoelastic. The third sequence (c), opposite, shows still another aspect of viscoelastic behavior.

A simple model for such materials would be a series combination of the contributions to the shear rate of a Newtonian liquid and an elastic solid; in other words, a simple Newtonian fluid that can exhibit elasticity. Equations (3–33) and (15–3) can be combined to give

$$d\gamma/dt = \dot{\gamma} = -\tau/\mu - (1/G)(d\tau/dt),$$

which can be rewritten as

$$\tau + (\mu/G)(d\tau/dt) = -\mu(d\gamma/dt). \tag{15–8}$$

This is the equation for the Maxwell liquid [109] or viscoelastic fluid. A mechanical

(c)

Fig. 15-13. (Continued)

model of Eq. (15-8) is given in Fig. 15-14. From either the equation or the figure, the following characteristics are apparent: (1) any stress will cause a continuous flow; thus the material is fluidlike; (2) μ/G is the time constant for the decay or relaxation of stress at a constant strain; i.e., when the forcing motion is stopped, the spring will relax as $\exp(-tG/\mu)$; and (3) the system will exhibit recoil when the stress is suddenly released.

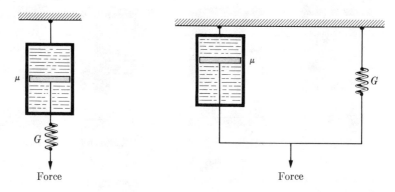

Fig. 15-14. Maxwell or viscoelastic fluid.

Fig. 15-15. Voigt or viscoelastic solid.

Another simple mechanical model that can be used to describe solidlike materials (a Voigt body or an elasticoviscous solid) is shown in Fig. 15-15. The resulting stress for this parallel system is the sum of the contributing parts:

$$\tau = -G\gamma - \mu(d\gamma/dt). \tag{15-9}$$

Such a model exhibits creep; that is, under a constant stress, the spring extends until equilibrium is obtained. Likewise, the system returns to its starting position when the stress is released.

If the deformation is kept very small, equations such as (15-8) and (15-9) can be used as crude approximations to real materials. In actuality, far more complex models are usually needed, but the simplest models just described contain the elements necessary to provide a qualitative picture of the material. Such analyses as these are a part of linear viscoelasticity. In contrast, if the deformations are large or if the viscous contribution is non-Newtonian, one must resort to nonlinear theories of viscoelasticity [133, 166]. Finally, viscoelastic materials may have a range of properties which depend on temperature. Amorphous polymers are glassy at low temperatures, but as the temperature is increased, they become rubbery. Finally, at high temperatures, they begin to flow.

Viscoelasticity, or normal stress effects, can be observed in other ways. The most classic of these is the Weissenberg effect [161, 211, 212] shown in Fig. 15-16; a Newtonian material is in the middle and the viscoelastic material of about the same viscosity on the right. In Fig. 15-17, the same types of materials are shown in the form of a jet exiting from a tube. Instead of the normal contrac-

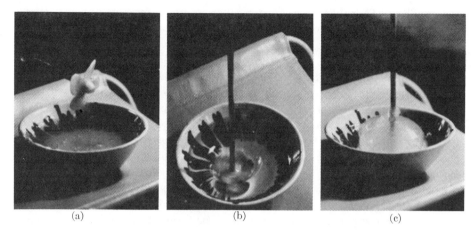

Fig. 15–16. (a) Mixer. (b) Purely viscous. (c) Viscoelastic.

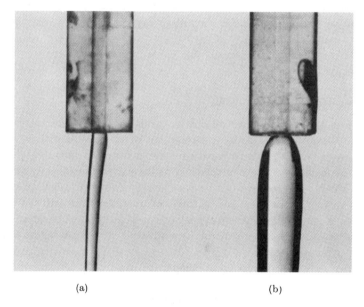

Fig. 15–17. (a) Purely viscous. (b) Viscoelastic. [From A.S. Lodge, Elastic Fluids, Academic Press, New York (1964). By permission.]

tion, there is a jet expansion, which is a result of normal stress effects [125]. Experiments have shown that when polymer melts, such as polyethylene, are extruded, the pressure loss in the contracting entrance to the capillary is in excess of that estimated for normal entrance losses [4]. Bagley and Birks [5] have demonstrated that the entrance loss can be attributed to viscoelastic effects associated with the sudden deformation of the melt as it contracts to pass into the capillary. They have measured relaxation times for polyethylene varying from about 30 to 600 seconds. Savins [174] has also measured relaxation times and has compared the results of the measurements with those obtained from jet expansion experiments.

The general field of viscoelasticity has been treated most extensively from a linear viewpoint, while relatively little work has been done on the nonlinear aspects. Furthermore, most of the work has been on viscoelastic solidlike materials. For information on the methods used and results obtained on a great variety of materials, such as metals, plastics, rubbers, raw elastomers, fibers, gelatins, asphalts, etc., see references 2, 3, 29, 54, 69, 86, 94, 95, 100, 160, 186, 194, 200, 218.

Before we conclude this section, a word of warning is in order. Although one can obtain the basic shear diagram without regard for normal stress effects, one cannot indiscriminately apply simple shear results to complex flow problems without also taking into account normal stress effects. Of course, if one's objective is the basic shear diagram (for correlation or elucidation of the chemistry of the system), then all the better; but if one wants to solve complex flow problems (such as those arising due to the boundary layer), one must exercise caution. Even more caution is necessary if very short-time processes are involved, for in such a case many materials can exhibit characteristics that are very much like those of solids.*

15-1.C Rheological Measurements

Viscometry is the measurement of the relations between shear stress and shear rate, while *rheogoniometry* is the measurement of all stresses within the material [88]. The latter, which encompasses the former, is far more difficult and to a degree is still in its formative stages, since there is little agreement between the results of normal stress measurements on various instruments [101, 174]. In this section, some measuring equipment and pitfalls of measurement will be considered. Later, several specific laminar-flow problems will be treated in detail to provide the means of obtaining the basic shear diagram and normal stress values from actual experimental results.

From the previous discussion it should be clear that single-point viscosity-measuring methods (i.e., methods involving only one shear rate) have only limited applicability to non-Newtonian systems. There are two major types of viscometers used to obtain multipoint curves: the capillary and the rotational instruments. In addition, there are many special types designed for specific measurements. In the capillary tube viscometer, the flow can be induced by either gravity or force. The Ostwald viscometer and its many modifications [206]

---

*See reference 235, in which the implications of the Weissenberg number (ratio of elastic to viscous forces) and the Deborah number (ratio of duration of fluid memory to duration of deformation process) are reviewed.

are of the gravitational type and are single-point instruments. However, by using a number of them with varying diameters and lengths, one may obtain a limited multipoint curve. This instrument is, of course, limited to fluids which flow under gravity. In the forced-flow capillary instrument the fluid is forced through various capillaries under pressure. It is one of the few units capable of very high shear rates. The rotational viscometer is usually of the concentric cylinder type (in which either cylinder can be rotated) or of a plate-and-cone design. The actual mechanical details of these various instruments will not be discussed in this text. However, Van Wazer et al. [206], in their book on viscometers and their use, have compiled an extensive list of the commercially available units. They have presented detailed descriptions, advantages and disadvantages, possible modifications, and the recommended experimental procedures. Rotational, capillary, and miscellaneous commercial viscometers are covered, and there is a chapter on viscoelastic systems.

A tabulation and limited evaluation of 56 industrial viscometers (including many single-point instruments) have been made by Bates [6]. A general article on viscosity measurements of all types, with emphasis on methods, has been presented by Peter [145]. Wilkinson [218] has considered several of the better multipoint instruments in some detail, including the Roberts-Weissenberg rheogoniometer and the Shirly plate-and-cone device. The former has the advantage of being able to measure normal stress effects. Further information on this has been given by Roberts [167] and Jobling and Roberts [88]. There is also the possibility of using the jet of fluid issuing from a long capillary to obtain normal stress measurements [57, 78, 125, 174, 180]. McKennell [113, 114] has discussed some of the advantages of rotational instruments, with emphasis on the plate-and-cone units. Details of an instrument capable of a response of 0.01 second have been given by Pawlowski [144]. Descriptions and ideas as to the use of a number of attachments for the versatile system by Haake [71] are available [53, 83]. Gabrysh et al. [61] and Harper [77] have described rotational viscometers they have built. These units are not available commercially, but could be reproduced with the help of detailed drawings from the authors. Techniques for experimenting with viscoelastic materials have received extensive review by Ferry [52]. Finally, a considerable amount of information is available from the various manufacturers of instruments. Some of this information not specifically referred to above is given in [27, 55, 87, 150, 206].

Of considerable interest in any measurement system are the various problems that can introduce unexpected complications [198]. Among these are: the existence of plug flow, wall slip, temperature heating effects, end effects, laminar instability, and turbulence. In most cases the effect can be eliminated or corrections can be made. These will not be mentioned at this time, but will be treated later when the specific flow geometries typical of viscometric work are solved. However, a general discussion at this point should give an idea of the multitude of problems that must be considered.

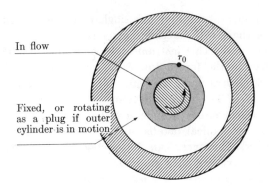

Fig. 15–18. A material with a yield in a rotational device.

Plug flow occurs if the material has a yield stress. Consider first the rotational instrument for which the stress will be higher at the inner cylinder and will decrease toward the outer cylinder. The yield stress τ_0 is reached first at the inner cylinder as the torque is increased, and flow will begin in this area. As the torque is further increased, there will be an annular ring next to the inner cylinder, which will be in flow, and an annular plug toward the outer cylinder. This is illustrated in Fig. 15–18. Eventually the torque will reach a point at which the stress is greater than τ_0 at the outer cylinder, and the entire field will be in flow. Figure 15–19 shows the qualitative effect that would be expected for an ideal Bingham plastic material.

The problem is more extreme when capillary instruments are involved, since the shear is always zero at the center line of the pipe. Thus $\tau_\mathfrak{C} < \tau_0$ and there will always be a plug near the center. The flow curve, shown in Fig. 15–20, can be seen to approach the ideal Bingham material only asymptotically, and to reach it only at infinite shear. With an analytical treatment it is possible for one to interpret the experimental results that are given in Figs. 15–19 and 15–20, and thus to obtain the characteristic quantities (τ_0 and μ') that describe the Bingham plastic material.

Many two-phase systems can be more easily studied as non-Newtonian materials than as complex two-phase materials. However, in such systems it is possible to have an effective slip at the wall. One can visualize this in the following manner: on account of the no-slip boundary condition at the wall, the fluid in contact with the wall will have zero velocity. Any particle suspended in the stream will be of finite size and will have a slightly higher velocity on the side that is away from the wall than on the side that is closer to the wall. This difference in velocity will be accentuated as the particle approaches the wall, where the velocity gradient is greater. The difference in velocity gives rise to a difference in pressure (Magnus effect, page 76), which acts in the direction that will move the particle toward the center of the flow stream. Thus, in the immediate area of the wall,

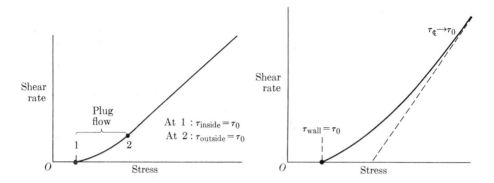

Fig. 15–19. Shear diagram for material with a yield in a rotational device.

Fig. 15–20. Shear diagram for material with a yield in a capillary system.

the stream can have a dearth of solid particles. This gives the appearance of a slip, since the non-Newtonian slurry can slide over the thin film of fluid next to the wall. The solid particles closest to the wall will have a velocity greater than zero. The possibility of the effective slip must be recognized and accounted for if satisfactory data on such materials are to be obtained.

In any viscous system under shear, there is heating from viscous dissipation of energy. On most viscometers of advanced design, there is an attempt at temperature control, but it is a very difficult experimental problem to solve. The effects of a temperature increase must be minimized or corrections must be made, since the apparent viscosity of most systems is a strong function of the temperature. Clearly, if it were not recognized, a drop in viscosity because of a temperature rise would be interpreted as a non-Newtonian effect, and it would not be possible for one to obtain meaningful data. Fortunately, however, estimates for the temperature rise can be made, and these will be presented with the analytical treatment.

End effects must be eliminated by reducing their contribution; otherwise corrections will have to be made. Since corrections are difficult, the easiest procedure is to eliminate the problem, but tests still have to be made to assure that this has been done.

As the shear rate is increased, laminar instability and turbulence become important. In a rotational viscometer, the effect depends on whether the inner or the outer cylinder is rotated. The work of Taylor has already been described in some detail on pages 228–231. Though Taylor's work was restricted to Newtonian materials, one would expect the same general trend to exist for non-Newtonian materials. When the inner cylinder is rotated, the vortices as described by Taylor form at some critical speed; at still higher speeds the fluid motion breaks down into turbulence. Laminar instability always precedes turbulence, and thus marks the point at which viscometer results are no longer valid. The flow pattern no longer conforms to a simple solution of the basic equations, and thus cannot be accurately interpreted in terms of the basic shear diagram. When the outer cylinder is rotated, the system is stable to the Taylor-type vortices, and the first deviation from the simple rotational flow form is the onset of turbulence.

Clay [37] studied a number of non-Newtonian materials in a large-gap inner-cylinder rotating unit (he used the large gap so that a large range of shear rate would exist in each experiment). He found that he could estimate the onset of the laminar instability to an accuracy within about 25% by using Taylor's theory and the apparent viscosity at the shear rate that would exist if the material were Newtonian. The accuracy of the estimate could have been increased had Clay had more reliable measurements of the non-Newtonian nature of the materials he considered.

Taylor [188] experimentally measured the onset of turbulence for the case of the outer cylinder in rotation. All the effects considered here require more energy than that necessary to maintain the simple laminar flow, and thus these effects would cause an apparent increase in viscosity. For non-Newtonian materials this increase could easily be misinterpreted as a characteristic of the materials, since the viscosity would already be varying with shear and the additional change might not be distinguishable from that already occurring. For Newtonian materials the change is quite apparent.

15-1.D Rheological Equations of State

To facilitate the representation of the many possible flow curves previously discussed, a number of equations have been proposed. Most of these are designed to describe the basic shear diagram, and for the most part are empirical or semi-empirical. Theoretically based equations are few in number; they are generally based on molecular considerations and on some specific assumptions about the mechanism which give rise to the observed non-Newtonian behavior. However, there are two theories, based on different aspects of kinetics, that are more general, and these will be presented here. These theories have not been developed as far as one might desire, but they do hold promise for the future. Finally, for those materials that exhibit normal stress effects, the equations just discussed are of use only if one is interested in the basic shear diagram *per se*. If one is interested in a more general three-dimensional representation (which would be necessary if one were to describe and solve for the flow in complex geometries involving more than simple shear), then one must turn to the constitutive equations of continuum mechanics. For those cases in which normal stresses do not exist, the simpler equations can be extended to a three-dimensional form, as was done for Newton's law [compare Eqs. (3-32) and (3-42)]. However, if normal stresses exist, then a reasonable description of the material requires far more complex equations.

(a) Empirical and semi-empirical equations. The simplest empirical relation describes the Newtonian fluid (Eqs. 3–32 or 15–5). The three-dimensional counterpart is Eq. (3–42). The ideal Bingham plastic relation has already been given by Eq. (15–7), but is more accurately written as

$$dU_x/dy = 0, \qquad \bar{\tau}_{yx} \leqslant \tau_0,$$
$$\bar{\tau}_{yx} - \tau_0 = -\mu'(dU_x/dy), \qquad \bar{\tau}_{yx} \geqslant \tau_0. \tag{15-10}$$

The power, or Ostwald-deWaele, law [142] is a two-constant empirical equation, which for simple shear is

$$\bar{\tau}_{yx} = -K |dU_x/dy|^{n-1}(dU_x/dy). \tag{15-11}$$

For dimensional consistency, either one needs a nondimensional form or one must assume that the absolute-value term is numerical only, and that it does not have units [158]. An alternate or inverse form is often used, in which the shear rate is expressed as a function of a power of the stress [158, 215]. The form used here (Eq. 15–11) is more common in the engineering literature [12, 121]. If Eq. (15–11) is valid (n and K constant), a log-log plot of the variables will give a straight line. Clearly, $n < 1$ for shear-thinning materials, $n = 1$ for Newtonians, and $n > 1$ for shear-thickening materials. The main objection to the power law is that it is not derived from any physical concept, but only from an empirical formula. Another objection is that the law breaks down for both small and large values of the variables (i.e., it cannot adequately describe the entire Ostwald curve of Fig. 15–6). Consequently the equation should be considered only as an interpolation formula. However, the law is often quite valid over a shear-rate range of as much as 100- to 1000-fold.

To describe shear-thinning materials more adequately, we need more than a two-constant equation. The Ellis [41, 65] and Sisko [181] models are designed to fit the low and high shear ranges respectively. They are

Ellis: $\qquad dU_x/dy = -(1/\mu_0 + K\bar{\tau}_{yx}^{(1/n)-1})\bar{\tau}_{yx}. \tag{15-12}$

Sisko: $\qquad \bar{\tau}_{yx} = -(\mu_\infty + K |dU_x/dy|^{n-1})(dU_x/dy). \tag{15-13}$

The Ellis model contains Newton's law and the power law as limiting forms. The Sisko model has similar limiting forms, and has found application in describing greases at high shear rates. If $n < 1$, the Ellis model approaches the limiting lower Newtonian range as $\bar{\tau}_{yx} \to 0$, and the Sisko model approaches the upper Newtonian range as $dU_x/dy \to \infty$. At the other extremes (that is, $\bar{\tau}_{yx} \to \infty$ for the Ellis model and $\bar{\tau}_{yx} \to 0$ for the Sisko), the equations fail.

EXAMPLE 15–1. The following basic shear data were obtained by Brodnyan and Kelly [26], partly by correcting capillary flow data and partly from a low-shear rotational instrument. Estimate the constants for the Ellis and Sisko models and compare the calculated curve with the experimental results.

dU_x/dy, sec^{-1}	$\bar{\tau}_{yx}$, dynes/cm^2	dU_x/dy, sec^{-1}	$\bar{\tau}_{yx}$, dynes/cm^2
3.52×10^0	2.21×10^1	2.05×10^4	3.42×10^4
6.92	4.42	2.81	4.16
1.45×10^1	8.84	3.79	4.89
3.68	2.21×10^2	2.58	3.68
6.82	4.42	4.96	5.49
1.84×10^2	8.80	8.50	7.30
2.40	1.33×10^3	1.14×10^5	8.34
3.81	1.83	1.64	1.01×10^5
5.52	2.88	2.70	1.38
8.95	4.18	4.10	1.62
1.47×10^3	6.37	4.46	2.14
2.23	8.29	6.55	2.65
3.10	1.06×10^4	8.08	3.17
5.04	1.50	9.34	3.69
7.27	2.04	1.12×10^6	4.25
1.12×10^4	2.47	1.60	6.60
1.45	2.89	1.91	7.76

Answer. From the data, an average value for μ_0 is 6.25 poises, and for μ_∞, 0.40 poise. For the Ellis model (15–12), a log-log plot of

$$1/\mu_a - 1/\mu_0 \quad \text{or} \quad (\mu_0 - \mu_a)/\mu_0 \mu_a$$

versus $\bar{\tau}_{yx}$ will have a slope of $(1/n) - 1$ and an intercept of K at $\bar{\tau}_{yx} = 1$. Reasonable values are $n = 0.5$ and $K = 1.35 \times 10^{-5}$ (see Fig. 15–21a). The single values for n and K are satisfactory, as shown by the calculation of the basic shear diagram (Fig. 15–21b). The curve is somewhat high at the low range and low at the high range (compare the data with the solid line). Over the range of the fit, the deviation from the experimental curve varies by up to 30% (remember, this is a log-log plot). As μ_∞ is approached, the equation becomes inadequate. This can also be seen from Fig. 15–21a. For the Sisko model (15–13), a log-log plot of

$$\mu_a - \mu_\infty \quad \text{versus} \quad dU_x/dy$$

will have a slope of $n - 1$ and an intercept of K at $dU_x/dy = 1$. Reasonable values for the high shear range are $n = 0$ and $K = 3.05 \times 10^4$ (see Fig. 15–21a). The fit was made to this range because the model was designed to fit such data. The inadequacy at lower shear rates is apparent from a comparison of the data with the dashed curve of Fig. 15–21b or from Fig. 15–21a. The deviations in the range of the fit are similar to those of the Ellis model. However, Eq. (15–12) provides a good fit over a much wider range than Eq. (15–13).

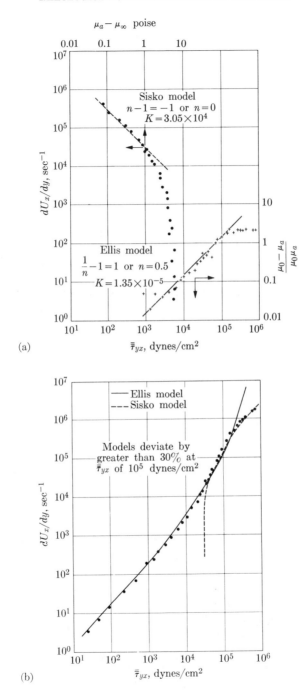

Fig. 15-21. (a) Determination of parameters for Sisko and Ellis models. (b) Comparison of data with Ellis and Sisko models.

Three constant equations can be formulated so as to fit the entire S-shaped Ostwald curve of Fig. 15–6. For example, the Reiner-Philippoff model [146] is

$$\bar{\bar{\tau}}_{yx} = -\left(\mu_\infty + \frac{\mu_0 - \mu_\infty}{1 + \bar{\bar{\tau}}_{yx}^2/A}\right)\frac{dU_x}{dy}, \tag{15-14}$$

and the Powell-Eyring model [153] (to be derived later) can be expressed as

$$\bar{\bar{\tau}}_{yx} = -\left[\mu_\infty + \frac{\mu_0 - \mu_\infty}{\beta(dU_x/dy)} \sinh^{-1}\beta\frac{dU_x}{dy}\right]\frac{dU_x}{dy}, \tag{15-15}$$

where β is a constant. A simple empirical formula has been suggested by Seely [229]:

$$\mu_a = \mu_\infty + (\mu_0 - \mu_\infty)\exp(-A\bar{\bar{\tau}}_{yx}).$$

Although the proper shape is obtained by these equations, they are not exact when detailed comparisons are made. Some suggestions have been made for equations with more than three constants; two of these are by Chang and Ramanaiah [35] and by Slibar and Paslay [184]. The first is a complex series equation which involves three universal constants to determine a master curve and five material constants to describe individual materials. The second is an integral representation involving five material constants. The equation contains a time-dependent term, and thus may be able to describe thixotropic systems. The solutions to the equations are quite complex, but have been worked out for both pipe and concentric cylinder flow.

EXAMPLE 15–2. Repeat Example 15–1 for the Reiner-Philippoff, Powell-Eyring, and Seely models.

Answer. The easiest means of evaluating the one unknown constant of these models is to use the inflection-point values in conjunction with the known values of μ_0 and μ_∞. At the inflection point,

$$dU_x/dy = 6 \times 10^4 \text{ sec}^{-1},$$
$$\bar{\bar{\tau}}_{yx} = 6 \times 10^4 \text{ dynes/cm}^2,$$
$$\mu_a = 1 \text{ poise}.$$

For the Reiner-Philippoff model, the constant A can be calculated as 4.12×10^8, and for the Powell-Eyring model, by trial and error, β is 7.28×10^{-4}. A comparison is given in Fig. 15–22. Both models (solid line) give the same result; they are quite good in the high range, but off by greater than 30% in the low range. An alternate value of the constant would give a better fit for the low range, but then the high range would be off. For example, Meter [119] has suggested evaluation of the constant at the point where $\mu_a = \frac{1}{2}\mu_0$. For the Reiner-Philippoff model the constant becomes 1.37×10^8, and the modified curve is shown as a dashed line. The fit is not greatly improved. The model by Seely is also shown

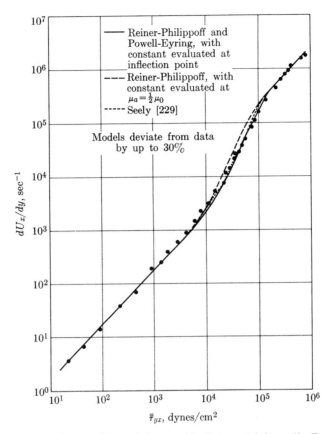

Fig. 15-22. Comparison of data with Reiner-Philippoff, Powell-Eyring, and Seely models.

on the figure for the constant evaluated at the inflection point. Its value is 3.8×10^{-5}. This simple empirical law gives a fairly good fit to the curve over the entire range. Although deviations outside the range of the experimental data do occur, it appears to be the best of the three constant models.

The Herschel-Bulkley law is an empirical combination of the Bingham plastic and power laws:

$$dU_x/dy = 0, \quad \bar{\bar{\tau}}_{yx} \leqslant \tau_0,$$
$$\bar{\bar{\tau}}_{yx} - \tau_0 = -K\,|dU_x/dy|^{n-1}(dU_x/dy). \quad (15\text{-}16)$$

A somewhat similar law has been suggested by Casson [33]:

$$\bar{\bar{\tau}}_{yx}^{1/2} - \tau_0^{1/2} = -K\,|dU_x/dy|^{1/2}, \quad \bar{\bar{\tau}}_{yx} \geqslant \tau_0. \quad (15\text{-}17)$$

Heinz [84] has shown agreement of this equation with data for such materials as suspensions in linseed oil and a polyester lacquer; however, a further modification was necessary to treat certain other suspensions in oil. The modified form was

$$\bar{\tau}_{yx}^{2/3} - \tau_0^{2/3} = -K\,|dU_x/dy|^{2/3}, \qquad \bar{\tau}_{yx} \geqslant \tau_0. \tag{15-18}$$

A general form of these has been suggested by Bruss [28]:

$$\bar{\tau}_{yx}^n - \tau_0^n = -K\,|dU_x/dy|^{n-1}(dU_x/dy). \tag{15-19}$$

(b) Theoretical approaches. One of the major problems in the study of non-Newtonian materials has been the lack of adequate theoretical equations for the description of the flow characteristics. The need for equations with *meaningful constants* arises from the necessity of correlating the rheological properties of the material under consideration with its fundamental structure. The variation of these constants with molecular weight, temperature, and structure could provide considerable insight into the mechanism of the flow and the reasons for the nonlinear behavior.

One general nonempirical approach has been that of Eyring and his coworkers. Their work is based on the application of the theory of rate processes to the relaxation processes that are believed to play an important part in determining the nature of the flow. The rate theory is presented in detail in [68, 85, 185]. Very briefly, the theory postulates an activated complex as an intermediate unstable state which would be formed from the reactants and which would decompose into the products. The assumption is made that the decomposition of the complex is the rate-controlling step, and that there is an activation energy associated with this. The form of the specific rate constant k' for the reaction is obtained. It is

$$k' = \kappa(kT/h)\exp(-\Delta G^*/RT), \tag{15-20}$$

where κ is a transmission coefficient used to correct for the fact that not all molecules that arrive at the activated state continue to complete the reaction. Here k is the Boltzmann constant, h is Planck's constant, and ΔG^* is the change in the molal free energy of activation.

The application of Eq. (15-20) to the relaxation theory of liquids is suggested by the model given in Fig. 15-23, which is a cross section of an idealized liquid. The flow is pictured as taking place by a unimolecular process [49]. A shearing force F is applied to the fluid, and flow occurs when the shaded molecule squeezes past the neighboring particles and moves into an unoccupied hole. The distance traveled by the particle is denoted by λ. The net velocity of flow would be the net number of jumps times the distance per jump:

$$U_\lambda = \lambda(k'_f - k'_r), \tag{15-21}$$

where U_λ is the net velocity and k'_f and k'_r are the specific jumping rates in the forward and reverse directions, respectively. If no force is acting, the forward

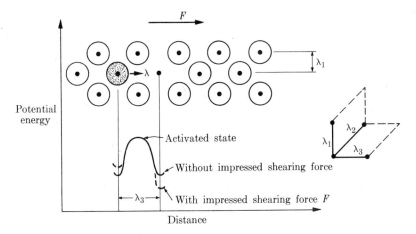

Fig. 15-23. Model for the relaxation theory of liquids. [From J. A. Hirschfelder, C. F. Curtiss, and R. B. Bird, Molecular Theory of Gases and Liquids, John Wiley & Sons, New York, (1954). By permission.]

and reverse rates will be equal, and from Eq. (15-20),

$$k'_f = k'_r = k'_0 = \kappa(kT/h) \exp(-\Delta G^*/RT), \quad (15\text{-}22)$$

where k'_0 is the specific rate for the flow process (the jumping process of a molecule from an equilibrium position into the neighboring vacant site at zero stress). If a force is acting, then the forward process is increased and the reverse decreased. The amount of change per mole is the force per unit area times the area, acting over a distance $\frac{1}{2}\lambda$, and all times Avogadro's number N; that is,

$$\text{change in free energy} = F(\lambda_2 \lambda_3)(\tfrac{1}{2}\lambda)N.$$

Since $R = kN$, the forward and reverse specific rates become

$$k'_f = k'_0 e^{\alpha F}, \qquad k'_r = k'_0 e^{-\alpha F},$$

where $\alpha = \lambda \lambda_2 \lambda_3 / 2kT$. Equation (15-21) then becomes

$$U_\lambda = \lambda k'_0 [e^{\alpha F} - e^{-\alpha F}] = 2\lambda k'_0 \sinh \alpha F. \quad (15\text{-}23)$$

The shear rate can be estimated as

$$dU_x/dy = U_x/y = U_\lambda/\lambda_1, \quad (15\text{-}24)$$

and the shear stress would be $\bar{\bar{\tau}}_{yx} = -F$. Combining Eqs. (15-23) and (15-24) gives us

$$\mu_a = -\frac{\bar{\bar{\tau}}_{yx}}{dU_x/dy} = -\frac{\bar{\bar{\tau}}_{yx} \lambda_1}{U_\lambda} = \frac{-\beta \bar{\bar{\tau}}_{yx}}{\sinh(-\alpha \bar{\bar{\tau}}_{yx})}, \quad (15\text{-}25)$$

where $\beta = \lambda_1/2k'_0\lambda$, a relaxation-type term. For small external forces, $\alpha \bar{\bar{\tau}}_{yx}$ is small, and the sinh term can be expanded in a Taylor's series. Using the first term gives

$$\mu \cong \beta/\alpha. \tag{15-26}$$

Equation (15-25) describes non-Newtonian behavior, since the viscosity is dependent on the stress $\bar{\tau}_{yx}$. In contrast, the expansion given in Eq. (15-26) describes Newtonian flow. Equation (15-25) does work satisfactorily for some materials over a limited range of shear rates.

The analysis was extended by Ree and Eyring [155] to cover a wider range of materials and shear rates. Rearrangement of Eq. (15-25) gives

$$\mu_a = (1/\alpha \mathbf{d}_{yx}) \sinh^{-1}(\beta \mathbf{d}_{yx}), \tag{15-27}$$

where $\mathbf{d}_{yx} = dU_x/dy$. If we assume that the material is composed of more than one type of flow unit (heterogeneity of flow units), we obtain

$$\mu_a = \sum_{i=1}^{n} \frac{x_i \beta_i}{\alpha_i} \frac{1}{\beta_i \mathbf{d}_{yx}} \sinh^{-1}(\beta_i \mathbf{d}_{yx}), \tag{15-28}$$

where the constants are the same, but now refer to the ith flow type. The term x_i is the fraction of area occupied by the ith type on the shear surface. Usually n is 2 or 3. Eyring's theory has been described in detail [91, 155, 156], including the molecular meaning of the constants and the methods of estimating their value from rheological data. In general, the constants were obtained by trial and error; however, a method has been devised which is based on the inflection point in the curve μ_a versus log $\mathbf{d}_{yx}$, and which gives unique and exact results [91]. For the more complicated systems involving more than one non-Newtonian flow unit, we must assume that their non-Newtonian contributions occur at different levels of shear rate. The results of the application of the relaxation theory are reviewed in the references cited. In addition, Maron and his coauthors [104, 105, 106] have used the theory for the interpretation of data on synthetic rubber–latex suspensions. The data obtained were extensive enough to establish the dependency of the constants in the equation on concentration (volume fraction) and temperature.

The Powell-Eyring equation (15-15) consists of the first two terms of Eq. (15-28) written in terms of shear stress and shear rate. Some interpretation of the constants $x_1\beta_1/\alpha_1$ and $x_2\beta_2/\alpha_2$ is possible, since the properties of the function $(1/\beta_i \mathbf{d}_{yx}) \sinh^{-1}(\beta_i \mathbf{d}_{yx})$ are known; that is,

$$\lim_{\beta_i \mathbf{d}_{yx} \to 0} [(1/\beta_i \mathbf{d}_{yx}) \sinh^{-1}(\beta_i \mathbf{d}_{yx})] = 1,$$
and
$$\lim_{\beta_i \mathbf{d}_{yx} \to \infty} [(1/\beta_i \mathbf{d}_{yx}) \sinh^{-1}(\beta_i \mathbf{d}_{yx})] = 0. \tag{15-29}$$

When the shear rate approaches infinity, terms containing $(1/\beta_i \mathbf{d}_{yx}) \sinh^{-1}(\beta_i \mathbf{d}_{yx})$ approach zero; therefore the first term must be Newtonian and associated with the viscosity at very high shear rates. The other limit can be used to show that the coefficient of the second term must be $\mu_0 - \mu_\infty$. Thus Eq. (15-15) is composed of two terms, one a Newtonian flow unit at μ_∞ and the other non-

Newtonian. In Eq. (15–15) the adjustable constant β (besides μ_0 and μ_∞) is β_2. In order to improve the Powell-Eyring equation's representation of the Ostwald curve, it was necessary to introduce a transition between groups; five constants were needed and were determined from the experimental data [157].

The application of the theory of rate processes to high polymers [50] was a necessary step toward the application to thixotropic substances [72, 73]. The polymer is pictured as an entangled molecule in a three-dimensional network. In this form its flow behavior is non-Newtonian. The entangled units are slowly converted to disentangled molecules, which are considered Newtonian. Thus Eq. (15–28) is used with only two terms, one Newtonian and one non-Newtonian. A net transition rate for the structure change is derived [68], and is combined with the experimental condition that dU_x/dy is varied linearly with time (increasing on the up-curve and decreasing on the down-curve). When the information concerning the transition and variation of shear rate is combined with Eq. (15–28) for two flow units, we obtain the final time-dependent basic shear relation as

$$\bar{\tau}_{yx} = [1 - (x_2)_0 \, e^{-k'_f \gamma I}] \frac{\beta_1}{\alpha_1} \mathbf{d}_{yx} + \frac{1}{\alpha_2} (x_2)_0 \, e^{-k'_f \gamma I} \sinh^{-1}(\beta_2 \mathbf{d}_{yx}), \qquad (15\text{--}30)$$

where the subscript 0 denotes initial time, k'_f is the specific rate for the forward reaction at $\bar{\tau}_{yx} = 0$, and γ and I are parameters. In this case, γ contains the proportionality constant between the shear rate and time, and thus makes the equation time-dependent. The method of obtaining the various constants is essentially a trial-and-error procedure designed to give the best fit to the data, and has been reviewed by Hahn, Ree, and Eyring [72, 73].

A development by Denny and Brodkey [40], based on a homogeneous reaction kinetic approach, can be used for thixotropic substances as well as time-independent materials. The analysis is based on the idea that the nonlinear characteristics are caused by a structural change, which is influenced by the level of stress applied to the material. The first form of the theory is given in the reference cited, and more recent work [225] has led to refinements that will be incorporated here.

Consider the thixotropic material shown in Fig. 15–10. The maximum viscosity that can be observed at any rate of shear is μ_0. There is also a minimum viscosity given by the equilibrium curve and dependent on the rate of shear (i.e., point B). It is hypothesized that a structural breakdown, weakening of forces between particles, or in fact any change, is the reason for the deviation from Newtonian behavior. One assumes that some function involving a difference between μ_0 and μ_a is proportional to the amount of material changed, and that a similar function of μ_a and μ_∞ is proportional to the amount of material remaining unchanged. The reasoning is as follows: When the viscosity equals a constant, then no change in structure occurs, i.e., there is no breakdown or buildup [72]. The viscosity of a non-Newtonian material approaches a value of μ_∞, which is constant at high shear rates. This is taken to mean that no further change can occur, no matter how high the shear rate. Therefore the limiting indications of no change and maximum change at any shear rate are μ_0 and μ_∞, respectively. The viscosity μ_a falls somewhere between these two limiting values. Applying the inverse lever principle to the function, the unchanged fraction is taken as

$$F = \frac{f(\mu_a) - f(\mu_\infty)}{f(\mu_0) - f(\mu_\infty)}, \qquad (15\text{-}31)$$

and the changed fraction would be $1 - F$. The function $f(\mu)$ relates the concentration of the structure undergoing change to the viscosity. For example, for polymer melts or polymers in solution, the proper empirical form might be

$$f(\mu) = A\mu^{1/3.5}.$$

Note that the constant A conveniently cancels out of the expression for the fraction in Eq. (15-31). In contrast to the foregoing, the authors in the original reference assumed that $f(\mu) = A\mu$, which should be good for suspensions and might still be adequate for polymer solutions.

Now consider the reaction,

$$\text{unchanged} \underset{k_2'}{\overset{k_1'}{\rightleftarrows}} \text{changed},$$

and the kinetic equation for such a reaction,

$$-d(\text{unchanged})/dt = k_1'(\text{unchanged})^m - k_2'(\text{changed})^n,$$

where m and n must be integers. The use of fractions gives us

$$-d(F)/dt = k_1'(F)^m - k_2'(1 - F)^n. \qquad (15\text{-}32)$$

The rate constants k_1' and k_2' must contain the effect of shear stress, which can be pictured as being analogous to the action of a catalyst; i.e., higher stresses produce a faster rate of breakdown, which is comparable to saying that the speed of a chemical reaction increases with the amount of catalyst used. Thus

$$k_i' = k_i(\bar{\tau}_{yx})^{p_i} \qquad (15\text{-}33)$$

is used, where i in this case is either 1 or 2. The use of shear stress rather than shear rate is dictated by the nature of the effect of temperature on viscosity. Whatever the mechanism is that causes a change in viscosity with temperature (for Newtonian materials as well), it follows closely the law,

$$\mu = A' e^{-B/RT}. \qquad (15\text{-}34)$$

For non-Newtonians, A' and B are constant with changes in temperature only if considered at constant $\bar{\tau}_{yx}$ [89, 115]. Thus a combination of Eq. (15-31) (with the function defined) and Eq. (15-34) gives an expression for F which is a function of the level of $\bar{\tau}_{xy}$ and T only, and in which the constants are not functions of temperature. For F in Eq. (15-32) (i.e., integrated and solved for F) to be similarly a function of $\bar{\bar{\tau}}_{yx}$ and T only, k_1' and k_2' are defined by Eq. (15-33). In the combination of Eqs. (15-32) and (15-33), a change in T at constant $\bar{\bar{\tau}}_{yx}$ causes a change in the rate, because

$$k_1 = k_0\, e^{-E^*/RT}, \qquad (15\text{-}35)$$

and because the p_i's may also be a function of T (E^* is the activation energy of the non-Newtonian reaction). Equation (15-33) is thus written as a function of $(\bar{\bar{\tau}}_{yx})^{p_i}$ rather than $\mathbf{d}_{yx}^{p_i}$:

$$-d(F)/dt = k_1(F)^m(\bar{\bar{\tau}}_{yx})^{p_1} - k_2(1-F)^n(\bar{\bar{\tau}}_{yx})^{p_2}. \qquad (15\text{-}36)$$

In reference 40, $\mathbf{d}_{yx}^p$ was used, but was always considered to be under isothermal conditions, so that the temperature-consistency problem did not arise. One can use Eq. (15-36), as one normally does in homogeneous-reaction kinetics. For example, initial-rate data could be used to evaluate k_1, m, and p_1 [225]. However, such data are not yet available, and so there will be no further discussion of this point. When such measurements can be made, they can be used to prove the internal consistency of the approach, and will offer an independent evaluation of the constants necessary to describe the steady-state data to be discussed next.

At equilibrium or steady state (curve $OFBDG$ of Fig. 15-10), $-d(F)/dt = 0$, and Eq. (15-36) becomes

$$K(\bar{\bar{\tau}}_{yx})^p = (k_1/k_2)(\bar{\bar{\tau}}_{yx})^{p_1-p_2} = (1-F)^n/F^m, \qquad (15\text{-}37)$$

where K is the ratio of forward-to-reverse reaction-rate constants, and is of the form of an equilibrium constant, and where p is $p_1 - p_2$. Clearly, from Eq. (15-37), a log-log plot of $(1-F)^n/F^m$ versus $\bar{\bar{\tau}}_{yx}$ should be a straight line of slope p, and an intercept of K at $\bar{\bar{\tau}}_{yx} = 1$.

In dealing with a range of concentrations of the material under test, the concentration can be included in the rate term as $(CF)^m$ and $(C(1-F))^n$, where C is the concentration. This particular form is that which would be expected from the concentration effect in homogeneous reaction kinetics. A demonstration of the validity of this is given in reference 225.

The theory makes no specific assumptions as to the exact nature of the fundamental mechanism involved. Furthermore, the constants p, m, n, K, μ_0, and μ_∞ are easy to determine and are readily interpretable. Several actual applications are given in [40, 225] as well as in Example 15-3. Ultimately all the parameters should be related to the suggested structural change reaction mechanism. This in itself may vary widely for the materials considered. Polymer entanglement-disentanglement may be one mechanism. Particle-packing arrangements and changes of alignment with shear may be applicable to clays, emulsions, polymers, etc. Each system must be studied with a view to the possible correlation of the parameters with system properties.

Finally, in contrast to Eyring's theory, the present theory takes the view that materials that are not time dependent are non-Newtonian because of a structural change rather than because of a multiplicity of flow types. The difference between such materials and thixotropic substances is that the latter have a much slower forward reaction rate and thus require considerable time to arrive at equilibrium.

EXAMPLE 15-3. Repeat Example 15-1, using the kinetic approach suggested by Denny and Brodkey.

Answer. For polymer solutions, Brodnyan and Kelly [26] have indicated that $f(\mu) = A\mu^{1/3.5}$ is valid; thus, from Eq. (15-31), the fraction broken can be calculated for each data point. A log-log plot of Eq. (15-37) can be made. For this calculation, $m = 1$ and $n = 2$ will be assumed, as suggested by Denny and Brodkey [40].

Figure 15-24 shows this plot, as well as that calculated for the basic shear diagram (with the constants $p = 1.76$ and $K = 1 \times 10^{-8}$). The fit is well within the experimental error over the entire range of the data. Scatter at the low-shear range on the right-hand curve is caused by inaccuracies in the viscometric measurements. These in turn cause large errors, since differences in nearly equal numbers are used. Fortunately, the back calculation of the basic shear diagram curve is insensitive to this. The back calculation is tedious, and simple empirical equations may be easier to use for specific needs; however, it is important to note that the constants obtained in this analysis have specific meaning, e.g., that K is a type of equilibrium constant. Values of $m = 1$ and $n = 3$ give as good a fit as those just calculated. For these, $p = 2.2$ and $K = 5 \times 10^{-11}$. Denny *et al.* [225] have shown that if we consider four different concentrations in the manner described, these values for p and K bring all the data together. Furthermore, the results conclusively show that $\bar{\tau}_{yx}$ and not $\mathbf{d}_{yx}$ is the correct form.

Microrheology can be defined as those investigations directed toward elucidation of the basic reasons why specific systems are non-Newtonian. In general, they are beyond the scope of this chapter because of their specific nature. They must at least be cited, however, since they are an important aspect of rheology. It is hoped that, in time, these advances will allow characteristic parameters of materials to be estimated and intelligently varied at the will of the investigator.

Two molecular-based theories for polymer systems are often referred to in the literature. These are the necklace model of Rouse [170] and the entangled ropelike model of Bueche [30]. Both are based on master curves that are designed to represent polymer flow data when plotted in the correct nondimensional coordinates (μ_0 must be known). The curves can be found in the references cited and in McKelvey [112], where there is a comparison with data on polyethylene melt. In both cases, the theories give rise to a lower Newtonian range, and an approach to an upper Newtonian limit. The literature on experimental polymer work (with the intent to explain the basic molecular structure) is extensive. Reference can be made to books by Ferry [52] and Tobolsky [195], to the review articles in

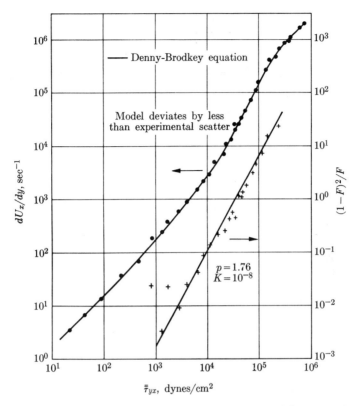

Fig. 15-24. Comparison of data with kinetic-model approach.

various series [8, 45, 80], and to the many articles in journals such as the *Transactions of the Society of Rheology*.

Another major area of microrheology is in the flow of suspensions. The work of Michaels and his coworkers [128, 129, 130] on aqueous kaolinite dispersions and of Mason and his coworkers (see the extensive review given in [107] and [108]) on variously shaped particles and their interaction during flow are of particular note. These studies have contributed much to the understanding of why certain fluids act in the manner observed.

(c) **Constitutive equations.** By *constitutive equations* of continuum mechanics, we mean those rheological equations of state that have been developed with due consideration to the three-dimensional tensoral nature of space and that are properly invariant to coordinate transformation. Of prime interest is the *principle of material indifference*, which states that the response of a material is the same for all observers, no matter what their coordinate system of reference. Truesdell [201–204] has commented extensively on this (see also [56–58, 136]) and has given the following example: One end of a spring is attached to the center of a rotating table and the other end to a free weight. Centrifugal force causes the spring to stretch. An observer standing on the ground, who can see the spring rotate with the table, measures a stretch corresponding to the applied force. An observer standing on the table does not see the spring rotate, but the amount of stretch he measures is exactly the same as that measured by the observer on the ground. The amount of stretch is independent of the frame of reference, or, in a more general sense, the response of a material is independent of the viewpoint of the observer. Although this principle seems trivial on first inspection, it does restrict the form of constitutive equations.

Oldroyd [136] has shown that the principle can be satisfied if the rheological equation is formulated in a frame of reference that moves, rotates, and is deformed with the substance. When partial differentiation is performed with respect to time, it must be observed from this frame of reference, and the quantity being differentiated must be referred to this path (called the *convected derivative*). The problem is to find the form that the derivative will take when referred to a fixed-coordinate system. A similar problem has already been considered. The *substantial derivative* (in rheology and continuum mechanics often called the material derivative) is a partial derivative with respect to time observed from a frame of reference that is being translated but not rotated or deformed. Equation (3–20), $D/Dt = (\partial/\partial t) + (\boldsymbol{U}\cdot\boldsymbol{\nabla})$, gives the form that this derivative assumes. Oldroyd [136, 138] has shown that the convective derivative can be expressed as

$$\frac{\delta A}{\delta t} = \frac{DA}{Dt} - \frac{1}{2}(\Omega \times A) \pm \frac{1}{2}(\mathbf{d}\cdot A), \qquad (15\text{–}38)$$

$$\frac{\delta \mathbf{T}}{\delta t} = \frac{D\mathbf{T}}{Dt} - \frac{1}{2}(\Omega \times \mathbf{T} - \mathbf{T} \times \Omega) \pm (\mathbf{d}\cdot\mathbf{T}), \qquad (15\text{–}39)$$

where

$$\mathbf{d} = (\boldsymbol{\nabla} U) + (\widetilde{\boldsymbol{\nabla} U}) \quad \text{or} \quad d_{ij} = (\partial U_i/\partial x_j) + (\partial U_j/\partial x_i), \qquad (15\text{–}40)$$

and is simply the rate of strain or shear-rate tensor previously given as a part of Eqs. (3–42) and (3–43). The term $\Omega = (\boldsymbol{\nabla} \times \boldsymbol{U})$ is given by Eq. (4–13). The $\pm$ signs in Eqs. (15–38) and (15–39) occur because vectors and higher-order tensors can themselves be transformed in two ways [110]. The cross product of a vector and a second-order tensor [for example, $(\Omega \times \mathbf{T})$] is given in Eq. (2–46), and the dot or inner product of two second-order tensors [for example, $(\mathbf{d}\cdot\mathbf{T})$] is given in Eq. (2–45). Clearly the substantial derivative is a part of the convective derivative. A scalar has the same value in any coordinate system, and because of this is often called an invariant. The time derivative of a scalar is also independent of the frame of reference, so that Eq. (15–38) reduces to

$$\delta A/\delta t = DA/Dt = \partial A/\partial t + (\boldsymbol{U}\cdot\boldsymbol{\nabla})A. \qquad (15\text{–}41)$$

Still another derivative is possible; it is denoted by $\mathscr{D}/\mathscr{D}t$, and is called the *Jaumann derivative*. It is a partial derivative with respect to time, and is observed from a frame of reference that is translated and rotated but not deformed. This derivative takes the same form as Eqs. (15–38) and (15–39), but without the $\pm$ deformation term. Further discussion of these derivatives can be found in the references cited and in [57, 165, 219].

An important implication [204] of the principle of material indifference is that for a tensoral quantity such as **d**, any vorticity or net velocity in the system can be transformed away; that is, by the selection of a frame of reference that is moving and spinning in an appropriate manner, one can observe a fluid under conditions of no spin and no velocity. This is merely what is done when one stands on the rotating table in the example cited, and is a simple restatement of Oldroyd's point, that the response remains invariant even if one chooses a frame of reference that moves with the flow. In the coordinate system for which the vorticity and mean motion are absent, the rate-of-strain tensor can be expressed in terms of a volumetric extension coefficient:

$$(\mathbf{d} \cdot d\mathbf{x}) = \lambda d\mathbf{x}. \tag{15-42}$$

By Eq. (2–49), we have

$$(\mathbf{d} \cdot \mathbf{n})\, dx = \lambda \mathbf{n}\, dx \quad \text{or} \quad (\mathbf{d} \cdot \mathbf{n}) = \lambda \mathbf{n}.$$

This can be rearranged by introducing the unit tensor:

$$((\mathbf{d} - \lambda \mathbf{U}) \cdot \mathbf{n}) = 0. \tag{15-43}$$

Equation (15–43) will have a nontrivial solution if and only if the determinant of the coefficients is equal to zero [223]; that is,

$$|\mathbf{d} - \lambda \mathbf{U}| = 0. \tag{15-44}$$

In cartesian-coordinate notation, this would be

$$|d_{ij} - \lambda \delta_{ij}| = \begin{vmatrix} d_{11} - \lambda & d_{12} & d_{13} \\ d_{21} & d_{22} - \lambda & d_{23} \\ d_{31} & d_{32} & d_{33} - \lambda \end{vmatrix} = 0. \tag{15-45}$$

The determinant can be expanded; after we group terms, it becomes

$$\lambda^3 - \mathrm{I}\lambda^2 + \mathrm{II}\lambda - \mathrm{III} = 0, \tag{15-46}$$

where

$$\begin{aligned} \mathrm{I} &= (\mathbf{d} : \mathbf{U}) = 2(\mathbf{\nabla} \cdot \mathbf{U}) && \text{[see Eq. (2-47)]}, \\ \mathrm{II} &= \tfrac{1}{2}(\mathrm{I}^2 - \mathbf{d} : \mathbf{d}) && \text{[see Eq. (2-47)]}, \\ \mathrm{III} &= |\mathbf{d}| && \text{[the determinant of } \mathbf{d}\text{].} \end{aligned} \tag{15-47}$$

These are the principal invariants of the shear-rate tensor. These scalar functions will be invariant during transformation. For incompressible flow $\mathrm{I} = 0$, and for simple flows such as that through pipes and between concentric cylinders, $\mathrm{III} = 0$.

For an isotropic viscous fluid, Reiner [159] and Rivlin [164] derived a rigorous constitutive equation based in part on an argument by Stokes that $\bar{\bar{\tau}}$ should not depend on the vorticity. Truesdell [204] has shown that this argument can be demonstrated from the principle of material indifference. If one takes the shear stress tensor as

$$\bar{\bar{\tau}} = \mathbf{P} - p\mathbf{U} = f(\mathbf{d}, \mathbf{\Omega}, U), \tag{15-48}$$

one can, by the principle of material indifference, transform away the vorticity and mean-motion dependencies by the same arguments just presented, so that Eq. (15-42) can be established. That is,

$$\bar{\bar{\tau}}^* = \mathbf{P}^* - p^*\mathbf{U} = f(\mathbf{d}^*, 0, 0),$$

where the starred terms are in the new coordinate system. Truesdell further shows that if this is true in any given frame of reference it must be true in all frames. Thus Eq. (15-48) reduces to Stokes' assumption that

$$\bar{\bar{\tau}} = f(\mathbf{d}). \tag{15-49}$$

The simplest form for this equation in three-dimensional space is

$$\bar{\bar{\tau}} = \alpha \mathbf{U} + \beta \mathbf{d} + \gamma(\mathbf{d}\cdot\mathbf{d}) \quad \text{or} \quad \bar{\tau}_{ij} = \alpha\delta_{ij} + \beta d_{ij} + \gamma d_{ik}d_{kj}. \tag{15-50}$$

All higher terms in $\mathbf{d}$ can be expressed as functions of lower ones [57]. Each of the scalar coefficients α, β, and γ are functions of the three principal invariants of $\mathbf{d}$ as given by Eq. (15-47). [Thus Eq. (15-49) is satisfied.] Equation (15-50) describes the Reiner-Rivlin fluid in its most general form. For a Newtonian,

$$\alpha = \tfrac{1}{2}(\tfrac{2}{3}\mu - \kappa)\mathbf{I}, \qquad \beta = -\mu, \qquad \gamma = 0.$$

A simple geometry, convenient for illustrative purposes, is the flow in the x-direction between two infinite walls separated by a distance y. The coordinates are x, y, z and the corresponding velocities are $U_x(y), 0, 0$. The pressure tensor, which is symmetric, is written as

$$\mathbf{P} = p\mathbf{U} + \bar{\bar{\tau}} = \begin{pmatrix} p + \bar{\tau}_{xx} & \bar{\tau}_{yx} & 0 \\ \bar{\tau}_{yx} & p + \bar{\tau}_{yy} & 0 \\ 0 & 0 & p + \bar{\tau}_{zz} \end{pmatrix}.$$

Actually $\bar{\tau}_{zz}$ must be zero if there is to be no flow in the z-direction.

For an incompressible non-Newtonian flow in this system [$\alpha = 0$ since $\mathrm{I} = 0$, $\beta = -\mu_a(\mathrm{II})$, and $\gamma = \eta_c(\mathrm{II})$], the pressure tensor [from Eq. (15-50)] becomes

$$\mathbf{P} = p\mathbf{U} + \bar{\bar{\tau}} = \begin{pmatrix} p - \eta_c(\mathrm{II}) d_{yx}^2 & -\mu_a(\mathrm{II}) d_{yx} & 0 \\ -\mu_a(\mathrm{II}) d_{yx} & p - \eta_c(\mathrm{II}) d_{yx}^2 & 0 \\ 0 & 0 & p \end{pmatrix}, \tag{15-51}$$

where $\mu_a(\mathrm{II})$ is the *apparent viscosity* and $\eta_c(\mathrm{II})$ is the *cross viscosity*. Both are functions of II only, since $\mathrm{III} = 0$ for this simple geometry; however, even if

III $\neq 0$, it has been suggested that μ_a can be only a very weak function of III [17, 182, 183]. Under the simple shear geometries normally used, such a material would conform to the equation,

$$\bar{\tau}_{yx} = \bar{\tau}_{xy} = -\mu_a(\text{II})\mathbf{d}_{yx} = -\mu_a(\text{II})(dU_x/dy).$$

By simple viscometric observations, therefore, this would not be distinguishable from any other non-Newtonian material. However, the inequality of the normal stresses has been suggested as the cause of the Weissenberg effect [139, 159, 161, 164]. In the present case, $\mathsf{P}_{xx} = \mathsf{P}_{yy} \neq \mathsf{P}_{zz}$. This is in contrast to the experimental evidence obtained by Roberts [88, 167] and Pollett [151], which has shown that for such materials as they studied, $\mathsf{P}_{xx} \neq \mathsf{P}_{yy} = \mathsf{P}_{zz}$. It is also in contrast to other experimental observations [81, 102] that imply $\mathsf{P}_{xx} \neq \mathsf{P}_{yy} \neq \mathsf{P}_{zz}$. However, this does not rule out the possibility that materials might be found that can be described by the Reiner-Rivlin equation.

Equation (15-50) also contains, as special cases, the empirical models previously cited [17, 182, 208]. For example, given an incompressible fluid ($\text{I} = 0$), and assuming any effect of III to be negligible, the power law of Eq. (15-11) can be expressed as

$$\bar{\tau} = -(K|\text{II}|^{(n-1)/2})\mathbf{d}, \tag{15-52}$$

where, for this case, $\text{II} = -\frac{1}{2}(\mathbf{d}:\mathbf{d})$, since $\text{I} = 0$. The Sisko model (15-13) can be expressed as

$$\bar{\tau} = -(\mu_\infty + K|\text{II}|^{(n-1)/2})\mathbf{d}, \tag{15-53}$$

and so on, simply by replacing dU_x/dy by $\sqrt{\text{II}}$ in each case. The ease of conversion to any coordinate system has been emphasized by Bird, Lightfoot, and Stewart [17]. The term in parentheses simply replaces μ in the Newtonian expressions. Of course, the proper transformation of II must also be made.

For several reasons, Eq. (15-50) is clearly inadequate for the representation of normal stresses. It is a constitutive equation of a purely viscous fluid and does not allow for viscoelasticity (i.e., there are no time derivatives), the predicted normal stress pattern is contrary to the available experimental evidence, and it has been observed that fluids that exhibit normal stress effects are also viscoelastic.

The observed normal stress effects might be caused by the Poynting effect, as discussed by Reiner [159]. In a purely elastic material, in order to maintain a simple shear, it is necessary to have a pressure p and a tension $\bar{\tau}_{xx}$. The pressure tensor for a simple shear (a rod in torsion) with no bulk effects is

$$\mathsf{P} = \begin{pmatrix} p - \gamma^2 G & -\gamma G & 0 \\ -\gamma G & p & 0 \\ 0 & 0 & p \end{pmatrix}. \tag{15-54}$$

In contrast to Eq. (15–51), $P_{xx} \neq P_{yy} = P_{zz}$. This particular analysis considers the solid characteristics only and ignores the viscous aspects; thus, as it stands, it is not directly applicable to viscoelastic materials.

Fröhlich and Sack [60] and Oldroyd [137] have derived an equation for viscoelastic fluids:

$$\bar{\bar{\tau}} + \lambda_1(\partial \bar{\bar{\tau}}/\partial t) = -\mu(\mathbf{d} + \lambda_2 \partial \mathbf{d}/\partial t), \tag{15-55}$$

where μ, λ_1, and λ_2 are constants. Here μ can be identified as the viscosity in steady-state experiments, λ_1 is a relaxation time [if motion is stopped, $\bar{\bar{\tau}}$ will decay as $\exp(-t/\lambda_1)$], and λ_2 is a retardation time [if stresses are removed, $\mathbf{d}$ will decay as $\exp(-t/\lambda_2)$]. Fröhlich and Sack derived the equation by considering a dilute suspension of elastic solid spheres in a viscous liquid, while Oldroyd considered a dilute emulsion containing liquid drops. Interfacial tension causes the drops to resist changes of shape, and thus they act in a way that is similar to that of elastic solid spheres. Equation (15–55) is an equation of linear viscoelasticity which shows a Newtonian viscosity in simple shear and does not predict normal stresses. However, if the partial derivatives are properly replaced with the convective derivatives of Eq. (15–39), we obtain

$$\bar{\bar{\tau}} + \lambda_1 \mathfrak{d}\bar{\bar{\tau}}/\mathfrak{d}t = -\mu(\mathbf{d} + \lambda_2 \mathfrak{d}\mathbf{d}/\mathfrak{d}t), \tag{15-56}$$

which can predict normal stresses, but still has a Newtonian viscosity in steady shear. There will be two forms of this equation because of the $\pm$ sign in Eq. (15–39). The plus sign gives rise to Oldroyd's fluid A, which does not predict the Weissenberg effect, and the negative sign gives fluid B, which does. In simple shear, for fluid B, the stress tensor takes the form

$$\mathbf{P} = p\mathbf{U} + \bar{\bar{\tau}} = \begin{pmatrix} p - 2\mu(\lambda_1 - \lambda_2)\,\mathbf{d}_{yx}^2 & -\mu\,\mathbf{d}_{yx} & 0 \\ -\mu\,\mathbf{d}_{yx} & p & 0 \\ 0 & 0 & p \end{pmatrix}, \tag{15-57}$$

which has $P_{xx} \neq P_{yy} = P_{zz}$, but still predicts a Newtonian viscosity under simple shear, and is therefore unsatisfactory as it stands. There is also a problem of having two forms (fluids A and B); however, this could be eliminated by using the Jaumann derivative rather than the convective. To rectify these shortcomings, Oldroyd [138] modified the derivative and added a number of nonlinear terms. The resulting expression for the non-Newtonian viscosity [57, 124, 138, 219] was

$$\bar{\bar{\tau}}_{yx} = -\mu \left[\frac{1 + a\,\mathbf{d}_{yx}^2}{1 + b\,\mathbf{d}_{yx}^2}\right] \mathbf{d}_{yx}, \tag{15-58}$$

where a and b are functions of the other constants. Williams and Bird [219] showed that for pipe-flow data the model was good for either low or high shear rates, but not for both at once. Tanner [187] found similar results for the same equation, using data from a helical flow. To be able to prove that such an equation of state is valid, one should show that the same constants can be applied to data for both shear and normal stress, or at least show that the constants can be applied to other flows. Tanner showed that the constants a and b of Eq. (15–58), obtained from the experiments on helical flow, were completely inadequate to describe the drag on a sphere, which should also be a function of the same two constants. One must conclude that this equation is inadequate for this task. However, improvements can be made, and some further modifications have been suggested by Spriggs and Bird [230]. Kapoor et al. [226] and Vela et al. [231] have also investigated the use of similar equations and found that the results obtained are quite good, but certainly not perfect. In particular, the fit with a three-constant model and a five-constant model were essentially the same, so that with the data available, one could not distinguish which model was better.

Rivlin and Ericksen [165] have developed a constitutive equation designed to avoid the inadequacies of the Oldroyd model. They assumed that the shear-stress tensor could be expressed as a polynomial in terms of the rate-of-strain tensor and its convective derivatives. The final result for an isotropic material without a yield value requires eight material functions. White and Metzner [232] have provided a general discussion of such constitutive equations for viscoelastic materials, and have indicated some possible simplifications and applications to certain specific flows for slightly viscoelastic non-Newtonian materials. Ericksen has extended this general approach even further and has developed a theory of anisotropic fluids. Several of the references on this analysis are listed under reference 165. The complete equation of Rivlin and Ericksen reduces to Eq. (15–50) if the time derivatives are dropped. For certain laminar flows (those common to viscometric measurement where $\mathrm{I} = \mathrm{III} = 0$), Markovitz [101] has shown that for all of the various theories only three material functions are needed. For the anisotropic fluid, Kaloni [165] has shown the same thing. Coleman and Noll [38] were able to show that this was true for a very general class of fluids, the simple fluid of Noll [134]. Since this latter theory includes the preceding isotropic theories, we shall present it briefly here. However, let us anticipate the result for our case of simple shear:

$$\bar{\bar{\tau}}_{xy} = \bar{\bar{\tau}}_{yx} = -\mu_a(\mathrm{II})\, \mathsf{d}_{yx},$$

$$\mathsf{P}_{xx} - \mathsf{P}_{zz} = \bar{\bar{\tau}}_{xx} - \bar{\bar{\tau}}_{zz} = \bar{\bar{\tau}}_{xx} = \sigma_1(\mathrm{II}), \qquad (15\text{–}59)^*$$

$$\mathsf{P}_{yy} - \mathsf{P}_{zz} = \bar{\bar{\tau}}_{yy} - \bar{\bar{\tau}}_{zz} = \bar{\bar{\tau}}_{yy} = \sigma_2(\mathrm{II}).$$

For $\sigma_1 = \sigma_2$, the result is the same as for the Reiner-Rivlin fluid (15–51).

Before we say anything further about the Coleman-Noll theory, let us review the implications of Eq. (15–59). First, there are three material functions dependent

---

*Another convention often used is $\bar{\bar{\tau}}_{11} - \bar{\bar{\tau}}_{33} = \sigma_1(\mathrm{II})$ and $\bar{\bar{\tau}}_{22} - \bar{\bar{\tau}}_{33} = \sigma_2(\mathrm{II})$, in which 1 is in the direction of the streamline, 2 is perpendicular to the shearing plane, and 3 is the remaining direction.

on II (that is, on d_{yx}^2 for these flows). One of these is the apparent viscosity, which can be represented on the basic shear diagram by several of the empirical or theoretical equations already given. The normal stresses (which are all that remain) are defined by two additional functions of the rate of shear ($\sqrt{\text{II}}$). In other words three curves, the basic shear diagram and two basic normal stress diagrams, are all that is needed for a complete definition of the stress tensor. As already noted, simple equations are available for the first of these, and it is suspected that somewhat similar equations can represent the latter two.

Let us now turn to the theory which is expressed in terms of functionals. We shall follow Truesdell's review of Coleman's and Noll's work [204]. A functional is a function whose value depends on *all* possible values of its argument. A function depends on *specific* values of its argument. For an incompressible fluid, the constitutive equation of Coleman and Noll is

$$\bar{\boldsymbol{\tau}} = \mathscr{F}[\mathbf{C}_{(t)}(t+s)]; \tag{15-60}$$

that is, $\bar{\boldsymbol{\tau}}$ is a functional ($\mathscr{F}$) of $\mathbf{C}_{(t)}(t+s)$, where $\mathbf{C}_{(t)}$ is the Cauchy-Green finite deformation tensor for the fluid at time t with respect to that at time $t+s$. Note that s is negative, since only the history is important (i.e., viscoelasticity). Any dependency of $\bar{\boldsymbol{\tau}}$ on vorticity or mean motion has been transformed away by the principle of material indifference. Without going into details, we can show the Cauchy-Green deformation tensor to be a function of the metric tensor **g** and the derivatives of the past coordinates. The metric tensor is simply the diagonal tensor composed of the h_i's of the general orthogonal coordinates discussed on page 19; that is,

$$\mathbf{g} = g_{ij} = \begin{pmatrix} h_i^2 & 0 & 0 \\ 0 & h_j^2 & 0 \\ 0 & 0 & h_k^2 \end{pmatrix}. \tag{15-61}$$

These are given for cylindrical coordinates in Eq. (2–69), and for spherical coordinates, in Eq. (2–75). For viscometric flows (I = III = 0), the net result is

$$\mathbf{C}_{(t)}(t) = \begin{pmatrix} 1 + \text{II}\, s^2 & \sqrt{\text{II}}\, s & 0 \\ \sqrt{\text{II}}\, s & 1 & 0 \\ 0 & 0 & 1 \end{pmatrix}. \tag{15-62}$$

Equation (15–62) can also be written as

$$\mathbf{C}_{(t)}(t) = \mathbf{U} + s\,\mathbf{A} + s^2\,\mathbf{B}, \tag{15-63}$$

where

$$\mathbf{A} = \sqrt{\text{II}} \begin{pmatrix} 0 & 1 & 0 \\ 1 & 0 & 0 \\ 0 & 0 & 0 \end{pmatrix} \quad \text{and} \quad \mathbf{B} = \text{II} \begin{pmatrix} 1 & 0 & 0 \\ 0 & 0 & 0 \\ 0 & 0 & 0 \end{pmatrix}.$$

Viscoelastic characteristics would appear in Eq. (15–63), since the deformation tensor is dependent on past time; the form of this dependency is quite specific. Of major importance is the result that the functional of Eq. (15–60) can only depend on specific values of **A** and **B**, since by Eq. (15–63) these determine $\mathbf{C}_{(t)}$ uniquely for all past time. Therefore

$$\bar{\bar{\tau}} = \mathscr{F}[\mathbf{C}_{(t)}(t+s)] = f(\mathbf{A}, \mathbf{B}), \tag{15-64}$$

where f is a simple function and not a functional. The function $f(\mathbf{A}, \mathbf{B})$ is tensoral, and by application of the principle of material indifference must transform as such; thus its components are functions of the principal invariant II (since I = III = 0). Since $\bar{\bar{\tau}}$ is symmetric, it has six components. In addition,

$$\bar{\bar{\tau}}_{yz} = \bar{\bar{\tau}}_{xz} = 0 \quad \text{and} \quad \bar{\bar{\tau}}_{xx} + \bar{\bar{\tau}}_{yy} + \bar{\bar{\tau}}_{zz} = 0,$$

constitute three additional conditions. Consequently, only three of the six components are independent. These can be taken as $\bar{\bar{\tau}}_{xy}, \bar{\bar{\tau}}_{xx} - \bar{\bar{\tau}}_{zz}, \bar{\bar{\tau}}_{yy} - \bar{\bar{\tau}}_{zz}$, and have already been given in Eq. (15–59); that is, μ_a, σ_1, and σ_2. Again, these are material functions and define the entire stress tensor. They should be valid for any viscometric flow in which I = III = 0. Note that $\bar{\bar{\tau}}_{yx}$ is an odd function of the shear rate, while μ_a, σ_1, and σ_2 are even. This is so because $\bar{\bar{\tau}}_{yx}$ must change sign with $\mathbf{d}_{yx}$, and the normal stress and viscosity itself will not change sign if one changes the direction of shearing.

Finally it should be pointed out that although, on the basis of certain experiments, it has been implied that $\bar{\bar{\tau}}_{yy} - \bar{\bar{\tau}}_{zz} = \sigma_2(\text{II}) = 0$, the last word on this has not been said. Lodge [97] criticized most of the measurements. Hayes and Tanner [81], using annular flow, found $\sigma_2 > 0$ and obtained values up to 30% of $\bar{\bar{\tau}}_{yx}$. Likewise Markowitz [102] found that $\sigma_2 > 0$ (about $\frac{1}{2}$ to $\frac{1}{3}$ of σ_1). In contrast to this, Ginn and Metzner [67] found equality to within 5% over the range of shear rates from 70 to 500 sec^{-1} for similar materials. These authors have criticized other work that has shown that an inequality exists; however, still more recent work by these authors has indicated that their conclusions in reference 67 may well be in error and that the equality does not exist.

15-2 NON-NEWTONIAN FLUID FLOW

Laminar flows that occur in geometries simple enough so that exact solutions can be obtained are of prime interest in the field of rheology. These are the viscometric flows previously mentioned. These problems must be solved so that the basic shear diagram, independent of the instrument, can be obtained. Undesirable sources of error, such as end effects, slip at the wall, heating effects, and instability must be considered. Very often the same flow situations occur in practical engineering problems, but more often complications of a three-dimensional nature, heat transfer, or turbulence may arise. These are some of the additional aspects of non-Newtonian flow that are to be considered.

15-2.A Viscometric Flows

Several possible viscometric flows will be considered in this section. In some of them an attempt will be made to first set up the basic equations without regard to any specific rheological model. Following this, several specific laws will be considered, and the means of obtaining the basic shear diagram will be given. Sources of error already cited will be developed, as well as the means of eliminating them. Finally, there will be some comments on normal force measurements in these flows.

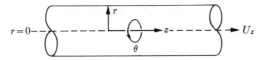

Fig. 15–25. *Coordinates for the capillary flow system.*

(a) Capillary flow. Steady pipe flow described in cylindrical coordinates (r, θ, z) has the velocity components $0, 0, U_z(r)$ (see Fig. 15–25). The stress tensor takes the form

$$\mathbf{P} = p\mathbf{U} + \bar{\bar{\tau}} = \begin{pmatrix} p + \bar{\bar{\tau}}_{rr} & 0 & \bar{\bar{\tau}}_{rz} \\ 0 & p + \bar{\bar{\tau}}_{\theta\theta} & 0 \\ \bar{\bar{\tau}}_{rz} & 0 & p + \bar{\bar{\tau}}_{zz} \end{pmatrix},$$

where $\bar{\bar{\tau}}_{\theta\theta}$ must be zero if there is to be no helical flow. The shear rate dU_z/dr is always negative, since r is measured from the center line, and $\bar{\bar{\tau}}_{rz}$ is always positive; i.e., momentum is transferred from the center line toward the wall. Equation (15–6) becomes

$$\bar{\bar{\tau}}_{rz} = -\mu_a(dU_z/dr), \tag{15-65}$$

and the equation of motion (4–1) reduces to [17]

$$\begin{aligned} 0 &= -\frac{\partial p}{\partial r} - \frac{1}{r}\frac{\partial}{\partial r}(r\bar{\bar{\tau}}_{rr}) - \frac{\bar{\bar{\tau}}_{\theta\theta}}{r}, \\ 0 &= -\frac{\partial p}{\partial z} - \frac{1}{r}\frac{\partial}{\partial r}(r\bar{\bar{\tau}}_{rz}), \end{aligned} \tag{15-66}$$

where it was assumed that no forces were acting in the radial or flow directions, and that the flow was steady. From the second equation of (15–66) or from a simple force balance,

$$\bar{\bar{\tau}}_{rz} = -\frac{r}{2}\frac{dp}{dz} = \frac{r}{r_0}\tau_w, \tag{15-67}$$

where r_0 is r at the wall, and $\tau_w = -(r_0/2)(dp/dz)$. The boundary condition is $U_z = 0$ at $r = r_0$, and the volume flow rate [given by Eq. (7–22)] becomes

$$Q = \int_0^{r_0} 2\pi r U_z\, dr = 2\pi\left(\frac{r_0}{\tau_w}\right)^2 \int_0^{\tau_w} \bar{\bar{\tau}}_{rz} U_z\, d\bar{\bar{\tau}}_{rz}. \tag{15-68}$$

The velocity would be

$$U_z = \int_r^{r_0}\left(-\frac{dU_z}{dr}\right)dr = \frac{r_0}{\tau_w}\int_{\bar{\tau}_{rz}}^{\tau_w}\left(-\frac{dU_z}{dr}\right)d\bar{\tau}_{rz}. \quad (15\text{-}69)$$

Combining Eqs. (15–68) and (15–69) gives

$$\frac{4Q}{\pi r_0^3} = \frac{8}{\tau_w^3}\int_0^{\tau_w}\bar{\tau}_{rz}\int_{\bar{\tau}_{rz}}^{\tau_w}\left(-\frac{dU_z}{dr}\right)d\bar{\tau}_{rz}\,d\bar{\tau}_{rz}, \quad (15\text{-}70)$$

and when we integrate by parts [139], we obtain

$$\frac{4Q}{\pi r_0^3} = \frac{4}{\tau_w^3}\int_0^{\tau_w}\bar{\tau}_{rz}^2\left(-\frac{dU_z}{dr}\right)d\bar{\tau}_{rz}, \quad (15\text{-}71)$$

which is the general equation that relates the flow to the velocity gradient in the system.

For a Newtonian fluid $(-dU_z/dr) = \bar{\tau}_{rz}/\mu$, and Eqs. (15–69) and (15–70) integrate to

$$U_z = -\frac{r_0^2}{4\mu}\frac{dp}{dz}\left(1 - \frac{r^2}{r_0^2}\right) \quad (7\text{-}21)$$

and

$$\frac{4Q}{\pi r_0^3} = \frac{\tau_w}{\mu} = -\frac{r_0}{2\mu}\frac{dp}{dz}. \quad (7\text{-}22)$$

This latter equation can be used to define a capillary shear diagram as $4Q/\pi r_0^3$ or $4U_{z,\text{ave}}/r_0$ versus τ_w. For non-Newtonian fluids this in turn defines the pipe apparent viscosity as

$$\frac{1}{\mu_{\text{ap}}} = \frac{4Q/\pi r_0^3}{\tau_w} = \frac{4}{\tau_w^4}\int_0^{\tau_w}\bar{\tau}_{rz}^2\left(-\frac{dU_z}{dr}\right)d\bar{\tau}_{rz}, \quad (15\text{-}72)$$

where μ_{ap} has been used to distinguish it from μ_a defined by Eq. (15–65); that is,

$$\tau_w = -\mu_a\left(\frac{dU_z}{dr}\right)_w = \mu_{\text{ap}}\frac{4Q}{\pi r_0^3}. \quad (15\text{-}73)$$

The term $4Q/\pi r_0^3$ is sometimes called the *pseudo-shear rate*.

A simple relation between the flow rate and the wall shear rate (or between μ_a and μ_{ap}) was obtained by Weissenberg [46], Rabinowitsch [154, 210], and Mooney [132]. Differentiation of Eq. (15–71) gives

$$\frac{d[\tau_w^3(4Q/\pi r_0^3)]}{d\tau_w} = 4\tau_w^2\left(-\frac{dU_z}{dr}\right)_w.$$

Further differentiating the left side results in

$$\left(-\frac{dU_z}{dr}\right)_w = \frac{3}{4}\frac{4Q}{\pi r_0^3} + \frac{\tau_w}{4}\frac{d(4Q/\pi r_0^3)}{d\tau_w}. \quad (15\text{-}74)$$

The history of this relation and the many alternate forms that have appeared in the literature are given by Savins et al. [176]. This equation can be used directly to reduce the capillary shear diagram to the basic shear diagram as follows: (1) For a given value of τ_w, the value of $4Q/\pi r_0^3$ can be obtained from the data. (2) The slope of the curve can also be obtained at that point. (3) These are the only terms that appear on the right-hand side of the equation; thus $(-dU_z/dr)_w$ can be calculated. (4) Both τ_w and $(-dU_z/dr)_w$ are measured at the same point and thus are the terms of the basic shear diagram. No rheological law of any kind need be assumed for this calculation.

(i) The power law. One definition of a power law can be taken from the capillary shear diagram, and is

$$\tau_w = -\frac{r_0}{2}\frac{dp}{dz} = K'\left(\frac{4Q}{\pi r_0^3}\right)^{n'}, \tag{15-75}$$

where the primes are used to indicate that these terms are based on the capillary shear diagram, and not on the basic shear diagram as they are in Eq. (15-11). In this definition, n' is not necessarily a constant. An alternate and better definition [121] would be

$$n' = \frac{d[\ln\{-(r_0/2)(dp/dz)\}]}{d[\ln(4Q/\pi r_0^3)]}, \tag{15-76}$$

which is a form defined from a further manipulation of Eq. (15-74); that is,

$$\left(-\frac{dU_z}{dr}\right)_w = \frac{3}{4}\frac{4Q}{\pi r_0^3} + \frac{1}{4}\left(\frac{4Q}{\pi r_0^3}\right)\frac{d[\ln(4Q/\pi r_0^3)]}{d[\ln \tau_w]}. \tag{15-77}$$

Using Eq. (15-76) in (15-77) gives

$$\left(-\frac{dU_z}{dr}\right)_w = \frac{3n'+1}{4n'}\left(\frac{4Q}{\pi r_0^3}\right). \tag{15-78}$$

Inspection of Eq. (15-78) shows that for $n' = 1$, $4Q/\pi r_0^3$ is equal to the shear rate at the wall, and Eq. (15-75) reduces to Newton's law with $K' = \mu$. This alternate form of the Weissenberg-Rabinowitsch-Mooney equation (15-78) can also be used to obtain the basic shear diagram from capillary data: (1) At a given value of τ_w, n' is obtained from the slope of a log-log plot of τ_w versus $4Q/\pi r_0^3$ (Eq. 15-76). (2) With this value of n' and $4Q/\pi r_0^3$, Eq. (15-78) is used to obtain $(-dU_z/dr)_w$.

It is important to note that the capillary flow experiment cannot give the basic shear diagram directly, but can be corrected to it exactly. Equation (15-74) or (15-78) can be used for this conversion.

EXAMPLE 15-4. The following data on a commercial high-pressure polyethylene melt (190°C) have been obtained on a capillary unit and have been corrected for end effects [120]. Obtain the basic shear diagram.

$4Q/\pi r_0^3$, sec^{-1}	τ_w, dynes/cm^2
10	2.24×10^5
20	3.10
50	4.35
100	5.77
200	7.50
400	9.73
600	11.10
1000	13.52
2000	16.40

Answer. The data as given are first plotted on log-log paper, and the slope n' is evaluated. One can then correct $4Q/\pi r_0^3$ to $(-dU_z/dr)_w$ by Eq. (15–78). The results are given in the following table, and, as can be seen, the correction varies from 26% to 53%.

$4Q/\pi r_0^3$, sec^{-1}	n'	$\dfrac{3n'+1}{4n'}$	$(-dU_z/dr)_w$, sec^{-1}	μ_{ap}, poise	μ_a, poise
10	0.49	1.26	12.6	22,400	17,800
20	0.45	1.31	26.2	15,500	11,800
50	0.44	1.32	66	8,700	6,600
100	0.41	1.36	136	5,770	4,250
200	0.39	1.39	278	3,750	2,700
400	0.35	1.46	584	2,430	1,660
600	0.34	1.49	894	1,850	1,240
1000	0.32	1.53	1530	1,352	885
2000	0.32	1.53	3060	820	535

(ii) Bingham ideal plastic. Equation (15–72) can be integrated for the Bingham ideal plastic material, since it is known that

$$-\frac{dU_z}{dr} = 0, \quad 0 \leqslant \bar{\bar{\tau}}_{rz} \leqslant \tau_0, \quad \text{and} \quad -\frac{dU_z}{dr} = \frac{\bar{\bar{\tau}}_{rz} - \tau_0}{\mu'}, \quad \bar{\bar{\tau}}_{rz} \geqslant \tau_0. \qquad (15\text{–}79)$$

The integration must be done in two parts, from 0 to τ_0 and from τ_0 to τ_w. For the pipe apparent viscosity, the result is

$$\frac{1}{\mu_{ap}} = \frac{1}{\mu'}\left[1 - \frac{4}{3}\frac{\tau_0}{\tau_w} + \frac{1}{3}\left(\frac{\tau_0}{\tau_w}\right)^4\right] \qquad (15\text{–}80)$$

for $\bar{\bar{\tau}}_{rz} \geqslant \tau_0$, and infinite for $\bar{\bar{\tau}}_{rz} \leqslant \tau_0$. In terms of the volume flow rate (15–71), the equation becomes

$$Q = -\frac{\pi r_0^4}{8\mu'}\frac{dp}{dz}\left[1 - \frac{4}{3}\frac{\tau_0}{\tau_w} + \frac{1}{3}\left(\frac{\tau_0}{\tau_w}\right)^4\right], \qquad (15\text{–}81)$$

which is called the *Buckingham-Reiner equation*. A slight rearrangement of Eq. (15-81) describes the Bingham ideal plastic material on the capillary shear diagram. It is shown in Fig. 15-26, and is the same as Fig. 15-20.

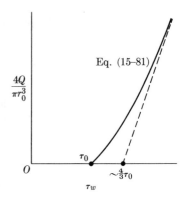

Fig. 15-26. Bingham ideal plastic material on the capillary shear diagram.

(iii) **End effects.** If end effects cause an appreciable amount of pressure drop, the estimation of τ_w by Eq. (15-67) is in error. Our accuracy in calculating the end effects for pipe flow is far from satisfactory, even for Newtonian materials. This is true because of the large number of factors contributing to the loss and the difficulty of separating the effects so that either we can check theoretical estimates or we can obtain general empirical correlations. The treatment is complicated by the fact that pressure-drop measurements are usually made between two reservoirs, and thus there may be both an upstream entrance effect and a downstream exit effect. These in turn involve frictional losses and kinetic energy corrections associated with the change in velocity and the development of the velocity profile. For Newtonians, enough theoretical and empirical information is available (for an example, see pages 129-134) so that a reasonable estimate can be made of the end effects. Bogue [18] and Collins and Schowalter [39] have made analyses for the entrance region of a pipe for power-law non-Newtonians, and these check well with the limited amount of data available. As yet, however, no information exists which would enable treatment of either the more complex entrance effects caused by viscoelastic materials or of exit effects.

Empirical correlations are generally a combination of a number of possible effects, and are made as an effective length to be added to the actual length, as a kinetic energy correction factor, or both. The general agreement between the results reported by various investigators is poor [135], one reason being that the kinetic energy correction coefficient has usually been assumed constant. However, Cannon *et al.* [32] have clearly shown that this is not true by demonstrating that the correction is a function of the Reynolds number. Their correlation is restricted to the flow of Newtonians in systems that have trumpet-shaped entrances and exits. Correlations for other geometries and for non-Newtonians are limited or simply do not exist.

It is preferable to design a capillary instrument so that end effects are negligible. That this is not always possible, however, can be shown by the case of molten polymers. Although such cases can be treated, additional experimentation and calculation are needed. The method of Metzger and Brodkey [120] provides a convenient means of making these corrections.

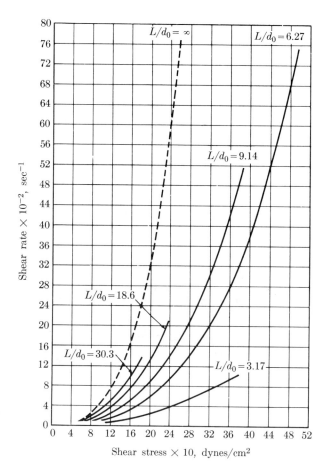

Fig. 15–27. Linear plot of flow behavior for a commercial low-density polyethylene. [From A. P. Metzger and R. S. Brodkey, J. Appl. Polymer Sci. 7, 399–410 (1963).]

The effect of capillary geometry on a polymer flow is illustrated in Fig. 15–27. It has been reported [118] that the L/d_0 ratio of polyethylene should be around 30 in order for the viscosity to be unaffected by length. Other investigators [125] and the measurements presented in Fig. 15–27 suggest that results may be influenced by L/d_0 ratios as high as 130. Because of the high viscosity of polymer melts it is difficult to work with long capillaries. Some attempts have been made to obtain corrections from the data for flow through sharp-edged orifices (no length) of the same diameter [216]. The difference between the pressure drop through the capillary and that through the orifice is assumed to be the driving force for the capillary alone. This method may not be valid, however [122], and a more effective approach has been suggested by Bagley [4]. In Bagley's

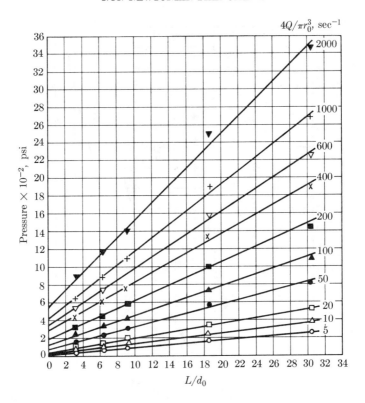

Fig. 15–28. Curves showing the relationship between pressure and capillary dimensions at constant shear rates. [From A. P. Metzger and R. S. Brodkey, J. Appl. Polymer Sci. 7, 399–410 (1963).]

technique, the equation for shear stress is modified by assuming that the entrance effect is a function of the capillary dimensions. If experimental data from a series of capillaries are available, the correction term can be obtained by extrapolating to zero pressure the linear plot of pressure versus L/d_0 (at constant shear rate; see Fig. 15–28). It was demonstrated that this correction term is strongly dependent on the shear rate and that it also varies from one polymer to another. If the correction terms are determined and the shear stresses recalculated, flow data from different capillaries reduce to a single curve. Thus not only does the technique illustrate the importance of correcting for entrance effects but it also permits comparison of data obtained with different capillaries.

A simpler, though similar, approach [120] is suggested by examination of the expression for shear stress (15–67), which can be rearranged to

$$\tau_w = \frac{1}{4} \frac{dp}{d(L/d_0)}. \qquad (15\text{–}82)$$

It follows that the wall shear stress can be obtained from the slope of the pressure versus L/d_0 without the necessity of determining the end correction. Thus experimental data from short capillaries can be used to calculate shear stresses which are independent of the geometry and are characteristic of the polymer. For the plot to be linear, however, fully developed flow must exist in the tubes. This causes no trouble for polymers because of their high viscosity. As already discussed in some detail, the expression $4Q/\pi r_0^3$ will have to be corrected by Eq. (15-74) or (15-78) to the wall conditions. The sequence of calculations is: (1) The values of $4Q/\pi r_0^3$ are plotted versus the corresponding pressure for each capillary. (2) At arbitrarily chosen values of $4Q/\pi r_0^3$, the corresponding pressures are obtained and cross-plotted versus the L/d_0 ratio, as in Fig. (15-28). (3) The slopes of these lines are used to calculate the corrected shear stresses for a capillary of infinite length. The linearity of the lines demonstrates the validity of Eq. (15-82), since in the data reported both L and d_0 were varied. (4) From a log-log plot of the corrected shear stress versus $4Q/\pi r_0^3$, one can proceed as already outlined, following Eq. (15-78) or as shown in Example 15-4.

EXAMPLE 15-5. Figure 15-28 shows a cross plot of the raw data for a high-pressure commercial polyethylene melt which were obtained on a capillary flow unit [120]. Obtain first the capillary shear diagram, then the basic shear diagram.

Answer. The raw data consisted of measured flow rates for the specific pressures necessary to force the polymer through five different capillaries. Figure 15-28 is a cross plot of this data, giving the pressure at specific levels of the parameter $4Q/\pi r_0^3$. The true shear stress at the wall is determined from one-fourth of the slope of the straight lines of this figure [Eq. (15-82) with the proper conversion factors]. These are the values tabulated at the beginning of Example 15-4. The reduction of these results to the basic shear diagram is given in that example.

(iv) Slip at the wall. We have already cited the possibility of an effective slip at the wall for two-phase systems. The equations we derived can easily be modified to allow for a slip velocity U_s by changing the boundary condition used in the derivations. The new boundary condition at the wall becomes $U_z = U_s$, rather than $U_z = 0$. It may be that U_s and terms depending on U_s are a function of τ_w. When we follow through the derivation as presented by Mooney [132] and modified by Oldroyd [139], Eq. (15-69) becomes

$$U_z = U_s + \int_r^{r_0}\left(-\frac{dU_z}{dr}\right)dr = U_s + \frac{r_0}{\tau_w}\int_{\bar{\tau}_{rz}}^{\tau_w}\left(-\frac{dU_z}{dr}\right)d\bar{\tau}_{rz}. \tag{15-83}$$

When we combine this with Eq. (15-68), we find that the relation similar to (15-70) is

$$\frac{4Q}{\pi r_0^3} = \frac{4U_s}{r_0} + \frac{8}{\tau_w^3}\int_0^{\tau_w}\bar{\tau}_{rz}\int_{\bar{\tau}_{rz}}^{\tau_w}\left(-\frac{dU_z}{dr}\right)d\bar{\tau}_{rz}\,d\bar{\tau}_{rz}. \tag{15-84}$$

An integration by parts gives the form analogous to Eq. (15–71):

$$\frac{4Q}{\pi r_0^3} = \frac{4U_s}{r_0} + \frac{4}{\tau_w^3} \int_0^{\tau_w} \bar{\bar{\tau}}_{rz}^2 \left(-\frac{dU_z}{dr}\right) d\bar{\bar{\tau}}_{rz}. \qquad (15\text{–}85)$$

Oldroyd defines

$$\xi = U_s/\tau_w \quad \text{and} \quad \phi = \frac{4}{\tau_w^4} \int_0^{\tau_w} \bar{\bar{\tau}}_{rz}^2 \left(-\frac{dU_z}{dr}\right) d\bar{\bar{\tau}}_{rz},$$

which can be combined with Eq. (15–85) to give

$$4Q/\pi r_0^3 = (4\xi/r_0 + \phi)\tau_w. \qquad (15\text{–}86)$$

Data obtained on a series of pipe diameters can be used with Eq. (15–86) to determine both ξ and ϕ (for any given τ_w) from a plot of $4Q/\pi r_0^3$ versus $1/r_0$. With this information we can then obtain the true basic shear diagram; however, we must be extremely careful to avoid misinterpreting turbulence and thixotropic characteristics as effective slip. Toms [197] has given an example for a polymeric system and Thomas [189] has demonstrated slip for a thorium-oxide suspension.

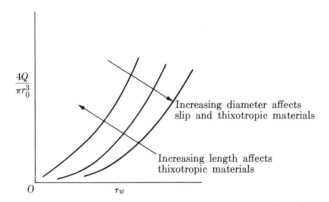

Fig. 15–29. *Effects of slip and thixotropic materials on the capillary shear diagram.*

Both slip at the wall and thixotropic behavior cause complications in the measurement of the capillary shear diagram. The effects are shown in Fig. 15–29 and can be explained as follows: At a constant $4Q/\pi r_0^3$, an increase in length or a decrease in r_0 (that is, a decrease in $U_{z,\text{ave}}$ to keep $4Q/\pi r_0^3$ constant) will increase the time a fluid element is in the capillary, and thus will increase the breakdown of a thixotropic substance. At constant τ_w, Eq. (15–86) shows that for a system with slip, $4Q/\pi r_0^3$ must decrease with increasing diameter, but there is no effect of length if steady state is obtained. To distinguish between thixotropic behavior and slip flow, it is most convenient to have a rotational viscometer available so that the change in viscosity with time can be observed at a given rate of shear.

If no time dependency exists, the noncoincidence of the curves of Fig. 15-29 can be attributed to slip. Turbulence effects can usually be established by calculation of the Reynolds number for each test pipe and the experimental observation that when one goes from laminar to turbulent flow the change in the calculated viscosity for a given pipe is quite large.

(v) **Heat effects.** Viscous heating effects in capillary flow are usually unimportant, since new fluid at reservoir temperature is continually being introduced and recycle is absent. This can be contrasted to the continuous testing of the total initial charge of material in rotational instruments. However, for very viscous materials (e.g., polymers and their solutions) the heating effect might be appreciable in all viscometers, including capillary tubes. Since such effects are to be avoided whenever possible, one must know when they are liable to be important. Bird [13] solved for the temperature distribution by using the momentum and energy equations for a power-law material. He calculated the distribution and the average and maximum temperature rises for an isothermal and an adiabatic wall. The numerical work was restricted to the case in which $n = 0.5$. Fredrickson [57] has reviewed the general approach used by Bird. In this analysis, he defines a dimensionless capillary length by

$$\zeta = \frac{2[(n+1)/(3n+1)](z/r_0)}{N_{\text{Pe}}}, \tag{15-87}$$

where

$$N_{\text{Pe}} = \frac{2r_0 U_{z,\text{ave}} \rho c_v}{k} = \frac{2Q\rho c_v}{\pi r_0 k},$$

and he defines a dimensionless temperature rise by

$$\theta = \frac{[(3n+1)/2n]^2 \Delta T k}{r_0^2 K (4Q/\pi r_0^3)^{n+1}}. \tag{15-88}$$

Figure 15-30 gives values of θ_{ave} and θ_{max} for the two cases as a function of ζ. The true temperature value probably lies somewhere between the two extremes. For a typical polyethylene melt, the calculated maximum rise was 7°C at a τ_w of 10^6 dynes/cm^2.

Gerrard and Philippoff [66] have shown that the average temperature rise is unimportant, as it is insensitive to changes that can occur because of gradients that exist across the flow stream. In the extreme case, a radial gradient greater than 150°F existed under conditions in which the bulk increase was less than 14°F. At a τ_w of 10^5 to 10^6, because of heating effects, there was as much as a fivefold decrease in apparent viscosity.

Toor [199] solved a number of similar problems, of which McKelvey [112] has given a brief review. In general, Toor included the effect of thermal expansion, but since this is always a cooling effect, Bird's analysis should still provide an upper limit to the temperature rise.

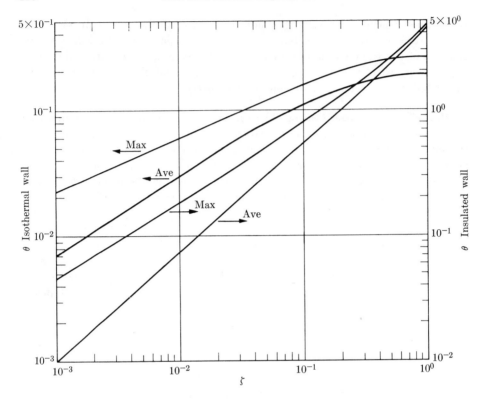

Fig. 15–30. Heat effects for the capillary system (after Bird).

EXAMPLE 15–6. Calculate the average and maximum temperature rises expected for the isothermal and adiabatic wall cases for the polyethylene melt discussed in Examples 15–4 and 15–5. Consider the highest rate of shear in the longest capillary for which $d_0 = 0.0191$ in. Assume for the calculation that $k = 0.0003$ cal/sec-cm-°C, and $c_v = 0.3$ cal/gm-°C. From the raw data, the flow rate was measured as 0.215 gm/min.

Answer. From the data given, N_{Pe} can be calculated as being 94. From Example 15–4, $n = 0.32$. From Fig. 15–28, the L/d_0 for the longest capillary is 30.3. From the information cited, ζ can be calculated from Eq. (15–87), and is 0.87. The corresponding values of θ can be read from Fig. 15–30. The actual temperature rise can be calculated from Eq. (15–88), with the proper conversion factors. Note that the term

$$K(4Q/\pi r_0^3)^{n+1} = \tau_w 4Q/\pi r_0^3,$$

which can be obtained from Example 15–4. Solving for the ΔT, one obtains $\Delta T = 16.4\,\theta$. The actual temperature rise estimates are: isothermal average, 3°C, isothermal maximum, 4°C, adiabatic average, 61°C, adiabatic maximum, 64°C.

One can make a crude estimate to establish which of the estimates might be more meaningful for the specific case under consideration. From the known geometry and heat capacity, about 0.034 cal would have to be removed during the transit of the volume of polymer in the capillary. This is equivalent to 0.26 cal/cm^2-sec. Even if a heat transfer coefficient of 100 Btu/hr-ft^2-°F could be obtained, a driving temperature difference of 19°C would be needed to transfer the heat away. Since this seems unlikely, the rise in the capillary would probably be closer to the higher value given above than to the lower.

(vi) Normal stress effects. For capillary flow (Fig. 15–25), the stress pattern (Eq. 15–59) becomes

$$\bar{\bar{\tau}}_{rz} = \bar{\bar{\tau}}_{zr} = -\mu_a(\mathrm{II})\,\mathbf{d}_{zr},$$
$$\mathbf{P}_{zz} - \mathbf{P}_{\theta\theta} = \bar{\bar{\tau}}_{zz} - \bar{\bar{\tau}}_{\theta\theta} = \bar{\bar{\tau}}_{zz} = \sigma_1(\mathrm{II}), \qquad (15\text{--}89)$$
$$\mathbf{P}_{rr} - \mathbf{P}_{\theta\theta} = \bar{\bar{\tau}}_{rr} - \bar{\bar{\tau}}_{\theta\theta} = \bar{\bar{\tau}}_{rr} = \sigma_2(\mathrm{II}),$$

where all stresses are a function of both r and z, that is, $\mathbf{P}_{zz}(r, z)$. It is by no means easy to evaluate the normal stress terms, and there is considerable disagreement on how measured variables are to be translated into meaningful results [64, 131, 174, 217]. The first capillary jet experiments reported were by Philippoff and Gaskins [62, 63, 148].

The term $\dot{\sigma}_2$ can be evaluated from a detailed knowledge of the conditions within the capillary tube. From the first equation of (15–66), we obtain

$$0 = -(\partial p/\partial r) - (\partial \bar{\bar{\tau}}_{rr}/\partial r) - \bar{\bar{\tau}}_{rr}/r - \bar{\bar{\tau}}_{\theta\theta}/r$$

or

$$\partial \mathbf{P}_{rr}/\partial r = -(\bar{\bar{\tau}}_{rr} - \bar{\bar{\tau}}_{\theta\theta})/r = -\sigma_2/r. \qquad (15\text{--}90)$$

Integration of this between 0 and r gives

$$\mathbf{P}_{rr}(r, z) - p(0, z) = -\int_0^r \sigma_2\,d\ln r, \qquad (15\text{--}91)$$

since by symmetry at the tube axis, $\bar{\bar{\tau}}_{rr}(0, z) = 0$. At the exit of the tube $z = L$, and at the wall $r = r_0$. Thus

$$\mathbf{P}_{rr}(r_0, L) = p(r_0, L) = p(0, L) - \int_0^{r_0} \sigma_2\,d\ln r. \qquad (15\text{--}92)$$

Differentiating with respect to $\ln \tau_w$ and using the linear variation of τ_w over the radius (15–67) gives

$$\frac{dp(r_0, L)}{d\ln \tau_w} = -\sigma_2(\mathbf{d}_w^2) + \frac{dp(0, L)}{d\ln \tau_w}. \qquad (15\text{--}93)$$

The expression $\sigma_2(\mathbf{d}_w^2)$ means that σ_2 will be a unique function of the square of the shear rate at the wall, $(\mathbf{d}_{rz})_w^2$. Sakiadis [172] assumed that the last term was zero because the system was open to the atmosphere; that is, at the center line, the stresses are zero by symmetry and only the pressure exists. This is taken as the ambient pressure p_0 and does not change:

$$\frac{dp(0, L)}{d \ln \tau_w} = \frac{dp_0}{d \ln \tau_w} = 0.$$

Sakiadis then obtained a nonzero σ_2. White [217] has critized Sakiadis' use of Eq. (15–93) for his data, so these results are open to question.

The term σ_1 can be evaluated from an integral momentum balance written for the jet issuing from a capillary tube. This takes the form

$$\int_0^{r_0} 2\pi r \rho U_z^2 \, dr - \int_0^{r_0} 2\pi r \mathsf{P}_{zz} \, dr + 2\pi\sigma(r_j - r_0) - \pi r_0^2 p_0 = \pi r_j^2 \rho U_j^2 = \rho Q^2/\pi r_j^2, \quad (15\text{–}94)$$

where the subscript j refers to the jet and p_0 is the ambient pressure, usually taken as zero, without loss of generality. The third term accounts for surface tension effects, and is probably small at high exit velocities [64]. This assumption is usually made in data evaluation [124, 127, 217]. It is easiest to assume a power law for U_z and differentiate Eq. (15–94) to obtain an expression for $\mathsf{P}_{zz}(r_0, L)$ in terms of the surface tension contribution and terms involving the jet contraction or expansion ratio [127]. An equivalent approach [172, 217] is to use Eq. (15–89) to give

$$\mathsf{P}_{zz}(r, z) - \mathsf{P}_{rr}(r, z) = \sigma_1 - \sigma_2,$$

then combine this with Eq. (15–91) to obtain

$$\mathsf{P}_{zz}(r, z) = p(0, z) - \int_0^r \sigma_2 \, d \ln r + \sigma_1 - \sigma_2. \quad (15\text{–}95)$$

At the wall, Eq. (15–95) becomes

$$\mathsf{P}_{zz}(r_0, L) = p(0, L) + \sigma_1(\mathsf{d}_w^2) - \sigma_2(\mathsf{d}_w^2) - \int_0^{r_0} \sigma_2 \, d \ln r, \quad (15\text{–}96)$$

or, if we use Eq. (15–92), we obtain

$$\mathsf{P}_{zz}(r_0, L) = p(r_0, L) + \sigma_1(\mathsf{d}_w^2) - \sigma_2(\mathsf{d}_w^2). \quad (15\text{–}97)$$

Sakiadis [172] in effect neglects $\mathsf{P}_{zz}(r_0, L)$ in Eq. (15–97) and calculates σ_1 from σ_2 obtained from Eq. (15–93) and the pressure $p(r_0, L)$. Because of existing questions of this measurement, White also questioned the use of Eq. (15–97) to interpret the data as measured.

Gavis and Middleman [64] have considered the source for the axial normal stress P_{zz}, concluding that there is, in addition to the internal normal stress, a profile relaxation stress and an external normal stress occurring outside the tube. The profile relaxation stress occurs in any fluid and is associated with the change in the velocity profile from that within the tube to the flat profile in the jet. The external normal stress develops in viscoelastic materials because of

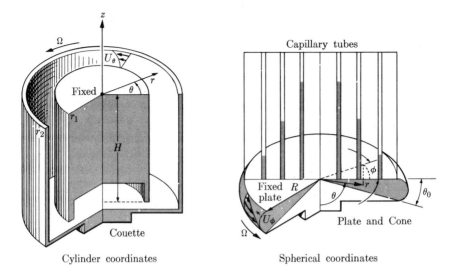

Fig. 15–31. Rotational Couette system and plate-and-cone system.

the change in rate of shear during the velocity profile relaxation (an elastic reaction to the change). Of course there is also the effect of surface tension or surface traction. Clearly, the simple jet is a complicated tool and measurements should be made with caution; it is not surprising that there are differences reported in the literature [67, 172, 203]. Savins [174], as still another alternative, has suggested the use of a pitot tube within the pipe to obtain a measure of

$$\mathsf{P}_{zz}(r, z) + \tfrac{1}{2}\rho U_z^2$$

and thus, by Eq. (15–95), to obtain information about σ_1 and σ_2.

(b) Rotational flow. Two designs in common use for viscometric measurements are the Couette (concentric cylinders) and the plate and cone. These can be designed to give a rate of shear that is essentially constant in the gap and can thus approximate the basic shear diagram directly. It must be emphasized, however, that to accomplish this a proper design is necessary. The two systems are shown in Fig. 15–31, and will be developed in parallel. For the Couette, we shall consider the case of the outer cylinder in rotation and the inner cylinder fixed. For the plate-and-cone design, the cone will be rotated and the plate fixed. In this unit there is a sequence of capillary tubes which can give the radial pressure gradient. This in turn can be used to establish normal stress effects. Once again it will be assumed that end effects and slip do not exist (these will be discussed in some detail in a later section). It will also be assumed that the rotational speed is below the point of laminar instability (Taylor-type vortices or turbulence). A number of other specialized rotational units have been built, among which are the rotating disk, rotating concentric spheres, double cone, conicylindrical, etc. Further reference to these more specific instruments can be found in the review by Oka [135].

The equations for the apparent viscosity (15–6) are:

Couette, $\quad \bar{\bar{\tau}}_{r\theta} = -\mu_a r \dfrac{d\omega}{dr} = -\mu_a \left[r \dfrac{d(U_\theta/r)}{dr} \right] = -\mu_a \left(\dfrac{dU_\theta}{dr} - \dfrac{U_\theta}{r} \right);$

Plate and cone, $\quad \bar{\bar{\tau}}_{\theta\phi} = -\mu_a \sin\theta \dfrac{d\omega}{d\theta} = -\mu_a \left[\sin\theta \dfrac{d(U_\phi/r \sin\theta)}{d\theta} \right].$
(15–98)

The stress tensor takes the form:

Couette, $\quad \mathbf{P} = \begin{pmatrix} p + \bar{\bar{\tau}}_{rr} & \bar{\bar{\tau}}_{r\theta} & 0 \\ \bar{\bar{\tau}}_{r\theta} & p + \bar{\bar{\tau}}_{\theta\theta} & 0 \\ 0 & 0 & p + \bar{\bar{\tau}}_{zz} \end{pmatrix},$ (15–99a)

where $\bar{\bar{\tau}}_{zz}$ should be zero for no vertical flow, and

Plate and cone, $\quad \mathbf{P} = \begin{pmatrix} p + \bar{\bar{\tau}}_{rr} & 0 & 0 \\ 0 & p + \bar{\bar{\tau}}_{\theta\theta} & \bar{\bar{\tau}}_{\theta\phi} \\ 0 & \bar{\bar{\tau}}_{\theta\phi} & p + \bar{\bar{\tau}}_{\phi\phi} \end{pmatrix},$ (15–99b)

where $\bar{\bar{\tau}}_{rr}$ should be zero for no radial flow. The equation of motion (4–1) reduces to (see Eq. 7–30):

$$-\rho U_\theta^2/r = -(\partial p/\partial r) - (1/r)(\partial r \bar{\bar{\tau}}_{rr}/\partial r) + \bar{\bar{\tau}}_{\theta\theta}/r,$$
$$\partial(r^2 \bar{\bar{\tau}}_{r\theta})/\partial r = 0,$$ (15–100)
$$\partial p/\partial z = -\rho g,$$

for Couette flow, and to

$$-\rho \dfrac{U_\phi^2}{r} = -\dfrac{\partial p}{\partial r} - \dfrac{\partial \bar{\bar{\tau}}_{rr}}{\partial r} - \dfrac{2\bar{\bar{\tau}}_{rr}}{r} + \dfrac{\bar{\bar{\tau}}_{\theta\theta} + \bar{\bar{\tau}}_{\phi\phi}}{r},$$

$$-\rho (\cot\theta) \dfrac{U_\phi^2}{r} = -\dfrac{1}{r} \dfrac{\partial p}{\partial \theta} - \dfrac{1}{r \sin\theta} \dfrac{\partial(\bar{\bar{\tau}}_{\theta\theta} \sin\theta)}{\partial \theta} + \dfrac{1}{r} (\cot\theta) \bar{\bar{\tau}}_{\phi\phi},$$ (15–101)

$$0 = \dfrac{1}{r} \dfrac{\partial \bar{\bar{\tau}}_{\theta\phi}}{\partial \theta} + \dfrac{2}{r} (\cot\theta) \bar{\bar{\tau}}_{\theta\phi},$$

for the plate-and-cone unit with no external forces acting. From the second equation of (15–100) and the last of (15–101), or from a simple force balance, one obtains for the stress:

Couette, $\quad \bar{\bar{\tau}}_{r\theta} = -T/2\pi r^2,$

Plate and cone, $\quad \bar{\bar{\tau}}_{\theta\phi} = 3T/2R \sin^2\theta,$
(15–102)

where in the first case T is the torque per unit height and in the second case T is the torque per unit area. If the angle of the gap (θ_0) for the plate-and-cone unit is small, the equation can be simplified, since $\sin\theta \cong 1$.

The boundary conditions for the two cases are:

Couette, $U_\theta = 0$ at $r = r_1$ and $U_\theta = r_2 \Omega$ at $r = r_2$, (15-103)

Plate and cone,

$U_\phi = 0$ at $\theta = \frac{1}{2}\pi$ and $U_\phi = r\Omega \cos \theta_0$ at $\theta = \frac{1}{2}\pi - \theta_0$.

The velocity distribution for the Couette flow is

$$\frac{U_\theta}{r} = \int_{r_1}^{r} \frac{d(U_\theta/r)}{dr} dr = \int_{\bar{\bar{\tau}}_{r\theta}}^{\tau_1} r \frac{d(U_\theta/r)}{dr} \frac{1}{2\bar{\bar{\tau}}_{r\theta}} d\bar{\bar{\tau}}_{r\theta}, \quad (15\text{-}104)$$

and for the flow in the plate-and-cone unit,

$$\frac{U_\phi}{r \sin \theta} = \int_{(1/2)\pi}^{\theta} \frac{d(U_\phi/r \sin \theta)}{d\theta} d\theta = -\int_{\bar{\bar{\tau}}_{\theta\phi}}^{\tau_{\text{plate}}} \sin \theta \frac{d(U_\phi/r \sin \theta)}{d\theta} \frac{1}{2\bar{\bar{\tau}}_{\theta\phi} \cos \theta} d\bar{\bar{\tau}}_{\theta\phi}. \quad (15\text{-}105)$$

If the inner cylinder is rotated, the problem remains the same except for the change in the boundary conditions in Eq. (15-103). In this case it is more convenient to express Eq. (15-104) in terms of the outer radius.

For a Newtonian fluid, Eq. (15-98) ($\mu = \mu_a$) can be substituted into Eqs. (15-104) and (15-105) and integrated. For Eq. (15-104),

$$\frac{U_\theta}{r} = -\int_{\bar{\bar{\tau}}_{r\theta}}^{\tau_1} \frac{1}{2\mu} d\bar{\bar{\tau}}_{r\theta} = -\frac{1}{2\mu}(\tau_1 - \bar{\bar{\tau}}_{r\theta}) = \frac{T}{4\pi\mu}\left(\frac{1}{r_1^2} - \frac{1}{r^2}\right).$$

Combining this equation, written at r_2, with the boundary condition, Eq. (15-103), gives

$$\Omega = \frac{T}{4\pi\mu}\left(\frac{1}{r_1^2} - \frac{1}{r_2^2}\right) \quad (15\text{-}106)$$

or

$$\tau_1 = -\frac{T}{2\pi r_1^2} = -\mu \frac{2\Omega r_2^2}{r_2^2 - r_1^2}, \quad (15\text{-}107)$$

which are forms of the Margules equation for the Newtonian viscosity in a Couette instrument. For the plate-and-cone design,

$$\frac{U_\phi}{r \sin \theta} = -\int_{(1/2)\pi}^{\theta} \frac{\bar{\bar{\tau}}_{\theta\phi}}{\mu \sin \theta} d\theta = -\frac{3T}{2R\mu}\int_{(1/2)\pi}^{\theta} \frac{d\theta}{\sin^3 \theta} = \frac{3T}{4R\mu}\left[\frac{\cos \theta}{\sin^2 \theta} - \ln(\tan \tfrac{1}{2}\theta)\right]$$

or

$$\frac{U_\phi}{r} = \frac{3T}{4R\mu}[\cot \theta - \ln(\tan \tfrac{1}{2}\theta) \sin \theta]. \quad (15\text{-}108)$$

Combining this, written at $\theta = \frac{1}{2}\pi - \theta_0$, with the boundary condition (15-103), we have

$$\Omega \cos \theta_0 = \frac{3T}{4R\mu}\left[\tan \theta_0 - \cos \theta_0 \ln\left(\frac{1 - \tan \tfrac{1}{2}\theta_0}{1 + \tan \tfrac{1}{2}\theta_0}\right)\right]. \quad (15\text{-}109)$$

For θ_0 very small, we have

$$\Omega = \frac{3T}{4R\mu}(\theta_0 + \theta_0) = \frac{3T\theta_0}{2R\mu} \qquad (15\text{--}110)$$

or

$$\tau_{\text{plate}} \simeq \frac{3T}{2R} \simeq \mu\frac{\Omega}{\theta_0}. \qquad (15\text{--}111)$$

Equation (15–109) or the simplified form (15–110) allows measurement of the viscosity from the geometry, angular velocity, and the torque per unit area. Equivalent equations were first derived by Braun [20].

Equations (15–107) and (15–111) can be used to define rotational shear diagrams, just as was done for the capillary tube case. For non-Newtonian materials, these in turn define rotational apparent viscosities for the Couette and plate-and-cone designs, respectively. For the Couette,

$$\tau_1 = -\mu_{\text{ar}}\left[\frac{2\Omega r_2^2}{(r_2^2 - r_1^2)}\right] = -\mu_a\left(\frac{r\,d\omega}{dr}\right)_1, \qquad (15\text{--}112)$$

where μ_{ar} has been used to distinguish this from μ_a, defined by Eq. (15–98) and given in (15–112). We wish to establish a relation between the pseudo-shear terms and the true shear rate at the inner cylinder or at the plate, that is, a relation between

Couette, $\quad \dfrac{2\Omega r_2^2}{r_2^2 - r_1^2} \quad$ and $\quad \left(\dfrac{r\,d\omega}{dr}\right)_1 = \left[r\dfrac{d(U_\theta/r)}{dr}\right]_1,$

and

plate and cone, $\quad \dfrac{\Omega}{\theta_0} \quad$ and $\quad \left(\dfrac{d\omega}{d\theta}\right)_{\text{plate}} = \left[\dfrac{d(U_\phi/r)}{d\theta}\right]_{\text{plate}},$

which would parallel the Weissenberg-Rabinowitsch-Mooney relation (15–74). For Couette flow with $\bar\tau_{r\theta} = \tau_2$, Eq. (15–104) can be differentiated (2–54) to give

$$2\tau_1\frac{d\Omega}{d\tau_1} = \left[r\frac{d(U_\theta/r)}{dr}\right]_1 - \left[r\frac{d(U_\theta/r)}{dr}\right]_2, \qquad (15\text{--}113)$$

which was first obtained by Mooney [132]. The terms on the left can be evaluated from a plot of Ω versus τ_1. But now the difficulty becomes apparent in the fact that $(r\,d\omega/dr)_1$ cannot be determined; instead we can determine only the difference between the true rate of shear at the inner and outer cylinders. For the special case of a rotating bob in an infinite medium, there is no motion away from the bob, so $(r\,d\omega/dr)_2 = 0$ and Eq. (15–113) can be used rigorously. Savins et al. [176] have discussed the various approaches taken in obtaining an approximate solution to Eq. (15–113). A similar solution exists when one starts with Eq. (15–105) for the plate-and-cone system.

(i) The power law. Equation (15–113) can be solved rigorously if a power law is assumed [22]:

$$\bar{\tau}_{r\theta} = -K\left[r\frac{d(U_\theta/r)}{dr}\right]^n, \tag{15-114}$$

where K and n have to be assumed constant. Combining this with the expression for shear stress (15–102) gives

$$\frac{T}{2\pi r^2} = K\left[r\frac{d(U_\theta/r)}{dr}\right]^n. \tag{15-115}$$

Integration gives

$$\frac{U_\theta}{r} = -\left(\frac{T}{2\pi K}\right)^{1/n}\frac{n}{2}r^{-2/n} + C. \tag{15-116}$$

By evaluating the constant at $r = r_1$ (Eq. 15–103), using the second boundary condition of (15–103), rearranging to the form of the shear rate, combining with Eq. (15–115) written at the inner cylinder, and finally rearranging once again, we get an equation somewhat analogous to (15–78):

$$\left[r\frac{d(U_\theta/r)}{dr}\right]_1 = \left\{\frac{1-(r_1/r_2)^2}{n[1-(r_1/r_2)^{2/n}]}\right\}\left[\frac{2\Omega\, r_2^2}{r_2^2 - r_1^2}\right]. \tag{15-117}$$

This equation, when contrasted to Eq. (15–78), can be seen to have a geometric dependency. As in Eq. (15–78), for $n = 1$, the correcting term becomes unity. For $n \neq 1$, the geometrical dependence can be used to advantage. The limit as r_1 approaches r_2 is

$$\lim_{r_1 \to r_2}\left\{\frac{1-(r_1/r_2)^2}{n[1-(r_1/r_2)^{2/n}]}\right\} = \lim_{r_1 \to r_2}\left[\frac{-2(r_1/r_2)}{-n(2/n)(r_1/r_2)^{(2/n)-1}}\right] = 1. \tag{15-118}$$

Thus as r_1 approaches r_2, the value of $2\Omega r_2^2/(r_2^2 - r_1^2)$ approaches the true shear rate at the inner cylinder for any value of n, and clearly one can obtain the basic shear diagram directly.

Corrections are necessary when the rotational viscometer has not been designed to take advantage of a narrow gap. Usually the variation in the shear rate across the gap is such that n can be considered as a constant over this limited range. According to Eq. (15–117), the true shear rate is directly proportional to the pseudo-shear rate, as given by the second bracketed term. Thus a log-log plot of the terms of Eq. (15–112) would still give the correct value of n, even though the graph would not be exactly the basic shear diagram. The necessary correction can now be made with the known n and Eq. (15–117). If n is constant in the gap, the method is rigorous.

For the special case of a bob in an infinite fluid, Eq. (15–117) reduces to

$$\left[r\frac{d(U_\theta/r)}{dr}\right]_1 = \frac{2\Omega}{n}. \tag{15-119}$$

Actually this equation is rigorous even if n is not constant; we can show this as follows: Define a power law as $\bar{\tau}_{r\theta} = -K''(2\Omega)^{n''}$, where n'' and K'' need not be constant. The term n'' is given by

$$n'' = d \ln \bar{\tau}_{r\theta}/d \ln 2\Omega. \tag{15-120}$$

Equation (15-113) can be rearranged to give

$$\left[r\frac{d(U_\theta/r)}{dr}\right]_1 = \frac{d \ln 2\Omega}{d \ln \tau_1} 2\Omega, \tag{15-121}$$

since for this case $(r\, d\omega/dr)_2 = 0$. Combining Eqs. (15-120) and (15-121) gives the form equivalent to (15-119), but now n'' is not restricted to a constant.

There have been a number of alternative methods proposed for obtaining the basic shear diagram if the gap is not narrow enough to allow it to be obtained directly. Krieger and Maron [93] and Krieger and Elrod [92] have suggested methods which are summarized by Metzner [124] and Savins et al. [176]. In these procedures, Eq. (15-113) is expanded in terms of an infinite series, and to obtain simpler forms one usually assumes a constant n. Thus for this case these methods can be only approximate when compared with Eq. (15-117). For example, Metzner has cited the values obtained by Calderbank and Moo-Young [31], who obtained them by calculating enough terms in the infinite series to obtain a one percent accuracy for constant n. Equation (15-117) gives more accurate values directly and thus avoids the need for using the series expansion. Since it is unusual for n not to be constant over the relatively small range of shear rates in a gap, Eq. (15-117) is recommended for correcting to the basic shear diagram. For $n = 0.25$ and a 10% gap, the change in shear stress is only a factor of 2.14. For higher n and narrower gaps, the difference is even less.

A similar analysis is possible for the plate-and-cone system; however, an analytical solution is possible only for certain specific values of n. Combining Eq. (15-102) with a power law based on Eq. (15-98) gives

$$\frac{3T}{2R \sin^2 \theta} = K\left[-\sin\theta \frac{d(U_\phi/r\sin\theta)}{d\theta}\right]^n.$$

Separation of variables and the assumption of n such that the integration can be made ($n = \frac{1}{2}, \frac{1}{3}$, etc.), followed by the same simplification as that used to convert Eq. (15-108) to (15-109), gives for $n = \frac{1}{2}$,

$$\Omega \cos\theta_0 = \left(\frac{3T}{4KR}\right)^2 \left\{\frac{\tan\theta_0}{\cos^2\theta_0} + \frac{3}{2}\left[\tan\theta_0 - \cos\theta_0 \ln\left(\frac{1 - \tan\frac{1}{2}\theta_0}{1 + \tan\frac{1}{2}\theta_0}\right)\right]\right\}, \tag{15-122}$$

which in turn, for θ_0 very small, gives

$$\Omega = (3T/4KR)^2[\theta_0 + \tfrac{3}{2}(\theta_0 + \theta_0)]$$

or

$$\tau_{\text{plate}} = 3T/2R \cong K(\Omega/\theta_0)^{1/2}. \tag{15-123}$$

When compared with Eq. (15-111), the above equations show that if the angle θ_0 is small, the non-Newtonian nature is immaterial, since the power law is recovered with the given n. The largest θ_0 normally used is 6°, and from Eq. (15-122) the error is less than 0.1 per cent for n of $\frac{1}{2}$. Thus the use of Eq. (15-123) is justified. However, as long as n can be considered constant, corrections can be made for larger cone angles by equations such as (15-122). With larger angles caution is necessary to avoid end and heating effects.

(ii) **Bingham ideal plastic.** Plastic flow occurs in the Couette instrument if $|\tau_1| > |\tau_0|$. If $|\tau_1| > |\tau_0| > |\tau_2|$, flow will take place in part of the material, and remain as a plug in the outer area. If $|\tau_2| > |\tau_0|$, plastic flow will occur everywhere. Equation (15-104), combined with the Bingham ideal plastic equation of state

$$-r\frac{d(U_\theta/r)}{dr} = 0, \qquad 0 \leqslant |\bar{\tau}_{r\theta}| \leqslant |\tau_0|,$$
$$-r\frac{d(U_\theta/r)}{dr} = (\bar{\tau}_{r\theta} - \tau_0)/\mu', \qquad |\bar{\tau}_{r\theta}| > |\tau_0|, \qquad (15\text{-}124)$$

gives

$$\frac{U_\theta}{r} = \frac{T}{4\pi\mu'}\left(\frac{1}{r_1^2} - \frac{1}{r^2}\right) + \frac{\tau_0}{\mu'} \ln \frac{r}{r_1}, \qquad (15\text{-}125)$$

where $r_1 \leqslant r \leqslant r_y$. The term r_y corresponds to τ_0, and is

$$\tau_0 = -T/2\pi r_y^2. \qquad (15\text{-}126)$$

Note that both $\bar{\tau}_{r\theta}$ and τ_0 are negative in this system. For the outer area where $r \geqslant r_y$,

$$U_\theta/r = (U_\theta)_{r_y}/r_y = \text{const.} \qquad (15\text{-}127)$$

The relation between Ω and T can be obtained from Eq. (15-125). For example, if the entire area is in flow ($r_2 \leqslant r_y$), then

$$\Omega = \frac{T}{4\pi\mu'}\left(\frac{1}{r_1^2} - \frac{1}{r_2^2}\right) + \frac{\tau_0}{\mu'} \ln \frac{r_2}{r_1}. \qquad (15\text{-}128)$$

This is the *Reiner-Rivlin equation* [158], which can be used to obtain the plastic viscosity from rotational data. The equation is the Couette counterpart to the Buckingham-Reiner equation (15-81) for capillary flow.

Equation (15-128) can be rearranged into a form similar to Eq. (15-117) by using Eqs. (15-102) and (15-124) written at the inner cylinder [22]. This is

$$\left[r\frac{d(U_\theta/r)}{dr}\right]_1 = \frac{2\Omega r_2^2}{r_2^2 - r_1^2} + \frac{\tau_0}{\mu'}\left[1 - \frac{r_2^2 \ln(r_2/r_1)^2}{r_2^2 - r_1^2}\right]. \qquad (15\text{-}129)$$

In the limit of r_1 approaching r_2, the last term goes to zero, and the basic shear diagram can be obtained directly. Equations (15-128) and (15-129) are both for $|\tau_2| > |\tau_0|$. Equation (15-128) is the equation for the curve shown in Fig. 15-19 when Ω or $2\Omega r_2^2/(r_2^2 - r_1^2)$ is plotted against $\bar{\bar{\tau}}_{r\theta}$. The similarity to the role of Eq. (15–81) in Figs. 15–20 and 15–26 is apparent. Equation (15–129) was derived earlier by Doherty and Hurd [44] and reported by McKennell [114].

(iii) End, wall, and heating effects. End effects in various types of rotational flow have been considered in some detail by Oka [135]. In some specially designed systems, the end effect can be estimated if the material is not too far removed from Newtonian characteristics. In other systems the design is such that the end effect is quite small. For the Couette system, this is generally accomplished by making the inner cylinder hollow and open on the bottom so that air will be trapped, as shown in Fig. 15–31. The air–fluid interface that replaces the metal–fluid interface offers considerably less resistance to flow, and thus contributes little to the measured torque. When corrections are necessary, or when very accurate results are needed, the equation for the torque at the inner cylinder wall (15–102) can be modified to

$$\tau_1 = T/2\pi r_1^2 = (M/H)/2\pi r_1^2, \qquad (15\text{–}130)$$

where M is the measured torque for a submersion depth of H, and T is the torque per unit of height. From this equation one sees that $(1/2\pi r_1^2)$ times the slope of a plot of M versus H gives the corrected value. This procedure is exactly analogous to that used for the pipe correction. A similar technique was used by McKennell [113] for the plate-and-cone system; he showed that the measured torque was proportional to R^3 for a Newtonian material. According to

$$\tau_{\text{plate}} = 3T/2R = 3M/2\pi R^3, \qquad (15\text{–}131)$$

this would be true only if the end effects are negligible over the range of radii used.

In a manner completely analogous to that of the development of capillary flow, Oldroyd [139] has presented the required modification of Eq. (15–104) to account for the slip in a rotational viscometer. In this case there are two walls to be considered; the slip at these may not be the same, since it can depend on the local value of the stress and may not be the same at both the inner and outer walls. In general, the experimental technique requires the use of a Couette system in which the ratio of inner to outer radii can be varied. Wall slip is an effect that is not necessarily time-independent, so one must be careful not to confuse wall slip with thixotropic effects. Time effects as long as one-half hour have been observed [228].

Of extreme importance in all viscometric systems are the viscous heating effects that can invalidate what might appear to be excellent measurements. For the Couette instrument, Fredrickson [57] has presented a simple analysis. A more general solution is possible if one assumes that the power law (15–114) is valid. For an incompressible fluid, with no external forces acting, the energy equation (3–29) at steady state is

$$0 = -(\nabla \cdot q) - (\bar{\bar{\tau}} : \nabla U). \qquad (15\text{–}132)$$

In cylindrical coordinates [(4–17) and (2–47)], the last term becomes

$$(\bar{\bar{\tau}} : \nabla U) = \bar{\tau}_{r\theta} r \frac{\partial \omega}{\partial r} = \tau_{r\theta} r \frac{\partial (U_\theta/r)}{\partial r}. \tag{15-133}$$

Combining Eqs. (3–45) and (15–133) into (15–132) gives

$$\bar{\tau}_{r\theta} r \frac{d\omega}{dr} = k(\nabla \cdot \nabla T) = k \frac{1}{r} \frac{d}{dr}\left(r \frac{dT}{dr}\right),$$

where T now stands for temperature and not torque per unit height. Expressing the shear stress by Eq. (15–102) (using M/H for the torque term to avoid confusion with temperature), and the shear rate by the power law (15–115) gives

$$-\frac{M/H}{2\pi k}\left(\frac{M/H}{2\pi K}\right)^{1/n} r^{-(n+2)/n}\, dr = d\left(r \frac{dT}{dr}\right).$$

The boundary condition can be taken as

$$T = T_0 \quad \text{at} \quad r = r_2 \quad \text{and} \quad dT/dr = 0 \quad \text{at} \quad r = r_1,$$

which should be valid for cooling of the cup and no transfer at the bob. When we integrate and use the second boundary condition, we get

$$-\frac{M/H}{2\pi k}\left(\frac{M/H}{2\pi K}\right)^{1/n}\left(-\frac{n}{2}\right)(r_1^{-2/n} - r^{-2/n}) = r\frac{dT}{dr}.$$

Separating variables, integrating again, and using the first boundary condition, we get

$$\frac{M/H}{2\pi k}\left(\frac{M/H}{2\pi K}\right)^{1/n}\left(\frac{n}{2}\right)^2\left(r_2^{-2/n} - r^{-2/n} + \frac{2}{n} r_1^{-2/n} \ln \frac{r_2}{r}\right) = T - T_0.$$

We rearrange the foregoing into a more convenient form by multiplying top and bottom by $r^{2+(2/n)}$ and by using Eq. (15–102). That gives us

$$\frac{\bar{\tau}_{r\theta}}{k}\left(\frac{\bar{\tau}_{r\theta}}{K}\right)^{1/n}\left(\frac{n}{2}\right)^2\left[r^2\left(\frac{r}{r_2}\right)^{2/n} - r^2 + \frac{2}{n} r^2\left(\frac{r}{r_1}\right)^{2/n} \ln \frac{r_2}{r}\right] = T - T_0. \tag{15-134}$$

The maximum temperature will occur at r_1, giving

$$\frac{\mu_a}{k}\left(r\frac{d\omega}{dr}\right)_1^2 r_1^2 \left(\frac{n}{2}\right)^2\left[\left(\frac{r_1}{r_2}\right)^{2/n} - 1 + \frac{2}{n}\ln\frac{r_2}{r_1}\right] = T_{\max} - T_0, \tag{15-135}$$

where $\tau_1(\tau_1/K)^{1/n} = \mu_a(r\, d\omega/dr)_1^2$. We can simplify further by combining Eqs. (15–117) and (15–135) to get

$$\left(\frac{\mu_a \Omega^2}{k}\right) r_1^2 \left\{\frac{(r_1/r_2)^{2/n} - 1 + (2/n)\ln(r_1/r_2)}{[1 - (r_1/r_2)^{2/n}]^2}\right\} = T_{\max} - T_0. \tag{15-136}$$

As r_1 approaches r_2, this equation reduces to

$$\mu_a \Omega^2 r_1^2 / 2k = T_{\max} - T_0, \tag{15-137}$$

and is independent of n. However, it must be recalled that k and K were assumed constant during the integration; thus the analysis can be valid only for small values of $T_{\max} - T_0$, say of the order of a degree or less. A more sophisticated analysis would include the temperature dependence of the viscosity constant K, possibly

$$K/K_0 = \exp\left[-B(T - T_0)/T_0\right], \tag{15-138}$$

and might also consider the temperature dependence of k. Turian and Bird [205] have made such an analysis for the plate-and-cone system and a Newtonian fluid. Bird and Turian [16] considered the constant-property case (as we have done here) by a variational method and arrived at an upper limit for any fluid:

$$T_{\max} - T_0 = \frac{\Omega R^2 \theta_0 \tau_{\text{plate}}}{8k} = \frac{3T\Omega R\theta_0}{16k} = \frac{3M\Omega\theta_0}{16\pi kR}, \tag{15-139}$$

where the unsubscripted T is the torque per unit area and $T_{\max}$ and T_0 are the maximum and reference temperatures, respectively. The analysis of Turian and Bird [205] can be used to make approximate corrections to data in which heating effects occur. These authors have considered two cases: one in which both the plate and the cone are maintained at a constant temperature by cooling, and another in which the plate is cooled and the cone insulated. For a small cone angle, the constant shear rate in the gap gives, in effect, a constant apparent viscosity which can be considered as Newtonian, and thus the analysis can be applied at each rate of shear to be considered. Correction factors are given as a function of BN_{Br}, where B is from Eq. (15–138) and

$$N_{\text{Br}} = \text{Brinkman number} = \mu_a \Omega^2 R^2 / IkT_0, \tag{15-140}$$

in which μ_a is the apparent viscosity as measured. The desired apparent viscosity at T_0 (i.e., if there is no heating effect) is

$$\mu_{a0} = \mu_a / I. \tag{15-141}$$

The correction involves a simple trial and error solution. No corrections are needed for $BN_{\text{Br}} \leqslant 0.05$. Some other representative values are shown below.

BN_{Br}	$I_{\text{case 1}}$	$I_{\text{case 2}}$
0.05	1.00	0.99
0.1	1.00	0.98
0.5	0.97	0.91
1.0	0.95	0.83
5.0	0.81	0.55
10.0	0.70	0.38

McKennell [113, 114] has made a rough comparison between the heating effects in a plate-and-cone unit and a typical Couette system.

EXAMPLE 15–7. Calculate the expected temperature rise for the polyethylene melt of Examples 15–4 and 15–6, if it were measured in a plate-and-cone system at the highest shear rate. Consider a $\frac{1}{4}°$ gap and a 1 cm radius.

Answer. In Example 15–4, the true shear rate of 3060 sec^{-1} was calculated for the stress of 16.40×10^5 dynes/cm^2. Since the gap is small, Eq. (15–111) or (15–123) can be used to calculate the angular velocity as the true shear rate times the gap angle. If one is to use Eq. (15–139), this velocity must be known; all other values are given in this or the preceding examples. Thus the upper limit for the temperature rise can easily be calculated as slightly less than 1°C. If one is to use Eqs. (15–140) and (15–141) to estimate the temperature rise, one must know the value of B. Let us assume a value of 60 for this calculation. As a first try, we use the apparent viscosity of 535 poises given in Example 15–4, and assume that $I = 1$. The temperature of 190°C was also given in Example 15–4. With these and the other values already given, we can calculate the Brinkman number (15–140) as 0.0164. With the assumed value for B, we can calculate $BN_{Br} = 1.0$. Thus $I = 0.95$ for case 1 and 0.83 for case 2, instead of the assumed value of 1. A second trial would increase N_{Br} and cause a slight further decrease in I. However, since the value of B was only a guess, further trial and error will not be attempted. The corresponding temperature rises can be calculated from Eq. (15–138) and the assumed value of B, the estimated values for I, and the reference temperature of 190°C. For case 1, this is 0.4°C, and for case 2 the rise is 1.5°C. Considering the assumptions necessary to arrive at an answer, these are not out of line with the estimate from Eq. (15–139). From Eq. (15–139), an increase in the gap size to $\frac{1}{2}°$ would increase the temperature rise to about 4°C; a 1° gap would give about 16°C, and a 2° gap would give 64°C.

EXAMPLE 15–8. Repeat Example 15–7 for a Couette system with $r_1 = 1$ cm, and a gap of 5%.

Answer. The angular velocity can be calculated from the geometry, the true shear rate cited in the previous problem, and Eq. (15–117). The equation gives $2\Omega r_2^2/(r_2^2 - r_1^2) = 2870$ sec^{-1}, or $\Omega = 133.5$ rad/sec. The apparent viscosity was determined in Example 15–4 as 535 poises. All other terms of Eq. (15–137) are known, so that the maximum temperature rise can be calculated as 38°C. If we use the more accurate equation (15–136), the geometrical term with $n = 0.32$ reduces the temperature rise estimate to 34°C. A comparison of Examples 15–6, 15–7, and 15–8 shows that the small-angle plate-and-cone system would probably have the smallest temperature rise, and the capillary tube the highest. The Couette lies between, and is about the same as a 1° to 2° gap for the plate-and-cone.

The actual data used were obtained on a capillary-flow system, and for comparison purposes the same conditions were used for the plate-and-cone and Couette systems. Such data could not actually have been obtained in these two systems, since normal stress effects would tend to cause the polyethylene melt to crawl out of the gap under these high shear conditions.

(iv) Normal stress effects. For rotational flow, the shear stress pattern (15–99) becomes, for the Couette* system,

$$\bar{\bar{\tau}}_{r\theta} = \bar{\bar{\tau}}_{\theta r} = -\mu_a(\mathrm{II})\,\mathbf{d}_{\theta r},$$

$$\mathbf{P}_{\theta\theta} - \mathbf{P}_{zz} = \bar{\bar{\tau}}_{\theta\theta} - \bar{\bar{\tau}}_{zz} = \bar{\bar{\tau}}_{\theta\theta} = \sigma_1(\mathrm{II}),$$

$$\mathbf{P}_{rr} - \mathbf{P}_{zz} = \bar{\bar{\tau}}_{rr} - \bar{\bar{\tau}}_{zz} = \bar{\bar{\tau}}_{rr} = \sigma_2(\mathrm{II}).$$

For the plate-and-cone system, it becomes

$$\bar{\bar{\tau}}_{\phi\theta} = \bar{\bar{\tau}}_{\theta\phi} = -\mu_a(\mathrm{II})\,\mathbf{d}_{\phi\theta},$$

$$\mathbf{P}_{\phi\phi} - \mathbf{P}_{rr} = \bar{\bar{\tau}}_{\phi\phi} - \bar{\bar{\tau}}_{rr} = \bar{\bar{\tau}}_{\phi\phi} = \sigma_1(\mathrm{II}),$$

$$\mathbf{P}_{\theta\theta} - \mathbf{P}_{rr} = \bar{\bar{\tau}}_{\theta\theta} - \bar{\bar{\tau}}_{rr} = \bar{\bar{\tau}}_{\theta\theta} = \sigma_2(\mathrm{II}).$$

For Couette flow: The Couette shear-stress pattern above, together with the first equation of (15–100) (neglecting its velocity-squared term), gives

$$\partial \mathbf{P}_{rr}/\partial \ln r = \sigma_2 - \sigma_1. \tag{15–142}$$

For the plate-and-cone instrument: The plate-and-cone shear-stress pattern above, together with the first equation of (15–101) (neglecting its velocity-squared term), gives

$$\partial \mathbf{P}_{rr}/\partial \ln r = \sigma_1 + \sigma_2. \tag{15–143}$$

Roberts [167] and Jobling and Roberts [88] have used the plate-and-cone system to show that $\mathbf{P}_{rr}(R) = p(R) = p_0$, where p_0 is the ambient pressure. The term $\mathbf{P}_{\theta\theta}$ is given by

$$\mathbf{P}_{\theta\theta} = p_0 + \rho g\, h(r), \tag{15–144}$$

where $h(r)$ is the height of fluid rise in the manometers shown in Fig. 15 31. Experimentally it was found that for many materials the rise at the rim was zero, which from the foregoing equations shows that $\mathbf{P}_{rr} = \mathbf{P}_{\theta\theta} = p_0$, or that $\sigma_2 = 0$ at R. Finding that $h(R) = 0$ when R is varied, and since σ_2 depends only on the shear rate, which in this constant-shear device would be constant, it was concluded that $\sigma_2 = 0$ everywhere. Under these conditions, Eq. (15–143) becomes

$$\partial \mathbf{P}_{\theta\theta}/\partial \ln r = \partial \mathbf{P}_{rr}/\partial \ln r = \sigma_1. \tag{15–145}$$

---

*In the alternative convention sometimes used, the relations

$$\bar{\bar{\tau}}_{11} - \bar{\bar{\tau}}_{33} = \sigma_1(\mathrm{II}) \quad \text{and} \quad \bar{\bar{\tau}}_{22} - \bar{\bar{\tau}}_{33} = \sigma_2(\mathrm{II})$$

are maintained. Thus, for Couette flow, the coordinates have been taken as (θ, r, z), and for the plate-and-cone system, as (ϕ, θ, r). This is equivalent to the convention of taking 1 as the direction of the streamline, 2 as perpendicular to the shearing plane, and 3 as perpendicular to both of these.

It was further found experimentally that the rise in the manometer tubes was logarithmic, and so Eq. (15-144) could be written as $\mathsf{P}_{\theta\theta} = p_0 + C \ln (R/r)$. Equation (15-145) and the above result combine to give

$$\partial \mathsf{P}_{\theta\theta}/\partial \ln r = -C(\text{II}) = \sigma_1(\text{II}). \tag{15-146}$$

It was also found that for low values of $\mathsf{d}_{\phi\theta}$, σ_1 tended to vary with $\mathsf{d}_{\phi\theta}^2$, which is in accord with the discussion of constitutive equations. However, at somewhat higher rates of shear, the variation was between the first and second power.

The above analysis can be criticized for several reasons [97]: The inertial term has been neglected, which restricts the analysis to low rates of shear at which the corrections will be small, and there is an implied idealization of the free surface. For $\mathsf{P}_{rr} = p_0$, one must assume an interface which is a spherical sector with no surface-tension effects, secondary motions, or momentum transfer across it. It is expected that the most serious of these effects will be secondary motions which take place when there is contact of the plate-and-cone surfaces with the fluid interface. Here large local effects could cause gross errors in manometer readings, although these local effects would scarcely affect integrated values averaged over the entire plate area. One might also question the assumption of $\sigma_2 = 0$ everywhere, since its measurement was only at this boundary at which the error would be greatest. In any event, Eq. (15-143) is still valid and Eq. (15-145) still results, with the modification that the right-hand side is $\sigma_1 + \sigma_2$. The slope of the semilogarithmic plot gives $\sigma_1 + \sigma_2$, rather than just σ_1, as indicated in Eq. (15-146). An additional experiment, as suggested by Eq. (15-142), is necessary to separate the two values. This has been done by Markovitz [101, 102], who found σ_2 to be of the order of one-half to one-third of σ_1. Williams [233] used the plate-and-cone system and confirmed the logarithmic variation of $\mathsf{P}_{\theta\theta}$. For the materials he used, the curves did not extrapolate to atmospheric pressure at the outer edge, as did those of Jobling and Roberts [88]. In the former case, this implies that $\sigma_2 \neq 0$, but unless another geometry is investigated, one cannot be sure that it is not just a boundary effect. As a further exception, Philippoff [147] has found that rigid particle suspensions and solutions have a $\sigma_2 \neq 0$.

An increasingly important means of obtaining normal stress information is by use of small-amplitude oscillatory motion. Two recent references [209, 220] have considered the plate-and-cone instrument for this purpose.

15-2.B Pipe and Other Flows

For laminar flow, it is easy to predict the pressure drop for pipeline systems, provided the necessary rheological data are available. If the data are obtained with a capillary instrument, the design is a simple scale-up. The capillary shear diagram gives the wall stress versus $4Q/\pi r_0^3$. The system to be designed usually has some known $4Q/\pi r_0^3$; thus one has only to turn to the capillary diagram, select the proper value of $(r_0/2)(dp/dz)$, and calculate the pressure drop. This, of course, assumes that the flow remains laminar, and that other complications

432 NON-NEWTONIAN PHENOMENA

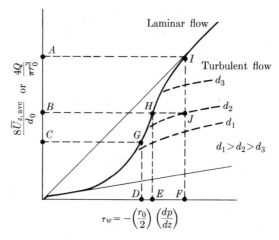

Fig. 15–32. *Turbulent flow on the capillary shear diagram.*

do not exist (slip, end effects, etc.). If the data are available as the basic shear diagram, then $(dU_z/dr)_w$ will have to be calculated from the known $4Q/\pi r_0^3$ and Eq. (15–78). From this, τ_w can be read from the diagram. If the radius is the unknown, and is to be selected so that a given flow will have a certain pressure drop, then the solution becomes one of trial and error.

For Newtonian pipe flow, both the laminar and turbulent regions have usually been expressed in terms of a friction factor–Reynolds number relation or plot. It will be useful to follow the same general method for non-Newtonians. The Fanning friction factor f, defined by Eq. (7–24), can be rewritten as

$$f = -\frac{(r_0/2)(dp/dz)}{\rho \bar{U}_{z,\text{ave}}^2/2}, \tag{15-147}$$

where $\bar{U}_{z,\text{ave}}$ is the average across the pipe of $\bar{U}_z$, which is the time average of the instantaneous velocity U_z. For laminar flow, $\bar{U}_{z,\text{ave}} = U_{z,\text{ave}}$. The overbar simply denotes the time average, so that turbulent flows can be treated (see pages 233–238). The definition of a Reynolds number presents the major problem, since there are several possible definitions of the viscosity term. Because of this variation, a number of different correlations have been published. However, each of the Reynolds numbers that has been used has a simple rheological definition, even if this was not recognized when the number was originally established [22].

(a) The Reynolds number. The Reynolds numbers for pipe flow can be defined with the aid of a capillary flow diagram (Fig. 15–32). If the flow in a pipe of diameter d_2 is laminar, say at point G, the pipe apparent viscosity would be a natural selection for the viscosity term:

$$N_{\text{Re},\mu_{\text{ap}}} = \frac{d_0 U_{z,\text{ave}} \rho}{\mu_{\text{ap}}}, \tag{15-148}$$

where μ_{ap} is the ratio of the value of τ_w at point D to $4Q/\pi r_0^3$ at point C. If the flow is increased to point B, so that it is now turbulent (point J), the same Reynolds number can be used, with the viscosity being evaluated at point H (i.e., from the laminar-flow curve at the same $4Q/\pi r_0^3$ that occurs in the turbulent-flow case). Under these conditions the velocity would now be $\bar{U}_{z,\text{ave}}$. Meter [119] has introduced a turbulent-flow Reynolds number, which is based on point I. This will be denoted as

$$N_{\text{Re},\mu_{apw}} = \frac{d_0 \bar{U}_{z,\text{ave}} \rho}{\mu_{apw}}, \qquad (15\text{-}149)$$

where μ_{apw} is evaluated at point I; that is, at the same *wall shear stress* as in the turbulent flow rather than at the same flow parameter as for μ_{ap}.

For laminar flow, the Reynolds numbers (15-148) and (15-149) based on the pipe apparent viscosity are identical to the modified Reynolds number used by Metzner and Reed [121], which is

$$N_{\text{Re}'} = \frac{d_0^{n'} \bar{U}_{z,\text{ave}}^{2-n'} \rho}{K' 8^{n'-1}}. \qquad (15\text{-}150)$$

This can be shown by combining Eqs. (15-73) and (15-75) to give

$$\mu_{ap} = K' \left(\frac{8\bar{U}_{z,\text{ave}}}{d_0} \right)^{n'-1} \qquad (15\text{-}151)$$

or

$$\mu_{ap} = \frac{K' 8^{n'-1} \bar{U}_{z,\text{ave}}^{n'-1}}{d_0^{n'-1}}. \qquad (15\text{-}152)$$

Substituting this into the Reynolds number of Eq. (15-148) gives exactly Eq. (15-150).

For turbulent flow, the Reynolds numbers of Eqs. (15-148) and (15-150) are the same if n' and K' are evaluated at point H. However, Metzner and Reed obtained a better correlation if they used the velocity equivalent to point B, but evaluated n' and K' at point I (i.e., at the same wall conditions, rather than at the same flow). This latter Reynolds number is denoted as $N_{\text{Re}'}$ and, as can be seen, is an empirical mixture not using any exactly defined viscosity. The actual difference between $N_{\text{Re},\mu_{ap}}$ and $N_{\text{Re}'}$ for turbulent flow is slight, since n' and K' do not vary much over small ranges of shear stress. If n' and K' are constants, the Reynolds numbers are of course exactly equivalent over the entire range. The Reynolds number based on μ_{ap} (or $N_{\text{Re}'}$) has been used by Metzner and Reed [121], Shaver and Merrill [179], Weltman [215], and Winning [222]. Thomas [189] has recently referred to this as the *effective viscosity*.

A simple relation can be established between $N_{\text{Re},\mu_{apw}}$ and $N_{\text{Re}'}$ [119]. Equations (15-151) and (15-152) written for μ_{apw} (that is, using n' and K' evaluated at I) can be combined with Eqs. (15-149) and (15-150) to give

$$N_{\text{Re},\mu_{apw}} = N_{\text{Re}'} (\bar{U}_{z,\text{ave}}/\bar{U}'_{z,\text{ave}})^{n'-1}, \qquad (15\text{-}153)$$

where $\bar{U}'_{z,\text{ave}}$ is the velocity corresponding to point A, and n' is evaluated at point I. This equation can be expressed in terms of the friction factor by using a modification of Eq. (15–147):

$$\frac{f}{16} = \frac{\tau_w}{8\rho \bar{U}^2_{z,\text{ave}}} = \frac{\mu_{\text{apw}}(8\bar{U}'_{z,\text{ave}}/d_0)}{8\rho \bar{U}^2_{z,\text{ave}}} = \frac{\mu_{\text{apw}} \bar{U}'_{z,\text{ave}}}{\rho d_0 \bar{U}^2_{z,\text{ave}}} = \frac{\bar{U}'_{z,\text{ave}}}{\bar{U}_{z,\text{ave}} N_{\text{Re},\mu_{\text{apw}}}},$$

which, when combined with Eq. (15–153), gives

$$N^{n'}_{\text{Re},\mu_{\text{apw}}} = (f/16)^{1-n'} N_{\text{Re}'}. \tag{15-154}$$

Still other Reynolds numbers can be defined if Fig. 15–32 is replaced with the basic shear diagram, that is, if $(dU_z/dr)_w$ replaces $4Q/\pi r_0^3$. For laminar flow,

$$N_{\text{Re},\mu_a} = d_0 U_{z,\text{ave}} \rho / \mu_a \tag{15-155}$$

can be used (say at point G). For turbulent flow, the same Reynolds number can be used in which μ_a is defined by point H (this clearly parallels $N_{\text{Re},\mu_{\text{ap}}}$). Bogue and Metzner [19] have defined the parallel to $N_{\text{Re},\mu_{\text{apw}}}$, which will be denoted as $N_{\text{Re},\mu_{aw}}$, where again the subscript w means that the viscosity is to be evaluated at the wall stress (i.e., at point I). Some further simple relations exist between the Reynolds numbers as defined. Equations (15–73) and (15–78) can be combined to give

$$N_{\text{Re},\mu_{aw}} = \frac{d_0 \bar{U}_{z,\text{ave}} \rho}{\mu_{aw}} = N_{\text{Re},\mu_{\text{apw}}} \frac{3n'+1}{4n'}. \tag{15-156}$$

And of course Eqs. (15–154) and (15–156) can be combined to give

$$N_{\text{Re},\mu_{aw}} = N^{1/n'}_{\text{Re}'}(f/16)^{(1-n')/n'}(3n'+1)/4n'. \tag{15-157}$$

Equation (15–156) was given in reference 22, but only for laminar flow in which $\mu_{\text{ap}} = \mu_{\text{apw}}$. The expression $N_{\text{Re},\mu_{aw}}$ has been used by Bogue and Metzner, as already cited, and by Eissenberg [47].

In turbulent-flow problems, it is usually the pressure drop (that is τ_w) that is the quantity which must be calculated. Since τ_w is unknown at the start, so is point I; thus the Reynolds number based on this point cannot be directly obtained, but rather will require a trial-and-error solution. In contrast, point H is usually known, so that the Reynolds numbers based on this point can be directly calculated. The only advantage gained by using numbers based on point I is that correlations of the available data have shown a smaller dependency on n'.

All the Reynolds-number correlations for turbulent flow are completely inadequate when normal stress effects are present [119]. Harris [79] has suggested a modification for finite values of $\sigma_1 - \sigma_2$, but his suggestion has not been thoroughly tested with data. In effect, the viscosity in the Reynolds number and τ_w appearing in the friction factor are modified by a factor

$$\sqrt{1 + (\sigma_1 - \sigma_2)^2/(2\bar{\bar{\tau}}_{rz})^2}, \tag{15-158}$$

which is meant to account for the rotation of the principal axes of stress. Harris pointed out that when $N_{\text{Re}'}$ is used, the actual values of the factor are difficult to choose. This results from the problem as already cited; i.e., whether the factor should be evaluated at point H or point I. There is no reason why the factor could not be used with any of the Reynolds numbers cited. If it is used with $N_{\text{Re},\mu_{\text{ap}}}$ or $N_{\text{Re},\mu_{\text{apw}}}$, the point of evaluation becomes unambiguous.

For the Bingham ideal plastic, the Reynolds number is defined as

$$N_{\text{Re},\mu'} = d_0 \bar{U}_{z,\text{ave}} \rho / \mu'. \tag{15-159}$$

Since Eq. (15-80) gives a relation between μ' and μ_{ap}, it can be used to relate the Reynolds numbers:

$$N_{\text{Re},\mu_{\text{ap}}} = N_{\text{Re},\mu'}\left[1 - \frac{4}{3}\frac{\tau_0}{\tau_w} + \frac{1}{3}\left(\frac{\tau_0}{\tau_w}\right)^4\right]. \tag{15-160}$$

Equation (15-79) can be used to relate μ' to μ_a, to give

$$\bar{\tau}_{rz}/\mu_a = (1/\mu')(\bar{\tau}_{rz} - \tau_0) \quad \text{or} \quad 1/\mu_a = (1/\mu')(1 - \tau_0/\bar{\tau}_{rz}).$$

Thus

$$N_{\text{Re},\mu_a} = N_{\text{Re},\mu'}(1 - \tau_0/\bar{\tau}_{rz}). \tag{15-161}$$

The Reynolds number based on plastic viscosity has been used by Hedstrom [82], McMillen [117], Thomas [189], Weltman [215], and Winning [222].

A parallel analysis of the Reynolds number involving a geometric correction factor can be made for annular flow. Savins [175] has reviewed the history of the various Reynolds numbers used, and Brodkey [22] has suggested the simple rheological basis, which is analogous to that just given for pipe flow. McEachern [111] has recently obtained shear-thinning annular flow data for both laminar and turbulent flow.

The Reynolds number can have many values, depending on the geometry and viscosity description used. However, any are usable so long as one appreciates the limitations, and realizes that each number can be related to the basic shear diagram. This basic diagram provides point relations, as contrasted to the average relations obtained in pipe flows or other gross flows. The Reynolds number so obtained is the only completely general one; the others are specific to some special case, and reduce to the basic number in the limit of a Newtonian material. Of course, these various numbers are often more convenient to use in their specific cases, and do have a relatively simple form when interpreted in terms of basic rheological terminology.

(b) Correlations for pressure drop. Both laminar and turbulent pressure-drop correlations have been presented in the literature [42, 82, 117, 119, 121, 179, 190, 214, 215]. For laminar flow the correlations are equivalent, since they are based on the same fundamental theory. For turbulent flow, however, Thomas [191] has suggested that the available correlations are in reality different correlations

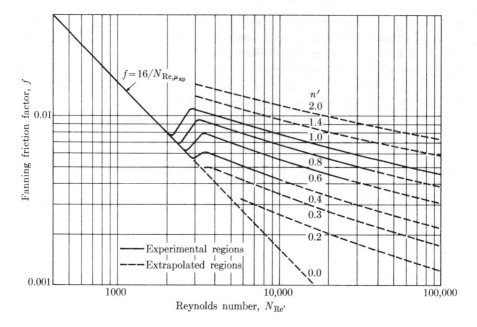

Fig. 15–33. Fanning friction factor for shear-thinning materials. [From D. W. Dodge, and A. B. Metzner, AIChEJ 5, No. 2, 198 (1959). By permission.]

for different types of materials. The correlation of Dodge and Metzner [42] and the modification by Meter [119] are for asymmetrically shaped particles and nonviscoelastic polymers; that of Shaver and Merrill [179] and the more recent one of Meter [119] are for viscoelastic polymer solutions; and that of Thomas [190] is for symmetrically shaped particles. In time and with more knowledge, it is expected that the differences can be resolved, and that a simple comprehensive correlation can be developed which will be good for Newtonians and for non-Newtonians with and without normal stress effects. However, since this is not yet a reality, several of the individual correlations will be discussed briefly.

(i) Shear-thinning materials. The Fanning friction factor for non-Newtonians corresponds to a relation which is the same as that for Newtonians (for laminar flow) if $N_{Re,\mu_{ap}}$, $N_{Re,\mu_{apw}}$, or $N_{Re'}$ are used; that is,

$$f = 16/N_{Re,\mu_{ap}}. \tag{15-162}$$

Figure 15–33 is the conventional plot used by Metzner and Reed [121] and modified for turbulence by Dodge and Metzner [42, 43]. For laminar flow, Eq. (15–162) is valid. The transition region was established by experiments, and was found to be a function of n'. For the turbulent region, a modified von Kármán equation (see Eq. 14–86) was used:

$$1/\sqrt{f} = (4.0/n'^{0.75}) \log [N_{Re'} f^{(1-n')/2}] - 0.4/n'^{1.2}. \tag{15-163}$$

This equation, with the empirical terms, correlated the pseudoplastic data with a mean deviation of 1.9%. The maximum deviation was 8.5%. In this correlation n' should be constant or evaluated at the existing value of τ_w. In Fig. 15-33, the lines were established from Eq. (15-163); the heavy lines have experimental confirmation.

Bogue and Metzner [19] were able to predict the plot of the friction factor versus the Reynolds number for very high Reynolds numbers by using the assumption that the flow is characterized by the apparent viscosity at the wall. Thus they used $N_{\text{Re},\mu_{aw}}$ and obtained a single curve for the higher Reynolds-number range. Meter [119] reworked Eq. (15-163) into terms of $N_{\text{Re},\mu_{apw}}$ and found that the lines obtained were not parallel lines, as in Fig. 15-33, but blended together at higher Reynolds numbers, just as found by Bogue and Metzner. However, in both cases the correlation was still a function of n'. Since $N_{\text{Re},\mu_{aw}}$ or $N_{\text{Re},\mu_{apw}}$ require a trial-and-error solution (evaluated at τ_w, which is obtained from the unknown f), the correlation of Fig. 15-33 is easier to use. If n' is not constant, $N_{\text{Re}'}$ requires a trial-and-error solution; otherwise $N_{\text{Re}'} = N_{\text{Re},\mu_{ap}}$, and is usually known immediately.)

(ii) **Viscoelastic materials.** Shaver and Merrill [179] have studied a series of shear-thinning materials (such as carboxymethylcellulose, or CMC), and obtained a correlation which differs considerably from that obtained by Dodge and Metzner. Dodge and Metzner were also unable to correlate their results on CMC, and suggested that it was viscoelastic. Thus the correlation presented by Shaver and Merrill may apply to viscoelastic materials, although this is only speculation. If it is true, however, the correlation must be limited, since the degree of viscoelasticity was not a parameter and was not measured. More recently Meter [119] investigated both CMC and hydroxyethylcellulose and found approximate agreement with the correlation of Shaver and Merrill. This latter material was clearly viscoelastic. However, he found better agreement when the friction factor was plotted as a function of $\tau_w/\tau_{1/2}$ (where $\tau_{1/2}$ is the shear stress at which $\mu_a = \frac{1}{2}\mu_0$). In this specific form the correlation was not a function of $N_{\text{Re},\mu_{apw}}$. Bird [15] and Meter [119] give the correlating equations as

$$f = 0.0064 - 0.00425 \log (\tau_w/\tau_{1/2}), \quad 0.2 < \tau_w/\tau_{1/2} < 10,$$
$$f = 0.0117(\tau_w/\tau_{1/2})^{-0.73}, \quad 10 < \tau_w/\tau_{1/2} < 40. \tag{15-164}$$

Possibly $\tau_{1/2}$ somehow accounts for the viscoelastic nature of these materials, but this seems unlikely, since it is a laminar-flow parameter independent of the viscoleastic nature. As pointed out by the authors, the validity of this correlation (15-164) will have to await further experimental information on other materials showing normal stress effects.

We should mention once again the suggestion made by Harris [79]. It is quite possible that, if the correction of Eq. (15-158) is used, $N_{\text{Re},\mu_{ap}}$ or $N_{\text{Re},\mu_{apw}}$ will be adequate for viscoelastic materials as well as for other materials. Harris has

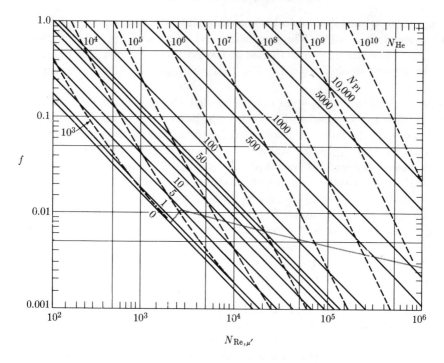

Fig. 15–34. *Fanning friction factor for plastic materials.* [*From* B. O. A. Hedstrom, *Ind. Eng. Chem.* **44**, 651–656 (1952). *By permission.*]

cited limited support for the idea, and further support can be obtained from the data of Meter and the extrapolations of Meter's normal stress measurements, as suggested by McEachern [111].

(iii) **Plastic materials.** Hedström [82] formulated a correlation based in part on the work of McMillen [117], who had put the Buckingham-Reiner pipe-flow equation (15–81) into dimensionless terms. The same basic correlation was used by Weltmann [214, 215] in her work. The correlation (Fig. 15–34) can be found in references 82, 214, and 215. Hedström used two groups,

$$N_{\text{Pl}} = \frac{\tau_0 d_0}{\mu' \bar{U}_{z,\text{ave}}} \quad \text{and} \quad N_{\text{He}} = N_{\text{Pl}} N_{\text{Re},\mu'} = \frac{\tau_0 d_0^2 \rho}{\mu'^2}, \quad (15\text{–}165)$$

either of which would determine the other on a friction factor–Reynolds number plot. At the time the correlation was presented, evaluation of the limited data available indicated that turbulence started at the Reynolds number where the line of the laminar flow curve intersected the turbulent branches of the curves for Newtonian flow. In the turbulent region, the curves for Newtonian flow were used for non-Newtonian fluids. It now appears that this simplification was optimistic, and a number of nearly parallel turbulent curves are necessary, which depend on the yield value and the plastic viscosity [190]. However, for a given concentration of one material, only one line is necessary, and it is independent of the pipe diameter.

The equations necessary to obtain the curves for the parameter N_{Pl} can be obtained from the Buckingham-Reiner equation. Combining Eq. (15–81) and the friction factor (15–147), and remembering $Q = \pi r_0^2 \bar{U}_{z,\text{ave}}$, gives

$$f = \frac{16\mu'}{d_0 \bar{U}_{z,\text{ave}} \rho [1 - \frac{4}{3}(\tau_0/\tau_w) + \frac{1}{3}(\tau_0/\tau_w)^4]} = \frac{16}{N_{\text{Re},\mu'}} \left[\frac{1}{1 - \frac{4}{3}(\tau_0/\tau_w) + \frac{1}{3}(\tau_0/\tau_w)^4} \right]. \tag{15-166}$$

Manipulating the bracketed term further, and using Eqs. (15–81) and (15–165), we get

$$f = \left(\frac{16}{N_{\text{Re},\mu'}} \right) \frac{N_{Pl}}{8(\tau_0/\tau_w)}. \tag{15-167}$$

For a given value of N_{Pl}, τ_0/τ_w is set, so the friction factor is set, once $N_{\text{Re},\mu'}$ has been selected. The value of the Hedström number N_{He} follows directly, since $N_{\text{Re},\mu'}$ is known.

Calculations of pressure drop from the various correlations will not be illustrated, since they would parallel the already well-known Newtonian calculations.

(c) The critical Reynolds number. In the preceding sections, several Reynolds numbers and pressure-drop correlations have been suggested. The critical Reynolds number is usually taken as the point of deviation from laminar flow on a friction factor–Reynolds number plot. The suggested Reynolds numbers ($N_{\text{Re},\mu_{ap}}$, $N_{\text{Re},\mu_{apw}}$, $N_{\text{Re}'}$) are not constant and vary slightly with n'. The concept of transition (pages 227–233) stresses the importance of local conditions, so a type of local Reynolds number might serve better as a criterion for transition than would the Reynolds number based on mean flow or on wall conditions [24]. Such a number can be defined for the position at which the maximum rate of change of kinetic energy occurs (see comment on page 233):

$$N_{\text{Re},R} = 2RU_z \rho/\mu_a, \tag{15-168}$$

where R is the radial position in the pipe at which the maximum occurs, U_z is the velocity at that point, and properties are taken as those existing at that position. In general, Eq. (15–65) is valid, so that

$$N_{\text{Re},R} = \frac{2U_z \rho R}{\bar{\tau}_{rz}} \frac{dU_z}{dr} = \frac{R\rho}{\bar{\tau}_{rz}} \frac{dU_z^2}{dr}. \tag{15-169}$$

Further equalities may be obtained from the linear variation of $\bar{\tau}_{rz}$ in a pipe, the definition of U^*, and the value of τ_w. These are

$$N_{\text{Re},R} = \frac{r_0}{\tau_w} \rho \frac{dU_z^2}{dr} = \frac{r_0}{U^{*2}} \frac{dU_z^2}{dr} = \frac{2\rho}{dp/dz} \frac{dU_z^2}{dr}. \tag{15-170}$$

Comparison of the first term on the right-hand side with the stability parameter proposed by Ryan and Johnson [171] shows that for laminar flow they are equivalent terms, differing only by one-half. Similarly, for pipe flow, the number suggested by Hanks [74] is different by a factor of one-quarter.

The local Reynolds number or its equivalent was considered by Ryan and Johnson [171], Sanghani [173], Hanks and Christiansen [76], and Hanks [74, 75]. These authors assumed that the materials considered followed certain specific rheological equations of state. The number calculated was to indicate the beginning of transition, and was hypothesized to be a constant. The resulting calculations showed qualitative agreement with the theory, the difference being attributed to the failure of the rheological law. More recent calculations [90], in which purely graphic methods were used on experimental data with no assumptions of rheological laws necessary, showed somewhat improved agreement between predicted and measured values. The major part of the disagreement appears to stem from the attempt to establish a local phenomenon of transition from a not-so-local measurement of pressure drop versus flow rate. Nevertheless, the best value for the prediction of the transition is $N_{\text{Re},R} = 1616$; it should be pointed out that this is not an easy number to evaluate, since the laminar velocity profile must be known in order to establish the point R or to use Eq. (15–169) or (15–170).

(d) **Velocity profiles.** Laminar velocity profiles for specific rheological models are easy to establish from Eq. (15–69). For example, the integration of the power law,

$$\bar{\bar{\tau}}_{rz} = -K(-dU_z/dr)^n, \qquad (15\text{–}171)$$

across the pipe gives

$$\frac{U_z}{U_{z,\text{ave}}} = \frac{3n+1}{n+1}\left[1 - \left(\frac{r}{r_0}\right)^{(n+1)/n}\right]. \qquad (15\text{–}172)$$

The laminar results are compared with experimental data [163, 179] in Fig. 15–35. Pitot tube data [179] shows good agreement for $n > 0.6$. The data of Richardson, McGinnis, and Beatty [163], obtained by a tracer-displacement method, shows excellent agreement with the predicted curves. For the Bingham ideal plastic material (15–79), the integration gives

$$U_z = \frac{1}{2\mu'}\frac{r_0}{\tau_w}(\tau_w - \tau_0)^2, \qquad r \leqslant r_0\frac{\tau_0}{\tau_w}, \qquad (15\text{–}173)$$

$$U_z = \frac{1}{2\mu'}\frac{r_0}{\tau_w}[\tau_w^2 - \bar{\bar{\tau}}_{rz}^2 - 2\tau_0(\tau_w - \bar{\bar{\tau}}_{rz})], \qquad r_0\frac{\tau_0}{\tau_w} \leqslant r \leqslant r_0. \qquad (15\text{–}174)$$

In terms of the radius, the latter equation becomes

$$U_z = (1/2\mu' r_0)[(r_0^2 - r^2)\tau_w - 2\tau_0 r_0(r_0 - r)]. \qquad (15\text{–}175)$$

In the range in which Eq. (15–173) is valid, the material is in plug flow, and remains as an elastic solid. This is shown in Fig. 15–36. Of course a graphic solu-

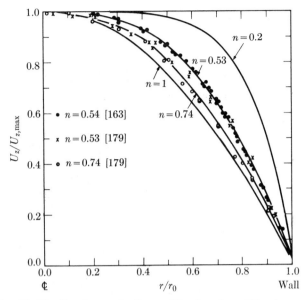

Fig. 15–35. Laminar velocity profiles for shear-thinning materials.

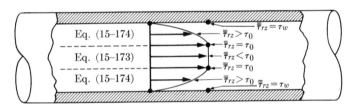

Fig. 15–36. Laminar velocity profile for a material with a yield.

tion of Eq. (15–69) is also possible [90]; one would not need to assume a rheological equation of state.

The desirability of a single equation for the description of the turbulent-flow velocity profile has long been recognized. The closest approach to this goal has been the semi-empirical method of Pai, which was discussed in detail beginning on page 250. Brodkey, Lee, and Chase [23] have extended this method to non-Newtonian fluids described by the power-law approximation. From an integration of the Reynolds equations it has been shown that the total shear is composed of a laminar and a turbulent component (14–14). For the power-law fluid, the laminar stress is given by Eq. (15–171). In general,

$$\tau_{\text{turbulent}} = \rho \overline{u_r u_z}, \qquad (15\text{–}176)$$

where u_z, the fluctuating component, has been defined from $U_z = \bar{U}_z + u_z$, in which U_z denotes the local point velocity, and $\bar{U}_z$ is the average velocity at the point. Combining equations (15–171) and (15–176) with Eq. (14–14), and using the linear variation of $\bar{\tau}_{rz}$ across the pipe, we get the non-Newtonian counterpart to Eq. (14-73):

$$(r/r_0)U^{*2} = (K/\rho)(-dU_z/dr)^n + \overline{u_r u_z}. \tag{15-177}$$

In analogy to Eq. (14-71), it is assumed that the velocity across the pipe is represented by a three-term power series of the form

$$\bar{U}_z/\bar{U}_{z,\max} = 1 + a_1(r/r_0)^{(n+1)/n} + a_2(r/r_0)^{2m}. \tag{15-178}$$

The coefficients a_1 and a_2 are obtained in exactly the same manner as before, giving

$$a_1 = \frac{(s-m)}{m - (n+1)/2n}, \quad a_2 = \frac{(n+1)/2n - s}{m - (n+1)/2n}, \tag{15-179}$$

where

$$s = \left(\frac{\rho U^{*2}}{K}\right)^{1/n} \frac{r_0}{2\bar{U}_{z,\max}} = \frac{(y_0^+)^{1/n}}{2U_0^+}, \tag{15-180}$$

in which

$$y_0^+ = \frac{r_0^n U^{*2-n} \rho}{K} \quad \text{and} \quad U_0^+ = \frac{\bar{U}_{z,\max}}{U^*}. \tag{15-181}$$

Further parallel development gives a result analogous to Eq. (14-78):

$$\frac{\bar{U}_{z,\text{ave}}}{\bar{U}_{z,\max}} = 1 + \frac{2a_1 n}{3n+1} + \frac{a_2}{m+1}. \tag{15-182}$$

The usefulness of Eq. (15-178), like that of its Newtonian counterpart, depends on the availability of the parameter m. Brodkey [21] has given a graphic correlation for nonviscoelastic shear-thinning materials. In this plot, m is a function of $N_{\text{Re},\mu_{\text{ap}}}$, with n as a parameter. However, extensive comparison of the actual velocity distributions with the theoretical ones has not yet been reported. Measurements of non-Newtonian turbulent velocity profiles have been reported in references 19, 48, and 179.

EXAMPLE 15-9. The following data were obtained by Bogue [19] on a 0.2% Carbopol solution at an $N_{\text{Re},\mu_{\text{ap}}} = 29{,}450$. Here $U^* = 0.835$ ft/sec, $\bar{U}_{z,\text{ave}} = 16.5$ ft/sec, and $n = 0.80$. Calculate and compare the velocity distribution.

r/r_0	$(y^+)^{1/n}$	$\bar{U}_z$, ft/sec	$\bar{U}_z/\bar{U}_{z,\max}$ ft/sec
0.0	1374	20.7	1.0
0.3	962	20.05	0.968
0.5	687	18.9	0.913
0.7	412	17.4	0.842
0.8	275	16.3	0.787
0.85	206	15.5	0.748
0.9	137	14.45	0.698

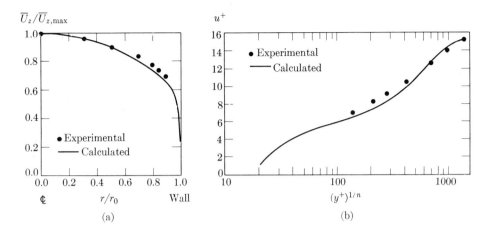

Fig. 15-37(a) Turbulent velocity profile for a shear-thinning material. (b) Universal turbulent velocity profile plot.

Answer. We can calculate s from Eq. (15–180), since $\bar{U}_{z,\max}$ from the table is 20.7, U^* is given, and y_0^+ from the table is 1374. Thus, $s = 27.5$. From Eq. (15–182), we can obtain m by trial and error, since the ratio on the left is known (0.797), on the right s and n are known, and only m is unknown. From the solution, $m = 47$, the value correlated by Brodkey [21]. Then we can calculate a_1, a_2, and thus the distribution from Eq. (15–178). A comparison is given in Fig. 15–37.

(e) Other problems. Drag reduction, experienced in the pumping of certain materials (usually viscoelastic ones), has received considerable attention because of the possible economic advantages to be gained by adding small amounts of these agents to the materials to be pumped. Drag reduction in some systems occurs because the laminar flow range is extended when the friction factor is expressed in terms of the Reynolds numbers already cited. It may be possible to predict the extent of the effect by using the factor suggested to account for the existence of normal stresses (15–158). Again, this has not been extensively checked because of the lack of data on normal stress. The exact mechanism responsible for the drag reduction is not known, but is believed to be associated with boundary-layer formation, damping of turbulence due to a viscosity gradient, or storage of kinetic energy in the elastic molecules of the viscoelastic material. A recent theoretical study on turbulence in a Reiner-Rivlin fluid (power-law considered) by Lumley [98] has cast doubt on the second possibility as a major contributing factor. Since it is known that once transition has occurred, the energy-containing range of wave numbers is almost independent of molecular viscosity, a time-dependent or viscoelastic mechanism (the third suggestion) is necessary to explain the observations. Although the analysis applied only to isotropic decaying turbulence, another argument can be advanced for turbulent shear flows. Figure 14–17 shows that by far the greater part of the production and dissipation of turbulence occurs in a narrow region near the wall. Over this region, the stress is nearly constant at the wall value, so that the production and dissipation of most of the turbulent energy is occurring in a region of constant apparent viscosity controlled by the value of τ_w. Thus one would not expect a non-Newtonian flow to differ from its Newtonian counterpart if the τ_w's were the same and $\mu = \mu_{aw}$. Under these conditions, Eqs. (15–73) and (15–78) combine to give

$$\tau_w = \mu\left(-\frac{dU_z}{dr}\right)_{w-N} = \mu\left(\frac{4Q}{\pi r_0^3}\right)_N = \mu_{aw}\left(-\frac{dU_z}{dr}\right)_{w-\text{non-N}} = \frac{3n'+1}{4n'}\left(\frac{4Q}{\pi r_0^3}\right)_{\text{non-N}}$$

or

$$Q_N = [(3n'+1)/4n']Q_{\text{non-N}}.$$

Since $(3n'+1)/4n'$ for shear-thinning materials is greater than unity, the result suggested $(Q_N > Q_{\text{non-N}})$ is the opposite of the result observed $(Q_{\text{non-N}} > Q_N)$. Thus another explanation, such as that of viscoelasticity, must be used. Additional references are 51, 140, 143, 177, 197, and 227.

Many other problems of engineering interest can be cited. There is a growing literature in non-Newtonian boundary layer theory, although normal stress effects have so far been neglected [1, 11, 57, 58, 70, 178] except in the recent analysis by White and Metzner [232]. Slow-motion flow in three-dimensional geometries, (such as the flow over a sphere) has been treated [34, 96, 182, 196, 207, 208]. Analyses of heat transfer in non-Newtonian flows, such as the Graetz-Nusselt problem for laminar flow, have also received recent attention [7, 13, 14, 15, 17, 32, 33, 36, 57, 58, 76, 99, 126, 193, 224, 234].

PROBLEMS

15–1. A series of points were obtained from capillary data. Derive the equation which would relate the apparent viscosity on the basic shear diagram to the apparent viscosity on the shear diagram obtained by plotting the capillary data. To what does this relation reduce for a Newtonian fluid?

15–2. The following capillary data have been obtained on a material. Obtain the basic shear diagram, and establish which rheological class would describe this material.

$4Q/\pi r_0^3$, sec^{-1}	τ_w, lbf/ft^2
11	1.46
24	1.97
42	2.49
75	3.30
125	4.20

15–3. We wish to determine the capacity of a 5-in Sch. 40 steel pipeline, 500 feet in length. The material to be pumped is a slurry (density is 80 lb/ft^3). The pressure drop in this line is to be 100 psi. A test was made on the same material in a $\frac{1}{2}$-in. Sch. 40 steel line, 10 ft in length. The flow in the test was found to be 1 gph for a pressure drop of 16.15. What is the capacity of the larger pipeline?

15-4. What would be the capacity of the pipeline of Problem 15-3 for the material of Problem 15-2. Assume the same density and pressure drop.

15-5. Determine the velocity distribution in the pipe of Problem 15-4.

15-6. Equation (15-57) is for Oldroyd's fluid B; repeat the derivation for fluid A.

15-7. Derive the equation for the flow of a Bingham plastic between infinite parallel plates. The equation will be analogous to the Buckingham equation for pipe flow.

15-8. Explain the difference between the different derivative operators considered and explain why they give the same result when operated on a scalar.

15-9. Describe the procedure necessary to obtain the basic shear diagram from a Brookfield instrument.

15-10. What would be the capacity of the pipeline of problem 15-3 for a material of the following properties (the data were obtained on a Brookfield viscometer):

Rev/min	Reading, centipoise
5	11,200
10	7,000
20	4,400
40	2,700
80	1,600

15-11. Derive the equations necessary to describe the flow between two flat plates. Consider the means to measure normal stresses in this configuration.

15-12. For the capillary shear diagram, provide a qualitative analysis of the effects of wall slip, heating effects, and thixotropic behavior. What might you expect to observe if entrance effects were superimposed on slip at the wall? if they were superimposed on thixotropic behavior?

NOTATION FOR CHAPTER 15

a	constant used in Eq. (15-58)
a_1, a_2	constants defined in Eq. (15-179)
A	constant
A	tensor defined in Eq. (15-63)
b	constant used in Eq. (15-58)
B	constant used in Eqs. (15-34) and (15-138)
B	tensor defined in Eq. (15-63)
c_v	heat capacity at constant volume
C$_{(t)}$	Cauchy-Green finite-deformation tensor (15-60)
d_0	diameter
d	rate-of-strain tensor
d$_w$	shear rate at the wall
e_c	contraction strain (15-4)
e_l	extensional strain (15-2)
e_v	dilational or volume strain (15-1)

e_τ	shear strain (15–3)
E^*	activation energy (15–35)
f	a function
f	Fanning friction factor
F	force; fraction unconverted (15–31)
g	metric tensor
G	shear modulus (15–3)
G^*	molar free energy of activation (15–20)
h	Planck's constant; height of fluid rise in manometer
H	submersion depth
I	parameter in Eq. (15–30); correction factor in Eq. (15–141)
k	bulk modulus; Boltzmann constant; thermal conductivity
k'	specific rate constant (15–20)
k'_f	forward specific rate constant (15–21)
k'_r	reverse specific rate constant (15–21)
k_1, k_2, k'_1, k'_2	forward and reverse rate constants for the non-Newtonian reaction (15–32) and (15–36)
K	equilibrium constant (15–37); power-law constant (15–11)
K'	power-law constant (15–75)
K''	power-law constant
L	length of pipe
l	length
n, n', n''	power-law constants (15–11), (15–75), (15–76), (15–120)
m	reverse-reaction order (15–32); parameter (15–178)
M	torque (15–130)
n	unit normal vector
N	Avogadro's number
N_{Re}	various Reynolds numbers defined in Eqs. (15–148), (15–149), (15–153), (15–155), (15–156), (15–159)
N_{Pe}	Peclet number (15–87)
N_{Br}	Brinkman number (15–140)
N_{He}	Hedström number (15–165)
N_{Pl}	plasticity number (15–165)
p_i	pressure; susceptibility to shear constant (15–33)
p_0	ambient pressure
P	pressure tensor
q	heat flux vector
Q	volumetric flow rate
r	radius
r_0	radius of pipe
R	radius of plate in plate-and-cone system; radial position used in Eq. (15–168)
s	parameter defined in Eq. (15–180)

T	tensile stress; temperature; torque per unit height in the Couette system; torque per unit area in the plate-and-cone system (15-102)
T_0	reference temperature
$\mathbf{U}$	velocity vector
U	unit tensor
U_λ	net velocity of flow
U_s	slip velocity
U_0^+	defined in Eq. (15-181)
x_i	fractional area term used in Eq. (15-28)
x, y, z	cartesian coordinates
y_0^+	defined in Eq. (15-181)

Greek letters

α	parameters in Eqs. (15-23) and (15-50)
β	parameters in Eqs. (15-15), (15-25), and (15-50)
β_i	parameter in Eq. (15-28)
γ	angle; parameters in Eqs. (15-30) and (15-50)
δ_{ij}	Kronecker delta or unit tensor
ζ	dimensionless length defined by Eq. (15-87)
η_c	cross viscosity (15-51)
θ	coordinate; dimensionless temperature (15-88)
θ_0	angle between the plate and the cone
κ	transmission coefficient (15-20); bulk viscosity
λ	distance traveled by a particle; volumetric extension coefficient (15-42)
$\lambda_1, \lambda_2, \lambda_3$	molecular distances; constants (15-55)
μ	viscosity
μ'	plastic viscosity
ξ	parameter in Eq. (15-86)
π	pi, 3.1416 ...
ρ	density
σ	Poisson's ratio (15-4); surface tension (15-94)
σ_1, σ_2	normal stress coefficients
$\boldsymbol{\tau}$	stress tensor
τ_0	yield stress
$\tau_{1/2}$	stress at point where $\mu_a = \frac{1}{2}\mu_0$
ϕ	coordinate; parameter in Eq. (15-86)
$\boldsymbol{\Omega}$	angular velocity vector
ω	angular velocity

Subscripts

a	apparent
ap	pipe apparent
apw	pipe apparent at the wall
aw	apparent at the wall

ar   rotational apparent
1, 2   inner, outer
¢   center line
0   initial
w   wall

Other symbols

I, II, III   principal invariants defined in Eq. (15–47)
𝔡   convective derivative defined by Eqs. (15–38) and (15–39)
𝒟   Jaumann derivative
ℱ   functional

REFERENCES

1. A. ACRIVOS, M. J. SHAH, and E. E. PETERSEN, *AIChEJ* **6**, 312–317 (1960); *Chem. Eng. Sci.* **20**, 101–105 (1965); see also N. HAYASI, *J. Fluid Mech.* **23**, 293–303 (1965).
2. T. ALFREY, JR. and E. F. GURNEE, *Rheology* (F. R. Eirich, editor), Vol. I, Chap. 11, Academic Press, New York (1956).
3. T. ALFREY, JR. *Mechanical Properties of High Polymers*, Interscience, New York (1948).
4. E. B. BAGLEY, *J. Appl. Phys.* **28**, 624–627 (1957).
5. E. B. BAGLEY and A. M. BIRKS, *J. Appl. Phys.* **31**, 556–561 (1960).
6. R. L. BATES, *Chem. Engr.* **67**, 145–148 (April 4, 1960).
7. W. J. BEEK and R. EGGINK, *De Ingen.* **74**, Chaps. 81–89 (1962).
8. J. T. BERGEN, *Viscoelasticity: Phenomenological Aspects*, Academic Press, New York (1960).
9. E. C. BINGHAM, *Fluidity and Plasticity*, McGraw-Hill, New York (1922).
10. E. C. BINGHAM and H. GREEN, *Proc. ASTM* **19**, 640 (1919).
11. G. C. BIZZELL and J. C. SLATTERY, *Chem. Eng. Sci.* **17**, 777–782 (1962).
12. R. B. BIRD, *AIChEJ* **2**, 6S and 8S (Sept. 1956).
13. R. B. BIRD, *SPEJ* **11** No. 7, 35 (1955); see also R. G. GRISKEY and I. A. WIEHE, *AIChEJ* **12**, 308–312 (1966).
14. R. B. BIRD, *Chem.-Ing.-Tech.* **31**, 569–572 (1959).
15. R. B. BIRD, *Report 22, Eng. Exp. Stat.* University of Wisconsin, Madison, Wis. (Oct. 1963).
16. R. B. BIRD and R. M. TURIAN, *Chem. Eng. Sci.* **17**, 331–334 (1962).
17. R. B. BIRD, W. E. STEWART, and E. N. LIGHTFOOT, *Transport Phenomena*, John Wiley & Sons, New York (1960).
18. D. C. BOGUE, *Ind. Eng. Chem.* **51**, 874–878 (1959).
19. D. C. BOGUE and A. B. METZNER, *Ind. Eng. Chem., Fund.* **2**, 143 (1963); D. C. BOGUE, Ph. D. thesis, University of Delaware, Newark, Del. (1960).
20. I. BRAUN, *Bull. Research Council Israel* **1**, 126 (1951).
21. R. S. BRODKEY, *AIChEJ* **9**, 448–451 (1963).

22. R. S. Brodkey, *Ind. Eng. Chem.* **54** No. 9, 44–48 (1962).
23. R. S. Brodkey, J. Lee, and R. C. Chase, *AIChEJ* **7**, 392–393 (1961).
24. R. S. Brodkey, E. R. Corino, and P. K. Sanghani, paper presented at AIChE meeting, Chicago (Dec. 1962).
25. J. G. Brodnyan, F. H. Gaskins, W. Philippoff, and E. G. Lendrat, *Trans. Soc. Rheology* **2**, 285 (1958).
26. J. G. Brodnyan and E. L. Kelly, *Trans. Soc. Rheology* **5**, 205–220 (1961).
27. Brookfield Engineering Lab. *Synchro-Lectric Viscometer*, Stoughton, Mass; *see also* Martin Sweets, Co., *Weissenberg Rheogoniometer*, Louisville, Ky. (1962) *and* Ferranti Electric, *Shirley Plate-and-cone Viscometer*, Plainview, N. Y. (1963).
28. H. Bruss, *Paint and Varnish Production* (July 1961).
29. R. Buchdahl, *Rheology* (F. R. Eirich, editor), Vol. II, Chap. 4, Academic Press, New York (1958).
30. F. Bueche, *J. Chem. Phys.* **22**, 1570 (1954).
31. P. H. Calderbank and M. B. Moo-Young, *Trans. Inst. Chem. Engrs.* (*London*) **37**, 26 (1959).
32. M. R. Cannon, R. E. Manning, and J. D. Bell, *Anal. Chem.* **32**, 355–358 (1960).
33. N. Casson, *Rheology of Dispersed Systems*, pp. 84–104, Pergamon Press, London (1957).
34. B. Caswell and W. H. Schwarz, *J. Fluid Mech.* **13**, 417 (1962).
35. C. C. Chang and P. Ramanaiah, *Phys. Fluids* **4**, 1179–1181 (1961).
36. R. M. Clapp, *Proc. 1961 Int. Heat Trans. Conf.*, University of Colorado, Boulder, Colorado (1961)
37. D. R. Clay, M. S. thesis, The Ohio State University, Columbus, Ohio (1960).
38. B. D. Coleman and W. Noll, *Arch. Rat'l. Mech. Anal.* **3**, 289–303 (1959); *see also* B. D. Coleman, H. Markovitz, and W. Noll, *Viscometric Flows of Non-Newtonian Fluids*, Springer-Verlag, Berlin (1966).
39. M. Collins and W. R. Schowalter, *AIChEJ* **9**, 804–809 (1963).
40. D. A. Denny and R. S. Brodkey, *J. Appl. Phys.* **33**, 2269–2274 (1962).
41. F. D. Dexter, *J. Appl. Phys.* **22**, 1124 (1954).
42. D. W. Dodge and A. B. Metzner, *AIChEJ* **5**, 189–204 (1959).
43. D. W. Dodge, Ph. D. thesis, University of Delaware, Newark, Del. (1957).
44. D. J. Doherty and R. Hurd, *J. Oil & Colour Chemists' Assoc.* **42**, 41 (1958).
45. F. R. Eirich, editor, *Rheology*, Vols. I–IV, Academic Press, New York (1956–1960, in press).
46. R. Eisenschitz, B. Rabinowitsch, and K. Weissenberg, *Mitt. der deut. Mat.* **9**, 21 (1929).
47. D. M. Eissenberg, *AIChEJ* **10**, 403–407 (1964).
48. D. M. Eissenberg and D. C. Bogue, *AIChEJ* **10**, 723–727 (1964); D. M. Eissenberg, M. S. thesis, University of Tennessee, Knoxville, Tenn. (1963).
49. H. J. Eyring, *Chem. Phys.* **4**, 283 (1936).
50. H. J. Eyring, T. Ree, and N. Hirai, *Proc. Natl. Acad. Sci. U.S.* **44**, 1213–1217 (1958).
51. A. G. Fabula, *Proc. Fourth Int. Congr. on Rheology* (E. H. Lee, editor), Part 3, pp. 455–479, Interscience, New York (1965).

52. J. D. FERRY, *Viscoelastic Properties of Polymers*, John Wiley & Sons, New York (1961); *Rheology* (F. R. Eirich, editor), Vol. II, Chap. 11, Academic Press, New York (1958).
53. A. FINCKE and W. HEINZ, *Rheol. Acta.* **1**, 530–538 (1961).
54. E. K. FISCHER, *Colloidal Dispersions*, John Wiley & Sons, New York (1950).
55. Fischer & Porter (Canada) Limited, *High Pressure Viscometer*, Toronto, Canada; see also Pressure Products Ind., *P. P. I. Rheometer*, Hatboro, Penn. (1966).
56. A. G. FREDRICKSON, *Chem. Eng. Sci.* **17**, 155–166 (1962).
57. A. G. FREDRICKSON, *Principles and Applications of Rheology*, Prentice-Hall, New York (1964).
58. A. G. FREDRICKSON, *Modern Chemical Engineering* (A. Acrivos, editor), Chap. 5, Reinhold, New York (1963).
59. H. FREUNDLICH and F. JULIUSBURGER, *Trans. Faraday Soc.* **31**, 920–922 (1935).
60. H. FRÖHLICH and R. SACK, *Proc. Roy. Soc. (London)* **185A**, 415 (1946).
61. A. F. GABRYSH, R. H. WOOLLEY, T. REE, H. EYRING, and C. J. CHRISTENSEN, *Bull. No. 106*, Utah Eng. Exp. Station, **51**, No. 23 (1960).
62. F. H. GASKINS and W. PHILIPPOFF, *Trans. Soc. Rheology* **3**, 181 (1959).
63. F. H. GASKINS and W. PHILIPPOFF, *J. Appl. Poly. Sci.* **2**, 143 (1959).
64. J. GAVIS and S. MIDDLEMAN, *J. Appl. Polymer Sci.* **7**, 493–506 (1963); *Phys. Fluids* **7**, 1097–1098 (1964).
65. R. E. GEE and J. B. Lyon, *Ind. Eng. Chem.* **49**, 956–960 (1957).
66. J. E. GERRARD and W. PHILIPPOFF, *Proc. Fourth Intern. Congr. on Rheology* (E. H. Lee, editor), Part 2, pp. 77–94, Interscience, New York (1965); see also J. E. GERRARD, F. E. STEIDLER, and J. K. APPELDOORN, *Ind. Eng. Chem., Fund.* **4**, 332–339 (1965); *ibid.* **5**, 260–263 (1966).
67. R. E. GINN, and A. B. METZNER, *Proc. Fourth Intern. Congr. on Rheology* (E. H. Lee, editor), Part 2, pp. 583–601, Interscience, New York (1965).
68. S. GLASSTONE, K. J. LEIDLER, and H. EYRING, *The Theory of Rate Processes*, McGraw-Hill, New York (1941).
69. B. GROSS, *Mathematical Structures of the Theory of Visco-Elasticity*, Hermann, Paris (1953).
70. C. GUTTINGER, and P. SINNAR, *AIChEJ* **10**, 631–639 (1964).
71. Haake Rotovisco, *Bulletin 93*, Poly Science Corp., Evanston, Ill. (1966).
72. S. J. HAHN, T. REE, and H. J. EYRING, *Ind. Eng. Chem.* **51**, 856–857 (1959).
73. S. J. HAHN, T. REE, and H. J. EYRING, *MLGI Spokesman* **23**, No. 3, 129–136 (1959).
74. R. W. HANKS, *AIChEJ* **9**, 45–48 (1963).
75. R. W. HANKS, *ibid.* 306–309.
76. R. W. HANKS and E. B. CHRISTIANSEN, *AIChEJ* **7**, 519–523 (1961).
77. J. C. HARPER, *Rev. Sci. Instr.* **32**, 425–428 (1961).
78. J. HARRIS, *Proc. Fourth, Intern. Congr. on Rheology* (E. H. Lee, editor), Part 3, pp. 417–428, Interscience, New York (1965).
79. J. HARRIS, *Brit. J. Appl. Phys.* **14**, 817–818 (1963).
80. V. G. W. HARRISON, *Proc. Second Intern. Congr. on Rheology*, Academic Press, New York (1954).

81. J. W. Hayes and R. I. Tanner, *Proc. Fourth Intern. Congr. on Rheology* (E. H. Lee, editor), Part 3, pp. 389–399, Interscience, New York (1965).
82. B. O. A. Hedström, *Ind. Eng. Chem.* **44**, 651–656 (1952).
83. W. Heinz, *Materialprüfung* **2**, 345–351 (1960); *Adhäsion* **6**, 1–4 (1957); *Kolloid-Zeit.* **145**, 119–125 (1956).
84. W. Heinz, *Materialprüfung* **1**, 311–316 (1959).
85. J. O. Hirschfelder, C. F. Curtiss, and R. B. Bird, *Molecular Theory of Gases and Liquids*, John Wiley & Sons, New York (1954).
86. I. L. Hopkins and W. O. Baker, *Rheology* (F. R. Eirich, editor), Vol. III, Chap. 10, Academic Press, New York (1960).
87. Instron Engineering Corp., *MCER Viscometer*, Canton, Mass. (1963); see also Tinius Olsen Co., *Semimicro Rheometer*, Willow Grove, Pa. (1964).
88. A. Jobling and J. E. Roberts, *Rheology* (F. R. Eirich, editor), Vol. II, Chap. 3, Academic Press, New York (1958); *J. Polymer Sci.* **36**, 433–441 (1959).
89. J. F. Johnson, M. J. R. Cantow, and R. S. Porter, *Proc. Fourth Intern. Congr. on Rheology* (E. H. Lee, editor), Part 2, pp. 479–489, Interscience, New York (1965).
90. H. T. Kim, M. S. thesis, The Ohio State University, Columbus, Ohio (1962).
91. W. K. Kim, N. Hirai, T. Ree, and H. J. Eyring, *J. Appl. Phys.* **31**, 358 (1960).
92. I. M. Krieger and H. Elrod, *J. Appl. Phys.* **24**, 134 (1953).
93. I. M. Krieger and S. H. Maron, *J. Appl. Phys.* **23**, 147 (1952); **25**, 72–75 (1954).
94. H. Leaderman, *Elastic and Creep Properties of Filamentous Materials and Other High Polymers*, Textile Foundation, Washington, D. C. (1943).
95. H. Leaderman, *Rheology* (F. R. Eirich, editor), Vol. II, Chap. 1, Academic Press, New York (1958).
96. F. M. Leslie, *Quart. J. Mech. Appl. Math.* **14**, 36 (1961).
97. A. S. Lodge, *Rheol. Abs.* **3** No. 3, 21–23 (1960); see also *Elastic Liquids*, Academic Press, New York (1964).
98. J. L. Lumley, *Phys. Fluids* **7**, 335–337 (1964).
99. R. N. Lyon, *Chem. Eng. Progr.* **47**, 75 (1951).
100. H. Mark and A. V. Tobolsky, *Physical Chemistry of High Polymer Systems*, Interscience, New York (1950).
101. H. Markovitz, *Trans. Soc. Rheol.* **1**, 37–52 (1957).
102. H. Markovitz, *Proc. Fourth Intern. Congr. on Rheology* (E. H. Lee, editor), Part 1, pp. 189–212, Interscience, New York (1965); *Phys. Fluids* **8**, 200 (1965).
103. S. H. Maron and R. J. Belner, *J. Colloid Sci.* **10**, 523–535 (1955).
104. S. H. Maron and P. E. Pierce, *J. Colloid Sci.* **11**, 80–95 (1956).
105. S. H. Maron and A. W. Sisko, *J. Colloid Sci.* **12**, 99–107 (1957).
106. S. H. Maron, N. Nakajima, and I. M. Krieger, *J. Polymer Sci.* **37**, 1–18 (1959).
107. S. G. Mason, *Proc. Fourth Intern. Congr. on Rheology* (E. H. Lee, editor), Part 1, pp. 367–369, Interscience, New York (1965).
108. S. G. Mason, *Rheology* (F. R. Eirich, editor), Vol. IV, Academic Press, New York (in press).
109. J. C. Maxwell, *Phil. Trans. Roy. Soc. London* **49**, 157 (1867).

110. A. J. McConnell, *Application of Tensor Analysis*, Dover Publications, New York (reprint of the 1931 edition).
111. D. W. McEachern, Ph. D. thesis, University of Wisconsin, Madison, Wis. (1963).
112. J. M. McKelvey, *Polymer Processing*, John Wiley & Sons, New York (1962).
113. R. McKennell, *Anal. Chem.* **28**, 1710–1714 (1956).
114. R. McKennell, *ibid.* **32**, 1458–1463 (1960).
115. J. Meissner, *Proc. Fourth Intern. Congr. on Rheology* (E. H. Lee, editor), Part 3, pp. 437–453, Interscience, New York (1965).
116. E. W. Merrill, *Modern Chemical Engineering* (A. Acrivos, editor), Chap. 4, Reinhold, New York (1963).
117. E. E. McMillen, *Chem. Eng. Prog.* **44**, 537–546 (1948).
118. E. H. Merz and R. E. Colwell, *ASTM Bull.* **63** (Sept. 1958).
119. D. M. Meter and R. B. Bird, *AIChEJ* **10**, 878–881 (1964); D. M. Meter, *ibid.* 881–884; Ph. D. thesis, University of Wisconsin, Madison, Wis. (1964).
120. A. P. Metzger and R. S. Brodkey, *J. Appl. Polymer Sci.* **7**, 399–410 (1963).
121. A. B. Metzner and J. C. Reed, *AIChEJ* **1**, 434–440 (1955).
122. A. B. Metzner, *Ind. Eng. Chem.* **50**, 1577 (1958).
123. A. B. Metzner, *Advances in Chem. Eng.* (T. B. Drew and J. W. Hoppes, editors), Vol. I, Academic Press, New York (1956).
124. A. B. Metzner, *Handbook of Fluid Dynamics* (V. L. Streeter, editor), Chap. 7, McGraw-Hill, New York (1961).
125. A. B. Metzner, E. L. Carley, and I. K. Park, *Mod. Plastics* **37**, 133 (July 1960).
126. A. B. Metzner and P. S. Friend, *Ind. Eng. Chem.* **51**, 879–882 (1959); *see also* A. W. Petersen and E. B. Christiansen, *AIChEJ* **12**, 221–232 (1966).
127. A. B. Metzner, W. T. Houghton, R. A. Sailor, and J. L. White, *Trans. Soc. Rheology* **5**, 133 (1961).
128. A. S. Michaels and J. C. Bolger, *Ind. Eng. Chem., Fund.* **1**, 24, 153 (1962); **3**, 14 (1964).
129. A. S. Michaels, *Ind. Eng. Chem.* **50**, 951 (1958).
130. A. S. Michaels and F. Tausch, *Ind. Eng. Chem.* **52**, 857 (1960).
131. S. Middleman and J. Gavis, *Phys. Fluids* **4**, 355–359, 963–969, 1450 (1961).
132. M. Mooney, *J. Rheol.* **2**, 210 (1931).
133. F. D. Murnaghan, *Finite Deformations of an Elastic Solid*, John Wiley & Sons, New York (1951).
134. W. Noll, *J. Rat'l. Mech. Anal.* **4**, 323–425 (1955).
135. S. Oka, *Rheology* (F. R. Eirich, editor), Vol. III, Chap. 2, Academic Press, New York (1960).
136. J. G. Oldroyd, *Proc. Roy. Soc. (London)* **200A**, 523–541 (1950).
137. J. G. Oldroyd, *ibid.* **218A**, 122–132 (1953).
138. J. G. Oldroyd, *ibid.* **245A**, 278 (1958).
139. J. G. Oldroyd, *Rheology* (F. R. Eirich, editor), Vol. I, Chap. 16, Academic Press, New York (1956).
140. J. G. Oldroyd, *Proc. First Intern. Congr. on Rheology*, p. 130, North-Holland, Schereningen (1949).
141. W. Ostwald, *Z. Physik. Chem.* **111A**, 62 (1924); *Kolloid-Zeit.* **36**, 99 (1925).

142. W. OSTWALD and R. AUERBACH, *Kolloid-Zeit.* **38**, 261 (1926).
143. R. S. OUSTERHOUT and C. D. HALL, JR., *J. Petrol. Technol.* **13**, 217 (1960).
144. J. PAWLOWSKI, *Chem.-Ing.-Tech.* **28**, 786–793 (1956).
145. S. PETER, *Chem.-Ing.-Tech.* **32**, 437–447 (1960).
146. W. PHILIPPOFF, *Kolloid-Zeit.* **71**, 1–16 (1935).
147. W. PHILIPPOFF, cited by D. M. Meter, *see* thesis reference 119.
148. W. PHILIPPOFF and F. H. GASKINS, *Trans. Soc. Rheol.* **2**, 263 (1958).
149. W. PHILIPPOFF and K. HESS, *Z. Physik. Chem.* **31B**, 237 (1936).
150. Polarad Electronics Corp., *Model RV2 Low shear viscometer* L. I., New York. (Note: low-shear instrument is no longer manufactured).
151. W. F. O. POLLETT, *Proc. Second Intern. Congr. on Rheology*, pp. 85–90, Academic Press, New York (1954).
152. R. S. PORTER and J. F. JOHNSON, *J. Appl. Polymer Sci.* **3**, 200 (1960).
153. R. E. POWELL and H. J. EYRING, *Nature* **154**, 427 (1944).
154. B. RABINOWITSCH, *Z. Physik. Chem.* **145A**, 1–26 (1929).
155. T. REE and H. J. EYRING, *J. Appl. Phys.* **26**, 793–800 (1955).
156. T. REE and H. J. EYRING, *Rheology* (F. R. EIRICH, editor), Vol. II, Chap. 3, Academic Press, New York (1958).
157. T. REE, S. J. HAHN, and H. J. EYRING, cited in R. C. POWELL, W. E. ROSEVEARE, and H. J. EYRING, *Ind. Eng. Chem.* **33**, 430 (1941).
158. M. REINER, *Deformation and Flow*, Lewis and Co., London (1949).
159. M. REINER, *Rheology* (F. R. Eirich, editor), Vol. I, Chap. 2, Academic Press, New York (1956).
160. M. REINER, *ibid.* Vol. III, Chap. 9 (1960).
161. M. REINER, *Am. J. Math.* **67**, 350–362 (1945).
162. O. REYNOLDS, *Phil. Mag.* **20**, 46–80 (1885).
163. F. M. RICHARDSON, P. H. MCGINNIS, JR., and K. O. BEATTY, JR., presented at AIChE meeting, Atlanta (Feb. 1960).
164. R. S. RIVLIN, *Proc. Roy. Soc. (London)* **193A**, 260 (1948); *Nature* **160**, 611 (1947); *Nature* **161**, 567 (1948); *Proc. Cambridge Phil. Soc.* **45**, 88 (1949).
165. R. S. RIVLIN and J. L. ERICKSEN, *J. Rat'l. Mech. Anal.* **4**, 323–425 (1955); J. L. ERICKSEN, *Kolloid. Z.* **173**, 117–122 (1960); *Arch. Rat'l. Mech. Anal.* **4**, 231–237 (1960); *ibid.* **8**, 1–8 (1961); *ibid.* **9**, 1–8 (1962); *Trans. Soc. Rheol.* **4**, 29–39 (1960); *ibid.* **6**, 275–292 (1962); P. N. KALONI, *Int. J. Eng. Sci.* **3**, 515–532 (1965).
166. R. S. RIVLIN, *Rheology* (F. R. Eirich, editor), Vol. I, Chap. 10, Academic Press, New York (1956).
167. J. E. ROBERTS, *Proc. Second Intern. Congr. on Rheology*, pp. 91–98, Academic Press, New York (1954).
168. P. S. ROLLER and C. K. STODDARD, *J. Phys. Chem.* **48**, 410–425 (1944).
169. I. R. ROSCOE, *Flow Properties of Dispersed Systems* (J. J. Hermans, editor), p. 31, North-Holland, Amsterdam (1953).
170. P. E. ROUSE, *J. Chem. Phys.* **21**, 1272 (1953).
171. N. W. RYAN and M. M. JOHNSON, *AIChEJ* **5**, 433 (1959).
172. B. C. SAKIADIS, *AIChEJ* **8**, 317–321 (1962).
173. P. K. SANGHANI, M. S. thesis, The Ohio State University, Columbus, Ohio (1959).

174. J. G. SAVINS, *J. Appl. Polymer Sci.* **6**, 567 (1962); *AIChEJ* **11**, 673–677 (1965).
175. J. G. SAVINS, *AIChEJ* **8**, 272 (1962).
176. J. G. SAVINS, G. C. WALLICK, and W. R. FOSTER, *Soc. Pet. Engrs. J.* **2**, 211–215, 309–316 (1962); **3**, 14–18, 177–184 (1963).
177. J. G. SAVINS, *Soc. Pet. Engrs. J.* **4**, 203–214 (1964).
178. W. R. SCHOWALTER, *AIChEJ* **6**, 24 (1960).
179. R. G. SHAVER and E. W. MERRILL, *AIChEJ* **5**, 181–188 (1959); R. G. SHAVER, Sc. D. thesis, Massachusetts Institute of Technology, Cambridge, Mass. (1957).
180. C. R. SHERTZER and A. B. METZNER, *Proc. Fourth Intern. Congr. on Rheology* (E. H. Lee, editor), Part 2, pp. 603–618, Interscience, New York (1965).
181. A. W. SISKO, *Ind. Eng. Chem.* **50**, 1789–1792 (1958).
182. J. C. SLATTERY, *AIChEJ* **8**, 663–667 (1962).
183. J. C. SLATTERY and R. B. BIRD, *Chem. Eng. Sci.* **16**, 231 (1961).
184. A. SLIBAR and P. R. PASLAY, *J. Appl. Mech.* **30**, 453–460 (1963).
185. J. M. SMITH, *Chemical Engineering Kinetics*, McGraw-Hill, New York (1956).
186. H. A. STUART, *Die Physik der Hochpolymeren*, Springer, Berlin (1956).
187. R. I. TANNER, *Chem. Eng. Sci.* **19**, 349–355 (1964).
188. G. I. TAYLOR, *Proc. Roy. Soc. (London)* **157A**, 546 (1936).
189. D. G. THOMAS, *AIChEJ* **6**, 631–639 (1960).
190. D. G. THOMAS, *ibid.* **8**, 266–271 (1962).
191. D. G. THOMAS, *Progr. Intern. Res. on Thermo. and Trans. Properties*, paper 61, ASME, New York (1962).
192. D. G. THOMAS, *Ind. Eng. Chem.* **55** No. 11, 18–29 (1963).
193. C. TIEN, *Can. J. Chem. Eng.* **39**, 45 (1961).
194. A. V. TOBOLSKY, *Rheology*, (F. R. Eirich, editor), Vol. II, Chap. 2, Academic Press, New York (1958).
195. A. V. TOBOLSKY, *Properties and Structure of Polymers*, John Wiley & Sons, New York (1960).
196. Y. TOMITA, *Bull. Japan Soc. Mech. Engrs.* **2**, 469 (1959).
197. B. A. TOMS, *Proc. First Intern. Congr. on Rheology*, p. 135, North-Holland, Schereningen (1949).
198. B. A. TOMS, *Rheology* (F. R. Eirich, editor), Vol. II, Chap. 12, Academic Press, New York (1958).
199. H. L. TOOR, *Trans. Soc. Rheol.* **1**, 177 (1957); *Ind. Eng. Chem.* **48**, 922 (1956); *AIChEJ* **4**, 319 (1958).
200. L. R. G. TRELOAR, *The Physics of Rubber Elasticity*, Oxford University Press, London (1949).
201. C. TRUESDELL, *J. Rat'l. Mech. Anal.* **1**, 125–300 (1952).
202. C. TRUESDELL, *ibid.* **2**, 593–616 (1953).
203. C. TRUESDELL, *Trans. Soc. Rheol.* **4**, 9 (1960).
204. C. TRUESDELL, *Principles of Continuum Mechanics*, Colloquium Lecture Series No. 5, Socony Mobil Oil Co., Dallas, Texas (1961); *The Elements of Continuum Mechanics*, Springer Verlag, New York (1967).
205. R. M. TURIAN and R. B. BIRD, *Chem. Eng. Sci.* **18**, 689–696 (1963).

206. J. R. VAN WAZER, J. W. LYONS, K. Y. KIM, and R. E. COLWELL, *Applied Rheology-Viscometers and Their Use*, Interscience, New York (1963).
207. G. C. WALLICK, J. G. SAVINS, and D. R. ARTERBURN, *Phys. Fluids* **5**, 367–368 (1962).
208. M. L. WASSERMAN and J. C. SLATTERY, *AIChEJ* **10**, 383–388 (1964).
209. K. WEISSENBERG, *The Testing of Materials by Means of the Rheogoniometer*, Farol Research Engineers Ltd., Bognor Regis, Sussex, England (1963).
210. B. RABINOWITSCH, *Z. Physik. Chem.* **166A**, 257–272 (1933).
211. K. WEISSENBERG, *Nature* **159**, 310 (1947).
212. K. WEISSENBERG, *Proc. First Intern. Congr. on Rheology*, pp. 1–29: North-Holland, Amsterdam (1949).
213. R. N. WELTMANN, *Ind. Eng. Chem., Anal. Ed.* **15**, 424–429 (1943).
214. R. N. WELTMANN, *NACA TN 3397*, Washington, D.C. (Feb. 1955).
215. R. N. WELTMANN, *Ind. Eng. Chem.* **48**, 386–387 (1956).
216. R. F. WESTOVER, and B. MAXWELL, *SPEJ* **13**, 27 (1957).
217. J. L. WHITE, *AIChEJ* **9**, 559–561 (1963).
218. W. L. WILKINSON, *Non-Newtonian Fluids*, Pergamon Press, London (1960).
219. M. C. WILLIAMS and R. B. BIRD, *AIChEJ* **8**, 378–382 (1962).
220. M. C. WILLIAMS and R. B. BIRD, *Ind. Eng. Chem., Fund.* **3**, 42–49 (1964).
221. R. V. WILLIAMSON, *Ind. Eng. Chem.* **21**, 1108–1111 (1929).
222. M. D. WINNING, M. S. thesis, University of Alberta, Edmonton, Alb. (1948).
223. C. R. WYLIE, JR., *Advanced Engineering Mathematics*, McGraw-Hill, New York (1960).
224. J. YAU and C. TIEN, *Can. J. Chem. Eng.* **41**, 139–145 (1963).
225. D. A. DENNY, H. T. KIM, and R. S. BRODKEY, paper presented at Soc. Rheology meeting, Pittsburgh (1965); H. T. KIM and R. S. BRODKEY, paper presented at AIChE meeting, Salt Lake City (1967).
226. N. N. KAPOOR, J. W. KALB, E. A. BRUMM, and A. G. FREDRICKSON, *Ind. Eng. Chem., Fund.* **4**, 186–194 (1965).
227. A. B. METZNER and M. G. PARK, *J. Fluid Mech.* **20**, 291–303 (1964).
228. S. R. MORRISON and J. C. HARPER, *Ind. Eng. Chem., Fund.* **4**, 176–181 (1965).
229. G. R. SEELY, *AIChEJ* **10**, 56–60 (1964); **11**, 364–365 (1965).
230. T. W. SPRIGGS and R. B. BIRD, *Ind. Eng. Chem., Fund.* **4**, 182–186 (1965).
231. S. VELA, J. W. KALB, and A. G. FREDRICKSON, *AIChEJ* **11**, 288–294 (1965).
232. J. L. WHITE and A. B. METZNER, *AIChEJ* **11**, 324–330, 989–995 (1965).
233. M. C. WILLIAMS, *AIChEJ* **11**, 467–473 (1965).
234. A. H. P. SKELLAND, *Non-Newtonian Flow and Heat Transfer*, John Wiley, & Sons, New York (1967),
235. A. B. METZNER, J. L. WHITE, and M. M. DENN, *Chem. Eng. Progr.* **62** No. 12, 81–92 (1966).

CHAPTER 16

MULTIPHASE PHENOMENA I: PIPE FLOW

The basic theory of homogeneous flow can be extended to multiphase systems if the assumption is made that the mass, momentum, and energy balances apply in each phase. The problem then resolves itself into an attempt to match the solutions at the interfaces. Knowledge of the mechanism of the various interphase transfers is far from complete, and so there is little hope at the present time for a general solution to problems of multiphase transport. The complexity of the problem has led mainly to an accumulation of empirical knowledge and, at the same time, to analysis of data by semi-empirical extensions of single-phase flow to simplified two-phase flow models. The literature in the field is quite extensive; for example, a recent annotated bibliography [136] cites over 2800 references for the period from 1950 to 1962, and a general index of 5253 references on two-phase flow has been compiled by Gouse [91]. A review by Dukler and Wicks [66] contains a summary of the available data in tabular form and some evaluation of the various means of data analysis. Earlier references can be found in several general literature reviews [48, 97, 122]. In this and subsequent chapters, some of these references will be reviewed in the hope that an understanding of them will provide a basis for a more detailed and improved analysis of the problem in the future.

The breakdown of two-phase fluid transport is quite arbitrary. One procedure is to use four divisions: (1) liquid–gas, (2) liquid–liquid, (3) solid–gas, and (4) solid–liquid flow. The first is common in combustion systems and in process equipment such as condensers, coolers, evaporators, absorbers, and pipe transport. The second, liquid–liquid flow, occurs in extractors, pipelines, and in many cases where chemical reaction is taking place. In this text, these two divisions will be combined and treated as pipe flow (in this chapter) and free flow (drops and bubbles, in Chapter 17). The third and fourth divisions, involving solids, which are used in solids transport and in fluidization operations, will also be combined and treated in Chapter 18 under the title solids–fluid flow.

For pipe flow, the problems of two-phase momentum transfer are quite complex in themselves, and are not often adaptable to a simple analytical solution. The addition of heat and/or mass transfer only adds complexities to any solution. Consequently, the case in which no mass or heat transfer occurs will be considered first, and then the effect of these on pipe flow will be considered.

16-1 TWO-COMPONENT ISOTHERMAL FLOW

The general direction of research on two-phase pipe flow has been toward the obtaining of analytical solutions for specific flow geometries rather than the development of gross overall correlations that encompass all flow regimes. The reason for this is simple; past experience has shown that the overall approach cannot provide a completely adequate correlation for all flow types. It is hoped, therefore, that specific solutions for specific flow types will prove to be satisfactory for correlation of data. Since each regime will ultimately be considered separately, one must be able to establish the conditions under which specific flow regimes will occur, and, more particularly, just what the possible flow configurations are.

16-1.A Flow Patterns

The initial problem of interest is that of establishing those fundamental variables which determine the nature of flow in liquid–gas systems without heat or mass transfer. Alves [6] considered the possible types of flow to be observed in horizontal concurrent flow of a liquid plus an increasing amount of gas. Figure 16–1 will aid in the following discussion of the types.

(1) *Bubble flow*: Bubbles of gas move along the upper part of the pipe at approximately the same velocity as the liquid.

(2) *Plug flow*: Alternate plugs of liquid and gas move along the upper part of the pipe.

(3) *Stratified flow*: Liquid flows along the bottom of the pipe and the gas flows above, over a smooth liquid–gas interface.

(4) *Wavy flow*: This is similar to stratified flow except that the gas moves at a higher velocity and the interface is disturbed by waves traveling in the direction of flow. The wavy flow may be subclassified [104, 105, 214] into:
   (i) two-dimensional waves
   (ii) squalls (pebbled appearance)
   (iii) roll waves (superimposed on squalls).

(5) *Slug flow*: In this case, the roll wave is picked up by the more rapidly moving gas to form a slug, which passes through the pipe at a velocity greater than the average liquid rate.

(6) *Annular flow*: Here the liquid flows in a thin film around the inside wall of the pipe and the gas flows at a high velocity as a central core. The surface is neither symmetrical nor smooth, but rather is similar to roll waves superimposed on squalls, as noted for wavy flow. More liquid flows on the bottom of the pipe than over the rest of the surface [128, 161].

(7) *Dispersed or spray flow*: Most of the liquid is entrained as spray by the gas. The spray appears to be produced by the high-velocity gas ripping liquid off the crests of the roll waves.

(8) *Foam flow*: Although not usually a part of the preceding sequence, foam flow can occur in systems in which materials such as soaps are present [140, 185].

Flow-pattern sketches

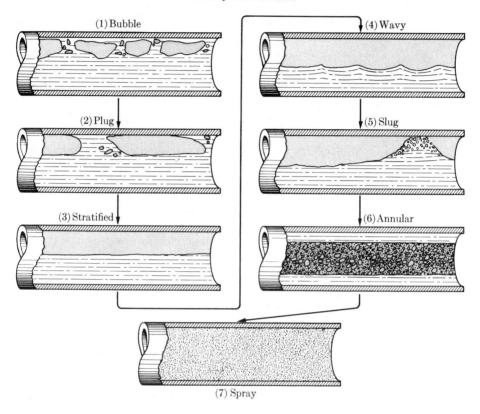

Fig. 16–1. *Flow pattern sketches.* [*From G. E. Alves,* Chem. Eng. Prog. **50**, 449–456 (1954). *By permission.*]

In many publications, the above flow patterns are not defined or utilized. As already noted, Dukler and Wicks [66] have summarized much of the available data and have indicated those for which flow patterns are known. Some work has been done to pinpoint the type of flow that occurs when the flow rates of the gas and liquid are known. (For reviews, see references 7, 66, 97, 122, 136, 216; for more specific articles, see references 1, 4, 6, 15, 29, 76, 79, 104, 111, 118, 125, 134, 139, 140, 142, 164, 172, 173, 177, 191, 192, 199, 214, 223.) Researchers in the field generally agree that, to date, there is no method in published literature which adequately describes or establishes just what type of flow will occur under the range of possible conditions. The information currently available has been presented in various papers as a plot of the liquid flow or related functions against the gas flow or related functions. In a plot of this nature, the flow types appear as certain areas. An example of such a plot is presented in Fig. 16–2, from Baker [15]. The terms G_g and G_l are the gas and liquid mass velocities in lb/hr-ft^2,

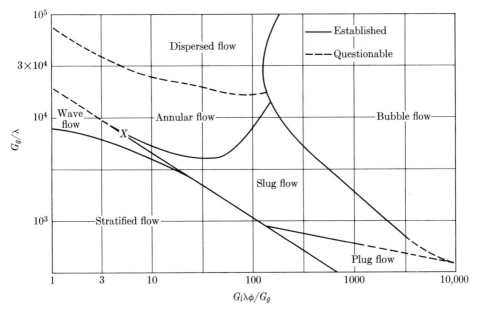

Fig. 16–2. An area plot for two-phase horizontal flow. [*From O. Baker*, Oil and Gas Journal **53**, 185 (July 26, 1954). *By permission.*]

and λ and ϕ are terms dependent on the physical properties only:

$$\lambda = [(\rho_g/0.075)(\rho_l/62.4)]^{1/2}, \tag{16-1}$$

$$\phi = (73/\sigma)\,[\mu_l(62.3/\rho_l)^2]^{1/3}, \tag{16-2}$$

where

σ = surface tension, dynes/cm,
ρ_l = density of the liquid, lb/ft^3,
μ_l = viscosity of the liquid, centipoise,
ρ_g = density of the gas, lb/ft.^{3}

Table 16-1 summarizes some of the area plots that have been used to analyze such data. Unfortunately, as just suggested, none of these are adequate for defining the flow regime for all experimental conditions [4, 66, 125, 192]. Al Sheikh [4] emphasized this by using the comprehensive data bank described by Dukler et al. [67] to test some 23 correlations, including those used in Table 16-1 and others based on fundamental dimensional reasoning. He found that no single correlation could sort all the available data into separate areas. However, he did find that the task could be accomplished by using a sequence of plots. Al Sheikh considered a number of methods in his analysis, including that by Lemaire and Grassman [148], as described in Chapter 11. He used certain normalizations and projections in an attempt to reduce the three-dimensional representation to

TABLE 16-1

Flow Pattern Correlations for Horizontal Flow

Reference	Phases	Conditions[†]	Diameter	Correlation parameters[‡]
Reviews [4, 7, 8, 97, 125, 216]				
Alves [6]	Air–water, oil	1, 2, 4, 5, 6	1.049 in.	V_l^* vs. V_g^*
Krasiakova [142]	Air–water	1, 3, 5, 6	30 mm	V_l^* vs. V_g^*
Martinelli et al. [164]	Air–water, oils	All	1 in.	W_l vs. W_g
Johnson and Abou-Sabe [134]	Air–water	1, 3, 4, 5, 6 and data of Bergelin et al. [29, 30, 84]		W_l vs. W_g
Schneider, White, and Huntington [199]	Air, natural gas–water, oils	3, 4, 5, 6	1 in., 1½ in., 2 in.	G_l vs. G_g
Baker [15]	Natural gas–oil	All, and data of Alves [6], Bergelin [29], and Kosterin [139]		G_g/λ vs. $G_l \lambda \phi/G_g$
Reichart [191]	Air–water	1, 3, 5	25 mm	Q_l vs. Q_g
Frazier [79]	Air–water	1, 3, 7	0.7 in.	Q_l vs. Q_g
Nesbit [177]	Air–water	3, 5, 7	19 mm	$N_{Re, l0}$ vs. $N_{Re, g0}$
Kosterin [139]	Air–water	1, 3, 5, 6	4 in.	V_g vs. $(V_l^* + V_g^*)$
Mologin [172, 173]	Air–water	All	25–100 mm	V_g vs. $(V_l^* + V_g^*)$
Hoogendoorn [111]	Air–water, oil	3, 4, 5, 6	24–140 mm	V_g vs. $(V_l^* + V_g^*)$
Abramson [1]	Air–water	6		u^+ vs. y^+

[†] 1 = bubble, 2 = plug, 3 = stratified, 4 = wavy, 5 = slug, 6 = annular, 7 = spray.
[‡] V^* = superficial velocity, W = mass flow rate, G = mass velocity, Q = volumetric flow rate, $N_{Re, l0}$ and $N_{Re, g0}$ = Reynolds numbers based on single-phase flow and pipe diameter, V = gas volume fraction.

a general two-dimensional one. The problem reduced to the selection of the proper variables and the possible dimensionless forms that these could take. Some idea of the type of groups useful for the description of the flow can be established from theoretical considerations. In particular, the formation of waves on the surface of a liquid over which a gas is flowing might be associated with a number of the transitions in pipe flow, i.e., stratified to wavy to slug, annular to mist, etc. Thus an examination of the problem of wave formation on a free surface could provide some insight into the question of which groups might be applicable to the more complex pipe-flow problem.

Analysis of a deep liquid layer is mainly concerned with the generation of surface waves, and thus finds its application to such bodies as the ocean. Thin layers or films are a common occurrence in film coolers, contacting devices, and two-phase fluid flow in pipes. Ursell [213] has critically reviewed most of the deep layer work on wave formation and concluded that at that time (1956) neither adequate data nor theoretical models were available. The details of these approaches can be found in Lamb [147] (for example, Jeffreys' sheltering model [130] is discussed in article 348). Some experimental results and a review of earlier work have been given by Hughes and Stewart [117]. The existing void has essentially been filled by the more recent theories of Miles [168], Brooke Benjamin [38] (based on the analysis by Miles [167]), and Phillips [183], together with the experimental confirmation by Longuet-Higgins [157]. The discussion by Phillips [184], the physical interpretation of the mathematical analysis by Lighthill [152], and the further comments by McGoldrick [240] have added much to the clarity of the analysis, which has put the theory of wave formation in deep water on a firm basis. Theoretical work on thin films has been presented by Brooke Benjamin [37], Feldman [75], Miles [169], Whitaker [244], Cohen and Hanratty [232], and Hanratty and Hershman [105]. Experimental efforts on thin films have been reported by Hanratty and Engen [104], van Rossum [214], Ellis and Gay [70], Hanratty and Hershman [105], Lilleleht and Hanratty [153, 154], and Cohen and Hanratty [232]. Fulford [235] has reviewed many aspects of thin-film flow and has provided an extensive literature review. Ostrach and Koestel [242] have considered a number of aspects of film stability, and have attempted to clarify the various sources of film breakup.

In terms of dimensionless groups, the following comments can be made from the various references cited:

(1) Miles' [169] work on thin films indicates that the important groups are:

$$N_{Re} = \frac{\lambda c \rho_l}{\mu_l}, \qquad N_{We} = \frac{\rho_l c^2 \lambda}{\sigma}, \qquad N_{Fr} = \frac{c^2}{g\lambda}, \qquad (16\text{-}3)$$

in which c is the wave velocity, λ is the wavelength, and the subscript l refers to the liquid or more dense phase. The numbers can be expressed in terms of the film thickness; however, a modified wave number is then necessary:

$$\alpha = 2\pi h/\lambda = kh, \qquad (16\text{-}4)$$

where k is the normally used wave number and h is the film thickness. The interface velocity or gas velocity can generally be substituted for the wave velocity in the above numbers, since these velocities are usually directly related. The specific form of the important term in the theory is

$$\frac{\alpha}{N_{We}} + \frac{1}{\alpha N_{Fr}} = \frac{1}{U_i^2}\left(\frac{\sigma k}{\rho_l} + \frac{g}{k}\right), \qquad (16\text{-}5)$$

in which the interface velocity U_i has been substituted for the wave velocity.

(2) The Kelvin-Helmholtz deep-layer model (Lamb, articles 267–268) can be expressed in terms of the same dimensionless groups as those given in Eq. (16-3).

The result is

$$\frac{2\pi}{N_{\text{We}}} + \frac{1}{2\pi N_{\text{Fr}}} = \frac{1}{c^2}\left(\frac{\sigma k}{\rho_l} + \frac{g}{k}\right) = \frac{1+s}{1-s}, \tag{16-6}$$

where s is the ratio ρ_g/ρ_l.

(3) Jeffreys' [130] sheltering model (Lamb, article 348) takes the form

$$N_{\text{Re}}N_{\text{Fr}}s = c^3 \, \rho_g/g \, \mu_l = 1/\beta, \tag{16-7}$$

where β is called the sheltering coefficient. In the theory, $c = \frac{1}{3} U_g$ and β is about 0.3, which gives a wind velocity of $U_g = 107$ cm/sec for wave formation. In an analysis by Miles [167], the corresponding value of the wind velocity was about 100 cm/sec, but could be as low as 80 cm/sec.

(4) In his series of articles, Miles [168] generalized the Kelvin-Helmholtz model and found the important groups to be a Reynolds number to establish the velocity profile of the upper phase,

$$\frac{N_{\text{We}}}{4\pi^2 N_{\text{Fr}}} = \frac{\rho_l g}{\sigma k^2}, \quad \text{and} \quad N_{\text{We}}N_{\text{Fr}}s^2 = \frac{U_2^4 \rho_g^2}{g\sigma\rho_l}, \tag{16-8}$$

where in the last expression U_2 has replaced c, and is a velocity related to the gas flow (defined as $U^*/0.4$).

For the various theories on wave formation discussed above, the dimensionless groups can be expressed as a combination of the following forces:

$$\text{Gravitation, } F_g = \lambda^3 \, \rho_l \, g,$$

$$\text{Inertia (of the liquid), } F_{\rho_l} = \lambda^2 \, c^2 \, \rho_l,$$

$$\text{Inertia (of the gas), } F_{\rho_g} = \lambda^2 \, c^2 \, \rho_g, \tag{16-9}$$

$$\text{Viscosity (of the liquid), } F_{\mu_l} = \lambda \mu_l c,$$

$$\text{Surface tension, } F_\sigma = \lambda \sigma.$$

A complete set (see Chapter 11) could be N_{Re}, N_{We}, N_{Fr}, s, and if necessary the geometric or modified wave number, α. From the theories of Miles [169] and Brooke Benjamin [37] on thin films, the density ratio s does not appear to be needed. Thus, for thin films, a representation such as that given in Fig. 11-1 should suffice for an area plot. However, if the film thickness is to be used in the groups, the modified wave number will probably have to be introduced. For very thin films (such as are found in annular flow), the Froude number approaches infinity, and the controlling groups should be the Reynolds, Weber, and modified wave numbers [see Eq. (16-5)]. Portalski [188] has made a similar suggestion, based on an analysis of film flow down a vertical plate. Further justification of the dimensional approach has been mentioned by Hughes et al. [118], who have considered the transition from annular to spray flow. (The analysis might also apply to the transition between wavy and slugging flow.)

The experimental work by the various authors already cited and by Keulegan [137] is in general agreement with the results of the theories. Hanratty and Engen [104] used a Reynolds number based on the film thickness, friction velocity, and liquid properties. Ellis and Gay [70] used a gas Reynolds number, but this could be related to the various numbers described above. Keulegan was able to describe his experiments by a modification of Jeffreys' analysis. Van Rossum [214] used a Weber number based on the film thickness, gas density, and gas velocity. This is easily related to the Weber number mentioned above. In addition he used a liquid Reynolds number and the ratio N_{We}/N_{Re}, substituting the gas velocity for the wave velocity. The theoretical work of Hanratty and Hershman [105] is an extension of Jeffreys' theory, and thus involves the same groups. The experimental check with the theory was quite good, being better when the roll waves moved over a squall surface (rough) than when they moved over a smooth one (surface active agent added in this case). Any of these experimental and theoretical results could be transferred as surfaces to the suggested area plot. A tabular summary of data on film thickness is given in reference 66. Al Sheikh [4] has also studied the transitions for the horizontal channel data just described and concluded that several possible plots could be used to describe the system.

Very little experimental work has been done on liquid–liquid pipe-flow systems. The first such work was by Russell *et al.* [197] on a mixture of mineral oil and water, followed by the work of Charles *et al.* [47] on an equal-density oil–water mixture. The oil density was increased by the addition of carbon tetrachloride. The flow patterns are very similar to those already described for liquid–gas flow; bubble, stratified, and mixed flows were observed. In the latter work, the bubble region could be subdivided into drops, bubbles, and slugs of oil in the water phase. Also, since the flow was of two equal-density fluids, the stratified region became a concentric annular flow of water surrounding a central core of oil, and the mixed-flow region was the flow of water drops in a continuous oil phase.

The situation in vertical flow is little different from that for the horizontal case. A similar flow-regime breakdown for a given liquid flow with increasing gas flow would be:

(1) *Bubble, aerated, or gas-dispersed flow*: Spherical and spherical cap bubbles, small compared with the tube diameter, are distributed across the pipe with some crowding toward the center.

(2) *Plug, gas-piston, or slug flow*: Large, approximately axially symmetric gas bubbles that almost bridge the tube cause the flow to alternate between being mainly liquid and mainly gas.

(3) *Churn, froth, or turbulent flow*: In this phase the gas is highly dispersed and interacting in a continuous liquid.

(4) *Developing slug or semi-annular flow*: Here, large slugs of gas have lost their axially symmetric characteristic. This is similar to plug flow, but the slug is now more of an interacting gas part. Generally the liquid contains gas and is similar to that described as churn flow.

(5) *Annular*: Annular flow consists of a gas core which usually contains liquid drops, with the main flow of liquid along the wall.

(6) *Dispersed, spray, or mist flow*: Most of the liquid is entrained as a spray in the gas.

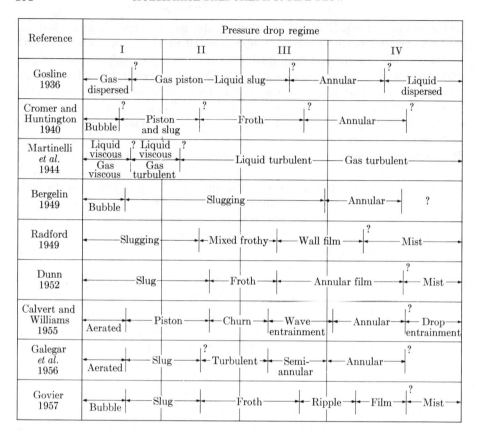

Fig. 16–3. Interrelationship of flow-pattern terminology. [*From G. W. Govier et al., Can. J. Chem. Eng.* **35,** *58 (1957). By permission.*]

Govier et al. [93] have made an attempt to bring some order to the terminology. They have compared the work of eight authors in terms of the types of flow (Fig. 16–3). The comparison is based on the shape of the pressure-drop curve versus the air–water volume ratio. Govier and Short [94] have extended this work to include the effect of tube diameter, and Brown et al. [40] have considered the density of the gas phase. In each of these works, the flow pattern, holdup, and pressure drop have been determined and empirically correlated. A general review of this work has been given by Govier [236]. Vertical flow regime data are also contained in references 43, 62, 141, 170, and 215. The reviews cited previously [48, 66, 97, 122, 136, 216] provide a useful compact source of information. Dukler and Wicks [66] also provide a review of vertical film flow on a flat plate. Moissis [170] has developed a simple theory for the breakup of developing slugs into a homogeneous flow, and has compared his result with the empirical correlations available. He has emphasized that a single line on an area plot will in general be inadequate, and that a band is necessary to separate the two regimes. The analysis appears to be most characteristic of the transition of slug flow (2) to a churn or froth flow (3). Griffith [99] experimentally examined the transition from developing slug to annular flow in detail, and was able to further define the stages of the transition.

With regard to vertical liquid–liquid systems, Govier et al. [95] and Wood [226] have extended the investigation to the upward vertical flow of oil–water mixtures in which the oil viscosity was varied. The flow patterns are essentially the same as in the air–water system. The pressure drop is independent of the oil viscosity so long as the water is the continuous phase. Brown and Govier [39] have provided details of the photographic study that was done in connection with the vertical oil–water study.

16-1.B Pressure-Drop and Void-Fraction Correlations

In view of our general inability to define the exact nature of two-phase flow without direct observation, it is not surprising to find a similar difficulty in predicting pressure drop. One way to avoid the problem of having to know the flow regime is simply not to incorporate it as a necessary part of the pressure drop correlation; in other words, to use an empirical or semi-empirical correlation, which treats all data in the same manner without regard to the particular flow regime. Such overall correlations are not expected to be perfect, but they can often provide an adequate estimate for a specific purpose. More accurate predictions of two-phase pressure drop is possible only after the type of flow has been established and correlations are available for each type. The first problem has already been discussed in some detail, and as noted, there is much work to be done. The second task involves many individual theoretical problems, and, of course, the checking of these with experimental data.

(a) Equations of two-phase flow. One must use the equation for the mechanical energy balance for two-phase flow in order to reduce the measured data to such a degree that one can separate the effects of acceleration and elevation change from the frictional contribution. In two-phase flow, just as in single-phase flow, it is the friction factor for which correlations are needed (i.e., the plot of friction factor versus Reynolds number). Obviously, the reverse is true; that is, in order to be able to calculate the expected drop in total pressure, one must be able to evaluate contributions from friction, from changes in acceleration or kinetic energy, and from changes in elevation. If one uses an integral approach, one need not consider the specific details of the flow regime. The correct form for the balance equation is obtained by taking Eq. (10–48) (which is on the basis of a unit mass of flowing fluid), writing it for the gas and liquid phases, then weighing them by their respective mass flow values, and summing. In all cases considered here, no mass transfer between phases occurs. That this is the correct procedure [146, 217] can be demonstrated by a complete rederivation of the equation. To illustrate, let us consider first the kinetic energy term, which was obtained from an integration of

$$\frac{1}{2} \int \rho U^3 \, dS, \tag{16-10}$$

which in turn was obtained from Eqs. (10-28) and (10-29) with the surfaces for integration (S_1 and S_2) taken perpendicular to the lines of flow. Integration over any configuration at Plane 1 and Plane 2 gives

$$\frac{1}{2}[\rho_{l1}(U_{l1}^3)_{\text{ave}} S_{l1} + \rho_{g1}(U_{g1}^3)_{\text{ave}} S_{g1} - \rho_{l2}(U_{l2}^3)_{\text{ave}} S_{l2} - \rho_{g2}(U_{g2}^3)_{\text{ave}} S_{g2}]. \tag{16-11}$$

In terms of the mass flow rate of each phase, this becomes

$$\frac{1}{2}\left[\frac{(U_{l1}^3)_{\text{ave}}}{U_{l1,\text{ave}}} w_{l1} + \frac{(U_{g1}^3)_{\text{ave}}}{U_{g1,\text{ave}}} w_{g1} - \frac{(U_{l2}^3)_{\text{ave}}}{U_{l2,\text{ave}}} w_{l2} - \frac{(U_{g2}^3)_{\text{ave}}}{U_{g2,\text{ave}}} w_{g2}\right]. \tag{16-12}$$

For steady state, $w_{l1} = w_{l2}$ and $w_{g1} = w_{g2}$, so that on the basis of a unit mass of flowing material (divided by $w_t = w_l + w_g$), the term becomes

$$\frac{1}{2} \frac{w_l}{w_t}\left[\frac{(U_{l1}^3)_{\text{ave}}}{U_{l1,\text{ave}}} - \frac{(U_{l2}^3)_{\text{ave}}}{U_{l2,\text{ave}}}\right] + \frac{1}{2} \frac{w_g}{w_t}\left[\frac{(U_{g1}^3)_{\text{ave}}}{U_{g1,\text{ave}}} - \frac{(U_{g2}^3)_{\text{ave}}}{U_{g2,\text{ave}}}\right],$$

or just the mass-flow-rate weighting of Eq. (10-48), written for each phase. The potential energy term from Eqs. (10-28) and (10-29) is

$$g \int \rho z U \, dS = g(\rho_{l1} z_1 U_{l1,\text{ave}} S_{l1} + \rho_{g1} z_1 U_{g1,\text{ave}} S_{g1} - \rho_{l2} z_2 U_{l2,\text{ave}} S_{l2} - \rho_{g2} z_2 U_{g2,\text{ave}} S_{g2})$$

$$= g(w_{l1} z_1 + w_{g1} z_1 - w_{l2} z_2 - w_{g2} z_2) = g(w_t z_1 - w_t z_2) = g w_t(z_1 - z_2),$$

or, on the basis of a unit mass of flowing fluid, it is $g(z_1 - z_2)$, which is unchanged from Eq. (10-48). The pressure term from Eq. (10-28) is

$$\int pU \, dS = (p_1 U_{l1,\text{ave}} S_{l1} + p_1 U_{g1,\text{ave}} S_{g1} - p_2 U_{l2,\text{ave}} S_{l2} - p_2 U_{g2,\text{ave}} S_{g2})$$

$$= p_1\left(\frac{w_{l1}}{\rho_{l1}} + \frac{w_{g1}}{\rho_{g1}}\right) - p_2\left(\frac{w_{l2}}{\rho_{l2}} + \frac{w_{g2}}{\rho_{g2}}\right). \tag{16-13}$$

For steady state, and on the basis of a unit mass of flowing fluid, Eq. (16-13) can be expressed as

$$\frac{w_l}{w_t} \frac{p_1 - p_2}{\rho_l} + \frac{w_g}{w_t}\left(\frac{p_1}{\rho_{g1}} - \frac{p_2}{\rho_{g2}}\right).$$

The final equation with all terms is

$$\frac{1}{2}\frac{w_l}{w_t}\left[\frac{(U_{l1}^3)_{\text{ave}}}{U_{l1,\text{ave}}} - \frac{(U_{l2}^3)_{\text{ave}}}{U_{l2,\text{ave}}}\right] + \frac{1}{2}\frac{w_g}{w_t}\left[\frac{(U_{g1}^3)_{\text{ave}}}{U_{g1,\text{ave}}} - \frac{(U_{g2}^3)_{\text{ave}}}{U_{g2,\text{ave}}}\right] + g(z_1 - z_2)$$

$$+ \frac{w_l}{w_t}\frac{p_1 - p_2}{\rho_l} + \frac{w_g}{w_t}\left(\frac{p_1}{\rho_{g1}} - \frac{p_2}{\rho_{g2}}\right) + \int_V p \, d\left(\frac{1}{\rho}\right) - F_{\text{Losses}} = 0, \tag{16-14}$$

where the work term cancels with no external work (since work done on the liquid by the gas is the negative of the work done on the gas by the liquid), where

ρ_l has been assumed constant, and where F_{Losses} represents the total viscous dissipation for both phases per unit mass of material flowing. The integral term requires a knowledge of the variation of pressure with density change. If the pressure ratio p_1/p_2 is no greater than 2, one can assume ρ constant (since an average gas density can be used), and the integral term is zero.

It should be emphasized that to obtain a good correlation for the frictional loss, one must eliminate kinetic energy effects and elevation changes. Unfortunately, the kinetic energy term requires that individual phase velocities at the entrance and exit of the pipe be known, and they usually are not known accurately.

An alternative approach is to start with the momentum equation (10–11 and 10–12) and derive the new relation by adding an equation for the gas to one for the liquid. For steady state, this is

$$w_l \left[\frac{(U_{l1}^2)_{\text{ave}}}{U_{l1,\text{ave}}}\right] + w_g \left[\frac{(U_{g1}^2)_{\text{ave}}}{U_{g1,\text{ave}}}\right] - w_l \left[\frac{(U_{l2}^2)_{\text{ave}}}{U_{l2,\text{ave}}}\right] - w_g \left[\frac{(U_{g2}^2)_{\text{ave}}}{U_{g2,\text{ave}}}\right] + (p_1 - p_2)S$$

$$- \tau_w A_w - g \int_V (\rho_l S_l + \rho_g S_g)\,dz = 0, \qquad (16\text{--}15)$$

where the drag at the wall is expressed in terms of the wall shear stress τ_w and a wall area A_w, where interactions between phases are equal and opposite and therefore cancel as before, and where the external force has been assumed to be gravity only.

Let us consider the case in which p_1/p_2 is less than 2, and an average density for the gas [defined as $1/\bar{\rho}_g = \frac{1}{2}(1/\rho_{g1} + 1/\rho_{g2})$] can be used, so that Eq. (16–14) can be rearranged to a pressure-drop form. Then

$$p_1 - p_2 = \underbrace{\Delta p_{\text{Fe}}}_{\text{Frictional}} + \underbrace{\frac{gw_l(z_2-z_1)}{(U_l S_l + \bar{U}_g S_g)}}_{\text{Head}} + \underbrace{\frac{w_l\left[\frac{(U_{l2}^3)_{\text{ave}}}{U_{l2,\text{ave}}} - \frac{(U_{l1}^3)_{\text{ave}}}{U_{l1,\text{ave}}}\right]}{2(U_l S_l + \bar{U}_g S_g)}}_{\text{Liquid kinetic}} + \underbrace{\frac{w_g\left[\frac{(U_{g2}^3)_{\text{ave}}}{U_{g2,\text{ave}}} - \frac{(U_{g1}^3)_{\text{ave}}}{U_{g1,\text{ave}}}\right]}{2(U_l S_l + \bar{U}_g S_g)}}_{\text{Gas kinetic}},$$

(Total)

$$(16\text{--}16)$$

where $U_l S_l + \bar{U}_g S_g = w_l/\rho_l + w_g/\bar{\rho}_g$ and where Δp_{Fe} is the frictional pressure drop contribution obtained via the energy equation; it is defined as

$$\Delta p_{\text{Fe}} \equiv F_{\text{Losses}} \frac{w_l}{w_l/\rho_l + w_g/\bar{\rho}_g} = F_{\text{Losses}} \frac{w_l}{U_l S_l + \bar{U}_g S_g} = F_{\text{Losses}} \bar{\rho}. \qquad (16\text{--}17)$$

From the definition, $\bar{\rho}$ above can be expressed as a mass weighted density:

$$1/\bar{\rho} = (1-x)/\rho_l + x/\bar{\rho}_g, \qquad (16\text{--}18)$$

where x is the weight fraction vapor flowing, or the quality. The density can also be expressed in terms of volume fractions:

$$\bar{\rho} = \overline{\rho_l R_l + \rho_g R_g} = \rho_l \bar{R}_l + \bar{\rho}_g \bar{R}_g.$$

In general, the area fractions within a pipe at any given cross section are defined as

$$R_l = S_l/S, \qquad R_g = 1 - R_l = S_g/S. \tag{16-19}$$

In the density equation, the overbar on $\bar{R}_l$ and $\bar{R}_g$ denotes an average over the pipe (this is the holdup or volume fraction as contrasted to an area fraction). In terms of the area fractions of Eq. (16–19), the momentum equation (16–15) becomes

$$\underbrace{p_1 - p_2}_{\text{Total}} = \underbrace{\Delta p_{\text{Fm}}}_{\text{Frictional}} + \underbrace{g\bar{\rho}(z_2 - z_1)}_{\text{Head}} + \underbrace{\left[\frac{G_l^2/\beta_{l2}}{\rho_{l2} R_{l2}} + \frac{G_g^2/\beta_{g2}}{\rho_{g2} R_{g2}} - \frac{G_l^2/\beta_{l1}}{\rho_{l1} R_{l1}} - \frac{G_g^2/\beta_{g1}}{\rho_{g1} R_{g1}}\right]}_{\text{Kinetic}}. \tag{16-20}$$

The term Δp_{Fm} represents the frictional pressure-drop contribution obtained via the momentum equation, and is defined as $\Delta p_{\text{Fm}} \equiv \tau_w A_w/S = \tau_w 2L/r_0$. The term $G_l = \rho_l R_l U_{l,\text{ave}}$, and is the superficial mass velocity of the liquid. (It is constant along the pipe by virtue of the assumption that there is no mass transfer between phases.) The β's are momentum-correction terms which are almost always assumed unity because their value is not known. Gravity acts on the total mass of fluid within the pipe, so an integrated average density is necessary. The equation can be expressed in terms of G_l and the quality, since $G_l = w_l/S = (w_l/w_t)(w_t/S) = (1-x)G_t$ and $G_g = x G_t$. One then obtains

$$p_1 - p_2 = \Delta p_{\text{Fm}} + g\bar{\rho}(z_2 - z_1)$$
$$+ G_t^2 \left[\frac{(1-x)^2}{\rho_{l2} R_{l2} \beta_{l2}} + \frac{x^2}{\rho_{g2} R_{g2} \beta_{g2}} - \frac{(1-x)^2}{\rho_{l1} R_{l1} \beta_{l1}} - \frac{x^2}{\rho_{g1} R_{g1} \beta_{g1}}\right]. \tag{16-21}$$

Equations (16–20) and (16–21) (with all β's equal to unity) are forms used by Martinelli and Nelson [166]. Equation (16–16) has been used to correct data by Hughmark and Pressberg [121] and Hughmark [120].

In the more restricted case of a *homogeneous mixture*, the mixture density can still vary because of pressure drop. For such a system, inlet and exit densities, ρ_1 and ρ_2, are defined similarly to $\bar{\rho}$ of Eq. (16–17), except that entrance (1) or exit (2) conditions are used; that is, for ρ_1,

$$\rho_1 = \frac{w_t}{w_l/\rho_l + w_g/\rho_{g1}} = \frac{1}{(1-x)/\rho_l + x/\rho_{g1}} = \rho_l R_{l1} + \rho_{g1} R_{g1}.$$

The total mass velocity is expressed as $G_t = w_t/S = \rho_1 U_{1,\text{ave}} = \rho_2 U_{2,\text{ave}}$, so that

$$\frac{(U^3)_{\text{ave}}}{U_{\text{ave}}} \equiv \frac{U^2_{\text{ave}}}{\alpha} = \frac{G_t^2}{\alpha \rho^2}.$$

With this, Eq. (16–16) reduces to

$$p_1 - p_2 = \Delta p_{\text{Fe}} + g\bar{\rho}(z_2 - z_1) + \frac{G_t^2}{\alpha} \frac{\bar{\rho}}{2}\left(\frac{1}{\rho_2^2} - \frac{1}{\rho_1^2}\right),$$

where it has been assumed that α does not vary between 1 and 2.

The term $\bar{\rho}$ has been defined by Eqs. (16–17) and (16–18) and can be combined with the above equation to evaluate the multiplier of (G_t^2/α):

$$\frac{1}{\rho_2^2} - \frac{1}{\rho_1^2} = \left(\frac{1-x}{\rho_l} + \frac{x}{\rho_{g2}}\right)^2 - \left(\frac{1-x}{\rho_l} + \frac{x}{\rho_{g1}}\right)^2 = x^2\left(\frac{1}{\rho_{g2}^2} - \frac{1}{\rho_{g1}^2}\right) + \frac{2x(1-x)}{\rho_l}\left(\frac{1}{\rho_{g2}} - \frac{1}{\rho_{g1}}\right)$$

$$= 2x\left(\frac{\rho_{g1} - \rho_{g2}}{\rho_{g1}\rho_{g2}}\right)\left[\frac{x(\rho_{g1} + \rho_{g2})}{2\rho_{g1}\rho_{g2}} + \frac{1-x}{\rho_l}\right].$$

Now, from Eq. (16–18), we have

$$\bar{\rho} = \frac{1}{\frac{1-x}{\rho_l} + \frac{x}{\bar{\rho}_g}} = \frac{1}{\frac{1-x}{\rho_l} + \frac{x(\rho_{g1} + \rho_{g2})}{2\rho_{g1}\rho_{g2}}};$$

therefore

$$\frac{\bar{\rho}}{2}\left(\frac{1}{\rho_2^2} - \frac{1}{\rho_1^2}\right) = x\left[\frac{\rho_{g1} - \rho_{g2}}{\rho_{g1}\rho_{g2}}\right] = x\left(\frac{1}{\rho_{g2}} - \frac{1}{\rho_{g1}}\right).$$

This is equivalent to $1/\rho_2 - 1/\rho_1$, if x and ρ_l are constant; that is,

$$\frac{1}{\rho_2} - \frac{1}{\rho_1} = \left(\frac{1-x}{\rho_l} + \frac{x}{\rho_{g2}}\right) - \left(\frac{1-x}{\rho_l} + \frac{x}{\rho_{g1}}\right) = x\left(\frac{1}{\rho_{g2}} - \frac{1}{\rho_{g1}}\right).$$

Thus, for homogeneous flow without mass transfer, $\bar{\rho}$ can be expressed as

$$1/\bar{\rho} = \tfrac{1}{2}(1/\rho_1 + 1/\rho_2), \tag{16-22}$$

since, when this is used, the multiplier of G_t^2/α is also $1/\rho_2 - 1/\rho_1$. The previous pressure drop equation can now be written as

$$p_1 - p_2 = \Delta p_{\text{Fe}} + g\bar{\rho}(z_2 - z_1) + (G_t^2/\alpha)(1/\rho_2 - 1/\rho_1). \tag{16-23}$$

For the momentum equation, the multiplier of G_t^2 is given in Eq. (16–21), but it is easier to work with Eq. (16–15), which, for homogeneous flow, reduces to

$$p_1 - p_2 = \Delta p_{\text{Fm}} + g\bar{\rho}(z_2 - z_1) + (w_t/S)(U_{2,\text{ave}}/\beta_2 - U_{1,\text{ave}}/\beta_1).$$

If β does not change between 1 and 2, it becomes

$$p_1 - p_2 = \Delta p_{\text{Fm}} + g\bar{\rho}(z_2 - z_1) + (G_t^2/\beta)(1/\rho_2 - 1/\rho_1). \tag{16-24}$$

In general it is assumed that $\alpha = \beta = 1$, and the energy and momentum analyses give the same result. For this rather restricted case, then, $\Delta p_{\text{Fe}} = \Delta p_{\text{Fm}}$. If $\alpha \neq \beta$, then Δp_{Fe} cannot equal Δp_{Fm}.

For the case of a nonhomogeneous horizontal flow with the momentum term negligible (which is often, but not always, the case for isothermal two-phase flow), Eqs. (16–16) and (16–20) reduce to

$$p_1 - p_2 = \Delta p_{\text{Fe}} = \Delta p_{\text{Fm}} = \tau_w 2L/r_0 = F_{\text{Losses}}\bar{\rho}. \tag{16-25}$$

Experiments performed under these conditions provide ideal data for correlating the drop in pressure due to friction.

For convenience, a friction factor is defined in terms of the frictional pressure drop:

$$\Delta p_F = 4f(L/d_0)(U_{ave}^2/2)\rho. \tag{16-26}$$

Since this is to be a frictional contribution only, it should be based on a location for which acceleration is not a factor. This is true for the pipe entrance. Thus, for the velocity, one can define U_{ave} as $U_{ave} = \sqrt{(U_1^2)_{ave}/U_{1,ave}} = G_t/\rho_1$. Using this in Eq. (16-26) gives us

$$\Delta p_F = 2f(L/d_0)(G_t^2/\rho_1^2)\rho. \tag{16-27}$$

One simple definition of ρ is ρ_1 [64, 92]; however, since the friction factor is usually correlated against a Reynolds number, the selection of ρ should be consistent with that selection used in the other dimensionless numbers [64]. It should be remembered that if the frictional pressure drop obtained from the energy equation is not the same as that obtained from the momentum equation, then friction factors based on these will be different.

(b) Overall correlations. One of the best correlations for pressure drop, a correlation requiring little information, was developed by Martinelli and his coworkers [156, 164, 165]. The correlation does not specifically include the concept of flow patterns, and is semi-empirical. In spite of the fact that the exact nature of the flow is neglected, pressure drops can be predicted to within 40% (standard deviation). (However, some individual predictions may be in error by as much as 200%, depending on the type of flow.) The work of Lockhart and Martinelli [156] will be considered first, because of its extensive historical use and the continued references to it in the literature (e.g., references 23 and 80). Shortcomings which have come to light as a result of more recent experimental data will then be pointed out and finally there will be a discussion of alternative methods necessary for more accurate estimates.

Martinelli et al. [164] suggested that two-phase flow could be divided into four flow regimes: (1) liquid and gas turbulent (tt), (2) liquid turbulent and gas viscous (tv), (3) liquid viscous and gas turbulent (vt), (4) liquid and gas viscous (vv).

The changeover points were selected to be consistent with single-phase flow and to give the best correlation of the experimental data. There are two basic postulates on which the analysis is based: (1) The static pressure drop for the liquid and gaseous phases must be equal regardless of the flow pattern. (2) The volume occupied by the liquid plus the gas at any instant (position) must equal the total volume of the pipe. These postulates imply that the flow pattern does not change along the tube length. In effect, they eliminate those flows which have large pressure fluctuations, such as slug flow, and those with radial pressure drop, such as stratified flow.

In the final correlation, the pressure drop per unit length during two-phase frictional flow, $(\Delta p/\Delta L)_{TPF}$, was expressed as

$$(\Delta p/\Delta L)_{TPF} = \phi_l^2 (\Delta p/\Delta L)_l, \tag{16-28}$$

where $(\Delta p/\Delta L)_l$ is the pressure drop per unit length that would exist if the liquid phase were assumed to flow alone. The term ϕ_l represents a parameter which is a function of a dimensionless variable X, which is in turn a function of the flow rates and physical properties of the fluids. In the following analysis, the equations for the liquid phase will be developed. A similar set of equations for the gas phase is needed, but these can be obtained simply by changing the appropriate subscripts.

The static pressure drop due to the liquid flow may be written in terms of a friction factor as

$$(\Delta p/\Delta L)_{\text{TPF}} = 4 f_l \, (\rho_l/d_l)(\bar{U}_l^2/2), \tag{16-29}$$

where d_l is defined as a "hydraulic diameter" of the liquid flow pattern. The area of the flow pattern for one phase can be related to the hydraulic diameter; for a complex cross section, one can use

$$A_l = \alpha \pi d_l^2/4. \tag{16-30}$$

For the corresponding gas phase, a constant β is used. So long as the shape of the individual phase is circular, α and β will be unity. Equation (16-30) can be used to obtain $\bar{U}_l$:

$$\bar{U}_l = w_l/\alpha(\pi d_l^2/4)\rho_l. \tag{16-31}$$

The friction factor f_l can be expressed in terms of the Blasius formula as

$$f_l = c_l/N_{\text{Re},\,l}^n, \tag{16-32}$$

where $N_{\text{Re},\,l} = 4w_l/\pi \alpha d_l \mu_l$. For the gas phase, the power in Eq. (16-32) is denoted as m. Combining Eqs. (16-28), (16-31), and (16-32) gives

$$\left(\frac{\Delta p}{\Delta L}\right)_{\text{TPF}} = \frac{2c_l w_l^2}{(4w_l/\pi \alpha d_l \mu_l)^n \alpha^2 d_l^5 (\pi/4)^2 \rho_l} = \left[\frac{2(4/\pi)^{2-n} c_l \mu_l^n w_l^{2-n}}{d_0^{5-n} \rho_l}\right] \alpha^{n-2} \left(\frac{d_0}{d_l}\right)^{5-n}.$$

The bracketed term is the pressure drop per unit length for only liquid flowing in the pipe; that is,

$$\left(\frac{\Delta p}{\Delta L}\right)_l = \frac{2 f_l \rho_l \bar{U}_l^2}{d_0} = 2 \frac{c_l \rho_l}{(4w_l/\pi d_0 \mu_l)^n} \frac{w_l^2}{(\pi/4)^2 d_0^4 \rho_l^2 d_0} = \frac{2(4/\pi)^{2-n} c_l \mu_l^n w_l^{2-n}}{d_0^{5-n} \rho_l}.$$

Thus

$$\left(\frac{\Delta p}{\Delta L}\right)_{\text{TPF}} = \left(\frac{\Delta p}{\Delta L}\right)_l \alpha^{n-2} \left(\frac{d_0}{d_l}\right)^{5-n}. \tag{16-33}$$

Equations (16-28) and (16-33) can be combined to give

$$\frac{(\Delta p/\Delta L)_{\text{TPF}}}{(\Delta p/\Delta L)_l} = \phi_l^2 = \alpha^{n-2}\left(\frac{d_0}{d_l}\right)^{5-n}. \tag{16-34}$$

A similar analysis of the gas phase leads to the equations corresponding to (16-33) and (16-34). However, these equations would be in terms of β, m, and ϕ_g.

The second postulate (i.e., that the liquid volume plus the gas volume equals the pipe volume) gives

$$\alpha d_l^2 + \beta d_g^2 = d_0^2. \quad (16\text{--}35)$$

The fraction of the tube area filled with liquid, R_l, would be

$$R_l = \alpha(d_l/d_0)^2 = 1 - \beta(d_g/d_0)^2 \quad (16\text{--}36)$$

or

$$\beta = (1 - R_l)(d_0/d_g)^2 = R_g(d_0/d_g)^2.$$

Cited in Eqs. (16–33) through (16–36) were four variables, d_l/d_0, d_g/d_0, α, and β, which can be expressed in terms of four experimental variables:

$$\phi_l, \quad \phi_g, \quad R_l, \quad \text{and} \quad R_g. \quad (16\text{--}37)$$

Another variable can be obtained by combining Eqs. (16–28) through (16–32) to give

$$\frac{d_l^{5-n}}{d_g^{5-m}} = \left(\frac{c_l w_l^{2-n} \mu_l^n \rho_g}{c_g w_g^{2-m} \mu_g^m \rho_l}\right) \frac{\beta^{2-m}}{\alpha^{2-n}} \left(\frac{\pi}{4}\right)^{n-m},$$

which can be rearranged to give

$$\left(\frac{4}{\pi d_0}\right)^{m-n} \frac{c_l w_l^{2-n} \mu_l^n \rho_g}{c_g w_g^{2-m} \mu_g^m \rho_l} = \left(\frac{d_l}{d_0}\right)^{5-n} \left(\frac{d_0}{d_g}\right)^{5-m} \frac{\alpha^{2-n}}{\beta^{2-m}} = X^2. \quad (16\text{--}38)$$

Using a new Reynolds number,

$$N_{\text{Re},l0} = 4w_l/\pi d_0 \mu_l, \quad (16\text{--}39)$$

which is simply the liquid Reynolds number (l) based on the inside pipe diameter (0), we can rewrite Eq. (16–38) as

$$\frac{N_{\text{Re},g0}^m}{N_{\text{Re},l0}^n} \frac{c_l}{c_g} \frac{w_l^2}{w_g^2} \frac{\rho_g}{\rho_l} = X^2. \quad (16\text{--}40)$$

This variable, X^2, is simply the ratio of the liquid to the gas pressure drop, if each phase flows separately; that is,

$$X^2 = (\Delta p/\Delta L)_l \big/ (\Delta p/\Delta L)_g. \quad (16\text{--}41)$$

We now make the assumption that the variables of Eq. (16–37) are functions of the parameter X. This parameter must be evaluated with the appropriate values of n, m, c_l, and c_g, as determined by the flow conditions. These values and the criteria for transition are tabulated in Table 16-2. Empirical correlations of X against ϕ_l, ϕ_g, $\bar{R}_l$, and $\bar{R}_g$ for the various flows are shown in Fig. 16-4, and the coordinates are given in Table 16-3. The overbar on $\bar{R}_l$ and $\bar{R}_g$ denotes that because of the experimental procedure used to obtain the data the correlation is really based on a volume fraction (holdup) in the pipe. If the original postulates are valid, then the area fraction at any cross section would be equal to the volume fraction. To calculate a $(\Delta p/\Delta L)_{\text{TPF}}$, we obtain X^2 and then ϕ_l from Fig. 16-4 or Table 16-3, calculate the value of $(\Delta p/\Delta L)_l$, and find that $(\Delta p/\Delta L)_{\text{TPF}}$ is simply

TABLE 16-2

Values of n, m, c_l, and c_g*

	t-t	v-t	t-v	v-v
n	0.2	1.0	0.2	1.0
m	0.2	0.2	1.0	1.0
c_l	0.046†	16	0.046†	16
c_g	0.046†	0.046†	16	16
$N_{\text{Re},l0}$	> 2000	< 1000	> 2000	< 1000
$N_{\text{Re},g0}$	> 2000	> 2000	< 1000	< 1000

*By permission from R. W. Lockhart and R. C. Martinelli [156].
†For smooth pipes

the product with the term ϕ_l^2 (16–34). The same procedure is valid for the corresponding gas equations.

EXAMPLE 16–1. A mixture of a gas and a liquid is to be carried 500 ft with a pressure drop of 1.5 psi. The rate of flow of the gas is 10 lb/min and of the liquid 1.5 ft³/min. Specify the nominal pipe size needed.

Answer. The solution will be a trial-and-error one. For the gas assume: $\rho_g = 0.0373$ lb/ft³, $\mu_g = 0.012$ centipoise. For the liquid assume: $\rho_l = 59.8$ lb/ft³, $\mu_l = 0.284$ centipoise.

	Velocity, ft/sec		Reynolds number		f	
Assume d_0	Gas	Liquid	Gas	Liquid	Gas	Liquid
3.068 in.	119	0.487	140,000	39,100	0.0053	0.0065
4.026	69.1	0.283	106,800	29,700	0.0054	0.0068
6.065	30.4	0.125	71,000	20,000	0.0058	0.0074

X	ϕ_g	$(\Delta p/\Delta L)_g = (2 f_g \bar{U}_g^2 \rho_g / d_0) \times 10^3$	$\left(\dfrac{\Delta p}{\Delta L}\right)_{\text{TPF}} \times 10^3$	
0.182	2.20	4.7 psi/ft	22.8 psi/ft	
0.184	2.20	1.238	6.0	
0.186	2.20	0.170	0.825	

The four-inch line provides the closest fit, but it would be advisable to use the next-larger size (six inch).

EXAMPLE 16–2. Use the Baker plot of Fig. 16–2 to estimate the flow type for the flow of Example 16–1. Assume that the gas is air, and that the liquid is water.

Answer. From Eq. (16–1), we have

$$\lambda = [(0.498)(0.96)]^{1/2} = \sqrt{0.478} = 0.692.$$

TABLE 16-3
Coordinates of ϕ and $\bar{R}$ vs. parameter X*

X	All mechanisms		Turbulent-turbulent		Viscous-turbulent		Turbulent-viscous		Viscous-viscous	
	$\bar{R}_l$	$\bar{R}_g$	ϕ_l	ϕ_g	ϕ_l	ϕ_g	ϕ_l	ϕ_g	ϕ_l	ϕ_g
0.01			128.0	1.28	120.0	1.20	112.0	1.12	105.0	1.05
0.02			68.4	1.37	64.0	1.28	58.0	1.16	53.5	1.07
0.04			38.5	1.54	34.0	1.36	31.0	1.24	38.0	1.12
0.07	0.04	0.96	24.4	1.71	20.7	1.45	19.3	1.35	17.0	1.19
0.10	0.05	0.95	18.5	1.85	15.2	1.52	14.5	1.45	12.4	1.24
0.2	0.09	0.91	11.2	2.23	8.90	1.78	8.70	1.74	7.00	1.40
0.4	0.14	0.86	7.05	2.83	5.62	2.25	5.50	2.20	4.25	1.70
0.7	0.19	0.81	5.04	3.53	4.07	2.85	4.07	2.85	3.08	2.16
1.0	0.23	0.77	4.20	4.20	3.48	3.48	3.48	3.48	2.61	2.61
2.0	0.31	0.69	3.10	6.20	2.62	5.25	2.62	5.24	2.06	4.12
4.0	0.40	0.60	2.38	9.50	2.05	8.20	2.15	8.60	1.76	7.00
7.0	0.48	0.52	1.96	13.7	1.73	12.1	1.83	12.8	1.60	11.2
10.0	0.53	0.47	1.75	17.5	1.59	15.9	1.66	16.6	1.50	15.0
20.0	0.66	0.34	1.48	29.5	1.40	28.0	1.44	28.8	1.36	27.3
40.0	0.76	0.24	1.29	51.5	1.25	50.0	1.25	50.0	1.25	50.0
70.0	0.84	0.16	1.17	82.0	1.17	82.0	1.17	82.0	1.17	82.0
100.0	0.90	0.10	1.11	111.0	1.11	111.0	1.11	111.0	1.11	111.0

*By permission from R. W. Lockhart and R. C. Martinelli [156].

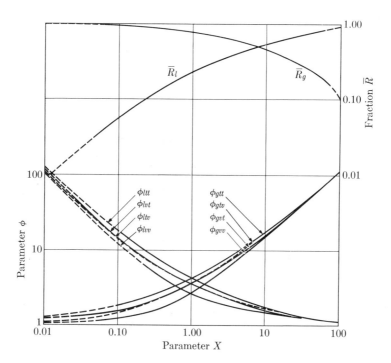

Fig. 16–4. *Faired curves showing relation between ϕ_l, ϕ_g, R_l and R_g for all flow mechanisms.* [From R. W. Lockhart, and R. C. Martinelli, Chem. Eng. Prog. **45**, 39–48 (1949). By permission.]

Let us assume $\sigma = 73$ dynes/cm for water, so that Eq. (16-2) gives

$$\phi = 1/(0.309)^{1/3} = 1/0.676 = 1.48,$$
$$G_l = (0.025)(59.8)(3600)/0.201 = 26{,}800 \text{ lb/hr-ft}^2,$$
$$G_g = (6.1)(0.0373)(3600)/0.201 = 4090,$$
$$G_g/\lambda = 5.91 \times 10^3, \qquad G_l \lambda \phi / G_g = 6.71.$$

The point is marked X in Fig. 16–2. One can see that it is in a poorly defined area and that the flow could be slug, annular, wavy, or even stratified. The large pipe diameter raises further doubts about the type, since there is some indication [66] that the shift with diameter would be in a direction which would result in stratified flow.

A number of investigators have reported two-phase pressure drop data [6, 15, 29, 30, 51, 63, 66, 80, 83, 84, 93, 111, 121, 134, 140, 156, 192, 219]. It would be well to review some of their specific findings, and then define as well as possible the limitations of the Lockhart-Martinelli correlation. Gazley and Bergelin [29, 83] pointed out that their data for a 1-in. copper pipe correlated to within about

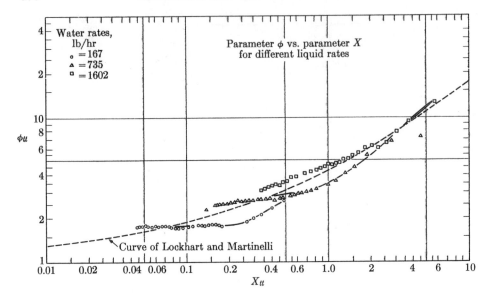

*Fig. 16–5. ϕ_{ll} versus X_{ll}. Data of Jenkins at three liquid rates compared with curve of Lockhart and Martinelli. [From discussion by C. Gazley Jr., and O. P. Bergelin, Chem. Eng. Prog. **45**, 45–46 (1949). By permission.]*

±40%; however, their data fell within ±5% of the best line through any one liquid flow rate (see Fig. 16–5). They also noted a definite change in slope, which corresponded to a change in the flow regime. In their results on stratified flow in which hydraulic gradients exist and the pressure drops in the two phases are not equal, predicted pressure drops were about twice the experimental values. Bergelin et al. [30] considered downward turbulent flow in a 1-in. vertical tube. The water was in annular flow and correlated with the Martinelli curve to within ±30%. The trend of the points for any one water rate was counter to the predicted trend. Kosterin and Rubanovich [140] reported that changes in surface tension are not important unless a stable foam is formed, in which case the pressure drop can increase by a factor of as much as three. An analysis of vertical up-flow by Hughmark and Pressburg [121] showed marked dependence on flow rate and poor agreement with the Lockhart-Martinelli analysis. The authors empirically correlated the data to an average error of less than 15%. Ward and DallaValle [219] considered a non-Newtonian system of kaolin clay in water. On all runs (120), a ±20% correlation with the Lockhart-Martinelli correlation was obtained. Fohrman [76] tested the rectangular channel flow of Newtonian materials with a viscosity range of 0.75 to 500 centipoises. The high viscosity data were only 15% below that predicted, and this could be attributed either to the use of a hydraulic radius or to the higher viscosity. The data of Chenoweth and Martin [51] at 100 psig show wide deviation from the correlation. The 1.5-in. line data were as low as one-half the predicted values. The data for the 3-in. line were also low, with the predicted values being up to 2.5 times those actually measured. Hoogendoorn [111] has compared his data with the correlation and found them valid for plug, slug, and froth flows at atmospheric pressure and for any pipe diameter. However, he found that the correlation was invalid for any of these flows if the gas density was different from air at atmospheric pressure and if the flow was stratified, wavy, or mist-annular. Hoogendoorn and Buitelaar [112] looked further into the effect of gas density.

One can conclude that the Lockhart-Martinelli correlation, because of its empirical nature, is capable of providing the level of pressure drop, but that it is not correct in detail. It is valid for horizontal flow in a pipe of any diameter, with predictions becoming increasingly higher as diameter increases. The effect is magnified at pressures greater than atmospheric. This error probably results from the fact that the original correlation was based on total pressure drop data obtained on small-diameter pipes, rather than on data for which the kinetic contribution was subtracted. As the pipe diameter is increased, the actual kinetic contribution becomes small. Thus, in these cases, if one used the correlation, one would expect to overestimate the pipe size required. The absolute pressure level is a major factor; the higher the pressure (and the larger the diameter at a given pressure), the higher the predicted pressure drop will be in comparison with experimental values [15, 51, 111]. Liquid flow rate does have a definite effect, but it lies within the limits of the correlation. The viscosity and surface tension of the fluid do not appear to affect the results to any major degree. The effect of the type of flow is marked; annular flow has pressure drops higher than those predicted, and stratified, wavy, and spray flows may have pressure drops considerably lower than those predicted, while bubble and slug flow check pretty well. The correlation is not overly accurate for vertical flow, being about one-half as accurate as for the horizontal case.

Dukler et al. [67] have tested the correlation with a set of data carefully culled from the data bank previously mentioned. In their test, the most extensive to date, nearly 2300 data points were checked. Where comparisons could be made, the effects already commented on were consistent with their results. For the entire set of data, a standard deviation of 36% was reported. The deviations are *not* normally distributed; it was found that 68% (one sigma) of the data were within 25% of the correlation. Unfortunately, the authors did not report the two-sigma (95.5%) and three-sigma (99.7%) limits, but one might expect as many as 100 points to be off by 50% or more and 10 points or so to be off by 75% or more. These rough estimates are consistent with the comments previously made on the level of accuracy that can be expected from this overall method, which requires no information other than the fluid properties, pipe diameter, and flow rates.

For horizontal flow, Dukler et al. [67] have suggested a correlation based on dynamic similarity (inertial, viscous, and pressure forces), with any effects of gravitational and surface forces neglected. For homogeneous flow, they obtained the friction factor (16–27) with ρ given by ρ_1. Under the assumptions made, the average viscosity needed for the corresponding Reynolds number, which was

$\rho U_{ave}d_0/\mu_{ave} = 4w_l/\pi d_0\mu_{ave}$, was no longer arbitrary, and was shown to be

$$\mu_{ave} = \mu_1 = \mu_l R_{l1} + \mu_g R_{g1} = \rho_1\left[\frac{\mu_l(1-x)}{\rho_l} + \frac{\mu_g x}{\rho_{g1}}\right], \quad (16\text{--}42)$$

which is a mass-weighted kinematic viscosity. The method is of importance, since one assumes that the numerical value of the friction factor is given by the normal single-phase correlation, so long as the proper density and viscosity are used. No two-phase flow information is needed. The authors compared the actual pressure drop for the culled horizontal-flow data with the data computed from the friction factor and corrected for acceleration by an equation equivalent to (16–24). The results were slightly better than those for the Lockhart-Martinelli correlation (68% within 21% versus 68% within 25%). This is both surprising and encouraging, since again no two-phase flow data were needed to establish the original correlation, as was necessary for the Lockhart-Martinelli work. A second correlation was tested, in which the friction factor was defined by Eq. (16–27), with ρ given by

$$\rho = \frac{\rho_l R_{l1}^2}{R_l} + \frac{\rho_{g1} R_{g1}^2}{R_g} = \rho_1^2\left[\frac{(1-x)^2}{\rho_l R_l} + \frac{x^2}{\rho_{g1}R_g}\right], \quad (16\text{--}43)$$

which gives

$$\frac{\rho}{\rho_1^2} = \frac{(1-x)^2}{\rho_l R_l} + \frac{x^2}{\rho_{g1}R_g}. \quad (16\text{--}44)$$

Note that this ratio is the same as that developed as a multiplier of G_l^2 in Eq. (16–21) (with the β's equal to unity), thus justifying the definition of ρ given in Eq. (16–43). The terms ρ and μ_{ave} for use in the Reynolds number are given by Eqs. (16–43) and (16–42), respectively. For 400 of the 2300 data points tested [for which Eq. (16–25) could be considered valid], reliable values of the area fraction were known, and were used to construct an empirical friction-factor curve. All the data were then tested by using this correlation, with corrections for kinetic energy given by Eq. (16–20) and an estimate of the area fraction from the empirical correlation by Hughmark [119]. The results were the best obtained; 68% of the data were within 18% of the correlation, as compared with 68% within 25% for the Lockhart-Martinelli work. Neither correlation procedure has been tested with vertical data. Other definitions of ρ and μ_{ave} can be generated, but none have been tested.

In addition to the above, a number of empirical correlations and corrections to existing correlations which are often used and referred to have been presented by the various authors already cited [51, 53, 92, 111, 112, 119, 120, 121, 228]. Several of these consider vertical flow.

There is usually a reduction in pressure drop when water is added to a flowing viscous oil. The references for liquid–liquid flow studies have already been cited [47, 197]; the pressure-drop data have been correlated on a plot of a friction factor versus superficial water velocity. The density differences in liquid–liquid flow are not large; thus it might be expected that in highly turbulent flow the system can be treated as a single phase, with average physical properties. The data of Russell et al. [197] indicate that this is true. This work has been extended by Charles et al. [47] to the horizontal flow of equal-density oil–water mixtures. Both holdup and pressure drop were considered and presented as plots against the input oil–water ratio. No comparisons with the various pressure-drop correlations were reported. It is interesting to note that, compared with the pressure gradient of oil alone, the pressure drop of the oil–water mixture was reduced, at maximum, by a factor of ten. Cengel et al. [44] have presented the flow of an unstable liquid–liquid emulsion considered as a homogeneous fluid with rheological properties.

In order to correct two-phase pressure-drop data for acceleration effects, and for many of the pressure-drop correlations, one must know the area fraction. One usually knows the inlet liquid fraction R_{l_1} from the inlet conditions, but because of the pressure drop and the resulting variation in the gas density, the area fraction R_l can change as the flow progresses down the pipe. Ideally, the area fraction should be measured at specific cross sections, but it is more often measured as an average over the entire test section by trapping the liquid and measuring the volume. This latter volume fraction or holdup is denoted by $\bar{R}_l$. The void fraction is often indicated by $\bar{R}_g$. The measurement of the area fraction is more difficult, and many procedures have been suggested. Neal and Bankoff [175] have cited such methods as gamma- and beta-ray absorption, radioactive tracers, photographic methods, and several direct-probe methods. Experimental methods are also considered in references 12, 19, 76, 110, 115, 124, 126, 181, and 182. In two-component isothermal flow, acceleration effects are not extreme because there is no mass transfer between phases; furthermore, if the pipe diameter is not too small, R_l will be essentially constant and equal to $\bar{R}_l$. For small-diameter pipes and one-component two-phase flow systems, this is not expected to be true.

A number of interrelated terms are in common use. The quality x has already been defined. The slip ratio is U_g/U_l and the relative or slip velocity is $U_g - U_l$. Useful relations can be obtained from the various definitions:

$$\frac{U_g}{U_l} = \frac{w_g}{w_l} \frac{\rho_l}{\rho_g} \frac{S_l}{S_g} = \frac{w_g}{w_l} \frac{\rho_l}{\rho_g} \frac{R_l}{R_g} = \frac{x}{1-x} \frac{\rho_l}{\rho_g} \frac{R_l}{1-R_l}, \qquad (16\text{-}45)$$

$$U_g - U_l = \frac{w_g}{\rho_g R_g S} - \frac{w_l}{\rho_l R_l S} = G_t \left[\frac{x}{\rho_g(1-R_l)} - \frac{1-x}{\rho_l R_l} \right]. \qquad (16\text{-}46)$$

For flows without mass transfer between phases, such as are being considered here, x is constant and the slip ratio or velocity can be established if the fluid properties and liquid area fraction are known.

The liquid volume fraction $\bar{R}_l$, for the Lockhart-Martinelli correlation, has already been given in Fig. 16–4 and in Table 16-3. Martinelli and Nelson [166] recognized the inadequacy of the correlation at high pressure, and developed additional lines on Fig. 16-4 for the steam-water one-component system. This correlation and similar ones will be cited later, when one-component systems are considered in more detail. Apparently the Lockhart-Martinelli volume-fraction curve and other void-fraction correlations are inadequate under the extreme density ratio of a gas and liquid metal [20, 206]. Viscosity also has a rather marked effect; the data used by Fohrman [76] for the liquid area fraction in a rectangular channel were much lower than predicted by the correlation.

Hughmark [119, 120] has presented a means of calculating holdup based on the analysis by Bankoff [16]. The basic equation is

$$1/x = 1 - (\rho_l/\rho_g)(1 - K/R_g), \qquad (16\text{--}47)$$

with K expressed as an empirical function of liquid viscosity, Reynolds number, Froude number, and R_{l1}. The correlation was applied to both horizontal and vertical flow data. Dukler et al. [67] compared the holdup correlations suggested by Hoogendoorn [111], Hughmark [119], and Lockhart-Martinelli with the culled data (about 700 points). The Hughmark method provided the best results: 68% of the data (one sigma) fell within 17.5% of the correlation. However, the other correlations gave better results under certain conditions; i.e., at high values of $\bar{R}_l$, in large-diameter pipes, and for annular and wavy flow.

(c) **Specific two-phase flow problems and analysis.** We can make an intelligent guess about the flow regime under a normal set of conditions, and we can make a reasonable estimate of the pressure drop, with no guarantee of accuracy. But under extreme conditions for which data are not available, a reliable estimate is not possible, and we must resort to experimentation. Furthermore, the overall correlations are unsatisfactory for accurate estimation of the pressure drop and area fraction. The approaches that can be used to define the flow regime, and the extent to which they are successful, have been discussed in some detail. Assuming for the moment that the type of flow is known, a more accurate and reliable correlation might be obtained by considering each flow regime separately. It is not proposed that the more analytical methods will eliminate empiricism, but rather that they will provide the insight which will enable one to do a more satisfactory job of treating the data. Remember that even for single-phase flow, the empirical plot of the friction factor versus the Reynolds number is still used. The analytical studies based on one-flow regime are relatively recent in two-phase flow research, and they have not yet been developed to the extent that they can be considered as a final answer. For practical problems, therefore, one is still forced to use the overall methods of the preceding sections, even though these might not be very satisfactory to the analytically minded. In the sequel, a number of the theoretical methods will be outlined and briefly discussed. None will be treated in detail, however, simply because it is recognized that these are not the final answers, and that further development of each is possible and (it is hoped) forthcoming.

(i) **Entrance effects.** The nature of the entrance has a profound effect on the subsequent two-phase flow. It is suspected that much of the disagreement found in the data on two-phase flow could be attributed to such effects, if they could be established. At this point in our discussion we shall present only the results of experiments. Theoretical efforts will be mentioned, but since each analysis of the entrance region assumes some general nature for the flow pattern, it will be appropriate to look at these in more detail when the particular flow regimes are considered.

Wicks and Dukler [224] obtained evidence that for horizontal flow in the annular regime, the configuration of the inlet has a significant effect on both entrainment and pressure drop. Jacowitz and Brodkey [128] found that for a small-diameter tube, the selection of the inlet dictated the type of flow in their test section. They observed annular flow and a slugging type of flow which was primarily annular but which had slugs of liquid bridging the pipe and traveling much faster than the main liquid flow. For vertical-upward annular flow, Gill et al. [86] found similar results for the flow injectors they used; although the flow was always annular with entrainment, different injectors gave different patterns of entrainment, film thickness, and pressure drop. Nedderman and Shearer [176] obtained results on the motion and frequency of waves in vertical annular flow which were different from those of Taylor et al. [211], and they attributed the difference to the calming by the inlet section length before the liquid injector (which was identical in both studies).

In a real sense, annular flow can be considered entirely as an entrance phenomenon. In looking at the horizontal case, Jacowitz and Brodkey assumed no entrainment, an assumption expected to be valid in the first part of the entrance region, or as long as the flow conditions are such that entrainment is nil [211]. Their calculations checked well with the extensive experimental data of McManus [161], who investigated pressure drop, mean film thickness, minimum film thickness, and maximum wave height for a wide range of conditions. McManus found that film thickness initially increased, then decreased steadily with distance and did not reach steady state, even 95 pipe diameters downstream from the entrance. There seem to be no actual entrainment data available for horizontal flow very close to the entrance. Wicks and Dukler [224] and Magiros and Dukler [162] have made measurements from about 9 to 19 ft from the entrance, and in each case the fraction entrained increased with distance from the inlet. Anderson et al. [10] obtained additional entrainment data which checked well with that just mentioned.

There is much more information available for vertical flow. Anderson and Mantzouranis [8] and the English Atomic Energy Research Group [56, 86, 87, 88, 145] have clearly shown that entrainment increases with distance. In reference 145, the authors show that with their injector there is no entrainment at the entrance, and that entrainment was still increasing 17 ft from the inlet. Reference 66 provides a tabular review of the experimental results. Thus if entrainment is considered as an integral part of annular flow, as it should be, then a steady-state annular flow has not really been observed. Of course, this conclusion would be subject to change in the light of additional experimental evidence.

Moissis and Griffith [171] considered entrance effects for vertical slug flow. The effects were marked, and for the ideal conditions considered, could persist for as much as 25 to 30 pipe diameters. The analytical results checked well with the carefully controlled experiments. Griffith and Wallis [101] have obtained data showing that for similar flows, entrance effects can be observed up to 300 pipe diameters from the inlet.

(ii) **Homogeneous models.** One approach to the two-phase flow problem is to assume that the gas and liquid can be considered as a homogeneous material, and to apply appropriately modified phenomenological methods of single-phase flow. It is not clear that such an analysis will result in better correlations or greater knowledge of the mechanism of flow, since the flow is not really homogeneous. In effect, the exact nature of the flow is not considered, and the possible interactions between phases are not considered in detail. Nevertheless, the methods must be cited, since in many cases they do predict the trends observed.

Bankoff [16] has suggested as a model a variable-density homogeneous fluid, based on a suspension of bubbles in the flowing liquid. Aoki et al. [13] have modified the analysis slightly. Since the method does not treat the actual bubble dynamics as such, it will be considered here. Simple power-law forms for the velocity (Eq. 14–61) and for the area-fraction profiles were assumed:

$$R_g(\eta)/R_g(1) = \eta^{1/n}, \tag{16-48}$$

$$U_l(\eta)/U_l(1) = \eta^{1/m}, \tag{16-49}$$

where $\eta = 1 - r/r_0$. In the notation used here, $R_g(\eta)$ is an area fraction of gas (or void fraction) at a specific position η, and $R_g(1)$ is the area fraction at the center line. The average across the cross section is denoted R_g, and the average over the entire pipe $\bar{R}_g$. In Eqs. (16–48) and (16–49), it is expected that m should be about 7 and n about 2 to 3. The volume flow rate of the gas is

$$Q_g = w_g/\rho_g = 2\pi r_0^2 \int_0^1 U_g(\eta) R_g(\eta)(1-\eta) d\eta. \tag{16-50}$$

A local slip velocity [see Eq. (16–46) for the average slip velocity] can be defined as

$$U_s(\eta) = U_g(\eta) - U_l(\eta). \tag{16-51}$$

For a true homogeneous flow, $U_s(\eta) = 0$ by definition. Equation (16–50) can be integrated for Q_g with $U_g(\eta)$ given by Eq. (16–51), $U_l(\eta)$ by Eq. (16–49), and $R_g(\eta)$ by Eq. (16–48). A corresponding equation for the liquid is

$$Q_l = 2\pi r_0^2 \int_0^1 U_l(\eta)[1 - R_g(\eta)](1-\eta) d\eta, \tag{16-52}$$

which can be integrated in the same manner. The average area fraction is given by

$$R_g = \frac{2\pi r_0^2 \int_0^1 R_g(\eta)(1-\eta)\,d\eta}{\pi r_0^2}. \tag{16-53}$$

Upon integration, we find R_g to be a function of n and the boundary conditions at the center line. Likewise Q_g and Q_l are functions of n and m, the local slip velocity, and the center-line boundary conditions. A ratio,

$$K = \frac{R_g}{Q_g/Q_l} = \frac{R_g}{R_{g1}}, \tag{16-54}$$

can be established from the results of the integration:

$$K = \kappa \left[\frac{1 + U_s(\eta)(\pi r_0^2/Q_l) R_g(1 - R_g/\kappa)}{1 + U_s(\eta)(\pi r_0^2/Q_l) \kappa(1 - R_g/\kappa)} \right], \tag{16-55}$$

where

$$\kappa = \frac{2(m+n+mn)(m+n+2mn)}{(m+1)(2m+1)(n+1)(2n+1)}.$$

Equation (16-55) was obtained by Aoki et al. and reduces to the form obtained by Bankoff for true homogeneous flow $[U_s(\eta) = 0]$. The quality or gas-weight fraction flowing can be expressed in terms of K:

$$\frac{1}{x} = \frac{w_t}{w_g} = 1 + \frac{w_l}{w_g} = 1 + \frac{\rho_l Q_l}{\rho_g Q_g} = 1 + \frac{\rho_l}{\rho_g}\left(\frac{Q_l}{Q_g} - 1\right) = 1 - \left(\frac{\rho_l}{\rho_g}\right)\left(1 - \frac{K}{R_g}\right) \tag{16-56}$$

or

$$R_g = \frac{K}{1 + (\rho_g/\rho_l)[(1-x)/x]}.$$

In terms of the slip ratio, Eq. (16-56) [combined with Eq. (16-45)] becomes $U_g/U_l = (1 - R_g)/(K - R_g)$. For this model there is no slip locally, but since the gas can concentrate in the high-velocity region, the average gas velocity can be greater than the average liquid velocity; thus K can be less than unity.

Bankoff determined the value of K empirically, since n and m are not exactly known. He used

$$K = 0.71 + 0.0001\,p \tag{16-57}$$

for the steam–water system he considered (atmospheric pressure to 2000 psia). Hughmark [119, 120] used the same approach for near-atmospheric two-phase flow data for which $K = 0.71$ was not valid; he empirically correlated K with certain dimensionless parameters, as already cited (Eq. 16–47). Aoki et al. further assumed that

$$U_s(\eta)(\pi r_0^2/Q_l) = CR_g/(\kappa - R_g),$$

and then assumed that C was unity. When the above equation is combined with Eq. (16-55), it gives

$$K = \kappa - R_g(\kappa - R_g). \tag{16-58}$$

From Eq. (16-56), for m of 7 and n of 2 or 3, $\kappa \cong 0.95$. This gives a variable K for use in Eq. (16-56).

Fujie [81] has derived an expression, restricted to the annular-flow regime, which is based on momentum considerations of each individual phase. [This is to be contrasted to the overall momentum equation given as Eq. (16-15) and will be discussed in more detail later, when mass transfer between phases is considered.] In addition, Fujie had to make an assumption about the interfacial forces. His theory can be interpreted as giving rise to a variable K. For large ρ_l/ρ_g, the result is

$$K = R_g + \left[(1 - R_g) \Big/ \left(1 + \frac{R_g^{1/4}}{1 - R_g^{1/2}} k\right)\right],$$

where k can be expressed in terms of the pressure (psia) as $\sqrt{10/p}$ [compare Eq. (16-57)]. The differences between the various approaches are small. To illustrate, at atmospheric pressure and $R_g = 0.4$, the K's are 0.71, 0.76, and 0.735, respectively, for Bankoff, Aoki et al., and Fujie. At 1000 psia, the values are 0.81, 0.76, and 0.89, respectively. A comparison of these results with the empirical correlation of Hughmark has not been made.

Zuber and Findlay [245] have suggested the general form

$$U_g = C_0 U_{\text{ave}} + \overline{R_g[U_g(\eta) - U_{\text{ave}}]}/R_g, \tag{16-59}$$

where C_0 is a function of the distribution, as given by Eqs. (16-48) and (16-49) or alternate distributions. The second term on the right-hand side is a weighted mean drift velocity. The equation suggests that a plot of U_g or its equivalent against U_{ave} would have a slope of C_0 and an intercept of the weighted mean drift velocity. In terms of the K of Eq. (16-54), the above equation becomes

$$1/K = C_0 + \overline{R_g[U_g(\eta) - U_{\text{ave}}]}/U_{\text{ave}} R_g.$$

Equation (16-55) can also be put in a similar form:

$$\frac{1}{K} = \frac{1}{\kappa} - \frac{U_s(\eta)(\pi r_0^2/Q_l)(1 - R_g/\kappa)^2}{1 + U_s(\eta)(\pi r_0^2/Q_l)(1 - R_g/\kappa)}.$$

Just as in the previous equation, κ, like C_0, depends on the distribution profiles. The second term involves the local slip velocity, and as such can vary across the cross section; in contrast, the previous equation involves the mean drift velocity and is fixed for a given cross section. The modified form for K, as given by Eq. (16-58), can be put in the same form, and is

$$\frac{1}{K} = \frac{1}{\kappa} + \frac{R_g(1 - R_g)/\kappa}{1 - R_g(1 - R_g/\kappa)}.$$

Because of the assumption that was made about the slip velocity in order to arrive at Eq. (16–58), this equation does not vary across the cross section.

Levy [151] used a mixing-length approach in a variable-density turbulent system. Assuming that the mixing length for velocity (Eq. 14–33) and a similarly defined mixing length for density were equal, he combined them with Eq. (14–16) [neglecting $\overline{\rho uu}$ and assuming $(\nabla \cdot \bar{U}) = 0$] to give

$$\bar{\tau}_{yx} = \rho l^2 \left(\frac{d\bar{U}_x}{dy}\right)^2 + l^2 \bar{U}_x \frac{d\bar{U}_x}{dy}\frac{d\bar{\rho}}{dy} - \mu \frac{d\bar{U}_x}{dy} = l^2 \frac{d\bar{U}_x}{dy}\frac{d\bar{\rho}\bar{U}_x}{dy} - \mu \frac{d\bar{U}_x}{dy}. \quad (16\text{–}60)$$

Levy defined the mixing length itself by Eq. (14–68), using the same constants (k and A). He then related the density to the velocity by the assumed relation

$$\bar{\rho} = B/(1 - A\bar{U}_x), \quad (16\text{–}61)$$

with the boundary conditions

$$\begin{aligned} r &= 0, & U_x &= U_{x,\max}, & \rho &= \rho_{\min}, \\ r &= r_0, & U_x &= 0, & \rho &= \rho_l, \end{aligned} \quad (16\text{–}62)$$

which he used to establish A and B of Eq. (16–61). He combined Eq. (16–60) with (14–68) and (16–61) to obtain a single expression for $\bar{\rho}/\rho_l$, an expression which could then be integrated to obtain the density distribution. The velocity distribution can be established by means of Eqs. (16–61) and (16–62). Levy presented the results of his numerical computation in the form of nondimensional plots. Further integration gave the mean density, which in turn could be used to obtain the area fraction. In Eq. (16–60), the dimensionless velocity and distance (in terms of u^+ and y^+) referred to the liquid viscosity rather than the mixture viscosity. Thus Levy had to define this viscosity μ; he selected as its meaning,

$$\mu/\mu_l = \mu^+ = 1 + 2.5[(\rho_l - \rho_g)/\rho_l]R_g, \quad (16\text{–}63)$$

which is a modification of the Einstein viscosity equation. The parameters of the analysis are the Reynolds number, $G_t d_0/\mu_l$, and the average fluid density. Once these are fixed, the dimensionless velocity and density distributions are set; these in turn fix the void fraction and pressure drop.

In general, the results of the various theories are in accord with both the magnitude and trends observed in the data. It should be emphasized that the results of these analyses, although often less accurate than the results of the empirical methods already cited, do not depend on extensive fits to two-phase flow data. One may surmise the reasons why these approaches are successful at all. First, it is known that the velocity distribution in single-phase flow can be predicated by a number of methods (i.e., based on assumptions of mixing length, the power law, etc.) so long as the proper boundary conditions are used. It may well be that the same is true for the density distribution. In Levy's analysis, the two parameters fix the velocity and density distributions between known limits. A similar situation exists in Bankoff's analysis. Since the distributions are fixed between known bounds, and since integrations are necessary to obtain the void fraction and pressure drop, these two results will not be sensitive to the details of the distributions. The important point is that the methods do not predict really accurate results, which leads one to suspect that the various assumptions (such as those based on mixing length, the power law, etc.) are not correct but are adequate for data correlation. The approach probably will not lead to a basic understanding of the mechanism because of the homogeneous fluid assumption. An extensive and valid comparison of the various homogeneous theories has not been made. Such a test would be useful in establishing the best constants for each method, the most fruitful approach, and the modifications that may lead to even better data correlation.

(iii) **Bubble, stratified, slug, and semi-annular flows.** The homogeneous models just discussed should apply best to vertical bubble flow. Levy [151] included the effect of a vertical orientation, and found that the additional parameter $(G_l/\rho_l)^2/d_0$ would be necessary. From his analysis, one finds that the effect of gravity is less at high flows or low mean fluid densities. Aoki *et al.* [13], in a separate part of their work, specifically considered vertical bubble dynamics and were able to arrive at an estimate of the distributions of the local slip ratio and the gas area fraction. Turbulent flow of the liquid was assumed. For the vertical direction a force balance included the drag and buoyancy effects; for the radial direction, drag (Stokes), interaction between bubbles, and the effect of a pressure distribution were considered. A range of bubble sizes was assumed, and the results were found to be consistent with those from the simpler homogeneous models. This lent support to the idea of using such models for data correlation.

Nicklin [178] has considered vertical bubble flow and has presented a clear discussion of the various measurable bubble velocities. The slip velocity is

$$U_g - U_l = Q_g/S_g - Q_l/S_l = U_0/R_l \qquad (16\text{-}64\text{a})$$

or as a slip ratio is

$$U_g/U_l = 1 + U_0/U_lR_l = U_0S/Q_l, \qquad (16\text{-}64\text{b})$$

where U_0 is the velocity of the bubbles relative to the liquid which is free of them. The slip velocity is always greater than the bubble velocity because there is a relative acceleration of the liquid, due to the decrease in flow area from that of the tube to R_lS. Nicklin has also included a discussion of the energy and momentum equations for vertical bubble flow.

The results of the work of Bergelin and Gazley [29] on stratified flow have already been cited. However, little theoretical work has been done on this problem. Gemmell and Epstein [85] have made a numerical analysis for stratified laminar flow of two immiscible liquids. Pressure-drop reduction occurs because of the addition of a less-viscous fluid. An exact solution is possible for the problem of flow between parallel plates, but recourse to numerical methods is necessary for pipe flow. Velocity profiles, holdup ratio, and pressure-reduction factors were determined which, where applicable, were in good agreement with the data of Russell et al. [197].

The slug-flow regime has recently received attention. Griffith and Wallis [101] have expressed the average density or void fraction in terms of volumetric flow rates, properties of pure materials, and bubble-rise velocity U_b (relative to the ground):

$$\bar{R}_g = Q_g/U_b S = Q_g/(Q_l + Q_g + U_0 S). \tag{16-65}$$

Behringer was apparently the first to suggest this relation [245]; he reported good agreement between the prediction of Eq. (16–65) and experimental results. Griffith and Wallis assumed U_b to be a modified form of the equation for the upward velocity of an infinitely long bubble in a tube filled with a stagnant fluid [60, 68]:

$$U_b = C_1 C_2 \sqrt{g d_0}. \tag{16-66}$$

For a stagnant fluid, $C_1 = 0.35$ and $C_2 = 1.00$. Later Nicklin et al. [179] found that in slug flow, the slug rose at a velocity given by

$$U_b = Q_g/S_g = Q_g/\bar{R}_g S = 1.2 U_{\text{ave}} + 0.35 \sqrt{g d_0}, \tag{16-67}$$

which implies that the slug rises as a bubble in a stagnant fluid relative to the center-line velocity, which is about 1.2 times the average fluid velocity for turbulent flow. A simple rearrangement gives an expression for $\bar{R}_g$ for vertical flow, which has been found valid for slug, semi-annular, and the beginnings of annular flow regimes. For horizontal flow the rise-velocity term (16–66) does not enter, and Eq. (16–67) reduces to

$$Q_g/\bar{R}_g S = 1.2 U_{\text{ave}} = 1.2(Q_g + Q_l)/S. \tag{16-68}$$

When this equation is expressed in terms of the quality, it is identical to the equation proposed by Bankoff (16–56), with $K = 1/1.2 = 0.83$. A direct comparison of Eq. (16–67) with Zuber and Findlay's proposal (16–59) suggests that $C_0 = 1.2$, and that if the weighted mean drift velocity is zero, $K = 1/C_0 = 1/1.2 = 0.83$. Scott [200] has proposed that this equation is adequate for slug, plug, bubble, or froth flows at high liquid Reynolds numbers ($U_{\text{ave}} \rho_l d_0/\mu_l > 8000$). The comparisons made to the mass of data currently available are limited. Hughmark [119] reported that a wide range of K values (0.15 to 0.98) is necessary to fit *all* the data, but it is not known if this is true for the specific flow regimes cited above. It is known that a constant value is inadequate for high-pressure steam–water data (Eq. 16–57). Scott made his one comparison at 1000 psia, for which Eq. (16–57) gives $K = 0.81$ versus the 0.83 used. Clearly, at any other pressure the constant value would not be correct.

Moissis and Griffith [171] have extended the analysis of Griffith and Wallis [101] to the developing slug flow or semi-annular regime. This is an entrance problem in which the pattern ultimately becomes the fully developed slug flow. Two effects exist (Moissis and Griffith have considered only the second). The first is the agglomeration of many small bubbles into a slug, which can be

488 MULTIPHASE PHENOMENA I: PIPE FLOW

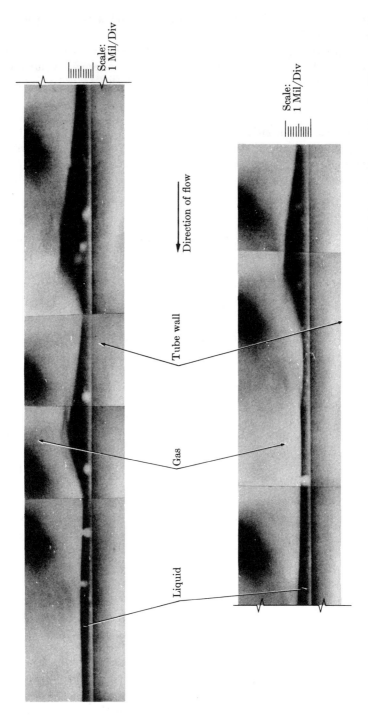

Fig. 16–6. Two-phase annular flow. [Gas: air, at 0.01 lb/sec. Liquid: turpentine, at 0.0012 lb/sec. Pressure: 40 psia. Temperature: 70°F. Inner diameter of glass tube: 9.5 mm.]

described by Eq. (16–67), and the second is the agglomeration of two or more large bubbles [each described by Eq. (16–67)] due to wake effects. The interaction of two large bubbles requires that two problems be solved: (1) The velocity profile behind a given bubble must be established. (2) The velocity of the trailing bubble (which is in the velocity profile now specified behind the leading bubble) must be established. The first problem was solved, checked, and used to establish the major variables that would be important in the second problem. For the second problem, an empirical approach was used to obtain

$$U_{bt}/U_b = 1 + 8 \exp(-1.06 \, L_s/d_0), \tag{16–69}$$

where U_{bt} is the velocity of the bubble's rise, U_b is given by Eq. (16–67), L_s is the separation distance, and d_0 must be greater than $\frac{1}{2}$ in. The first bubble injected will rise at a constant velocity U_b; subsequent bubbles will move faster than this, since L_s will be finite, and consequently many of the trailing bubbles will reach the first one. As the flow in the entrance continues, an oscillatory character develops. Moissis and Griffith have solved the problem by a finite-difference method for the average density at a given L/d_0. They have presented the results as plots (one for each L/d_0), giving a dimensionless form of $\bar{\rho}$ [Eq. (16–65), with U_{bt} replacing U_b] as a function of inlet conditions and initial bubble size. If one assumes further that friction and acceleration effects are small, then one can find the pressure drop per unit length by multiplying the force of gravity times the average density up to the point under consideration. The comparison of the results obtained by Moissis and Griffith with their carefully controlled experiments was quite satisfactory, and serves as an excellent example of what can be done from a fundamental basis in two-phase flow research. Griffith and Lee [237] have attempted to clarify the transition between annular and slug flow. By a potential flow theory, they showed that annular flow should be stable for liquid volume fractions less than about 6%.

(iv) **Annular flow and entrainment.** Essential to the formulation of a successful model are background experimentation aimed at visualization of the flow, establishment of important variables, and elucidation of the mechanism. Annular flow appears to be an entrance problem, and experimentation on this has been cited in a previous section. Probably the best visualizations of the flow are the direct high-speed movies made by Jacowitz and Brodkey [128]; a composite of one sequence is shown in Fig. 16-6. The duration of this sequence was 0.006 of a second. Clearly, the nature of the gas–liquid interface has much to do with the characteristics of this two-phase flow. The roll waves observed correspond closely to those suggested by Lacey et al. [145] for vertical upward flow, and by Hanratty and Engen [104] for horizontal channel flow. Detailed measurements of the interface have been reported in references 142 and 161 for horizontal pipe flow, in references 70, 104, 105, 153, 154 and 214 for horizontal channel flow, in references 176, 211, and 243 for vertical upward pipe flow, and in references 49 and 52 for vertical downward pipe flow. Reference 66 provides a tabular summary.

Reference 57 presents a comparison and evaluation of the many possible methods that can be used to measure film thickness and interface characteristics. An analytical treatment of the interface structure *per se* would be nearly impossible, considering the complexity of the action observed in Fig. 16-6. However, some progress has been made in correlating the interfacial stress with the pertinent variables by semi-empirical means [137, 154]. The basis for such work rests in part on the material treated in the discussion on wave motion (pages 460–463).

Entrainment is expected to be important in annular flow, and certainly in the extended region of transition to mist flow. For the horizontal and vertical upward cases it may well be, as already pointed out, that annular flow with negligible entrainment can exist only very near the entrance, or at very low liquid flow rates. Most of the work on entrainment was briefly discussed in the section on entrance effects; however, some additional comments are in order. Anderson *et al.* [10] have measured an interchange rate for horizontal annular pipe flow which they defined as the rate at which droplets are torn away from the liquid surface. If a steady state is assumed, this equals the rate at which droplets are deposited on the annular film. They found that the percentage of the liquid flow interchanged was approximately constant at 3% to 5% over a wide range of gas flows and superficial-liquid Reynolds numbers ($4w_l/\pi d_0 \mu_l$) in excess of about 2400. Below this figure, the percentage of interchange dropped to less than 0.5%. They hypothesized that the breakpoint corresponds to the point at which the wave heights of the liquid film just penetrated the highly turbulent region of the gas stream. For vertical upward flow, Gill *et al.* [86–88] have reported extensive measurements of mass-velocity distributions in the gas core. The effects of distance from the injector, air-flow rate, and liquid-flow rate were all considered.

The geometry of the annular flow regime in its ideal form is simple enough so that an analytical description can be made, and consequently this flow has been treated most extensively from a theoretical viewpoint. For horizontal flow, Levy [149] obtained a solution by assuming: (1) a concentric undistorted flow with no gravity effects, (2) no interfacial waves, (3) a steady-state system with no change along the pipe length, (4) no shear differential at the interface, (5) no transfer between phases, (6) constant gas and liquid properties, and (7) the velocity profiles are given by the one-seventh power law for turbulent flow and by a modified solution of the Navier-Stokes equation for laminar flow. He found that the pressure drop could be related to the parameter X of Eq. (16–40), which lends support to Martinelli's choice. Additional parameters would need to be introduced only for large μ_g/μ_l. The specific results of the analysis illustrate the fact that if there are inadequate assumptions, a model analysis can give poorer results than an empirical method. Experimental values for the pressure drop in annular-flow are greater than that predicted by the Martinelli correlation (up to a factor of two). Levy's theoretical prediction for the drop is always less than Martinelli's. There can easily be a factor-of-three difference between experimental results and theoretical prediction. Levy suggested that the deviation is probably due to the assumptions of ideal geometry, equal shear stresses at the interface,

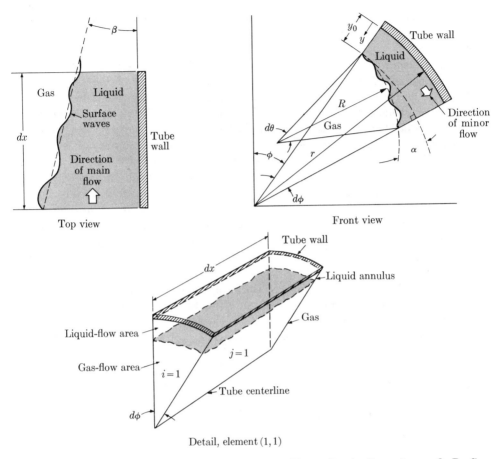

Fig. 16-7. Details of annular flow element. [*From* L. A. Jacowitz, and R. S. Brodkey, Chem. Eng. Sci. **19**, 261–274 (1964).]

constant properties, and neglect of such effects as unsteady state and varying flow patterns.

As already pointed out, the effect of gravity cannot be neglected. The liquid film is far from concentric, as has been showed by McManus [161], Isbin *et al.* [122], and Krasiakova [142]. As shown in Fig. 16-6, surface waves are usually present. A more realistic model was used by Jacowitz and Brodkey [128] and extended by Corder [58]. In this analysis, it was assumed that the flow could be divided into many circumferential and axial elements and that a momentum balance would apply over each element (see Fig. 16-7). The conditions at the beginning of the tube were set, and the numerical solution, based on the force balance and continuity, gave the flow conditions and geometry for each element around and along the tube, thus defining the geometry, flow, and pressure drop of the entire film. In this manner, the flow was defined and limited only by the various assumptions necessary to make possible a solution: e.g., there was no entrainment; the circumferential flow was finite but small, so that second-order effects could be neglected; only the axial and circumferential force balances were needed to define the flow; and the various forces could be estimated from work done on channel flow. In order to obtain complete stability of the calculation, Corder found it necessary to iterate over each individual element in addition to stepping around the tube (usually in elements of 5°) and then down it (usually in elements of 1/8 ft). A reasonably complete description of the flow was obtained, which is encouraging. However, some of the remaining assumptions limit the applicability of the solution, and will require further work.

Fujie's momentum approach for annular flow [81] has already been mentioned. It resulted in an equation like (16–56), but with K now defined as already cited. The analysis was based on a differential balance (in the direction of flow) on each individual phase. In contrast, Jacowitz and Brodkey attempted to use a differential balance on small angular segments of the liquid phase in addition to small elements in the direction of flow. Thus they were able to study the changes around the tube as well as along it.

Zivi [230], in an approximate approach, has applied the principle of minimum entropy production in order to obtain an estimate of the effect of entrainment and wall friction. He took a very idealized view so as to establish the kinetic energy flux, which he in turn minimized:

$$\mathrm{E} = (G_t/2)[U_g^2 x + U_l^2(1-x)]. \tag{16-70}$$

Differentiating yields

$$\frac{d\mathrm{E}}{dR_g} = \frac{G_t}{2}\left[2U_g \frac{dU_g}{dR_g} x + 2U_l \frac{dU_l}{dR_g}(1-x)\right] = 0. \tag{16-71}$$

In terms of G_l, G_g, x, and R_g, the mass velocities are

$$G_l = \frac{G_t(1-x)}{\rho_l(1-R_g)} \quad \text{and} \quad G_g = \frac{G_t x}{\rho_g R_g}. \tag{16-72}$$

Differentiating these with respect to R_g, and combining with Eq. (16–71), one obtains

$$\frac{1}{x} = 1 - \left(\frac{\rho_l}{\rho_g}\right)^{2/3}\left(1 - \frac{1}{R_g}\right) \quad \text{or} \quad R_g = \frac{1}{1 + (\rho_g/\rho_l)^{2/3}[(1-x)/x]}, \tag{16-73}$$

which can be compared to the similar result given in Eq. (16–56). As expected, the two developments are quite different in terms of the slip ratio; Eq. (16–56) gave

$$U_g/U_l = (1 - R_g)/(K - R_g),$$

while Eq. (16–73) gives

$$U_g/U_l = (\rho_l/\rho_g)^{1/3}.$$

Comparing Eqs. (16–56) and (16–73), one gets

$$K = R_g + (1 - R_g)(\rho_g/\rho_l)^{1/3}. \tag{16-74}$$

This is expected to be a lower bound, and for atmospheric pressure one obtains $K = 0.45$ [compared with 0.71 from Eq. (16–57)]. At 1000 psia, one finds $K = 0.64$, as compared with 0.81 previously cited. Zivi found that wall friction lowered the void fraction, but that entrainment increased it. To account for entrainment, Zivi assumed that a certain fraction was removed from the film moving at the liquid velocity and placed in the gas stream moving at the gas velocity. The solution is similar to that just presented, except that a new expression is necessary for U_l. The results bracket the experimental data; i.e., complete entrainment or homogeneous flow [Eq. (16–56) with K equal to unity] is an upper limit, and no entrainment (Eq. 16–73) is a lower limit. Various levels of entrainment give results that lie between these extremes and qualitatively check the data.

Calvert and Williams [42] studied upward, co-current annular flow of an air–water mixture. Briefly, the steps of the analysis can be described as follows: A force balance was made on the liquid film, and on the gas-filled pipe. The shear at the interface was assumed to be the same as would be found in single-phase gas flow at the same radial position. The film was assumed to flow in the same manner as would a liquid in single-phase flow. Thus the velocity profile could be obtained from known principles. The film was assumed to be composed of a laminar sublayer and a turbulent outer layer. From the assumptions, the shear distribution, interface velocity, and film thickness could be obtained. The interfacial shear or pressure drop was then expressed as a function of the film thickness and flow rates. In the analysis, the two-phase pressure drop was assumed to be composed of two parts: first, the contribution that would exist if the gas flowed alone through the dry pipe, and second, the drop due to the drag of the gas over the surface of the liquid. The first contribution was obtainable in the conventional manner from the gas Reynolds number and a friction-factor plot. The second contribution was estimated from an equation involving a drag coefficient and an area of drag. These latter two terms had to be evaluated from experimental data.

A similar derivation and similar results have been obtained by Anderson and Mantzouranis [8]. Their results are presented in terms of dimensionless groups, and thus can be used for systems other than air–water. Qualitative experimental confirmation has been obtained by Bennett and Thornton [28]. The agreement obtained by Calvert and Williams was excellent at lower flow rates; they believed that the discrepancy at high water flow rates was due to wave formation and entrainment. It is of interest to note that they chose to use a drag coefficient in their analysis rather than an equivalent roughness, because the roughness required to fit the data was greater than the tube radius. Further comments on the distinction between wave drag and flow over a rough solid surface have been given by Stewart [209]. Besides those authors already mentioned, a number of others [81, 123, 146, 155, 186, 217] have analyzed the annular flow problem by applying the energy, momentum, and continuity equations to each phase. Westmoreland [220] has performed an analysis similar to that made by Levy [149]. All these are directed primarily toward problems in which mass transfer can occur, such as in the adiabatic flow of a saturated liquid. Consequently, these will be discussed in a later section, when mass transfer is considered in more detail.

Bashforth et al. [21] have considered the upward vertical flow in a tube at the flooding point, i.e., no net liquid flow. The general flow structure is quite similar to that studied by Keulegan [137] for horizontal flow with no net liquid motion. Another area of vertical flow that is quite similar, and that has received considerable attention, is falling films. The work started with the classic analysis of Nusselt [180] on film condensation. Jeffreys [131, 132] made a further extension of this which was then modified by Green [96] and Wyllie [227]. The problem is of course closely related to horizontal film flow, and the reader is directed to the review by Ursell [213], the recent references 103 and 188, and to those references cited therein.

Concerning liquid–liquid systems, the possible pressure-drop reduction obtained by injection of a less viscous liquid has already been mentioned. A comparison between stratified flow and a concentric annular flow has been made by Russell and Charles [196] and has been given in reference 85. Other references have already been cited. The annular core analysis suggests that the pressure drop can be reduced by a factor as great as 500. However, the model flow is difficult to obtain in practice. Russell et al. [197] have reported on the flow of a mineral oil and water.

16-2. ADIABATIC, EVAPORATING, ONE-COMPONENT FLOW

The adiabatic flow of a one-component, two-phase system with mass transfer between phases (flow with flashing) is important in liquid–gas contactors, separators, pipelines, and chemical reactors. Considering that the more elementary problem of flow without interphase transfer has not been solved satisfactorily, it is not surprising that much work still remains to be done on the problem of flow *with* interphase transfer. The flashing process of a mixture of steam and water flowing through a pipe can be described as follows [25]: As the pressure decreases along the pipe, the saturation temperature decreases, the liquid becomes superheated, and the enthalpy of the fluid is reduced in proportion to the drop in temperature. The heat liberated by the reduction in enthalpy of the water is absorbed as latent heat, and evaporates part of the water. The specific volume of the mixture increases rapidly as steam is produced, and the energy which becomes available with the decrease in pressure is expended in accelerating the mixture of steam and water, thus increasing its kinetic energy. In the foregoing process, the mass flow of each individual phase is not constant (as it is for the isothermal case) and thus the quality, and all terms associated with it will vary as the flow progresses along the pipe. As a first result, the overall equations of momentum and energy previously developed cannot be used, since it was assumed that there would be no mass transfer between phases; that is, that $w_{l1} = w_{l2}$ and $w_{g1} = w_{g2}$.

16-2.A Equations of Flow with Interphase Transfer

The derivation of the momentum equation is similar to that described in the obtaining of Eq. (16–15), except that w_l and w_g are not constant along the pipe and so will require subscripts. In addition, the interphase transfer must be considered. For the liquid phase, the momentum balance equation is

$$\left[w_{l1}\frac{(U_{l1}^2)_{\text{ave}}}{U_{l1,\text{ave}}} - w_{l2}\frac{(U_{l2}^2)_{\text{ave}}}{U_{l2,\text{ave}}}\right] - U_{l,\text{ave}}(w_{l1} - w_{l2}) + (p_1 S_{l1} - p_2 S_{l2}) + \bar{p}(S_{l2} - S_{l1})$$

$$- \tau_w A_w + \tau_i A_i - g\int \rho_l S_l \, dz = 0, \qquad (16\text{--}75)$$

where τ_i and A_i are the interface stress and area, respectively. The second term is the momentum associated with the liquid that is converted to vapor at some average velocity denoted by $U_{l,\text{ave}}$. This term was absent in the derivation of Eq. (16–15) because $w_{l1} = w_{l2}$. The term $\bar{p}(S_{l2} - S_{l1})$ is the effect of the average pressure within the pipe on the change in area occurring within the pipe. The term cancels out in the summation to obtain the overall momentum balance because $\bar{p}(S_{g2} - S_{g1}) = \bar{p}(S_{l2} - S_{l1})$. This term and the $\tau_i A_i$ term were simply called interactions between phases in the discussion of Eq. (16–15). Equation (16–75) and the corresponding equation for the gas, when summed, give Eq. (16–15), with the modification that subscripts are retained on w_l and w_g [as was done in Eq. (16–12)]. Note that $U_{l,\text{ave}}(w_{l1} - w_{l2}) = U_{l,\text{ave}}(w_{g1} - w_{g2})$, so that this term, as well as the average pressure and interface shear terms, cancels out in the summation. In other words, it is also an interaction between phases, and the momentum transfer from the liquid (which is caused by the mass transfer) is equal and opposite to that received by the gas. Corresponding subscripts would also have to be retained on G_l and G_g in Eq. (16–20), and on x in Eq. (16–21). If the flow is homogeneous, the slightly modified Eq. (16–15) still results in Eq. (16–24). This in turn can be combined with the friction factor of Eq. (16–27) (with $\rho = \rho_1$) to give

$$G_t^2 = \beta\left[\frac{(p_1 - p_2)\rho_1 + g\rho_1\bar{p}(z_1 - z_2)}{2f_m(L/d_0) + (\rho_1/\rho_2) - 1}\right]. \qquad (16\text{--}76)$$

A similar approach can be used for the mechanical energy balance (10–48), which was used to derive Eq. (16–14). When one follows through the analysis, all terms are the same, except that one cannot assume steady state in the kinetic and pressure terms, and one must consider the energy associated with mass transfer. The subscripts will have to be retained in Eqs. (16–12) and (16–13). On the basis of a unit mass of flowing fluid, Eq. (16–13) can also be expressed as

$$(p_1/\rho_1) - (p_2/\rho_2), \qquad (16\text{--}77)$$

where the definitions of ρ_1 and ρ_2 are identical to those used in Eq. (16–22), expect that the subscripts are now retained on all terms. The overall mechanical energy balance equation now becomes

$$\frac{1}{2w_t}\left[w_{l1}\frac{(U_{l1}^3)_{ave}}{U_{l1,ave}} + w_{g1}\frac{(U_{g1}^3)_{ave}}{U_{g1,ave}} - w_{l2}\frac{(U_{l2}^3)_{ave}}{U_{l2,ave}} - w_{g2}\frac{(U_{g2}^3)_{ave}}{U_{g2,ave}}\right] + \left(\frac{p_1}{\rho_1} - \frac{p_2}{\rho_2}\right)$$
$$+ g(z_1 - z_2) + \int p\,d\left(\frac{1}{\rho}\right) - F_{\text{Losses}} = 0, \qquad (16\text{-}78)$$

where once again the work term cancels, as do the interactions between phases caused by the mass transfer within the pipe. Since the changes in pressure and density will not be small, the integral term must be retained and an equation like (16–16) does not follow from Eq. (16–78). However, the integral term can be combined with Eq. (16–77) [as was done in obtaining Eq. (10–49) from (10–48)] to give the energy equation as

$$\frac{1}{2w_t}\left[w_{l1}\frac{(U_{l1}^3)_{ave}}{U_{l1,ave}} + w_{g1}\frac{(U_{g1}^3)_{ave}}{U_{g1,ave}} - w_{l2}\frac{(U_{l2}^3)_{ave}}{U_{l2,ave}} - w_{g2}\frac{(U_{g2}^3)_{ave}}{U_{g2,ave}}\right] + g(z_1 - z_2)$$
$$- \int \frac{1}{\rho}\,dp - F_{\text{Losses}} = 0. \qquad (16\text{-}79)$$

If the flow is homogeneous, Eq. (16–79) becomes

$$\frac{G_t^2}{2\alpha}\left(\frac{1}{\rho_1^2} - \frac{1}{\rho_2^2}\right) + g(z_1 - z_2) - \int \frac{1}{\rho}\,dp - F_{\text{Losses}} = 0, \qquad (16\text{-}80)$$

and if $\bar{\rho}$ is defined by Eq. (16–22), we obtain

$$\frac{G_t^2}{\alpha}\left(\frac{1}{\rho_2} - \frac{1}{\rho_1}\right) + g\bar{\rho}(z_2 - z_1) + \bar{\rho}\int \frac{1}{\rho}\,dp + \Delta p_{Fe}' = 0. \qquad (16\text{-}81)$$

This in turn can be combined with the friction factor of Eq. (16–27) (with $\rho = \rho_1$) to give

$$G_t^2 = \alpha\left[\frac{-\rho_1\bar{\rho}\int(1/\rho)\,dp + g\rho_1\bar{\rho}(z_1 - z_2)}{2f_e(L/d_0) + (\rho_1/\rho_2) - 1}\right]. \qquad (16\text{-}82)$$

Equations (16–76) and (16–82) [or Eqs. (16–24) and (16–81)] will be the same only if $\alpha = \beta$ (which can occur only at unity) and if $-\int (1/\rho)dp = (p_1 - p_2)/\bar{\rho}$. Since the integral term depends on an assumed path, the two sides are not expected to be equal. Thus the pressure contributions, Δp_{Fe} and Δp_{Fm}, or the corresponding friction factors, f_e and f_m, cannot be the same. This is to be contrasted to the case for no mass transfer, where (for $\alpha = \beta$) Eqs. (16–23) and (16–24) are the same. Of course, a relation can be obtained between them by simply equating the equations being considered.

The homogeneous assumption is known to be inadequate for much of the available experimental evidence, and consequently more exact equations are necessary. The two momentum balances on the individual phases and their sum are available, as are the mechanical energy balances on the individual phases and on the total flow. One can also write an energy balance on each phase [e.g.,

Eq. (10-32), as contrasted to the mechanical energy balance], but it is easier to use the enthalpy ($H + U + p/\rho$). For the liquid,

$$(w_{l1}H_{l1} - w_{l2}H_{l2}) - H_l(w_{l1} - w_{l2}) + \frac{1}{2}\left[w_{l1}\frac{(U_{l1}^3)_{\text{ave}}}{U_{l1,\text{ave}}} - w_{l2}\frac{(U_{l2}^3)_{\text{ave}}}{U_{l2,\text{ave}}}\right]$$
$$- \tfrac{1}{2}U_{l,\text{ave}}^2(w_{l1} - w_{l2}) + g(w_{l1}z_1 - w_{l2}z_2) + \tau_i U_i A_i - \Delta H_v(w_{l1} - w_{l2}) = 0, \tag{16-83}$$

in which H is the enthalpy, U_i is the interface velocity, and ΔH_v is the latent heat of vaporization. The second term (including H_l) is simply the rate of removal of enthalpy from the liquid phase by virtue of the loss of liquid by mass transfer. For the gas,

$$(w_{g1}H_{g1} - w_{g2}H_{g2}) - H_l(w_{g1} - w_{g2}) + \frac{1}{2}\left[w_{g1}\frac{(U_{g1}^3)_{\text{ave}}}{U_{g1,\text{ave}}} - w_{g2}\frac{(U_{g2}^3)_{\text{ave}}}{U_{g2,\text{ave}}}\right]$$
$$- \tfrac{1}{2}U_{l,\text{ave}}^2(w_{g1} - w_{g2}) + g(w_{g1}z_1 - w_{g2}z_2) - \tau_i U_i A_i = 0. \tag{16-84}$$

The second term involving H_l and the fourth term involving $U_{l,\text{ave}}$ are based on the liquid, since the rates of enthalpy and kinetic energy transfer from the liquid (by the loss of liquid through mass transfer) are gained by the gas. Of course, $w_{l1} - w_{l2} = w_{g1} - w_{g2}$. The total energy balance is the sum of Eqs. (16-83) and (16-84), but it can also be obtained from Eq. (16-78):

$$\frac{1}{2w_t}\left[w_{l1}\frac{(U_{l1}^3)_{\text{ave}}}{U_{l1,\text{ave}}} + w_{g1}\frac{(U_{g1}^3)_{\text{ave}}}{U_{g1,\text{ave}}} - w_{l2}\frac{(U_{l2}^3)_{\text{ave}}}{U_{l2,\text{ave}}} - w_{g2}\frac{(U_{g2}^3)_{\text{ave}}}{U_{g2,\text{ave}}}\right] + g(z_1 - z_2)$$
$$+ (1-x_1)H_{l1} + x_1 H_{g1} - (1-x_2)H_{l2} - x_2 H_{g2} - (x_2 - x_1)\Delta H_v = 0, \tag{16-85}$$

where x_1 and x_2 are the qualities at 1 and 2, respectively. The various phase equations are difficult to use, since such terms as the wall area in contact with an individual phase, the interface velocity, and the interface area are unknown unless some specific flow geometry and flow characteristics are assumed. Nevertheless, further progress in the field will no doubt rely to a great extent on these equations.

In a number of cases in the literature [123, 155, 186] the equations have been used in differential form and step integrations or differential relations have been used in their solutions. For a small element of flow, such differential forms are easily generated from the preceding equations by replacing differences with differentials. To illustrate, Eq. (16-75) in differential form is

$$d\int \rho_l U_l^2 \, dS - U_{l,\text{ave}} \, dw_l + S_l \, dp + \tau_w \, dA_w - \tau_i \, dA_i - g\rho_l S_l \, dz = 0, \tag{16-86}$$

where $d(S_l p) - p\, dS_l = S_l\, dp$ has been assumed for a small element. In a somewhat less exact form, Eq. (16-86) has been written as

$$d(w_l U_l) - U_l\, dw_g + S_l\, dp + \tau_w\, dA_w - \tau_i\, dA_i + g\rho_l S_l\, dz = 0$$

or
$$w_l\,dU_l + s_l\,dp + \tau_w\,dA_w - \tau_i\,dA_i + g\rho_l S_l\,dz = 0. \tag{16-87}$$

As another example, the total mechanical energy balance equation (16–78) becomes

$$\tfrac{1}{2}d\left(\int \rho_l U_l^3\,dS_l + \int \rho_g U_g^3\,dS_g\right) + gw_l\,dz + (U_l S_l + U_g S_g)\,dp + w_l F_{\text{Losses}} = 0. \tag{16-88}$$

Equation (16–87) and others like it are usually derived by considering the liquid and vapor to be in annular flow and using each phase separately as a control volume in the differential element.

16-2.B Overall Methods

The use of an overall momentum or energy balance on a homogeneous flow model neglects the true nature of real flows, and so this approach will be treated as an overall method. In this model, the two phases flow as a finely divided mixture with equal average velocities. Because of this, it is often referred to as the *fog model*. Since the overall equations require the use of a friction factor, the term *friction-factor model* [122] is also often applied and will be used here. If, instead of using the overall balance equations, one were to try to treat the details of a homogeneous flow (e.g., as described previously under homogeneous models for two-component isothermal flow on pages 482–486), then these would more appropriately be considered in the next section, which deals with specific flow areas. Another overall approach was used by Martinelli and Nelson [166], in which they modified the two-component isothermal correlation of Lockhart and Martinelli, so that it applied to flashing steam–water flow. Once again, we shall discuss these methods in some detail because of their historical importance and continued use in the literature.

(a) Friction-factor models. The homogeneous-flow equations involving friction factors have been given as (16–76) and (16–82). The equation from momentum considerations (16–76) is preferred because it is simpler to use; i.e., no stepwise integration is necessary. However, this is not the equation that has been extensively used in the previous literature [25, 207]. Rather, the approximate differential relation (16–87) and a corresponding one for the gas are summed and combined with the friction-factor definition to give

$$U\,dU + (1/\rho)\,dp + g\,dz + 4f(1/d_0)(U^2/2)\,dL = 0. \tag{16-89}$$

Since $d(\rho U) = dG_t = 0$, this is rearranged to

$$-(G_t^2/\rho)\,d\rho + \rho\,dp + g\rho^2\,dz + G_t^2(2f/d_0)\,dL = 0, \tag{16-90}$$

which, when one integrates between 1 and 2 and solves for G_t^2, gives

$$G_t^2 = \frac{-\int_{p_1}^{p_2} \rho\,dp - \int_{z_1}^{z_2} g\rho^2\,dz}{(2fL/d_0) - \ln(\rho_2/\rho_1)}. \tag{16-91}$$

An equation quite similar to (16-76) (with β equal to unity) can be obtained from Eq. (16-90) by dividing by ρ and integrating. Here one must assume that $g\rho\,dz$ integrates to $g\bar{\rho}(z_2 - z_1)$ and that the ρ associated with the friction factor is ρ_1.

Equations (16-76), (16-82), or (16-91) can be solved for any system by obtaining an equivalent length of pipe, assuming a friction factor, and evaluating the integral terms. The change in density with pressure can be evaluated by isentropic, isenthalpic, constant-stagnation enthalpy, or by allowing for heat transfer. This latter possibility will be treated further in the section on two-phase flow with heat transfer, and is mentioned here to indicate that the equations are not restricted to adiabatic flow. If friction losses are present, the isentropic expansion is not exact. Even if the flow were adiabatic, the isenthalpic expansion is not exact either, because of the decrease in enthalpy due to the increase in kinetic energy. The most correct evaluation would be constant stagnation enthalpy; however, even this would not be exact if there were heat losses. Probably the most helpful point is that, in most practical problems, it does not appear to make much difference which expansion method is selected. For example, Haubenreich [107] obtained less than 1% variation in any of the assumptions for his calculations on a half-inch line. For his one-inch system, the calculated values were 6% higher than experimental when he accounted for the kinetic energy effect and the heat losses from the system. For the water–steam system, the steam tables of Keenan and Keyes [135] are used. Several workers have published the results of their calculations [36, 69, 107], thus enabling the rest of us to avoid the calculation for this system.

Equations (16-76), (16-82), or (16-91) can, in some ways, be compared to the compressible flow equation (12-25); that is, if for some given upstream pressure, calculations are made for a number of exit pressures, it will be found that the flow rate passes through a maximum. In experiments, the flow rate increases to the maximum, but the decrease is not observed. This is called the *critical flow* or *critical pressure*, and the flow is said to be *choked*. It may be that in some way this critical phenomenon in two-phase flow corresponds to the propagation of a small pressure wave in compressible flow; however, this has not yet been established. It is clear from the equations that the critical phenomenon is obtained when the increase in energy due to a small pressure drop just equals the increase in kinetic energy and friction.

EXAMPLE 16-3. Benjamin and Miller [25] considered a horizontal steam–water system at an initial saturation temperature of 269.3°F (41.4 psia). The initial pressure of the system was 35 psia. The line was equivalent to 90.3 ft. of 4-in. pipe. Establish the manner in which the mass flow varies as the exit pressure is reduced.

Answer. One needs to find the density as a function of pressure. An example of this calculation, if one assumes an isentropic expansion, is, for $p = 24$ psia,

$$s_{41.4} = 0.3948, \quad s_{24} = 0.3500, \quad \Delta s = 0.0448,$$

$$s_{lg} = 1.3672, \quad x = \Delta s/s_{lg} = 0.0328, \quad \text{and} \quad 1 - x = 0.9672.$$

The average density is obtained from $1/\rho = (1-x)/\rho_l + x/\rho_g$, where, for the present conditions,

$$\rho_l = 60.9 \text{ lb/ft}^3 \quad \text{and} \quad \rho_g = 0.061 \text{ lb/ft}^3,$$

so that one can calculate

$$1/\rho = 0.571 \text{ ft}^3/\text{lb} \quad \text{or} \quad \rho = 1.75 \text{ lb/ft}^3.$$

Other densities can be calculated in a similar manner. The final tabulation of this computation is as follows.

p, psia	ρ, lb/ft^3	p, psia	ρ, lb/ft^3
41.4	58.3	24	1.75
36	8.25	20	1.15
32	4.39	16	0.75
28	2.69	8.4	0.27

Three equations can be used, (16–76), (16–82), and (16–91). The friction factors, which are different for each equation, are unknown; therefore, so that a comparison can be made, assume $f_e = f_m = f = 0.003$. Where integrations are necessary, they can be performed incrementally with the decreasing exit pressure. In the present case, increments of 1 psia were used, with averages of the ends of the increment being considered representative of the increment. A tabular solution for Eq. (16–91), as shown in Fig. 16–8, is available in reference 25. The results of the three equations do not differ greatly; each has a critical point in the range of 22 to 25 psia exit pressure. The friction factors in the three equations used in the calculation were assumed the same, even though they would be different if they were properly established by comparison with experimental data. It is clear from the equations and the figure that, in order to bring the curves together in the range of $G_t < G_c$ to represent an experimental set of conditions, f_e will have to be greater than f_m, which will in turn have to be greater than f.

What is of importance here is not so much the details of the curves or even their representation of data, but rather that a maximum occurs which can limit the flow. (This maximum is also observed in real systems.) This critical flow and pressure are important, and will be considered in more detail later.

In each of these overall methods it is necessary to estimate the friction factor. The values to be used in calculations based on Eq. (16–91) have been suggested as 0.003 for steam–water systems [25, 35] and 0.005 for oil–oil vapor systems [34, 64]. These figures are for high Reynolds numbers. As an approximation, one may use the standard f–N_{Re} curves by using a Reynolds number calculated from average properties. One suggestion is to use an average specific volume of the two-phase mixture and the viscosity of the liquid [34, 59]. McAdams et al.

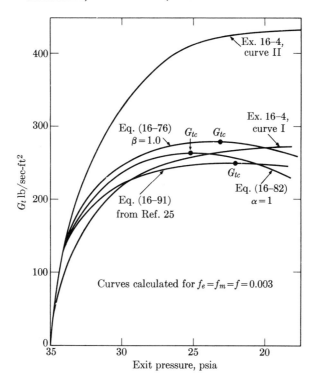

Fig. 16-8. Critical phenomenon for two-phase flow.

[159] suggested a weighted reciprocal of the viscosity of the two phases rather than the liquid viscosity. Isbin et al. [122] summarized a number of investigations and concluded that this viscosity of the mixture gave values of f within about -50% of that observed. However, Isbin et al. [125] in a later study, with new data, have found that the observed two-phase frictional pressure drop is from 0.5 to 2.3 times that calculated. Qualitatively, the use of the liquid viscosity leads to better agreement, but in many cases gives an overcorrection. Haubenreich [107] measured several friction factors, and found a variation of 0.004 to 0.006 for a half-inch pipe and 0.002 to 0.005 for a one-inch tube. This is a relatively large variation, but it was not correlated with other data against the Reynolds number. Note that, of the various suggestions, none correspond to ρ_l and μ_l as given by Eq. (16-42). However, even when these were used and were tested [67] on over 300 experimental steam–water systems, the results were far from satisfactory. One is left with the conclusion that an overall balance on a homogeneous flow model is at best only a fair representation of most real flows. Apparently the homogeneous model can be applied to long, complicated horizontal systems [3, 25, 35, 107] but fails for short, simple systems such as condensers, boilers, and evaporators [123, 190].

(b) Martinelli-Nelson overall approach. The Lockhart-Martinelli correlation was developed from extensive studies of two-component, two-phase flow with no mass transfer between phases. The adaptation of this to the flow of flashing steam–water mixtures and flow with heat transfer was done by Martinelli and Nelson [166]; they used the data of Davidson et al. [59] to help establish the empirical curves, and those of McAdams et al. [159] to test the method. Martinelli and Nelson postulated that the pressure drop resulting from the flow of a boiling or flashing mixture is made up of two parts: that which is due to the action of frictional forces during two-phase flow, and that resulting from the rate at which the momentum of the mixture increases as it flows through the tube and vaporizes. They further assumed that each could be determined separately, as indicated by the previously developed equations; i.e., that the frictional contribution could be calculated by means of a correlation such as the friction factor, and that the accelerative or kinetic energy part could be separately calculated by a term such as $(G_t^2/\beta)(1/\rho_2 - 1/\rho_1)$ of Eq. (16-24).

The frictional pressure drop was obtained in a manner analogous to the isothermal case; however, the experimental results of Davidson et al. showed that Fig. 16-4 overestimated the high-pressure data, with increasing error as the critical pressure of the fluid was approached. At the critical pressure, the error was fivefold. The curve from Fig. 16-4, which was found to be good for atmospheric-pressure runs, was used as one limit. The other limit was the curve at the critical pressure, and was obtained as follows: A term $\mathcal{X}_{tt}$ [closely related to X of Eq. (16-40)] is expressed as a function of the quality:

$$\mathcal{X}_{tt} = (\rho_g/\rho_l)^{0.571}(\mu_l/\mu_g)^{0.143}[(1-x)/x]. \tag{16-92}$$

It is assumed that a linear relationship exists between x and L. Thus, once the exit quality is established, the value of $\mathcal{X}_{tt}$ is known at any point along the tube. From the equation just above Eq. (16-33),

$$(dp/dL)_l = \phi(w_l^{2-n}) = \phi(w_l^{1.75}), \tag{16-93}$$

where n has been taken as 0.25. The term $(dp/dL)_0$ is defined as the pressure gradient for the flow of 100% liquid. Therefore

$$(dp/dL)_l = (dp/dL)_0(w_l/w_t)^{1.75} = (dp/dL)_0(1-x)^{1.75}. \tag{16-94}$$

The two-phase drop due to friction can be obtained from Eq. (16-28). At the critical point, since there is no distinction between phases,

$$(\Delta p/\Delta L)_{\text{TPF}} = (\Delta p/\Delta L)_0. \tag{16-95}$$

Combining Eqs. (16-28), (16-94), and (16-95) gives

$$\phi_{lt} = 1/(1-x)^{0.875}. \tag{16-96}$$

At the critical point, Eq. (16-92) becomes

$$\mathcal{X}_{tt} = (1-x)/x. \tag{16-97}$$

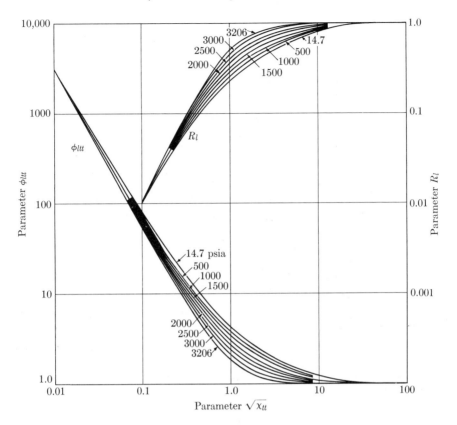

Fig. 16-9. Plot of parameters ϕ_{ltt} and R_l versus parameter $\sqrt{\chi_{tt}}$ for various pressures from one atmosphere absolute pressure to critical pressure for water and water vapor. [From the American Society of Mechanical Engineers, Trans. ASME **70**, 695–702 (1948). By permission.]

Combining Eqs. (16–96) and (16–97) gives the desired curve for ϕ_{ltt} and χ_{tt} at the critical pressure:

$$\phi_{ltt} = [(\chi_{tt} + 1)/\chi_{tt}]^{0.875}. \tag{16-98}$$

The remaining curves at intermediate pressures were established by trial and error, using the data of Davidson *et al.* The resulting plot is similar to Fig. 16–4 and is given in Fig. 16–9.

If we use the curves established for ϕ_{ltt} versus χ_{tt} (with system pressure as a parameter), the ratio

$$\frac{(dp/dL)_{\mathrm{TPF}}}{(dp/dL)_0} = (1 - x)^{1.75} \phi_{ltt}^2 \tag{16-99}$$

from Eqs. (16–28) and (16–94) can be plotted as a function of x for any specific pressure. This is possible because the variables in Eq. (16–99) are functions of

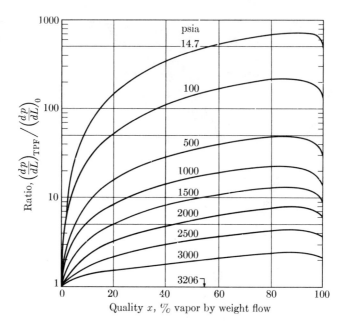

*Fig. 16–10. Ratio of local two-phase pressure gradient to pressure gradient for 100% liquid flow as a function of quality and pressure. [From the American Society of Mechanical Engineers, Trans. ASME **70**, 695–702 (1948). By permission.]*

TABLE 16-4

Values of $\Delta p_{\text{TPF}}/\Delta p_0$ as a Function of x_2 and p*

Steam quality, percent by weight flow	Pressure, psia							
	14.7	100	500	1000	1500	2000	2500	3206
1	4.7	2.8	1.79	1.45	1.28	1.17	1.08	1.00
5	16.5	7.4	3.42	2.30	1.77	1.43	1.21	1.00
10	33.0	13.8	5.13	3.10	2.23	1.72	1.39	1.00
20	68.5	25.9	8.90	4.92	3.18	2.27	1.70	1.00
30	108.	39.5	12.1	6.40	4.00	2.68	1.90	1.00
40	152.	52.5	15.1	7.70	4.74	3.15	2.20	1.00
50	198.	65.0	18.0	8.96	5.40	3.55	2.48	1.00
60	246.	76.0	21.2	10.3	6.10	3.92	2.62	1.00
70	293.	87.2	24.2	11.9	6.93	4.36	2.82	1.00
80	337.	97.	27.3	13.2	7.62	4.70	2.95	1.00
90	377.	108.	29.2	14.0	8.00	4.98	3.10	1.00
100	408.	118.	31.0	14.9	8.50	5.26	3.22	1.00

*By permission of the American Society of Mechanical Engineers [166].

ADIABATIC, EVAPORATING, ONE-COMPONENT FLOW

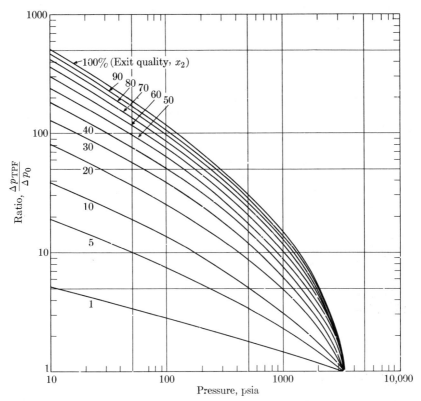

Fig. 16–11. Ratio Δp_{TPF} to Δp_0 as a function of exit quality and absolute pressure. [From *The American Society of Mechanical Engineers*, Trans. ASME **70**, 695–702 (1948). By permission.]

p and x only. This plot is shown in Fig. 16–10. If we use it, we can calculate the pressure drop for any given pipe at any given mean pressure p from

$$\frac{\Delta p_{\text{TPF}}}{\Delta p_0} = \frac{1}{x_2} \int_0^{x_2} \frac{(dp/dL)_{\text{TPF}}}{(dp/dL)_0} \, dx, \qquad (16\text{-}100)$$

where Δp_{TPF} is the two-phase friction pressure drop in the pipe with mean pressure p and exit quality x_2. The pressure drop across the same tube is Δp_0, assuming that no vaporization occurs. Integration of the curves of Fig. 16–10 gives the final curves shown in Fig. 16–11. Table 16-4 lists the numerical values. The curves must be considered tentative, since they are based only on the data of Davidson *et al*. The pressure drop due to friction during forced-circulation evaporation is obtained in the following manner: By the usual method, we make a calculation of the pressure drop which assumes no evaporation (Δp_0); then, using the average pressure in the evaporator section and the exit quality, we determine the pressure drop due to friction from Fig. 16–11.

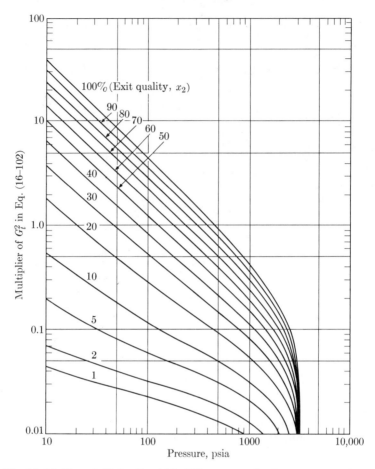

*Fig. 16–12. Multiplier of G_t^2 in Eq. (16–102) versus absolute pressure for various exit qualities. [From The American Society of Mechanical Engineers, Trans. ASME, **70**, 695–702 (1948). By permission.]*

In order to obtain the total pressure drop, we must calculate not only the frictional contribution but also the accelerative contribution. For the homogeneous model, we can use Eq. (16–24) (with $\beta = 1$) to give, for this contribution,

$$\Delta p_{a\,I} = G_t^2 \left(\frac{1}{\rho_2} - \frac{1}{\rho_1} \right) = G_t^2 \left(\frac{x_2}{\rho_{g2}} + \frac{1 - x_2}{\rho_l} - \frac{1}{\rho_l} \right),$$

where we have assumed that $x_1 = 0$ (that is, all liquid). The term ρ_l is assumed constant, but ρ_g can vary. Rearranged, this gives

$$\Delta p_{a\,I} = G_t^2 \left[(1 - x_2) + \frac{x_2 \rho_l}{\rho_{g2}} - 1 \right] \frac{1}{\rho_l}. \tag{16-101}$$

For a less restrictive flow, where R_l and R_g are known, Eq. (16–21) (with $\beta = 1$)

TABLE 16-5

Values of Multipler of G_t^2 in Eq. (16–102) as a Function of x_2 and p*

Steam quality, percent by weight flow	Pressure, psia							
	14.7	100	500	1000	1500	2000	2500	3206
1	0.04	0.023	0.0130	0.0093	0.0067	0.0041	0.0021	0
5	0.16	0.068	0.0316	0.0211	0.0153	0.0100	0.0064	0
10	0.41	0.118	0.0498	0.0313	0.0227	0.0163	0.0100	0
20	1.30	0.280	0.093	0.054	0.037	0.026	0.017	0
30	2.70	0.522	0.146	0.084	0.054	0.037	0.025	0
40	4.60	0.850	0.209	0.112	0.074	0.049	0.033	0
50	7.03	1.250	0.291	0.150	0.096	0.064	0.043	0
60	9.95	1.730	0.384	0.193	0.121	0.080	0.053	0
70	13.40	2.27	0.494	0.241	0.149	0.098	0.064	0
80	17.20	2.90	0.618	0.296	0.180	0.117	0.076	0
90	21.80	3.63	0.762	0.360	0.215	0.139	0.088	0
100	26.7	4.40	0.903	0.420	0.253	0.162	0.102	0

*By permission of the American Society of Mechanical Engineers [166].

can be used to give the accelerative contribution:

$$\Delta p_{a\text{II}} = G_t^2 \left[\frac{(1 - x_2)^2}{R_{l2}} + \frac{x_2^2}{R_{g2}} \frac{\rho_l}{\rho_{g2}} - 1 \right] \frac{1}{\rho_l}, \qquad (16\text{--}102)$$

where in addition it has been assumed that $R_{l1} = 1$, since the entrance has been assumed to be entirely liquid. At the critical point, $\Delta p_a = 0$ and $\rho_l/\rho_{g2} = 1$; thus, from the equation, $R_{l2} = 1 - x_2$ and $R_{g2} = x_2$. Since, from Eq. (16–97), we know the relation between X_{tt} and x at the critical point, we can obtain the curve of R_l versus X_{tt} at that point. These values are shown in Fig. 16–9. The values at the intermediate pressure were arbitrarily interpolated, since no data were available. Values of R_{g2} were plotted as a function of x with pressure as a parameter. From this plot and Eq. (16–102), a plot was made of the multiplier of G_t^2 versus pressure with a parameter of x_2. This is shown in Fig. 16–12, and the values are tabulated in Table 16-5. The final value of Δp_a will probably lie between the limits set by Eqs. (16–101) and (16–102), since the water–vapor mixture leaving the heating section will be partially in the form of fog and partially in the form of separate liquid and vapor. Begell and Hoopes [23] have presented a chart and nomograph for a rapid solution of X_{tt} and the acceleration pressure drop.

Martinelli noted certain assumptions on which the entire analysis is based. Those assumptions were:

(1) That the work of Lockhart and Martinelli [156] on air and various liquids could be extended to the flashing-water case.

(2) That the point-to-point evaluation of $(dp/dL)_t$ and the subsequent integration were valid.
(3) That extrapolation of the Lockhart-Martinelli curves to the critical point was valid.
(4) That there was a linear relation between x and L.
(5) That the tubes used were horizontal and of constant diameter.

The correlation procedure, as outlined above was applied to the data of Davidson et al. [59] and McAdams et al. [159]. One must recall that the data of Davidson et al. were used to establish the correlation. On these data, the correlation was good ($\pm 30\%$) for data from 6 to 100% exit quality and 500 to 3000 psia. In this case, the pressure drop due to acceleration of the fluid was a small fraction of the pressure drop due to friction, so no significant difference was noted between calculations based on Eqs. (16–101) and (16–102). The data of McAdams et al. were obtained under the conditions of 4 to 95% quality and 18 to 25 psia. The pressure drop due to acceleration was appreciable for these runs, being of the same order of magnitude as the pressure drop due to friction. It was noted that, on the average, the measured pressure drops lay between the two limits obtained by using Eqs. (16–101) and (16–102). It was further noted that, in the runs in which the vaporization was more gradual, the correction of Eq. (16–102) gave the best correlation, while for runs in which vaporization was more rapid, somewhat better correlation was obtained by using the correction of Eq. (16–101). Overall, the correlation is probably good to about $\pm 30\%$.

Although this overall method is often used to calculate the frictional contribution, it should be noted that Isbin et al. [125] tried to use the method for data in the range from 400 to 800 psig, without satisfactory results. Consequently, some brief comments are in order on further use of this method in the literature. Altman et al. [5] have reworked the method for refrigerant-22 and have presented figures analogous to 16–11 and 16–12. They calculated the pressure drop for 14 runs, and found that the data were predicted to within 10%; however, the extreme error was a plus 30%. Baroczy [20] has presented a somewhat similar analysis and an acceleration-correction figure similar to 16–12 for mercury. Hoopes [113] used the Martinelli-Nelson method for flow in an annulus. According to the Baker flow-area chart (Fig. 16–2), the flow changed from slug to annular to dispersed in 2 ft. The results of the correlation were satisfactory, being $+30\%$ to -11% at higher qualities, and within 45% at lower qualities at which the actual quality was in doubt by as much as 30%. Harvey and Foust [106] tried to correlate their data obtained in a vertical tube boiler by the method, but the results were not satisfactory. It was believed that this was due to the use of high-pressure data in the correlation, while their data were for pressure below atmospheric. As an alternative they used the Lockhart-Martinelli correlation to establish the relation between ϕ_{tt} and $\mathcal{X}_{tt}$, then applied a graphical method, obtaining excellent results (4%) for subcritical flow. When critical flow occurred, the error was plus 20%. Choking was accounted for by a development based on Eq. (16–91). Further discussion of this can be found in references 106 and 198.

Rogers [195] has reviewed and applied the work of Harvey and Foust to hydrogen flow in horizontal tubes. The analysis leads to rather complicated equations which were solved on a numerical basis and presented as a series of graphs. Isbin et al. [122] point out that these procedures are really mixed models, since they use homogeneous flow in the assumptions of velocity and density, and the Martinelli model for the two-phase pressure drop. Since some of the foregoing also considered flow with heat transfer, further comments will be made in the section on boiling.

Isbin et al. [124] applied the Martinelli work to their studies of void fractions in two-phase steam–water flow. In their work, they made a comparison of the voids with those values obtained from the homogeneous model or from the Martinelli-Nelson correlation. In a more recent and extensive study, Isbin et al. [126] have presented data for void fractions at higher pressures. In their paper they provided a comprehensive survey of the void data available, and methods of correlation. In general, none of the possible methods are completely satisfactory, since the various methods check one set of data and not another. Isbin et al. concluded that the systems are too complex, and the data are neither extensive nor precise enough to allow an overall correlation. Their article provides a good review of the status of research on the problem. Shear and Green [202] have emphasized the need for void measurements, along with pressure-drop measurements, to aid in the interpretation of data.

EXAMPLE 16–4. For the system of Example 16–3, determine, by means of the Martinelli-Nelson method, the curve for mass velocity versus exit pressure, and compare the result with Fig. 16–8.

Answer. Since the Martinelli-Nelson method requires a knowledge of exit conditions, it is easier to assume these and calculate G_t by trial and error, rather than the reverse. For the calculation of quality from pressure, we shall assume isentropic expansion consistent with Example 16–3. To illustrate the calculation for just one pressure, let us consider an exit pressure of 20 psia.

(1) Exit pressure is 20 psia (given).

(2) Quality is 0.0424 (isentropic assumption).

(3) Average pressure is 27.5 psia (average of 20 and 35).

(4) $(\Delta p_{\text{TPF}}/\Delta p_0)$ is 11 (Fig. 16–11).

(5) $\Delta p_{a\,\text{I}}/G_t^2$ is 0.000151 [Eq. (16–101), psia/G_t^2].

(6) $\Delta p_{a\,\text{II}}/G_t^2$ is 0.0000195 (Fig. 16–12, psia/G_t^2).

(7) $\Delta p_0/G_t^2$ is 0.001923f [Eq. (16–27), psia/G_t^2].

Assume G_t, and calculate the total two-phase pressure drop until a match with the known drop of 15 psia is obtained. For example, let us take the following.

(8) G_t is assumed to be 300 lb/sec-ft².

(9) μ_l is 0.20 cp (at about 250°F).

(10) N_{Re} is 560,000 ($N_{\text{Re}} = d_0 G_t/\mu_l$).

(11) f is 0.0031 (smooth pipe assumed).

(12) Δp_0 is 0.54 psia (from step 7).

(13) Δp_{TPF} is 5.9 psia (from steps 4 and 12).

(14) Δp_{aI} is 13.6 psia (from step 5).

(15) Δp_{aII} is 1.8 psia (from step 6).

(16) Δp_I is 19.5 psia (sum of steps 13 and 14).

(17) Δp_{II} is 7.7 psia (sum of steps 13 and 15).

If the homogeneous, or fog-acceleration, correction is used (I), then the drop in step 16 is too large and a lower G_t is necessary. The correct value is near 270 lb/sec-ft^2. If separate flow is assumed (II), the drop in step 17 is too small, and a larger G_t is necessary. The correct value for this case is near 430 lb/sec/-ft^2. In Fig. 16-8 for Example 16-3, the two cases (labeled I and II) are shown. Within the accuracy of using the graphical method, the curves do not show a maximum value. It is interesting to note the close accord of the results using the homogeneous assumption for the acceleration correction (I) with the curve based on homogeneous flow.

16-2.C Specific Problems and Analyses

Overall methods are necessary for design purposes; friction-factor models appear good for long complicated lines and, of course, in systems in which highly dispersed conditions exist. The Lockhart-Martinelli approach, with its modifications by Martinelli and Nelson and by others, is more generally useful. However, as pointed out, limitations exist and caution must be exercised in any design. In order to obtain more accurate and reliable means to predict the flow characteristics, we must consider the various flow regimes separately, and some steps, though not many, have already been taken along these lines. Some of these problems, and others of importance in evaporating flows, will be briefly considered in this section.

(a) **Homogeneous models.** The homogeneous models considered by Bankoff [16] and by Levy [150, 151] can be applied to flows with evaporation. The equations and curves for void fraction as a function of quality are directly applicable and are analogous to Fig. 16-9 for the Martinelli-Nelson approach. As in this method, one must integrate local values along the length of the tube to obtain the total frictional pressure drop (analogous to obtaining Fig. 16-11 from Fig. 16-10). Usually these methods are compared with the data and the predictions made by the Martinelli-Nelson procedure, and usually the comparison is good. The advantage of such methods lies in their analytical nature, which does not require the use of graphical representations. For example, the variable-density model by Bankoff requires a single-parameter K to provide a reasonable representation. The disadvantage is that when the Martinelli-Nelson method is inadequate, the homogeneous models are usually inadequate also.

(b) Annular-flow models. Linning [155] made a detailed analysis based on the annular flow pattern. In this, he assumed that each phase had a uniform velocity ($\alpha = \beta = 1$), that the pressure and temperature were uniform over any given cross section, that equilibrium existed, and that the film was symmetrical. The momentum balance in differential form for the total flow, the momentum balance for the liquid (16–87), and the energy balance for the vapor phase [Eq. (16–84) in differential form], were combined to eliminate τ_w and τ_i:

$$w_g(dH_g + U_g dU_g) - U_i[w_g dU_g + (U_g - U_l)dw_g + S_g dp]$$
$$+ \tfrac{1}{2}(U_g^2 - U_l^2)dw_g = 0. \quad (16\text{--}103)$$

To put the equation in a more workable form, one must also eliminate U_i. By applying the equation of continuity to the two phases at a given cross section, one can obtain the following equations:

$$U_g = B(k+a), \qquad U_l = \frac{B(k+a)}{k}, \qquad S_l = \frac{Sk}{k+a}, \qquad S_g = \frac{Sa}{k+a} = R_g S, \quad (16\text{--}104)$$

where

$$k = \frac{U_g}{U_l}, \qquad a = \frac{y\rho_l}{(1-y)\rho_g}, \qquad B = \frac{G_l(1-y)}{\rho_l},$$

and y is the mean radius of the interface. In addition, the relation $n = U_i/U_l$ is defined. Combining these in Eq. (16–103) gives a final result that is tedious to use, since it involves an iteration process:

$$\frac{k\, dH_g \rho_g}{n} - dp + \left[G_l^2 \frac{(k+a)(1-y)}{2a\rho_l} \right] \left\{ \frac{2y(k-n)(da+dk)}{n} \right.$$
$$\left. + \left(\frac{k+a}{kn} \right)[(k^2-1) - 2n(k-1)]\, dy \right\} = 0. \quad (16\text{--}105)$$

For the outlet conditions, Linning used 1.2 as the value of n. The analysis checks well with some of the experimental data for adiabatic flow at low pressures (8–70 psia), but not with all the available data. Actually, because of the complexity of the analysis, the extent of cross-checking with experimental data has not been as extensive as would be desirable.

A modification of Linning's analysis has been made by Isbin et al. [123]. The final equation required a knowledge of the liquid and gas fractions, which were obtained from the Lockhart-Martinelli correlation. Agreement with the experimental data was good at high qualities, with the predicted values becoming increasingly larger than the observed values as the quality was decreased. At 10% quality, the error was as high as 24%. Pike and Ward [186] used the same basic equations as Linning, but approached the problem in a different manner. A numerical integration was necessary and design charts were provided. The method predicted the pressure drop to within 10% for a limited number of data, provided that the assumption of a uniform temperature over a given cross section was valid. The experimental data were for low pressures, and as would be expected, the Martinelli-Nelson procedure was inadequate (with 46%). The friction-factor models were better, being within 24%. Other comparisons are given by the authors [186].

(c) **Critical flow.** Analysis of data on critical flow can be separated into two main divisions. First, there are those methods that predict the critical flow from a knowledge of the initial conditions and other parameters of the flow, and second, those that predict the relation between variables at the local critical flow point. For design use, this latter method would have to be coupled with a means of measuring pressure drop from the inlet conditions, as for example the Martinelli-Nelson method. The pressure drop would be obtained in the normal manner and then tested by one of the local critical-flow methods to see if choking flow conditions exist. To illustrate the two means of analysis, the overall homogeneous (friction-factor) model can be used. In Example 16–3, the model was used in a stepwise manner to obtain the mass velocity as a function of exit pressure. The results are shown in Fig. 16–8, where the critical condition is given at the point at which

$$dG_t/dp = 0. \tag{16-106}$$

In this first case, the critical-flow point is determined from the initial upstream pressure, friction in the system, and the conditions under which Eq. (16–106) was found to apply. The results of these calculations for the critical flow are reviewed in Table 16-6; G_{tc} and p_c are the critical points from the figure and x_c was obtained from the density calculation illustrated at the beginning of the example. For the local type of analysis, one can use a criterion other than Eq. (16–106). The homogeneous flow equation (16–76) can be differentiated and combined with Eq. (16–106) to give

$$G_{tc}^2 = \rho_c^2 \left(\frac{dp}{d\rho}\right)_c = -\left(\frac{dp}{d(1/\rho)}\right)_c \quad \text{or} \quad U_c^2 = \left(\frac{dp}{d\rho}\right)_c. \tag{16-107}$$

This is the same as Eq. (12–12), which was established for the propagation of a small pressure wave (the velocity of sound). Equation (12–12) was for single-phase compressible flow, however, and it is not known if the critical conditions actually correspond to the sound velocity for the complex two-phase flow. Fauske [234] has investigated this problem further and concluded that the implications for the two cases varied significantly, and that sonic conditions do not exist for the two-phase flow. In any event, Eq. (16–107) is used with thermodynamic data to establish the local critical flow point relation between p_c, x_c, and G_{tc}. Any two must be known to establish the third. The final results are usually presented in graphic form [73, 123, 229] or as a correlating equation [72, 90]. The main use for the homogeneous analysis has been for normalization of the data; i.e., the ratio of the actual critical flow to that obtained from the homogeneous relation for the same conditions. The various general plots just cited will not all be given

TABLE 16-6

Critical Values for Example 16–3

	G_{tc}, lb/sec-ft^2	p_c, psia	x_c
Friction-Factor Models			
Initial conditions			
Eq. (16–76)	278	23	0.0353
Eq. (16–82)	264	25	0.0305
Eq. (16–91)	248	22	0.0377
Local conditions			
Ref. 123	264	(23)*	(0.0353)
Ref. 73 (Fig. 16–13)	266	(23)	(0.0353)
Ref. 229	273	(23)	(0.0353)
Ref. 72 (Eq. 16–108)	324†	(23)	(0.0353)
Complex Models			
Initial conditions			
Ref. 186	500	21.3	
Local conditions			
Ref. 73 (Fig. 16–14)	815	(23)	(0.0353)
Ref. 123 (modified Linning)	825	(23)	(0.0353)
Experimental			
Ref. 72	about 730		
Ref. 122	about 660		

*Figures in parentheses are assumed values.
†Outside range of validity of Eq. (16–108).

here; the graphic method of reference 73 is given as Fig. 16–13 and the equation of Faletti and Moulton [72] is

$$G_{tc,\text{homogeneous}} = 96.2604 \, (6.1602 \, p_c^{-1.974} x_c + 17.198 \, p_c^{-1.684})^{-1/2} \quad (16\text{--}108)$$

and is valid (0.75%) for 28 psia $< p_c <$ 100 psia. The various solutions show good agreement, and to illustrate, several are tabulated in Table 16-6. The variation between results is of the order of one's ability to read the graphs presented in the various references.

Unfortunately, the overall homogeneous model is not an adequate representation of the extensive data that have been accumulated [72, 73, 123, 207], and so alternate methods are necessary. Again, the extensive review of critical flow by Smith [207] is helpful. Probably the best local method to date is that of Fauske [73, 74]. The condition used for his method is that the rate of pressure drop has a finite maximum value at the critical flow point, and that this is caused by variation in the slip-velocity as the flow progresses down the pipe. Equation (16–21) in differential form, with each $\beta = 1$ and for horizontal flow, can be combined with Eqs. (16–27) and (16–44) to give $dp = G_t^2[(2f/d_0)(\rho/\rho_1^2)dL + d(\rho/\rho_1^2)]$. Expressing

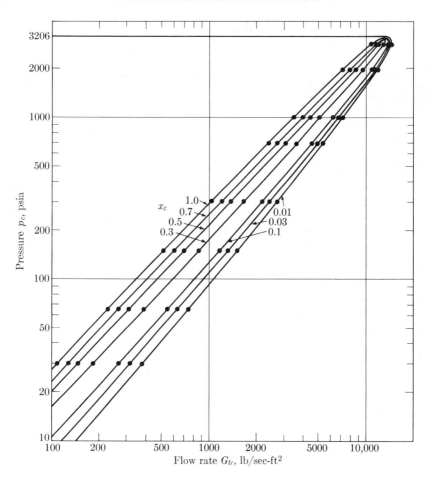

Fig. 16–13. Prediction of critical-flow rate from homogeneous-flow model. [From H. K. Fauske, ANL-6633 (Oct. 1962). By permission.]

this as pressure drop per foot (dp/dL) and differentiating with respect to the slip ratio $(k = U_g/U_l)$ gives us

$$\left[\frac{\partial(dp/dL)}{\partial k}\right]_c = G_{tc}^2 \left[\frac{2f}{d_0}\frac{\partial(\rho/\rho_1^2)}{\partial k} + \frac{2}{d_0}\frac{\rho}{\rho_1^2}\frac{\partial f}{\partial k} + \frac{d}{dL}\left(\frac{\partial(\rho/\rho_1^2)}{\partial k}\right)\right]_c = 0, \quad (16\text{–}109)$$

which equals zero because dp/dL is to be maximized with respect to variation in the slip ratio. It should be emphasized that this is true only at the critical flow point; hence the subscript c. Fauske suggests that the bracketed term can be zero for nonisentropic real flows only if each term is zero; i.e., if

$$\left(\frac{\partial f}{\partial k}\right)_c = \left[\frac{\partial(\rho/\rho_1^2)}{\partial k}\right]_c = 0. \quad (16\text{–}110)$$

Equation (16–44) can be combined with the slip-ratio expression of Eq. (16–45) to give

$$\frac{\rho}{\rho_1^2} = \frac{1}{k}\left[\frac{(1-x)k}{\rho_l} + \frac{x}{\rho_g}\right][1 + x(k-1)]. \tag{16-111}$$

Differentiating this and using the condition of Eq. (16–110) gives us $x_c(1 - x_c)[1/\rho_l - 1/\rho_g k_c^2] = 0$, which in turn can be zero only if

$$k_c = \left(\frac{U_g}{U_l}\right)_c = \left(\frac{\rho_l}{\rho_g}\right)^{1/2}. \tag{16-112}$$

In terms of the void fraction, this is

$$R_{gc} = \frac{1}{1 + (\rho_g/\rho_l)^{1/2}[(1-x_c)/x_c]}. \tag{16-113}$$

Note the similarity of Eqs. (16–112) and (16–113) (valid at the critical flow point only) to those for the condition of minimum entropy production as given by Eq. (16–73) and following.

Fauske determined the critical flow rate from a modification of Eq. (16–107). The term $1/\rho$ was replaced by ρ/ρ_1^2 [given by Eq. (16–44) or the equivalent (16–21)]; that is, the critical flow was determined from

$$G_{tc}^2 = -\left[\frac{dp}{d(\rho/\rho_1^2)}\right]_c^{1/2}. \tag{16-114}$$

The slip ratio is specified by Eq. (16–112), and the various thermodynamic derivatives $[d(1/\rho_g)/dp, dx/dp,$ and $d(1/\rho_l)/dp)]$ were evaluated by assuming steam–water isenthalpic flow. The results were presented in a general graphical form as shown in Fig. 16–14, and in addition as a series of graphs (with comparison to experimental data), in which p_c was plotted against G_{tc} for specific values of x_c. The main justification for the basic premise of Eq. (16–110) is the good comparison with the data of references 72, 73, 74, and 123. For easy comparison with the overall homogeneous results, Table 16-6 gives the estimation and approximate data values.

Several more recent proposals [233, 239, 241] are about the same in their agreement with the data. Levy [239] has extended his momentum model [150] to the prediction of critical flow rates, and for the case of no friction or head loss, obtained essentially the same agreement with the data as did Fauske, without making any additional assumptions about the void fraction. Moody [241] has used the energy equation to establish the slip ratio under critical flow conditions, whereas Fauske used the momentum equation. Moody obtained the result given following Eq. (16–73), in contrast to Eq. (16–112); i.e., the one-third power rather than the one-half. Because of the various assumptions, the two derivations are, in effect, the same. Cruver and Moulton [233] have attempted to show that the slip ratio should tend to the one-third-power law because of considerations of entropy production. The implication is that even though the momentum-derived one-half-power law implies a higher slip velocity, this velocity cannot

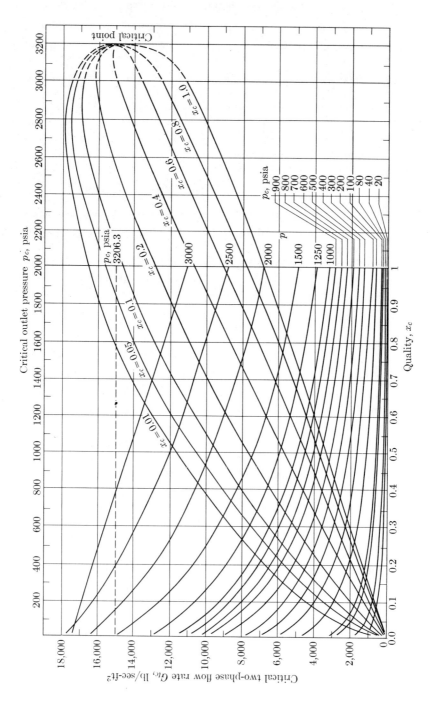

Fig. 16–14. Theoretical predictions of critical-flow rates. [From H. K. Fauske, ANL-6633 (Oct. 1962). By permission.]

be obtained because of entropy considerations that indicate the maximum to be given by the one-third-power law. The actual differences in the final predictions are not large, and it is difficult to decide whether one approach is better than another. It is recognized that the basic model is a crude approximation of the real case, and that a difference in the estimation of the slip velocity appears relatively unimportant when compared with the formulation of a better model.

As already mentioned, Isbin et al. [123] modified the approach of Linning and obtained a solution which was good at high qualities but relatively poor at low values. As presented, this was also a local method. The result calculated from their graph is shown in Table 16-6. For the specific case illustrated, both the estimation based on Fauske's analysis and the modified Linning approach are about 25% higher than experimental, which must be considered good for this type of analysis.

Pike and Ward [186] have presented a complex numerical solution based on initial conditions. As already noted, they used the equations presented by Linning as a basis, and performed a stepwise solution with distance down the tube. The critical flow point was assumed to be the maximum flow rate for a given pressure (at the initial vaporization point) and L/d_0 ratio. Several design charts were presented, and the result for our case is shown in Table 16-6. The comparison is quite good for the critical pressure, but underestimates the critical flow by as much as the other methods overestimate.

Before closing this section, we should mention the metastable flow conditions that often exist in variable area systems. Silver and Mitchell [203, 204] developed a theory based on the assumption that vaporization occurs only at the surface. This model is an annular core of water with a surrounding vapor ring. Bailey [14] used this approach to study the nature of saturated and nearly saturated water in orifices and nozzles, making the following assumptions: (1) Vapor begins to form when the saturation pressure is reached. (2) Vapor forms at the interface between the fluid and the nozzle. (3) The flowing fluid consists of a metastable liquid core surrounded by an annular ring of vapor. (4) The mass of vapor discharging from the nozzle is negligible in comparison with the mass of liquid. (5) There is no change in the temperature of the liquid core. (6) The velocity and pressure are uniform at any cross section of the fluid stream. From an analysis based on these assumptions, Bailey obtained an expression for the coefficient of contraction due to evaporation (ratio of area of liquid core from nozzle to area of nozzle exit):

$$C_e = 1 - 17.2 \times 10^3 \, K(L/d_0)\sqrt{\rho_l}\,(p_1)^{-0.435} Z. \tag{16-115}$$

The coefficient is C_e, L and d_0 are the length and diameter of the nozzle, p_1 is the initial pressure, K is the coefficient of evaporation, and Z is a complicated dimensionless parameter. Equation (16-115) is used with

$$G_t = C_e C \sqrt{2(p_1 - p_2)\rho_l}, \tag{16-116}$$

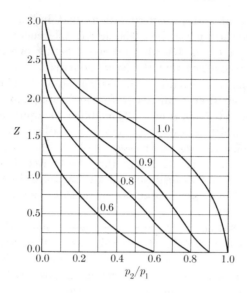

Fig. 16–15. The value of Z in Eq. (16–115). Parameter is p_s/p_1, where p_s is the saturation pressure. [From the American Society of Mechanical Engineers, Trans. ASME, **73**, 1109–1116 (1951). By permission.

which is the normal discharge equation with the normal coefficient C. Bailey derived an equation for K, and established values from a set of experimental runs. A single average value can be used (2.45×10^{-5}). The values of Z are given in Fig. 16–15.

The results for flow through an orifice are the same as for an incompressible fluid. This is the same conclusion reached by Silver [203] and by Benjamin and Miller [25]. For a nozzle (L of 2 in. and d_0 of $\frac{1}{8}$ in.), the fit is quite good; however, for different initial temperatures the error is as much as 4%. This can be reduced to 1% if the proper value of K is used rather than the average value mentioned above. Equation (16–116) reaches a maximum value, which is the critical flow condition. For metastable flow conditions, the data follow the incompressible equation and then drop down to the value given by Eq. (16–116). This was observed in the data of Danforth, which were used by Bailey. Two additional references [210] cover flow in two orifices in series. The review by Smith [207] summarizes and illustrates the use of the various available methods. In addition, Smith has included several working charts for hydrogen, nitrogen, oxygen, and certain refrigerants.

EXAMPLE 16–5. Bailey [14] suggested the problem of determining the mass velocity of water through a nozzle 0.125 in. in diameter and $\frac{1}{2}$ in. long. The conditions of flow are

$$p_1 = 45 \text{ psia}, \qquad p_2 = 25 \text{ psia}, \qquad p_{\text{saturation}} = p_s = 35 \text{ psia}, \qquad C = 0.99.$$

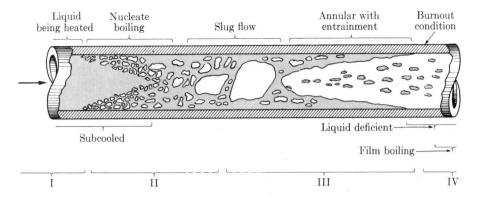

Fig. 16–16. Two-phase flow and boiling. (Drawn symmetrical for vertical flow; expected to be highly nonsymmetrical for horizontal flow.)

Answer. From Fig. 16–15, $Z = 0.45$ for $p_2/p_1 = 25/45 = 0.555$, and $p_s/p_1 = 35/45 = 0.778$. From the steam tables, $\rho_l = 58.6 \text{ lb/ft}^3$ at p_s. With $K = 2.45 \times 10^{-5}$ and the known values above, Eq. (16–115) gives $C_e = 0.873$. All terms are known in Eq. (16–116), which gives $G_t = 2850 \text{ lb/sec-ft}^2$.

16-3 ONE-COMPONENT TWO-PHASE FLOW WITH HEAT TRANSFER

Forced convection boiling and condensation are the major areas of two-phase flow with heat transfer. Unfortunately, because of the interactions of the heat transfer with the flow, these areas are the least understood of the general fields of boiling and condensation. Far more information is available on pool conditions where there is no force flow of the fluid. A few general references [160, 194, 201, 221] can be cited for these conditions, and, since the present discussion will emphasize the flow aspects of the fluid, no further comments will be offered here. The first step in solving the flow problem is to establish the rate of heat transfer for specific levels of the wall superheat, given by $T_w - T_s$ (T_w is the wall temperature and T_s is the saturation temperature). This is a problem of heat transfer, in which a correlating equation is desired for the heat-transfer coefficient in the familiar equation,

$$q/A_w = h(T_w - T_s). \tag{16–117}$$

One would expect that a number of correlations would be necessary because of the various regimes shown in Fig. 16–16. After one establishes the level of heat input along the length of the system, one can undertake the two-phase flow problem. If suitable methods are available, one can determine the pressure drop, voids, and other characteristics dependent on the flow. For the fluid-flow problem *per se*, the most obvious solution is to use an overall method, so that one does not consider the details of the flow, and thus one can avoid the complicated interactions occurring between the flow and the boiling. Little has been accomplished in the way of more sophisticated treatments of the details of the flow, but a beginning has been made in solving several specific problems, such as visualization of the flow, voids, burnout conditions, and unsteady oscillations. Before turning to a discussion of the problem, we should mention that a number of reviews may be found in references 27, 55, 97, 136, 158, 222, and 231.

A visualization of the flows during the boiling process will help us to understand the complex nature of the problem. Figure 16–16 will thus aid in our discussion. First, the liquid near the wall is heated to the boiling point. Local nucleate boiling will occur there, but since the total fluid is still subcooled there will be no net vaporization. It is assumed that the heat flux is below the critical level for which film boiling occurs. As the fluid approaches saturation temperature, the bubbles will remain, and will begin to coalesce as boiling progresses. Hsu and Graham [114] have studied the flow visually during heat addition and have found that this transition step toward slug flow is largely determined by the density of nucleation sites on the wall. If a large number of sites exist, the flow follows the process described, and eventually slugs are formed. However, if the number of sites is small, a metastable state apparently forms, so that the nucleate boiling range is small or nonexistent, and sudden slug-formation occurs. Eventually the slugs combine into annular flow. Near the end of the process, the wall becomes dry in places (film boiling occurs), and the heat cannot be conducted away through the vapor layer. The wall temperature rises to a point where structural failure can occur (the burnout condition).

16-3.A Heat Transfer During Forced-Convection Boiling

Because of the necessity of establishing the heat load from Eq. (16–117), and because h will depend on the nature of the flow, a brief review of the literature on this subject is in order. The general references already cited for pool boiling offer some information on the forced-convection problem. In Fig. (16–16), four zones have been noted (I to IV). The first is the single-phase, forced-convection section which precedes a second region involving forced convection nucleate boiling. The third zone is a two-phase convection area involving slug, annular, and annular-mist flow. The final zone is a liquid-deficient region associated with burnout.

For region I, standard equations such as those presented in McAdams [160] can be used; that is,

$$\left(\frac{hD}{k_l}\right)_{\mathrm{I}} = 0.023 \left(\frac{d_0 G_l}{\mu_l}\right)^{0.8} \left(\frac{c_p \mu}{k}\right)_l^{0.4}. \tag{16–118}$$

In region II, the flow is developing and can be pictured as a superimposed nucleate boiling on the single-phase convection of region I. The simplest recommendation was made by Rohsenow [193], who suggested that

$$\left(\frac{q}{A_w}\right)_{\mathrm{II}} = \left(\frac{q}{A_w}\right)_{\mathrm{I}} + \left(\frac{q}{A_w}\right)_{\mathrm{PB}}, \tag{16–119}$$

where, for example, the pool boiling flux, $(q/A_w)_{\mathrm{PB}}$, can be obtained from the equation by Forster and Zuber [78]:

$$\left[\frac{(q/A_w)_{PB} c_{pl} \rho_l}{k_l \Delta H_v \rho_g}\right] (\pi \alpha_l)^{1/2} \left(\frac{2\sigma}{\Delta p}\right)^{1/2} \left(\frac{\rho_l}{g_c \Delta p}\right)^{1/4} = 0.0015 \left[\frac{\rho_l}{\mu_l}\left(\frac{\Delta T c_{pl} \rho_l}{\rho_g \Delta H_v}\right)^2 \pi \alpha_l\right]^{5/8} \left(\frac{c_p \mu}{k}\right)_l^{1/3},$$
(16-120)

in which α_l is the thermal diffusivity of the liquid and σ the surface tension. Of particular importance is the study by Bergles and Rohsenow [31], which shows that the simplicity of Eq. (16-119) is not really valid. They suggest the combination equation,

$$\left(\frac{q}{A_w}\right)_{II} = \left(\frac{q}{A_w}\right)_{I} \left\{1 + \frac{(q/A_w)_{CB}}{(q/A_w)_{I}}\left[1 + \frac{(q/A_w)_{CBi}}{(q/A_w)_{CB}}\right]^2\right\}^{1/2}, \qquad (16\text{-}121)$$

in which $(q/A_w)_{CB}$ is measured under actual conditions of fully developed, forced-convection boiling, and may not be the same as $(q/A_w)_{PB}$. The expression $(q/A_w)_{CBi}$ is the extrapolation of this back to the value of $T_w - T_s$, corresponding to the incipient boiling point (i.e., it is the boiling contribution to the total flux at incipient boiling). Region II has a critical or maximum heat flux above which film boiling at the surface will occur. In cryogenic fluids, such as liquid hydrogen studied by Hendricks et al. [108], a stable forced-convection film boiling is difficult to avoid. Alternatives to Eqs. (16-119) and (16-121), means of estimating the pool-boiling heat flux, and the maximum heat flux have been reviewed by Seader et al. [201], and in part in the general references already cited.

Davis and David [61] have reviewed the available experimental data on steam-water systems for region III. Seader et al. [201] presented a corresponding review for cryogenic fluids. A summary of the many proposed correlating equations can be found in Davis and David. A comparison, showing that little agreement exists, has been made by Anderson et al. [9]. For this region, Davis and David used a modified form of Eq. (16-118) to obtain a correlation with an average absolute error of 17%:

$$\left(\frac{h d_0}{k_l}\right)_{III} = 0.060 \left(\frac{\rho_l}{\rho_g}\right)^{0.28} \left(\frac{d_0 G_l x}{\mu_l}\right)^{0.87} \left(\frac{c_p \mu}{k}\right)_l^{0.4}. \qquad (16\text{-}122)$$

Chen [50] suggested that Eq. (16-119), slightly modified, could be used for zone III:

$$(q/A_w)_{III} = (q/A_w)_{I} F + (q/A_w)_{PB} S. \qquad (16\text{-}123)$$

The single-phase forced-convection and the pool-boiling contributions are obtained as before. The terms F and S are empirical parameters that account for the presence of vapor and the suppression of bubble growth due to flow, respectively; F is a function of the Lockhart-Martinelli parameter given in Eq. (16-40):

$$X_{tt} = \left(\frac{w_l}{w_g}\right)^{0.9} \left(\frac{\rho_g}{\rho_l}\right)^{0.5} \left(\frac{\mu_l}{\mu_g}\right)^{0.1} = \left(\frac{1-x}{x}\right)^{0.9} \left(\frac{\rho_g}{\rho_l}\right)^{0.5} \left(\frac{\mu_l}{\mu_g}\right)^{0.1}, \qquad (16\text{-}124)$$

and S is a function of

$$[d_0 G_l (1-x)/\mu_l] F^{1.25}. \qquad (16\text{-}125)$$

The functions are presented graphically in references 50 and 201.

The final region (IV) is a complex series of steps in which, at the level of critical heat flux, called burnout, the surface becomes liquid-deficient and the vapor makes contact with the wall. The operation is unsteady, with large temperature oscillations occurring. Over a relatively short range the vapor blankets the surface, causing a drastic drop in the heat-transfer coefficient and a corresponding rise in the wall temperature. Usually the rise is more than the system can tolerate and structural failure occurs. If the vapor velocity is high enough, however, it is possible that the tube temperature will remain below the failure level, and that a subsequent stable film-boiling range can be obtained. Polomik et al. [187] have demonstrated that these conditions can exist, and suggested the following correlation equation for the film-boiling range:

$$\left(\frac{hd_0}{k_f}\right)_{\mathrm{IV}} = 0.00136 \left(\frac{d_0 G_t}{\mu_f} \frac{1-x}{x}\right)^{0.853} \left(\frac{c_p \mu}{k}\right)_f^{1/3} \frac{R_g}{1-R_g}$$

$$= 0.00136 \left(\frac{\rho_l}{\rho_g}\right)^{2/3} \left(\frac{d_0 G_t}{\mu_f}\right)^{0.853} \left(\frac{1-x}{x}\right)^{-0.147} \left(\frac{c_p \mu}{k}\right)_f^{1/3}, \quad (16\text{--}126)$$

where the latter form was obtained by using Eq. (16–73). The subscript f denotes properties of the *steam* (not the liquid) at a temperature which is the arithmetic average of the wall and the bulk.

16-3.B Overall Models

From the preceding section one can make a reasonable estimate of the heat load as a function of length. With this established, one can investigate the actual flow problem. The simplest model, homogeneous boiling, is similar to flashing, except that the heat transfer, not the pressure drop, controls the vaporization. The various overall analyses, involving two-phase flow with heat transfer, apply only to regions II and III. If a fluid is subcooled when it enters the system, then an estimate must be made for the length of region I before the two-phase pressure-drop calculation can be undertaken. Once this estimate has been made, it is generally assumed that the heat-absorption rate per unit area of tube surface is constant in regions II and III. This assumption implies that the heat absorbed by the liquid is directly proportional to the distance traveled by the fluid. Since the heat input up to any point is known, the heat content can be calculated. If one knows the pressure (so that one can establish the temperature), one can, from this, determine the fraction vaporized. Knowing the fraction vaporized, one can determine the mean density of the flowing system at any point along the tube. Equations (16–76), (16–82), or (16–91) can then be used as before to determine the pressure drop for a known flow. The average pressure must be known in order to establish, from the heat content, the fraction vaporized, but since this is the pressure that is being determined, a stepwise solution is usually necessary. For the sake of accuracy, the stepwise solution is done in small increments, and in terms of length rather than in terms of pressure, since it is the length that determines the heat content, and it is the heat content which eventually determines the density needed in the calculation. Thus, so long as some starting pressure is known, a point-by-point calculation along the heater is possible. The friction term usually makes up the main part of the total pressure drop in the system. The error from neglecting the velocity and static heads is usually small. Further, as the density decreases the static heat term becomes less and less important. In a horizontal system, this term is zero at all times. At the very start of the vaporization, the static head term may be of importance but when this is true the velocity head term can usually be neglected.

Various authors have used the overall homogeneous approach for a variety of flows. For example, Dittus and Hildebrand [64] used a combined mathematical and graphical method to arrive at the pressure drop in an oil furnace. Since this work several authors have suggested ways that are somewhat shorter [77, 163] and more accurate [41, 163]. Maker [163] presented a simplified modification worked out by Ludwig, which assumes that the velocity and potential heads are negligible, and that the pressure–density ratio varies linearly with distance. Foltz and Murray [77] presented a more detailed development along the same line, in which they assumed a linear relation for the specific volume. Maker also developed a method which used a plot of pressure versus enthalpy, with specific volume as the parameter. Buthod [41] and Doll-Steinberg [65] have made further contributions to this area of study.

The simplified method of Ludwig is quite easy to use, and can provide a good preliminary estimate of the pressure drop. In addition, it can be used to establish some idea of the magnitude of the velocity head term. The method is developed as follows: The friction-factor relation, in differential form, is

$$dp = -\rho(4f)(U_{\text{ave}}^2/2)(1/d_0)\,dL.$$

Multiplying both sides by p and re-expressing the velocity in terms of the mass-flow rate gives us

$$p\,dp = -p(4f)\frac{16w_t^2}{2\pi^2 d_0^5 \rho}dL = -0.811\frac{w_t^2}{d_0^5}4f\frac{p}{\rho}dL.$$

Since p/ρ is assumed linear with H or L, integration gives

$$\int_{L_1}^{L_2} \frac{p}{\rho}\,dL = \frac{1}{2}\left(\frac{p_1}{\rho_1} + \frac{p_2}{\rho_2}\right)(L_2 - L_1).$$

The pressure drop is

$$p_1^2 - p_2^2 = 0.811\frac{w_t^2}{d_0^5}\left(\frac{p_1}{\rho_1} + \frac{p_2}{\rho_2}\right)4f(L_2 - L_1). \qquad (16\text{--}127)$$

The velocity head term can be computed separately and then added to the value obtained from Eq. (16-127):

$$\Delta p_a = \frac{1}{2}\rho U_{\text{ave}}^2 = \frac{w_t^2}{2\rho S^2} = \frac{16w_t^2}{2\rho\pi^2 d_0^4} = \frac{0.811 w_t^2}{\rho d_0^4}. \qquad (16\text{--}128)$$

Foltz and Murray [77] have presented an alternative to the above which accounts for the velocity and static heads. A linear relation between the specific volume and heater length is assumed, which allows the conditions along the line to be represented in terms of the physical properties, heat input, flow rate, and heater length. However, Foltz and Murray reported that the predictions were about 25% low for a Freon-114 system. They believed that this was due to a length of nonboiling section and to the fact that the difference in velocity of the two phases was neglected. It is expected that, for such equipment as a vacuum furnace for a crude vacuum-distillation column, consideration of the velocity head will be necessary. For this case, the velocities may even be high enough to obtain critical flow. Finally, the Martinelli-Nelson method can also be applied to systems with heat transfer. Here the amount of transfer will determine the exit quality, which must be established before the method can be used.

16-3.C Specific Flow Problems

Although a great deal of effort has been expended on the boiling problem, only a qualitative insight into the mechanism has been obtained. The foregoing discussion, then, constitutes the means for design, while what now follows is the beginning of a better understanding of the process, from which, one hopes, improved designs will come. The breakdown suggested in Fig. 16-16 will be used to outline our discussion here. As usual, no attempt will be made at completeness, and only a few of the more recent and important efforts will be discussed. In each of the cases mentioned, rather complete literature reviews of the earlier works are available.

(a) Nucleate boiling region (II). The flow aspects of forced-convection nucleate boiling, such as pressure drop and void distribution, have received almost no attention. Of primary interest to researchers has been the mechanism by which the observed increase in heat transfer occurs. Most of these studies, though not all of them, have considered the slightly less complicated pool-boiling conditions, and have suggested that transfer is enhanced by: (1) microconvection, which is the local turbulent mixing caused by the stirring action of the bubbles [17, 18, 46, 71, 102, 129, 143], (2) vapor–liquid exchange, which is a quenching of the wall by the cold bulk fluid that has replaced the hot fluid removed from the surface by the vapor bubbles [17, 71, 143], (3) microlayer vaporization, which is the vaporization of a thin layer at the base of vapor bubbles [18, 174], and (4) latent heat transport within the bubbles [18, 46, 78, 102, 171]. There is little agreement as to which of the mechanisms is the most important. Arguments have been presented for all of them, and in all probability all are significant, depending on the conditions which exist during boiling.

So far as the two-phase flow problem is concerned, what is of greatest interest is a visualization of the flow process on which estimates of voids and pressure drop can eventually be based. To this end, Jiji and Clark [133] photographed the bubble boundary layer in vertical flow along a flat plate. In the one photograph shown (2 μ sec exposure), individual bubbles are clearly distinguishable in the layer, although temperature fluctuation measurements very near the surface indicate that the population of bubbles is high enough to cause partial blocking of the surface with vapor. A similar effect was observed by Kudirka [143] when he simulated boiling by air injection during heat transfer.

(b) Slug-flow boiling region (II-III). Slug flow is a condition which is unique to the forced-convection problem, and as such has not been treated extensively. However, nucleate boiling will still occur in slug flow, and nucleate boiling mechanisms would be expected to apply. In addition, the sweeping action of the large slugs would contribute to mixing by interchange, and probably would result in higher heat-transfer coefficients. This is only a premise, however, since the process is complex and other effects may be occurring at the same time. Specific to the fluid-flow problem is Griffith's application [100] of the slug-flow theory previously presented (pages 487–489) to the prediction of voids during boiling. For systems with heat transfer, he found that C_2 in Eq. (16–66) was 1.6, rather than unity. The value of C_1 and the coefficient of U_{ave} in Eq. (16–67) are specific to circular pipe systems. Griffith has estimated values for the alternative geometries of rectangular channels, annuli, and tube bundles. With this generalization of the constants, one can determine the void fraction from Eq. (16–65) for nearly any vertical geometry, so long as one knows that the flow is in the slug-flow regime. Griffith compared the results of the analysis with the available experimental data in terms of the void fraction or slip ratio (16–64). Considering that knowledge of the flow regime is uncertain, the comparison is quite good, providing a reasonable estimate for the voids and, indirectly, of the pressure drop.

(c) Annular-flow boiling region (III). The regime of annular flow during boiling has received practically no detailed analytical or visual consideration. Of course, experimental results are available for void fractions and pressure drops, and overall methods do apply. No doubt nucleation exists in the earlier part of the flow, but it is apparently suppressed in the latter stages. An estimate of the suppression can be made from the heat-transfer analysis of Chen [50] by estimating the contribution of S in Eq. (16–123). Annular flow should exist at an area fraction of about 0.9 (less than 1% quality for a steam–water flow). From Fig. 16–4, X_{tt} for this level of void is about 0.25. The plots in either reference 50 or 201 show that F is about 7. At high velocities, $(q/A_w)_I$ can be more important than $(q/A_w)_{PB}$, while at low velocities the opposite is true. For the sake of calculation, let us assume $(q/A_w)_{PB} = 2(q/A_w)_I$; then for a pool-boiling contribution of 5% or less, $S = 0.05(\frac{1}{2})7 = 0.175$. This value occurs when the parameter of Eq. (16–125) is 2.25×10^5. For the given F, the liquid Reynolds number $(d_0 G_l/\mu_l)$ would be about 20,000. Higher Reynolds numbers would mean a contribution of less than 5%, and as the area fraction increases above 0.9, the contribution would again decrease. It appears that a good part of the annular flow regime is not influenced by nucleate boiling, and so it may be that the flow analysis during flashing can be applied with little modification.

(d) Critical heat flux or burnout region (IV). For economy of area in heat-transfer systems, it is desirable to operate at as high a heat input as possible. Because of this, considerable research effort has been expended to determine the maximum or critical heat level for both pool boiling and forced-convection boiling. Reviews of such work have been given by Bernath [32] and Seader et al. [201]. Two critical locations are possible; the first is in the nucleate-boiling range (before slug flow forms) where the heat flux can be so high as to cause a continuous vapor layer along the wall. The second, which is of more interest here, occurs at the end of the annular-flow regime. For the former problem, several correlations have been suggested [32, 82, 98, 144, 201] in which the maximum (q/A_w) can be determined from the system properties and overall conditions of flow. For the latter case, one major problem has been that of defining just what is meant by the burnout condition. Actually there is reasonable agreement as to what happens: dry areas appear on the tube surface, are covered again by liquid, and then become dry again. This cycle is repeated again and again, and results in temperature oscillations. The disagreement is not so much with what happens, but with finding acceptable ways to measure the condition and interpret the experimental results [205]. In a short distance, which depends on the heat flux, the wall becomes dry and the temperature increases. This in turn gives rise to new mechanisms of transfer, such as film boiling [45, 187], vapor superheat [26], or spray evaporation [89], or it may result in tube failure. The exact details of the transition involved in going from the first appearance of dry patches to the new mechanism for transfer are essentially unknown, but the transition apparently depends on the quality of flow at burnout, flow rate, and pressure [26, 187, 205]. However, other factors, such as geometry, heat flux distribution, and compressibility [2, 22, 138] do enter the problem. Figure 16–17 illustrates the trend observed in the experimental data for a given geometry at a constant pressure and mass velocity [187]. Critical heat-flux rates decrease with increased mass velocity (constant p_b and x_b), and with increased pressure (constant G_{tb} and x_b). The subscript b is used to denote the critical heat-flux rate associated with burnout. In the area above point A in Fig. 16–17, the quality is relatively low and the critical flux high. Because of the high flux, the burnout is characterized by some initial oscillation, followed by a very rapid and probably destructive rise in temperature (fast or normal burnout). Beyond point A, the quality is higher, and critical heat fluxes much lower. This area is characterized by temperature oscillations (slow burnout).

Various means of correlation have been suggested, some of them empirical [22, 24, 208, 238] and others theoretical [54, 89, 127, 212]. If the phenomena are local in nature, a correlation such as that suggested in Fig. 16–17 should be generally valid [33]. In this correlation the burnout is a function of local conditions (critical-heat flux and x_b), with geometry, pressure, and mass flow being parameters. However, most of the recent experimental evidence fails to support this contention [109]. More common, and more recent, are a number of hydrodynamic models [54, 89, 127, 212] which are based on an assumed mechanism for the

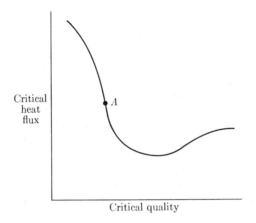

Fig. 16-17. Qualitative correlation for critical heat flux.

onset of burnout and the subsequent transition. Most of these involve an evaporating liquid film along the wall, with an exchange of liquid between the film and drops in the core (assumed as a diffusion mechanism). Such models are in part supported by observations [212]. In one case [212], it is assumed that critical conditions begin when the rate of evaporation exceeds the rate of deposition, and in another [127], when the liquid film disappears. One group of observers [89] has taken the opposite view that droplet diffusion from the core through a vapor boundary layer is the limiting mechanism for burnout. It is assumed that liquid flow through the film is negligible. When the heat flux just equals the ability of turbulent diffusion to supply droplets from the core for evaporation, the burnout condition occurs. These various theories all agree that the critical-heat flux decreases with x_b, and thus can be applied to the region above point A in Fig. 16-17. All give reasonable results, because the adjustable constants needed are obtained by comparison with actual burnout data. It is unfortunate that we cannot describe the burnout region on the basis of a real knowledge of what is occurring. Tippets [212], who took movies of the onset of burnout, has suggested that a more accurate correlation of the burnout problem requires basic information which can be provided only by an accurate visualization of the burnout region over the entire range of qualities and other parameters. It is doubtful that one mechanism can account for all the variation observed.

The possibility of oscillatory motions, and thus the necessity of considering transient behavior, is associated with boiling in both forced and natural circulation loops. Due partly to our lack of understanding of the mechanism of boiling, the problem is complex. Nevertheless, the various equations (continuity, motion, and energy) with an unsteady state contribution can be used to study the problem [11, 116, 189, 218, 225].

PROBLEMS

16-1. Water and air (saturated) are flowing in a 4-in. Sch. 40 steel pipeline. The weight ratio of air to water is 0.02. The line is at 90°F and atmospheric pressure. Calculate the pressure drop in the line for the assumptions of turbulent-turbulent flow and equal friction factors in both phases. Use both the Lockhart-Martinelli and Chenoweth-Martin methods. The pressure drop is 10 in. of water, when the water is flowing alone.

16-2. Consider the possibility of an analytical determination of ϕ and X for the assumed model of symmetrical annular flow. This is the Levy solution given in reference 149.

16-3. A vertical pipe 15 ft long (1-in. Sch. 40) is being used as an air lift for an oil of 0.9 specific gravity. Calculate the performance curve of liquid- versus gas-flow rate. What is the maximum oil-flow rate? Assume an average oil–air temperature of 70°F.

16-4. Outline the computation necessary to establish the initial pressure of a system similar to that in Example 16-3, but with a flow rate of 248 lb/sec-ft².

16-5. Apply the Ludwig method to Example 16-3. Make any assumptions necessary for the solution of the problem, and tabulate these assumptions.

NOTATION FOR CHAPTER 16

A, B	constants
A	area
c	wave velocity used in Eq. (16-3), constant
C	constant, coefficient
C_e	coefficient in Eq. (16-115)
d_0	diameter
E	kinetic energy flux
f	Fanning friction factor
F	force, parameter
g	gravity
G	mass velocity
h	heat-transfer coefficient, film thickness
H	enthalpy
ΔH_v	heat of vaporization
k	wave number, parameter, slip ratio, thermal conductivity
K	parameter used first in (16-47)
L	length
N_{Re}, N_{We}, N_{Fr}	dimensionless numbers (Reynolds, Weber, Froude)
n, m	constants used for power laws
p	pressure
Q	volumetric flow rate
R	area or void fraction
r_0	radius

NOTATION FOR CHAPTER 16

s	gas-to-liquid density ratio
S	cross-sectional area, parameter
s	entropy
T	temperature
U	velocity
u	internal energy
U^*	friction velocity
V	volume
w	mass flow rate
x	quality
X	parameter defined in Eq. (16–40)
Z	parameter defined in Fig. 16–15
z	height

Greek letters

α	wave number modified in Eq. (16–4), energy correction, thermal diffusivity, variable in Eq. (16–30)
β	momentum correction, sheltering coefficient in Eq. (16–7), variable in Eq. (16–35)
Δ	difference
η	$1 - r/r_0$
κ	defined in Eq. (16–55)
λ	wavelength used in Eq. (16–13), parameter defined in Eq. (16–1)
μ	viscosity
π	3.1416...
ρ	density
$\bar{\rho}$	density given in Eq. (16–18)
σ	surface tension
τ	stress
ϕ	parameter defined in Eqs. (16–2) and (16–28)
χ	parameter defined in Eq. (16–92)

Subscripts

a	acceleration
ave	average
b	bubble
c	critical
CB	fully developed convective boiling
e	from energy considerations
f	film
g	gas, gravitational
i	interface, incipient
l	liquid
m	from momentum considerations
0	relative to free liquid in Eq. (16–64), 100% liquid in Eq. (16–94)
PB	pool boiling

s	slip, separation, saturation
t	total
w	wall
Fe	friction from energy equation used in Eq. (16-20)
Fm	friction from momentum equation used in Eq. (16-21)
Losses	losses
TPF	two-phase frictional in Eq. (16-28)
1	initial
2	final
σ	surface
μ	viscous
ρ	inertial

Superscripts

$+$	nondimensional form
overbar	average over the pipe

REFERENCES

1. A. E. ABRAMSON, *J. Appl. Mech.* **19**, 267–272 (1952).
2. I. T. ALADYEV, Z. L. MIRPOLSKY, V. E. DOROSHCHUK, and M. A. STYRIKOVICH, *International Developments in Heat Transfer*, paper 28, 237–243, ASME (1961).
3. W. F. ALLEN, JR., *Trans. ASME* **73**, 257–265 (1951).
4. J. AL SHEIKH, Ph. D. thesis, The Ohio State University, Columbus, Ohio (1963).
5. M. ALTMAN, F. W. STAUB, and R. H. NORRIS, *Chem. Eng. Progr. Symposium Series* No. 30 **56**, 151–159 (1960).
6. G. E. ALVES, *Chem. Eng. Progr.* **50**, 449–456 (1954).
7. T. W. AMBROSE, *Doc. HW-52927, OTS*, Dept. Commerce (Oct. 1957).
8. G. H. ANDERSON and B. G. MANTZOURANIS, *Chem. Eng. Sci.* **12**, 109–126, 233–242 (1960).
9. G. H. ANDERSON, G. G. HASELDEN, and B. G. MANTZOURANIS, *Chem. Eng. Sci.* **16**, 222–230 (1961).
10. J. D. ANDERSON, R. E. BOLLINGER, and D. E. LAMB, *AIChEJ* **10**, 640–645 (1964); see also T. W. F. RUSSELL and D. E. LAMB, *Can. J. Chem. Eng.* **43**, 237–245 (1965).
11. R. P. ANDERSON, L. T. BRYANT, J. C. CARTER, and J. F. MARCHATERRE, *Chem. Eng. Progr. Symposium Series* No. 41 **59**, 96–103 (1963).
12. T. T. ANDERSON, *AIChEJ* **10**, 776 (1964).
13. S. AOKI, T. ICHIKI, and T. TAKAHASHI, *Tokyo Inst. Tech. Bulletin* **49**, 127–139 (1962).
14. J. F. BAILEY, *Trans. ASME* **73**, 1109–1116 (1951).
15. O. BAKER, *Oil and Gas J.* **53** No. 12, 185–195 (July 26, 1954).
16. S. G. BANKOFF, *J. Heat Transfer* **82C**, 265–272 (1960); see also N. ZUBER, *ibid.*, 255–258.
17. S. G. BANKOFF, *Chem. Eng. Progr. Symposium Series* No. 32 **57**, 156–172 (1961).
18. S. G. BANKOFF, *AIChEJ* **8**, 63–65 (1962).
19. S. G. BANKOFF, *AIChEJ* **10**, 776 (1964).

20. C. J. BAROCZY, *Chem. Eng. Progr. Symposium Series* No. 57 **61**, 179–191 (1965); see also L. R. SMITH, M. R. TEK, and R. E. BALZHISER, *AIChEJ* **12**, 50–58 (1966).
21. E. Q. BASHFORTH, J. B. P. FRASER, H. P. HUTCHISON, and R. M. NEDDERMAN, *Chem. Eng. Sci.* **18**, 41–46 (1963).
22. K. M. BECKER and G. HERNBORG, *J. Heat Transfer* **86C**, 393–404 (1964).
23. W. BEGELL and J. W. HOOPES, JR., *CU-18-54-At-dP-Ch.E.*, Columbia University, New York (April 1954).
24. D. S. BELL, *Nucl. Sci. and Eng.* **7**, 245–251 (1960).
25. M. W. BENJAMIN and J. G. MILLER, *Trans. ASME* **64**, 657–669 (1942).
26. A. W. BENNETT, H. A. KEARSEY, and R. K. F. KEEYS, *AERE-R 4352*, Atomic Energy Research Estab., Harwell, England (June 1964).
27. J. A. R. BENNETT, *AERE CE/R-2497*, Atomic Energy Research Estab., Harwell, England (1958).
28. J. A. R. BENNETT and J. D. THORNTON, *Trans. Inst. Chem. Engrs. (London)* **39**, 101–126 (1961).
29. O. P. BERGELIN and C. GAZLEY, JR., *Proc. Heat Transfer and Fluid Mech. Inst.*, pp. 5–18, Berkeley, California (June 1949).
30. O. P. BERGELIN, P. K. KEGEL, F. G. CARPENTER, and C. GAZLEY, JR., *ibid*, pp. 19–28.
31. A. E. BERGLES and W. M. ROHSENOW, *J. Heat Transfer*, **86C**, 365–372 (1964).
32. L. BERNATH, *Chem. Eng. Progr. Symposium Series* No. 50 **56**, 95–116 (1960).
33. S. BERTOLETTI, C. LOMBARDI, and G. PETERLONGO, *Energia Nucleare* **11**, 269–272 (1964).
34. L. M. K. BOELTER and R. H. KEPNER, *Ind. Eng. Chem.* **31**, 426–434 (1939).
35. W. T. BOTTOMLEY, *Trans. Northeast Coast Institution of Engineers and Shipbuilders*, No. 8, Session 53, 65–100 (1936–1937).
36. T. E. BRIDGE, *Heating, Piping and Air Conditioning*, 69–73 (March 1949); 92–96 (April 1949); 98–100 (May 1949).
37. T. BROOKE BENJAMIN, *J. Fluid Mech.* **2**, 554–574 (1957); **3**, 657 (1957); **10**, 401–419 (1961).
38. T. BROOKE BENJAMIN, *J. Fluid Mech.* **6**, 161–205 (1959).
39. R. A. S. BROWN and G. W. GOVIER, *Can. J. Chem. Engr.* **39**, 159–164 (1961).
40. R. A. S. BROWN, G. A. SULLIVAN and G. W. GOVIER, *Can. J. Chem. Engr.* **38**, 62–66 (1960).
41. P. BUTHOD, *Oil and Gas J.*, **55** No. 26, 111–118 (July 1, 1957).
42. S. CALVERT and B. WILLIAMS, *AIChEJ* **1**, 78–86 (1955).
43. C. O. CARTER and R. L. HUNTINGTON, *Can. J. Chem. Engr.* **39**, 248–251 (1961).
44. J. A. CENGEL, A. A. FARUGUI, J. W. FINNIGAN, C. H. WRIGHT, and J. G. KNUDSEN, *AIChEJ* **8**, 335–339 (1962).
45. R. D. CESS and E. M. SPARROW, *J. Heat Transfer* **83C**, 370–379 (1961).
46. Y. P. CHANG and N. W. SNYDER, *Chem. Eng. Progr. Symposium Series* No. 30 **56**, 25–38 (1960).
47. M. E. CHARLES, G. W. GOVIER, and G. W. HODGSON, *Can. J. Chem. Engr.* **39**, 27–36 (1961); see also M. E. CHARLES and L. U. LILLELEHT, *ibid.* **43**, 110–117 (1965); **44**, 47–49 (1966); *J. Fluid Mech.* **22**, 217–224 (1965).

48. D. A. Charronia, *TM-58-1*, Jet Prop. Center, Purdue University, Lafayette, Ind. (1958).
49. D. A. Charronia, *Interim Report 50-1, JPC I-59-1*, Jet Prop. Center, Purdue University, Lafayette, Ind. (May 1959).
50. J. C. Chen, *Ind. Eng. Chem., Proc. Design and Develop.* **5**, 322–329 (1966).
51. J. M. Chenoweth and M. W. Martin, *Pet. Ref.* **34**, No. 10, 151–155 (1955).
52. S. F. Chien and W. Ibele, *J. Heat Transfer* **86C**, 89–96 (1964).
53. D. Chisholm and A. D. K. Laird, *Trans. ASME* **80**, 276–286 (1958).
54. A. Cicchitti, M. Silvestri, G. Soldaini, and R. Zarattarelli, *Energia Nucleare* **6**, 637–660 (1959).
55. J. G. Collier, *AERE CE/R-2496*, Atomic Energy Research Estab., Harwell, England (1958).
56. J. G. Collier and G. F. Hewitt, *Trans. Inst. Chem. Engrs. (London)* **39**, 127–136 (1961).
57. J. G. Collier and G. F. Hewitt, *AERE-R-4684*, Atomic Energy Research Estab., Harwell, England (July 1964).
58. W. C. Corder, M. S. thesis, The Ohio State University, Columbus, Ohio (1963); see also L. S. Lowe, *ibid.* (1966).
59. W. F. Davidson, P. H. Hardie, C. G. R. Humphreys, A. A. Markson, A. R. Mumford, and T. Ravese, *Trans. ASME* **65**, 553–591 (1943).
60. R. M. Davies and G. I. Taylor, *Proc. Roy. Soc. (London)* **200A**, 375–390 (1950).
61. E. J. Davis and M. M. David, *Ind. Eng. Chem., Fund.* **3**, 111–118 (1964).
62. R. T. P. Derbyshire, G. E. Hewitt, and B. Nicholls, *AERE-M 1321*, Atomic Energy Research Estab., Harwell, England (1964).
63. J. E. Diehl, *Pet. Ref.* **36**, No. 10, 147–153 (1957).
64. F. W. Dittus and A. Hildebrand, *Trans. ASME* **64**, 185–192 (1942).
65. A. Doll-Steinberg, *Pet. Ref.* **38**, No. 1, 217 (1959).
66. A. E. Dukler and M. Wicks III, *Modern Chemical Engineering* (A. Acrivos, editor), Chap. 8, Reinhold, New York (1963).
67. A. E. Dukler, M. Wicks, III, and R. G. Cleveland, *AIChEJ* **10**, 38–51 (1964).
68. D. T. Dumitrescu, *ZAMM* **23**, No. 3, 139–149 (1943).
69. G. M. Dusinberre, *Trans. ASME* **64**, 666 (1942).
70. S. R. M. Ellis and B. Gay, *Trans. Inst. Chem. Engrs. (London)* **37**, 206 (1959).
71. K. Engelberg-Forster and R. Greif, *J. Heat Transfer* **81C**, 43–53 (1959).
72. D. W. Faletti and R. W. Moulton, *AIChEJ* **9**, 247–253 (1963).
73. H. K. Fauske, *ANL-6633*, Argonne National Lab., Argonne, Ill. (Oct. 1962).
74. H. K. Fauske, *ANL-6779*, Argonne National Lab., Argonne, Ill. (Oct. 1963).
75. S. Feldman, *J. Fluid Mech.* **2**, 343–370 (1957).
76. M. J. Fohrman, *ANL-6256*, Argonne National Lab., Argonne, Ill. (Nov. 1960).
77. H. L. Foltz and R. G. Murray, *Chem. Eng. Progr. Symposium Series* No. 30 **56**, 83–94 (1960).
78. H. K. Forster and N. Zuber, *AIChEJ* **1**, 532–535 (1955).
79. A. W. Frazier, B. S. thesis, Carnegie Institute, Pittsburgh, Pa. (1942).
80. L. Fried, *Chem. Eng. Progr. Symposium Series* No. 9 **50**, 47–51 (1954).

81. H. FUJIE, *AIChEJ* **10**, 227–232 (1964).
82. W. R. GAMBILL *Chem. Eng. Progr. Symposium Series* No. 41 **59**, 71–87 (1963).
83. C. GAZLEY, JR. and O. P. BERGELIN, *Chem. Eng. Progr.* **45**, 45–46 (1949).
84. C. GAZLEY, JR., *Proc. Heat Transfer and Fluid Mech. Inst.*, pp. 29–40, Berkeley, California (June 1949).
85. A. R. GEMMELL and N. EPSTEIN, *Can. J. Chem. Engrs.* **40**, 215–224 (1962).
86. L. E. GILL, G. F. HEWITT, and P. M. C. LACEY, *Chem. Eng. Sci.* **20**, 71–88 (1965).
87. L. E. GILL, G. F. HEWITT, J. W. HITCHON, and P. M. C. LACEY, *Chem. Eng. Sci.* **18**, 525–535 (1963).
88. L. E. GILL, G. F. HEWITT, and P. M. C. LACEY, *Chem. Eng. Sci.* **19**, 665–682 (1964).
89. K. GOLDMANN, H. FIRSTENBERG, and C. LOMBARDI, *J. Heat Transfer* **83C**, 158–162 (1961).
90. K. GOLDMANN, R. HANKEL, and R. P. STEIN, *J. Appl. Mech.* **31**, 380–382 (1964).
91. S. W. GOUSE, JR., *Index to the Two-Phase Gas–Liquid Flow Literature*, *MIT Rept. No. 9*, The MIT Press, Cambridge, Mass. (1966).
92. G. W. GOVIER and M. M. OMER, *Can. J. Chem. Engrs.* **40**, 93–104 (1962).
93. G. W. GOVIER, B. A. RADFORD, and J. S. C. DUNN, *Can. J. Chem. Engrs.* **35**, 58–70 (1957).
94. G. W. GOVIER and W. L. SHORT, *Can. J. Chem. Engrs.* **36**, 195–202 (1958).
95. G. W. GOVIER, G. A. SULLIVAN, and R. K. WOOD, *Can. J. Chem. Engrs.* **39**, 67–75 (1961).
96. G. GREEN, *Phil. Mag.* **22**, 730 (1936).
97. W. A. GRESHAM, Jr., P. A. FOSTER, JR., and R. J. KYLE, *WADC-TR-55-422*, Eng. Exp. Stat., Georgia Institute of Technology, Atlanta, Ga. (June 1955).
98. P. GRIFFITH, *ASME Paper 57-HT-21* (Aug. 1957).
99. P. GRIFFITH, *ANL-6796*, Argonne National Laboratory, Argonne, Ill. (Nov. 1963).
100. P. GRIFFITH, *J. Heat Transfer* **86C**, 327–333 (1964).
101. P. GRIFFITH and G. B. WALLIS, *J. Heat Transfer* **83C**, 307–320 (1961); see also E. S. KORDYBAN, *J. Basic Eng.* **83D**, 613–618 (1961); R. A. S. BROWN and G. W. GOVIER, *Can. J. Chem. Eng.* **43**, 217–230 (1965).
102. F. C. GUNTHER and F. KREITH, *Heat Transfer and Fluid Mechs. Inst.* pp. 113–126, Berkeley, Calif. (1949); *JPL-Prog. Rept.-4-120*, Jet Propulsion Laboratory, Pasadena, Calif. (1950).
103. C. GUTFINGER and J. A. TALLMADGE, *AIChEJ* **10**, 774 (1964).
104. T. J. HANRATTY and J. M. ENGEN, *AIChEJ* **3**, 299–304 (1957).
105. T. J. HANRATTY and A. HERSHMAN, *AIChEJ* **7**, 488–497 (1961).
106. B. F. HARVEY and A. S. FOUST, *Chem. Eng. Progr. Symposium Series* No. 5 **49**, 91–106 (1953).
107. P. N. HAUBENREICH, *AEC CF-55-5-200*, TIS, Oak Ridge, Tenn. (May 1955).
108. R. C. HENDRICKS, R. W. GRAHAM, Y. Y. HSU, and R. FRIEDMAN, *NASA TN D-765*, Nat. Aero. and Space Adm., Washington, D. C. (May 1961).
109. G. F. HEWITT, H. A. KEARSEY, P. M. C. LACEY, and D. J. PULLING, *Int. J. Heat and Mass Trans.* **8**, 793 (1965).
110. G. F. HEWITT and P. C. LOVEGROVE, *AERE-M 1203*, Atomic Energy Research Estab., Harwell, England (1963).

111. C. J. Hoogendoorn, *Chem. Eng. Sci.* **9**, 205–217 (1959).
112. C. J. Hoogendoorn and A. A. Buitelaar, *Chem. Eng. Sci.* **16**, 208–221 (1961).
113. J. W. Hoopes, Jr., *AIChEJ* **3**, 268–275 (1957).
114. Y. Y. Hsu and R. W. Graham, *NASA TN D-1564*, Nat. Aero. and Space Adm., Washington, D. C. (Jan. 1963).
115. Y. Y. Hsu, F. F. Simon, and R. W. Graham, paper presented ASME Winter Meeting, Philadelphia (Nov. 1963).
116. J. L. Hudson, K. M. Atit, and S. G. Bankoff, *Chem. Eng. Sci.* **19**, 387–402 (1964).
117. B. A. Hughes and R. W. Stewart, *J. Fluid Mech.* **10**, 385 (1961).
118. R. R. Hughes, H. D. Evans, and C. V. Sternling, *Chem. Eng. Progr.* **49**, 78–87 (1953).
119. G. A. Hughmark, *Chem. Eng. Progr.* **58**, No. 4, 62–65 (1962).
120. G. A. Hughmark, *Ind. Eng. Chem., Fund.* **2**, 315–321 (1963).
121. G. A. Hughmark and B. S. Pressburg, *AIChEJ* **7**, 677–682 (1961).
122. H. S. Isbin, R. H. Moen, and D. R. Mosher, *AEC Pub. AECU-2994*, TIS, Oak Ridge, Tenn. (1954).
123. H. S. Isbin, J. E. Moy, and A. J. R. DaCruz, *AIChEJ* **3**, 361–365 (1957).
124. H. S. Isbin, N. C. Sher, and K. C. Eddy, *AIChEJ* **3**, 136–342 (1957).
125. H. S. Isbin, R. H. Moen, R. O. Wickey, D. R. Mosher, H. C. Larson, *Chem. Eng. Progr. Symposium Series* No. 23 **55**, 75 (1959).
126. H. S. Isbin, H. A. Rodriguez, H. C. Larson, and B. D. Pattie, *AIChEJ* **5**, 427–432 (1959).
127. H. S. Isbin, R. Vanderwater, H. Fauske, and S. Singh, *J. Heat Transfer* **83C**, 149–157 (1961).
128. L. A. Jacowitz, and R. S. Brodkey, *Chem. Eng. Sci.* **19**, 261–274 (1964).
129. M. Jakob, *Heat Transfer*, John Wiley & Sons, New York (1949).
130. H. Jeffreys, *Proc. Roy. Soc.* **107A**, 186–204 (1925); **110A**, 241–247 (1926).
131. H. Jeffreys *Phil. Mag.* **49**, 793 (1925).
132. H. Jeffreys, *Proc. Cambridge Phil. Soc.* **26**, 204 (1930).
133. L. M. Jiji and J. A. Clark, *J. Heat Transfer* **86C**, 50–58 (1964).
134. H. A. Johnson and A. H. Abou-Sabe, *Trans. ASME* **74**, 977–987 (1952).
135. J. H. Keenan and F. G. Keyes, *Thermodynamic Properties of Steam*, John Wiley & Sons, New York (1936).
136. R. R. Kepple and T. V. Tung, *ANL-6734*, Argonne National Laboratory, Argonne, Ill. (July 1963).
137. G. H. Keulegan, *J. Res. Natl. Bur. Std.* (U.S.) **46**, 358 (1951).
138. S. P. Kezios, T. S. Kim, and F. M. Rafchiek, *Inter. Dev. in Heat Transfer*, paper 31, 262–269, ASME, New York (1961).
139. S. I. Kosterin, *Izvest. Akad. Nauk SSSR Otdel. Tekh. Nauk*, 1824 (1949).
140. S. I. Kosterin and M. N. Rubanovich, *Izvest. Akad. Nauk SSSR Otdel. Tekh. Nauk*, 1085–1093 (1949).
141. B. K. Kozlov, *Zh. Techn. Fiz.* **24**, 2285–2288 (1954); *AEC-tr-2258* translation.
142. L. Y. Krasiakova, *Zh. Techn. Fiz.* **22**, 656–669 (1952); *AERE Lib/Tr 695* translation.

143. A. A. KUDIRKA, *ANL-6862*, Argonne National Laboratory, Argonne, Ill. (March 1964).
144. S. S. KUTATELADZE, *Energetika* **2**, 229–239 (1959).
145. P. M. C. LACEY, G. F. HEWITT, and J. G. COLLIER, paper presented to the Thermodynamics and Fluid Mechanics Group of the Inst. Mech. Engrs. (Feb. 1962); *AERE-R-3962*, Atomic Energy Research Establishment, Harwell, England (1962).
146. D. E. LAMB and J. L. WHITE, *AIChEJ* **8**, 281–283 (1962).
147. H. LAMB, *Hydrodynamics* (Sixth edition), Dover Publications, New York (reprint of the 1932 edition).
148. L. H. LEMAIRE and P. GRASSMAN, *Compt. Rend.* **246**, 1378, 3152 (1958).
149. S. LEVY, *Proc. Second Midwest. Conf. on Fluid Mech.*, pp. 337–348, The Ohio State University, Columbus, Ohio (1952).
150. S. LEVY, *J. Heat Transfer* **82C**, 113–124 (1960).
151. S. LEVY, *J. Heat Transfer* **85C**, 137–152 (1963).
152. M. J. LIGHTHILL, *J. Fluid Mech.* **14**, 385–398 (1962).
153. L. V. LILLELEHT, and T. J. HANRATTY, *J. Fluid Mech.* **11**, 65–81 (1961).
154. L. V. LILLELEHT, and T. J. HANRATTY, *AIChEJ* **7**, 548–550 (1961).
155. D. L. LINNING, *Inst. Mech. Engrs. (London)*, Proc. B **18**, No. 2, 64–75 (1952).
156. R. W. LOCKHART, and R. C. MARTINELLI, *Chem. Eng. Progr.* **45**, 39–48 (1949).
157. M. S. LONGUET-HIGGINS, *Proc. Roy. Soc. (London)* **265A**, 286 (1962); see also A. W. R. GILCHRIST, *J. Fluid Mech.* **25**, 795–816 (1966).
158. P. A. LOTTES, R. P. ANDERSON, B. M. HOGLUND, J. F. MARCHATERRE, M. PETRICK, G. F. POPPER, and R. J. WEATHERHEAD, *ANL-6561*, Argonne National Laboratory, Argonne, Ill. (Feb. 1962).
159. W. H. MCADAMS, W. K. WOODS, and L. C. HEROMAN, *Trans. ASME* **70**, 695–702 (1948).
160. W. H. MCADAMS, *Heat Transmission* (third edition), McGraw-Hill, New York (1954).
161. H. M. MCMANUS, JR., Ph. D. thesis, University of Minnesota, Minneapolis, Minn. (1956); *OOR Proj. 2117*, Cornell University, Ithaca, New York (1959); *ASME Paper 61-HYD-20* (May 1961).
162. P. G. MAGIROS, and A. E. DUKLER, *Developments in Mechanics*, Vol. 1, p. 532, Plenum, New York (1961).
163. F. L. MAKER, *Pet. Ref.* **34**, No. 11, 140–152 (1955).
164. R. C. MARTINELLI, L. M. K. BOELTER, T. H. M. TAYLOR, E. G. THOMSEN, and E. H. MORRIN, *Trans. ASME* **66**, 139–151 (1944).
165. R. C. MARTINELLI, J. A. PUTNAM, and R. W. LOCKHART, *Trans. AIChE* **42**, 681–705 (1946).
166. R. C. MARTINELLI and D. B. NELSON, *Trans. ASME* **70**, 695–702 (1948).
167. J. W. MILES, *J. Fluid Mech.* **3**, 185–202 (1957).
168. J. W. MILES, *J. Fluid Mech.* **6**, 568, 583 (1959); **7**, 469 (1960); **13**, 433 (1962).
169. J. W. MILES, *J. Fluid Mech.* **8**, 593 (1960).
170. R. MOISSIS, *J. Heat Transfer* **85C**, 366–370 (1963).
171. R. MOISSIS and P. GRIFFITH, *J. Heat Transfer* **84C**, 29–39 (1962).

172. M. A. Mologin, *Dokl. Akad. Nauk SSSR* **94**, 807–810 (1954); *AERE-Lib/Tr. 479* translation.
173. M. A. Mologin, *Zh. Techn. Fiz.* **26**, 1823–1835 (1956).
174. F. D. Moore and P. B. Mesler, *AIChEJ* **7**, 620–624 (1961).
175. L. G. Neal and S. G. Bankoff, *AIChEJ* **9**, 490–494 (1963).
176. R. M. Nedderman and C. J. Shearer, *Chem. Eng. Sci.* **18**, 661–670 (1963).
177. J. R. Nesbit, B. S. thesis, Carnegie Institute of Technology, Pittsburgh, Pa. (1940).
178. D. J. Nicklin, *Chem. Eng. Sci.* **17**, 693–702 (1962).
179. D. J. Nicklin, J. O. Wilkes, and J. F. Davidson, *Trans. Inst. Chem. Engrs. (London)* **40**, 61–68 (1962).
180. W. Nusselt, *Z. ver. dtsch. Ing.* **60**, 541–546, 569–575 (1916).
181. M. Petrick, *ANL-5787*; Argonne National Laboratory, Argonne, Ill. (1958); see also M. Petrick and B. S. Swanson, *Rev. Sci. Inst.* **29**, 1079–1085 (1958).
182. M. Petrick, *ANL-6581*, Argonne National Laboratory, Argonne, Ill. (1962).
183. O. M. Phillips, *J. Fluid Mech.* **2**, 417–445 (1957); **4**, 426–434 (1958); **5**, 177 (1959); **9**, 193 (1960).
184. O. M. Phillips, *J. Geophys. Res.* **67**, 3135–3141 (1962).
185. R. Phinney, *J. Appl. Mech.* **30**, 448–452 (1963).
186. R. W. Pike and H. C. Ward, *AIChEJ* **10**, 206–213 (1964).
187. E. E. Polomik, S. Levy, and S. G. Sawochka, *J. Heat Transfer* **86C**, 81–88 (1964).
188. S. Portalski, *Chem. Eng. Sci.* **18**, 787–804 (1963); **19**, 575–582 (1964); *AIChEJ* **10**, 584 (1964); **11**, 369 (1965).
189. E. R. Quandt, *Chem. Eng. Progr. Symposium Series* No. 32 **57**, 111–126 (1961).
190. W. E. Ranz, *On Spray and Spraying: A Survey of Spray Technology for Research and Development Engineers*, Dept. of Eng. Res., Pennsylvania State University, University Park, Pa. (March 1956).
191. H. L. Reichart, Jr., B. S. thesis, Carnegie Institute of Technology, Pittsburgh, Pa. (1934).
192. R. C. Reid, A. B. Reynolds, A. J. Diglio, I. Spiewak, and D. H. Klipstein, *AIChEJ* **3**, 321–324 (1957).
193. W. M. Rohsenow, *Trans. ASME* **74**, 969–976 (1952).
194. W. M. Rohsenow, in *Modern Developments in Heat Transfer* (W. Ibele, editor), Academic Press, New York (1963).
195. J. D. Rogers, *AIChEJ* **2**, 536–538 (1956).
196. T. W. F. Russell and M. E. Charles, *Can. J. Chem. Engrs.* **37**, 18–24 (1959).
197. T. W. F. Russell, G. W. Hodgson, and G. W. Govier, *Can. J. Chem. Engrs.* **37**, 9–12 (1959).
198. J. L. Scheppe and A. S. Foust, *Chem. Eng. Progr. Symposium Series* No. 5 **49**, 77–89 (1953).
199. F. N. Schneider, P. D. White, and R. L. Huntington, *Pipe Line Industry* **1**, No. 4, 47–51 (Oct. 1954).
200. D. S. Scott, *Can. J. Chem. Engrs.* **40**, 224–225 (1962).
201. J. D. Seader, W. S. Miller, and L. A. Kalvinskas, *R-5598*; Rocketdyne Division of North American Aviation (May 1964).

202. N. C. SHER and S. J. GREEN, *Chem. Eng. Progr. Symposium Series* No. 23 **55**, 61–73 (1959).
203. R. S. SILVER, *Proc. Roy. Soc. (London)* **194A**, 464–480 (1948).
204. R. S. SILVER and J. A. MITCHELL, *Trans. North East Coast Inst. of Engrs. and Shipbuilders* **62**, 51–72, D15–30 (1945–1946).
205. M. SILVESTRI, *Inter. Dev. in Heat Transfer*, paper 39, pp. 341–353, ASME, New York (1961).
206. C. R. SMITH, Y. S. TANG, and C. L. WALKER, *AIChEJ* **10**, 586–589 (1964).
207. R. V. SMITH, *Natl. Bur. Std. Tech. Note 179*, U.S. Dept. Comm. (Aug. 1963).
208. G. SONNEMAN, *Nucl. Sci. and Eng.* **5**, 237–241 (1959).
209. R. W. STEWART, *J. Fluid Mech.* **10**, 189–194 (1961).
210. M. D. STUART and D. R. YARNALL, *Mech. Eng.* **58**, 479–484 (1936); *Trans. ASME* **66**, 387–397 (1944).
211. N. H. TAYLOR, G. F. HEWITT, and P. M. C. LACEY, *Chem. Eng. Sci.* **18**, 537–552 (1963).
212. F. E. TIPPETS, *J. Heat Transfer* **86C**, 12–38 (1964).
213. F. URSELL, in *Surveys in Mechanics* (G. K. Batchelor and R. M. Davies, editors), pp. 216–249, Cambridge University Press, Cambridge (1956).
214. J. J. VAN ROSSUM, *Chem. Eng. Sci.* **11**, 35–52 (1959).
215. T. VEDA, *J. Japan Soc. Mech. Engrs.* **1**, 139–144 (1958).
216. J. H. VOHR, *TID-11514* Dept. of Commerce, Washington, D. C. (Dec. 15, 1960).
217. J. H. VOHR, *AIChEJ* **8**, 280–281 (1962).
218. G. B. WALLIS and J. H. HEASLEY, *J. Heat Transfer* **83C**, 363–369 (1961).
219. H. C. WARD and J. M. DALLAVALLE, *Chem. Eng. Progr. Symposium Series* No. 10 **50**, 1–14 (1954).
220. J. C. WESTMORELAND, *AEC Pub. KAPL-1792, OTS*, Washington, D. C. (Feb. 1957).
221. J. W. WESTWATER, in *Advances in Chemical Engineering* (T. B. Drew, and J. W. Hooper, editors), Vols. I and II, Academic Press, New York (1956, 1958).
222. J. W. WESTWATER, in *Research in Heat Transfer*, pp. 61–72, Pergamon Press, Oxford (1963).
223. P. D. WHITE, Ph. D. thesis, University of Oklahoma, Norman, Okla. (1954).
224. M. WICKS, III and A. E. DUKLER, *AIChEJ* **6**, 463–468 (1960).
225. E. H. WISSLER, H. S. ISBIN, and N. R. AMUNDSON, *AIChEJ* **2**, 157–162 (1956).
226. R. K. WOOD, M. S. thesis, University of Alberta, Edmonton, Alb. (March 1960).
227. G. WYLLIE, *Phil. Mag.* **36**, 581 (1945).
228. S. YAGI, *Chem. Eng. (Japan)* **18**, 2 (1954).
229. F. R. ZALOUDEK, *HW-68934*, Hanford Atomic Works, Hanford, Washington (1961).
230. S. M. ZIVI, *J. Heat Transfer* **86C**, 247–252 (1964); *see also* E. E. POLOMIK, *ibid.* **88C**, 10–18 (1966).
231. N. ZUBER and E. FRIED, *Am. Rocket Soc. J.* **32**, 1332–1341 (1962).
232. L. S. COHEN and T. J. HANRATTY, *AIChEJ* **11**, 138–144 (1965).
233. J. E. CRUVER and R. W. MOULTON, paper presented at AIChE meeting, Houston, Texas (Feb. 1965).

234. H. K. Fauske, *Appl. Sci. Res.* **13A**, 149–160 (1964).
235. G. D. Fulford, in *Advances in Chemical Engineering*, (T. B. Drew, J. W. Hoopes, Jr., and T. Vermeulen, editors), Vol. V, pp. 151–236, Academic Press, New York (1964).
236. G. W. Govier, *Can. J. Chem. Eng.* **43**, 3–10 (1965).
237. P. Griffith and K. S. Lee, *J. Basic Eng.* **86D**, 666–668 (1964).
238. G. F. Hewitt, *AERE-R 4613*, Atomic Energy Research Estab., Harwell, England (1964).
239. S. Levy, *J. Heat Transfer* **87C**, 53–58 (1965).
240. L. F. McGoldrick, *J. Fluid Mech.* **21**, 305–331 (1965).
241. F. J. Moody, *J. Heat Transfer* **87C**, 134–142 (1965).
242. S. Ostrach and A. Koestel, *AIChEJ* **11**, 294–303 (1965).
243. E. R. Quandt, *AIChEJ* **11**, 311–318 (1965).
244. S. Whitaker, *Ind. Eng. Chem., Fund.* **3**, 132–142 (1964).
245. N. Zuber and J. A. Findlay, *J. Heat Transfer* **87C**, 453–468 (1965).

CHAPTER 17

MULTIPHASE PHENOMENA II: FREE FLOW

A knowledge of the formation and subsequent motion of drops and bubbles is essential to the eventual understanding of the transfer mechanisms occurring in many unit operations, such as absorption, extraction, spray-drying, distillation, foam fractionation, and dispersion in mixers. Our treatment of such complex phenomena to date has been semi-empirical, and has usually been based on correlations of overall transfer coefficients. To approach these subjects from a more basic viewpoint, we must have an understanding of hydrodynamics, and within this framework we can also acquire an understanding of heat- and mass-transfer mechanisms. Much work is being done along these lines, and although direct application to the unit operations is limited, we shall concentrate our effort on the more fundamental aspects.

We shall follow the general convention that a drop falls and a bubble rises relative to the continuous phase. So far as possible we shall attempt to integrate, rather than discuss separately, the various aspects of drops and bubbles in both liquid–gas and liquid–liquid systems. There are many areas that must be considered, as well as many questions that must be proposed and, if possible, answered. For example, what is the nature of drops and bubbles during their formation? What is the mean of the sizes formed; what is the distribution of the sizes, and how do we treat it? What determines velocity of rise? What determines the shape of a drop or bubble? Is there circulation in the bubbles? Does the bubble oscillate, and, if it does, what are the controlling factors? How are bubbles and drops further dispersed or coalesced? How do mass and heat transfer affect the dynamics, and vice versa? And what are the interaction effects when concentrated populations of drops or bubbles exist? These and other questions have been asked. The answers to some of the questions are known, and answers to others will result from current research efforts. Before we leave this introduction, we should mention that many reviews of this area of study may be found in references 10, 18, 24, 34, 48, 84, 92, 95, 96, 106, 114, 134, 137, 150, 152, 157, 164, 168, and 183.

17-1 FORMATION OF DROPS AND BUBBLES

The ideal pendant drop (a drop hanging from a flat plate) has been described accurately enough to allow evaluation of the interfacial tension [3] by a method based on measurement of the shape. The geometric form was first investigated by Bashforth and Adams [6], and the calculations were later extended by Fordham

[45]. The equation for the profile of such a drop is (see Fig. 17–1)

$$\frac{1}{R/b} + \frac{\sin \alpha}{x/b} = 2 + \frac{z}{b}\beta, \quad (17\text{--}1)$$

where R = the radius of curvature at the point (x, z), α = the slope of the drop surface at point (x, z), b = the radius of curvature at point $(0, 0)$, x = the horizontal distance from center of drop, z = the vertical distance from the bottom of the drop, and $\beta = -b^2 \rho / \sigma$, in which σ is the surface tension.

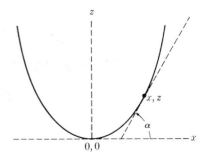

Fig. 17–1. Ideal pendant drop.

The quantity β describes the shape of the drop, and the value of b determines its size. References 6 and 45 tabulate values of x/b and z/b for values of α and β.

For a drop on a horizontal circular tip, with liquid being added, the visual changes which occur as the flow rate through a nozzle is increased can be described as follows:* "At low flow rates, drops form individually at the nozzle tip and grow in size until the weight or buoyance force overcomes the interfacial tension and the drop is released. At increased flow rate, a point is reached where a very short continuous neck of liquid exists between the nozzle tip and the point of drop detachment; this velocity is called the jetting point. Further increases rapidly lengthen the jet, which appears as a smooth column of liquid with occasional transient lumps (Rayleigh jet). Finally, the jet takes on a ruffled appearance at its outer end and the drops formed are less uniform than in the earlier stages; this occurs at or near the maximum length of the jet and will be called the critical velocity. Increasing the flow further decreases the jet length and increases drop nonuniformity until the jet breakup point retreats to the nozzle tip and a nonuniform spray of rather small drops results; this last point is called the disruptive velocity or point of atomization." Figure 17–2 is an area plot of jet length versus flow rate, as described above. Merrington and Richardson [135] referred to the two areas as *varicose*, because of occasional lumps in the jet, and *sinuous*, because of irregular weaving of the jet.

There are two devices commonly used for forming bubbles: a submerged horizontal orifice and a submerged vertical slot. The qualitative picture is approximately the same for both, although the open area in the latter case is not constant with increasing gas-flow rate. Leibson *et al.* [109] and Davidson and Amick [32] describe the various modes of bubble formation at an orifice as the gas rate is varied over a wide range.

As the gas-flow rate is increased, three distinct regions of bubble formation are found [206]. At very low rates the bubble volume remains essentially constant for a given orifice diameter, and hence the frequency of bubble production is

---

*By permission from F. W. Keith and A. N. Hixson *Ind. Eng. Chem.* **47**, 258–267 (1955).

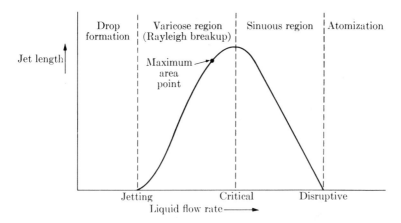

Fig. 17-2. Jet length as a function of liquid-flow rate.

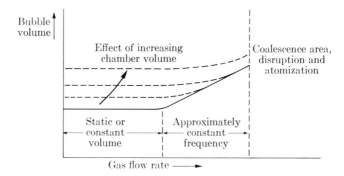

Fig. 17-3. Bubble volume as a function of gas-flow rate.

proportional to the gas-flow rate. This type of formation is called the *constant-volume* or *static* bubble. As the gas flow is increased past a transition region, the bubble frequency levels off to a constant value and the volume increases in proportion to the gas-flow rate. This second type is called the *constant-frequency* bubble. At higher rates of flow, breakup and coalescence occur; these are the conditions of operation of most industrial equipment. The various regions are qualitatively described in Fig. 17-3.

In the material to follow, each of these types of generation will be treated in more detail.

17-1.A Detachment of Drops and Bubbles

(a) Formation of drops from tips. For low flow rates, at which individual drops are produced, Hauser *et al.* [78] established the steps of the formation and detachment of drops from tips in air. These steps can be described as follows: a waist forms first, and rapidly necks down into a stem, which then breaks off close to the drop. The stem detaches from the liquid still on the tip and then

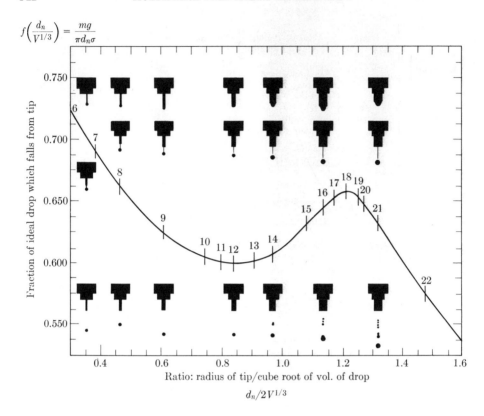

*Fig. 17–4. Fraction of an ideal drop which falls from a tip. [From H. E. Edgerton, E. A. Hauser, and W. B. Tucker, J. Phys. Chem. **41**, 1017 (1937). By permission.]*

breaks into one or more satellite droplets; these then fall behind the main drop. Harkins and Brown [71] developed the following expression from a force balance of the counteracting effects of gravity and surface tension:

$$mg = \pi \, d_n \, \sigma \, f(d_n/V^{1/3}), \qquad (17\text{--}2)$$

where m is the mass of the drop and its satellites, V is the volume of the detached drop, d_n is the orifice diameter, and $f(d_n/V^{1/3})$ is an empirical correction function with the value of unity for an ideal drop (one formed with no satellites and no material left on the tip). Using materials of known surface tension, the authors plotted $mg/\pi \, d_n \sigma$ as a function of $d_n/2V^{1/3}$. Figure 17–4, from Edgerton et al. [42], shows the curve, in which the ordinate can be considered as the fraction of an ideal drop which falls from the tip. The initial decrease of the curve is caused by additional material which is left on the tip without satellite formation. The curve then increases to a maximum because of satellite formation and little material being left on the tip. The curve finally decreases again because the effect of an increasing amount of material on the tip becomes the controlling factor.

Because of the complexity of the formation just described, a detailed theoretical analysis has not yet been made. The production of satellites from the stem is due to the instability of a long liquid cylinder. As will be seen, this breakup is similar to that of a jet, the main cause being that surface tension forces the cylinder into a spherical pattern. It is the viscosity of the material, however, that determines the length and diameter of the stem, and thus the size and number of the satellites.

Hayworth and Treybal [77] considered individual drop production in liquid–liquid systems, and extended their experiments slightly beyond the jetting velocity into the varicose region; however, the characteristics of the varicose region were not as obvious as for a liquid jet in a gas. Actually, a less-dense material is injected in most liquid–liquid systems, and thus we are dealing with bubbles. The injected material behaves more like a liquid than a gas, however, and so the somewhat paradoxical title "rising drop" has come into use. Hayworth and Treybal observed that for low nozzle velocities the drop size was uniform and increased with nozzle velocity, with the drop size reaching its upper limit at very near the jetting velocity. For somewhat higher velocities, the size decreased and became less uniform (varicose region). At even higher levels, the operation was erratic and nonreproducible. Hayworth and Treybal had little success with the use of dimensional analysis and several equations proposed by others. They obtained a reasonable correlation of the data by using a force balance on the drop during formation. They assumed that the drop would be stable (would not separate from the nozzle) if the velocity of rise of the drop (if free) in the liquid were less than the velocity of the fluid in the nozzle, and if the buoyant force acting on the drop were less than the force of interfacial tension. They assumed that the drop would break away when both these conditions were reversed. Izard et al. [93], using a comparable system, could not obtain a single uniform drop. Their frequency-distribution curves were bimodal, with the second peak associated with satellite formation. As a result they investigated liquid drop production by discontinuous injection (from a sharp-edged orifice) and found that equal-sized drops could be obtained if the injected liquid did not wet the nozzle material. When the material was wetted, one or more satellites always formed.

(b) Formation of bubbles from orifice

(i) The constant-volume region. For gas injection, we can use an approach paralleling the development of Eq. (17–2). If we assume that an equivalent spherical bubble exists, and that the upward force due to buoyancy just equals the surface tension force, we obtain

$$Vg\,\Delta\rho = (\pi d^3/6)g(\rho_l - \rho_g) = \pi\, d_n \sigma \cos\theta f(d_n/a), \qquad (17\text{–}3)$$

where V is the bubble volume, θ is the angle of contact at the liquid-solid-gas interface, and $f(d_n/a)$ is a shape factor equal to unity for a spherical bubble. If the orifice is wetted by the liquid, $\theta = 0$, and Eq. (17–3) becomes [131]

$Vg = \pi d_n \sigma/(\rho_l - \rho_g)$, or

$$d[g(\rho_l - \rho_g)/d_n\sigma]^{1/3} = 6^{1/3} = 1.82. \tag{17-4}$$

For an air bubble in water at 20°C, this becomes

$$V/d_n = 0.231 \text{ cm.}^2 \tag{17-5}$$

Lane and Green [106] reported that Datta et al. [31] obtained an experimental value of 0.33, while Newman and Whelan [144] obtained 0.19. Benzing and Meyers [9] have surveyed the literature and found values varying from 0.04 to 1.0 for the same system. They have indicated that Eq. (17–4) could be empirically modified and used to correlate constant-volume bubble data. Their equation used the same numerical constant, with a modified power of $\frac{1}{4}$ rather than $\frac{1}{3}$. There is a small tendency for the bubble volume to increase with bubble frequency over the entire range. This was observed by Davidson and Amick [32], and a good review has also been presented by Siemes [183].

(ii) **Effect of chamber volume.** As suggested by Hughes et al. [91], the discrepancy in the data just cited [which is indicated by the wide range of experimental values for the expected theoretical constant of Eq. (17–5)] can be attributed to the fact that the effect of the volume of the chamber upstream from the orifice has been neglected. As a bubble forms at an orifice at a low flow rate, it causes the liquid above it to be accelerated away from the orifice area. If the volume of the upstream chamber is small, acceleration of the liquid will cause the pressure to drop, and the bubble growth will be arrested. If the rate at which the gas flows to the chamber is constant, the pressure will again rise and the growth will continue at an increasing rate. This oscillation of pressure causes what appears to be a delayed release of the bubble. The bubble grows slowly and, when its buoyancy becomes too great, separates. At slightly higher chamber volumes, the effect of the pressure fluctuation is dampened out and the release is immediate, corresponding to an abnormally large flow just prior to breakoff. Due to the extra flow, the bubble volume is greater than it would be in a chamber with less volume. For even larger volumes, Davidson and Amick [32] observed that the bubbles form in pairs because of the excess gas accumulated in the chamber between the normal bubble release times. However, the accumulation can also appear as an approximate doubling of the bubble volume (i.e., escape without having time to form a pair). It should be emphasized that two interacting factors are involved. First there is the rate of gas flow (see Fig. 17–3) and second there is the volume of the chamber. Thus one would expect a whole sequence of curves on Fig. 17–3 with the parameter descriptive of chamber volume.

The analysis by Hughes et al. obtained this parameter, called a *capacitance number*, which related the acoustics of the chamber to its volume:

$$N_c = \frac{4(\rho_l - \rho_g)V_c g}{\pi d_n^2 \rho_g c^2} \simeq \frac{4\rho_g V_c g}{\pi d_n^2 p\gamma}, \tag{17-6}$$

where V_c is the chamber volume, c is the velocity of sound in the gas, p is the total pressure at the orifice, and γ is the ratio of specific heats. The latter form [without the γ, so that there is approximately a 1.4 factor difference in the number from Eq. (17-6)] was used by Tadaki and Maeda [196, 197], and is obtained from the former by using Eq. (12-14) and the ideal-gas law. The low-chamber-volume system exists for values of N_c less than 0.7, although there is a small dependence on the rate of gas flow. For very low rates of gas flow within this range, the bubble volume is independent of chamber volume [32, 196], and Tadaki and Maeda [196] obtained an excellent check with Eq. (17-4) for a wide range of densities, viscosities, and surface tension values. For higher gas-flow rates, they modified Eq. (17-4) to

$$\bar{d}^3(g\rho_l/d_n\sigma) = 6 + 2.5\, N_{\mathrm{We}} N_{\mathrm{Fr}}^{-1/2}, \qquad (17\text{-}7)$$

where $N_{\mathrm{We}} = d_n U_n^2 \rho_l / \sigma$, $N_{\mathrm{Fr}} = U_n^2/gd_n$, U_n is the gas velocity in the orifice, and $\bar{d}$ is a mean bubble diameter. The parameter $N_{\mathrm{We}} N_{\mathrm{Fr}}^{-1/2}$ can be expressed as $4Q\rho_l g^{1/2}/\pi\sigma \bar{d}^{1/2}$. When this parameter is small, one obtains bubbles of constant volume, and Eq. (17-7) reduces to (17-4). For larger values of the parameter, $Q/\bar{d}^3 \sim Q/V = $ (frequency) approaches a constant, and one is in the range of transition to bubbles of constant frequency. Equation (17-7) is descriptive of most of the solid curve of Fig. 17-3, and is an adequate representation of the data for $N_{\mathrm{We}} N_{\mathrm{Fr}}^{-1/2} < 18$, for any $N_c \gamma < 1$. Furthermore, Tadaki and Maeda found that the bubbles were quite uniform for a parameter less than 6 (that is, $\bar{d} = d$). In the range of $1 < N_c \gamma < 9$, chamber volume had great effect on bubble volume. This effect corresponds to the range of intermediate chamber volumes where, just prior to breakoff, the bubble volume was increased because of the abnormally large flow. In this range, the authors found that

$$\bar{d}^3(g\rho_l/d_n\sigma) = 6N_c\gamma, \qquad (17\text{-}8)$$

for $N_{\mathrm{We}} N_{\mathrm{Fr}}^{-1/2} < (6/2.5)(1.4N_c - 1)$. For the parameter greater than this, but less than 18, Eq. (17-7) gave good results. The effect of increasing N_c is shown by the dashed lines in Fig. (17-3). Finally, for the range $N_c \gamma > 9$, where the bubbles were essentially twice the size given by Eq. (17-7), they found that

$$\bar{d}^3(g\rho_l/d_n\sigma) = 55 \qquad (17\text{-}9)$$

for any $N_{\mathrm{We}} N_{\mathrm{Fr}}^{-1/2} < 18$. For exactly twice the size, one would expect approximately $2^3 \cdot 6 = 48$, rather than the 55 observed; however, both are within the scatter of the experimental data. As already mentioned, Davidson and Amick observed pair formation under these same conditions. That factor of the system which causes the development of either bubble pairs or single bubbles of twice the basic size is still unknown. Differences in wetting of the orifice material or in the shape and size of the orifice could contribute to the disagreement.

Mahoney and Wenzel [130] considered a finite chamber above the liquid surface (all at large values of N_c) in addition to the entrance chamber. An effect did exist, but could be expressed in terms of an equivalent lower chamber; i.e., the relation

$$C_1/V_{c'} + C_2/V_c = 1 \qquad (17\text{--}10)$$

appeared valid, where $V_{c'}$ is the upper chamber volume. Constants for the system are C_1 and C_2. For specific values of these, an infinite number of combinations of $V_{c'}$ and V_c will give the same curve for bubble frequency versus gas flow rate. For a lower chamber only, $V_{c'} = \infty$ and $V_c = C_2$; thus C_2 is the desired equivalent volume of the lower chamber. Sullivan *et al.* [195] also obtained data at large values of N_c. The frequency of formation was correlated by a semi-empirical equation.

One would expect commercial equipment to have large chamber volumes, and thus large values of N_c, but this could be misleading. For example, a perforated-plate column, with many orifices connected in parallel between a common chamber and a common liquid pool, might have a relatively low value of N_c. Tadaki and Maeda [197] have shown that, for their range of orifice sizes, the use of V_c/m (where m is the number of holes) in place of just V_c will give the same correlations that would apply for a single orifice. The average bubble diameter was predicted to within 20%, but the distribution of sizes was somewhat greater than for the case of a single orifice.

(iii) **Further observations.** Mahoney and Wenzel and Sullivan *et al.* used larger orifice sizes than did the experimenters in the other references cited. Mahoney and Wenzel used 0.32 cm and Sullivan *et al.* used 0.16 to 0.39 cm, compared with 0.04 to 0.17 cm for reference 196. Apparently as a result of this, the different experimenters found differences in bubble frequency or equivalent bubble volume. For $N_{\text{We}} N_{\text{Fr}}^{-1/2} > 18$, Tadaki and Maeda observed that the average diameter decreased when there was an increase in flow. Accompanying this was a considerable spread in the bubble distribution. This area of flow is beyond the range of constant frequency and is the area of strong disruption or atomization, as shown in Fig. 17–3. In the case of the larger bubbles from the larger orifices, the onset of the disruptive region is retarded considerably beyond a value of $N_{\text{We}} N_{\text{Fr}}^{-1/2}$ of 18. The entire flow picture is shifted toward higher values of this parameter. The proper critical form or value is unknown, but $N_{\text{We}} N_{\text{Fr}}^{-1/2}$ is clearly inadequate. A nozzle Reynolds number might be applicable, but has not been extensively tested. In any event, there is an extended delay in these systems while the chamber pressure builds up, then a sudden release gives the larger bubble. These are the conditions under which Eq. (17–8) should be valid, but the equation is off by a factor of at least two. It may be of importance that as the orifice size increases, a point is reached where $d = d_n$. From Eq. (17–4), this occurs at $d^2 = d_n^2 = 6\sigma/g(\rho_l - \rho_g)$, and is the point at which the interface is horizontal and leads to some sort of instability. For water, this is 0.33 cm, which is close to the larger orifice diameters cited above.

There is a transition area (see Fig. 17–3) between the regions of constant volume and constant frequency. This area apparently involves an alternate single- and double-bubble formation [32]. The frequency of formation has increased so that bubble interaction now exists. A pair formation occurs in which one bubble hovers above the orifice until the next bubble emerges, after which the two bubbles rise together. This pattern is repeated for awhile, then the flow reverts to single-bubble formation. The reason for the hovering has not been established, but it may well be that a circulation causes a velocity increase in the continuous phase between the upper bubble and the one just forming. Such a velocity increase would cause a decrease in pressure in the area, analogous to that experienced between ships in a moving stream. The additional pressure force acting downward would tend to hold the upper bubble in place and accelerate the production of the one forming. When the second bubble has formed, they both move off together, since no appreciable retaining force exists once the lower one breaks away. As the flow is further increased, the bubbles come into contact. The lower bubble often momentarily protrudes into the top bubble. When the rate of flow increases further, coalescence occurs in any of several different ways. This point also marks the maximum frequency of bubble formation: Davidson and Amick also observed that the second bubble sometimes shoots completely through the first bubble, rising through the inner column of liquid in the bubble and emerging on top. Under other conditions, the first bubble absorbs part of the second and leaves a small satellite as residue. This may be a simple breakoff, or it may be a throwout or reject of the combined drop. In still other cases, the second bubble is completely swallowed by the first, although their interface remains unruptured. In most cases, at these flow rates, two bubbles completely coalesce at some distance above the orifice, yielding a single large, irregular bubble. As the flow rate increases, coalescence takes place closer and closer to the orifice. This is the beginning of the area of industrial importance, for here turbulence increases and shatters the large bubbles into clouds of fine bubbles.

17-1.B Breakup of Jets

(a) The Rayleigh jet. Lord Rayleigh [160–162] made an analytical study of the collapse of a liquid jet in the varicose region of jet flow (see Fig. 17–2). Assuming axial symmetry, irrotational flow, and a nonviscous liquid, he showed that a small disturbance symmetrical about the axis of the jet would cause breakup when the amplitude of the disturbance grew to one-half the diameter of the undisturbed liquid jet. Lamb [104] gives the derivation as based on a method of velocity potential. When the jet is considered to be symmetrical and to possess inertia (liquid jet), the term q (the phase velocity) is given by

$$q^2 = \frac{\sigma k r_j (k^2 r_j^2 - 1)}{\rho_l r_j^3 [I_0(kr_j)/I_0'(kr_j)]}, \qquad (17\text{--}11)$$

where r_j is the jet local radius, usually taken as $d_n/2$ (the initial radius),
k is the period of oscillation $2\pi/\lambda$,
I_0 and I_0' are modified Bessel functions of the first kind of zero order and its derivative, and
ρ_l is the density of the liquid jet or the phase to be dispersed.

When the phase into which the jet issues is considered to possess inertia (bubble jet), the value of q is

$$q^2 = -\frac{\sigma k r_j(k^2 r_j^2 - 1)}{\rho_l r_j^3 [K_0(kr_j)/K_0'(kr_j)]}. \tag{17-12}$$

Here K_0 and K_0' are modified Bessel functions of the second kind of zero order and its derivative. Christiansen and Hixson [24] report that Christiansen [23], using Lamb's method of velocity potential, modified the solution to account for the inertial effects of both the jet and the surrounding medium (liquid jet in liquid). In this work, the restriction of axial symmetry was not made. The result was

$$q^2 = \frac{\sigma k r_j(k^2 r_j^2 + s^2 - 1)}{r_j^3\{\rho_d[I_s(kr_j)/I_s'(kr_j)] - \rho_c[K_s(kr_j)/K_s'(kr_j)]\}}, \tag{17-13}$$

where the subscripts d and c refer to the discontinuous and continuous phases, and where s is the order of the Bessel function. The term s is 0 if symmetrical, 2 if an ellipse, 3 if an equilateral triangle, and 4 if a square.

Equation (17-11) for the breakup of a liquid jet in air can be plotted as a function of $\lambda/2r_j$; a maximum exists at 4.508. If Eq. (17-12) is plotted, a maximum occurs at 6.48. Drops that form at the maximum point *ideally* contain the same volume as a cylinder with a diameter of the jet and a length of λ. That is,

$$\text{Volume} = \pi d^3/6 = \pi r_j^2 \lambda = \pi r_j^2 (4.508) 2 r_j. \tag{17-14}$$

Thus $d = 3.78 r_j$ or $1.89 d_j$. The drop will have a diameter nearly twice that of the undisturbed jet. For the gas jet, a diameter 2.13 times the jet diameter is obtained. For the case of Eq. (17-13) with axial symmetry ($s = 0$), the densities of both phases enter into computation of the most unstable wavelength, so kr_j must be calculated for each system. One would think that the value for the entire density range would lie between 1.89 and 2.13. The check of Christiansen and Hixson is good for 5 systems in water, being 2.07 ± 0.123. Tyler [202] measured the frequency with which drops were formed, N, during jet disintegration. The velocity of the jet was

$$U_j = N\lambda. \tag{17-15}$$

Equating the volumes of the cylindrical jet and the drop, and using Eq. (17-15), Tyler obtained

$$\lambda/d_j = U_j/Nd_j = \frac{2}{3}\left(\frac{d}{d_j}\right)^3.$$

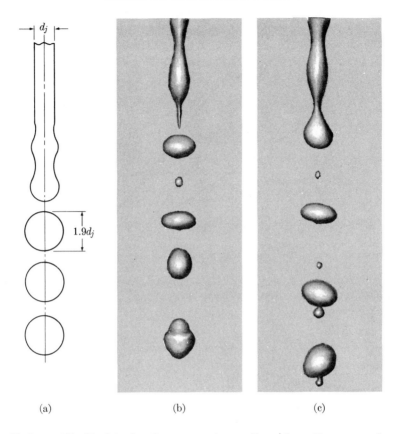

Fig. 17–5. (a) *Idealized jet breakup suggesting uniform drop diameter and no satellites.* (b *and* c) *Actual breakup of a water jet as shown by high-speed photographs.* [*From* W. R. Marshall, Atomization and Spray Drying, Chem. Eng. Progr. Monograph Series, *No. 2 (1954). By permission.*]

Tyler measured λ/d_j by three methods (by measurement of λ, d_j, d, U_j, and N). The values he obtained were nearly the same, and averaged 4.69, or an equivalent drop diameter of 1.91 d_j. Figure 17–5 shows an idealization of the breakup according to Rayleigh's model, and breakup as it actually occurs. The Rayleigh analysis does not predict the occurrence of the satellite drops at the point of necking down. The gas jet of Eq. (17–12) has not been observed.

(b) Areas of flow. Several regions of drop formation were defined in Fig. 17–2. In general, the many references that have considered the various possible flows have used dimensionless groups. Grassman and Lemaire [58] have reviewed these references and tabulated the results. The Reynolds, Weber, and Froude numbers are generally used, individually and in combination. Also used are the density and viscosity ratios and the ratio of drop size to vessel size. The jet Reynolds

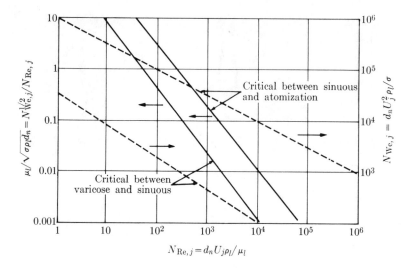

Fig. 17–6. Flow-area plot suggested by Ohnesorge. [Modified from W. R. Marshall, Atomization and Spray Drying, Chem. Eng. Progr. Monograph Series, No. 2 (1954). By permission.]

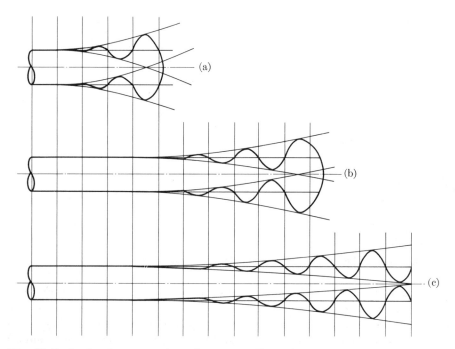

*Fig. 17–7. Jet breakup at various flow rates. [From R. M. Christiansen and A. N. Hixson, Ind. Eng. Chem. **49**, 1017–1024 (1957). By permission.]*

number is defined as $N_{\text{Re},j} = d_n U_j \rho_l / \mu_l$, where U_j is the jet velocity and d_n is still the nozzle diameter. Ohnesorge [147] used the Reynolds number and the group

$$\mu_l / \sqrt{\sigma \rho_l d_n} = N_{\text{We},j}^{1/2} / N_{\text{Re},j}, \qquad (17\text{--}16)$$

where $N_{\text{We},j} = d_n U_j^2 \rho_l / \sigma$ is a jet Weber number. Tyler and Watkin [203], for example, used $N_{\text{Re},j}/N_{\text{We},j}^{1/2}$ and $N_{\text{We},j}^{1/2}$. The flow-area plot of Ohnesorge is given in Fig. 17–6. Since the Weber number is a common term, a plot of $N_{\text{We},j}$ versus $N_{\text{Re},j}$ is also given. The upper critical point can be expressed as

$$\log(N_{\text{We},j}/N_{\text{Re},j}^2) = 6 - 2.45 \log N_{\text{Re},j},$$

or more simply as $N_{\text{We},j} = 10^6 N_{\text{Re},j}^{-0.45}$.

(c) **Drop distributions.** Drop distribution for liquids into air in the Rayleigh region was determined by Duffie and Marshall [40], who found a small but definite distribution around the expected single diameter from a Rayleigh breakup. One run had 100% within $\pm d$, while another had 90% within $\pm d$ and 99.8% within $\pm 2d$. Those particles larger than the main diameter were generally a multiple of the fundamental size. Those smaller were probably produced from the satellites. An empirical equation, good to 7.3% average deviation, was given as

$$\bar{d} = 36 \, d_n^{0.56} N_{\text{Re},j}^{-0.10}, \qquad (17\text{--}17)$$

where $\bar{d}$ is the geometric mass mean drop diameter. Christiansen and Hixson [24] have suggested a simple model which offers considerable insight into the mechanism of jet breakup about the maximum area point in the Rayleigh region (see Fig. 17–2). Part (b) of Fig. 17–7 is a representation of the flow at the maximum area point. Christiansen and Hixson note that in the absence of transients, a node pinches off when the inertial forces of the drop forming beyond it just balance the net force of surface tension at the node. If the flow rate is greater than the maximum area condition, but still less than the critical velocity, as shown in Fig. 17–7(c), a number of nodes along the jet approach instability at the same time, and one or more may separate. A separation upstream produces a drop larger than the ideal one-wavelength drop, and simultaneous separation of two or more nodes can cause both large and small drops, resulting in a system which produces a wide size distribution. For a flow rate below the maximum area point, as in part (a) of Fig. 17–7, the nodes form frequently. Christiansen and Hixson made experimental observations which showed that drops formed in two sizes: one equal to the ideal Rayleigh size and one double that size. They describe their observations as follows.* "High-speed pictures showed that the double-size drops were created by oscillations in the jet length. While at the maximum area point the jet length was approximately constant, at lower flow rates, the jet appeared to shorten with each successive drop split off. Finally, at its shortest length, it

---

*By permission from R. M. Christiansen and A. N. Hixson, *Ind. Eng. Chem.* **49**, 1017–1024 (1957).

appeared to omit splitting off a drop and grew in length to its longest stable position before the next drop was released. The drop which was now released in this extended jet position had traveled some distance with the jet and at the jet velocity, while the drop broken off at the shortest jet length slowed in the continuous phase to its equilibrium velocity. Thus, the former overtook the latter. This phenomenon was called 'twinning' and was observed only below the maximum area point. When the velocity was further decreased, jet oscillation increased, producing a corresponding increase in the number of twins."

Keith and Hixson observed that at the lowest flow rate, just above the jetting point, the drop size was quite uniform. Near the maximum area condition, the drop size was greatly decreased without any appreciable decrease in uniformity. As the rate was increased, the effect of large drops was noted. The highest flow rate indicated an increase in the proportion of large drops, which corresponded to a decrease in uniformity. It is the sudden appearance of large drops near the critical velocity that ends the constant increase of surface area found in the varicose region. As the flow rate is increased further, the frequency with which these large drops occur also increases, but at a faster rate than the increase in flow. This accounts for the sudden drop in surface area beyond the maximum value.

The degree of uniformity can be measured by the standard deviation (see Fig. 17-8). Duffie measured distributions for drops in air, while Keith obtained data on liquid–liquid systems. For Keith's data, the flow rate at which the maximum area point occurs is noted. The Rayleigh breakup theory would predict one drop size with no distribution. The maximum area flow rate of Keith and Hixson would not necessarily occur at the point of the most uniform drop size, but rather would occur where a balance existed between the high-area satellites and the low-area large drops. For the data shown in Fig. 17-8, this occurs at a flow rate somewhat greater than the most uniform size. One would expect Rayleigh's theoretical drop diameter to correspond to the minimum deviation. The one measurement available does not, and the difference noted in Fig. 17-8 might be accounted for by the tapering of the jet as it leaves the nozzle; thus d_n is not strictly equal to d_j. This tapering is high in continuous phases of high viscosity [95].

(d) Further observations. To avoid the problem of dependence of data on size of nozzle and properties of the system, Christiansen and Hixson defined a system-length parameter which depends only on the properties of the system (but not on its viscosity), and has the dimensions of length,

$$l = \pi(\sigma/\Delta\rho)^{1/2}. \tag{17-18}$$

This group of properties was used previously by Bond and Newton [13] to define the critical radius of a drop of liquid. The critical radius is the largest radius that is stable when the drop is rising or falling in a second liquid. Lewis

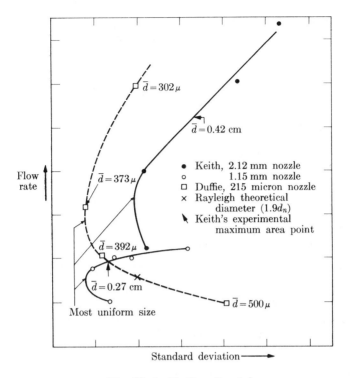

Fig. 17-8. *Uniformity of drops.*

et al. [118] used this group in their analysis of the size of the average drop that leaves a packed section of a column. Each author has in effect defined the radius of the maximum stable drop or minimum jet diameter as a fraction of the critical wavelength for stability, which is given by Lamb [104] as

$$\lambda_c = 2\pi(\sigma/\Delta\rho)^{1/2}. \tag{17-19}$$

Thus $l = \frac{1}{2}\lambda_c$. Equations (17-18) and (17-19) can be expressed as a dimensionless group, which is the ratio of the Weber to the Froude numbers. The term l was used by Christiansen and Hixson to obtain an empirical relation (Fig. 17-9) for the jet contraction at the flow rate for maximum area in a liquid–liquid system. For the liquid–air data of Duffie, $l \cong 5.5$ mm; however, d_n is so small that the ratio of d_n/d_j is 1.0; consequently, correction of the Rayleigh point for jet contraction would not be necessary. For the toluene–water data of Keith in Fig. 17–8, l from Eq. (17-18) is about 1.4 cm; thus d_n/l is 0.15 for the 2-mm nozzle and 0.08 for the 1-mm nozzle. The corrections from Fig. 17–9 are 1.09 and 1.00, respectively. The corrected Rayleigh theoretical drop size for the 2-mm nozzle would be 0.37 cm (versus 0.4 cm), and that of the 1-mm nozzle would remain at 0.22 cm. Both these values are slightly less than the minimum average diam-

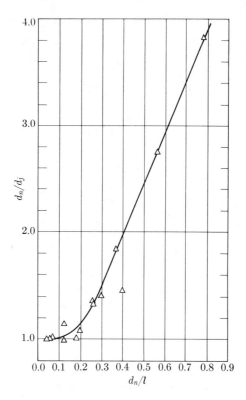

Fig. 17–9. Jet contraction. [*From R. M. Christiansen and A. N. Hixson,* Ind. Eng. Chem. **49,** *1017–1024 (1957). By permission.*]

eter (maximum area point) observed by Keith and Hixson, these being 0.42 and 0.27 cm, respectively.

Studies of the effect of viscosity on breakup [77, 188, 202, 203] have shown that modest viscous forces are negligible in the Rayleigh breakup or varicose region. At the critical velocity, however, all viscous force becomes important as the flow enters the sinuous region. A correction factor for viscosity was obtained by Weber [216], who extended the work of Rayleigh to include both viscous and inertial forces; for a jet in air, the correction factor is

$$\lambda/d_j = \pi\sqrt{2}\,[1 + (3\mu_l/\sqrt{\sigma\rho_l d_j})]^{1/2}. \tag{17-20}$$

When $\mu_l = 0$, λ/d_j is 4.44, which is close to Rayleigh's 4.508 (that is, $d = 1.88d_j$, rather than $1.89d_j$). Haenlein [66] showed that, for very viscous liquids (860 cps), the λ/d_j was as much as nine times greater than for the nonviscous case. According to Eq. (17–20), an increase in viscosity increases the Rayleigh diameter. However, the experimental increase is greater than that expected from the equation. The work of Ranz and Dreier [158] has demonstrated the difficulty of obtaining an analytical treatment to account for the instability of a viscous fluid interface under the action of a surface-tension force. The major problem is probably the complexity of the flow as it approaches the sinuous region, which is the area in which viscosity plays an important role.

2-1.C Atomization

For commercial applications, the simple orifice or nozzle is an inefficient producer of surface area. Commercial users want atomizers which will disintegrate and disperse a liquid over the maximum area at the least cost. The three types of atomizers in general use are: pressure, centrifugal, and pneumatic. Lane and Green [106] have reviewed the production of very uniform drops from tips and spinning disks. Excellent reviews of most phases of atomization have been given by Fraser and Eisenklam [48] and by Marshall [134]. The common types of atomizers can be briefly described as follows: The swirl-type or centrifugal-pressure nozzle atomizes by imparting a swirling or spinning motion to the liquid before it is sprayed. The spinning motion can be provided by a spin chamber within the nozzle, much like the action in a cyclone. These units produce a conical spray pattern with a hollow cone, although changes in design permit any shape to be obtained, with or without the hollow cone. The spinning-disk and other rotating atomizers produce a centrifugal acceleration of the fluid. The liquid accelerates until it reaches the outer edge, where it is dispersed into the gas stream. In this method of atomization, the liquid attains its velocity without high pressures and frequently with a free liquid surface exposed. In some designs, the liquid is confined and forced to flow through radial tubes arranged as spokes in a wheel. An important difference between the rotating and pressure atomizers is that the liquid feed rate can be independently controlled with the former, but can be changed only by changing the pressure or orifice diameter of the latter. Pneumatic atomization uses the action of a high-velocity gas stream on a liquid jet to produce the desired liquid breakup. The simplest arrangement is to discharge the liquid into a high-velocity gas stream. This type of atomizer will carry the spray and gas for a considerable distance.

The discussion to follow will treat several aspects of the mechanism of drop formation, but no attempt will be made to describe the detailed practical aspects of atomization units. For this, the reader is referred to the reviews and the references cited therein. Such units are extremely complex systems and for the most part do not lend themselves to rigorous theoretical treatment.

(a) Breakup of a liquid sheet. Liquid sheets are formed under certain conditions in many atomizing systems; for example, in fan- and swirl-type nozzles [36, 37, 39, 49, 134, 199], from spinning disks and cups [43, 47, 85, 106], and between impinging jets [38, 79, 198]. Probably the simplest system, and the one that has received the most attention, is the formation of a liquid sheet from a fan nozzle.

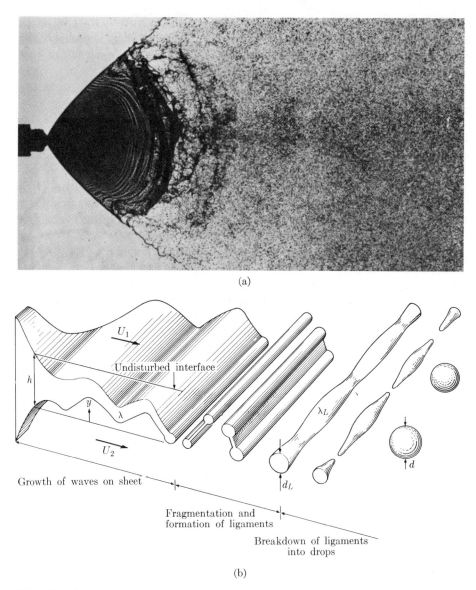

Fig. 17–10. (a) Fan-nozzle spray. (b) Idealization of breakup process. [From N. Dombrowski and W. R. Johns, Chem. Eng. Sci. **18**, 203–214 (1963). By permission.]

Figure 17–10 shows this and an idealization of the breakup process. The waves are a result of aerodynamic instability [36], and completely disappear at low subatmospheric pressures. Under these conditions, a smooth elliptically shaped film with a thick rim is obtained, and relatively large drops are formed along the edges directly from the rim.

FORMATION OF DROPS AND BUBBLES

Dombrowski and Johns [39] have extended the inviscid uniform-thickness solutions for aerodynamic instability made by Squire [192] and by Hagerty and Shea [67] to the viscous-fluid case, in which thinning of the film could be specified. They applied a force balance to the disturbed film and considered the effects of pressure, surface tension, inertial forces, and viscous forces. After some simplification, their balance can be modified to

$$N_{gr}^2 + \alpha^2 N_{gr}/N_{Re,s} + 2s\alpha^2(1/N_{We,s} - 1/\alpha) = 0, \qquad (17\text{-}21)$$

where

$$N_{gr} = \beta h/U, \qquad N_{Re,s} = hU\rho_l/\mu_l, \qquad N_{We,s} = hU^2\rho_g/\sigma.$$

The term α is defined by Eq. (16–4):

$$\alpha = 2\pi h/\lambda = kh, \qquad (17\text{-}22)$$

and β is defined as the amplitude growth factor:

$$y = y_0 e^{\beta t}. \qquad (17\text{-}23)$$

Here s is the ratio of gas densities to liquid densities as used in Eq. (16–6), and U is the mean relative air-wave velocity defined by

$$U^2 = \tfrac{1}{2}(U_1^2 + U_2^2), \qquad (17\text{-}24)$$

where U_1 and U_2 are the relative velocities over the upper and lower surfaces, as noted in Fig. 17–10. Finally, h is the sheet thickness. The various terms in Eq. (17–21) are associated with the forces: the first stems from the inertial force, or the rate at which the momentum of the liquid element changes; the second is associated with the viscous forces, and is zero for zero viscosity (that is, $N_{Re,s} \to \infty$); the first part of the third term is a result of surface tension forces and would be zero for zero surface tension ($N_{We,s} \to \infty$), and the last part of the third term results from the force caused by the air pressure. The wave that has the maximum growth is the one for which the growth parameter (N_{gr}) is a maximum with respect to wavelength (α). For constant h, from Eq. (17–21), we obtain

$$2N_{gr}(dN_{gr}/d\alpha) + (1/N_{Re,s})[2N_{gr}\alpha + \alpha^2(dN_{gr}/d\alpha)] + 4s\alpha/N_{We,s} - 2s = 0.$$

Since $(dN_{gr}/d\alpha) = 0$ for the maximum, this reduces to

$$N_{gr}\alpha/N_{Re,s} + 2s\alpha/N_{We,s} - s = 0, \qquad (17\text{-}25)$$

where N_{gr} and α in this and subsequent equations refer to conditions of maximum growth rate. Combining Eq. (17–25) with (17–21) gives us

$$N_{gr}^2 = \alpha s. \qquad (17\text{-}26)$$

Using Eq. (17–26) in (17–25), we obtain an expression for α, which is

$$1/\alpha = 2/N_{We,s} + (\alpha/s)^{1/2}/N_{Re,s}. \qquad (17\text{-}27)$$

The inviscid solution (with constant thickness) of references 67 and 192 is obtained for $N_{\text{Re},s} = \infty$; thus

$$\alpha = \tfrac{1}{2} N_{\text{We},s} \quad \text{and} \quad N_{\text{gr}}^2 = \alpha s = \tfrac{1}{2} s N_{\text{We},s}. \tag{17-28}$$

An alternative condition not previously considered is the viscous controlled sheet. Here the surface tension would be zero, or $N_{\text{We},s} = \infty$. Equation (17-25) becomes

$$\alpha = s N_{\text{Re},s}/N_{\text{gr}}, \tag{17-29}$$

which, when it is combined with Eq. (17-21), still gives Eq. (17-26). Combining Eqs. (17-26) and (17-29) gives us

$$\alpha^3 = s N_{\text{Re},s}^2 \quad \text{and} \quad N_{\text{gr}}^3 = s^2 N_{\text{Re},s}. \tag{17-30}$$

Dombrowski and Johns considered an attenuating sheet, for which the thickness was assumed to decrease in proportion to the time of transit of the fluid element from the nozzle.

To translate these results into a drop size, we must make assumptions about the breakup process. It is assumed that, at some distance from the nozzle, the waves separate into ligaments, which are initially $\lambda/2$ in width. These then contract into cylinders, which by continuity would be described by $\pi d_L^2/4 = h(\lambda/2)$ or

$$d_L^2 = 4h/k. \tag{17-31}$$

It is further assumed that these ligaments break up according to Weber's modification of Rayleigh's jet-breakup theory; that is, Eq. (17-20), with d_j replaced by d_L and λ replaced by λ_L. The subscript L indicates that these are the waves along the ligament and not along the sheet. The actual drop diameter is obtained from Eq. (17-14); that is,

$$d^3 = 3 d_L^2 \lambda_L / 2 = 3\pi d_L^2 / k_L. \tag{17-32}$$

Equations (17-20) and (17-32) can be combined to give

$$d^3 = (3\pi/\sqrt{2}) d_L^3 (1 + 3\mu_l/\sqrt{\sigma \rho_l d_L})^{1/2}, \tag{17-33}$$

as presented by Dombrowski and Johns.

The term d_L must be obtained from Eq. (17-31). The solution (Eqs. 17-26 and 17-27) can be used in part. Combining Eqs. (17-26) and (17-31) gives us

$$d_L^2 = 4(h/k) = 4s U^2/\beta^2. \tag{17-34}$$

Experimental observations [37] show that at the breakup point, r, for liquid sheets, βt is a constant equal to 12, where t is the time for an element to travel to this point. The term Ut is the distance traveled, r, if U is now associated with the sheet velocity. Further, the actual film thickness at any point can be expressed [49] as $hr = K$. Even though the film thickness has been assumed constant in the analysis, it is applied at some point where h will now refer to the breakup thickness. We must account for the error in this assumption either by using empirical constants or by using the more refined and difficult analysis of an attenuating liquid film thickness. In any event, Eq. (17–34) becomes

$$d_L^2 = 4s(Ut)^2/(\beta t)^2 = 4sr^2/12^2 = sK^2/36h^2. \qquad (17\text{--}35)$$

For the inviscid solution of constant film thickness, Eq. (17–28) can be used to obtain

$$(\beta t)^2 = 12^2 = \tfrac{1}{2}sN_{\text{We},s}U^2t^2/h^2 = \tfrac{1}{2}sN_{\text{We},s}r^2/h^2 = \tfrac{1}{2}sN_{\text{We},s}K^2/h^4$$

or

$$h^3 = sK^2\rho_g U^2/2\cdot 12^2\sigma. \qquad (17\text{--}36)$$

After some rearrangement, the combination of Eqs. (17–35) and (17–36) gives us

$$d_L^3 = 4s^{1/2}K\sigma/3\rho_g U^2 = \tfrac{4}{3}(K\sigma/\rho_l U^2 s^{1/2}). \qquad (17\text{--}37)$$

For the case of an attenuating sheet with viscosity considered, Dombrowski and Johns showed that this equation would be modified to

$$d_L^3 = 0.8885(K\sigma/\rho_l U^2 s^{1/2})[1 + 2.60\,\mu_l(\rho_l^2 Ks^4 U^7/72\sigma^5)^{1/3}]^{3/5}. \qquad (17\text{--}38)$$

For the case of viscosity controlling with constant film thickness, Eq. (17–30) can be used to give the parallel to Eq. (17–36):

$$(\beta t)^3 = 12^3 = s^2 N_{\text{Re},s} U^3 t^3/h^3 = s^2 N_{\text{Re},s} r^3/h^3 = s^2 N_{\text{Re},s} K^3/h^6$$

or

$$h^5 = s^2 K^3 U \rho_l / 12^3 \mu_l. \qquad (17\text{--}39)$$

Equation (17–35) becomes

$$d_L^5 = \tfrac{2}{9}(s^{1/2}K^2\mu_l/\rho_l U). \qquad (17\text{--}40)$$

In the limit of a high viscosity [unity negligible inside the bracket of Eq. (17–38)], only the numerical coefficient of Eq. (17–40) is different from that obtained from Eq. (17–38).

The complete solution for the constant-thickness case can be obtained by solving Eq. (17–27) for α (a form of a cubic equation) in terms of $N_{\text{Re},s}$, s, and $N_{\text{We},s}$. This solution can then be combined with an alternative to Eq. (17–35) which is obtained as follows: Eq. (17–26) can be expressed as

$$\beta t = 12 = (\alpha s)^{1/2}Ut/h = (\alpha s)^{1/2}r/h = (\alpha s)^{1/2}K/h^2,$$

which can be combined with Eqs. (17–34) or (17–35) to give

$$d_L^2 = (K/3)(s/\alpha)^{1/2}. \qquad (17\text{--}41)$$

The solution for $\alpha^{1/2}$ depends on whether the parameter

$$(3N_{\text{We},s})^3/s(4N_{\text{Re},s})^2 = 27hU^4\rho_g^2\mu_l^2/16\sigma^3\rho_l$$

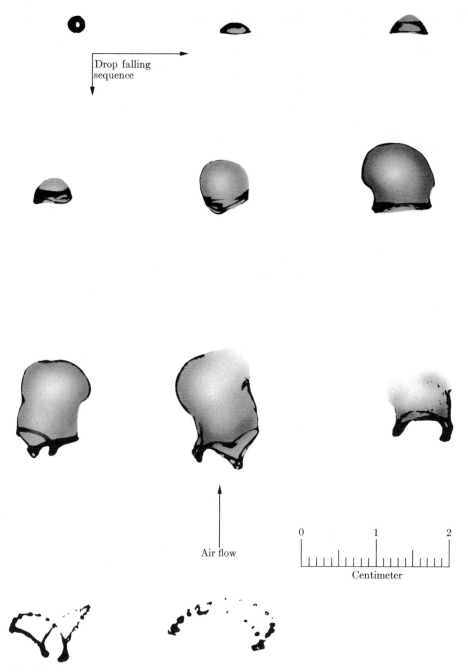

Fig. 17-11. *Stages of breakup of drop of water. Diameter of drops: 2.2 mm. Velocity of air stream: 24 m/sec. [Crown copyright reserved. By permission of the Controller of Her Britannic Majesty's Stationery Office.]*

is less than or greater than two. For the wide range of conditions tested in the references already cited, the value is less than two, and there are three real roots for $\alpha^{1/2}$. If the parameter is much less than unity, two of these roots are zero and the third is

$$\alpha^{1/2} = 2s^{1/2}N_{\text{Re},s}/N_{\text{We},s} = 2\sigma/\mu_l U s^{1/2}. \qquad (17\text{--}42)$$

When combined with Eq. (17–41), this gives

$$d_L^2 = Ks\mu_l U/6\sigma, \qquad (17\text{--}43)$$

which is also the ratio of Eqs. (17–37) and (17–40). Even though this condition (i.e., very small value of the parameter) is not often met, it is only the numerical coefficient of Eq. (17–43) that changes for the exact solution, and not the form. The actual drop size is obtained by the combination of Eq. (17–33) and the appropriate case from (17–37), (17–38), (17–40), and (17–43). The data of Dombrowski and Hooper [37] suggest that Eq. (17–37) is valid at subatmospheric pressures. The situation above atmospheric pressure is not clear. Equations (17–37) and (17–40), with modified numerical coefficients, are the limits of Eq. (17–38) for low and high viscosity. Equation (17–43) defines a diameter for which both viscous and surface forces are important in the case of a constant film thickness. However, the meaning of this equation is not clear when it is compared with Eq. (17–38). Further experimentation over a wider range of viscosities would help clarify the situation.

(b) Breakup of drops. A moderately large drop falling in air accelerates to a critical speed and becomes unstable. Instability also occurs in small drops in a rapid airstream. Drop breakup is important in any operation in which surface area must be created, and is basic to pneumatic atomization. The mode by which the breakup occurs depends on the conditions of experimentation. The large free-falling drop in still air, or the somewhat smaller drop in a steady stream of air, was first considered by Lenard [110, 111] and by Hochschwender [86]. They allowed water drops to fall into a free upward current of air and noted that some of the drops were blown inside out. Figure 17–11 shows this phenomenon [106], which can be described as follows: At the critical velocity of breakup, the drop passes through stages of increasing flattening, formation of a torus (a roughly circular rim with an attached film shaped like a hollow bag), increasing bag size, bursting of the film to produce a shower of very fine droplets, and breakup of the torus rim into larger droplets which contain a large fraction of the original drop mass (about 75%). Another mode of breakup, very different from the previous one, was first observed by Lane [105],who subjected a drop to a transient pulse of air. In this case the drop was not blown out into a thin hollow bag anchored to a rim; instead, it was deformed in the opposite direction and presented a convex surface to the flow of air (see Fig. 17–12). The edges of the saucer shape were drawn out into a thin sheet, then into fine filaments which in turn

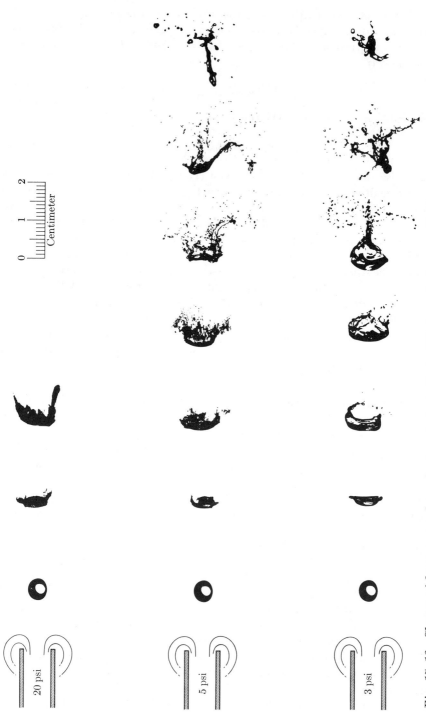

Fig. 17–12. Shatter of large water drops by transient air blast. (Photographed perpendicular to direction of motion of drop.) [Crown copyright reserved. By permission of the Controller of Her Britannic Majesty's Stationery office.]

broke up into droplets. This breakup was a much more chaotic disintegration process than that of a drop falling in a free upward current of air.

Hinze [84] has reviewed the work on breakup. This review and references 63, 70, 105, 129, and 166 comprise the bulk of the literature on the subject. Hinze has described three basic types of deformation: the flattening (or lenticular) type already discussed; a cigar-shaped deformation, like a jet; and "bulgy" deformation, in which local bulges and protuberances occur. He has considered these in combination with six types of flow. The breakup just considered would be lenticular deformation in parallel flow.

Taylor [200] determined that a drop, when accelerated in an air stream, would be plano-convex lenticular, with the diameter being about twice that of the original spherical drop. This prediction has been verified by photographic measurements.

Lane [105, 106], when considering the first mode of breakup, used water drops varying from $\frac{1}{2}$ to 5 mm in diameter. Measurements were made of the critical velocity by measuring the difference between the velocity of the air stream required to break drops and the velocity of the drop at the instant of breaking. The results for the water were:

$$U_c^2 d_{\max} = 612 \times 10^3 \text{ cm}^3/\text{sec}^2, \tag{17-44}$$

where U_c is the difference in velocity between the drop and its surroundings at the breakup conditions, and $d_{\max}$ is the maximum stable drop size. A relation of this form occurs when the force of drag over the drop exceeds the force of surface tension:

$$C_D \tfrac{1}{2} \rho_g U_c^2 = 4\sigma/d_{\max}, \qquad U_c^2 d_{\max} = 8\sigma/C_D \rho_g,$$

or

$$N_{\text{We},d} = \rho_g U_c^2 d_{\max}/\sigma = 8/C_D. \tag{17-45}$$

The value of C_D is unknown, but is probably between the values for a disk and for a sphere. Lane confirmed the relation by checking a range of surface tensions from 28 to 475 dynes/cm. Using Eq. (17-44) and *assuming* that it would hold well below the value of $\frac{1}{2}$ mm, we would find that a 5-μ water drop would remain just intact at the sonic velocity of air. This does not mean that smaller drops cannot exist, nor does it mean that larger drops cannot exist in an atomizer, since sonic velocities do not occur everywhere. Lane also notes that the extrapolation probably does not hold, since drops larger than 5μ are observed under controlled conditions.

Hinze [83] proposed the combination of the forces of pressure and surface tension as a critical measurement of drop breakup. The greater the value of the Weber number, the greater are the deforming external pressure forces when compared with the reforming surface tension force. Hinze also related the flow velocity force to the surface tension force and obtained $\mu_l/\sqrt{\rho_l d_{\max} \sigma}$ as a controlling group, which is of the same form as Eq. (17-16) except that $d_{\max}$ replaces d_n. Two critical points were defined; the lower critical would be the bag-forming breakup, which would be expected under most common applications. The upper critical can occur only if the experiment has been designed to get past the lower critical (as in a shock tube with a high gas velocity). From the data of Merrington and Richardson [135], Hinze has estimated the lower critical to be 13 for low-viscosity liquids. This may be compared with the value of 10.6 from Eq. (17–44), 10.6 to 14 from the data of Volynski [208], 10.3 for mercury drops in air, obtained by Haas [63], and 7.2 to 16.8 (with an average of about 13) for water, methyl alcohol, and a low-viscosity silicone oil, obtained by Hanson et al. [70], whose detailed photographs are excellent. Hinze [84] also considered the effect of viscosity and the way in which the relative velocity varies with time. The change in the critical Weber number should be some function of the second dimensionless group; for example, Hinze suggested

$$N_{\text{We},d} = (N_{\text{We},d})_{\mu_l=0}[1 + f(\mu_l/\sqrt{\rho_l d_{\max}\sigma})]. \tag{17-46}$$

The data of Hanson et al. give qualitative support to the effect, but do not agree in detail. A simple empirical relation can be expressed as

$$N_{\text{We},d} = (N_{\text{We},d})_{\mu_l=0} + 14(\mu_l/\sqrt{\rho_l d_{\max}\sigma})^{1.6}, \tag{17-47}$$

which is good to a maximum deviation of approximately 20% at the high-viscosity end.

The foregoing material implies a gross relative motion between the drop and its surroundings. Such a difference might not always exist or be definable; e.g., as in pipeline transport, in stirred mixing vessels used for extraction, absorption, or chemical reaction, and in stirred spray columns. In such cases one may surmise that the parameters of the local turbulence might be more important and controlling, and that the large-scale motions will have little effect on the process. Kolmogoroff [99] and Hinze [84] took this view, and further assumed that since the breakup was to be considered local, the principles of local isotropic turbulence would be valid. Equation (14–169) expresses the characteristic length and velocity of the equilibrium range in terms of the energy input ϵ and the kinematic viscosity. The inertial subrange is that range of eddy sizes, within the equilibrium range, which is independent of viscous dissipation. The only factors of importance for these eddies are their size and the rate of energy transfer through them from the very large energy-containing eddies to the small dissipation eddies. For this range, the correlation of velocities can be obtained from the inverse transform of Eq. (14–173) or simply by dimensional reasoning, and is

$$\overline{u_i(\boldsymbol{x})u_i(\boldsymbol{x}+\boldsymbol{r})} = C_1(\epsilon r)^{2/3}. \tag{17-48}$$

Assuming that a constant critical Weber number still applies, we can write

$$N_{\text{We},d} = \rho_c \overline{u_i(\boldsymbol{x})u_i(\boldsymbol{x}+\boldsymbol{d})}d_{\max}/\sigma = \rho_c C_1(\epsilon d_{\max})^{2/3} d_{\max}/\sigma = C_1\rho_c\epsilon^{2/3}d_{\max}^{5/3}/\sigma = \text{const}, \tag{17-49}$$

where the variable distance r has been replaced by $d_{\max}$ so as to apply across the drop. Equation (17–49) can be rearranged to

$$d_{\max} = C_2(\sigma/\rho_c)^{3/5}\epsilon^{-2/5}, \tag{17-50}$$

as given by Kolmogoroff and Hinze. In effect, it has been assumed that the inertial-force term is proportional to $\rho_c \overline{u_i(\boldsymbol{x})u_i(\boldsymbol{x} + \boldsymbol{d})}\, d_{\max}^2$ and the surface-force term to $\sigma d_{\max}$. [Of course, energies could be used instead of forces, in which case the inertial force would become the kinetic energy mu^2, $\rho_c d_{\max}^3 \overline{u_i(\boldsymbol{x})u_i(\boldsymbol{x} + \boldsymbol{d})}$. See Example 11–1, page 167.] The ratio of the forces is the Weber number, or Eq. (17–49). Experimental data on breakup in an isotropic field are nonexistent, so direct verification of the equations is not possible. Some indirect support [180, 207] will be cited when interaction effects are considered. Sleicher [187] has shown that Eq. (17–50) is not valid for pipe flow. The breakup occurs in the vicinity of the wall, where the conditions are the farthest from the approximate isotropic conditions at the center line. The breakup for the pipe system is probably a result of a balance between surface forces, velocity fluctuations, pressure fluctuations, and the steep velocity gradient.

(c) **Atomizing systems.** The various equations discussed in the previous sections form a reasonable beginning for a description of the breakup processes of sheets and drops. For example, for the liquid sheet issuing from a fan spray nozzle, Eqs. (17–33) and (17–37) show the correct dependency for the air density and other variables at subatmospheric pressures [37], and Eqs. (17–33) and (17–38) predict the mean drop size [39] if the numerical coefficient in Eq. (17–33) is lowered by a factor of 0.878. Further refinement of the theory and more extensive data will certainly lead to more satisfactory results. One shortcoming of the theory is that it predicts a uniform single-particle size, while in reality a complete distribution of sizes is obtained. At present there is no way to predict this distribution. However, the representation of distributions will be discussed in a later section.

Drops that are uniform in size can be formed from a spinning disk if the liquid wets the disk material and the flow rate is controlled within certain limits. The drop size is dependent on the speed with which the disk rotates. For a given disk profile, $\omega d(d_0 \rho_l/\sigma)^{1/2} = 3.3$, for $0.2 < d < 3.0$ mm, where d is the drop diameter, d_0 is the disk diameter, and ω is the rotational speed of the disk. Outside this range, the drop-size distribution depends on the rotational speed or centrifugal force [106]. In a series of papers, Fraser et al. [47] have made a comprehensive investigation of all aspects of atomization from a spinning cup. The photographs used to support the work are most impressive (for example, see Fig. 17–13). Fraser et al. first considered the cup design which would be needed to provide a uniform liquid sheet from the lip. They found that uniformity was dependent on the manner of feed distribution, liquid flow rate, and viscosity. A uniform rotating feed into a dampening reservoir was necessary to provide a uniform sheet over a wide range of conditions. With final cup design established, they next investigated the film characteristics in detail. They defined conditions for establishing

*Fig. 17–13. Mechanisms of drop formation from rotating cups. (Clockwise rotation) 2 in. dia, 165 cS viscosity. (a) Direct drop formation. (b) Break-down of threads produced from cup lip. (c) Sheet disintegration. [From R. P. Fraser, N. Dombrowski, and J. H. Routley, Chem. Eng. Sci. **18**, 315–353 (1963). By permission.]*

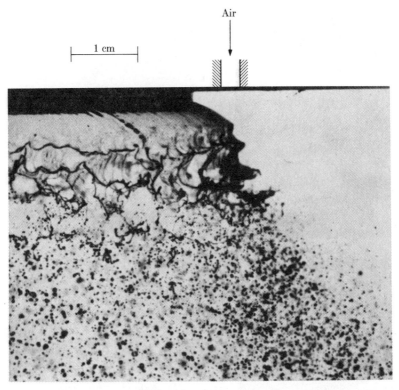

Fig. 17–14. Disintegration of sheet in air stream. Rotary speed 1500 rpm; liquid flow rate 250 lb/hr; liquid viscosity 40 cS; air velocity 100 ft/sec; air-flow rate 72 lb/hr. [From R. P. Fraser, N. Dombrowski, and J. H. Routley, Chem. Eng. Sci. **18**, *313–353 (1963). By permission.]*

a sheet from thread production (see Figs. 17–13b and 17–13c) and found that they occurred at values of

$$(\rho_l \omega^2 d_0^3/\sigma)(Q/\pi d_0^3)^{4/3}(\nu_l d_0/Q)^{0.19} > 0.363,$$

where ω is the rotational speed of the cup, d_0 is the inside cup diameter at the lip, and Q is the liquid volumetric flow rate. Some comments on thread production have been given by Eisenklam [43], and one should also refer to Hinze and Milborn [85, 134]. The variation of the film thickness with distance from the cup was predicted by Fraser *et al.* to within the experimental error of their measurements by geometrical considerations, continuity, and the neglect of the various forces acting. The extent of the sheet was based on the instability analysis previously presented [starting with Eq. (17–28)]. Semi-empirical equations were established for the distance of breakup and for film thickness at breakup. The final phase of the study considered atomization of the sheet by an impinging air stream (see Fig. 17–14). Again the instability analysis was used as a basis for a semi-empirical correlation.

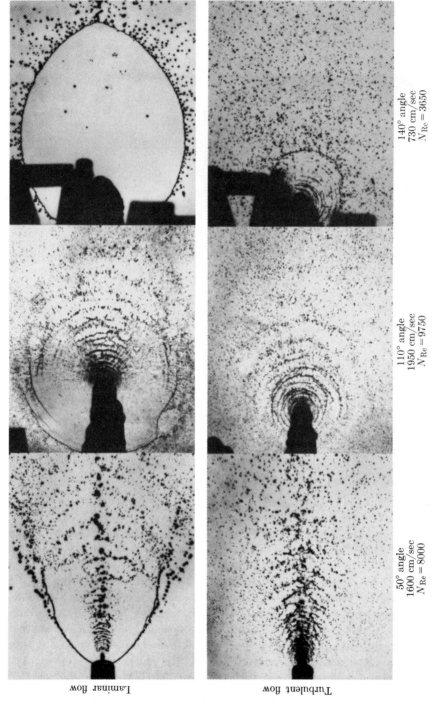

Fig. 17-15. Breakdown of a liquid sheet. [From N. Dombrowski and P. C. Hooper, J. Fluid Mech. **18**, 392–400 (1964), published by the Cambridge University Press, New York. By permission.]

Several investigations have considered the breakdown of sheets formed by impinging jets at various angles [38, 79, 80, 198]. The breakdown occurs by several mechanisms (see Fig. 17-15): by the aerodynamic instability and directly from the rim, as previously considered, and by a hydrodynamic instability which originates at the point of impact. Although aerodynamic instability disappears at low subatmospheric pressures, hydrodynamic instability does not. Little is really known about the actual mechanism involved, and the amount of experimental data is not extensive. The evidence indicates that the same variables are involved as for the fan jet, but that the relations are quite complex. Velocity and angle of jets are the main variables, and under various conditions the drops can be formed directly from the sheet rim, from aerodynamic instability, from hydrodynamic instability, or from combinations thereof.

Popov [154] has considered scaling for atomization. For jets under the range of conditions he tested, the viscosity ratio had little effect on the drop-size distribution and thus could be neglected. Of importance were the $N_{Re,j}$, $N_{We,j}$ or $N_{We,j}^{1/2}/N_{Re,j}$, density ratio, and complex geometrical factors.

Much more analytical work is needed on ligament and drop breakup before anything approaching a satisfactory analysis of something like pneumatic atomization can be obtained. Clearly there is much work to be done.

The foregoing discussion has concentrated on drops. Atomization also occurs in bubble formation, although the mechanisms must be quite different. Leibson et al. [109] used an orifice Reynolds number (same as $N_{Re,j}$, but with the gas velocity) to describe the flow areas they observed in single-orifice bubble formation. The point of atomization (see Fig. 17-16) occurred very near a Reynolds number of 2100. Coalescence occurred very near the orifice, and large bubbles rose three to four inches before shattering into a wide distribution of smaller bubbles. The description of the complex breakup is best taken from the authors:*

"As gas turbulence becomes fully developed ($N_{Re,j} > 10,000$), stroboscopic examination and high-speed motion pictures show that what appears to be a continuous jet is actually a series of closely spaced, irregular bubbles rising with a very rapid counterclockwise swirling motion. This may be a function of orifice design. A large number of very fine bubbles are formed by being torn from the swirling air stream. Rough portions of the air bubble surface are pinched or torn off to form very minute bubbles. The formation of bubble distribution is accelerated by the virtual explosion of the large irregular bubbles at a point approximately 4 in. above the orifice. The liquid circulates in a large eddy near the orifice, flowing toward the jet at the level of the orifice, rotating very rapidly around the central core of bubbles, and flowing away from the top of the forced air vortex as a gas-liquid mixing occurs with the shattering of the large bubbles." When a normal butyl alcohol system was used the mechanism was the same; however, the effect of lowering the surface tension was to increase the formation

---

*By permission from I. Leibson, E. G. Holcomb, A. G. Cacaso, and J. J. Jacmic, *AIChEJ* **2**, 296–306 (1956).

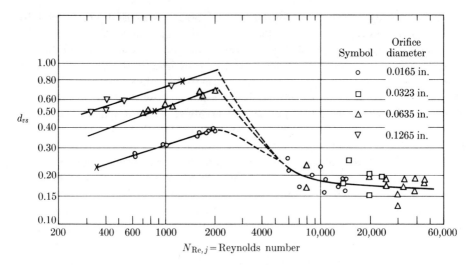

Fig. 17–16. *Bottom-entrance-flow mean bubble diameter d_{vs} versus Reynolds number, where × is the transition to the constant frequency range and d_{vs} is in inches.* [*From I. Leibson, E. G. Holcomb, A. G. Cacoso, and J. J. Jacmic, AIChEJ* **2**, *296 (1956). By permission.*]

of very small bubbles, apparently by causing a reduction in the amount of energy required to form a unit of interfacial area.

In the laminar range, which is of very nearly constant frequency, the bubble diameter increases with flow rate and with orifice diameter. Both references 32 and 109 suggest that the effect of orifice diameter and Reynolds number could be expressed as

$$d = 0.18\, d_n^{1/2} N_{\text{Re},j}^{1/3} \text{ in.} \tag{17-51}$$

Since the drop size is uniform, the values of d are equivalent to mean diameter values. In the turbulent region, the drops are not uniform and an analysis of bubble size distribution was necessary. Siemes and Borchers [184] obtained somewhat similar results for a gas distribution plate rather than a single orifice. The mean bubble diameter was a very weak function of pore size of the distribution plate and of the column diameter. The mean bubble diameter increased to a limiting value with the gas flow, and increased with height above the plate. The distribution of bubble sizes in these various studies was considerable, and will receive further attention at the end of the next section.

Chu *et al.* [25] studied bubble formation at vertical slots. They critically surveyed the literature in the field, noting in particular the two articles by Spells and Bakowski [189, 190]. On the basis of this work and its applications to distillation by Geddes, Chu *et al.* set up a program to consider the effects on interfacial area and contact time of the following variables: gas rate, liquid rate, slot and cap design, liquid submergence, surface tension, and liquid viscosity. The review indicated that bubbles formed at vertical slots have characteristics which are less dependent on physical and mechanical variables of the system than those formed at horizontal orifices. Experimental measurement and correlation of the above variables with the measured interfacial area and contact time showed that the static slot submergency or liquid seal on the slot was the most important variable considered.

17-1.D Drop and Particle Size Distribution

The various theories on the formation of drops and bubbles usually predict a single uniform size, with no distribution. However, such distributions do exist, for example, in spray units for dryers or spray columns, in stirred tanks, and for the solid particles in a fluidized bed system. These distributions are fundamental characteristics which are important factors in determining the physical nature of the overall system. Methods of sampling for distribution measurement are important, since obviously the correct sample is necessary for valid results. The sampling of liquid sprays has been discussed in some detail in the literature [134, 157]. The main methods that have been considered are: (1) collection of droplets in sampling cells containing an immiscible solvent, (2) impaction on slides coated with fine solids, (3) optical techniques that give mean drop diameter only, (4) conversion of molten liquid to solid by cooling, or conversion by freezing of materials with a low boiling point, (5) direct photography, (6) special automatic and electronic devices, and (7) encapsulation by a rapid interfacial polycondensation reaction [128]. Methods for the measurement of solids distribution are easier, since deformation, breakup, and evaporation are usually minor items. The main methods that have been used are [46, 107]: (1) sieving, (2) microscopic examination, (3) measurement of settling velocity, (4) measurement of surface area, (5) light scattering and other optical methods, (6) use of filters, and (7) use of electronic devices. The book by Orr and DallaValle [152] is an excellent starting reference for information on the measurements associated with fine particles, since it treats most of the methods mentioned above, and others on the measurement of diameter, surface area, pore size, etc.

(a) **Distribution analysis.** Particle-size distribution can be represented by a plot of the number of particles of a given diameter versus the diameter, as in Fig. 17–17. This plot is called a *histogram*. A smooth curve through the center points of the maxima give the distribution curve, and if this curve can be represented by some mathematical function $f(x)$, it is called the *distribution function*. If the distribution function is known, then one needs only a limited number of parameters (usually two or three) to define a given distribution. One is the mean diameter and another is a measure of the distribution about this mean. The additional parameters that are sometimes used are often associated with a minimum or maximum particle size. In distributions one sometimes uses the surface area or volume corresponding to a diameter plotted against the diameter; these curves would skew further to the right because of the weighting effect of the large

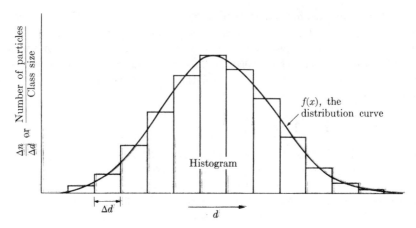

Fig. 17–17. Histogram.

particles. The ordinate values can be expressed in a number of ways: the number of particles of a given diameter (as in Fig. 17–17), relative fraction of the total, or the fraction of the total number per size class; in this case, the area under the curve is equal to 1.0. The last method is a common procedure (normalizing) and the ordinate used is $(\Delta n/n)/\Delta d$.

A second (and generally more convenient) method of representing particle-size distributions is the cumulative curve. In this method, the fraction of the total number (by diameter, area, etc.), which has a diameter smaller than a given size, is plotted against that given size. This is essentially an integrated form of the frequency curve. Figure 17–18 is a cumulative plot of the data given in Fig. 17–17.

In order to facilitate the use of distribution plots, various analytical relationships have been used to represent both frequency and cumulative particle size distributions. These functions are both analytical and empirical [8, 220] (normal, log-normal, error function, Cauchy, exponential, Rosin-Rammler volume, Nukiyama-Tanasawa, Tate-Marshall, etc.). Since no one has a complete understanding of particle production, and no single distribution function can represent all particle-size data, it is necessary to approach the problem by testing each of the various distributions. The normal distribution function (Gaussian distribution) has proved useful in the study of random occurrence on which it is based. It is quite often used (and misused), since it is comparatively simple and has been highly developed. Its misuse arises when it is applied to systems (such as those producing particles) which have a specific bias, and as such, are not random in nature, nor are they normally distributed. If one appreciates this limitation, it is worth while to consider such distributions further. A source of detailed information on statistics can be found in Bennett and Franklin [8].

The function $f(x)$ is a number-distribution function which gives the number of particles $f(x)$ of a given diameter x. The total number is

$$n = \int_{-\infty}^{\infty} f(x)\,dx, \qquad (17\text{–}52)$$

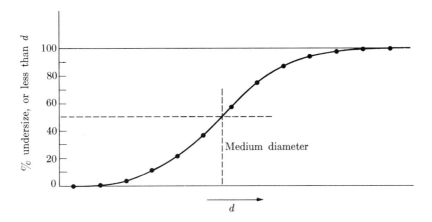

Fig. 17–18. *Cumulative plot.*

and if this is set equal to 1, the distribution is normalized and $f(x)$ becomes the fraction of particles as determined by x. The average of a length or a diameter can be defined by

$$\bar{x}_d = \int_{-\infty}^{\infty} xf(x)\,dx \bigg/ \int_{-\infty}^{\infty} f(x)\,dx. \tag{17-53}$$

In general

$$(\bar{x}_{ab})^{a-b} = \int_{-\infty}^{\infty} x^a f(x)\,dx \bigg/ \int_{-\infty}^{\infty} x^b f(x)\,dx. \tag{17-54}$$

If b is zero, the denominator is given by Eq. (17–52) and is n. Table 17-1, modified from Marshall and attributed to Mugele and Evans [142], reviews the various mean diameters as defined by Eq. (17–54). The geometric mean diameter is given by

$$\ln \bar{x}_g = \int_{-\infty}^{\infty} \ln x f(x)\,dx \bigg/ \int_{-\infty}^{\infty} f(x)\,dx. \tag{17-55}$$

Again the denominator is simply n. This diameter can also be defined in a manner similar to Eq. (17–54). When using a statistical average diameter to correlate data, one must take care to use an average diameter that has some meaning. For example, a simple number average diameter, $\bar{x}_d$, for a surface kinetic reaction would not have as much meaning as a surface average diameter. Such oversight in kinetics has at times prevented satisfactory correlation of good experimental data.

For a variable that is normally distributed, the *normal number distribution* function $f(x)$ is

$$dn/dx = f(x) = (1/\sqrt{2\pi}\sigma)\exp[-(1/2\sigma^2)(x - \bar{x}_d)^2], \tag{17-56}$$

where $\int_{-\infty}^{\infty} f(x)\,dx = 1$ (that is, the distribution is normalized), σ is the standard deviation (σ^2 is the variance), and $\bar{x}_d$ is the mean value of the variable x, such as

TABLE 17-1*

Table of Mean Diameters

a	b	Mean diameter	Symbol	Field of application
1	0	Length	$\bar{x}_d$	Evaporation
2	0	Surface	$\bar{x}_s$	Absorption, crushing
3	0	Volume	$\bar{x}_v$	Mass distribution
2	1	Surface-diameter	$\bar{x}_{sd}$	Adsorption
3	1	Volume-diameter	$\bar{x}_{vd}$	Evaporation, diffusion
3	2	Volume-surface (Sauter)	$\bar{x}_{vs}$	Mass transfer, kinetics

*By permission from W. R. Marshall, Jr., *Atomization and Spray Drying*, Chem. Eng. Prog. Monograph Series No. 2, **50** (1954).

particle diameter, etc. A graph of this function for $\bar{x}_d = 0$ and $\sigma^2 = 1$ is shown in Fig. 17–19. Values of x exist from $-\infty$ to $+\infty$, 99.99% of the curve lies in the range $\bar{x}_d \pm 4\sigma$, and 99.74% between $\bar{x}_d \pm 3\sigma$. The conflict between the infinite range of the normal distribution and a real finite range is not of concern if the range of possible values is greater than 3 or 4σ from the mean $\bar{x}_d$ [8].

The integral of the standard normal curve is the cumulative number distribution function $F(x)$:

$$\int_0^n dn' = F(x) = \int_{-\infty}^x f(x')\,dx' = (1/\sqrt{2\pi}\sigma)\int_{-\infty}^x \exp[-(1/2\sigma^2)(x' - \bar{x}_d)^2]dx', \quad (17\text{–}57)$$

where the prime denotes the dummy variable. A change of variable,

$$t = (x - \bar{x}_d)/\sigma, \quad (17\text{–}58)$$

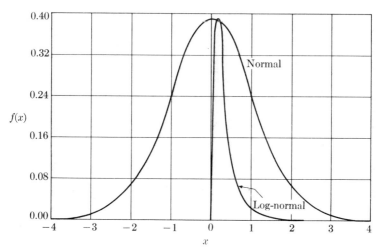

Fig. 17–19. Normal and log-normal distributions.

FORMATION OF DROPS AND BUBBLES

TABLE 17-2

t	$F(t)$	t	$F(t)$
0	0.50000 00000	2	0.97724 98680
0.5	0.69146 24613	3	0.99865 01019
1	0.84134 47461	4	0.99996 83288
1.5	0.93319 27987	5	0.99999 97133

gives $dx = \sigma\, dt$, and Eq. (17–57) becomes

$$F(x) = (1/\sqrt{2\pi}) \int_{-\infty}^{x} e^{-t^2/2}\, dt. \qquad (17\text{–}59)$$

This integral is tabulated in most handbooks. The normal distribution is symmetrical, so that $F(-x) = 1 - F(x)$. The area in an interval from x_1 to x_2 will be $F(x_2) - F(x_1)$, where Eq. (17–58) is used to establish x_1 and x_2. Some typical values are given in Table 17-2 [220].

Equation (17–59) is the basis for the arithmetic-probability graph paper. Figure 17–20 shows this with the data of Fig. 17–17 plotted. If a material follows a normal distribution, it will plot as a straight line. The distribution can be defined completely by the mean diameter ($\bar{x}_d = x_{50}$) and the standard deviation, which can be obtained by applying Eq. (17–58) to the figure. The term $F(t)$ from Table 17-2 for $t = +1$ is 84.13%, or for $t = -1$ is 15.87%; therefore, the standard deviation σ is $\sigma = (x - \bar{x}_d)/t = x_{84.13} - x_{50} = x_{50} - x_{15.87}$.

If $\ln(x/\bar{x}_{ng})$ is a variable to be normally distributed, then from Eqs. (17–55) and (17–56), the *log-normal number distribution* is

$$dn/d\ln x = x(dn/dx) = xf(x) = (1/\sqrt{2\pi}\sigma_g) \exp[-(1/2\sigma_g^2)(\ln x - \ln \bar{x}_{ng})^2], \qquad (17\text{–}60)$$

where $\bar{x}_{ng}$ is the number geometric mean diameter and σ_g is the geometric standard deviation. The terms σ_g and $\bar{x}_{ng}$ possess the same significance on an ln plot as σ

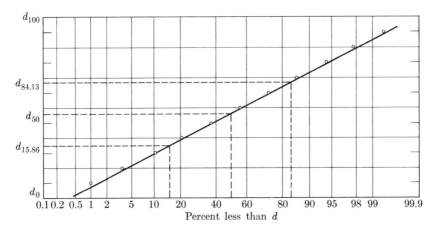

Fig. 17–20. Normal probability paper.

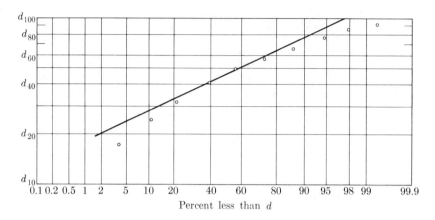

Fig. 17–21. Log-normal probability paper.

and $\bar{x}_d$ do on an arithmetic plot (that is, $\sigma_g = \ln \sigma$). In Fig. 17–19 the log-normal distribution is shown, and Fig. 17–21 shows log-normal probability paper with the data of Fig. 17–17 plotted. This type of distribution has some theoretical basis in particle size work, since it can be derived by assuming formation is a result of pulses proportional to drop size [100].

Equation (17–60) can be written on a volume basis also:

$$dV/d \ln x = x(dV/dx) = xf(x^3) = (1/\sqrt{2\pi}\sigma_g) \exp[-(1/2\sigma_g^2)(\ln x - \ln \bar{x}_{vg})^2], \quad (17\text{–}61)$$

where $\bar{x}_{vg}$ is the geometric volume mean diameter. The relationship between mean diameters can be established as follows: The volume can be expressed in terms of the number by $V = \int_0^n Kx^3 \, dn$, so that

$$dV/d \ln x = K \, x^3 (dn/d \ln x) = Kx^4 (dn/dx) = Kx^4 f(x). \quad (17\text{–}62)$$

Since $x^3 = e^{3(\ln x)}$, Eq. (17–62) can be combined with (17–60) to give

$$dV/d \ln x = Kx(1/\sqrt{2\pi}\sigma_g) \exp\left[-(1/2\sigma_g^2)(\ln x - \ln \bar{x}_{ng})^2\right] \times \exp(3 \ln x)$$

$$= Kx(1/\sqrt{2\pi}\sigma_g) \exp\left[-(1/2\sigma_g^2)(\ln^2 x - 2\ln x \ln \bar{x}_{ng} \ln^2 \bar{x}_{ng} - 6\sigma_g^2 \ln x)\right].$$

Completing the square in the exponent gives

$$dV/d \ln x = (Kx/\sqrt{2\pi}\sigma_g) \exp\{-(1/2\sigma_g^2)[\ln x - (\ln \bar{x}_{ng} + 3\sigma_g^2)]^2\}$$
$$\times \exp(3 \ln \bar{x}_{ng} + 9/2\sigma_g^2).$$

Comparing this with Eq. (17–61), we get the useful relation

$$\ln \bar{x}_{vg} = \ln \bar{x}_{ng} + 3\sigma_g^2, \quad (17\text{–}63)$$

and in this case, the term $K \exp (3 \ln \bar{x}_{ng} + 9/2\sigma_g^2)$ is taken as unity. If Eq. (17–61) were written on a surface fraction basis, the result would be

$$\ln \bar{x}_{sg} = \ln \bar{x}_{ng} + 2\sigma_g^2, \qquad (17\text{-}64)$$

and for the volume-surface or Sauter mean, the relation is

$$\ln \bar{x}_{vsg} = \ln \bar{x}_{ng} + 2\tfrac{1}{2}\sigma_g^2. \qquad (17\text{-}65)$$

There are a number of *empirical distributions* in use: The Rosin-Rammler volume-distribution function [167] is

$$\phi_v = 1 - e^{-(x/\bar{x}')\delta}, \qquad (17\text{-}66)$$

where ϕ_v is the volume fraction undersize (fraction contained in drops of diameter less than x), $\bar{x}'$ is the Rosin-Rammler mean, and δ is a dispersion coefficient. If this equation is applicable to the distribution, a graph of $\log[1/(1-\phi_v)]$ versus x should give a straight line of slope δ. The value of $\bar{x}'$ can be obtained from the curve, since it will be the particle diameter below which 63.2% of the volume lies. The values of δ will be between 2 and 4, the higher value indicating a more uniform distribution. *The Nukiyama-Tanasawa number distribution* [146] has been extensively used in pneumatic atomization and other drop-breakup phenomena. Its algebraic form is:

$$\Delta N = dn/dx = f(x) = ax^m \exp(-cx^\delta), \qquad (17\text{-}67)$$

where ΔN is the number of particles in the size group $x \pm \Delta x/2$, a and c are constants, m is usually taken as 2, and δ is the dispersion coefficient, which is a constant for a given nozzle over a wide range of conditions, and is determined by trial and error. A graph of $\log(\Delta N/x^2)$ against x^δ gives a straight line whose slope is a measure of the surface mean diameter. Here δ varies from $\tfrac{1}{6}$ to 2, high values indicating a narrow distribution. The *log number distribution* [26] is

$$\Delta N/\Delta x = aNe^{-ax}. \qquad (17\text{-}68)$$

A graph of $\log(\Delta N/N\Delta x)$ against x would give a straight line if this distribution is obeyed. The *root normal volume distribution* has been used by Tate and Marshall [199], and is

$$dn/dx = f(x) = (1/2\sqrt{2\sigma_r \pi x}) \exp[-(1/\sigma_r)(\sqrt{x} - \sqrt{\bar{x}_r})^2]. \qquad (17\text{-}69)$$

The $\sqrt{x}$ distributes normally in this type of distribution, and can be obtained from Eq. (17-56), since $2\sqrt{x}(dn/dx) = dn/d(\sqrt{x})$ and $\sigma_r = \sigma^2$. Mugele and Evans [142] have critically analyzed the above functions for a number of drop-size distributions. Some were found to be quite good in some areas, although poor in others. They developed a distribution which is called the *upper-limit log normal distribution*, in which

$$\ln[ax/(x_{\max} - x)] \qquad (17\text{-}70)$$

is normally distributed. In this case, a is called the skewness parameter and $x_{\max}$ is the maximum stable droplet diameter; $x_{\max}$ is determined by trial and

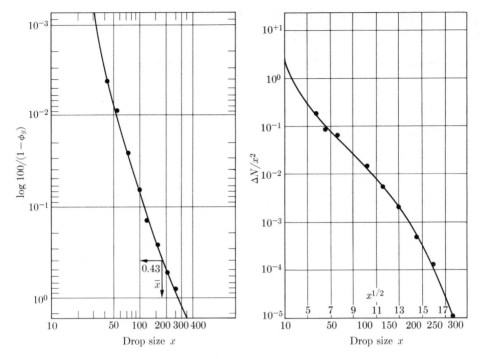

Fig. 17-22. Rosin-Rammler distribution. [From R. P. Fraser and P. Eisenklam, Trans. Inst. Chem. Engrs., London **34**, 294 (1956). By permission.]

Fig. 17-23. Nukiyama-Tanasawa distribution. [From R. P. Fraser and P. Eisenklam, Trans. Inst. Chem. Engrs., London **34**, 294 (1956). By permission.]

error to give the best alignment of data points when $x/(x_{\max} - x)$ is plotted against ϕ_v on a log-probability plot. From this figure,

$$a = (x_{\max} - x_{50})/x_{50}, \tag{17-71}$$

and σ would be determined in the normal manner.

(b) Some experimental results. Because methods of reporting results vary widely, there has been little comparison between the various distributions from different types of drop production equipment. However, some preliminary comments can be made about the methods of distribution representation. In general, Mugele and Evans considered the Rosin-Rammler distribution the poorest, the Nukiyama-Tanasawa distribution next, and the log-probability distribution nearly the best, in fact, superior to all if modified to include the maximum diameter value. Figures 17-22 through 17-25 give some of the distribution plots [49]. Fraser *et al.* [47] found that the Rosin-Rammler distribution was the best for their data on drops formed by an air stream impinging on a liquid sheet from a spinning cup. None would be directly useful for the impinging-jet data of Heidmann and Foster [79], since this showed a bimodal distribution under all operating conditions. However, with the hypothesis that two distinct

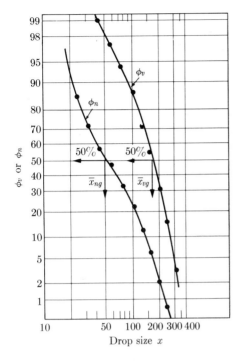

Fig. 17–24. *Log-normal distribution.* [From R. P. Fraser and P. Eisenklam, Trans. Inst. Chem. Engrs., London **34**, 294 (1956). By permission.]

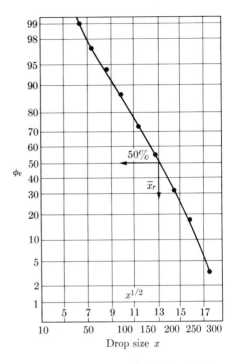

Fig. 17–25. *Root-normal distribution.* [From R. P. Fraser and P. Eisenklam, Trans. Inst. Chem. Engrs., London **34**, 294 (1956). By permission.]

factors contributed to the overall distribution, a qualitative match could be obtained by using two log-normal distributions. For a liquid–liquid dispersion, Olney [149] found the upper-limit log-normal distribution to be the most satisfactory. And finally, for the bubble formation of Leibson et al. [109], the log-normal distribution was found to be satisfactory. Here, standard statistical methods were used to obtain $\bar{x}_{ng}$ and σ_g from the plots, and from these the volume-surface or Sauter mean was obtained from Eq. (17–65). It is the Sauter mean which is plotted in Fig. 17–16 for the turbulent-flow area, and which would be the diameter of a hypothetical bubble whose ratio of volume to surface area is equivalent to that of the entire bubble size distribution. When these various distributions are used, the distribution should always be replotted on the original data to ascertain the validity of the $f(x)$ used.

The results of many atomization studies are summarized by Marshall [134], who points out that there is much to be desired, much confusion of conflicting data, and a wide variability in methods of reporting results. It is difficult to conclude that any one method is superior to all others; such a conclusion must be postponed until an understanding of the mechanism allows a reasonable theoretical determination of the proper distribution. At the present time, it may

TABLE 17-3

Range, mm	Average, d	d^2	d^3	Number, n	Number, %	Relative mass, nd^3	Weight, %	Relative surface, nd^2	Larger than bottom size, in cumulative %	
									Count	Weight
0–1	0.5	0.25	0.13	400	51.2	52	0.04	100	100	100
1–2	1.5	2.25	3.4	190	24.3	646	0.49	427	48.8	99.96
2–3	2.5	6.25	18.13	80	10.25	1456	1.11	500	24.5	99.47
3–4	3.5	12.25	42.88	55	7.04	2357	1.80	688	14.25	98.36
4–6	5	25	125	20	2.56	2500	1.91	500	7.21	96.56
6–8	7	49	343	13	1.66	4460	3.41	638	4.65	94.65
8–10	9	81	729	10	1.28	7290	5.56	810	2.99	91.29
10–15	12.5	156	1953.1	6	0.77	11719	8.96	936	1.71	85.68
15–20	17.5	306	5359.4	4	0.51	21438	16.40	1224	0.94	76.72
20–30	25	625	18125	2	0.26	36250	27.70	1250	0.43	60.32
30–40	35	1225	42875	1	0.13	42875	32.75	1225	0.17	32.62
40				0	0.00	0	0.00	0	0.00	0.00
Totals				781	99.96	131043	100.13	8298		

be well to report [134] in any work the cumulative distribution curve, a mean value (surface, volume, or number), and a value of the dispersion (such as the fraction range of drop sizes between 5 and 95% oversize divided by the mean diameter used).

EXAMPLE 17-1. The following size distribution has been obtained:

Particle size range, mm	0–1	1–2	2–3	3–4	4–6	6–8	8–10	10–15	15–20	20–30	30–40	40
Number in range	400	190	80	55	20	13	10	6	4	2	1	0

Construct a log-normal probability-distribution plot on a count basis and on a weight basis. Assume that the material follows this distribution. What is the best geometric mean diameter and standard deviation? From the information on a count basis, determine the curve for the weight distribution and compare it with that calculated. Determine the specific surface from the data and from the mean diameter and standard deviation. The particles can be assumed to be spherical and to have a specific gravity of 6.0.

Answer. Table 17-3 gives the calculations needed to construct the plots that are shown in Fig. 17-26. From this figure, $\bar{d}_{ng}$ is 0.90 mm and $\bar{d}_{vg}$ is about 37.5 mm. The standard deviation is obtained from Eq. (17-56), that is, the ratio of the

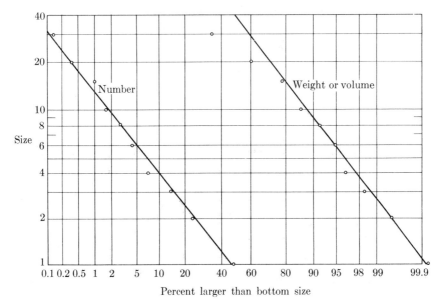

Fig. 17-26. Log-normal probability plot for Example 17-1.

84.13 or 15.87% level to the 50% level:

$$\sigma_n = 2.88/0.90 = 3.20, \qquad \sigma_v = 37.5/11.7 = 3.20.$$

Since these should be the same, and by chance are, 3.20 can be used. The geometric deviation is $\sigma_g = \ln 3.20 = 1.16$. Equation (17–63) can now be used to obtain the geometric volume diameter from the number mean diameter:

$$\ln \bar{d}_{vg} = \ln \bar{d}_{ng} + 3\sigma_g^2 = \ln 0.90 + 3(1.16)^2 = 3.92, \qquad \bar{d}_{vg} = 50 \text{ mm},$$

which can be compared with the estimated 37.5 mm above. The Sauter mean diameter can also be obtained, and is $\bar{d}_{svg} = 25.8$ mm. The surface-to-mass ratio can be obtained as follows:

$$\frac{s}{m} = \frac{s}{\rho V} = \frac{\pi d^2}{\rho(\pi d^3/6)} = \frac{6}{d\rho} = \frac{6}{25.8(0.1)6.0} = 0.388 \text{ cm}^2/\text{gm}.$$

The same ratio from the tabulation of data is

$$\frac{s}{m} = \frac{s}{\rho V} = \frac{\Sigma \pi d^2}{\Sigma (\rho \pi d^3/6)} = \frac{6}{\rho} \frac{\Sigma d^2}{\Sigma d^3} = \frac{6}{6.0} \frac{(8298)(0.01)}{(131043)(0.001)} = 0.633 \text{ cm}^2/\text{gm},$$

which is only a fair comparison.

A part of the discrepancy can be attributed to the bias that can occur in analyses of particle size made by the count method; i.e., a finite number of particles are counted and as a result a finite upper limit is obtained. In contrast, the distribution used here (log normal) is infinite in extent. Gwyn et al. [223] have considered this problem in some detail and have provided a good illustration of the type of bias that can occur. For a count of the size used here, the bias introduced into the number distribution is small, and as a result the number distribution shown in Fig. 17–26 is well fitted by the log-normal distribution. For the weight or volume distribution, the bias in the large-size particles can be large, because of the heavy weighting of particles in this range. As a result of this bias the curve deviates from the log-normal distribution as shown in Fig. 17–26. Fortunately, as shown by Gwyn et al., a reasonable correction for this bias can be made. The improvement in fit made by the bias correction suggests that one should approach the assumption of a finite upper-limit size with caution.

17-2 MOTION OF SINGLE DROPS AND BUBBLES

Once a drop or bubble distribution has been established, the subsequent motion in the process system will, to a great extent, determine the degree of heat and mass transfer, and thus the effectiveness with which the equipment accomplishes the desired ends. It is necessary to understand the motions before one can understand many of the unit operations and such problems as kinetics in heterogeneous gas–liquid or liquid–liquid systems. The characteristics of individual drops and bubbles will be considered in this section, and interactions will be treated in the next section. It should be emphasized at this point that, for an accurate solution to problems involving complex flow, the entire distribution of sizes will have to be taken into account, unless the proper average and effect of the measure of the distribution can be correctly established by theory or experimental means.

As in most areas of two-phase flow, photographic techniques have played an important role in establishing the motion of drops and bubbles. Kintner et al. [97] have provided the working details of many successful methods and have presented a review of the use of photography in bubble and drop research. Sideman [182] has worked out another interesting photographic method, which emphasizes the interface and gives a three-dimensional effect. A review by Kintner [96] is also worthy of note.

17-2.A The Effect of Circulation

The slow motion (low Reynolds number) of a solid sphere was treated in Chapter 8, and the resulting equation for the terminal velocity (in the current notation) is

$$U_t = d^2(\rho_d - \rho_c)g/18\mu_c, \quad (17\text{-}72)$$

where the subscript t denotes terminal (for solid sphere in this case), the subscript d denotes discontinuous (solid), and the subscript c denotes continuous (liquid). Stokes' analysis was modified to account for circulation and surface effects, and the resulting Eq. (8-28) was given as

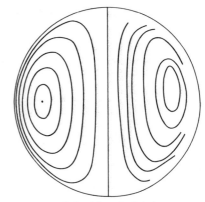

Left side calculated
Right side observed

Fig. 17-27. Drop circulation. [From W. R. Lane and H. L. Green, Surveys in Mechanics (G. K. Batchelor and R. M. Davies, editors), Cambridge University Press, New York (1956). By permission.]

$$U_{tc} = U_t[(3\mu_d + 3\mu_c + \gamma)/(3\mu_d + 2\mu_c + \gamma)], \quad (17\text{-}73)$$

where the subscript tc denotes terminal with the effect of circulation considered. A retardation coefficient, γ, is associated with the retarding effect which surface-active materials have on the motion of the surface of the drop or bubble. This form is due to Levich [115], where the complete derivation can be found (see also references 90 and 104), and reduces for $\gamma = 0$ to the form derived by Rybczyński [173] and Hadamard [65]. Streamlines for circulatory motion are shown in Fig. 17-27; on the left are the streamlines calculated, and on the right are those observed by Spells [191]. For higher Reynolds numbers, other circulation patterns have been suggested [68]. A number of ways of determining γ, as well as a number of alternative forms, have been suggested [12, 13, 59, 115, 170, 176, 177]. No definitive answer to the problem has yet been obtained, but apparently $\gamma \sim 1/d$, so that it becomes the controlling term for very small drops and bubbles (< 0.01 cm or $N_{Re,d} = dU_t\rho_c/\mu_c < 1$). Thus, for any such system, $U_{tc} = U_t$ by Eq.

(17-73), and since $N_{\text{Re},d} < 1$, U_t is given by Eq. (17-72). If $\gamma = 0$, the correction of Eq. (17-73) is small for drops of water in air (1.006), but large for air bubbles in water (1.49). For liquid–liquid systems (γ still zero), intermediate values would exist. For example, for drops of carbon tetrachloride in water, the correction is about 1.2. The retardation coefficient γ also depends on materials present that act as surface-active agents [59]. As a result, it may be increased (to reduce the correction factor toward unity) far beyond that predicted from the $1/d$ dependency. Thus the corrections cited above are upper limits.

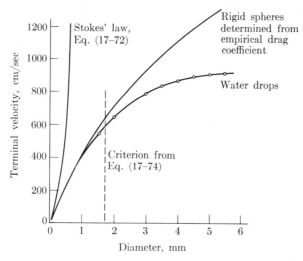

Fig. 17-28. *Terminal velocities of spheres of unit density and of water drops falling in air at 20°C and 75° mm Hg pressure. [From W. R. Lane and H. L. Green, Surveys in Mechanics (G. K. Batchelor and R. M. Davies, editors), Cambridge University Press, New York (1956). By permission.]*

17-2.B Some Experimental Evidence

Figures 17–28 through 17–31 contain selected examples of data. Figure 17–28 shows data for water drops in air [62, 106, 108], Stokes' law, and the curve to be expected for rigid spheres (which was obtained by using the empirical drag-coefficient curve for spheres). The data curve deviates from Stokes' law at a very small diameter (extrapolated), but still follows the solid sphere curve to nearly 0.1 cm ($N_{\text{Re},d} \sim 300$). A rough criterion for the deviation from solidlike behavior is shown in the figure, and was obtained by Bond and Newton [13] from dimensional arguments:

$$(\rho_d - \rho_c)d^2 g/\sigma \cong 0.4. \tag{17-74}$$

This criterion should be associated with distortion and swerving motion of the drops, rather than any effect of circulation. The latter is unimportant for drops in air, as shown by the low correction of 1.006. Bond and Newton originally

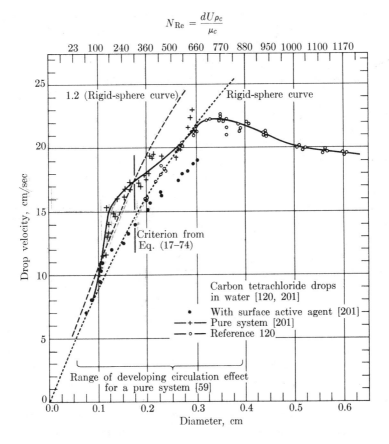

Fig. 17–29. Terminal velocities of carbon tetrachloride drops in water.

proposed this as a criterion for circulation, but as will be seen, this is not justified [59, 176]. The swerving motion occurs at a diameter of about 0.1 cm, even though both larger and smaller drops move vertically. Gunn [61] has indicated that this may be due to the detachment of eddies which have just the right frequency to excite self-oscillation of the falling drops [106]. Mills [225] and Schroeder and Kintner [228] have established the necessity of a vortex trail for the existence of drop oscillation. A correlation for the frequency of oscillation has been given in reference 228. In air, water drops flatten on the bottom, while the upper half maintains a hemispherical shape. Above 0.4 cm, the drops becomes flatter and flatter.

Figure 17–29 shows data for the carbon tetrachloride–water system [120, 201]. Values for small diameters where γ should be controlling (as a result of the $1/d$ effect) are obtained at about 0.1 cm and below. For larger diameters (up to 0.2 or 0.3 cm), the factors for circulation would predict that the data should lie above the rigid-sphere curve (1.2 for carbon tetrachloride). As can be seen in the figure, the data of Thorsen and Terjesen [201] for a pure system verify this. When surface active material was added, circulation was suppressed and the data followed the rigid-sphere curve. Griffith [59] showed similar increases for the systems he investigated and also made an extensive study of the effect produced by the addition of varying amounts of surface-active agent. He developed a criterion similar to Eq. (17–74) for circulation. Here the important factor is not σ itself, but $\Delta\sigma$, the reduction in surface tension caused by the added surface-active agent. Circulation begins to exert an effect at a value of the modified group of about 4, and should be completely effective in increasing the velocity at an estimated value of about 50. The range, based on these values and a guess at possible values of $\Delta\sigma$, is shown on Fig. 17–29.

Careful observations and measurements of other liquid–liquid systems were made by Satapathy and Smith [175]. The drop sizes were such that circulation would be expected to occur, and Eq. (17–73) was found valid for a drop having a Reynolds number less than 4. A ring vortex existed between 4 and 10, and the drag became somewhat greater than Eq. (17–73) would predict (see also reference 68). In the range from 10 to 45, considerable distortion occurred; however, the vortex remained attached to the drop, and the drag coefficient was the same as that for solid spheres up to a Reynolds number of 40. Between 40 and 45 the coefficient increased above the solid-sphere line. Above 45 the vortex began to break up, with little or no apparent internal circulation. Beyond a Reynolds number of 100, the vortices shed from alternate sides, causing the drop to oscillate. Again no internal circulation was apparent. The deviation from a straight path to a helical one occurs above a Reynolds number of 300. One of the most important conclusions made was that the drag coefficient for liquid–liquid systems is not a unique function of the Reynolds number. Some of the effects suggested might be attributed to an interesting observation by Elzinga and Banchero [44]; namely, that the drop circulation induces a shift in the point of separation, which in turn causes a reduction in the induced drag [230]. This is in addition to the reduction in drag caused by the slip effect. It should be pointed out that when the drag coefficient is based on an equivalent spherical drop diameter (even when the drops are highly distorted and possibly oscillate), the drag coefficients may have values less than solid spheres [44].

The criterion for the onset of distortion is given reasonably well by Eq. (17–74). For the pure system, it is at this point that the decrease in drag caused by circulation is counteracted by the increase in drag caused by distortion and swerving motion. The data of Licht and Narasimhamurty [120] continue the trend and eventually cross the rigid-sphere line, go through a maximum, and level off. For the system containing a surface-active agent, there is no decrease in drag because of circulation, and the criterion given by Eq. (17–74) marks the point at which distortion and swerving motion cause deviation from the rigid-sphere curve. The path is usually helical and the amplitude goes through a maximum,

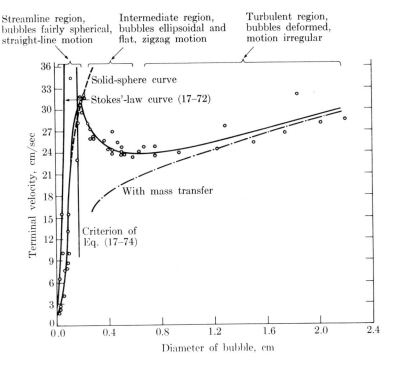

Fig. 17–30. *Terminal velocities of air bubbles in water.* [*From R. L. Datta, D. H. Napier, and D. M. Newitt,* Trans. Inst. Chem. Engrs., London **28**, *14 (1950). By permission.*]

which occurs at a diameter of from 0.2 to 0.8 cm. The amplitude decreases with increasing drop size, but does not die down completely.

The decrease in velocity from the rigid-sphere curve is mainly due to a deviation of the drop from a spherical shape. The change in shape is toward an increase in cross section, which results in a higher drag or lower velocity. Any circulation of fluid inside the drop induces a slip effect, which results in a reduction of drag. A second effect is also possible; a flattening of the drop due to centrifugal forces set up by the circulation. In addition, vortex formation at the rear of the drop [175] and shifting of the point of boundary-layer separation [44] can occur at Reynolds numbers beyond the Stokes region [230]. Still other effects can occur when the continuous phase is non-Newtonian; however, an apparent viscosity method allows comparison with the normal drag curves [136]. All these effects might be of importance if the entire range of drop sizes is to be investigated, but such an investigation is so complex that only limited analytical effort has been made and empirical methods are usually used.

Comparable data for air bubbles in water are given in Fig. 17–30, with more detail for the smaller diameters in Fig. 17–31. In the very low range (< 0.01 cm),

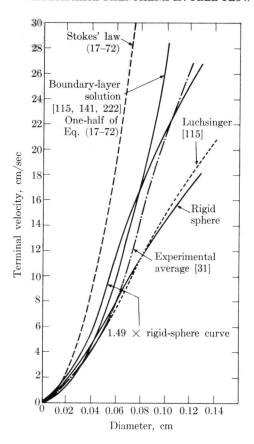

Fig. 17–31. Terminal velocities of air bubbles in water.

Stokes' law for rigid spheres applies, and the circulation correction suggested by Eq. (17–73) is not obeyed (1.49 correction). Here $\gamma \sim 1/d$, and is controlling. From 0.01 to 0.06 cm, the experimental data approximate the rigid-sphere curve, and are probably a result of contamination by a surface-active agent. Above 0.06 cm the data spread, some following the rigid-sphere curve and others (which were carefully purified to remove as much surface-active agent as possible) following close to curves calculated from 1.49 times the rigid-sphere or boundary-layer solutions [115, 141, 222]. The criterion of Eq. (17–74) is good for the description of the onset of bubble deformation. This is noted in Fig. 17–30, and is beyond the range of Fig. 17–31. However, circulation can cause deviation from solid-sphere behavior for lower values of d, if the system is clean. Bubbles that rise during simultaneous mass transfer (of major importance in process equipment) apparently do not pass through the maximum shown in Fig. 17–30. Thus their rise velocity is considerably below the curve in the range $d < 0.8$ cm [20, 112, 227].

All the previous work was for motion in calm liquids. Baker and Chao [221] have considered the problem of a bubble moving in a turbulent stream, and have provided what is perhaps the only reference on this subject.

17-2.C Analytical Representations

The equations available for the description of the flows just discussed have for the most part been empirical. For example, Davies [33] obtained an equation for the terminal velocities of drops in air at pressures down to a half atmosphere, which was accurate to within 3%. The analysis was for any liquid, so long as the parameter of Eq. (17–74) was in the range of 0.4 to 1.4. The empirical equation involved the Reynolds number, drag coefficient, and three constants that were different empirical functions of the group in Eq. (17–74). Klee and Treybal [98] considered values of liquid–liquid drop velocity over a wide range of variables. They established correlations by dimensional analysis, and found that two equations were necessary to describe the regions on either side of the maximum (see Fig. 17–29). The critical diameter for the changeover between regions was established by solving the empirical equations simultaneously. References 55, 89, 120, and 184 should be cited for other empirical correlations based on dimensional analysis, further data, and comparisons with equations previously used. Warshay et al. [212] investigated the effect of viscosity. Previous correlations just cited were satisfactory for a low-viscosity continuous phase, such as water, but not for high values. Warshay et al. suggested that for high-viscosity continuous-phase systems, one should use the standard drag versus Reynolds number curve. For non-Newtonian continuous phase, the work of Mhatre and Kintner [136] should be consulted. Olney and Miller [150] have reviewed this area and other aspects of liquid–liquid extraction, such as drop circulation, breakup, etc.

Datta et al. [31] described the motions and shapes of bubbles as a function of orifice size. Saffman [174] has given the corresponding bubble sizes. Grassman and Lemaire [58] have reviewed the literature on flow types and have presented a description in terms of dimensionless groups. The areas can be described as follows:

(1) For $d_n < 0.04$ cm or $d < 0.015$ cm, the bubbles are approximately spherical and move straight up at a uniform velocity, as dictated by Stokes' law ($N_{Re,d}/N_{Fr,d} = 18$). The boundary between this and the next flow type is $N_{Re,d} = 2$.

(2) For spherical bubbles with a slight internal circulation, the equation that describes the motion is $N_{Re,d} = 47\ N_{Fr,d}^{3/2}$ (for fluids not too different from water). The boundary between this and the next flow type is given by $N_{We,d}^3 = 667\ N_{Fr,d} N_{Re,d}^{-0.643}$.

(3) For 0.04 cm $< d_n <$ 0.4 cm, the bubbles are spherical or paraboloid at the orifice, but on release assume an ellipsoidal shape, with the longer axis horizontal. For 0.14 cm $< d <$ 0.2 cm, the bubbles zigzag, while between 0.2 cm and 0.46 cm they show a tendency to spiral or follow a helical path.

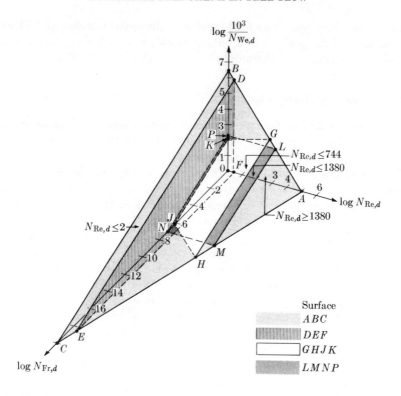

*Fig. 17–32. Three-dimensional area plot for water at 20°C. [From P. Grassman and L. H. Lemaire, Chem.-Eng.-Tech. **30**, 450–454 (1958). By permission.]*

A constant Weber number of 3.46 describes this type, and the equation for the boundary between this type and the next is $N_{\text{We},d} = 92\, N_{\text{Fr},d}$.

(4) For $d_n > 0.4$ cm, 0.4 cm $< d < 1.5$ cm, the bubbles are less stable [20], have quite variable shape, and may break up under certain conditions. This might be described by $N_{\text{Fr},d} = 0.51$.

(5) Still larger bubbles ($d > 1.5$ cm) will be discussed in more detail later. However, it can be noted that these usually have a spherical capped shape.

If we assume that the liquid properties control (probably not true with regard to internal circulation), then a simple right-angle coordinate system in space can be used to represent the various types of flow; in this system, the coordinates are the logarithms of the dimensionless groups being considered. Figure 17–32 is such a representation. The surface ABC depends only on the properties of the fluid in question, and is given by $N_{\text{Re},d}^4 N_{\text{Fr},d} N_{\text{We},d}^{-3} = \rho_c \sigma^3 / g \mu_c^4 = 3.9 \times 10^{10}$ for the water–air system. This particular combination was discussed in Chapter 11. The various boundary planes are shown; for example, the boundary between (1) and (2) is given by $N_{\text{Re},d} = 2$ and thus is the plane DEF; (2) occurs to the right of this plane to the plane $GHJK$. The area from this plane to the right and below to plane $LMNP$ ($N_{\text{We},d} = 92\, N_{\text{Fr},d}$) is the area of flow (3); beyond this plane lies area (4). Different liquids would simply shift the plane ABC in a parallel fashion. Caution, however, must be taken in any extrapolation, since little experimental confirmation has been obtained with alternative liquids.

The effect of viscosity on the general flow has been investigated by Hartunian and Sears [75]. They used two critical conditions, one for high- and one for low-viscosity fluids. Siemes [183] showed that the shapes of medium-size bubbles can be predicted by assuming an oblate spheroid, whose shape is determined by potential flow. The drag cannot be obtained if any vortex trail forms. Circulation within the bubble appears to be critical in determining trail formation. If viscous forces suppress circulation, vortex shedding, with lower bubble velocities and higher drag [54, 64], may be observed.

17-2.D Large Bubbles

The rise of large, free bubbles, shaped like spherical caps, and the rise of similar bubbles restricted in a vertical cylinder have received considerable attention both experimentally and theoretically. Walters and Davidson [209] have dealt with the motion in time of a two-dimensional bubble that started from rest and was originally in the form of a horizontal cylinder. The bubble (see Fig. 17–33) rapidly distorted by flatting on the bottom, and continued this motion until a tongue of liquid was projected well into the interior. The ∩-shaped bubble, now rising, pinched off a small bubble (vortex center) from each end of the ∩ to develop the vorticity necessary to maintain the steady motion of the now crescent-shaped, spherical-capped, two-dimensional Taylor bubble. For the free, spherical-capped bubble, Davies and Taylor [35] neglected surface tension and viscous forces, and assumed that the drop motion was determined by the free fall of the liquid along the upper surface of the bubble. A Bernoulli balance between the velocity term and the gravity force causing the free fall results in

$$U_\theta^2 = 2gh, \qquad (17\text{-}75)$$

where U_θ is the relative liquid velocity along the surface, and h is the vertical distance from the apex of the bubble (see Fig. 17–34). Since the viscous forces are assumed to be zero, the flow about the spherical cap will be potential. The flow over a cylinder or two-dimensional bubble was derived as Eq. (6–27) for a unit radius, and the velocity was obtained as shown following Eq. (6–28). For any bubble radius r_0, the functions become

$$\phi = U_b(r + r_0^2/r)\cos\theta, \qquad \psi = U_b(r - r_0^2/r)\sin\theta,$$

where U_∞ has been replaced by the bubble velocity U_b, since the velocity at infinity in a moving field must be the same as the bubble velocity in a stationary field. For a three-dimensional sphere the correct form is

$$\phi = U_b(r + r_0^3/2r^2)\cos\theta, \qquad \psi = \tfrac{1}{2}U_b(r^2 - r_0^3/r)\sin^2\theta.$$

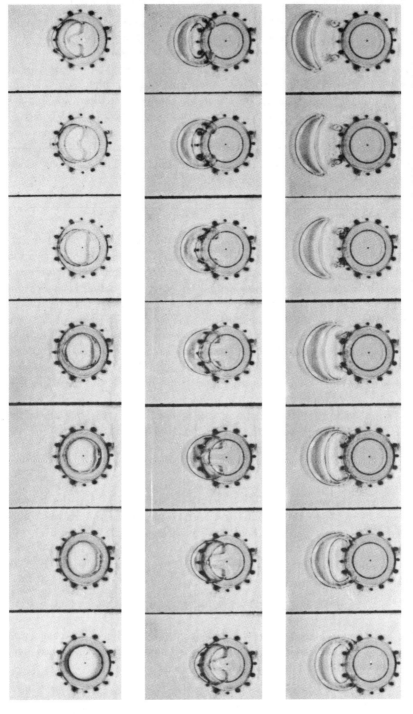

Fig. 17-33. Motion pictures, at about 80 frames per second, of the initial motion of a bubble two in. in diameter. [From J. K. Walters and J. F. Davidson, J. Fluid Mech. **12**, 408 (1962), published by the Cambridge University Press, New York. By permission.]

The surface velocity (at $r = r_0$) can be determined by combining either of these with Eq. (6–28) to give $U_\theta = \tfrac{3}{2} U_b \sin \theta$. Combining this with Eq. (17–75) and using geometry gives us

$$U_\theta^2 = 2gh = 2gr_0(1 - \cos \theta) = \tfrac{9}{4} U_b^2 \sin^2 \theta$$

or

$$U_b^2 = \tfrac{8}{9} gr_0 (1 - \cos \theta)/\sin^2 \theta. \qquad (17\text{–}76)$$

The term r_0 is the radius of the spherical-capped bubble; however, the bubble is not totally spherical, so r_0 is not the equivalent spherical radius. An integration from $\theta = 0$ to θ will give the actual volume as $V = \pi r_0^3(\tfrac{2}{3} - \cos \theta + \tfrac{1}{3} \cos^3 \theta)$. Combining this with Eq. (17–76) gives

$$U_b = g^{1/2} V^{1/6} \left[\frac{2\sqrt{2}}{3\pi^{1/6}} \left(\frac{1 - \cos \theta}{\sin^2 \theta} \right)^{1/2} \left(\frac{2}{3} - \cos \theta + \frac{1}{3} \cos^3 \theta \right)^{-1/6} \right].$$

Experimentally it was observed by Davies and Taylor that

$$U_b = 0.792 \, g^{1/2} V^{1/6}$$

or $\qquad (17\text{–}77)$

$$U_b = 0.711 \sqrt{g \, d_e},$$

where d_e is the equivalent diameter defined from $V = \pi d_e^3/6$. This can also be expressed as $N_{\text{Fr},d} = 0.503$. For bubbles with a diameter of 2 cm, this gives a velocity of 31.5 cm/sec ($N_{\text{Re},d} \cong 5000$), which is a good check with Fig. 17–30.

More extensive treatment has been made for the restricted bubble in a tube. Davies and Taylor obtained

$$U_b = 0.464 \sqrt{gr_0}$$

or $\qquad (17\text{–}78)$

$$N_{\text{Fr},d} = 0.1078,$$

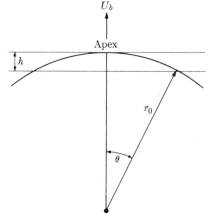

Fig. 17–34. Notation for the large-bubble problem.

and Dumitrescu [41] obtained similar equations with the coefficients 0.496 and 0.1225, respectively. For this case, r_0 is the radius of both the tube and the bubble. Laird and Chrisholm [103] have checked this equation closely. Dumitrescu's coefficient gives a velocity of 0.81 ft/sec for a tube with a 2-in. diameter, the observed values being from 0.78 to 0.88 ft/sec. Equation (17–78) is also based on viscous and surface-tension forces that are negligible, and fails when these become important. Thus it is valid only for values of [72, 215]

$$N_{\text{We},d}/N_{\text{Fr},d} = \rho_c g \, d^2/\sigma > 70 \quad \text{and [215]} \quad N_{\text{Re},d}^4 N_{\text{Fr},d} N_{\text{We},d}^{-3} = \rho_c \sigma^3/g \mu_c^4 > 10^6.$$

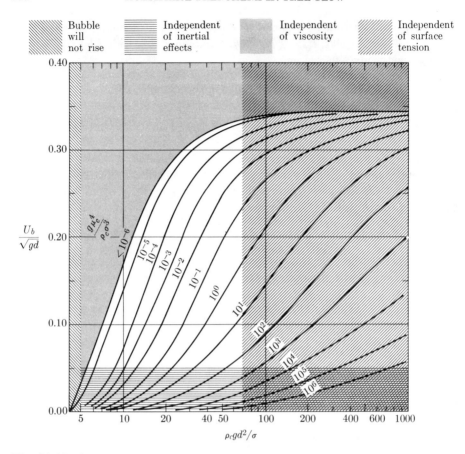

Fig. 17–35. General correlation for velocity of rise of cylindrical air bubbles in liquids. [From E. T. White and R. H. Beardmore, Chem. Eng. Sci. **17**, 351–361 (1962). By permission.]

White and Beardmore [215] have obtained extensive data and developed the general correlation shown in Fig. 17–35, which is in terms of the groups cited above. The various areas indicate when the effects of certain forces can be safely neglected.

Goldsmith and Mason [57] have considered positions along the bubble in which end effects were negligible and parallel incompressible creeping flow ($N_{\text{Re},d} < 1$) could be assumed. The Navier-Stokes equation became

$$-\frac{\partial p_c}{\partial x} + \rho_c g + \frac{\mu_c}{r}\frac{\partial}{\partial r}\left(r\frac{\partial U_c}{\partial r}\right) = 0 \qquad (17\text{–}79)$$

for the continuous phase, and a comparable equation for the bubble or discontinuous phase. The pressure on the two sides can be related through the surface tension (Laplace's capillary equation):

$$p_d = p_c + \sigma/(r_0 - h), \qquad (17\text{–}80)$$

where h is the film thickness along the wall. Equation (17–80) shows that

$$\partial p_c/\partial x = \partial p_d/\partial x = C.$$

The general solution to Eq. (17–79) was next obtained as

$$U_c(r) = (-\rho_c g + H)r^2/4\mu_c + A_c \ln r + B_c, \tag{17-81}$$

and a comparable equation was obtained for the bubble or discontinuous phase. Five boundary conditions were used to evaluate H, A_c, B_c, A_d, and B_d; that is, (i) $U_d(r)$ is regular or $A_d = 0$, (ii) $U_c(r_0) = 0$, (iii) the bubble velocity U_b can be obtained by an integration of Eq. (17–81) or its counterpart for the discontinuous phase, (iv) $U_c(r_0 - h) = U_d(r_0 - h)$, and (v) $\mu_c[\partial U_c(r_0 - h)/\partial r] = \mu_d[\partial U_d(r_0 - h)/\partial r]$. The end result is that we have explicit expressions for $U_c(r)$, $U_d(r)$, and U_b in terms of ρ_c, ρ_d, μ_c, μ_d, r_0, and h. The limit of an inviscid bubble ($\mu_d = 0$) was also obtained, and for the bubble velocity is

$$U_b = \left[\frac{(\rho_d - \rho_c)g}{2\mu_c}\right]\left[\frac{r_0^4}{4(r_0 - h)^2} - r_0^2 + \tfrac{3}{4}(r_0 - h)^2 + (r_0 - h)^2 \ln \frac{r_0}{r_0 - h}\right]. \tag{17-82}$$

This equation was applied to the data for low-viscosity bubbles for $N_{\text{Re},d} < 1$ by calculating h from the known values for ρ_c, μ_c, ρ_d, and the measured U_b. The mean experimental value of all runs calculated varied by less than 1% for gas bubbles and less than 2% for low-viscosity liquid bubbles. Some individual deviations were much greater. The more general equation was applied to viscous-bubble data (μ_d/μ_c as high as 40), and the agreement for individual measurements was better than 1%. The profiles, $U_c(r)$ for gas bubbles, and $U_c(r)$ and $U_d(r)$ for liquid bubbles, were measured and found to be in excellent agreement with the values calculated from the respective equations. One must remember that the entire analysis and data discussed here are for very slow motion and are parallel to Stokes' law for the free sphere. The conditions are far different from those in which Eq. (17–78) would apply. Uno and Kintner [204] empirically related bubble velocity in the proximity of a wall to the system variables. They concluded from their work that in order to neglect wall effects, the minimum tube size that should be used is 10 to 15 times the equivalent diameter of the largest bubble to be studied. Articles by Bretherton [15] and Cox [28] contain references to the motion of bubbles in tubes and the amount of fluid that remains along the wall during the passage of a long bubble.

17-2.E The Interface

The hydrodynamics of drops and bubbles have been treated in some detail in the preceding material. Mass or heat transfer in such systems clearly depends on the motion, distortion, and circulation that exists. However, before discussing the effect of these, one additional complexity must be cited. It has often been observed that spontaneous contractions, eruptions, localized agitation, interface oscillation,

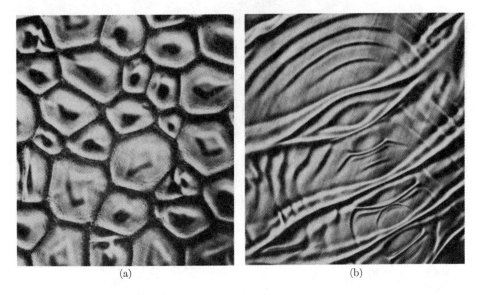

Fig. 17–36. (a) *Polygonal cells.* (b) *Ripples and stripes.* [*From A. Orell and J. W. Westwater, AIChEJ* **8**, *350–356 (1962). By permission.*]

and even spontaneous emulsification can occur during mass transfer. Such occurrences give rise to a great deal of motion at the interface and induce motion in the adjacent fluid [16, 34, 53, 76, 94, 101, 116, 117, 119, 122, 124, 133, 151, 179, 185, 211]. Clearly this phenomenon also has a considerable effect on the degree of transfer. As a group, the various forms of interfacial agitation are called interfacial turbulence [116] and often are chaotic and much like regular turbulence [53, 76, 101, 119, 179, 185], although organized cellular patterns have also been predicted [193] and observed [122, 151] (e.g., see Fig. 17–36). Interfacial turbulence has been observed at flat interfaces [16, 101, 133, 151, 211], along pendant droplets [53, 76, 119, 185], and at rising or falling drops [116, 179, 185, 205], and thus could well be a contributing factor in any system with liquid–liquid contact in which transfer occurs. It is apparent that the strange behaviors observed are a result of variations in the interfacial tension (Marangoni effect) as caused by variations in the interfacial concentration of the material diffusing between the phases [76, 101, 185, 193, 205]. A somewhat analogous situation can be found in the Bénard cell [7], in which cellular convection patterns are formed as a result of heating a fluid from below. The variation in temperature causes variations in surface tension, which in turn induces the patterns [11, 153]. A rather complete literature survey has been given by Pearson [153] and by Scriven and Sternling [178, 193], who have also treated a simplified model from a theoretical approach [193].

A knowledge of the dynamics of the interface is inherent in any detailed mathematical analysis of a two-phase flow system. A balance of the tangential stresses on an element of the surface must exist; thus, from Fig. 17-37, we obtain for the simple one-dimensional case,

$$-\mathsf{T}_{xx} + \mathsf{T}_{xx} + (\partial \mathsf{T}_{xx}/\partial x)\, dx + \bar{\bar{\tau}}_{yxc}\, dx - \bar{\bar{\tau}}_{yxd}\, dx = 0$$

or

$$\bar{\bar{\tau}}_{yxd} - \bar{\bar{\tau}}_{yxc} = \partial \mathsf{T}_{xx}/\partial x. \quad (17\text{-}83)$$

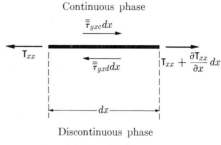

Fig. 17-37. Surface stress.

The term T_{xx} represents the longitudinal surface stress. This would be exactly the same as σ, the equilibrium interfacial tension, only in the restricted case of an isotropic interface with no viscous stresses peculiar to the interface itself. In the limiting case of an isothermal system, without gradients in concentration and in the absence of surface active agents, $\mathsf{T}_{xx} = \sigma = $ const. Thus, from Eq. (17-83), we see that $\bar{\bar{\tau}}_{yxd} = \bar{\bar{\tau}}_{yxc}$, which is the familiar boundary condition of equal tangential stress at an interface used in our previous analyses under what it was hoped were valid conditions. For more complex systems, the expression for $\bar{\bar{\tau}}_{yxd} - \bar{\bar{\tau}}_{yxc}$ can be established by an analysis of the dynamics of a fluid interface, as has been done by Scriven [177] in a very general manner.

This analysis is roughly parallel to the Navier-Stokes equation previously derived; however, complications exist because of the possible curvature of the surface and the change of shape with time. The interface equation relates the variable components lying in the interface and gives no information about the characteristics of the accompanying bulk-phase flow; thus it is also necessary to establish the connection between them. The final detailed equation will not be given here; the following is an abbreviated form intended to illustrate the various contributions discussed by Scriven:

$$\rho_s(D\boldsymbol{U}/Dt) - \boldsymbol{F} = \boldsymbol{\nabla}_s \sigma + (\kappa + \epsilon)\boldsymbol{\nabla}_s(\boldsymbol{\nabla}_s \cdot \boldsymbol{U}) + \epsilon f_1(\boldsymbol{U}) + N f_2(\sigma, \kappa, \epsilon, \boldsymbol{U}). \quad (17\text{-}84)$$

The net external force on the interface, $\boldsymbol{F}$, is composed of the regular external forces already discussed in the derivation of the Navier-Stokes equation and the stresses exerted on the interface by the bulk phases on either side. That is to say,

$$\boldsymbol{F} = \sum_s \rho_s \boldsymbol{F}_s + \boldsymbol{\tau}_c - \boldsymbol{\tau}_d,$$

where $\boldsymbol{\tau} = (\boldsymbol{N} \cdot \bar{\bar{\tau}})$, which can be obtained from Eq. (2-40) or (2-42), and where $\bar{\bar{\tau}}$ can be obtained from Eq. (3-43) evaluated in the bulk fluid about the interface. The terms κ and ϵ are Boussinesq's [14] surface coefficients of viscosity (dilational and shear, respectively). These coefficients play a part in Eq. (17-84) analogous to the part played by μ in the Navier-Stokes equation; however, this should not be taken to mean that they are related in any way, since they reflect different molecular constituents. From surface chemistry it is known that adsorption can produce large interface concentrations of components that are quite dilute in the bulk phases. The molecular viscosity is an empirical bulk characteristic, whereas the surface viscosities are empirical coefficients that characterize the interface. All of these depend on composition, temperature, and density. The existence of surface coefficients is believed to be the result of the presence of surface-active agents. The two-dimensional operator ∇_s is analogous to ∇ but operates only tangential to the surface, N is the unit vector normal to the surface, and ρ_s is the surface density (mass per unit area), sometimes referred to as the surface excess mass density. The first term on the right is the gradient of the interfacial tension as a result of variations in temperature, composition, or density along the surface. The second term is a result of the gradient of the surface divergence, that is, viscous resistance to area expansion or contraction of surface material. The third term is a result of viscous resistance to the shear of surface material, where the vector function f_1 depends on the motion (uniform motion in a surface of finite total curvature, rotational motion in the surface, or nonhomogeneous normal displacement of the surface). The final term on the right is the normal stress resultant, which can occur if surface curvature exists. A part of this is the familiar term for the pressure across an interface as a result of surface-tension effects. For any curved surface, the surface-tension contribution is given by $\sigma(1/r_1 + 1/r_2)$, where r_1 and r_2 are the principal radii of curvature. For a spherical surface of radius r_0, this reduces to $2\sigma/r_0$.

Scriven gives several examples of simplified forms of the equation for specific geometries. For a flat horizontal surface in the xz-plane with no inertial or external force terms, the equation reduces to

$$\tau_{xd} - \tau_{xc} = \frac{\partial \sigma}{\partial x} + (\kappa + \epsilon)\frac{\partial}{\partial x}\left(\frac{\partial U_x}{\partial x} + \frac{\partial U_z}{\partial z}\right) + \epsilon\frac{\partial}{\partial z}\left(\frac{\partial U_x}{\partial z} - \frac{\partial U_z}{\partial x}\right),$$

$$\tau_{yd} = \tau_{yc}, \tag{17-85}$$

$$\tau_{zd} - \tau_{zc} = \frac{\partial \sigma}{\partial z} + (\kappa + \epsilon)\frac{\partial}{\partial z}\left(\frac{\partial U_x}{\partial x} + \frac{\partial U_z}{\partial z}\right) + \epsilon\frac{\partial}{\partial x}\left(\frac{\partial U_z}{\partial x} - \frac{\partial U_x}{\partial z}\right).$$

For a cylindrical jet, the z and θ directions are

$$\tau_{zd} - \tau_{zc} = \frac{\partial \sigma}{\partial z} + (\kappa + \epsilon)\frac{\partial}{\partial z}\left(\frac{1}{r_0}\frac{\partial U_\theta}{\partial \theta} + \frac{\partial U_z}{\partial z}\right) + \left(\frac{\epsilon}{r_0}\right)\frac{\partial}{\partial \theta}\left(\frac{1}{r_0}\frac{\partial U_z}{\partial \theta} - \frac{\partial U_\theta}{\partial z}\right),$$

$$\tau_{\theta d} - \tau_{\theta c} = \frac{1}{r_0}\left(\frac{\partial \sigma}{\partial \theta}\right) + \frac{(k+\epsilon)}{r_0}\frac{\partial}{\partial \theta}\left(\frac{1}{r_0}\frac{\partial U_\theta}{\partial \theta} + \frac{\partial U_z}{\partial z}\right) + \epsilon\frac{\partial}{\partial z}\left(\frac{\partial U_\theta}{\partial z} - \frac{1}{r_0}\frac{\partial U_z}{\partial \theta}\right). \tag{17-86}$$

The radial or normal direction is across a curved surface and is different from that given in Eq. (17–85). The various terms derive from the f_2 term, and the result is

$$\tau_{rd} - \tau_{rc} = -\frac{\sigma}{r_0} - \frac{k+\epsilon}{r_0}\left(\frac{1}{r_0}\frac{\partial U_\theta}{\partial \theta} + \frac{\partial U_z}{\partial z}\right) + \frac{2\epsilon}{r_0}\frac{\partial U_z}{\partial z}. \tag{17-87}$$

Equations can also be derived for a spherical drop; they can be found in the paper cited, but will not be presented here.

Sternling and Scriven [193] have treated the isothermal, two-dimensional problem of the hydrodynamic instability of a flat horizontal (xz-plane) interface during steady mass transfer (the same analysis actually is valid for three-dimensional disturbances). The solution is ideal from a pedagogical viewpoint, since it deals with the onset of motion in an initially quiescent system, rather than with the appearance of secondary flow in a disturbed laminar flow system, as in the chapter on turbulence. Just as in the discussion in that chapter, we here write the Navier-Stokes equation and linearize it by neglecting the inertial terms. In addition, field forces are assumed to be absent:

$$\begin{aligned}\frac{\partial U_x}{\partial t} &= -\frac{1}{\rho}\frac{\partial p}{\partial x} + \nu\left(\frac{\partial^2 U_x}{\partial x^2} + \frac{\partial^2 U_x}{\partial y^2}\right), \\ \frac{\partial U_y}{\partial t} &= -\frac{1}{\rho}\frac{\partial p}{\partial y} + \nu\left(\frac{\partial^2 U_y}{\partial x^2} + \frac{\partial^2 U_y}{\partial y^2}\right).\end{aligned} \quad (17\text{-}88)$$

The pressure term is eliminated by cross-differentiation and subtraction. The velocities (or an equivalent stream function) are then expressed as a mean (zero) and fluctuating part, after which they are expressed as a complex exponential function. After simplification this becomes the equation for a two-dimensional flow disturbance in an initially quiescent fluid which can be solved subject to the boundary conditions. Its counterpart for the disturbance in an initially laminar flow is the Orr-Sommerfeld equation [121]. One of the boundary conditions involves the interface (Eq. 17–85), and for the flat, horizontal, two-dimensional problem becomes

$$\tau_{xd} - \tau_{xc} = \bar{\bar{\tau}}_{yxd} - \bar{\bar{\tau}}_{yxc} = \partial\sigma/\partial x + \mu_s(\partial^2 U_x/\partial x^2), \quad (17\text{-}89)$$

where $\mu_s = \kappa + \epsilon$ and variation in the z-direction vanishes. The variation of surface tension with position was approximated by

$$\partial\sigma/\partial x = (\partial\sigma/\partial c)(\partial c/\partial x), \quad (17\text{-}90)$$

which still leaves the problem of evaluating the surface concentration. For this one can use the diffusion equation written for the two-dimensional system:

$$\partial c/\partial t + U_x(\partial c/\partial x) + U_y(\partial c/\partial y) = D(\partial^2 c/\partial x^2 + \partial^2 c/\partial y^2). \quad (17\text{-}91)$$

The velocities are defined by the previous analysis, and will be perturbed according to the complex exponential function. A similar expression is used for the concentration fluctuation. The resulting stability equation is to diffusion what the two-dimensional flow-disturbance equation is to the flow, and can be solved subject to the boundary conditions. At this point, the equations are combined, and we attempt to solve for the growth-rate constant (real part of the exponential time factor), which determines whether the small disturbance will grow or be damped. Of course, this depends on all the parameters of the system: D_c, D_d, μ_c, μ_d, κ, ϵ,

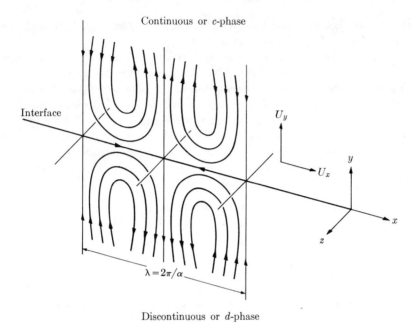

Fig. 17-38. *Two-dimensional roll waves.*

$\partial\sigma/\partial c$, σ, etc. The complete solution is understandably complex, and so a qualitative analysis of the behavior of the growth-rate constant was made by using certain limiting forms. The system studied (which again is an isothermal, two-dimensional, semi-infinite, flat interface between two quiescent fluids, with a single component at very low concentration being steadily transferred) is found to be always unstable for transfer in one direction or the other, and so waves should form, possibly as suggested in Fig. 17-38. The roll waves shown, however, are only one of the many possible forms that might exist with the dominant wavelength obtained from the solution (i.e., hexagonal cells or other three-dimensional forms could exist and have the same wavelength). If a system is stable from phase c to phase d, it will be unstable in the opposite direction. Under certain conditions, it may be unstable to transfer in either direction. The expression $\partial\sigma/\partial c$ is commonly negative; for this condition, the system is unstable for transfer out of the higher-viscosity phase or for transfer out of the phase characterized by the lower diffusivity. (For $\partial\sigma/\partial c$ positive, the reverse is true.) Other factors contributing to the instability are large differences in ν and D between phases, large values of $\partial c/\partial y$, large $\partial\sigma/\partial c$, low μ and D in both phases, absence of viscous or film-forming surface-active agents, and large interfaces. All of these factors promote smaller cell size and more rapid amplification. A numerical calculation of four examples, covering typical situations with no surface-active agent present ($\mu_s = 0$), confirmed the analysis. For flow-limited modes (in which $D_c \sim D_d$), the wavelength is of the order of 2 mm and the intensity doubles approximately every second. For diffusion-limited cases (in which $\mu_c \sim \mu_d$), the wavelength is very small (of the order of 0.03 mm), and the amplification rate is very rapid. It would be well to emphasize Sternling's and Scriven's point that the model is quite idealized, and that other factors could well negate the results for specific systems. For example, no one has yet considered the effects of temperature variation at the interface, movement and curvature of the interface rather than the maintenance of a fixed plane, nonsteady conditions of transfer, and nonconstant properties such as diffusivities.

In some cases the stability-direction effect has been confirmed [148, 151], while in others it has not [133, 151]. In all cases, however, other factors could have been present. The criteria are probably valid enough to establish when (in the absence of surface-active agents) interfacial activity will occur, but not valid enough to establish when it will not.

Ruckenstein and Berbente [171] have extended the preceding analysis to the case in which a first-order chemical reaction with no heat of reaction is also occurring, and found that even small values of the rate constant could alter the criteria for instability. In fact, one set of conditions gave rise to a system which was stable for transfer in either direction.

The parallel problem of the instability of a liquid-gas interface was discussed in Chapter 16 (wave formation). Whitaker [214] treated the problem of falling film in a manner somewhat analogous to the interface problem just discussed, and considered the effects of surface tension, surface viscosity, and the surface-tension gradient with concentration. He found that increasing surface tension or viscosity can lower the growth rate of a small disturbance, but does not stabilize the film; for finite surface viscosity, waves still form, but at some distance further from the top of the plate (about 10 times that with no surface viscosity). The surface-concentration gradient can cause film stabilization when $\partial\sigma/\partial c$ is negative, and is probably the reason why surface-active agents can prevent wave formation. Whitaker found that a critical Reynolds number for stabilization in the absence of diffusion was

$$N_{\text{Re},c} = [-16\rho_l(\partial\sigma/\partial c)^3/g\mu_l^4]^{1/5}, \tag{17-92}$$

and that if he approximated $\partial\sigma/\partial c$ by σ [115] and used the properties of water, $N_{\text{Re},c} \cong 200$. Further calculation showed that in the presence of a reasonable degree of diffusion, the critical Reynolds number increased to about 500. In this and all the preceding work, plane interfaces were treated; the extension to drops has been made by Bupara [231], but no work has been reported for jets.

Foams constitute still another gas-liquid interface problem. For the most part, interest has centered around the reasons for foam stability, with the aim of eventually eliminating troublesome foam problems [1, 4, 10, 132]. Foams can be divided into two types [1, 132]: wet foams with high liquid content and spherical bubbles (*kugelschaums*), and dry foams with low liquid content and polyhedral bubbles (*polyederschaums*). The latter type is the more important of the the two, since the former exists for only one bubble diameter or so above the liquid surface. A good approximation for the dry foam is the regular dodecahedra [1, 10, 132]. The hydrodynamics of the interstitial flow of liquid in foams has been studied [17, 113, 168], and its possible application to a separation process called foam fractionation has been reviewed [168] and considered in some detail [73, 113]. Separation by foam fractionation is caused by the adsorption of the solute at the surfaces of bubbles formed by sparging a liquid mixture. The foam then carries the solute overhead. The introductory analysis by Brunner and Lemlich [17, 168] does not consider foam density or the overhead flow-rate. Leonard and Lemlich [113] have extended the analysis and presented the results of a computer integration in the form of dimensionless plots. They considered the interstitial liquid flow in a stationary or moving foam and found that it depends on the foam structure and movement. These, of course, depend on the operating variables and the physical properties.

17-2.F Mass and Heat Transfer

From the preceding discussion it is apparent that in systems involving drops and bubbles, or in fact any interface, the solution to heat and mass transfer problems is quite complex, and empirical or semi-empirical approaches will continue to be used for some time. The material clearly demonstrates the need for caution in design. Consider the danger in designing a mass-transfer device from data in which the flow patterns, contact area, etc., have been determined from literature correlations obtained in the absence of mass transfer. What interfacial effects might be induced by the surface-tension gradient that would cause even greater rates of transfer? Consider also the possible consequences if a surface-active-agent contaminant is present in the experiments upon which the design correlation is based, in the pilot plant, or in the final process system. Unfortunately we cannot deal with mass or heat transfer in any great detail, but we must elaborate further on the effects of circulation, distortion, and interface turbulence, and we must cite some of the extensive literature on the subject (i.e., the reviews 18, 34, and 150).

For transfer from a continuous liquid phase to a body, the existence or absence of internal circulation implies that a mobile or rigid interface is presented to the flow. If the flow is laminar, the continuity equation can be written for the fluid as a whole and for the diffusing species. If we make some assumption about the velocity profile in the fluid, we can obtain a numerical solution for various shapes, and we can also obtain analytical solutions under the limiting conditions of whether or not circulation exists and whether one has slow motion ($N_{\text{Re},d} < 1$) or boundary-layer flow ($N_{\text{Re},d} \gg 1$). For the more chaotic flows of turbulence, these theories will be inadequate, and one must turn to: the film theory, in which transfer is controlled by molecular diffusion through a nearly stagnant laminar sublayer or by a combined molecular and eddy diffusion mechanism; to the penetration theory, which extends the film concept to one in which concentration eddies penetrate the film and transfer occurs through an unsteady-state, molecular-diffusion mechanism; or to one of the more refined recent models.

Mass transfer to a sphere suspended in a stagnant fluid follows the expected basic solution, in which the mass transfer is proportional to the diffusivity. For the film theory, this would be the limiting case of an infinite film, but direct proportionality to the diffusivity has not been found valid for any other situation [74, 138]. In laminar-flow problems where analytical solutions can be obtained, the dependency is on whether the interface is mobile or not, and on the nature of the assumption that is made about the velocity distribution. The usual assumption is that the Schmidt number is high (liquid systems) so that the major change in concentration occurs over a region much thinner than the velocity change, which in turn allows one to assume a linear velocity gradient in the region important to mass transfer. For the immobile interface the continuous-phase mass transfer is proportional to $D^{2/3}$ [50, 51, 88, 123, 140, 159, 224]. For mobile interfaces the transfer is proportional to $D^{1/2}$ [21, 59, 123, 169, 210, 224, 227]. Of course, if the mobility lies between the rigid and the completely mobile cases, one would expect the power to lie between $\frac{1}{2}$ and $\frac{2}{3}$ [123, 156, 176].

Other factors of importance have received some consideration. The penetration theory [30, 81] also leads to a $D^{1/2}$ dependency. Although appropriate to turbulent conditions, the resulting forms can also be applied to laminar flow in which a specific contact time [81] is known or a fractional rate of surface renewal (a distribution of contact times) [30] can be defined. Distortion will of course have a considerable effect on the transfer. Lochiel and Calderbank [123] have considered the boundary layer solution for oblate and prolate spheroids and the potential flow for spherical caps. The general form of the equation and the dependency on diffusivity remain the same; however, they introduce the additional factor of an eccentricity measurement. For oblate spheroids, the actual effect of eccentricity on the mass-transfer coefficient is small; but the product of the coefficient and the transfer area is important in mass transfer, and because of it, the rate of transfer is quite dependent on the shape [20]. For the spherical cap, the transfer is also proportional to $D^{1/2}$ [5, 20, 123]. The effect of a surface-tension gradient alters the velocity distribution in the vicinity of a body, and thus can also affect the transfer coefficient in the continuous phase [170]. Popovich et al. [155] have obtained mass-transfer data during the formation stage of a drop and have compared their results with several unsteady-state diffusion theories available in the literature. Experimental support and details of analytical development can be found in the various references, and in the further references cited therein. In a review of mass transfer in mixing, Calderbank [18] has brought together nearly all the available data into two empirical correlations which are quite consistent with the derived equations cited above (i.e., small-bubble, rigid, or immobile correlation, and large-bubble or mobile correlation).

The foregoing material considers transfer external to drops or bubbles. It is also necessary to consider the mechanism of internal transfer [18, 74, 138, 150, 186]. The simplest model is Newman's unsteady-state molecular diffusion in a stagnant fluid sphere [143]. Kronig and Brink [102] extended this model to include circulation of the type described by Hadamard [65]. The modification results in an effective diffusion coefficient $2\frac{1}{4}$ times the molecular value [18, 19]. Still other suggestions [19, 69, 186] have been made to help explain the results. McDowell and Myers [125] made the following experimental observations: (1) High-viscosity oil in water gave heat-transfer rates which showed negligible circulation. (2) High-viscosity oil in water was closely approximated by no circulation, but even closer by conduction plus a small surface heat-transfer resistance. (3) If the viscosity of the drop was low in comparison with the oil–water system, then circulation occurred, and higher heat transfer rates were obtained, in good accord with the results of Kronig and Brink. Calderbank and Korchinski [19] reported similar results when drops were circulating but not oscillating. Very high rates (from 7 to 70 times that for molecular diffusivity alone [18]) are experienced when oscillation occurs.

17-3 INTERACTION EFFECTS FOR DROPS AND BUBBLES

It can be said that there exists a reasonable understanding and working knowledge of the fundamentals of the behavior of single drops and single bubbles. However, the additional factors introduced when a swarm of drops or bubbles interact preclude a quantitative solution of the problem at this time. The complications of collision, distortion, coalescence, internal mixing, breakup, distribution in the vessel, altered residence time in the equipment, and modified interfacial area all contribute to the complexity of the problem. Even so, advances are being made in current research efforts that, it is hoped, will shed light on the entire problem. The problem of interaction is probably at its worst in the very important stirred-vessel contactor, and so this particular contacting device has received considerable attention.

Of the individual items contributing to the nature of highly interacting systems, only drop and filament breakup have been considered in detail earlier in this chapter. The competing process is, of course, coalescence, and fundamental studies in this area, together with the work on breakup, will in time enable us to establish the factors that control the dynamic balance between these two.

If we are ever to understand the entire process, we must understand the mechanism that causes coalescence of a drop at an interface, or of two drops together. MacKay and Mason [126] considered such problems and cited most of the previous work on the subject [2, 25, 27, 56, 145, 213]. As a fluid drop approaches the interface, the fluid between the drop and the surface is forced out. The drop rests for a period of time before the actual coalescence takes place. During this period the residual film between the drop and the surface is thinned until it becomes thin enough to rupture. The length of this time period is a function of the initial approach velocity and the final rupture thickness. In many cases the coalescence is not complete. The drop drains into the lower fluid and, as it does so, forms a cylindrical column which pinches off to form a secondary drop [22]. This secondary drop goes through the same cycle, which may be repeated a number of times. Coalescence at an interface is one geometric limit of interaction between drops, and the other limit is drops of equal size. MacKay and Mason observed that complete coalescence occurred if the ratio of the diameters of the drops was about 3.5 or less. Above a ratio of 12, the mechanism was identical to that for a flat interface (a ratio of ∞). Park and Crosby [226] have developed a device for producing controlled collisions between pairs of drops in a gaseous medium. In their experiments the drops are uniform in size and can be varied between 0.2 and 1.6 mm. In addition, their velocity, direction of motion, and position at impact can be independently controlled.

Only a first step has been taken in relating the coalescence parameters to the operating variables of specific systems. Shinnar and Church [181] and Shinnar [180] have suggested that coalescence of droplets can be both hindered and aided by turbulent velocity fluctuations. Fluctuations increase the rate of collision and thus aid coalescence; however, if the time necessary for coalescence to occur is long, velocity fluctuations can cause a separation of the droplets before coalescence can occur. The time for coalescence can be increased by the addition of a protective colloid which will hinder coalescence. The equation that describes the minimum diameter for the absence of coalescence (i.e., the diameter below which coalescence will occur) can be obtained in the same way as the breakup equation (17–49). The energy of adhesion of two droplets of equal diameter can be assumed to be $C_3 d_{\min}$. The ratio of kinetic energy to adhesion energy gives the group that is expected to be constant, that is, $\rho_c \overline{u_i(\boldsymbol{x}) u_i(\boldsymbol{x} + \boldsymbol{d})}\, d_{\min}^2 / C_3 = \text{const}$. Using Eq. (17–48), we obtain

$$C_1 \rho_c \epsilon^{2/3} d_{\min}^{8/3} / C_3 = \text{const} \tag{17-93}$$

or

$$d_{\min} = C_4 \rho_c^{-3/8} \epsilon^{-1/4}. \tag{17-94}$$

Howarth [87] has used a different approach to obtain an expression for the collisions per unit time for each drop in a system of uniform drop size. In terms of the volume fraction of the dispersed phase V_f and the rms Lagrangian turbulent velocity fluctuation v', the equation is

$$\omega_c = (24 V_f \sigma v'^2 / d_{\min}^3)^{1/2}.$$

The expression is then modified to give the coalescence frequency:

$$\omega = (24 V_f \sigma v'^2 / d_{\min}^3)^{1/2} \exp(-3 U_c^2 / 4 v'^2), \tag{17-95}$$

where U_c is a critical velocity along the lines of centers between colliding drops. Rietema [164] has raised some question as to the applicability of this equation.

In some cases the parameters descriptive of coalescence have been expressed as a power per unit volume, which can be related to ϵ by modification of Eq. (14-269), and in other cases as functions of the agitator rotational speed N, which can be related to ϵ by the equation [172]

$$\epsilon = KN^3 d_0^2, \qquad (17\text{-}96)$$

where d_0 is the agitator diameter and K is a geometric constant dependent on the vessel and agitator design. Equation (17-96) can be substituted into Eq. (17-49) for the maximum drop size as a result of breakup, to give

$$\rho_c N^2 d_0^{4/3} d_{\max}^{5/3} \sigma^{-1} = C_5, \qquad (17\text{-}97)$$

which should be valid under conditions for which the maximum diameter of the drop is larger than the characteristic length given by Eq. (14-169). Equations (17-93) and (17-96) can be combined to give

$$\rho_c N^2 d_0^{4/3} d_{\min}^{8/3} = C_6. \qquad (17\text{-}98)$$

The value of $d_{\min}$ can be important only if it is less than the $d_{\max}$ given by Eq. (17-97). By equating Eqs. (14-269) and (17-96), one can show that $\rho_c^{2/3} N^2 d_0^{4/3} \sim (P/V)^{2/3}$, which can also be used in either Eqs. (17-97) or (17-98). Equations (14-197) and (14-206) can be combined to show that for isotropic turbulence, $v' \sim \epsilon^{1/3}$ or $v' \sim N$ from Eq. (17-96). By Eq. (17-97), $d_{\max} \sim N^{-6/5}$, so that with these, Eq. (17-95) becomes

$$\omega = C_7 V_f^{1/2} N^{2.8} e^{C_8 N^{-2}}. \qquad (17\text{-}99)$$

A coalescence frequency in a system of uniform drop size is ω. The term ω/n (n is the number of drops in the system) is equal to the volume of dispersed phase coalescing per unit time per volume of dispersed phase. The frequency was used in reference 127, and the latter in references 60 and 139.

Shinnar and Church have performed mixing experiments using protective colloids, and have shown that conditions can be obtained in which no coalescence occurs. They introduced colored drops, and observed that these did not disappear or intermix even after 12 hours of agitation. The data support Eqs. (17-97) and (17-98). The critical speed at which the minimum coalescence diameter becomes greater than the maximum breakup diameter is the break point for the validity of the two equations [164]. Shinnar [180] has further shown that in two geometrically similar mixing tanks, Eq. (17-98) does predict the effect of agitator diameter. Hill [82] derived a relation which was based partly on dimensional analysis and partly on the various empirical and semi-empirical analyses of others. For the case of a fully baffled mixer, where the Froude number can be neglected, the equation takes the form

$$d/d_0 = k(N^2 d_0^3 \rho_c/\sigma)^{-0.6}(P/d_0^2 N^3 \rho_c V)^{-1}(d_0^2 N \rho_c/\mu)^{-m} = k(N_{\text{We},a})^{-0.6}(N_P)^{-1}(N_{\text{Re},a})^{-m}, \qquad (17\text{-}100)$$

where N_P is the power number as defined, P is the power input to the system, and V is the system volume. The power $-m$ depends on the Reynolds-number range, and is zero for high Reynolds numbers ($> 10^4$). In a form analogous to Eq. (17–97) (for the case of high Reynolds numbers) this becomes

$$\rho_c^{-2/5} N^{-3} d_0^{-2} d^{5/3} \sigma^{-1} P^{5/3} V^{5/3} = C_9. \qquad (17\text{–}101)$$

If the restrictions of geometric and dynamic similarity are introduced ($V \sim d_0^3$ and $P \sim N^3 d_0^5 \rho_c$), Eq. (17–101) is identical to (17–97). Thus Eq. (17–101) is more general and its only restriction is to systems that are fully baffled and operate at a high Reynolds number. Rodger et al. [165] measured contact area in a liquid–liquid system of standard mixer design. Their correlation was based on dimensional analysis and was of a rather complicated form. A somewhat less accurate correlation, which was, however, easier to use, was also presented. For a single set of liquids and similar agitators, the equation reduces to one quite similar to (17–98) ($N^{-19/25}$ versus $N^{-3/4}$). Although this would imply that coalescence was controlling, Shinnar and Church point out that such a conclusion would be questionable, since the equation is not likely to be valid for the high concentrations used. Vermeulen et al. [207] made a range of measurements on a liquid–liquid system. An empirical equation was used which was identical to (17–97). The constant C_5 was 0.016. The comparison implies that breakup was controlling. Other information which supports Eq. (17–97) [and thus (17–49)] in agitated systems can be found in the review by Calderbank [18]. References to both liquid–liquid and gas–liquid systems can be found there.

The data of Madden and Damerell [127] showed the $\omega \sim V_f^{1/2}$ dependency of Eq. (17–94) (the range of N was quite small); however, the data of Miller et al. [139] showed $\omega \sim V_f$ to $V_f^{1/2}$, depending on the power input. In all cases the dependency on N (or ϵ^3) could be adequately represented by a 2.5 power [164].

Groothuis and Zuiderweg [60] have dramatically demonstrated the effect of mass transfer on coalescence. The direction of transfer is of utmost importance. When a material that reduces σ is transferred out of the dispersed phase, the rate of coalescence can be increased manyfold (for example a 1.5% addition to the dispersed phase of an agent to be transferred increased the rate by a factor of 20). There is no enhancement of transfer in the other direction.

The analyses and experimental confirmations cited above are concerned with what are essentially steady-state conditions. For a flow system, such conditions may not exist; for example, Rietema [163] found that, even after six hours of vigorous mixing, the size of the particles in a specific liquid–liquid emulsion he studied was not at an equilibrium. Vermeulen et al. [207] found similar results, in that up to two hours were required to obtain steady state for some systems. Finally, the previously cited analysis was concerned with minimum, average, or maximum drop sizes and not with the distribution of sizes that must exist. Sullivan and Lindsey [194] used a light-scattering technique and found the distribution to be binodal. They found that the size associated with the principal peak was apparently of the same order as the Kolmogoroff scale, and thus application of isotropic turbulence would not be valid; however, they found it adequate for scale-up. These authors also found that as much as one hour was necessary to obtain steady state in their system. An interesting direct measurement of the distribution by encapsulation methods has been suggested [128].

A good tabulation and discussion of interfacial areas in gas–liquid systems can be found in Westerterp et al. [217]. Other references of interest in this general area of study are 29, 52, 166, 218, 219, and 229, and the reviews on mixing and agitation [92], drop phenomena and extraction [96], and segregation in liquid-liquid dispersions [164].

PROBLEMS

17-1. Explain why the Rayleigh breakup mechanism is good for the varicose region of drop breakup, and not good for the breakup of bubbles. Compare the expected and observed effect of orifice diameter in all cases.

17-2. The following distribution of particle sizes is typical for fluid catalytic cracking:

Size, μ	Average, %
0–20	25
20–40	30
40–80	35
40+	10
	100

Prepare a histogram and a cumulative plot of the data. Prepare both an arithmetic and a log-normal probability plot of the data. Fit the data with the best straight line and calculate the mean diameter and standard deviation.

17-3. The following size distribution was obtained on a fluid-bed product.

Screen analysis

Mesh	% On
48	45
65	9
150	6
100	10
200	7
270	6
Pan	17
	100

Prepare a histogram and a cumulative plot of the data. Prepare both an arithmetic and a log-normal probability plot of the data. Fit the data with the best straight line and calculate the mean diameter and standard deviation.

17-4. A material is found to have a standard geometric deviation of 3.0 and a geometric mean weight diameter of 0.5 mm. What would be the geometric mean weight diameter of another material of the same specific surface, and a standard geometric deviation of 2.0?

NOTATION FOR CHAPTER 17

b	radius of curvature in Eq. (17-1)
c	velocity of sound, concentration
C_n	constants
C_D	drag coefficient
d	diameter, drop or bubble diameter
d_0	cup, disk, or impeller diameter
D	diffusivity
e	natural base
$f(x)$	distribution function
F	force
F_s	force on the species s
g	gravity
h	film thickness
I_0, I_0'	Bessel functions of the first kind of zero order and its derivative
k	constant, $2\pi/\lambda$ as used in Eqs. (17-11) to (17-13)
K	constant
K_0, K_0'	Bessel functions of the second kind of zero order and its derivative
l	defined in Eq. (17-18)
m	mass
n	number
N	frequency, number, agitator rotational speed
N_c	capacitance number defined by Eq. (17-6)
$N_{\text{Re}}, N_{\text{Fr}}, N_{\text{We}}$	Reynolds, Froude, and Weber numbers, because of the variability of definition, usually defined as used in the text. For example, $N_{\text{We},j}$ is defined by Eq. (17-16), N_{gr}, $N_{\text{Re},s}$, and $N_{\text{We},s}$ are defined in Eq. (17-21), $N_{\text{We},d}$ is defined by Eq. (17-45), $N_{\text{Re},c}$ is defined by Eq. (17-92), etc.
p	pressure
P	power
q	phase velocity used in Eqs. (17-11) through (17-13)
Q	volumetric flow rate
r	radius
r_0	tube radius
R	radius of curvature in Eq. (17-1)
s	order of the Bessel function, ρ_g/ρ_l, surface
t	transformation defined in Eq. (17-58)
T_{xx}	longitudinal surface stress
u_i	velocity fluctuation
U	velocity
U_c	critical velocity
v'	rms Lagrangian velocity fluctuation
V	volume
V_c	chamber volume used in Eq. (17-6)

V_f	volume fraction
x	item
x, y, z	coordinates
$\tilde{x}_d$	defined in Eq. (17-53)

Greek letters

α	angle, wave number defined in Eq. (17-22)
β	growth factor, parameter in Eq. (17-1)
γ	ratio of specific heats, retardation coefficient used in Eq. (17-73)
δ	dispersion coefficient used in Eq. (17-66) and (17-67)
Δ	difference
ϵ	turbulent dissipation, surface shear viscosity used in Eq. (17-84)
θ	angle of contact, coordinate
κ	surface dilational viscosity used in Eq. (17-84)
λ	wavelength
λ_c	critical wavelength
μ	viscosity
μ_s	$\kappa + \epsilon$
ν	kinematic viscosity
π	3.1416 ...
ρ	density
σ	surface tension, standard deviation
σ_g	geometric standard deviation as defined in Eq. (17-60)
$\bar{\bar{\tau}}$	stress tensor
$\bar{\bar{\tau}}_{yx}$	stress tensor component
$\boldsymbol{\tau}$	$(\boldsymbol{N} \cdot \bar{\bar{\tau}})$
ϕ_v	defined in Eq. (17-66)
ω	frequency, rotational speed

Subscripts

b	bubble
c	continuous
d	discontinuous
g	gas
j	jet
l	liquid
L	ligament
max	maximum
min	minimum
n	nozzle or orifice
s	surface
t	terminal

Superscripts

overbar	average

REFERENCES

1. A. W. ADAMSON, *Physical Chemistry of Surfaces*, Interscience, New York (1960).
2. R. S. ALLAN, G. E. CHARLES, and S. G. MASON, *J. Colloid Sci.* **16**, 150 (1961).
3. J. M. ANDREAS, E. A. HAUSER, and W. B. TUCKER, *J. Phys. Chem.* **42**, 1001 (1938).
4. S. P. S. ANDREW, *Intern. Symp. on Distillation, Inst. of Chem. Engrs. (London)*, p. 73 (1960).
5. M. H. I. BAIRD and J. F. DAVIDSON, *Chem. Eng. Sci.* **17**, 87–93 (1962).
6. F. BASHFORTH and H. ADAMS, *The Theories of Capillary Action*, Cambridge University Press, Cambridge (1883).
7. H. BÉNARD, *Rev. gen. Sci. pur. appl.* **11**, 1261, 1309 (1900); see also S. CHANDRASEKHAR, *Hydrodynamic and Hydromagnetic Stability*, Oxford University Press, London (1961).
8. C. A. BENNETT and N. L. FRANKLIN, *Statistical Analysis*, John Wiley & Sons, New York (1954).
9. R. J. BENZING and J. E. MEYERS, *Ind. Eng. Chem.* **47**, 2087–2090 (1955).
10. J. J. BIKERMAN, *Foams: Theory and Industrial Applications*, Reinhold, New York (1953).
11. M. J. BLOCK, *Nature* **178**, 650 (1956).
12. W. N. BOND, *Phil. Mag.* **4**, 889–898 (1927).
13. W. N. BOND and D. A. NEWTON, *Phil. Mag.* **5**, 794–800 (1928).
14. J. BOUSSINESQ, *Ann. chim. Phys.* **29**, 349 (1913).
15. F. P. BRETHERTON, *J. Fluid Mech.* **10**, 166–188 (1961).
16. R. BRÜCHNER, *Naturwissenschaften* **47**, 372 (1960).
17. C. A. BRUNNER and R. LEMLICH, *Ind. Eng. Chem., Fund.* **2**, 297 (1963).
18. P. H. CALDERBANK, *Mixing: Theory and Practice, Vol. 2* (V. Uhl and J. Gray, editors), Chap. 6, Academic Press, New York (1967).
19. P. H. CALDERBANK and I. J. O. KORCHINSKI, *Chem. Eng. Sci.* **6**, 65 (1957).
20. P. H. CALDERBANK and A. C. LOCHIEL, *Chem. Eng. Sci.* **19**, 485–503 (1964).
21. B. T. CHAO, *Phys. Fluids* **5**, 69–79 (1962).
22. G. E. CHARLES and S. G. MASON, *J. Colloid Sci.* **15**, 105, 236 (1960).
23. R. M. CHRISTIANSEN, Ph. D. thesis, University of Pennsylvania, Philadelphia, Pa. (1955).
24. R. M. CHRISTIANSEN and A. N. HIXON, *Ind. Eng. Chem.* **49**, 1017–1024 (1957).
25. J. C. CHU, J. FORGRIEVE, R. GROSSO, S. M. SHAH, and D. F. OTHMER, *AIChEJ* **3**, 16–28 (1957).
26. H. CLARE and A. RADCLIFFE, *J. Inst. Fuel* **27**, 510 (1954).
27. E. G. COCKBAIN and T. S. McROBERTS, *J. Colloid Sci.* **8**, 440 (1953).
28. B. G. COX, *J. Fluid Mech.* **14**, 81–96 (1962).
29. R. L. CURL, *AIChEJ* **9**, 175–181 (1963).
30. P. V. DANCKWERTS, *Ind. Eng. Chem.* **43**, 1460 (1951).
31. R. L. DATTA, D. H. NAPIER, and D. M. NEWITT, *Trans. Inst. Chem. Engrs. (London)* **28**, 14–26 (1950).
32. L. DAVIDSON and E. H. AMICK, JR., *AIChEJ* **2**, 337–342 (1956).

33. C. N. DAVIES, quoted by O. G. SUTTON, *Met. Res. Comm.* **MRP 40** (1942).
34. J. T. DAVIES, *Advances in Chemical Engineering* (T. B. Drew, J. W. Hoopes, Jr., and T. Vermeulen, editors), Vol. IV, pp. 1–50, Academic Press, New York (1963).
35. R. M. DAVIES and G. I. TAYLOR, *Proc. Roy. Soc. (London)* **200A**, 375–390 (1950).
36. N. DOMBROWSKI and R. P. FRASER, *Phil. Trans. Roy. Soc. London* **247A**, 101 (1954).
37. N. DOMBROWSKI and P. C. HOOPER, *Chem. Eng. Sci.* **17**, 291–305 (1962).
38. N. DOMBROWSKI and P. C. HOOPER, *J. Fluid Mech.* **18**, 392–400 (1964).
39. N. DOMBROWSKI and W. R. JOHNS, *Chem. Eng. Sci.* **18**, 203–214 (1963).
40. J. A. DUFFIE and W. R. MARSHALL, JR., *Chem. Eng. Progr.* **49**, 417, 480 (1953).
41. D. T. DUMITRESCU. *ZAMM* **23**, No. 3, 139–149 (1943).
42. H. E. EDGERTON, E. A. HAUSER, and W. B. TUCKER, *J. Phys. Chem.* **41**, 1017 (1937).
43. P. EISENKLAM, *Chem. Eng. Sci.* **19**, 693–694 (1964).
44. E. R. ELZINGA, JR. and J. T. BANCHERO, *AIChEJ* **7**, 394–399 (1961).
45. S. FORDHAM, *Proc. Roy. Soc. (London)* **194A**, 1 (1948).
46. A. S. FOUST, L. A. WENZEL, C. W. CLUMP, L. MAUS, and L. B. ANDERSON, *Principles of Unit Operations*, Appendix B, John Wiley & Sons, New York (1960).
47. R. P. FRASER, N. DOMBROWSKI, and J. H. ROUTLEY, *Chem. Eng. Sci.* **18**, 315–321, 323–337, 339–353 (1963).
48. R. P. FRASER and P. EISENKLAM, *Trans. Inst. Chem. Engrs. (London)* **34**, 294–319 (1956).
49. R. P. FRASER, P. EISENKLAM, N. DOMBROWSKI, and D. HASSON, *AIChEJ* **8**, 672–680 (1962).
50. S. K. FRIEDLANDER, *AIChEJ* **7**, 347 (1961).
51. N. FROESSLING, *Gerlands. Beitr. Geophys.* **52**, 170 (1938).
52. B. GAL-OR and W. RESNICK, *Chem. Eng. Sci.* **19**, 653–661 (1964).
53. F. H. GARNER, C. W. NUTT, and M. F. MOHTADI, *Nature* **175**, 603 (1955).
54. F. H. GARNER and A. H. P. SKELLAND, *Ind. Eng. Chem.* **46**, 1255–1264 (1954).
55. J. H. GIBBONS, G. HOUGHTON, and J. COULL, *AIChEJ* **8**, 274–276 (1962).
56. T. GILLESPIE and E. K. RIDEAL, *Trans. Faraday Soc.* **52**, 173 (1956).
57. H. L. GOLDSMITH and S. G. MASON, *J. Fluid Mech.* **14**, 42–58 (1962).
58. P. GRASSMAN and L. H. LEMAIRE, *Chem.-Ingr.-Tech.* **30**, 450–454 (1958).
59. R. M. GRIFFITH, *Chem. Eng. Sci.* **12**, 198 (1960); **17**, 1057–1070 (1962).
60. H. GROOTHUIS and F. J. ZUIDERWEG, *Chem. Eng. Sci.* **12**, 288 (1960); **19**, 63–66 (1964).
61. R. GUNN, *J. Geophys. Res.* **54**, 383 (1949).
62. R. GUNN and G. D. KINZER, *J. Meteorol.* **6**, 243 (1949).
63. F. C. HAAS, *AIChEJ* **10**, 920–924 (1964).
64. W. L. HABERMAN and R. K. MORTON, *D. W. Taylor Model Basin, R. P. 802* (9/50); *Proc. Am. Soc. Civil Engrs.* **80**, Separ. No. 387 (Jan. 1954).
65. M. J. HADAMARD, *Compt. Rend.* **152**, 1735 (1911); **154**, 107 (1912).
66. A. HAENLEIN, *Forschung Gebiete Ingeniurw.* **2**, 139–149 (1931); translation, *NACA TM 659* (1932).

REFERENCES

67. W. W. HAGERTY and J. F. SHEA, *J. Appl. Mech.* **22**, 509 (1955).
68. A. E. HAMIELEC and A. I. JOHNSON, *Can. J. Chem. Eng.* **40**, 41–45 (1962); *see also* T. J. HORTON, T. R. FRITSCH, and R. C. KINTNER, *ibid.* **43**, 143–146 (1965).
69. A. E. HANDLOS and R. BARON, *AIChEJ* **3**, 127–136 (1957).
70. A. R. HANSON, E. G. DOMICH, and H. S. ADAMS, *Phys. Fluids* **6**, 1070–1080 (1963).
71. W. D. HARKINS and F. E. BROWN, *J. Am. Chem. Soc.* **41**, 499 (1919).
72. T. Z. HARMATHY, *AIChEJ* **6**, 281–288 (1960).
73. D. A. HARPER and R. LEMLICH, *Ind. Eng. Chem., Proc. Res. and Dev.* **4**, 13–16 (1965).
74. P. HARRIOTT, *Can. J. Chem. Eng.* **40**, 60–69 (1962).
75. R. A. HARTUNIAN and W. R. SEARS, *Heat Trans. and Fluid Mechs. Inst. Proc.*, 23, Pasadena, Calif. (1957); *J. Fluid Mech.* **3**, 27–47 (1957).
76. D. A. HAYDON, *Nature* **176**, 839 (1955); *Proc. Roy. Soc. (London)* **243A**, 483 (1958).
77. C. B. HAYWORTH and R. E. TREYBAL, *Ind. Eng. Chem.* **42**, 1174–1181 (1950).
78. E. A. HAUSER, H. E. EDGERTON, B. M. HOLT, and J. T. COX, JR., *J. Phys. Chem.* **40**, 973 (1936).
79. M. F. HEIDMANN and H. H. FOSTER, *NASA TN-D-872* (July 1961).
80. M. F. HEIDMANN, R. J. PRIEM, and J. C. HUMPHREY, *NACA TN-3835* (1957).
81. R. HIGBIE, *Trans. AIChE* **31**, 365–389 (1935).
82. B. A. HILL, *British Chem. Engr.* **6**, 104–106 (1961).
83. J. O. HINZE, *Appl. Sci. Res.* **A1**, 263, 273 (1948).
84. J. O. HINZE, *AIChEJ* **1**, 289–295 (1955).
85. J. O. HINZE and H. MILBORN, *J. Appl. Mech.* **17**, 145 (1950).
86. Z. HOCHSCHWENDER, dissertation, University of Heidelberg, Germany (1919).
87. W. J. HOWARTH, *Chem. Eng. Sci.* **19**, 33–38 (1964).
88. N. T. HSU and B. H. SAGE, *AIChEJ* **3**, 405 (1957).
89. S. HU and R. C. KINTNER, *AIChEJ* **1**, 42–48 (1955).
90. R. R. HUGHES and E. R. GILLILAND, *Chem. Eng. Progr.* **48**, 497–504 (1952).
91. R. R. HUGHES, A. E. HANDLOS, H. D. EVANS, and R. L. MAYCOCK, *Chem. Eng. Progr.* **51**, 557–563 (1955).
92. D. HYMAN, *Advances in Chemical Engineering* (T. B. Drew, J. W. Hoopes, Jr., and T. Vermeulen, editors), Vol. III, pp. 119–202, Academic Press, New York (1962).
93. J. A. W. IZARD, S. D. CAVERS, and J. S. FORSYTH, *Chem. Eng. Sci.* **18**, 467–468 (1963).
94. A. KAMINSKY and J. W. MCBAIN, *Proc. Roy. Soc. (London)* **198A**, 447 (1949).
95. F. W. KEITH and A. N. HIXSON, *Ind. Eng. Chem.* **47**, 258–267 (1955).
96. R. C. KINTNER, *Advances in Chemical Engineering* (T. B. Drew, J. W. Hoopes, Jr., and T. Vermeulen, editors), Vol. IV, pp. 51–94, Academic Press, New York (1963).
97. R. C. KINTNER, J. T. HORTON, R. E. GRAUMANN, and S. AMBERKAR, *Can. J. Chem. Eng.* **39**, 235–241 (1961).
98. A. J. KLEE and R. E. TREYBAL, *AIChEJ* **2**, 444–447 (1956).
99. A. N. KOLMOGOROFF, *Dokl. Akad. Nauk SSSR (NS)* **66**, 825–828 (1949).

100. J. KOTTLER, *J. Franklin Inst.* **250**, 339, 419 (1950).
101. H. KROEPELIN and H. J. NEUMANN, *Naturwissenschaften* **43**, 347 (1956); **44**, 304 (1957).
102. R. KRONIG and J. C. BRINK, *Appl. Sci. Res.* **A2**, 142 (1951).
103. A. D. K. LAIRD and D. CHRISHOLM, *Ind. Eng. Chem.* **48**, 1361–1364 (1956).
104. H. LAMB, *Hydrodynamics* (sixth edition), Art. 274, Cambridge University Press; reprinted by Dover Publications, New York (1945).
105. W. R. LANE, *Ind. Eng. Chem.* **43**, 1312–1317 (1951).
106. W. R. LANE and H. L. GREEN, *Surveys in Mechanics* (G. K. Batchelor and R. M. Davies, editors), pp. 162–215, Cambridge University Press, New York (1956).
107. C. LAPPLE, *Fluid and Particle Mechanics* University of Delaware, Newark, Del. (1956).
108. J. O. LAWS, *Trans. Am. Geophys. Un.* **22**, 709 (1941).
109. I. LEIBSON, E. G. HOLCOMB, A. G. CACASO, and J. J. JACMIC, *AIChEJ* **2**, 296–306 (1956).
110. P. LENARD, *Meteorol. Zeits.* **21**, 249 (1904).
111. P. LENARD, *Ann. Physik* **65**, 629 (1921).
112. J. H. LEONARD and G. HOUGHTON, *Chem. Eng. Sci.* **18**, 133–142 (1963).
113. R. A. LEONARD and R. LEMLICH, *AIChEJ* **11**, 18–29 (1965).
114. M. LEVA, *Tower Packing and Packed Tower Design*, U. S. Stoneware Co. Akron, Ohio (1951).
115. V. G. LEVICH, *Physicochemical Hydrodynamics*, Prentice-Hall, Englewood Cliffs, New Jersey (1962).
116. J. B. LEWIS, *Trans. Inst. Chem. Engrs. (London)* **31**, 323, 325 (1953).
117. J. B. LEWIS, *Chem. Eng. Sci.* **3**, 248, 260 (1954); **8**, 295 (1958).
118. J. B. LEWIS, I. JONES, and H. C. R. PRATT, *Trans. Inst. Chem. Engrs. (London)* **29**, 144 (1951).
119. J. B. LEWIS and H. R. C. PRATT, *Nature* **171** 1155 (1953).
120. W. LICHT and G. S. R. NARASIMHAMURTY, *AIChEJ* **1**, 366–373 (1955).
121. C. C. LIN, *Theory of Hydrodynamic Stability*, Cambridge University Press, Cambridge (1955).
122. H. LINDE, *Fette u. Seifen, Anstrichmittel* **60**, 826, 1053 (1958); *Monatsber. Deutsch. Akad. Wiss*, **1**, 699 (1959); *Colloid-Zh.* **22**, 333 (1960).
123. A. C. LOCHIEL and P. H. CALDERBANK, *Chem. Eng. Sci.* **19**, 471–484 (1964); A. C. LOCHIEL, *Can. J. Chem. Eng.* **43**, 40–44 (1965).
124. J. W. MCBAIN and T. M. WOO, *Proc. Roy. Soc. (London)* **163A**, 182 (1937).
125. R. V. MCDOWELL and J. E. MEYERS, *AIChEJ* **2**, 384–388 (1956).
126. G. D. M. MACKAY and S. G. MASON, *Can. J. Chem. Eng.* **41**, 203–212 (1963).
127. A. J. MADDEN and G. L. DAMERELL, *AIChEJ* **8**, 233–239 (1962).
128. A. J. MADDEN and B. J. MCCOY, *Chem. Eng. Sci.* **19**, 506–507 (1964).
129. R. H. MAGARVEY and B. W. TAYLOR, *J. Appl. Phys.* **27**, 1129–1135 (1956).
130. J. F. MAHONEY, JR. and L. A. WENZEL *AIChEJ* **9**, 641–645 (1963).
131. C. G. MAIER, *U.S. Bur. Mines Bull. 260* (1927).

132. E. MANEGOLD, *Schaum*, Strassenbau, Chemie, and Technik Verlagsgesellschaft m.b.H., Heidelberg (1953).
133. N. G. MAROUDAS and H. SAWISTOWSKI, *Chem. Eng. Sci.* **19**, 919–931 (1964); *Nature* **188**, 1186 (1960).
134. W. R. MARSHALL, JR., *Atomization and Spray Drying, Chem. Eng. Progr. Monograph Series No. 2*, **50** (1954).
135. A. C. MERRINGTON and E. G. RICHARDSON, *Proc. Phys. Soc. (London)* **59**, 1 (1947).
136. M. V. MHATRE and R. C. KINTNER, *Ind. Eng. Chem.* **51**, 865 (1959); see also G. ASTARITA and G. APUZZO, *AIChEJ* **11**, 815–820 (1965).
137. C. C. MIESSE, *Proc. Gas Dynamics Symp. on Aerothermo-Chem*, pp. 7–26 (1956).
138. D. N. MILLER, *Ind. Eng. Chem.* **56** No. 10, 18–27 (1964).
139. R. S. MILLER, J. L. RALPH, R. L. CURL, and G. D. TOWELL, *AIChEJ* **9**, 196–202 (1963).
140. F. O. MIXON and J. J. CARBERRY, *Chem. Eng. Sci.* **13**, 30 (1960).
141. D. W. MOORE, *J. Fluid Mech.* **16**, 161–176 (1963); **6**, 113–130 (1959); **23**, 749–766 (1965).
142. R. A. MUGELE and H. D. EVANS, *Ind. Eng. Chem.* **43**, 1317–1324 (1951).
143. A. B. NEWMAN, *Trans. AIChE* **27**, 203 (1931).
144. P. C. NEWMAN and P. F. WHELAN, *Nature* **169**, 326–327 (1952).
145. L. E. NIELSON, R. WALL, and G. ADAMS, *J. Colloid Sci.* **13**, 441 (1958).
146. S. NUKIYAMA and Y. TANASAWA, *Trans. Soc. Mech. Engrs. (Japan)* **4**, No. 14, 86, No. 15, 138 (1938); **5**, No. 18, 63–75 (1939); **6**, No. 22, II 7, No. 23, II 8 (1940).
147. G. OHNESORGE, *Z. angew. Math. Mech.* **16**, 355 (1936).
148. D. R. OLANDER and L. B. REDDY, *Chem. Eng. Sci.* **19**, 67–73 (1964).
149. R. B. OLNEY, *AIChEJ* **10**, 827–835 (1964).
150. R. B. OLNEY and R. S. MILLER, *Modern Chemical Engineering* (A. Acrivos, editor), Chap. 3, Reinhold, New York (1963).
151. A. ORELL and J. W. WESTWATER, *Chem. Eng. Sci.* **16**, 127 (1961); *AIChEJ* **8**, 350–356 (1962).
152. C. ORR, JR. and J. M. DALLAVALLE, *Fine Particle Measurement*, Macmillan, New York (1959).
153. J. R. A. PEARSON, *J. Fluid Mech.* **4**, 489 (1958).
154. M. POPOV, *Acad. Rép. Populare Romane, Rev. de Mécanique Appliquée (Bucarest)* **1**, No. 1, 71–88 (1956); translation, *NASA-TT-F-65* (1965).
155. A. T. POPOVICH, R. E. JERVIS, and O. TRASS, *Chem. Eng. Sci.* **19**, 357–365 (1964).
156. O. E. POTTER, *Chem. Eng. Sci.* **6**, 170 (1957).
157. W. E. RANZ, *On Spray and Spraying: A Survey of Spray Technology*, Dept. Chem. Eng., Pennsylvania State University, University Park, Pa. (March, 1956).
158. W. E. RANZ and W. M. DREIER, JR., *Ind. Eng. Chem., Fund.* **3**, 53–60 (1964).
159. W. E. RANZ and W. R. MARSHALL, JR., *Chem. Eng. Progr.* **48**, 141, 173, 247 (1952).
160. LORD RAYLEIGH, *Proc. London Math. Soc.* **10**, 4 (1878/9).
161. LORD RAYLEIGH, *Phil. Mag.* **34**, 177 (1892).
162. LORD RAYLEIGH, *Phil. Mag.* **48**, 321 (1899).
163. K. RIETEMA, *Chem. Eng. Sci.* **8**, 103–112 (1958).

164. K. RIETEMA, *Advances in Chemical Engineering* (T. B. Drew, J. W. Hoopes, Jr., and T. Vermeulen, editors), Vol. V, pp. 237–302, Academic Press, New York (1964).
165. W. A. RODGER, V. G. TRICE, JR., and J. H. RUSHTON, *Chem. Eng. Progr.* **52**, 515–520 (1956); **53**, 153M (1957).
166. F. RODRIGUEZ, L. C. GROTZ, and D. L. ENGLE, *AIChEJ* **7**, 663–665 (1961).
167. P. ROSIN and E. RAMMLER, *Z. Ver. deut. Ing.* **71**, 1 (1927); *J. Inst. Fuel* **7**, 29 (1933).
168. E. E. RUBIN and E. L. GADEN, *New Chemical Engineering Separation Techniques* (H. M. Schoen, editor), Chap. 5, Interscience, New York (1962).
169. E. RUCKENSTEIN, *Rev. Chim.* **6**, 221 (1961).
170. E. RUCKENSTEIN, *Chem. Eng. Sci.* **19**, 505–506 (1964); *Inzh. Fiz. Zhur.* **7**, No.7, 116–120 (1964); translation, *Inter. Chem. Eng.* **5**, 88–90 (1965).
171. E. RUCKENSTEIN and C. BERBENTE, *Chem. Eng. Sci.* **19**, 329–347 (1964).
172. J. H. RUSHTON, E. W. COSTICH, and H. J. EVERETT, *Chem. Eng. Progr.* **46**, 395–404, 467–476 (1950).
173. D. P. RYBCZYŃSKI, *Bull. intern. acad. sci., Cracovie* **A403**, 40 (1911).
174. P. G. SAFFMAN, *J. Fluid Mech.* **1**, 249–275 (1956).
175. R. SATAPATHY and W. SMITH, *J. Fluid Mech.* **10**, 561–570 (1961).
176. R. S. SCHECHTER and R. W. FARLEY, *Can. J. Chem. Eng.* **41**, 103–107 (1963); *Brit. Chem. Eng.* **8**, 37–42 (1963).
177. L. E. SCRIVEN, *Chem. Eng. Sci.* **12**, 98–108 (1960).
178. L. E. SCRIVEN, and C. V. STERNLING, *Nature* **187**, 186–188 (1960).
179. T. K. SHERWOOD and J. C. WEI, *Ind. Eng. Chem.* **49**, 1030–1034 (1957).
180. R. SHINNAR, *J. Fluid Mech.* **10**, 259–275 (1961).
181. R. SHINNAR and J. M. CHURCH, *Ind. Eng. Chem.* **52**, 253 (1960).
182. S. SIDEMAN *Chem. Eng. Sci.* **19**, 426 (1964).
183. W. SIEMES, *Chem. Ing. Tech.* **26**, 479–496, 614–630 (1954).
184. W. SIEMES and E. BORCHERS, *Chem. Eng. Sci.* **12**, 77–87 (1960).
185. K. SIGWART and H. NASSENSTEIN, *Naturewissenschaften* **42**, 458 (1955); *Ver. Deutsch. Ing. Zeit.* **98**, 453 (1956).
186. A. H. P. SKELLAND and R. M. WELLEK, *AIChEJ* **10**, 491–496 (1964).
187. C. A. SLEICHER, JR., *AIChEJ* **8**, 471–477 (1962); H. I. PAUL and C. A. SLEICHER, JR., *Chem. Eng. Sci.* **20**, 57–59 (1965).
188. S. W. J. SMITH and H. MOSS, *Proc. Roy. Soc. (London)* **93A**, 373 (1917).
189. K. E. SPELLS and S. BAKOWSKI, *Trans. Inst. Chem. Engrs. (London)* **28**, 38–51 (1950).
190. K. E. SPELLS and S. BAKOWSKI, *Trans. Inst. Chem. Engrs. (London)* **30**, 189–196 (1952).
191. K. E. SPELLS, *Proc. Roy. Soc. (London)* **65B**, 541 (1952).
192. H. B. SQUIRE, *Brit. J. Appl. Phys.* **4**, 167–169 (1953).
193. C. V. STERNLING and L. E. SCRIVEN, *AIChEJ* **5**, 514–523 (1959); *J. Fluid Mech.* **19**, 321–340 (1964); see also K. A. SMITH, *ibid.* **24**, 401–414 (1966).
194. D. M. SULLIVAN and E. E. LINDSEY, *Ind. Eng. Chem., Fund.* **1**, 87–93 (1962).

195. S. L. SULLIVAN, JR., B. W. HARDY, and C. D. HOLLAND, *AIChEJ* **10**, 848–854 (1964).
196. T. TADAKI and S. MAEDA, *Kagaku Kōgaku*, **27**, 147 (1963); translation, *Chem. Eng. (Japan)* **1**, 55–60 (1963).
197. T. TADAKI and S. MAEDA, *Kagaku Kōgaku*, **27**, 402 (1963); translation, *Chem. Eng. (Japan)* **1**, 106–109 (1963).
198. Y. TANASAWA, S. SASAKI, and N. NAGAI, *Tech. Rep. Tohoko Univ.* **22**, 73 (1957).
199. R. TATE and W. R. MARSHALL, JR., *Chem. Eng. Progr.* **49**, 169, 226 (1953).
200. G. I. TAYLOR, Ministry of Supply Rept. *AC 10647/Phys. C69* (1949).
201. G. THORSEN and S. G. TERJESEN, *Chem. Eng. Sci.* **17**, 137–148 (1962).
202. E. TYLER, *Phil. Mag.* **16**, 504 (1933).
203. E. TYLER and F. WATKIN, *Phil. Mag.* **14**, 849 (1932).
204. S. UNO and R. C. KINTNER, *AIChEJ* **2**, 420–425 (1956).
205. R. S. VALENTINE and W. J. HEIDEGER, *Ind. Eng. Chem., Fund.* **12**, 242–244 (1963).
206. D. W. VAN KREVELEN and P. J. HOFTIJZER, *Chem. Eng. Progr.* **46**, 29–35 (1950).
207. T. VERMEULEN, G. M. WILLIAMS, and G. E. LANGLOIS, *Chem. Eng. Progr.* **51**, 85F (1955).
208. M. S. VOLYNSKI, *Compt. Rend. (Doklady) Acad. Sci. USSR* **62**, 301 (1948).
209. J. K. WALTERS and J. F. DAVIDSON, *J. Fluid Mech.* **12**, 408–416 (1962); **17**, 321–336 (1963).
210. D. M. WARD, O. TRASS, and A. I. JOHNSON, *Can. J. Chem. Eng.* **40**, 164 (1962).
211. F. H. WARD and L. H. BROOKS, *Trans. Faraday Soc.* **48**, 1124 (1952).
212. M. W. WARSHAY, E. BOGUSZ, M. JOHNSON, and R. C. KINTNER, *Can. J. Chem. Eng.* **37**, 29–36 (1959).
213. T. WATANABE and M. KUSUI, *Bull. Chem. Soc. Japan* **31**, 236 (1958).
214. S. WHITAKER, *Ind. Eng. Chem., Fund.* **3**, 132–142 (1964).
215. E. T. WHITE and R. H. BEARDMORE, *Chem. Eng. Sci.* **17**, 351–361 (1962).
216. C. WEBER, *Z. angew. Math. Mech.* **11**, 136–154 (1931).
217. K. R. WESTERTERP, L. L. VAN DIERENDONCK, and J. A. DE KRAA, *Chem. Eng. Sci.* **18**, 157–176 (1963).
218. I. YAMAGUCHI, S. YABUTA, and S. NAGATA, *Kagaku Kōgaku* **27**, 576 (1963); translation, *Chem. Eng. (Japan)* **2**, 26–30 (1964).
219. F. YOSHIDA and Y. MIURA, *Ind. Eng. Chem., Proc. Res. Dev.* **2**, 263–268 (1963).
220. M. ZELEN and N. C. SEVERO in *Handbook of Mathematical Functions* (M. Abramowitz and I. A. Stegun, editors), National Bureau of Standards, Washington, D.C. (1964).
221. J. L. L. BAKER and B. T. CHAO, *AIChEJ* **11**, 268–273 (1965).
222. B. K. C. CHAN and R. G. H. PRINCE, *AIChEJ* **11**, 176 (1965).
223. J. E. GWYN, E. J. CROSBY, and W. R. MARSHALL, JR., *Ind. Eng. Chem., Fund.* **4**, 204–208 (1965).
224. B. D. MARSH and W. J. HEIDEGER, *Ind. Eng. Chem., Fund.* **4**, 129–133 (1965).
225. J. W. MILLS, M. S. thesis in Chem. Eng., The Ohio State University, Columbus, Ohio (1965).
226. R. W. PARK and E. J. CROSBY, *Chem. Eng. Sci.* **20**, 39–45 (1965).

227. J. A. REDFIELD and G. HOUGHTON, *Chem. Eng. Sci.* **20**, 131–139 (1965).
228. R. R. SCHROEDER and R. C. KINTNER, *AIChEJ* **11**, 5–8 (1965).
229. L. A. SPIELMAN, and O. LEVENSPIEL, *Chem. Eng. Sci.* **20**, 247–254 (1965).
230. S. WINNILOW and B. T. CHAO, *Phys. Fluids* **9**, 50–61 (1966).
231. S. S. BUPARA, PH. D. thesis in Chem. Eng., University of Minnesota, Minneapolis (1964).

CHAPTER 18

MULTIPHASE PHENOMENA III: SOLIDS-FLUID FLOW

18-1 INTRODUCTION

Operations such as slurry transport, settling, fluidization, pneumatic transport, etc., involve the simultaneous flow of a gas and a solid or a liquid and a solid. In these various operations, the superimposed problems of mass transfer, heat transfer, and kinetics are often of paramount interest. The field of solids–fluid flows covers a number of important practical problems, but before we deal with the subject in detail, a few introductory comments will be helpful.

When the solid and the fluid move together as a homogeneous mass, the system is usually treated as a problem in non-Newtonian flow (Chapter 15). However, the solid and fluid do not always do this, and the result is a net slip of the two phases. This does not mean that there is slip at any boundary, but rather that the average velocity of the fluid through the system is different from the average velocity of the solids. The packed bed is a good illustration of this condition.

Broadly speaking, multiparticle systems are either homogeneous or nonhomogeneous. For vertical systems in which there is no forced feed of the solids, the limits of homogeneous conditions are formed by flow through a packed bed on the one hand, and dilute phase transport on the other. (Dilute phase transport can be described as a low concentration of noninteracting solids carried along by a gas or liquid stream.) For a solid–gas system, in which the density difference is large, homogeneous conditions do not normally exist between these limits. However, if a liquid flows upward through a solid whose density does not differ greatly from its own, smooth motion, called *particulate fluidization*, can occur over the entire range. As the liquid flows between the particles, a pressure drop occurs. The drop in pressure increases with the velocity of the liquid, and a critical point is eventually reached at which the upward drag just equals the force of gravity on the particles. (This point is called the *minimum fluidization velocity*, U_{mf}.) A further increase puts the system out of balance and the particles rise, thus increasing the voids, which in turn cause the velocity to decrease. A new balance position is obtained. In this particulate type of fluidization, further increase in the velocity will cause a further bed expansion, and the mode will continue until the particles are all lifted out of the system container. If a distribution of particle sizes were involved, considerable classification would occur. Settling is a particulate type of system in which, at one time, the existing conditions range from the completely dilute form to solids with stagnant fluid in the void spaces. Homogeneous flow for horizontal systems is limited by the capacity of the turbulent motion of the fluid to prevent settling of solids to the bottom of the pipe by gravity. It is thus enhanced by lower weight loadings of solids and by higher fluid velocities.

When a gas is passed upward through a solid whose density is quite different from its own, the bed initially expands in the same manner as in particulate motion; however, a point is reached at which the bed becomes very mobile. The upper surface is analogous to a liquid surface, in that wave motion can occur, heavy objects can sink, and light objects can float. This is called *aggregate motion*, and is nonuniform when compared with the particulate counterpart. As the gas velocity is increased in aggregate fluidization, the volume of the dense phase remains about the same, with the excess gas passing through the fluid bed as large bubbles.

Fluidization has several advantages over static bed systems. The small size of the particles used in fluidized beds results in a large surface area, which in turn gives high rates of surface reaction, heat transfer, and mass transfer. The smallness of the particles also means that there is little resistance to diffusion within the particle. The rapid mixing in a fluidized bed results in a uniform temperature and in a high rate of heat transfer from a solid surface to the system materials. This is particularly advantageous when highly exothermic or endothermic reactions are being used. The ease with which solids are transported is important when there are cases in which catalyst regeneration is necessary, as in fluid catalytic cracking. The available methods of solids transport make it easy to introduce fresh solids into a system or to remove solid products from it. There are difficulties, however. The fluid bed may approximate a well-mixed tank, and thus all the advantages of countercurrent or stagewise operation could be lost. This may be quite important in a kinetic analysis. Several multistage fluid bed systems have been built, but their expense can be justified only if other methods are inadequate, as in the recovery of uranium by the fluorine process. Attrition of the solid due to the violent mixing motion is important, and recovery of the fines must be made. As in pipe flow, erosion by the moving solid can be a major problem.

Nonhomogeneous flow often occurs in horizontal systems, and is characterized by an accumulation of solids along the bottom of the pipe. It is very important that we know what type of solids–fluid flow will occur under specific conditions, especially in the transport of solids, and considerable work has been done in this area. In vertical systems of solids and fluids, either type of fluidization (aggregate or particulate) may exist; under some conditions, for example, aggregate fluidization is observed with lead shot and water and the area above an aggregate fluidized bed can be considered as particulate. Light plastic beads fluidize particulately in high-pressure (and thus relatively high-density) gases. The type of fluidization depends on the properties of the particles and the fluid, and of course on the operating conditions. An early suggestion was made by Wilhelm and Kwauk [315], who thought that the Froude number might be the controlling group. This has been found to be qualitatively correct, but later experimental work has shown that modifications are necessary. This and other aspects of the transition problem will be taken up in the section on stability, or the transition analysis.

18-2 PARTICLE BEHAVIOR IN DILUTE SYSTEMS

A very dilute system is one in which the particles are far enough removed from one another so that they may be treated as individual particles, each individually contributing to the overall character of the flow. The nature of the fluid surrounding the particles, either laminar or turbulent, has a marked influence on the resulting motion. There is of course a whole range of dilutions; at slightly more concentrated conditions, simple particle-to-particle interactions occur, and it is hoped that they might be treated in an analytical manner.

18-2. A Motion Which is, in Effect, Single-Particle Motion

(a) **Laminar flow conditions.** Individual particle motions in very slow flow and under steady-state conditions were already considered in some detail in Chapter 8.

As discussed by Hinze [124], the equation of motion for the Stokes flow of a single *accelerating* particle in a fluid at rest was obtained by Basset [17], Boussinesq [28], and Oseen [213], and subsequently modified by Tchen [285] to account for a variable velocity in the fluid. This was accomplished by rederiving the equation for a particle whose velocity is $U_d - U_c$ in a fluid at rest, and then superposing on the system a uniform velocity U_c. The final form, valid for the slow motion of a spherical particle, is

$$\frac{\pi}{6} d^3 \rho_d \frac{dU_d}{dt} = 3\pi \mu_c d(U_d - U_c) - \frac{\pi}{6} d^3 \frac{dp}{dz} + \frac{1}{2} \frac{\pi}{6} d^3 \rho_c \left(\frac{dU_d}{dt} - \frac{dU_c}{dt} \right)$$
$$+ 3\pi \mu_c d \left(\frac{d}{2\sqrt{\pi \nu_c}} \int_{t_0}^{t} \frac{dU_d/dt' - dU_c/dt'}{\sqrt{t - t'}} dt' \right) - \frac{\pi}{6} d^3 g(\rho_d - \rho_c), \quad (18\text{-}1)$$

where a downward velocity is taken as positive, and U_c is the fluid velocity in the area of the particle but not close enough to be affected by the particle. For steady state in an infinite medium at rest with no pressure gradient in the fluid, the equation reduces to Stokes' law, which, in the notation being used here, is

$$U_t = d^2(\rho_d - \rho_c)g/18\mu_c, \quad (18\text{-}2)$$

where U_t is of course the terminal velocity. In terms of a drag coefficient, this becomes

$$U_t^2 = 4d(\rho_d - \rho_c)g/3\rho_c C_D, \quad (18\text{-}3)$$

where $C_D = 24/N_{\text{Re},d}$ for Stokes' law. The various terms in Eq. (18-1) are: on the left, the force required to accelerate the particle; on the right, (1) the viscous drag from Stokes' law, (2) the effect of a pressure gradient in the fluid, (3) the force associated with the acceleration of the apparent mass of the particle relative to the fluid, (4) the effect of acceleration on the viscous drag, and (5) the force of gravity. Equation (18-1) is the starting point for a number of analyses of solids–fluid flow, as will be discussed later.

For the fluid under uniform laminar conditions, but with a particle Reynolds number greater than for the slow-flow region, the standard drag curves for spheres, disks, etc., can provide much of the necessary information. However, Christiansen and Barker [45] have pointed out that at higher Reynolds numbers where rotation and oscillation occur, the drag coefficient is a function not only of the particle Reynolds number, but also of the density ratio and a shape factor, or parameter. The extensive review by Torobin and Gauvin [296] covers most of the detailed problems one might encounter: Single particles, ideal model solutions, the nature of the wake, the effect of an accelerative motion, rotation, roughness, shape, and fluid turbulence are all treated. Zenz and Othmer [322] devote a chapter to similar problems.

Eichorn and Small [66] considered the effect of a single spherical particle ($80 < N_{\text{Re},d} = dU_c\rho_c/\mu_c < 250$) in laminar pipe flow. Here both drag forces and lateral or lift forces exist in addition to particle rotation, because of the velocity gradient in the system and the resulting asymmetric distribution of surface pressure. Standard drag-coefficient plots would not be expected to apply because of the possible proximity of the particle to the wall and the gradients encountered. Because of the difficulty of obtaining accurate data, adequate correlations for these conditions are not available; however, these preliminary experimental results help to indicate the expected dependency of the drag on the particle Reynolds number, on the ratio of particle to pipe diameter, and on a shear parameter.

Segré and Silberberg [249] considered the Poiseuille flow of an initially uniform dilute suspension of spheres in an equal-density fluid at low pipe Reynolds numbers (< 30). They found that the lateral forces (or in this case the radial forces) caused the particles to tend toward an equilibrium position at about r/r_0 of 0.6, resulting in an inhomogeneous distribution of particles. This means that particles have a tendency to move away from the region of the center line, as well as away from the region of the wall. While it is believed that the radial force is associated with inertial effects in the fluid, a clear mechanistic picture is not available.

Happel and Brenner [112] have considered, in theory, the possibility of particle location anywhere in a laminar flow field. Their work also extended to the problems of noninteracting, and thus dilute, multiparticle systems. In their analysis, the equations of motion are simplified and the method of reflection is applied, the latter involving a series of approximate solutions which satisfy the equations and partially satisfy the boundary conditions. The drag is given by Stokes' law, and the pressure drop due to each particle depends solely on its location and velocity. Each particle moves in an axial direction without collisions and without encountering radial forces. This assumption comes from neglecting

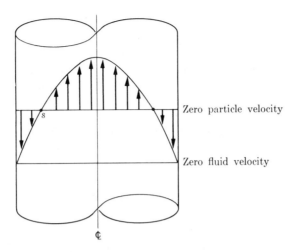

Fig. 18–1. *A single particle under vertical laminar-flow conditions.*

the inertial terms when using the slow-motion approximation. For this model the variables are the distribution of the particles, the terminal velocity of the particles, and the fluid velocity. A balance may be set up when the spheres are free to move in an axial direction, but the fluid velocity is laminar and varies across the tube. The result may be pictured as in Fig. 18–1. At some point s on the radius, the particle velocity is zero. At $r < s$, the particles have an upward motion, and at $r > s$, they have a downward motion. At the outside, gravitational forces exceed those due to friction, since the velocity is low in the neighborhood of the wall. There is no net particle transport for particulate fluidization; the upflow balances the downflow. Thus the analysis describes down-moving particles near the wall, which of course results in back-mixing. The frequency of recirculation for a particle is

$$f = 1/t = U_t(U_\mathbb{C} - U_t)/2LU_\mathbb{C}, \qquad (18\text{--}4)$$

where L is the bed length, U_t is the terminal velocity of the particles, and $U_\mathbb{C}$ is the velocity at the center line. Holdup time can be expressed as $\theta = L/U_{\text{ave}} = 2L/U_\mathbb{C}$, where U_{ave} is the average fluid velocity. The product

$$f\theta = (U_t/U_\mathbb{C})(1 - U_t/U_\mathbb{C}) = \phi(U_t/U_\mathbb{C}) \qquad (18\text{--}5)$$

would be a dimensionless parameter to correlate back-mixing. Although this idealized model is probably valid for very dilute systems, it is not achieved in cases of real particulate fluidization for several reasons: For one thing, the fluid velocity distribution is greatly modified by the presence of the solids and is generally quite flat [37] (in fact, in some cases a higher velocity is observed at the wall than at the center of the column); for another, circulation patterns are set up which are not symmetrical [39].

(b) **Turbulent flow conditions.** When stream conditions are such that the flow is inherently turbulent, the interactions between the particles and the fluid are far more complex than in the case of laminar flow. Consequently, this situation is generally less amenable to detailed analysis. The starting point for such an analysis is the single particle in a turbulent field. Torobin and Gauvin have provided a thorough review of the effects of turbulence on the drag coefficient [296]. In addition, they have described a rather unique facility for making such studies [297]. Both the relative intensity of turbulence and the particle Reynolds number are important factors in determining the drag level. Both reduction and increase of drag have been reported.

For very dilute turbulent flow systems, the character of the particle motion depends on its size relative to the scale of turbulence of the flow. If the particle is large compared with this scale of turbulence, its motion is not greatly affected by fluctuations in velocity, and the particle acts as a disturbing element in the flow field. Depending on its size and the relative density, it may well respond to some of the slower, larger-scale fluid motions. Friedlander [82] has suggested that an Eulerian formulation would be most appropriate for such a case. At the other extreme, if the particle is very small compared with the smallest eddy size (i.e., if the system is composed of colloidal or small, equal-density particles), the particle follows the detailed motion of the fluid, and a Lagrangian formulation would be appropriate. Between these two extremes lies the region in which the particle is most likely to have a mean velocity relative to the fluid and a turbulent velocity in response to the turbulence of the fluid. Here some sort of mixed Lagrangian and Eulerian formulation might be used.

In extending the theory to deal with the effect of turbulence on a particle in a very dilute system, we must make certain assumptions and modifications of Eq. (18–1). Tchen has suggested some simplifications which could allow a solution to be obtained. Corrsin and Lumley [50] have further clarified the nature of the assumptions that would be appropriate to the turbulence problem. First of all, Tchen assumed that: (1) the turbulence itself was homogeneous, infinite in extent, and steady, (2) the particle was small enough so that its motion relative to the fluid could be given by Stokes' law, (3) the particle was smaller than the shortest wavelength of the turbulence, and (4) the same fluid remained with the solid particle during its motion. Hinze [124] suggests that the last assumption is the weakest and could be valid only if the densities of the particle and the fluid were not too different.

Using the Lagrangian view, we assume that the coordinate system is fixed with the particle; however, the fluid velocity associated with the turbulence is a vector quantity relative to the particle (or the coordinate system), and Eq. (18–1), when it is rewritten, will have to reflect this change. The pressure gradient in the second term on the right can be obtained from the Navier-Stokes equation for slow motion, Eq. (8–1) [50]. In the present notation, this is

$$-\boldsymbol{\nabla} p = \rho_c (D\boldsymbol{U}_c/Dt) - \mu_c \nabla^2 \boldsymbol{U}_c. \tag{18–6}$$

The difference between the time derivatives in the third and fourth terms on the right must also be modified [50], since a Lagrangian view is being used. The term dU_d/dt stays the same, since it is fixed with respect to our coordinate system which is moving with the particle. The term dU_c/dt must be modified to reflect the movement with the particle, and thus is replaced by the substantial derivative moving with the coordinate system:

$$\partial U_c/\partial t + (U_d \cdot \nabla)U_c. \tag{18-7}$$

Note that this is not the same as

$$DU_c/Dt = \partial U_c/\partial t + (U_c \cdot \nabla)U_c \tag{18-8}$$

used in Eq. (18-6), which must reflect the fluid motion in the vicinity of the particle. With the modifications of Eqs. (18-6) through (18-8), Eq. (18-1) becomes

$$\frac{\pi}{6} d^3 \rho_d \frac{dU_d}{dt} = 3\pi \mu_c d(U_d - U_c) - \frac{\pi}{6} d^3 \rho_c \left(\frac{DU_c}{Dt} - \nu_c \nabla^2 U_c \right)$$
$$+ \frac{1}{2} \frac{\pi}{6} d^3 \rho_c \left[\frac{dU_d}{dt} - \frac{\partial U_c}{\partial t} - (U_d \cdot \nabla)U_c \right]$$
$$+ 3\pi \mu_c d \left[\frac{d}{2\sqrt{\pi \nu_c}} \int_{t_0}^{t} \frac{dU_d/dt' - \partial U_c/\partial t' - (U_d \cdot \nabla)U_c}{\sqrt{t - t'}} dt' \right] - \frac{\pi}{6} d^3 g(\rho_d - \rho_c), \tag{18-9}$$

which is now the particle motion equation from a Lagranian view. This can be expressed in cartesian tensor notation by replacing U_d and U_c by U_{di} and U_{ci}, respectively, noting that

$$(U_d \cdot \nabla)U_c = U_{dj}(\partial U_{ci}/\partial x_j), \tag{18-10}$$

and that g is g_i, the acceleration associated with external forces (usually only gravity and usually neglected). Corrsin and Lumley further pointed out that under certain conditions it might be acceptable to approximate $(U_d \cdot \nabla)U_c$ by $(U_c \cdot \nabla)U_c$, and thus make Eqs. (18-7) and (18-8) identical. Hinze rearranged Eq. (18-9) to establish the conditions necessary for the nonlinear terms to be neglected and for the equation to remain first order and to depend only on time (that is, $\nu_c \nabla^2 U_c = 0$). He found that the nonlinear terms could be neglected if

$$\partial (dU_c/\nu_c)/\partial (x/d) \ll 1, \tag{18-11}$$

and that the equation will remain first order and depend only on time if

$$\frac{dU_d/\nu_c}{\partial^2 (dU_c/\nu_c)/\partial (x/d)^2} \ggg 1. \tag{18-12}$$

Under these conditions, which would be a considerable idealization for turbulent flow, Tchen assumed a stationary situation in which the particle was suspended in a sinusoidally oscillating infinite element of fluid. In this case, the mean velocity is zero and the instantaneous velocities are the same as the velocity fluctuation. With these various assumptions and approximations, Eq. (18–9) can be reduced to an ordinary second-order differential equation, and for long diffusion times, it can be shown that the fluid and particle eddy diffusivities are the same. For shorter times some assumption must be made about the Lagrangian correlation coefficient; for example, one can assume, as in Eq. (14–227), that

$$R_{Lc}(t) = \exp(-t/T_{Lc}), \tag{18-13}$$

where the subscript Lc denotes the fact that the correlation is for the fluid or continuous phase. The results given by Hinze for the particle-to-fluid mean square velocity fluctuations are

$$\overline{u_d^2}/\overline{u_c^2} = (\alpha T_{Lc} + \beta^2)/(\alpha T_{Lc} + 1), \tag{18-14}$$

where $\alpha = 36\mu_c/(2\rho_d + \rho_c)d^2$ and $\beta = 3\rho_c/(2\rho_d + \rho_c)$, and for the diffusivities,

$$\frac{\epsilon_d}{\epsilon_c} = 1 + \left(\frac{1-\beta^2}{\alpha^2 T_{Lc}^2 - 1}\right)\left[\frac{\exp(-\alpha t) - \exp(-t/T_{Lc})}{1 - \exp(-t/T_{Lc})}\right]. \tag{18-15}$$

For long times the ratio is unity. If $\rho_c = \rho_d$, $\beta = 1$ and the velocities and diffusivities are the same; i.e., the particles and fluid move together. For solids in air, β can be assumed zero, and α reduces to $18\mu_c/\rho_d d^2$, which is the Stokes resistance per particle; that is,

$$\alpha = 3\pi\mu_c d/m_p = 3\pi\mu_c d/\rho_c(\pi d^3/6) = 18\mu_c/\rho_d d^2.$$

For these conditions the particle fluctuation is less than that of the fluid, unless $\alpha T_{Lc} \gg 1$, which should be true for d very small. Hinze has plotted Eq. (18–15) for selected values of αT_{Lc}. The particle spread can be much less than the fluid if αT_{Lc} is not much greater than unity. For example, at $t/T_{Lc} = 0$ (the initial point), Eq. (18–15) reduces to $\epsilon_d/\epsilon_c = \alpha T_{Lc}/(1 + \alpha T_{Lc})$, which means that at $\alpha T_{Lc} = 1$, $\epsilon_d/\epsilon_c = \frac{1}{2}$, and that ϵ_d/ϵ_c does not equal 0.99 until $\alpha T_{Lc} = 99$. Notice, however, that even at $\alpha T_{Lc} = 1$, ϵ_d/ϵ_c is essentially unity by the time t/T_{Lc} is equal to 6 [124]. Finally, the velocity fluctuation ratio given in Eq. (18–14) is for a Lagrangian view. However, because at the homogeneity assumption, this will be the same as an Eulerian formulation. Thus no special notation distinction has been made.

Friedlander [82] used a different approach, which paralleled the mathematics used in the calculation of Brownian motion and obtained essentially the same results. For very small particles he derived

$$\overline{u_r^2}/\overline{u_c^2} = 2(1 - \rho_c/\rho_d)^2(1/\lambda_L^2\alpha^2), \tag{18-16}$$

where $U_r = u_r = U_d - U_c$ and λ_L is the Lagrangian microscale defined analogously to Eq. (14–192). To follow the relative velocity fluctuation as a function of time, Friedlander further assumed a parabolic approximation to the exponential form given by Eq. (18–13). The final result was

$$\overline{u_r^2}/\overline{u_c^2} = (1 - \rho_c/\rho_d)^2[1/(1 + \alpha T_{Lc})]. \tag{18-17}$$

This is essentially the same as Eq. (18–14) for $\rho_c \ll \rho_d$ or $\beta = 0$, that is, for $\alpha T_{\text{Lc}} = 1$, Eq. (18–14) gives $\overline{u_d^2}/\overline{u_c^2} = \frac{1}{2}$ and Eq. (18–17) gives $\overline{u_r^2}/\overline{u_c^2} = \frac{1}{2}$. For $\alpha T_{\text{Lc}} = 99$, the values are 0.99 and 0.01, respectively. It should be emphasized that this is for a small particle. Friedlander pointed out that the correlation to be used is that of the *fluid velocities over the path of the particles*, and only for small particles would Eq. (18–13) be applicable. For larger particles, an Eulerian formulation of the correlation should be used. The correlation of Eq. (18–13) is for the fluid, and that for the particle is much more complex, as shown by Friedlander and by Hinze. Of course it reduces to Eq. (18–13) if $\beta = 1$ (that is, if $\rho_c = \rho_d$) or if $\alpha T_{\text{Lc}} \gg 1$ when $\beta = 0$.

Friedlander also considered the problem of the initial spread of particles *originally at rest*. The time was assumed short enough so that the correlation could be taken as unity, as was done in Eq. (14–224). The net result is that the spread varied as

$$\overline{Y_d^2} = \alpha^2 \overline{u_c^2}(t^4/4), \tag{18–18}$$

which is in contrast to $\overline{Y_c^2} = \overline{v_y'^2} t^2 = \overline{u_c^2} t^2$ given for the fluid by Eq. (14–224). Thus one must be sure that the particles being used to trace a turbulent motion do indeed follow that motion. To ensure this, the particles should be properly distributed into the flow and a reasonable time should be allowed for their adjustment to the flow conditions.

Several authors have assumed that the very small particles follow the fluid, which gives the immediate result that the ratio of ϵ_d/ϵ_c is unity for all time. In contrast, Eq. (18–15) gives a ratio of unity for long time only. In neither case, however, does this mean that the fluctuating velocity ratio must also be unity, as can be seen from Eq. (18–14), or from Eq. (18–17), which is for the long time limit. Soo [259] and Soo and Tien [267] in their analysis made the assumption cited at the beginning of this paragraph. They further assumed that Stokes' law would be valid for the relative motion, that the average fluid and particle velocities were the same, and that Eq. (18–13) was a valid form for the Lagrangian correlation. The solution was obtained in terms of a parameter K defined as

$$K = \tfrac{1}{9} N_{\text{Re},d} (d/\Lambda_L)(\rho_d/\rho_c + \tfrac{1}{2}), \tag{18–19}$$

where $N_{\text{Re},d}$ is given by du_r'/ν_c, in which $u_r' = \sqrt{\overline{u_r^2}} = \sqrt{\overline{(U_d - U_c)^2}}$ and $\Lambda_L = u_r' T_{\text{Lc}}$. The parameter K can be expressed in terms of αT_{Lc} by

$$K = 2/\alpha T_{\text{Lc}}. \tag{18–20}$$

Specifically, Soo and Tien obtained

$$\overline{u_d^2}/\overline{u_c^2} = 1 - \tfrac{1}{2}K^2 + (1 \cdot 3/2^2)K^4 - (1 \cdot 3 \cdot 5/2^3)K^6 + \ldots \tag{18–21}$$

for $K < 0.05$, and for the complete range they offered Fig. 18–2. The comparison between this and the previous results is good; at $\alpha T_{\text{Lc}} = \frac{1}{2}$, $K = 4$, and $\overline{u_d^2}/\overline{u_c^2}$ varies from 0.2 to 0.3 from the figure, depending on a parameter which accounts for the vicinity of the wall. This should be compared with the previous value

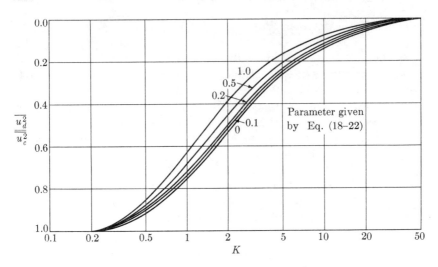

Fig. 18-2. Deviation of local intensity of motion of a particle from that of the stream at various distances from the wall. [From J. Appl. Mech. **27E**, 8 (1960). By permission.]

of 0.5. At $\alpha T_{Lc} = 99$, $K \cong 0.02$ and $\overline{u_d^2}/\overline{u_c^2}$ is 0.9998 by Eq. (18–21), which can be compared with 0.99. In Fig. 18–2, the parameter for the wall effect is

$$3\rho_c(d/y)^3/32(2\rho_d + \rho_c) = (\beta/32)(d/y)^3, \qquad (18\text{--}22)$$

where y is the distance of the particle from the wall. Calculations using Laufer's data for air in a pipe show that the intensity of the particle motion increases toward the wall and may be greater than the intensity of the local fluid stream. The smaller the value of K, and the lower the Reynolds number of the main flow, the more likely such an increase becomes. For other studies, see references 126, 173, 218, 265, and 268.

18-2.B Elementary Particle-to-Particle Interactions

Elementary particle-to-particle interactions are those interactions between particles for which we might reasonably expect to develop an analytical solution based on the particle motions. At the present time this would involve only a few particles at most. As a first step towards obtaining basic information on sedimentation we may observe the interactions of two or more spherical particles falling in a viscous fluid. Such flows are interesting in themselves, and Jayaweera et al. [137] have experimentally determined the nature of the motion of small clusters of uniform spheres ($10^{-4} < dU_\infty/\nu_c < 10$). The settling velocity of the cluster is always greater than that of an individual sphere, and it becomes still greater as the cluster becomes more compact. For two spheres side by side, rotation and gradual separation occur for a Reynolds number > 0.05, but not for values < 0.03. Spheres in a vertical line interact since one sphere moves faster than the other. Two equal spheres inclined to the horizontal will fall vertically and, at the same time, slide along the line of their centers. There is also lateral divergence. Three uniform spheres, equally spaced in a horizontal line, will interchange positions ($0.06 < N_{Re,d} < 0.16$); if they are unequally spaced, one sphere will be left behind. The motions of three to six uniform spheres ($0.06 < N_{Re,d} < 7$) that start as a compact cluster are most interesting; they eventually draw level and arrange themselves into a slowly expanding regular polygon. No tendency to form the polygons was observed when the Reynolds number was greater than seven, or when there were seven or more uniform spheres. Using slow-motion theory, Hocking [125] was able to prove the actions observed for three horizontal spheres, but he was unable to explain why clusters of from three to six spheres form a regular horizontal polygon. He was able to show that a steady configuration of seven or more spheres was unstable, and thus would not be expected to exist, which was in agreement with the experiments. Before Hocking, Kaye and Boardman [147] had investigated cluster formation in dilute suspensions, and had observed the increase in settling velocity for concentrations of from 0.1 to 3%. At between 3 and 10%, the induced upward return flow of the continuous liquid phase becomes important. Above 10%, normal hindered settling is the predominant mechanism. One important implication of this work is that, in the analysis of particle size by dilute suspension sedimentation, the concentration should not exceed 0.05%.

Other efforts that should be cited are the laminar flow studies of Mason and his coworkers, which have been mentioned in the chapter on non-Newtonian flow; the two-particle analytical solutions of settling spheres found in references 76, 159, 257, and 274; and the discussion by Peskin [218] on interactions in turbulent flow.

18-3 MULTIPARTICLE SYSTEMS IN HOMOGENEOUS FLOW

Because of the nonlinearity of the descriptive equations, detailed theoretical analysis of particle systems beyond the Stokes region are most difficult to obtain. For multiparticle systems, the motions are most certainly far too complex to solve. Consequently, effort along these lines has been restricted to the slow-motion flow of single particles or of a few interacting particles. One approach for a homogeneous system has been to assume that the mixture acts as a fluid with modified properties. This is the familiar method used for two-phase systems treated as non-Newtonians, or as homogeneous gas–liquid flow. Such an approach cannot give much insight into the detailed mechanism, since it ignores all the real characteristics of the two-phase motion; nevertheless, it is useful and will be discussed. The limitations, however, should be clearly recognized.

The more fruitful, but less developed, method of treating these two-phase flow systems is to ask somewhat more detailed questions about the motions of the particles, and about their interactions with other particles and the fluid itself. This extends the discussion in the previous section to far more complex systems, but such an extension must of necessity be approximate.

18-3.A The Integral Approach for Vertical Systems

If we are to expand our analysis to cover the flow of many particles in a medium of restricted extent, we must account for the effect of both the motion and the simple presence, even without motion, of all the other particles on the one we wish to consider [36, 326]. The effect of the motion of other particles is accounted for by use of the integral equations of motion and continuity for the two-phase media. We account for the effect of the simple presence of other particles by using the overall approach of an effective viscosity of the two-phase mixture in the equation that specifies the drag of a single particle moving in the two-phase system. [This is analogous to Eq. (18-3), which specifies the drag for a particle in a uniform infinite medium.]

(a) The equations. The integral momentum equation was formulated in Chapter 16, and for steady state was presented as Eq. (16–20). For the simplest case (steady state, completely homogeneous, and a vertical system), this equation reduces to the hydrostatic equation of the mixture:

$$p_1 - p_2 = g(\rho_c R_c + \rho_d R_d)(z_2 - z_1) = g[\rho_c(1 - R_d) + \rho_d R_d](z_2 - z_1)$$

or

$$dp/dz = -g[\rho_c(1 - R_d) + \rho_d R_d], \tag{18-23}$$

where it has been assumed that τ_w, and thus Δp_{Fm}, is zero, and that conditions at 1 and 2 are identical because of the completely homogeneous requirement. The subscripts have been changed to correspond to the present notation. In the general literature, alternate notations are also in common usage:

$$R_d = 1 - R_c = \alpha = 1 - \epsilon \quad \text{and} \quad R_c = 1 - R_d = \epsilon = 1 - \alpha.$$

The average velocity of this uniform density system can be obtained from identities previously obtained [following Eqs. (16–16) and (16–17)]:

$$U_{\text{ave}} = \frac{G_t}{\bar{\rho}} = \frac{w_t}{\bar{\rho} S} = \frac{1}{S}\left(\frac{w_c}{\rho_c} + \frac{w_d}{\rho_d}\right) = \frac{1}{S}(U_c S_c + U_d S_d)$$

$$= U_c R_c + U_d R_d = U_c(1 - R_d) + U_d R_d, \tag{18-24}$$

or directly from continuity considerations [326].

The general equation for drag on a single sphere in an infinite medium has been given as Eq. (18–1), which Zuber [326] has modified to apply to the motion of a single particle in an infinite two-phase mixture. The first modification replaces the continuous phase viscosity μ_c with the effective viscosity of the system μ_e. The second modification changes the third term (which is the force associated with the acceleration of the apparent mass of the particle relative to the fluid) to account for the effect of the concentration of particles.

As will be shown, the effective viscosity of the system μ_e can be expressed as

$$\mu_e = \mu_c f^{-1}(R_d). \tag{18-25}$$

That is, the effective viscosity experienced by the particle depends only on the viscosity of the medium and the presence of other particles, as given by some function of R_d. The concentration effect is obtained by considering a fluid sphere about the solid sphere. The diameter of the fluid sphere is taken as the average center-to-center spacing between the particles, and thus can be related to the number and size of the uniform solid spheres. The acceleration of the apparent mass of the particle relative to the fluid is obtained by considering the motion of the inner solid sphere relative to the now-specified outer fluid sphere. The correction factor is

$$(1 + 2R_d)/(1 - R_d), \qquad (18\text{–}26)$$

which, as expected, goes to unity for a very dilute system. The final modified form of Eq. (18–1) becomes

$$\frac{\pi}{6} d^3 \rho_d \frac{dU_d}{dt} = 3\pi \mu_c f^{-1}(R_d) d(U_d - U_c) - \frac{\pi}{6} d^3 \frac{dp}{dz} + \frac{1}{2} \frac{\pi}{6} d^3 \rho_c \left(\frac{1 + 2R_d}{1 - R_d}\right)$$
$$\times \left(\frac{dU_d}{dt} - \frac{dU_c}{dt}\right) + 3\pi \mu_c f^{-1}(R_d) d \left[\frac{d}{2\sqrt{\pi \nu_c f^{-1}(R_d)}} \int_{t_0}^{t} \frac{dU_d/dt' - dU_c/dt'}{t - t'} dt'\right]$$
$$- \frac{\pi}{6} d^3 g(\rho_d - \rho_c). \qquad (18\text{–}27)$$

For steady state, this reduces to

$$3\pi \mu_c f^{-1}(R_d) d(U_d - U_c) - (\pi/6) d^3 (dp/dz) - (\pi/6) d^3 g(\rho_d - \rho_c) = 0. \qquad (18\text{–}28)$$

Combining Eqs. (18–2), (18–23), and (18–28), we get

$$U_r = U_d - U_c = U_t \left(1 - R_d - \frac{\rho_c}{\rho_d - \rho_c}\right) f(R_d), \qquad (18\text{–}29)$$

where the function $f(R_d)$ is yet to be defined. Equation (18–29) suggests that to account for the presence of other particles, the relative velocity can be expressed as a modification of Stokes' law. Equation (18–29) with $\rho_c \ll \rho_d$, or an equivalent form, has been suggested by a number of investigators as the proper means of generalizing most two-phase flow systems [31, 161, 193, 221, 225, 308, 326]. One such form is to express the equation in terms of superficial velocities,

$$U_{sc} = U_c R_c = U_c(1 - R_d), \qquad U_{sd} = U_d R_d. \qquad (18\text{–}30)$$

In these terms, Eq. (18–29) becomes

$$U_r = U_d - U_c = U_{sd}/R_d - U_{sc}/(1 - R_d) = U_t \left(1 - R_d - \frac{\rho_c}{\rho_d - \rho_c}\right) f(R_d),$$

or

$$R_d(1 - R_d) U_r = U_{sd}(1 - R_d) - U_{sc} R_d = U_t R_d(1 - R_d) f(R_d) \qquad (18\text{–}31)$$

where $\rho_c \ll \rho_d$ has been used in Eq. (18–31).

We must establish $f(R_d)$, as defined in Eq. (18–25), so that specific results can be obtained. We usually obtain the relation by considering the initial rate of settling of a homogeneous mixture or by observing bed expansion during particulate fluidization. Equations (18–24) and (18–29) can be combined to give [326]

$$-U_{\text{ave}}/(1-R_d) + U_d R_d/(1-R_d) + U_d = U_t(1-R_d)f(R_d), \qquad (18\text{–}32)$$

which for sedimentation reduces to

$$U_d = U_t(1-R_d)^2 f(R_d), \qquad (18\text{–}33)$$

since in sedimentation there is no net average velocity ($U_{\text{ave}} = 0$). Equations (18–30) and (18–33) combine to give us

$$U_{sd}/U_t = R_d(1-R_d)^2 f(R_d). \qquad (18\text{–}34)$$

Thus, from the initial solids fraction, rate of solids settling, and terminal velocity, the function can be evaluated. Equation (18–29), for batch particulate fluidization, reduces to

$$U_c = -U_t(1-R_d)f(R_d), \qquad (18\text{–}35)$$

since there is no net solids flow, and thus $U_{sd} = U_d = 0$. The minus sign indicates the upward countergravity flow of the fluid. In terms of the superficial velocity, this becomes

$$U_{sc}/U_t = -(1-R_d)^2 f(R_d). \qquad (18\text{–}36)$$

A knowledge of the superficial liquid velocity, terminal velocity, and solids fraction allows us to determine $f(R_d)$. The sign on $f(R_d)$ depends on the coordinates used. For a positive sign, as used below, the vertical direction is positive downward.

(b) The viscosity function. Lapidus and Elgin [161] suggest that the relation $f(R_d)$ be obtained by empirical methods for each system; i.e., the relative velocity would be expressed as a function of R_d for any given particle system

$$U_r = F(R_d), \qquad (18\text{–}37)$$

and would be determined empirically. This of course eliminates the problems of possibly restricting the relation to spherical particles or of using various analytical representations that are inadequate. It will be seen that this latter factor is of some consequence, since there is considerable variation in available equations and all those proposed to date are inadequate at high solids fractions.

The simplest analytical relation is Einstein's [68] viscosity law for a suspension of dilute rigid spheres, in which

$$f(R_d) = (1 + 2.5R_d)^{-1}. \qquad (18\text{–}38)$$

More complex equations have been proposed; for example, Hawksley's [119] simplification of Vand's [303] viscosity equation can be expressed as

$$f(R_d) = \exp\left[\frac{-2.5R_d}{1-(39/64)R_d}\right], \qquad (18\text{–}39)$$

which neglects particle collisions and which has been checked by the experimental data of Hanratty and Bandukwala [109] for $N_{\text{Re},d} < 0.07$. It was approximately correct up to a Reynolds number of 58.5. However, visual observations indicated particulate fluidization for $N_{Re,d} < 0.8$ and nonuniform conditions for $N_{Re,d} > 2$. Brinkman [32] extended Einstein's analysis to get

$$f(R_d) = \frac{1}{(1-R_d)^2}\left[1 + \frac{3R_d}{4}\left(1 - \sqrt{(8/R_d) - 3}\right)\right]. \quad (18\text{-}40)$$

The empirical law by Richardson and Zaki [234] can be expressed as

$$f(R_d) = (1 - R_d)^{n-2}, \quad (18\text{-}41)$$

where n is an empirical parameter with a value of

$n = 4.65$	for	$N_{Re,d} < 0.2$,
$n = 4.35 N_{Re,d}^{-0.03}$	for	$0.2 < N_{Re,d} < 1$,
$n = 4.45 N_{Re,d}^{-0.1}$	for	$1 < N_{Re,d} < 500$,
$n = 2.39$	for	$N_{Re,d} > 500$.

Oliver [209] has obtained some new data and found that the Brinkman equation (18–40) was a good fit for low values of $R_d (< 0.03)$. The Vand-Hawksley equation (18–39) was a better fit in the intermediate range of R_d (0.1 to 0.4). Oliver has proposed still another semi-empirical equation ($R_d < 0.3$), which gives a good fit to the data, and is, in terms of $f(R_d)$,

$$f(R_d) = \frac{(1 - 0.75 R_d^{1/3})(1 - 2.15 R_d)}{(1 - R_d)^2}, \quad (18\text{-}42)$$

where the two constants were obtained from the best fit to a considerable amount of experimental data. Other methods have been proposed by Zenz [320, 322], Jahnig [136], Andersson [4], Happel [111], Steinour [272], and others. Reviews and further equations have been given by Zenz and Othmer [322], Frisch and Simha [84], and Shannon et al. [251]. Data on nonspherical coal particles have been obtained by Moreland [200], who found Eq. (18–41) adequate with n varying from 6.5 to 7.5.

A great many experimental data are available, and selected parts can be found to match any of the curves. The major problem in analytical representations is that they do not match the necessary boundary conditions. The usual representation is a plot of U_d/U_t or U_{sc}/U_t versus R_d, as shown in Fig. 18-3. The diversity of results is quite apparent. At the extreme of infinite dilution $[R_d = 0, f(R_d) = 1]$, the particle may not fall at the Stokes terminal velocity U_t calculated from average properties (i.e., variations caused by particle size distributions and deviations from spherical shape for real particles are not taken into account). This problem can be rectified in part by using an experimental value for U_t, thus losing some of the generality. At the other extreme, most representations predict the velocity as zero at $R_d = 1$. Exceptions to this are

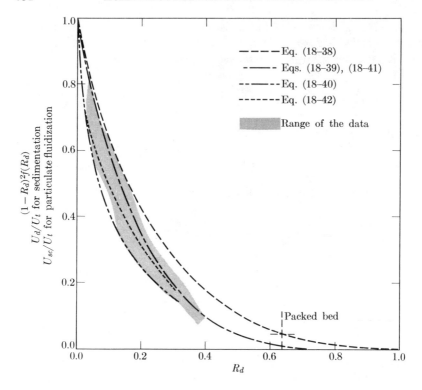

Fig. 18–3. Reduced velocity versus solids fraction.

Eqs. (18–40) and (18–42), which predict zero at an R_d of 0.667 and 0.465, respectively. As pointed out by Shannon et al. [251], the velocity should be zero when the particles are resting on each other and are at the bottom of the column, i.e., when they form a fixed bed. The maximum solids level for fixed beds is always less than one, no matter what the shape or distribution. Thus those representations that require R_d to approach unity as U_d or U_{sc} approach zero cannot be used as such at high solids levels, which are of major importance in sedimentation. Wallis [308] suggested a discontinuity at the point corresponding to the packed bed solids content, R_d, because an initial mixture at this concentration would settle uniformly until it arrived at the bottom of the container and then abruptly stopped. For data obtained from batch particulate fluidization experiments, the cutoff velocity would correspond to the point of minimum fluidization. Shannon et al. [250] suggested the somewhat less abrupt shape shown in Fig. 18–4, since particle-to-particle effects could retard the settling at the higher solids level. The exact nature of the curve in the region of the high solids apparently depends primarily on the nature of the material and on the manner in which the data are obtained.

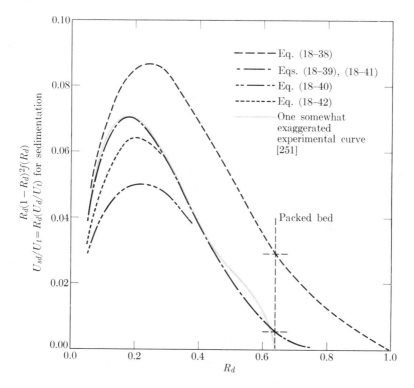

Fig. 18-4. Solids flux versus solids fraction.

Another common representation is the solids flux plot suggested by Kynch [158]. This is a plot of U_{sd}/U_t versus R_d, shown in Fig. 18-4, and is just $R_d(U_d/U_t)$ versus R_d. Again the diversity of results is apparent. All the equations predict singly concaved flux plots, whereas the experimental results are usually doubly concaved (see [250] through [252] and references cited therein). Clearly, work is needed to develop a more rational means of presenting data on initial hindered settling and particulate fluidization. We shall discuss the viscosity of a non-homogeneous fluidized bed (gas–solids) later; in contrast to the present results, which imply that μ_e is Newtonian, the aggregately fluidized bed exhibits non-Newtonian characteristics.

(c) **Operational diagrams.** With $f(R_d)$ defined approximately by the analytical relationships, or more exactly from experimental measurements, we can turn our attention to general representations of the data in operational diagrams. Several types of plots have been suggested [161, 193, 308, 326], one of which is to use U_{sc}/U_t versus R_d with U_{sd}/U_t as the parameter. Equation (18–36) for batch fluidization would define one line on the plot ($U_{sd} = 0$). Other curves can be calculated directly from Eq. (18–31), so long as due regard is given to the sign on the velocity (again, as used here, positive is down). Certain limitations on ranges do exist, becoming effective when the system floods or has various physical restraints designed to keep the solids within the system. Figure 18-5 shows such a generalized diagram. The areas of operation are indicated. The following example illustrates the construction and further meaning of the areas of operation.

EXAMPLE 18-1. For all values of R_d and U_{sc}/U_t, calculate the general operating lines for $U_{sd}/U_t = \pm 0.1$. Use Eq. (18-41) for $f(R_d)$, with $n = 4.65$.

Answer. Table 18-1 gives the values obtained by combining Eqs. (18-31) and (18-41), plotted in Fig. 18-5 as lines *abc* and *def*. For the first case, *abc*, both signs are negative, implying cocurrent-countergravity flow. From *a* to *b*, the flow is simply the transport of the solids. At *b*, U_{sc} becomes equal to the terminal velocity for a single particle; if free, the particle would tend to go down, i.e., become a positive U_{sd}. But the curve was determined for U_{sd} negative, and thus describes a system in which a vertical upward flow of the solids is forced. Such a flow can be created by using a bottom restraint, then forcing solids in at the bottom and removing them at the top (part *bc* of the line). For the second case, both velocities are positive, implying cocurrent-cogravity flow. At the point *e*, however, the line crosses the locus of flooding points, which defines the maximum value of U_{sc} for the given U_{sd} in an unrestrained system. Since the solids flow would have to be slowed down by a bottom restraint in order to maintain the low U_{sd}, a restrained cocurrent-cogravity feed is defined between *e* and *f*. Otherwise the liquid flow would cause an increase in U_{sd}.

Wallis [308] has suggested a very simple and rapid single-line operational diagram which is based on Fig. 18-4 and shown in Fig. 18-6. The various types are noted once again, and can be compared with Fig. 18-5.

Finally, for flow systems in which the slip velocity is negligible and gravity effects need not be considered (for either very small particles or very small density differences), the system can be treated as homogeneous non-Newtonian material.

18-3.B Specific Analyses for Vertical Systems

The operational diagram based on the integral approach can help establish the type and overall nature of vertical homogeneous solids-fluid flow. The details of such flows must come from more specific analyses, which are necessary for the design of the various transfer operations in such systems.

(a) **Particulate fluidization.** At any given liquid velocity, the expansion (or equivalent, solids fraction) is the most important parameter of the particulately fluidized bed, and is reasonably described by the material presented in the preceding sections. The subject is also considered in the reference books by Leva [164] and Zenz and Othmer [322]. The particulately fluidized bed is relatively uniform, but there are variations and gross circulation patterns [37, 39] that

TABLE 18-1

U_{sd}/U_t	R_d	$R_d(1 - R_d)^2 f(R_d)$	$R_d U_{sc}/U_t$	U_{sc}/U_t
−0.1	0.05	0.038	−0.133	−2.7
	0.1	0.061	−0.151	−1.51
	0.2	0.070	−0.151	−0.76
	0.3	0.058	−0.128	−0.42
	0.4	0.038	−0.098	−0.245
	0.5	0.023	−0.073	−0.146
+0.1	0.05	0.038	0.057	1.14
	0.1	0.061	0.029	0.29
	0.2	0.070	0.010	0.05
	0.3	0.058	0.012	0.04
	0.4	0.038	0.022	0.055
	0.5	0.023	0.027	0.054

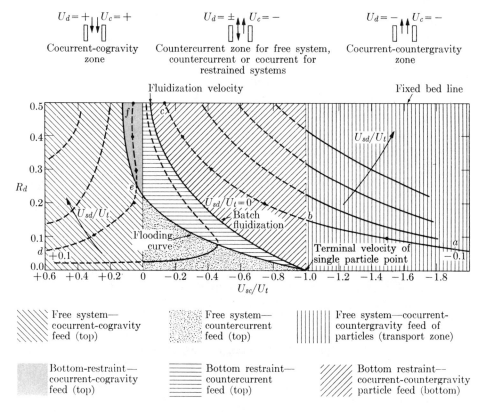

Fig. 18–5. *Operational diagram for vertically-moving fluidized systems* ($\rho_d > \rho_c$). [Modified from L. Lapidus and J. C. Elgin, AIChEJ **3**, 63 (1957). By permission.]

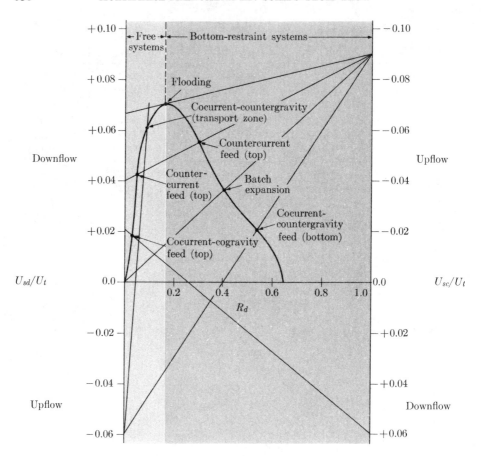

Fig. 18–6. Operational diagram. [After G. B. Wallis, Interaction Between Fluids and Particles, Inst. Chem. Engrs., London (1962). By permission.]

depend to a great extent on the exact nature of the inlet liquid distributor [102, 185]. As a result, circulation of solids is quite specific to given system geometries and cannot be easily generalized.

The *radial* dispersion of mass in the liquid phase of a particulately fluidized bed has been measured by Hanratty et al. [110] and Cairns and Prausnitz [39]. They obtained experimental results that showed that for diffusion from a point along the center line, the Taylor relationships for homogeneous isotropic turbulence are obeyed. A Gaussian distribution of C/C_{ave} existed with radial position (C being the concentration of the material in the fluid being diffused). There was a marked dependence of the radial dispersion on the fraction of solids. At $R_d = 0.3$, the scale of turbulence was a maximum and the radial Peclet number dU_c/E_r was a minimum (where E_r is the radial eddy coefficient of diffusion). Cairns and Prausnitz made a detailed visual study of the motions within the bed and concluded that at this fraction of solids ($R_d = 0.3$) there was a drastic change in the flow pattern. The bed characteristics depended mainly on the ratio of the diameters of tube and particle and the ratio of the densities of particle and fluid. Higher radial diffusion was obtained with particles which had greater density, and it was also observed that the heavier particles (lead) always gave more violent motion than did the lighter particles (glass).

Cairns and Prausnitz [38, 40] also obtained experimental measurements of the *axial* diffusion in a particulately fluidized bed. Measurement of the radial distribution of the local axial Peclet number, $dU_c(r)/E_a(r)$, indicated that at any given flow conditions (one average solids fraction), the local axial Peclet number was greater near the wall than at the center, which paralleled the results found for packed beds [70]. Although both $U_c(r)$ and $E_a(r)$ could vary, the greater local Peclet number corresponded to less axial diffusion in the wall area. In general, the distribution for lead spheres was quite flat across the bed when the tube-to-particle-diameter ratio was 78. At a ratio of 39, considerable variation occurred. In contrast, the distribution in a packed bed is quite flat for a ratio of 25 or greater. The variation across the bed is probably a result of the velocity profile in the particulately fluidized bed being less flat than the velocity profile of the packed bed [37]. This is apparently a direct function of the nature of the inlet liquid distributor [102]. Bruinzeel et al. [34] reported a similar study, which used larger vessels and smaller particles. The porosity effect was not observed, and it was found that the particle Peclet number increased slowly with the particle Reynolds number. Additional data on axial dispersion have been obtained by Kramers et al. [157], who suggested that the axial dispersion coefficient is composed of two parts, the first associated with the wakes of the individual particles and the second associated with the fluctuations in local porosity that travel upward through the bed. Variation of the solids fraction gave results for axial diffusion similar to those for radial diffusion. A minimum in the axial Peclet number also occurred at 0.3 solids fraction. The axial Peclet number was less for higher-density materials, and decreased with the tube-to-particle ratio.

Ruckenstein [243] has proposed a model for the homogeneous particulately fluidized bed, and although it is only approximate, it is a starting point for the eventual prediction of bed characteristics. The weakest point of the analysis lies in Ruckenstein's assumption that solid particles are immobile, that they remain in fixed locations and only move vertically about certain points that might be pictured as nodes of a lattice through whose voids the liquid flows. This motion, Ruckenstein feels, results in a local variation in the voids from which a mean-squared deviation about the mean void value can be defined (variance of local void fraction). This is related to the mean-squared velocity (velocity variance) of the particles about their node point by using a simplification of Eq. (18–1). The effects of the pressure gradient in the fluid and of acceleration on the viscous drag (the second and fourth terms on the right) were neglected. The continuous phase velocity U_c was modified to be the velocity through the minimum cross section, and the assumption was made that no further interaction effects occurred. Because of this, no alteration of the viscosity on account of the concentration effect need be considered, as was necessary in order to obtain

Eq. (18–27) from (18–1). Several further assumptions and simplifications were necessary in order to obtain an equation for the velocity variance which could be used to obtain the final equation for the variance of the local void fraction. With this variance established, it was possible to study the effect on axial diffusion of the particle movement, or of the variation in local void fraction. Still further approximations and estimates were needed to obtain the final expression for the axial diffusion coefficient, which has a single adjustable constant and is a function of the system parameters. The final equation shows the same general tendencies as the data concerning particle size and void fraction, but because of the rather large number of assumptions necessary to obtain the solution and the questionable accuracy of some of them, we cannot yet establish the validity of the model. The discussion by Beek [21] should also be cited as a means of representing the same data, although again a number of adjustable constants exist in the analysis.

In particulately fluidized systems, most of the research has been restricted to particles of uniform size, and cannot, without certain reservations, be applied to systems involving a distribution of sizes. Stratification by size will usually occur, and certain aspects of this have been discussed in references 142 and 222.

(b) Hindered settling. The hindered settling of a homogeneous slurry of solid particles of uniform size is a close parallel to particulate fluidization. An approximate settling process from some initial uniform concentration is shown in Fig. 18–7. At $t = 0$, the uniform suspension (I) begins to settle; at t_2, clear liquid (II) appears at the top, and a dense phase (III) at the bottom. Depending on the nature of the material (i.e., its compressibility), the dense phase may involve

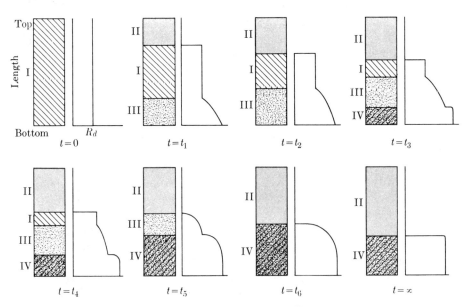

Fig. 18–7. Hindered settling.

a concentration gradient [194]. As settling progresses, phase III grows and phase I decreases. The two interfaces, or concentration waves, approach each other and finally merge at $t = t_5$. In the meantime, an even more dense phase (IV), nearly at the densest packing, begins to form (t_3), and continues to grow. Eventually the variable-density phase (III) is settled to the final concentration (IV). However, given enough time, vibration, etc., some further compaction can occur. Again, depending on the nature of the material, phase III may not even exist, and then phase I will meet the upward-traveling interface wave of phase IV directly without the intermediate phase III.

The settling curve of height versus time is usually desired, and can be obtained from the solids flux plot of Fig. 18–4 [158, 250, 252, 308]. The exact nature of the resulting settling curve depends on the initial concentration and the shape of the flux curve. In Fig. 18–8 (after Wallis [308], whose analysis paralelled that of Kynch [158]) three possible cases are considered. If the initial concentration lies in region A or D, the variable-density region (III) does not exist and the height changes linearly from the initial to the final values. The settling velocity U_d/U_t (or the velocity of the upper interface I–II) is constant and can be obtained from the figure; that is,

$$(U_d R_d/U_t)/R_{d,t=0} = U_d/U_t, \quad \text{or upper interface velocity.} \qquad (18\text{--}43)$$

This is simply the slope of the line from the origin to the point $R_{d,t=0}$ on the curve (line i in Fig. 18–8). The velocity of the lower interface (I–IV) is found from the slope of the line from $R_{d,t=\infty}$ to $R_{d,t=0}$ (line ii); that is,

$$(U_d R_d/U_t)/(R_{d,t=\infty} - R_{d,t=0}) = \text{lower interface velocity.} \qquad (18\text{--}44)$$

More complex settling curves are obtained if $R_{d,t=0}$ lies in regions B or C. For both regions, the upper interface (I–II) falls as in cases A and D, and is similarly determined (line i). The velocity at the lower interface (III–IV) has an upper limit given by the slope from $R_{d,t=\infty}$ to the tangent point (line iii). When upward-moving interface I–III meets downward-moving interface I–II at $t = t_5$, phase I disappears and the settling slows down (new interface II–III) until upward-moving interface III–IV meets downward-moving interface II–III, at $t = t_6$ and the settling is essentially complete. If the settling starts in region C, the interface between I and III is continuous and is given by the slope of the tangent line at $R_{d,t=0}$ (line iv); if in region B, a limit on the tangent line exists and is given by the line that connects $R_{d,t=0}$ to the inflection point (line v). This would cause a rather abrupt change and the interface I–III should be sharp. This second sharp interface was clearly shown in the photographs presented by Shannon et al. [250]. Tory [298] used Eq. (18–41) to establish an expression for the inflection point in terms of n. He set the second derivative equal to zero (see Fig. 18–8), getting

$$R_{d,i} = 1 - (n-1)/(n+1) \qquad (18\text{--}45)$$

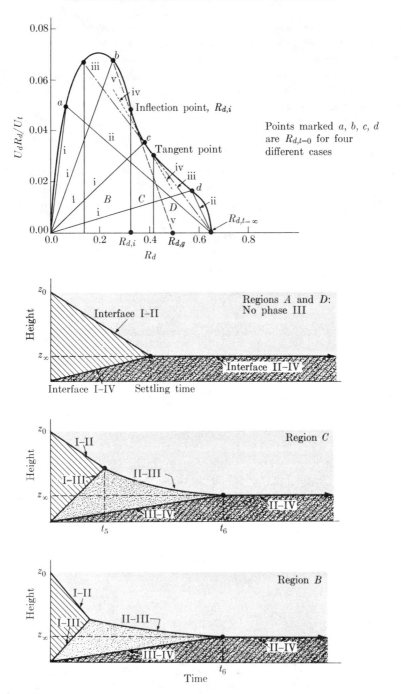

Fig. 18-8. Flux plot and settling curves. [After G. B. Wallis, Interaction Between Fluids and Particles, *Inst. Chem. Engrs.*, London (1962). By permission.]

and an expression for the intercept $R_{d,g}$, which would be the final solids fraction just large enough to keep the sharp concentration gradient (I–III) from forming;

$$R_{d,g} = 1 - [(n-1)/(n+1)]^2. \tag{18-46}$$

From Eqs. (18–41) and (18–46), for regular close packing of $R_d = 0.74$, the gradient cannot form above a Reynolds number of 40. For close random packing of spheres ($R_d \sim 0.61$), the sharp interface should form for Reynolds numbers less than one. The total settling time in both cases B and C is the same (t_6) and is controlled by the tangent curve iii.

Shannon et al. [250] have provided a detailed description of the types of flows to be expected. The only thing that remains to be established is the actual curve describing the movement of the interface II–III. It is here that the various authors differ, but their work is equivalent. Shannon and Tory [252] suggested a rapid graphical method and Wallis [308] used a generalized transformed curve of $Y = z/z_0 R_{d,t=0}$ versus $X = t/z_0 R_{d,t=0}$, which has the effect of causing all curves to pass through the common point $1/R_{d,t=\infty}$ when sedimentation is completed. In all cases the calculated curves compare well with the experimental results, with the limitation that the final consolidation period shown in Fig. 18–7 is absent [331].

Further problems of a more specific nature associated with settling have been treated. Particle size distributions and stratification have been discussed in references 148 and 222, and others cited therein. Reference 194 gives a treatment of the settling rates and sediment characteristics of a flocculated kaolin suspension in which the larger floc structure, rather than the individual clay particles, is the settling unit. At low rates, the flocs form weak clusters, which can in turn form networks. An automatic device for measuring the upper interface has also been described [79]. Extensions to continuous thickening have been given in references 117, 251, 252, 319, and those cited therein. The enhancement of settling by inclination of the settling system has been considered in [206, 210, 217], and references cited therein.

18-3.C Solids Transport Systems

Over certain narrow ranges of variables, we can approximate real flows by assuming the material to be homogeneous. The main problem is to establish when such an assumption is valid enough to allow a reasonable estimation of the flow characteristics. Any computational method that takes this approach and ignores the details of the flow is a homogeneous model, and its usefulness depends on the availability of empirical constants that adequately account for deviations from the model. Of course, there are many factors that complicate such a treatment, and one should be aware of their influence. In this section, we shall discuss the adequacy of the homogeneous assumption and the complicating factors that go with it. References will be given to literature that is readily available and that provides design information for such flows. The problem of the transition between nonhomogeneous and homogeneous conditions is for the most part avoided, since it, as well as nonhomogeneous conditions *per se*, will be taken up in more detail in later sections.

The homogeneous assumption for slurries would be valid for colloidal particles in the submicron size range; these should be maintained in a true homogeneous suspension by Brownian motion, and as such can be treated as Newtonian or non-Newtonian materials. They need not be considered further here. The distribution of particles in flow systems above this lower limit is probably never truly homogeneous, although in highly turbulent flow the approximation in many cases is quite good. Concentration distributions for liquid systems have been given in references 22, 41, 64, 202, and 208.

One rule of thumb is that slurries can be considered homogeneous if the settling velocity is less than 0.002 to 0.005 ft/sec [105], although considerable variation from this has been reported [22, 64]. Thomas, in a series of articles [288–290], investigated the minimum transport velocity of suspensions. The minimum transport velocity is defined as the average fluid velocity which just prevents the accumulation of a layer of particles (either stationary or moving) on the bottom of the pipe. Thomas [288] and others [41, 64] found that if this velocity was exceeded, the flow could be considered homogeneous for micron-size particles that are small enough to be in the Stokes region ($N_{\text{Re},d} < 1$) and, at the same time, smaller than a size equivalent to a y^+ of 5 for the flowing system. Homogeneous slurries are generally non-Newtonian and can often be approximated by the law of Bingham's ideal plastic [289]. Data for homogeneous flow of larger particles are meager [22, 202, 208, 290] and it is questionable whether one can treat such slurries as simple homogeneous materials and obtain any reasonable estimate of the flow characteristics. This doubt apparently does not arise in the case of coal in oil suspensions [22]; the data from all sources show that rather drastic concentration gradients exist.

Pneumatic transport generally involves particles of such size that a homogeneous flow is never obtained. This has been verified by the solids concentration distributions that have been measured and reported in references 65, 261, 266, 269, 305, and 311. Of course, in spite of the lack of uniformity, we can assume that the flow is effectively homogeneous, and hope that the approximation is adequate for the purpose or, if it is not, that any adjustable constants can empirically account for the deviation from homogeneity [104, 293, 320, 322]. Experimental pressure drop and solids transport results and their interpretation by such overall methods (which constitute the main methods of design) will not be discussed here because of the adequate coverage in references cited above, but a few complicating factors and extensions should be mentioned briefly.

For multiparticle flow, such as the turbulent flow of a dusty gas, Sproull [271] observed a reduction in pressure drop which apparently resulted from the dampening of the turbulence level by the dust because of the dissipation associated with the relative motion of the particles and the stream. The dampening could be quite large if the particles did not follow the small turbulent fluctuations [247].

The distribution of solids across the flow system depends not only on gravity effects but also on the presence of electrostatic forces. It has been known for a long time that such forces do exist and are of major consequence [296]. Since electrostatic forces are nearly always present, recent efforts have been made to establish their effect [233, 262, 263, 265, 268, 269]. Measurements of concentration distributions generally show an increase in solids in the vicinity of the wall [65, 269, 305], which is in accordance with what we might expect from consideration of electrostatic forces [262, 263, 296].

Dispersion measurements have been made for both the fluid elements and the solid particles in flow systems [83, 143, 174, 305]. For example, Kada and Hanratty [143] studied the *radial* dispersion of fluid elements in slurry flows, and found that the solid particles had no effect unless there was an appreciable average slip velocity and relatively high concentration of solids. The interdependence of concentration, slip velocity, and Reynolds number was illustrated, but no final correlation was obtained. In their system, the point at which radial diffusion was affected appeared to be associated with agglomeration and the associated large fluctuations in solid particle concentration. Van Zoonen [305] studied *radial* and *axial* dispersions of both the fluid elements (air) and the solid particles (cracking catalyst) in a vertical riser. The radial dispersion coefficients were of the same order of magnitude for both gas and solid, and were much larger than the axial values. The axial dispersion of the gas elements exceeded that of the solid particles.

Under flow conditions in which nonequilibrium, slip, or compressibility factors can be of importance, still another approach to the problem of cocurrent solid–gas flow is to make the solution of the problem parallel to the solutions for single-phase flow. For the most part, these solutions involve gas-dynamic problems of the nature discussed in Chapter 12. Such problems have been investigated primarily because of interest in rocket propulsion systems, in which part of the combustion product is a solid phase dispersed in a gas. For simplicity, the solutions are usually restricted to the one-dimensional case [42, 99, 131, 245, 246, 260, 263, 264], although the axial-symmetric three-dimensional case of a nozzle has been considered [8, 154]. As in the corresponding single-phase problem, the velocity is assumed uniform, so the viscous force terms involving velocity gradients do not enter the problem. The fluid viscosity does enter the problem, however, in that particle drag exists and is taken into account. In such compressible flow problems, in which rapid changes in velocity and temperature occur (as in a nozzle or in a shock wave), the particles cannot follow the changes rapidly enough to prevent slip and relaxation processes from occurring [42, 245, 246, 260] and giving rise to the deviations from single-phase flow. Other assumptions usually made are as follows: (1) The equation of the perfect-gas law is valid for the gas, but apparently not valid for systems in which the loadings are high enough to invalidate the fourth assumption below. (2) Uniform spherical particles are uniformly distributed and no loss of particle mass occurs; that is, there is no reaction, no mass transfer, etc. (This assumption has been removed in reference 260.)

(3) The drag is associated with the slip velocity only, and Brownian motion, etc., make no contributions. (4) The volume occupied by the solid particles is neglected. (This assumption has been removed in reference 245). (5) Heat transfer is caused by a mean temperature difference between the gas and the solid. (6) The temperature of a solid particle is uniform because of its small size and relatively high thermal conductivity. (7) Steady-state conditions exist (not used in reference 246). With these assumptions, six equations (two each for continuity, momentum, and energy) can be written and solved for certain specific flow cases (usually by numerical methods).

Nozzle flow has been treated in references 1, 8, 92, 98, 99, 131, 154, 178, and 260, and normal shock conditions in references 42, 244, 245, 246, 260, and 263. Because of the dissipative nature of particles in gas flow (due to drag), sonic velocity does not occur at the throat, but rather at some point beyond [131, 263]. The effect of solids volume [245] can be important in systems that have high loadings, and errors can be introduced [264] by assuming a uniform particle size rather than a distribution of sizes. Although most analytical solutions are for laminar conditions in the fluid, these theories can be extended to turbulent flow conditions [263].

18-4 TRANSITION BETWEEN FLOWS

The difference between an operation that is uniformly homogeneous (e.g., particulate fluidization) and one that is not (e.g., aggregate fluidization) is fairly easy to establish experimentally, but most difficult to define analytically. As with any analysis of transition or stability, the complexity of the problem forces us to approximate the physical reality before a solution is possible. We must make empirical correlations in order to define the transition region, although mathematical analyses do provide insight into the role played by the variables.

18-4.A Transition from Particulate to Aggregate Fluidization

Wilhelm and Kwauk [315] observed that solids–liquid systems generally fluidize particulately and that solids–gas systems generally fluidize aggregately, which led them to suggest the Froude number

$$N_{\text{Fr,mf}} = U_{\text{mf}}^2/g\,d \tag{18-47}$$

as a criterion. Here U_{mf} is the minimum fluidization velocity, previously defined, and d is the particle diameter. It was suggested that particulate fluidization occurred for values less than unity and aggregate fluidization occurred for values greater than unity. Later results [55, 113, 235, 254, 299, 313] showed that there is a continuous transformation between types, and that a large area of uncertainty exists in the region $0.003 < N_{\text{Fr,mf}} < 0.3$, with particulate fluidization occurring below the lower limit, and aggregate fluidization occurring above the upper limit. The overlap occurs when such systems as phenolic microballoons or very small particles are fluidized with gas [113] and systems of large metal shot are fluidized with liquids [113, 235]. Because of this rather wide range of overlap, alternate and modified criteria have been investigated. Romero and Johanson [235] have suggested and tested several possibilities which depend on the Froude number, the Reynolds number (dU_{mf}/ν_c), relative density ($\rho_d/\rho_c - 1$), and a length group. The various numbers were suggested by the stability analysis of Rice and Wilhelm [231], while their critical value was determined empirically. Harrison et al. [113] and Davidson and Harrison [55] have attempted to establish the conditions necessary for the stability of an existing void. They have postulated as a criterion the value of the maximum upward velocity in the interior of the bubble (U_{bm}) relative to the free-fall velocity of the particles in the fluid of the system (U_t). The bubble is stable if $U_t > U_{bm}$, since any particle in the interior will fall back into the wake, and the bubble will be maintained. The estimation of the velocity of a rising bubble will be discussed in a later section, and the values for U_t have already been discussed, experimental values being the most reliable. Both U_{bm} and U_t can be expressed in terms of the Froude number [the former by a modification of Eq. (17–77), and the latter in terms of Eq. (18–3)]. Thus this criterion can also be pictured as a modification of the Froude number.

Rather than investigate the stability of a bubble once formed, one can study the hydrodynamic instability of a homogeneous bed to determine the conditions necessary for the fastest growth of small disturbances. Analyses of this nature have been presented by Baron and Mugele [14], Jackson [3, 134], Murray [204], Pigford and Baron [220], Rice and Wilhelm [231], and Ruckenstein [242]. In general, these references show that fluidized beds are always unstable and that the growth rate is slow in liquid systems and fast in gas systems. These findings can generally be expressed also in terms of various groups, including the Froude number. There is still another view: that the nonhomogeneous nature is caused by interparticle forces (Teoreanu [286, 287]), which are known to be important in gas fluidized systems [47, 314].

A comparison of the various criteria and all the available data has not as yet been made, so that the adequacy of any given method is not as yet established. Of the modifications suggested by Romero and Johanson, the simplest is

$$N_{\mathrm{Fr,mf}} N_{\mathrm{Re,mf}}(\rho_d/\rho_c - 1) = U_{\mathrm{mf}}^3(\rho_d - \rho_c)/g\mu_c, \tag{18–48}$$

in which, for the experiments tested, the dividing line was between 65 and 125. This group and the critical value cited have received further theoretical support from the work of Pigford and Baron [220]. The reciprocal of the parameter of their analysis (which is called a dissipation factor, owing to the viscosity of the bed) is in terms of Eqs. (18–25) and (18–48):

$$[U_{\mathrm{mf}}^3(\rho_d - \rho_c)/g\mu_c]f(R_d). \tag{18–49}$$

At the point of minimum fluidization, R_d is bracketed by the range 0.3 to 0.55, with a good average being 0.4. For particulate fluidization, the highest value of Eq. (18–48) was about 65, which occurred at $N_{\mathrm{Re,mf}} \cong 50$. Using this value to estimate $f(R_d)$ from Eq. (18–41), we get a value of 0.6, which in turn gives us about 40 for Eq. (18–49). The lowest value of Eq. (18–48) for aggregate fluidization was about 125, which occurred at $N_{\text{Re,mf}} \cong 1500$. Likewise, the value for Eq. (18–49) is about 105. The analysis by Pigford and Baron indicated that, for a given disturbance, the larger the value of the parameter, the higher the growth rate. Because of the effect of viscous dampening, particulate beds with values of Eq. (18–49) less than 10 are quite stable to very small disturbances, while, to these same disturbances, beds with the parameter at 100 or more are far less stable. The difference between these two levels is quite dramatic, and although calculations have not been made, it is expected that there is a continuous transition between the essentially stable and unstable states. Thus we see that this theoretical analysis of hydrodynamic instability is in quantitative accord with the experimental observations previously stated. We have said that instability will increase with an increase in the parameter of Eq. (18–49); it is also true that the rate of growth will increase linearly with an increase in

$$(\rho_d - \rho_c)(1 - R_d)/R_d \rho_c + (1 - R_d)\rho_d. \tag{18-50}$$

For gas systems, this is essentially unity.

A comparison between the analysis of stability by Davidson and Harrison [55] and the analysis of hydrodynamic instability by Jackson [134] has been given by Anderson and Jackson [3]. The results correspond almost exactly. Systems requiring a long growth distance (i.e., relatively stable systems) have bubbles whose maximum stable size is correspondingly small, while those that require a short growth distance have bubbles whose maximum stable size is quite large. Similar correspondences exist between all the approaches. There are of course deficiencies in all the methods proposed to date (we cannot predict all cases), and further work in this area can be expected.

18-4.B Minimum Transport Velocity

The transport of solids in horizontal or inclined tubes is closely related to fluidization in vertical systems. For both slurry and pneumatic transport, the velocity at which deposition of solids occurs is clearly of importance. Thomas [288–290] has defined the transition point for slurries in a phenomenological manner by considering the nature of the particle settling and the size of the particle relative to the turbulent characteristics near the boundary (pipe bottom). We have already mentioned the homogeneous condition which would exist for particles small enough to (1) be in the Stokes flow region, (2) have a size equivalent to a y^+ which is less than 5 (that is, be of such a size that they could be buried in the laminar sublayer), and (3) be in a flow field in which the fluid velocity exceeds the minimum transport velocity as given by

$$U_t/U_{om}^* = 0.010(dU_{om}^*/\nu_c)^{2.71}, \tag{18-51}$$

where U_{om}^* is the friction velocity at minimum transport and infinite dilution.

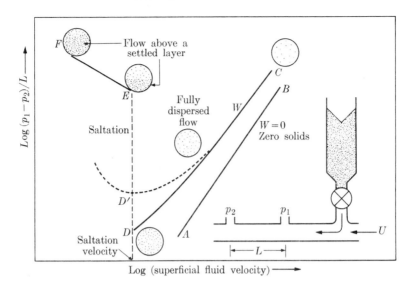

Fig. 18–9. Saltation velocity. [From F. A. Zenz, Petroleum Refiner 36, No. 6, 134, Gulf Publishing Co., Houston, Texas (1957). By permission.]

Equation (18–51) is for nonflocculated glass beads; for systems that can flocculate, Thomas [288] found the coefficient to be 0.0083 and the power to be 2.61. For particles larger than an equivalent of a y^+ of 5 (larger than the laminar sublayer), the correlation equation was

$$U_t/U_{om}^* = 4.90(dU_{om}^*/\nu_c)(\nu_c/d_0 U_{om}^*)^{0.60}(\rho_d/\rho_c - 1)^{0.23}, \qquad (18\text{–}52)$$

where d_0 is the pipe diameter. A map of the general flow regime has been suggested [290] and can be constructed for any $(\rho_d/\rho_c - 1)$. The use of this map to obtain the friction velocity, pressure drop, and mean stream velocity is discussed in the same reference. Besides homogeneous and heterogeneous flows, transverse and longitudinal wave formation are considered. Other references that can be cited are the review by Ellis et al. [69] and references 22, 105, 129, 202, 270, 320, and 322.

In pneumatic transport, the minimum transport velocity is often called the *saltation velocity*. Zenz and Othmer [320, 322] have given a rather complete discussion of saltation velocity, illustrating it by means of a plot of the log pressure drop versus the log superficial velocity (Fig. 18–9). The curve AB is the pressure drop for no solids in the pipe. As solids are added, the pressure drop is increased; for example, from B to C. If the rate of gas flow is now decreased, the pressure drop will decrease along the line CD; at D (the saltation velocity), solids start to settle out. The minimum transport velocity varies for different rates of solids flow. This is a time-dependent settling which, when equilibrium is obtained, will have a pressure drop corresponding to E. As the gas flow further decreases, the pressure drop changes along the line EF, each step of which would be an

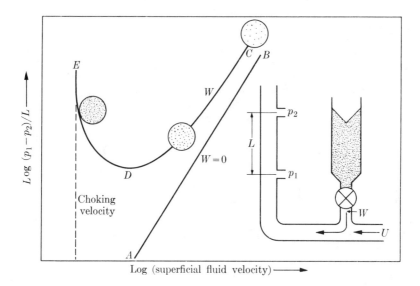

Fig. 18–10. *Choking velocity.* [From F. A. Zenz, Petroleum Refiner **36**, No. 6, 134, Gulf Publishing Co., Houston, Texas (1957). By permission.]

equilibrium process. It appears that some authors have reported curves as indicated by the dashed line in the figure. Apparently during the approach to equilibrium along the line DE, the point D' was obtained and resulted in the dashed line, rather than the correct curve $CDEF$. A correlation for the minimum transport velocity is difficult to obtain, and an early one by Zenz [320, 322] has been shown by both Leva [164] and Zenz [321] to be inadequate. Zenz showed by further experimentation that the groups originally proposed did not account for particle density and size distribution. He then developed the concept of a drag coefficient for the horizontal flow of a single particle which is analogous to the regular drag coefficient for vertical free fall. These coefficients were determined experimentally and are best represented graphically; however, the following equation is an approximate expression for solid spheres:

$$(C_D)^{1/2} = 1 - (1/N_{\text{Re},d}). \tag{18-53}$$

For particle Reynolds numbers of less than 50, the curve deviates markedly from the drag coefficient curve of vertical free fall, which explains why it is more difficult to transport small particles in a horizontal system than in a vertical one, and why it is more difficult to transport small particles than large ones.

For vertical pipe flow systems, the *choking velocity* is the flow rate at which slug formation occurs (see Fig. 18–10). The flow is similar to the horizontal case; as the gas velocity is decreased from C to D to E, the drag on the particles decreases until, at E, the drag is no longer able to support the solids and the suspension collapses. This is the point of slug formation, or choking velocity. Zenz observed that the minimum transport and choking velocities were the

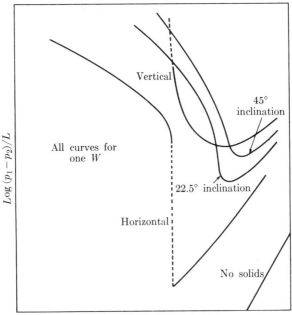

Fig. 18–11. Effect of tube inclination. [From F. A. Zenz, Petroleum Refiner **36**, No. 7, 177, Gulf Publishing Co., Houston, Texas (1957). By permission.]

same for a uniform particle size, but that for a distribution of particles, the minimum transport velocity was from three to six times the choking velocity.

Zenz has also considered velocities in an inclined tube. Figure 18–11 shows this changeover as a function of inclination. All the data curves were qualitatively the same.

The status of this field of research does not differ much from that of two-phase liquid–gas flow, on which comment has already been given. Much work is still needed if we are to arrive at the fundamentals of solids transport, and the data now available must be brought into some accord. Leva [164] has reviewed the literature on the movement of gas streams with a high solid content (25 to 100 weight ratio). Leva [164], Zenz [320], and Zenz and Othmer [322] have considered the nature of the gravity flow of solids in some detail. The literature surveys by both these authors are excellent, and should provide the basis for the definitive study still so badly needed.

18–5 MULTIPARTICLE SYSTEMS IN NONHOMOGENEOUS FLOW

The terms nonhomogeneous and heterogeneous are used here in the context to identify a solids–fluid system which is in direct contrast to the homogeneous solids–fluid system. Here we shall deal with systems in which concentrations of solids vary considerably from point to point. The discussion will center about the bed which is aggregately fluidized by gas, since this is the most practical application of the solids–gas technique.

18-5.A General Considerations

Before turning to fluidized beds *per se*, we should make reference to the empirical design methods for solids transport. Most of the information is contained in the references which have already been cited for homogeneous conditions and the transition between types. For slurry transport, references 22, 69, 105, 129, 202, 270, 288, 289, 290, 320, 322, 328, and 330 contain most of the available information, as well as further references. For pneumatic transport, references 61, 164, 192, 226, 320, 321, 322, and 329 provide parallel information, with Zenz and Othmer [322] being an excellent starting point because of their extensive review of the available experimental data.

At this point we must make some comments on, and references to, the literature for the design of fluidized bed systems. The minimum fluidization velocity, bed expansion, and the bed pressure drop are of particular importance. These are overall bed characteristics and do not reflect, except indirectly, the nature of the details within the bed, e.g., the existence of bubbles, their velocity, shape, etc., fluctuations in voidage, the circulation patterns of solids, and gradients in the concentration of solids with distance from the gas distributor. These details are considered a part of the fundamental analyses of bubble and solid dynamics, which will be treated in subsequent sections. The reference books on fluidization [164, 322] are the best sources of design information. To determine the point of incipient fluidization, we must estimate the minimum fluid bed voidage. Leva [164] has summarized the experimental measurements on a wide variety of materials. He has also summarized and compared six correlations for the minimum fluidization velocity suggested in the literature. One of these is based on data obtained by ten authors [165]:

$$G_{mf} = 1.40 \times 10^5 \, d^{1.82}[\rho_c(\rho_d - \rho_c)]^{0.94}/\mu_c^{0.88}, \qquad (18\text{--}54)$$

where consistent English engineering units are used, and G_{mf} is the superficial gas rate in lb/hr-ft^2. An alternate approach has been suggested by Zenz and Othmer [322], and Frantz [81] has given more recent data and a slightly modified correlation of the same nature as Eq. (18–54). There is only fair agreement between predictions from the various correlations. Work on the bed expansion at fluidization velocities above the minimum has also been reported. The methods suggested are empirical in nature and have been adequately covered and criticized in the reference books cited and in references 57, 72, 88, and 258. The pressure drop is often considered to be equal to the weight of bed per unit area; however, this is generally found to be only approximate [81, 168]. Comments on possible errors in measurements have been made by Sutherland [276]. Besides the reference books and the references cited in them, the effect of the distributor [43, 107, 121,

212], entrainment [57, 107, 171], particle size distribution [48, 57, 291], and electrostatic effects [47, 314] have been considered further.

In spite of the nonhomogeneous nature of the aggregately fluidized bed, it is a useful design approach to treat the system as homogeneous and to measure such properties as viscosity, surface tension, and miscibility. This approach has been taken in references 59, 85, 156, 187, and 248. The basic idea is that there is an analogy to a fluid, and that some property of the fluidized bed can be found for each property of the fluid. Furukawa and Ohmae [85] suggested that the temperature of a fluid would correspond to a term of the form of the kinetic energy as given by the product of the fluid viscosity and the fluid velocity in the fluidized system. With this term, a number of equations from liquid theory were used to correlate the properties of the fluidized bed. Wilhelm [314] discussed some of these problems in general terms, and noted that a fluidized bed should be non-Newtonian and should depend on the bulk density and powder dielectric constant; this latter variable is evidence that electrical forces will contribute significantly to particle interaction. With regard to the viscosity of the bed, Schügerl et al. [248] found that the shear diagram could be described by a hyperbolic sine function, with the constants related to the particle size and gas velocity.

18-5.B Bubble Dynamics

Before we proceed with a discussion of the characteristics of bubbles in a gas fluidized bed, we should understand that there are no bubbles in the bed in the conventional sense. The system is a mixture of solids and gas, in some areas of which the concentration of solids is so low that these areas are effectively large pockets of gas (voids) in an otherwise homogeneous mixture of solids and gases. These voids appear to be analogous to bubbles of gas in a liquid. In fact the analogy is so good under some conditions that similar equations apply to both cases. Underlying the various attempts to represent the motion is the intuitive idea that a fluidized bed is composed of two phases. We must be careful at this point, because the two phases refer to the uniform solids–gas mixture (called the *emulsion*) and to *voids* essentially free of solids, rather than to solids as one phase and gas as the other. The idea, first suggested by Toomey and Johnstone [294], will be discussed in more detail later. At present, what is important is the assumption that the solids–gas mixture phase is the same as the entire bed was at the point of incipient fluidization; i.e., the gas flow in this solids–gas emulsion is the rate for minimum fluidization, and the solids concentration (or void fraction) within this phase has not changed from the minimum fluidization value. All the gas flow in excess of that needed to maintain the emulsion passes up the bed as the *bubble phase*, and it is the characteristics of these bubbles that we wish to study. Just as in the gas–liquid system studied under free flow, our first step in the understanding of bubble dynamics will be to study a single bubble in an otherwise homogeneous emulsion just at the point of incipient fluidization. This is accomplished by injecting a bubble into a fluidized bed maintained by a separate air supply just at the conditions at which no bubbles form naturally. Such experiments cannot suggest or even exactly match the nature of natural bubble formation (i.e., the transition problem previously discussed), but they can give some insight into the characteristics of a bubble once it is formed. Of course, the formation of such bubbles at submerged orifices can be considered.

(a) Bubble formation at a submerged orifice. In the constant frequency range of bubble formation in liquids, the bubble volume is directly proportional to the volumetric flow. The form can be obtained by inspection of the empirical relation given in Eq. (17–51): $d = (6V/\pi)^{1/3} \sim d_n^{1/2} N_{Re,j}^{1/3}$, or

$$V \sim d_n^{3/2} N_{Re,j} \sim d_n^{3/2} Q/\nu_c. \qquad (18\text{--}55)$$

This indicates the additional dependency of bubble volume on the orifice diameter and on the continuous phase viscosity. As a more theoretical mechanism, Davidson and Harrison [55] have suggested that the buoyancy of the bubble in a liquid balances the rate at which the upward momentum of the liquid surrounding the bubble changes. This neglects the inertia of the air within the bubble and the surface tension effects. From this suggestion of Davidson and Harrison, we can show that the growing bubble has an upward acceleration equal to the downward acceleration of gravity. When the displacement distance equals the radius, the bubble is assumed to detach, and a volume balance gives us

$$V = (6/\pi)^{1/5} Q^{6/5}/g^{3/5}. \qquad (18\text{--}56)$$

Here neither the orifice size or the continuous phase viscosity are involved. Equation (18–56) fits the data reasonably well, although the diameter and flow rate dependencies shown in Eq. (18–55) are definitely valid.

Bubbles injected into fluidized beds at incipient fluidization have been treated in a similar manner by Davidson and Harrison. The data are that of Harrison and Leung [114]. We would expect Eq. (18–56) to be valid, since surface tension effects should be absent and a satisfactory fit is obtained. However, the dependency on the flow rate is much closer to that given by Eq. (18–55) than to that given by Eq. (18–56). The dependency on the orifice diameter exists, but is small and lies somewhere between that indicated by the two equations. It is most difficult to translate these results to a fluidized bed considerably above the minimum fluidization level and in which natural bubble formation occurs. The only indication is the measurements of Yasui and Johanson [235, 318]. Here the bubble thickness at any one height above the distributor was found to vary approximately with the 0.54 power of the excess velocity over that required for minimum fluidization. This would imply that the bubble volume should vary with about a 1.6 power of the flow rate. Thus, for liquids or fluidized beds at incipient fluidization, the dependency is to the first power by Eq. (18–55) or to a power of 1.2 by Eq. (18–56). At higher flow levels, the dependency is nearer to the power of 1.6.

(b) **Bubble shape.** The general shape of the bubble in an incipiently fluidized bed is that of a large liquid bubble and resembles the shape shown in the fifth and sixth frames of Fig. 17-33. Since it is difficult to observe these in a three-dimensional system, most pictures have been obtained in two-dimensional beds in which the solids are sandwiched between two transparent plates set close together. The x-ray photographs made by Rowe [55, 238] are exceptions to this general rule. He took x-ray pictures of bubbles in a large column containing glass beads or sand at the point of incipient fluidization (Fig. 18-12a). Rowe's results agree with the results obtained in simple two-dimensional beds, several examples of which are shown in Fig. 18-12. The general shape is circular, with a rather large wake or accumulation of solids occupying the lower third of the volume. The actual void is similar to that shown in the frames already cited in Fig. 17-33.

(c) **Motion.**

(i) **The rise of bubbles.** The velocity with which the void or bubble actually rises after being introduced into an incipiently fluidized bed is of primary concern to us. We might suspect from the previous discussion that the analogy to liquids will remain valid, and Davidson et al. [56] found this to be true. Further corroboration was obtained by Harrison and Leung [115], all of which is summarized by Davidson and Harrison [55]. The results, analogous to Eq. (17-77), were found to be

$$U_b = (0.71 \pm 0.11)g^{1/2}V^{1/6}. \tag{18-57}$$

In the experiments, the bubble was injected at various heights without any significant variation in its volume or velocity of rise. The inherent inaccuracies in the measurements and the weak dependency of the velocity on the bubble volume make Eq. (18-57) hard to establish; however, all the evidence supports such a relation for the quiescent incipient fluidized bed. The implication of the analysis is that bubble velocity does not depend on the nature of the particles. The data [55] indicate that a number of quite different materials can be represented by the same correlation. Sutherland [279] did find a small particle-diameter effect for glass beads, but it was within the deviation indicated in Eq. (18-57); Sutherland's data supported the $V^{1/6}$ dependency and was well represented by the equation, after a correction for wall effect was made. The correction was appreciable and it was apparent that the larger bubbles were becoming wall controlled slugs, as given by Eq. (17-78). Apparently a similar problem occurred in the data of Angelino et al. [5], as pointed out in the discussion by Pyle and Stewart [224].

Yasui and Johanson [318] and Baumgarten and Pigford [18] obtained results at conditions well above incipient fluidization. Extrapolation of their data to minimum conditions shows no change in bubble volume. At conditions greater than the minimum, the bubbles clearly grow as they rise, and the volume change is dependent on particle density, particle diameter, and gas flow rate. The growth of the bubbles appears to be linear with distance up the bed.

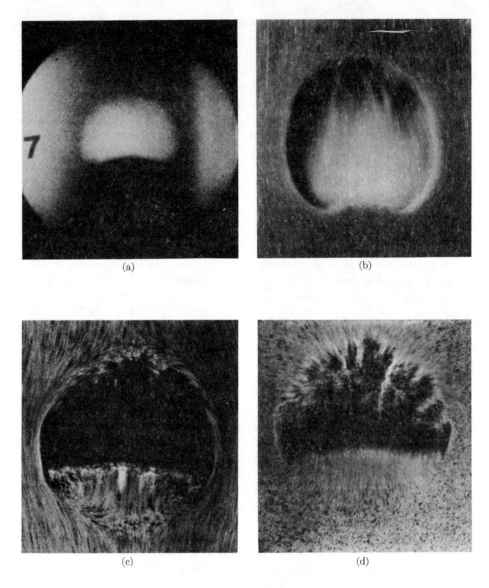

Fig. 18–12. Photographs of bubbles in a fluidized bed. (a) X-ray photograph of a bubble in soda-glass spheres carrying a wake of underlying lead-glass particles. [From P. N. Rowe, Chem. Eng. Progr. 60 No. 3, 75 (1964). By permission.] (b) 0.5 ballotini and air: [Both (a) and (b) from the Chemical Engineering Division, UKAEA, Research Group, Harwell, England. By permission.] (c) Shape of bubble and relative motion of solids (0.2–0.3 mm semolina fluidized in air). (d) Absolute motion of solids around rising bubble (0.2–0.3 mm semolina fluidized in air). [Both (c) and (d) from H. Reuter, Chem.-Ingr.-Tech. 35, 219–228 (1963). Verlag Chemie, Gmbh, Weinheim/Bergstr. By permission.]

At flow rates at which many bubbles appear, the bubble velocity relative to the ground is not the same as that relative to the emulsion. The difference is caused by the additional upward flow associated with all the other bubbles [see Eqs. (16–65) and (16–67)]. The rise relative to the ground is given by the sum of the rise relative to the fluid plus the upward induced velocity due to the inflow; that is,

$$U_b = U - U_{\mathrm{mf}} + 0.71 g^{1/2} V^{1/6}, \tag{18-58}$$

where $U - U_{\mathrm{mf}}$ is the flow that passes through as bubbles. Davidson *et al.* reported a constant bubble rise value, which was to be expected, since under conditions of minimum fluidization the bubble volume did not change [see Eq. (18–57)]. At higher flow rates, Yasui and Johanson reported a bubble velocity constant with height, even though the bubble volume increased. However, by Eq. (18–58), the bubble velocity is very weakly dependent on the volume. The actual dependency is dictated by the relative magnitude of $U - U_{\mathrm{mf}}$ and the term for the volume effect, and at a minimum, one has to nearly double the volume to see just a 10% change in the bubble velocity. In contrast, Baumgarten and Pigford deduced that the superficial bubble velocity increased with height from zero to nearly the superficial gas velocity. This apparent dilemma can be resolved by estimating, from the superficial bubble velocity and bubble size data, the true velocity of rise U_b. The calculations are crude, but for the higher flow rates, where the superficial velocity is known accurately, the values of U_b are in reasonable agreement with the measurements by others and are essentially constant. Further confirming information can be obtained from Bakker and Heertjes [13], who measured bed porosity. They found that this was independent of the amount of material in the bed for a given set of flow conditions, which can be true only if the bubble velocity is constant. Thus it appears that the bubble grows as it rises at an essentially constant velocity. The growth is probably associated with coalescence of the voids.

(ii) **Solids and gas flow patterns.** The velocity of a rising bubble can be adequately described by Eq. (18–57), which implies that the flow might be satisfactorily described by a potential motion. However, the differences between a bubble moving in a liquid and a void moving in a solids–gas mixture are quite marked, since, in the latter case, an exchange can occur between the two phases. Consequently, further experimental observations of the patterns of solids and gas flow are needed before an analytical representation will be possible. To an observer who is stationary relative to the ground, both the flow and the streamlines of the solids are unsteady, and photographs of the motion (Fig. 18–12d) must be interpreted very carefully. However, if one moves with the bubble, then the flow is steady and the streamlines and path lines of the solids are identical. Figures 18–12(b) and (c) are short time exposures moving with the bubbles. The path lines (or in this case streamlines) are similar to what one would expect for the potential flow over a cylinder or sphere; one such streamline is shown in

Fig. 6–7 and a number of them in Fig. 6–6. The wake, which is quite apparent, is for the most part internal to the sphere, and the flow external to the sphere is quite close to potential flow. There does not appear to be any viscous-type boundary layer in Figs. 18–12(b) or (c), since the particle velocity (as measured by length of streaks) does not seem to diminish as the hypothetical boundary is approached. Indeed, no viscous-type boundary layer should be expected, since there is no real interface on which it could form. This does not mean that interparticle forces, due to the interaction of particles, might not exist near the hypothetical surface, but rather that the interface is mobile and can move as a streamline with a finite velocity. The front and rear stagnation areas shown in Fig. 18–12(c) are somewhat clearer. Figure 18–12(d) is a photograph taken from a point of view which is stationary relative to the ground, and thus represents the instantaneous streamlines shown as dashed lines in Fig. 6–7. The correspondence is apparent. Rowe and Partridge pointed out [239] that the flow external to the void is probably potential. In the first place, it has a mobile boundary that prevents setting up a viscous boundary layer. In the second, its viscosity is low [Eq. (18–25) with (18–41)], i.e., only slightly greater than that of the gas, and it has a relatively high mixture density, both of which contribute to an extremely high Reynolds number.

Other predictable characteristics of the external flow, which are given in Fig. 6–8, are particle paths and the drift line. Comparable photographs of particle paths have not been made, but it is clear from Fig. 18–12(d) that the characteristic loops form. Drift lines have been reported in references 236 and 239. One drift line is reproduced as Fig. 18–13, and is of the expected shape. When the gas passes through the surface and out of the system, the particles (carried by the bubble as the internal vortex of solids) are left at the surface. On the left center of the drift profile are small trails, or barbs, probably a result of a rapid exchange by shedding from the internal vortex of solids to the external potential flow [239]. This is analogous to the vortex shedding from the wake in a real body in a liquid. The amount of exchange from both a slow continuous interchange and from sudden shedding is by no means small. The actual motion within the internal wake is difficult to establish from these photographs alone, but is apparently [203, 230, 239] a vortex ring with upflow at the axis and downflow near the boundary. Figures 18–12(c) and (d) infer that the wake might be for the most part like a small particulately fluidized bed lying at the bottom of the bubble [230]. The interchange occurs near the spherical surface and from the intermittent, but often present, raining of solids from the upper, inherently unstable surface; i.e., fresh solids enter from above, causing the wake to increase in size. When the wake is large enough, the potential flow shears away a part of the vortex, which becomes the sudden exchange cited above. The small internal fluidized bed is also in accord with the ideas of Harrison et al. [113], already presented, on the transition between flows. If the maximum upward velocity within the bubble is less than the free-fall velocity of the solids within the bed, then the particle will fall into the wake. If the opposite is true, the wake or particulate bed will expand and fill the void. A great deal is known

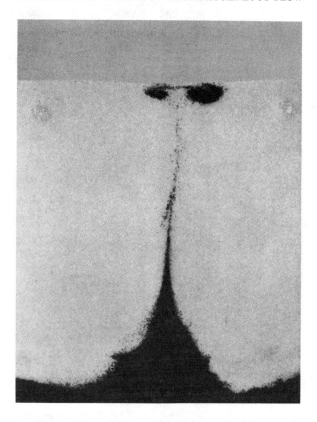

Fig. 18–13. Photograph of the solids displacement or drift line as caused by a single bubble. [From P. N. Rowe, and B. A. Partridge, Interactions Between Solids and Particles, *Inst. Chem. Engrs.*, London (1962), and the Chemical Engineering Division, UKAEA, Research Group, Harwell, England. By permission.]

about the motion of solids, but more information is needed on the interface between the void and the emulsion, and about the wake. It is not completely clear whether the interface is a surface of the emulsion maintained by gas flow, or whether particle interactions result in the formation of a denser layer or shell [228, 229, 230, 237, 239].

Little has been said about the flow of gas. The motion of solids approximates potential flow over a solid sphere, and we might suspect that gas flow could also be approximated by a potential motion. Specific experiments conducted to investigate this possibility [236, 240, 241, 307], have shown that under incipient fluidization conditions the gas streak lines were smooth, as might be expected for nonturbulent conditions. Because of the unsteady nature of a passing bubble (see Fig. 6–10), the exact interpretation of streak lines is difficult, but the analysis can be made and will be discussed in the next section. At this point some experimentally observed streak lines are shown in Fig. 18–14 (NO_2 streak

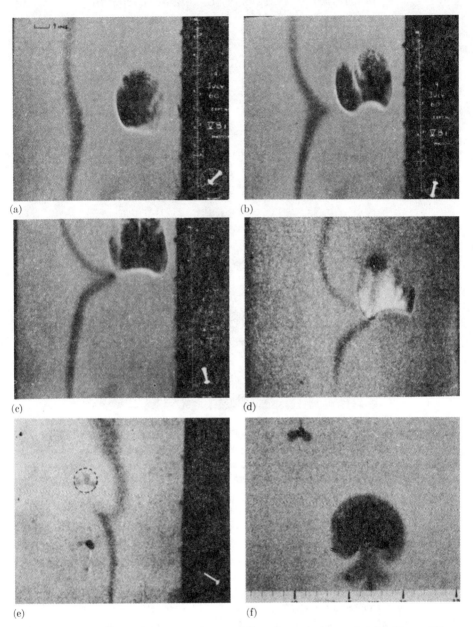

Fig. 18–14. (a), (b), (c) The drift of a streakline shown at intervals of 0.12 sec. [*From P. N. Rowe*, Chem. Eng. Progr. Symposium Series No. 38, **58**, 42 (1962). *By permission.*] (d) NO_2 streakline, $U_b/U_{mf} = 3/5$. (e) NO_2 streakline, $U_b/U_{mf} = 5/3$. [*Both (d) and (e) are from P. N. Rowe and B. A. Partridge*, Chem. Eng. Sci. **18**, 511 (1963). *By permission.*] (f) NO_2 cloud around two-dimensional bubble in 0.23 mm ballotini, $U_b/U_{mf} = 2.4$. [*From P. N. Rowe, B. A. Partridge, and E. Lyall*, Chem. Eng. Sci. **19**, 973 (1964). *By permission. All parts of Fig. 18–14 also by permission of The Chemical Engineering Division, UKEAA, Research Group, Harwell, England.*]

in a minimum fluidized system of glass microspheres with an air bubble injected). The sequence of Figs. 18–14(a), (b) and (c) show the smooth motion and the lateral movement of the entire streak line in time as the void progresses up the column. This is the same drift as was calculated in Example 6–4, and shown in Fig. 6–10. Figure 18–14(d) shows another streak line somewhat closer to the void. Here the gas from the streak is in part captured within the void. In Figs. 18–14(a), (b), (c), (d), the ratio, U_b/U_{mf} (bubble velocity to the minimum fluidization velocity) was less than unity. In Fig. 18–14(e), the ratio is greater than unity, and as will be seen, the upturn in the streak line [as contrasted to that in Figs. 18–14(a),(b),(c),(d)] can be predicted. Finally, Fig. 18–14(f) shows an NO_2 bubble injected, rather than an NO_2 streak line and air bubbles. In this case, U_b/U_{mf} is greater than unity. The rather sharp cloud around the void is typical of this type of injection, and once again can be reasonably well described by potential theory. Wace and Burnett [307] drew a very important conclusion from streak-line studies, observing that the gas flows into the bottom of the void and out through the sides and top. They further suggested at that time that potential flow theory might apply.

(d) The ideal bubble model. The following tentative model for the bubble rise can be made on the basis of experimental observations of the motion of solids and gases about a void in an incipiently fluidized bed.

(i) The model. A mixture of solid particles is fluidized by a well-distributed gas source moving at as high a velocity as possible without natural bubble formation. A sudden injection of gas through a separate source causes formation of a void, which then travels up the bed. The general shape of the void is circular, with the bottom third filled with a wake that has the characteristics of a little particulately fluidized bed. Circulation is quite apparent near the outer part and resembles a toroidal vortex, such as we often see when we start a circular object into motion in a fluid. The solids pass around the void in a potential motion. This external flow exchanges solids with the wake by raining solids down from the roof of the void and by an exchange along the surface between the emulsion and the wake. When the wake has increased in size because of the occlusion of solids, a part of the circulating motion is suddenly shed in a manner similar to the shedding of a vortex in a liquid. The gas also flows in a potential manner and is drawn toward the void. It enters at the bottom and exits from the sides and top. In some manner the gas flow and solids motion maintain the void as a discrete identity, even though it is continually exchanging both gas and solids with the emulsion.

The exact nature of the mechanism which supports the inherently unstable void is not yet known, and, as indicated previously, further experimental information is necessary. Even so, some comments are in order on the suggestions already made. An early suggestion, which can be rejected, is that the void progresses up the bed by a rapid and continuous raining of solids from the top to the bottom. The actual pictures of lateral solids motion can be cited as being inconsistent with this hypothesis. Further, the gas flow experiments show that the gas is drawn toward the void, which must result in a higher velocity of gas in the void, flowing from the bottom to the top. This flow must be a factor in retarding the raining of solids. The remaining mechanistic suggestions are that the void can be supported entirely by the gas flow just described [55, 237, 239], or that a shell of slightly more dense solids forms around the void [228–230]. Photographs of bubbles in a fluidized bed (see Fig. 18–12), particularly of bubbles in water [230], do not support the hypothesis of a shell structure. The analysis by Rowe [237] shows that the gas flow is sufficient to cause particle paths like those observed; however, this in itself does not rule out the existence of interparticle forces, and in fact it is hard to rationalize that the observed pressure gradient occurs without some interaction. It is entirely conceivable that solid-to-solid contact in combination with the gas flow patterns contributes to the existence of voids, and that each plays its part to varying degrees of importance at different positions and periods in the life of a void.

(ii) Potential motion representations. Using experimental evidence to justify the assumption of potential motion, we can make further analytical comparisons to help establish the adequacy of the assumption. For the motion of solids, the velocity potential and stream function have been given in Chapter 17 in the discussion of large bubble motion, i.e., for the two-dimensional case,

$$\phi_s = U_b(r + r_0^2/r) \cos \theta \quad \text{and} \quad \psi_s = U_b(r - r_0^2/r) \sin \theta, \quad (18\text{–}59)$$

and for the three-dimensional case,

$$\phi_s = U_b(r + r_0^3/2r^2) \cos \theta \quad \text{and} \quad \psi_s = \tfrac{1}{2}U_b(r^2 - r_0^3/r) \sin^2 \theta. \quad (18\text{–}60)$$

These are equations for the flow over a stationary sphere and can be compared directly with Fig. 18–12(c), since this was taken from a frame of reference moving with the void. This figure is reproduced again as Fig. 18 15, but this time with an overlay of streamlines from the potential motion. The agreement is quite satisfactory, although clearly there are local differences.

The gas flow is quite complex; a simple approximate form can be obtained from the flow in a uniform field with a void. For a stationary void, the stream function is quite similar to Eqs. (18–59) and (18–60):

$$\psi_g = U_{mf}(r + r_0^2/r) \sin \theta, \quad \text{two-dimensional,}$$
$$\psi_g = U_{mf}(r^2 + r_0^3/r) \sin^2 \theta, \quad \text{three-dimensional.} \quad (18\text{–}61)$$

Equation 18–61 for the two-dimensional case is shown as Fig. 18–16(a). The gas passing through the void doubles; i.e., all the flow within two diameters of the void center passes through the void. For the three-dimensional case, the flow inside triples. Since there is a change in void fraction, the velocity within the void is $2U_{mf}(1 - R_{g,mf})$ and $3U_{mf}(1 - R_{g,mf})$, respectively.

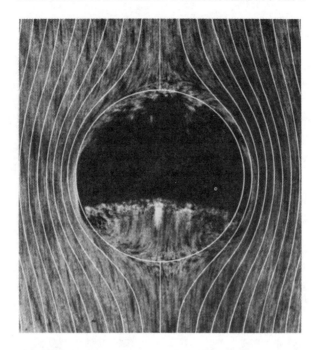

Fig. 18–15. Shape of bubble and relative motion of solids as in Fig. 18–12(c), but with addition of an overlay of the potential motion. [From paper presented by H. Reuter, at meeting of AIChE, Houston, Texas (Feb. 1965).

If this void is now translated at a velocity U_b, one must modify the stream function, just as was done in Example 6–2. This involves superimposing the motion given by

$$\psi = -U_b y = -U_b r \sin \theta \tag{18-62}$$

on Eq. (18–61). For the two-dimensional case, we obtain the instantaneous streamlines from the viewpoint of an observer watching the void pass:

$$\psi_g = U_{\mathrm{mf}}(1 - U_b/U_{\mathrm{mf}} + r_0^2/r^2) \, r \sin \theta. \tag{18-63}$$

This function, suggested by Woodrow (see Rowe [240]), ignores the particle motion and in effect maintains the particles stationary. They disappear when they encounter the top of the void and immediately reappear at the bottom. The equation does provide streak lines similar to those experimentally observed [240], but the internal velocity is now

$$(2U_{\mathrm{mf}} - U_b)(1 - R_{g,\mathrm{mf}})$$

for the two-dimensional case, which does not seem likely, especially for $U_b/U_{\mathrm{mf}} \geqslant 2$.

Davidson's analysis [54] accounted for the motion of the particles by adding the appropriate particle streamline to Eq. (18–63). Equation (18–59) is for the

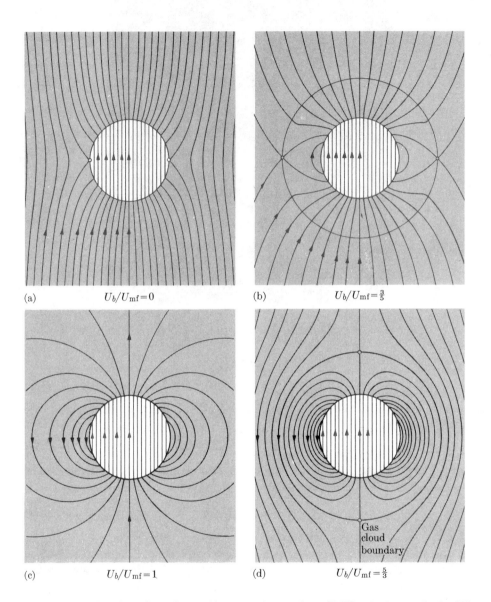

Fig. 18–16. Gas flow through a void in relative motion; U_b/U_{mf} is, in (a) 0, (b) 3/5, (c) 1.0, (d) 5/3. [From P. N. Rowe, Chem. Eng. Progr. Symposium Series No. 38 **58**, 42 (1962). By permission.]

view translated with the sphere; this must first be transformed into a form compatible with Eq. (18–63), that is, to a form which describes what is observed by a viewer who is stationary with respect to the ground, watching the sphere pass. This is again the problem of Example 6–2, and, for the streamline, a combination of Eqs. (18–59) and (18–62) gives

$$\psi_s = U_b(r - r_0^2/r) \sin \theta - U_b r \sin \theta$$

or

$$\psi_s = -U_b(r_0^2/r^2)r \sin \theta. \tag{18-64}$$

Combining Eqs. (18–63) and (18–64) ($-\psi_s$ is used, since the solids motion opposes the gas motion) gives

$$\psi_g = U_{mf}[(1 - U_b/U_{mf}) + (1 + U_b/U_{mf})(r_0^2/r^2)] r \sin \theta, \tag{18-65}$$

which has been rearranged to an alternate form [55]:

$$\psi_g = (U_b - U_{mf})(1 - A^2/r^2) r \sin \theta, \tag{18-66}$$

where

$$A^2/r_0^2 = (U_b + U_{mf})/(U_b - U_{mf}).$$

For the three-dimensional case, Davidson [55] gives

$$\psi_g = \tfrac{1}{2}U_{mf}[(1 - U_b/U_{mf}) + (2 + U_b/U_{mf})(r_0^3/r^3)] r^2 \sin^2 \theta$$
$$= \tfrac{1}{2}(U_b - U_{mf})(1 - A^3/r^3) r^2 \sin^2 \theta, \tag{18-67}$$

where

$$A^3/r_0^3 = (U_b + 2U_{mf})/(U_b - U_{mf}).$$

Equation (18–65) is plotted in Figs. 18–16 (b, c, d) for U_b/U_{mf} of $\tfrac{3}{5}$, 1, and $\tfrac{5}{3}$, respectively. It reduces to Fig. 18–16(a) for $U_b/U_{mf} = 0$.

These are the instantaneous streamlines associated with the upward movement of a void as observed from a frame of reference fixed in relation to the ground. These cannot be observed directly, but from them one can calculate the expected fluid path lines and streak lines. The method has been given in Examples 6–3 and 6–4, respectively. Figure 6–9 shows one example of this calculation for U_b/U_{mf} of $\tfrac{5}{3}$; other examples can be found in reference 240. Several specific calculated path lines are shown in Fig. 18–17(a), together with a comparison between them and experimental data (obtained by following a small element of injected NO_2 tracer). Such a comparison is difficult to make because the finite size of the element introduces experimental error and because the calculation is extremely sensitive to the starting point, as indicated by the variation of the path line with injection point (A, B, and C). Figure 18–17(b) shows a streak line constructed by connecting the equal time points on various path lines (a few are shown as dashed lines). Figure 18–17(c) is a repeat of Fig. 18–14(d), but with the calculated streak line superimposed. The comparison is quite reasonable. Figure 18–17(d) shows the development in time of a streak line calculated for the conditions of Fig. 18–14(e).

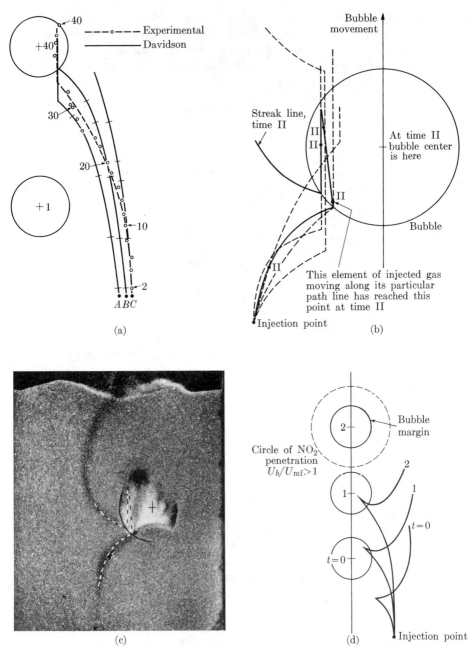

Fig. 18–17. (a) Comparison of observed and calculated pathlines. (b) Streakline constructed from a series of path lines. (c) Comparison of observed streakline with calculated streakline. (d) Series of streaklines calculated for $U_b/U_{mf} = 5/3$. [All from P. N. Rowe and B. A. Partridge, Chem. Eng. Sci. **18**, 511 (1963). By permission.]

The upturned segment is clear in both the experiment and the calculation. For $U_b/U_{mf} > 1$, the streamlines are shown in Fig. 18–16(d), and for these conditions there is, inside the line (called the gas cloud boundary), a circulation of fluid with no exchange between it and the external flow of emulsion gas. If a gas is injected outside this sphere, as was done in Figs. 18–14(e) and 18–17(d), it can never reach the bubble. If a gas is injected within this sphere, as shown with an NO_2 bubble in Fig. 18–14(f), then ideally it should remain with the void as observed. Of course, this condition is not exactly obtained in practice, since the wake, also containing some of the gas, can be shed [241]. This shedding can be seen to occur in the lower part of Fig. 18–14(f). Even in the ideal case, in which gas in the void is trapped within the cloud boundary, the gas comes into contact with new solids. Remember that fresh solids are passing all parts of the rising void, as shown in Fig. 18–15, and that these solid streamlines are superimposed on the fluid streamlines shown in Fig. 18–16(d). The actual ideal flow patterns within the bubble have been calculated by Pyle and Rose [223], who also estimated the fraction of fluid in contact with the particles as a function of the relative bubble velocity.

For any set of conditions, the theories can predict the ratio of the radius of the gas cloud to the radius of the void. A comparison of Eq. (18–59) with (18–66) [or (18–60) with (18–67)] shows that the flow external to the cloud boundary is the same as the flow over a circular shape, with the cloud radius being given by A. Thus the terms A/r_0 of Eqs. (18–66) and (18–67) are the desired ratios. For two-dimensional flow, Davidson's analysis gives

$$\frac{A}{r_0} = \sqrt{\frac{U_b + U_{mf}}{U_b - U_{mf}}} = \sqrt{\frac{1 + U_b/U_{mf}}{U_b/U_{mf} - 1}}. \tag{18-68}$$

Jackson [134] and Murray [203, 204] have presented more refined theories. Jackson's analysis is three-dimensional and requires a numerical solution. For the two-dimensional stream function, Murray's analysis gives

$$\psi_g = [1 - U_b/U_{mf} + (U_b/U_{mf})(1/r^2) + (1/2r^3) \cos \theta] \, r \sin \theta, \tag{18-69}$$

where it has been assumed that $r_0 = 1$. The ratio of the cloud radius to the void radius is given by $\psi_g = 0$ streamline, or

$$(1 - U_b/U_{mf})r^3 + (U_b/U_{mf})r + \tfrac{1}{2} \cos \theta = 0, \tag{18-70}$$

and since $r_0 = 1$, the value of r from this equation will be the ratio of the cloud radius to the void radius. Figure 18–18 shows a comparison at $\theta = 0°$, and although there is considerable scatter, the data support the analysis of Murray. Both Jackson and Murray predicted that the cloud would be displaced ahead of the bubble, as observed. Rowe et al. [241] have further discussed the cloud formation, including its three-dimensional aspects, loss of gas from the cloud, and the effects on the cloud of void splitting and coalescence. Figure 18–14(f) shows how one

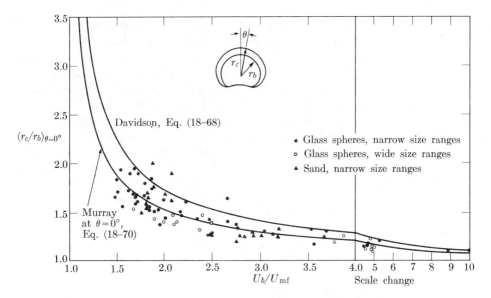

Fig. 18–18. $(r_c/r_b)_{\theta=0°}$ versus U_b/U_{mf} for two-dimensional case. [From P. N. Rowe, B. A. Partridge, and E. Lyall, Chem. Eng. Sci. **19**, 973 (1964). By permission.]

side of the cloud thickens at the bottom and becomes displaced into the surrounding flow, after which it is swept around toward the wake and is left in the trail. Reference 241 contains movie sequences and other illustrations of this action. This thickening and displacement must of course contribute considerably to the process of contact between the gas and solids in a fluidized bed. The subject is quite complex, and further work is needed before a more quantitative appraisal is possible.

Although it is quite apparent that the assumption of a nonviscous emulsion (under conditions of incipient fluidization) can offer a good description of the observed flow, such an assumption does not explain why the motions occur. For this, we must start from the equations of motion and continuity which are descriptive of the system, and predict the stream function from the reasonable basic assumptions that are necessary to obtain a solution. This is of course a formidable task, and in effect is the approach taken by the various authors cited. Davidson [54] assumed potential flow for the solids. He also assumed that the relative velocity between the fluid and the particles was proportional to the pressure gradient within the fluid. These assumptions, together with the equations of continuity of the solid and fluid, allowed him to prove that the fluid pressure obeys the Laplace equation. A form of the fluid pressure was selected to satisify certain boundary conditions; at the void it was assumed that the pressure was constant and far removed from the void it was assumed that the pressure gradient should be constant. With these assumptions, the previously cited streamlines for the fluid (Eqs. 18–66 and 18–67) can be derived. The pressure distribution has been measured

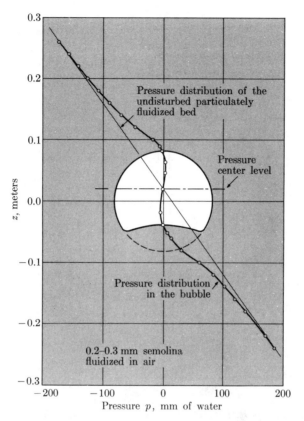

Fig. 18–19. Pressure distribution along the axis of symmetry of a bubble. [From paper presented by H. Reuter at meeting of AIChE, Houston, Texas (Feb. 1965). By permission.]

in one case [228–230] and is shown in Fig. 18–19. The form for the fluid pressure assumed by Davidson gives a total pressure in the vicinity of and above the void of the same shape, but is unsatisfactory below the void. Apparently we need only a weak assumption about the pressure distribution in order to show that potential flow is reasonable. Jackson [134] treated a more general set of equations and, although numerical methods were necessary, the similarity of the results implies that reasonable assumptions can lead to a potential-type motion. Finally Murray [203, 204] used a detailed and quite rigorous application of the basic equations. Unfortunately, a solution is impossible without further assumptions. For the two-dimensional case, it was found that the pressure (solved for) was not constant on the void surface, as assumed by Davidson; however, the vertical pressure gradient did not agree well with the observations cited in Fig. 18–19. Nevertheless, the streamlines given by Eq. (18–69) are quite reasonable, and Fig. 18–20 from Rowe et al. [241] offers a good summary for the conditions under which cloud formation occurs.

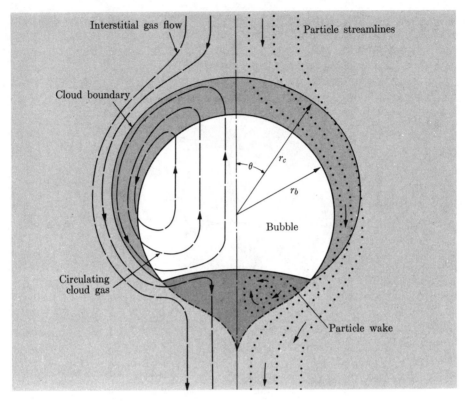

Fig. 18–20. Gas and particle flow patterns near a bubble, $U_b/U_{mf} \sim 2.5$. [From P. N. Rowe, B. A. Partridge, and E. Lyall, Chem. Eng. Sci. **19**, 973 (1964). By permission.]

(e) Bed characteristics at higher velocities

(i) **Bubble interactions.** Practical aggregate fluidization systems operate under conditions far removed from minimum fluidization and single bubble formation. Nevertheless, these latter studies do shed light on the fundamental processes that operate in fluidized beds. The natural extension of single bubble dynamics is the interaction of more than one void and the occurrence of channeling. Harrison and Leung [55, 116] first studied the coalescence of bubbles in a fluidized bed at a few percent above minimum fluidization conditions. We have already cited earlier studies [18, 318] made at fluidization conditions higher than minimum, which showed that bubble volume increased as the bubble rose. This was attributed to coalescence, but direct visual observation could not be made and the exact mechanism could not be established. The experiments made under conditions of essentially incipient fluidization indicated that the velocity of the lead bubble was unaltered and that the following bubble was unaffected if it was about one bubble diameter behind the lead void. Within this range, however, the following bubble appeared to be in a wake which was being transported with the lead bubble, and moved upward with its normal velocity relative to this wake. As a result, the rear bubble moved faster than the lead bubble and coalesced with it. So long as wall effects were absent, the main part of the wake was transported at essentially the same velocity as the lead bubble. Rowe et al. [241] considered the gas cloud surrounding two voids during coalescence, observing that the clouds also coalesced.

Chains of bubbles injected near the conditions of minimum fluidization have received attention by Botterill et al. [24, 26]. It is expected that the conditions used were comparable to those experienced in a bed operated at conditions considerably above minimum fluidization; the only apparent difference was that they studied a single chain of bubbles, rather than the interactions among a number of chains. In their earlier work, U_b/U_{mf} was less than unity because of the large particles used. Less than 10% of the gas volume injected ended up in the bubbles; apparently the gas could readily escape from the injector directly into the bed without being observed as a void. The actual gas velocity through the closed chain of bubbles near the injector was high and provided a low-pressure path for gas flow. There were indications of defluidization in this area. The chain of bubbles rapidly grew with height into separate bubbles and slowed down. Coalescence was a major factor. Much smaller particles were used in the later work, so that U_b/U_{mf} was greater than unity. Here the motion was followed by an elegant x-ray motion picture method. The results clearly showed that the bubbles grew as they rose, and that the equilibrium size for the material studied (depending on injection rate) was obtained within about 10 cm above the injector. The bubble velocity decreased as the bubble grew by coalescence, and leveled out when the bubble obtained its equilibrium size. The velocity of the bubbles did not correspond to the potential model, but neither did their shape. For all beds, most of the flow was through the bubble chain; in deep beds, however, where bubbles can develop to a nearly independent state, some gas appeared to leak into the emulsion. For a shallow bed, the flow was so high that the leakage was to the chain of bubbles and the emulsion could have become somewhat defluidized. The observations of Hassett [118], and comments thereon, illustrate some of the difficulties one can encounter in experimentation, such as partially fluidized beds, initial consolidation effects, and entrance design effects.

It is hoped that the work so far accomplished will prove to be the means of obtaining a better understanding of fluidized bed systems. But clearly much work needs to be done. For example, we have just discussed the observation that bubble size increases as a result of coalescence and that there is an accompanying decrease in bubble velocity; yet Yasui and Johanson [318] have observed that bubble size increases at a constant bubble velocity. How shall we rationalize this inconsistency? Does it result from injecting bubbles into fluidized beds which were at conditions of minimum fluidization, or perhaps from studying natural bubbles at conditions of greater than minimum fluidization? Or are other factors, such as a range of void sizes, gas leakage, etc., of importance? Such questions cannot be answered conclusively at the present time. For practical reasons, however, fully fluidized beds must be studied, although not always from the fundamental view of bubble dynamics. Secondary effects and statistical criteria, which are a direct result of bubble dynamics, can be used in this regard and will be discussed briefly.

(ii) **Voidage distributions.** We can ignore the local fluctuations of voidage in the fully fluidized bed and consider average values. It is expected that these averages will vary widely with location within the bed. Various means can be used to measure local porosities: light, acoustics [106], x-ray or gamma-ray attenuation [71], and local capacitance measurement [12]. The most successful and complete survey was obtained by Bakker and Heertjes [11] by the capacitance method, and is shown as Fig. 18–21. The fluidized bed can be separated into three major vertical zones, as can be seen in the figure. The lower zone is expected to be dependent on the nature of the distributor, and is one of lower solids. The main central zone is composed of most of the bed material, and is essentially at a constant void fraction. The upper dilute region of decreasing concentration of solids is caused by the entrainment of solids by the vigorous bubbling at the surface. Supporting evidence for the multizone model can be found in references 60, 71, and 106. Empirical correlations for the solids concentration in the various zones are given in the references cited.

(iii) **Voidage fluctuations.** The local variation in solids concentration or void fraction as a result of the rapid bubble movement can be divided in a statistical manner into an average and fluctuating part, as was done for velocity in turbulence. The preceding section considered the average part, and the fluctuating contribution (some statistical average) can be obtained in a number of ways: capacitance [62, 160, 180, 201], piezoelectric crystals to measure impulses in the bed [90], hot-wire methods [235], pressure difference [162, 253, 279, 287], gamma-ray absorption [10], and others [101, 284]. In each case the bed quality must be defined in some manner, usually as a deviation from an average divided by the frequency. This measure of the quality is obtained by dividing the fraction area of a recording of the fluctuations by the number of times the curve has intersected the average line. Sutherland [279] used a method equivalent to the length of chart line in a given time interval. Teoreanu [287] and Bailie et al. [10] used a more statistical approach, in which certain variances in solids concentration or voids were defined. At this stage, accurate comparisons between approaches are nearly impossible, since each method of measurement and eventual evaluation differs from the others. However, each analysis is internally consistent and a number of general conclusions can be reached: Bed quality decreases with an increase in velocity, size, height above the distributor, and solids density; bed quality is increased by the use of a wider distribution of solids at a given average particle size; and bed quality is unaffected by the height of the bed above the point being measured.

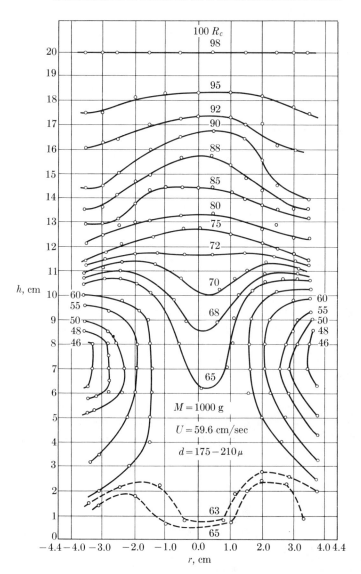

Fig. 18–21. *Voidage-distribution plot.* [From P. J. Bakker, and P. M. Heertjes, Chem. Eng. Sci. **12**, 260 (1960). By permission.]

(iv) Gas residence time distribution. Another overall characteristic of aggregately fluidized beds, which is a direct consequence of bubble action, is the gas residence time distribution. The interpretation of experimental results usually requires the formulation of models which have not yet been discussed here; therefore, we shall cite some of the experimental work at this point, but reserve the interpretation for a later section. There is little difference in principle between the gas residence

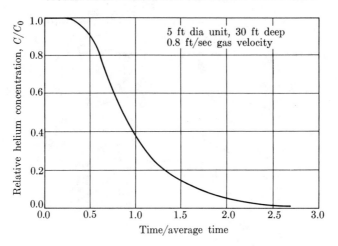

Fig. 18–22. Typical decay curve. [*From W. May*, Chem. Eng. Progr. **55**, *No. 12*, *49 (1959). By permission.*]

time distribution for the fluidized bed reactor and the previously mentioned liquid residence time distribution for a mixing-tank reactor system. They both serve the same purpose. In either case, there are several relatable procedures, such as pulse injection and step change, that can be used to determine the curve. The method used by a number of experimenters [53, 96, 97, 108, 132, 191] is to follow the time decay of the tracer concentration at the top of the bed when the injection of tracer into the feed is stopped, or to study the spread of an initial pulse [132]. A typical decay curve is shown plotted in two ways in Figs. 18–22 and 18–23. In Fig. 18–23, the uniform front of piston or plug flow and the semilogarithmic decay of complete mixing are shown for the sake of comparison. References on the subject of residence time distribution and its measurement are 52, 53, 58, 95, 96, 97, 108, 132, 191, 211, and 216. DeMaria and Longfield [58] reported experimental work on fluidized beds with diameters of from 4 in. to 13 ft. Their report is noteworthy because they also obtained and evaluated a number of local measurements within the beds, and thus could arrive at some conclusions about the detailed mixing patterns and residence time distribution within the vessels. It is of course impossible to present curves (such as Figs. 18–22 or 18–23) for all conditions, materials, and geometeries of fluidized beds. Therefore, so that simple correlations can be obtained, it is desirable to interpret measurements in terms of, for example, a diffusivity and/or cross flow between the emulsion and bubble phases. As noted earlier, we need a model or a statistical analysis of the results in order to do this. Thus this subject will be considered again after such models and procedures have been formulated.

18-5.C Solid Dynamics

The motion of solids in aggregately fluidized beds can be divided into three parts: the actual motions and patterns traveled by the solids, a description of the mixing

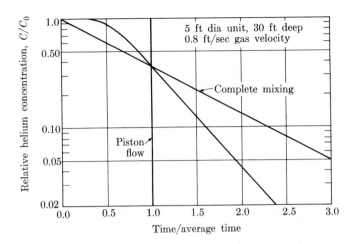

Fig. 18-23. Typical decay curve. [From W. May, Chem. Eng. Progr. **55**, No. 12, 49 (1959). By permission.]

in terms of some model, and the particle residence time distribution when there is an inflow and outflow of solids as might exist in a chemical reactor consisting of a fluidized bed. Compared with bubble dynamics, relatively little work has been done on solid dynamics.

Kondukov et al. [155] used a radioisotopic method to continuously follow the position of a radioactive particle. Like the corresponding work on a liquid fluidized system already mentioned, this method has the unique capability of following the particle motion anywhere in the fluidized bed. The particle clearly moved over the whole volume of the bed and made continuous upward and downward movements both at the wall and in the interior. It was observed that, during this oscillatory motion, the particle almost never reached the gas distributor grid, which is in accord with the previously noted lower density of this area, and is probably the result of the jetting action from the grid. The results of the work of Kondukov et al. suggest that large masses of material move downward with slow velocities, and smaller (or less dense) masses move upward at higher velocities. Considerable information is also available on the actual particle velocities within the bed; they are far from steady. Visual studies [186, 294] of the wall area have been made and suggest a general downward motion at the wall. Abrupt changes in velocity were noted as occurring in this area. Marsheck and Gomezplata [179] used a directional thermistor anemometer probe to measure the total mass velocity, and since the contribution of the solids was so much greater than the gas, it effectively measured the particle flow. Very rapid changes in particle velocity were observed, but these were averaged over a long period to obtain values for the bulk movement of solids rather than instantaneous values. One of the most interesting observations made by Marsheck and Gomezplata was that, although axial symmetry existed, the flow patterns were not the same along different radii at a given height. They discovered that in the system studied, the profiles were the same at 60° intervals, i.e., there were six similar wedge-shaped flow pattern cells arranged around the column axis. It is expected that the type of air distributor used was a factor in determining the exact nature and stability of the flow pattern. Variations of patterns with height and radial position were complex, but they were consistent with the hypothesis that bubbles grow in size through coalescence as they rise in the bed, and that the solids flow pattern is dictated by these bubbles. The solids motion observed was also in agreement with Kondukov et al. in that, to maintain a condition of no net solids transport, the particles must have moved upward at a higher velocity, and in less dense surroundings, than when they moved downward.

The detailed solids motion is clearly complex and apparently determined by the gas bubble flow. For their system, Marsheck and Gomezplata obtained a simple relation for the mass average velocity (lb/hr-ft^2) of circulation (the net flow of solids was zero; the circulation was obtained from averaging without regard to direction). The bulk circulation rate was found to be proportional to the square root of the product of the superficial air velocity and the nondimensional height above the air distributor. This implies that turnover of solids is probably better at the top of the bed than near the bottom, or at least that it is not the same everywhere.

A means of estimating the degree of solids mixing is to introduce tagged particles into the bed and follow their dispersion [29, 33, 120, 130, 144, 164, 171, 172, 207, 255, 278, 282, 283, 301, 302]. It is hoped that a simple model can be used to give a measure of mixing, such as a diffusivity, mixing time constant, or a mixing circulation rate. The last term does not mean the rate measured by Marsheck and Gomezplata; they measured the slow bulk turnover, or time-averaged circulation rate, of a randomly mixed bed. One would expect the actual mixing of solids to depend on both this and the details of the random mixing. Thus any measured mixing circulation rate would be much higher. The material to be followed has been tagged by widely varying means, of which colored particles [33, 144, 164, 207, 278, 282, 301] and radioactive isotopes [130, 172, 191, 255] are probably the most common. Chemical means have also been used; e.g., analysis of particles impregnated with NaCl, coke deposition on catalyst [302], and ion exchange resin types [283]. Still other particles are magnetically tagged [29], and particles with different resistivities are followed with a resistance probe [120]. The results have generally been reported by one of the three methods cited (diffusivity, time constant, circulation rate), but there has been little comparison between reports because of this and because of the widely varying conditions, geometry, and materials. However, in terms of diffusivity and circulation rate, the representations appear to be linearly dependent on U/U_{mf}. The circulation rate decreases with an increase in particle diameter. In some cases, the representation in terms of a diffusivity was not satisfactory, in that a single value could not be established [172, 191, 282].

If the tagged material is followed below the injection point as a function of time and distance, and if uniformity with cross section is assumed, then the

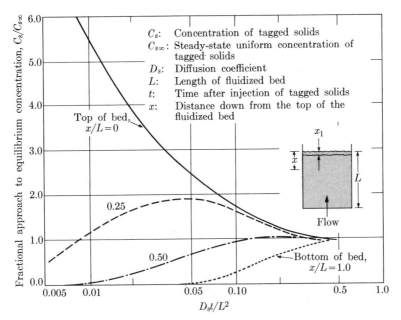

Fig. 18-24. Solids mixing. [From paper presented by W. May at meeting of AIChE, Chicago (Dec. 1958). By permission.]

differential equation is

$$D_s(\partial^2 C_s/\partial x^2) = \partial C_s/\partial t, \qquad (18\text{-}71)$$

where D_s is an average solids mixing coefficient or diffusivity, and C_s is the concentration of tagged solids. The boundary conditions for injection at the top are

$$\begin{aligned} t &= 0, & C_s &= C_{s0}, & 0 &\leqslant x \leqslant x_1, \\ & & C_s &= 0, & x_1 &< x \leqslant L, \\ t &= \infty, & C_s &= C_{s\infty}, & C_{s\infty} &= C_{s0}x_1/L, \qquad (18\text{-}72) \\ t &\geqslant 0, & \partial C_s/\partial x &= 0, & x &= 0 \text{ and } x = L, \end{aligned}$$

where x_1 is the thickness of the tagged solids layer at $t = 0$ and L is the height of the bed. The distance x is measured in the direction of flow of the solids (see Fig. 18-24). For extremely small amounts of tagged solids (x_1 small), the solution is

$$C_s/C_{s\infty} = 1 + 2 \sum_{n=1}^{\infty} e^{-D_s(n\pi/L)^2 t} \cos(n\pi x/L), \qquad (18\text{-}73)$$

which is plotted in Fig. 18-24 as a function of $D_s t/L^2$ and a position parameter x/L. May [191] determined D_s by matching experimental curves to this figure. Data for a column with a diameter of 15 in. gave a reasonably good fit to the curves, while a unit with a diameter of 5 ft gave results that were more erratic, and a single parameter was considered inadequate; even so, a single best-fit

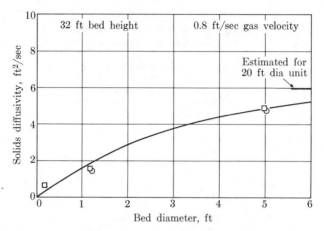

Fig. 18–25. Solids mixing-rate data. [From W. May, Chem. Eng. Progr. **55**, No. 12, 49 (1959). By permission.]

coefficient was used. Figure 18–25 gives the results from a very large number of tests made on a fluid catalyic cracking catalyst at about 50 times the minimum fluidization conditions. May noted that in the large-diameter column, the mixing is enormous; 50 grams of tagged solids were almost completely mixed into 15 tons of catalyst in less than one minute. Results for larger particles and conditions only several times minimum fluidization are much lower, being measured better in terms of ft^2/hr than in terms of ft^2/sec, as in Fig. 18–25. Although considerable data are now available, the form of presentation precludes a genuine comparison, and what is now needed in this area of research is a re-evaluation of the available information and an attempt to correlate this material in terms of more nearly universal mixing parameters.

When there is net solids flow, the residence time distribution of the solids is an important variable, especially if the fluidized bed is being used as a catalytic reactor. The data show that for all practical purposes the bed of solids can be treated as being perfectly mixed [141, 144, 216, 255, 282, 317]. This would imply that the residence time function $F(t)$, the fraction of material that is in the system less than the time t, would be

$$F(t) = 1 - e^{-t/\tau}, \tag{18-74}$$

where τ is the mean residence time. Wolf and Resnick [316] used the modified form

$$F(t) = 1 - e^{-\eta(t-\theta)/\tau}, \tag{18-75}$$

where η is an efficiency term and θ is a phase-shift term expressed in terms of time. The equation was found to be quite good for multistage systems, with η and θ varying as expected. Yagi and Kunii [317] obtained an expression for the mean residence time of solid particles for a fluidized system containing a wide range of sizes of solid particles. Zenz and Othmer [322] have reviewed and extended the work on entrainment from fluidized beds.

18-5.D Fluidized Bed Models

In order to interpret a large quantity of experimental results in a simple manner, we must assume a model for the system. Since all models are approximations of reality, they will succeed or fail to the extent to which they give logical and consistent results and predict new conditions accurately. Certainly, experimental results are correct if carefully obtained, but they are inconvenient, since they usually entail reams of data, graphs, and tables. Therefore we turn to a model to reduce the information to terms of some simple parameter or parameters. We do this all the time, of course, and for the most part the systems are simple, results which are quite reasonable are obtained, and we do not think about the problem further. But the fluidized bed system is one of extreme complexity, and simple models have proved to be totally inadequate. It is at this point that one can be led astray. The experimental data and the chosen model together may result in a set of parameters that nicely reduce the data to manageable proportions. But even though a correlation is obtained, there is no guarantee that the model is correct. The test of the model lies in the logic of the model itself, in the logic of its results, and in the prediction of new conditions. If we obtain a correlation from a model that is totally inadequate from these standpoints, that correlation will be useful so long as it is used *only* with that model. In extrapolating to conditions beyond the range of the data, one should be very careful, for an inadequate model can at times lead to worse results than a totally empirical approach. To cite one example as an illustration of this problem, we can consider the heat transfer from particle to gas in a fluidized bed. For very small particles, the particle Reynolds number is small because of the particle size and the relatively low velocity needed for fluidization. Under these conditions a great many heat transfer results have been obtained, models assumed where necessary, and heat transfer coefficients calculated. In all cases, these coefficients were one or two orders of magnitude less than the corresponding values for single particles or for packed beds [164, 322]. So long as we use these correlations with the appropriate model, we should be able to predict the heat transfer from fluid to particle; however, extrapolation to other conditions may lead to gross errors, since it is illogical to assume that heat-transfer coefficients will be lower than the corresponding packed-bed values. Some authors have assumed the calculated heat-transfer coefficients to be correct, and have attempted to explain the low values as being due to agglomeration effects that give an effectively larger diameter. This reasoning cannot be correct, however, since the agglomerate is in effect a small packed bed, and no matter what its size or motion relative to the fluid, it has a coefficient which is more than an order of magnitude greater than that suggested. It cannot, therefore, contribute to a lowering of the coefficient. The correlations giving low values of heat-transfer coefficients at low Reynolds numbers can be used with their corresponding model, but they are not in any way indicative of the true mechanism of heat transfer or of the level of the coefficients actually existing in the bed. Clearly, to be sure of the design, and especially if we are working outside the range of available data, we would like to have a better, more logical model. With such limitations of model analysis in mind, we can turn our attention to some of the models that have been proposed.

(a) The two-phase model. The treatment of an aggregately fluidized bed as a single phase has been useful for the interpretation of experimental results in certain specific cases, but for most purposes it has proved inadequate, and so will not be discussed here. Rather, we shall turn to the most complex two-phase model proposed [191, 304], which is a modification of that originally suggested by Toomey and Johnstone [294], and which reduces to the simpler single-phase model in one limit. Other examples of model analysis can be found in references 6, 58, 95, 96, 97, 108, 140, 170, and 188.

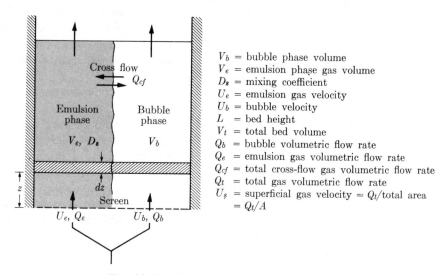

V_b = bubble phase volume
V_e = emulsion phase gas volume
D_s = mixing coefficient
U_e = emulsion gas velocity
U_b = bubble velocity
L = bed height
V_t = total bed volume
Q_b = bubble volumetric flow rate
Q_e = emulsion gas volumetric flow rate
Q_{cf} = total cross-flow gas volumetric flow rate
Q_t = total gas volumetric flow rate
U_s = superficial gas velocity = Q_t/total area
 = Q_t/A

Fig. 18–26. *Two-phase fluidized-bed model.*

The basic bed model is pictured as a two-phase system (Fig. 18–26) in which the emulsion phase consists of solid particles with the amount of gas necessary to provide fluidization without bubbles. The remainder of the gas passes up the bed as bubbles. May [191] suggested that the emulsion phase could be characterized with a mixing coefficient D_s, which is the same as for the solids. No mixing occurs in the bubble phase; however, a cross flow (mass transfer) exists between the emulsion and bubble phases, and can be characterized by an exchange coefficient or a cross-flow ratio (cross flow/bubble flow). The cross-flow ratio would indicate the number of times a bubble would exchange its contents with the emulsion phase during its rise through the bed.

The major factors controlling the velocity of rising bubbles appear to be material properties [318] and the value of $U_s - U_e$, where U_e is the emulsion gas velocity, which is the superficial gas velocity (U_s) at conditions of incipient fluidization (the bubble point). For a given material and superficial gas velocity, the bubble velocity should be constant, and can be estimated from the bed properties. There are no bubbles at minimum fluidization conditions, and the bed parameters are characteristic of the emulsion (dense phase); thus the emulsion gas velocity U_e, and volume $V_e + V_s$ (V_e = emulsion gas volume and V_s = solid volume), can be determined. In the two-phase model, these factors are assumed to remain constant with increasing superficial gas velocity; at a superficial gas velocity higher than minimum, the additional bed volume would be the total bubble volume; that is, $V_b = V_t - (V_e + V_s)$. The minimum fluidization velocity would be $U_e = Q_e/\text{emulsion gas area} = Q_e L/V_e$, where Q_e is the volumetric flow rate of gas in the emulsion. Likewise, $U_b = Q_b/\text{bubble area} = Q_b L/V_b$, and of course, $Q_t = Q_e + Q_b$. Combining these equations gives

$$U_b = (LQ_t - U_e V_e)/[(V_t - (V_e + V_s))]. \tag{18-76}$$

In terms of bed expansion ϵ,

$$L = \epsilon L_s = \epsilon(V_e + V_s)/A, \qquad V_t = \epsilon V_s = \epsilon(V_e + V_s), \qquad Q_t = U_s A,$$

where L_s is the height at conditions of minimum fluidization. Combining these with Eq. (18-76) gives

$$U_b = \frac{\epsilon(V_e + V_s) - U_e V_e}{\epsilon(V_e + V_s) - (V_e + V_s)} = \frac{(\epsilon U_s - x_e U_e)}{(\epsilon - 1)}, \tag{18-77}$$

where x_e is the void fraction at the bubble point, i.e., $V_e/(V_e + V_s)$, and would be a constant for any given material. The void fraction, or porosity of the bed under conditions above minimum fluidization, has been measured [11, 12, 13]; in terms of the present notation, it would be

$$x_b = (V_b + V_e)/V_t = (1/\epsilon)(V_e + V_b)/(V_e + V_s). \tag{18-78}$$

When plotted, the data of May and of Minet et al. [198] show that U_b is a function of about the one-third power of the superficial gas velocity. The bubble-to-emulsion gas volume ratio can be determined as

$$V_b/V_e = (\epsilon - 1)/x_e. \tag{18-79}$$

When one determines these parameters, the bed diameter may be an important factor; the work of Minet et al. shows its effect, which takes the form

$$\epsilon = 1 + U_s(k_1 + k_2/d_0), \tag{18-80}$$

where d_0 is the column diameter and k_1 and k_2 are correlating constants, different for each type of particle used. This work emphasizes the need for a cautious approach to scale-up and the application of small bed parameters in the model.

The main purpose of the models is to aid in the rational interpretation of experimental data, such as gas residence time distributions, heat and mass transfer results, and kinetics. Through such interpretation we hope to obtain some insight into the actual mechanism. For interpretation of the gas residence time data, May [191] made a material balance on each phase (refer to Fig. 18–26) over a distance dz, and obtained,

$$V_b(\partial C_b/\partial t) \qquad\qquad\qquad + Q_b(\partial C_b/\partial z) = Q_{cf}(C_e - C_b),$$
$$\underbrace{V_e(\partial C_e/\partial t)}_{\text{Accumulation}} - \underbrace{(D_s V_e/L^2)(\partial^2 C_e/\partial z^2)}_{\text{Mixing}} + \underbrace{Q_e(\partial C_e/\partial z)}_{\text{Convection}} = \underbrace{Q_{cf}(C_b - C_e)}_{\text{Exchange}}, \qquad (18\text{–}81)$$

where, as noted earlier, there is no mixing in the bubble phase (plug flow). The boundary conditions selected were

$$t < 0: \quad \text{all } C = 1;$$
$$t > 0: \quad \text{at } z = 0, \qquad C_b = 0, \qquad Q_e C_e = (D_s V_e/L^2)(\partial C_e/\partial z);$$
$$\qquad\qquad \text{at } z = 1, \qquad \partial C_e/\partial z = 0. \qquad\qquad (18\text{–}82)$$

An analytical solution to Eqs. (18–81) and (18–82) is not available; however, a numerical integration [191] or a statistical analysis [304] can be made. For either of the analyses, certain bed parameters must be known (emulsion and bubble volumes and velocities). These have already been discussed in some detail; however, it should be emphasized that in this model they are considered to be functions of the particle nature and superficial gas velocity only, and do not vary with distance up the bed.

In the numerical solution, general graphs are obtained which relate the tracer decay slope (on semilogarithmic paper, see Fig. 18–23) to the cross-flow ratio with the parameter as D_s/L. Each graph is for a different physical set of operating conditions. An analogous set of graphs (restricted to no flow in the dense phase) is available from the statistical solution. These graphs relate the relative standard deviation of the residence time distribution to the number of transfer units, with the eddy diffusivity as a parameter. Each graph is for a given bubble fraction.

May's residence time studies were made in a 3-in. column with the cracking catalyst at different total bed heights. Using the general graphs from the solution of Eq. (18–81), he plotted the cross-flow ratio against the bed height (see Fig. 18–27). The data for the bed of low height showed a small systematic variation with the superficial velocity. Van Deemter obtained the same values for the data on the 3-ft bed height, using his statistical approach. In addition he obtained a cross-flow ratio of 4.5 for 102-micron glass beads and a ratio of 6 for 155-micron glass beads; this can be compared with about 1.8 for the cracking catalyst.

An experiment to measure D_s by following gas back-mixing has been described and tested [6, 53, 181, 304]. In this experiment, tracer gas is injected near the top of the bed, rather than radioactive solids, and the composition is monitored at points below the injection plane. The drawback of this method is that it is quite dependent on the fraction of tracer injected into the dense and bubble phases. One advantage is that a steady-state experiment can be performed, and Eq. (18–81) is simplified enough so that an analytical solution is possible [304].

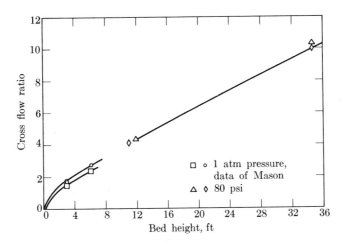

Fig. 18-27. *Cross-flow ratio results.* [*From* W. May, Chem. Eng. Progr. **55**, No. 12, 49, (1959). *By permission.*]

The model is not by any means a perfect description of the fluidized bed. There is clearly an effect of the air distributor plate [11–13, 122, 123] and of the transition from the dense to the dilute bed at the top [128]. The variation in porosity between these zones is completely ignored in this approach.

May [191] and van Deemter [304] have both considered a first-order catalytic kinetic reaction in conjunction with the two-phase model (no reaction in the bubble phase). May's analysis is the more detailed of the two, and contains comparisons with some actual pilot unit conversion data. Equation (18–81), for steady state and a first-order reaction, becomes

$$Q_b(\partial C_b/\partial z) = Q_{cf}(C_e - C_b),$$

$$(D_s V_e/L^2)(\partial^2 C_e/\partial z^2) - Q_e(\partial C_e/\partial z) + Q_{cf}(C_b - C_e) = kWC_e, \qquad (18\text{-}83)$$

$$(Q_b + Q_e)C_2 = Q_b C_{b2} + Q_e C_{e2},$$

where k is the reaction rate constant and W is the weight of catalyst in the bed. The boundary conditions are,

$$z = 0, \qquad C_b = C_1, \qquad Q_e(C_1 - C_e) = (D_s V_e/L^2)(\partial C_e/\partial z);$$

$$z = 1, \qquad \partial C_e/\partial z = 0.$$

May's solution of these equations indicates that, for low cross-flow ratios, the back-mixing has no effect (van Deemter also shows this). Further, the mixing levels given in Fig. 18-25 result in conversions that are closer to no mixing than to complete mixing. These theoretical results are encouraging, and further experimental confirmation and application of the models will be a helpful addition to the literature. A first step in this direction has been made by Orcutt *et al.* [211].

They used the reaction of ozone decomposition on a silica-alumina cracking catalyst impregnated with iron oxide. There was a finite amount of bypassing, and it was independent of the system variables. The results, based on the relative standard deviation of the contact time distribution, were in reasonably good agreement with the two-phase model.

(b) Multizone model. The two-phase model has found wide use in the interpretation of gas flow characteristics and the closely dependent experiments involving kinetics. A rational analysis of heat transfer data has not been possible with this and similar models, and some alternate description appears necessary. One important factor is that the heat of the gas per unit volume is several orders of magnitude less than that of the solid. Thus it takes little solid to cool a large gas flow. The rapid cooling of a gas stream has been well documented by Heertjes et al. [122, 123], who showed that the cooling was essentially complete in the first $\frac{1}{4}$-in. above the distributor. Kim [151] found similar results, using a frequency-response method. It should be recalled that this is the zone that is low in solids as a result of the jetting action of the grid [121] (see Fig. 18–21). Furthermore, the voids in this region are apparently small and well distributed, in that the fluctuations associated with bubbles have not been seen [318]. Through the process of coalescence, the bubbles grow to rather large size, but this is a function of distance from the grid, and there is apparently little if any coalescence of voids at the short distance of $\frac{1}{4}$-in. from the grid. In the region above the lowest zone (main bed zone and upper dilute zone), bubbles have formed and solids mixing is intense. For most practical problems, these regions are so intensely mixed that the assumption of complete mixing is adequate. Figure 18–28 is an illustration of this model. For the purpose of heat transfer, the two-phase aspect need not be considered, since the bubble phase acts as a mixing device in the upper regions

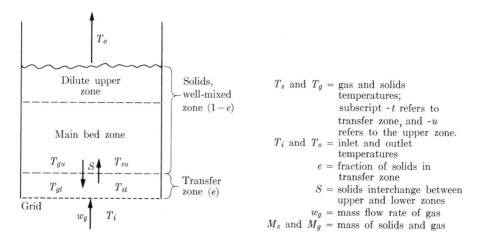

Fig. 18–28. Multizone model for heat transfer.

and does not exist in the transfer region [151]. This zone model for heat transfer shows that the cool solids come into contact with the entering hot gas in the transfer zone and are heated. By the solids mixing process, the now heated solids are returned to the upper zones and are extremely rapidly mixed. Further transfer from the solids to the gas can occur here; however, since the solids have not been heated up drastically (the heat content difference), and since those that are heated are mixed into a much larger body of solids, we would not expect to see a really measurable temperature difference in the upper zones. This does not mean that equilibrium exists in these areas, as has been assumed in most previous models for heat transfer. It means, rather, that finite heat transfer occurs and cannot be ignored, because even though the temperature driving force is very small, the area for transfer is very large. An overall heat balance on the system gives us

$$w_g c_{pg}(T_i - T_o) = M_{gt} c_{pg}\frac{dT_{gt}}{dt} + eM_s c_{ps}\frac{dT_{st}}{dt} + M_{gu} c_{pg}\frac{dT_{gu}}{dt} + (1-e)M_s c_{ps}\frac{dT_{su}}{dt}, \quad (18\text{-}84)$$

and, for the solids,

$$h_t A(T_{gt} - T_{st}) + (1-e)h_u A(T_{gu} - T_{su})$$
$$= eM_s c_{ps}(dT_{st}/dt) + (1-e)M_s c_{ps}(dT_{su}/dt), \quad (18\text{-}85)$$

where the notation is given in Fig. 18-28. To simplify the problem, Kim assumed: (1) The thermal capacitance of the gas was negligible in both zones. (2) Solids mixing was great so that $T_{st} = T_{su} = T_s$. (3) The gas temperature in the upper zone was equal to T_o. (4) The gas temperature in the lower zone was equal to T_i. The last assumption is the weakest, since it is known that there is a gradient in this area. Equations (18-84) and (18-85), with the additional assumption that e is very close to unity (so that the last term of (18-85) can be dropped), become

$$M_s c_{ps}(dT_s/dt) = -eAh(T_o - T_s) = w_g c_{pg}(T_i - T_o). \quad (18\text{-}86)$$

In terms of the frequency-response method used, phase angles very close to 90° had to be measured, which was difficult; nevertheless, measurements could be made, and the coefficients obtained were considerably greater than the results for single or packed beds at low Reynolds numbers. Since a wide range of variables was not investigated, no correlation was attempted, and so the results obtained cannot be used for design purposes. However, the work does show that logical results can be obtained and points the way for future research efforts in this area.

18-5.E Modified Systems

Since the advent of the fluidized-bed technique, a great number of attempts have been made to improve the general workability of the system. The effects of baffles, screens, or staged operation have been extensively investigated [9, 170, 182, 183, 186, 216, 235, 306]. Spouted beds are a modified solids–fluid contacting device, which is a partially fluidized bed. The fluid is injected through a small orifice at the center of a conical based column. The jet of fluid entrains some of the solids and carries them to the top of the bed, where they move laterally outward from the center core and slowly return toward the bottom as part of an angular ring. This system shows promise for coarse, relatively uniform solids that fluidize poorly, such as wheat, wood chips, coal, and certain food items. The first suggestion of this system was made by Mathur and Gishler [189], and was followed by a number of reports [19, 20, 35, 51, 91, 153, 175, 176, 177, 190, 219, 256, 292] on the operations, pressure drop, applications, and heat transfer in the system. Another variation is semifluidization [73, 74], which is the name given to a fluidized bed that has a porous screen at the top to restrict the passage of solids. Still another system, designed for large nonfluidized particles, is a packed-fluidized bed, which is a bed of large particles with fine particles fluidized in the void spaces [86, 87, 133, 277, 323, 325]. This modification was first investigated in a program aimed at the recovery of spent uranium dioxide reactor fuel. Finally, gas injection into a liquid fluidized bed can expand the usability of the fluidized-bed method [300]. Additional work in this area can be found in references 145, 146, 184, 214, 215, and 273.

18-5.F Mass and Heat Transfer and Kinetics

This short section, like the previous one, is meant to offer a number of references (including some already cited) on the further developments of heat and mass transfer and kinetics in fluidized systems. The number of heat-transfer references is large; there are several general references [11, 164, 322], and references on heat transfer either between the bed and the boundaries [7, 15, 16, 23, 25, 27, 63, 75, 89, 135, 150, 166, 169, 195, 196, 197, 295, 312, 324], or from the fluid to the solid particles [2, 30, 67, 78, 80, 121, 122, 123, 127, 149, 232, 275, 280, 309, 310, 327]. Many earlier and other current references can be found in those cited above. The references for mass transfer are fewer in number; some of them are 11, 30, 46, 121, 122, 123, 149, 163, 280, and others cited therein. References on the basic analysis of kinetic problems by the application of model analysis have been given in references 55, 138, 140, 170, 188, 191, 211, 227, 238, 255, 281, 294, 304, 317, and others cited therein. Various applications of the fluidized bed for reaction kinetics have been given in references 44, 49, 77, 93, 94, 100, 103, 133, 135, 139, 141, 152, 167, 198, 199, 205, 227, and 317.

PROBLEMS

18-1. The turbulent motion in the vicinity of a pipe wall is to be followed by visual observation of the motion of small particles. For the case of a pipe with an inner diameter of 3 in. and a flow of 3 ft/sec of water, suggest the way to determine whether or not the particles are following the motion. To ensure that the observations are correct within 1%, determine the largest particle that can be used.

18–2. Determine the settling velocity for spherical catalyst particles of 0.1 in. diameter being used in a water system at 70°F, and at a concentration of 40% solids. The solids density is 80 lb/ft³.

18–3. The catalyst of Problem 18–2 is to be conveyed in an air stream at a solids-to-air ratio of 4.0. What is the minimum air velocity that can be used to accomplish the desired transport?

18–4. For the catalyst of Problem 18–2, determine the minimum fluidization velocity both in air and in water.

18–5. Calculate one representative streamline for Fig. 18–16(a), (b), (c), and (d).

18–6. Derive Eq. (18–61).

18–7. Determine several points of the two curves of Fig. 18–18.

18–8. Derive Eq. (18–81) and solve for steady-state conditions.

NOTATION FOR CHAPTER 18

A	parameter in Eqs. (18–66) and (18–67)
C	concentration
C_D	drag coefficient
c_p	heat capacity
D_s	mixing coefficient
d	particle diameter
E	eddy coefficient of mixing
e	fraction solids in transfer zone, used in Eq. (18–84)
F	functions defined in Eqs. (18–37) and (18–74)
f	function defined in Eq. (18–25)
f	frequency
G	mass velocity
$g, \boldsymbol{g}, g_i$	gravity
K	defined in Eq. (18–19)
k	constant, reaction rate constant
L	length
m_p	mass of a particle
n	power, number
p	pressure
Q	volumetric flow rate
R	fraction
R_L	Lagrangian correlation
r	radius
S	area, solids exchange
s	point in Fig. 18–1
T	temperature
t	time
$U, \boldsymbol{U}, u$	velocity, velocity fluctuation

u'_r	relative rms velocity fluctuation
U^*_{om}	defined in Eq. (18–51)
V	volume
W	weight of catalyst
w	weight flow rate
X	parameter
x	distance, void fraction
Y	parameter
Y	instantaneous spread in the y-direction
y	defined in Eq. (18–22)
y^+	yU^*/ν_c
z	height

Greek letters

α	defined following Eq. (18–14)
β	defined following Eq. (18–14)
ϵ	diffusivity, bed expansion
θ	angle, holdup time, phase shift
Λ_L	Lagrangian macroscale (Eq. 18–19)
λ_L	Lagrangian microscale
μ	viscosity
ν	kinematic viscosity
π	3.1416...
ρ	density
τ	shear stress, residence time
T_L	Lagrangian time scale
ϕ	function, stream potential
ψ	stream function

Subscripts

a	axial
b	bubble
c	continuous or fluid
cf	cross flow
d	discontinuous or particle
e	emulsion, effective
i	initial
max	maximum
mf	minimum fluidization
n	nozzle
0	pipe, initial in time, exit
r	relative, radial
s	solids, superficial
t	terminal, total, transfer

u	upper
w	wall
1	one
2	two
₵	center line
ω	free stream

Other symbols

∇	del
∞	infinity

REFERENCES

1. D. ALTMAN and J. M. CURTIS, *High Speed Aerodynamics and Jet Propulsion*, Vol. II, Ser. C, Princeton University Press, Princeton, N.J. (1956).
2. N. R. AMUNDSON and R. ARIS, in Third Congr. Europ. Fed. Chem. Engrs. (London), *Interaction Between Fluids and Particles*, pp. 176–182, Inst. Chem. Engrs. (London) (1962).
3. T. B. ANDERSON and R. JACKSON, *Chem. Eng. Sci.* **19**, 509–511 (1964).
4. K. E. B. ANDERSSON, *Chem. Eng. Sci.* **15**, 276–297 (1961).
5. H. ANGELINO, C. CHARZAT, and R. WILLIAMS, *Chem. Eng. Sci.* **19**, 289–304 (1964).
6. J. W. ASKINS, G. P. HINDS, and J. KUNREUTHER, *Chem. Eng. Progr.* **47**, 401–404 (1951).
7. A. BAERG, J. KLASSEN, and P. E. GISHLER, *Can. J. Res.* **28F**, 287 (1950).
8. W. S. BAILEY, E. N. NIELSON, R. A. SERRA, and T. F. ZUPNIK, *ARSJ* **31**, 793–799 (1961).
9. R. C. BAILIE, D. S. CHUNG, and L. T. FAN, *Ind. Eng. Chem., Fund.* **2**, 245–246 (1963).
10. R. C. BAILIE, L. T. FAN, and J. J. STEWART, *Ind. Eng. Chem.* **53**, 567–569 (1961); *J. Chem. Eng. Data* **6**, 469–473 (1961).
11. P. J. BAKKER and P. M. HEERTJES, *Brit. Chem. Eng.* **3**, 240–246 (1958).
12. P. J. BAKKER and P. M. HEERTJES, *ibid.* **4**, 524–529 (1959).
13. P. J. BAKKER and P. M. HEERTJES, *Chem. Eng. Sci.* **12**, 260–271 (1960).
14. T. BARON and R. A. MUGELE, paper presented at AIChE meeting, Chicago (Dec. 1957).
15. R. N. BARTHOLOMEW and D. L. KATZ, *Chem. Eng. Progr. Symposium Series* No. 4 **48**, 3 (1952).
16. A. P. BASKAKOV, *Inzh. Fiz. Zh.* **6**, No. 11, 20–25 (1963); translation, *Int. Chem. Eng.* **4**, 320–324 (1964).
17. A. B. BASSET, *A Treatise on Hydrodynamics*, Vol. II, Chap. 5, Deighton, Bell and Co., Cambridge, England (1888); Dover Reprints (1961).
18. P. K. BAUMGARTEN and R. L. PIGFORD, *AIChEJ* **6**, 115–123 (1960).
19. H. A. BECKER, *Chem. Eng. Sci.* **13**, 245–262 (1961).
20. H. A. BECKER and H. R. SALLANS, *Chem. Eng. Sci.* **13**, 97–112 (1961).
21. W. J. BEEK, in Third Congr. Europ. Fed. Chem. Engrs. (London), *Interaction Between Fluids and Particles*, pp. 163–165, Inst. Chem. Engrs. (London) (1962).

22. N. Berkowitz, C. Moreland, and G. F. Round, *Can. J. Chem. Engrs.* **41**, 116–121 (1963).
23. J. S. M. Botterill, *Brit. Chem. Eng.* **8**, 21–26 (1963).
24. J. S. M. Botterill and P. D. Bloore, *Can. J. Chem. Eng.* **41**, 111–115 (1963).
25. J. S. M. Botterill, G. L. Cain, G. W. Brundrett, and D. E. Elliott, paper presented at AIChE meeting, Boston (Dec. 1964).
26. J. S. M. Botterill, J. J. George, and H. Besford, paper presented at AIChE meeting, Boston (Dec. 1964).
27. J. S. M. Botterill, K. A. Redish, D. K. Ross, and J. R. Williams, in Third Congr. Europ. Fed. Chem. Engrs. (London), *Interaction Between Fluids and Particles*, pp. 183–189, Inst. Chem. Engrs. (London) (1962).
28. J. Boussinesq, *Théorie analytique de la chaleur*, Vol. II, p. 224, Gauthier-Villars, Paris (1903).
29. K. McG. Bowling and A. Watts, *Australian J. Appl. Sci.* **12**, 413 (1962); see also P. N. Rowe and K. S. Sutherland, *Trans. Inst. Chem. Engrs.* (London) **42**, 55–62 (1964); P. N. Rowe, B. A. Partridge, A. G. Cheney, G. A. Henwood, and E. Lyall, *ibid.* **43**, 271–286 (1965).
30. R. D. Bradshaw and J. E. Myers, *AIChEJ* **9**, 590–595 (1963).
31. A. G. Bridge, L. Lapidus, and J. C. Elgin, *AIChEJ* **10**, 819–826 (1964).
32. H. C. Brinkman, *Appl. Sci. Res.* **1A**, 27 (1947).
33. W. Brötz, *Chem.-Ingr.-Tech.* **28**, 165–174 (1956).
34. C. Bruinzeel, G. H. Reman, and E. Th. van der Laan, in Third. Congr. Europ. Fed. Chem. Engrs. (London), *Interaction Between Fluids and Particles*, pp. 120–126, Inst. Chem. Engrs. (London) (1962).
35. R. H. Buchanan and F. Manurung, *Brit. Chem. Eng.* **6**, 402–407 (1961).
36. J. M. Burgers, *Proc. Acad. Sci. Amsterdam.* **44**, 1045–1051, 1177–1184 (1941); **45**, 9–16, 126–127 (1942).
37. E. J. Cairns and J. M. Prausnitz, *Ind. Eng. Chem.* **51**, 1441–1444 (1949).
38. E. J. Cairns and J. M. Prausnitz, *AIChEJ* **6**, 400–405 (1960).
39. E. J. Cairns and J. M. Prausnitz, *ibid.*, 554–560.
40. E. J. Cairns and J. M. Prausnitz, *Chem. Eng. Sci.* **12**, 20–34 (1960).
41. R. C. Cairns, K. R. Lawther, and K. S. Turner, *Brit. Chem. Eng.* **5**, 849–856 (1960).
42. G. F. Carrier, *J. Fluid Mech.* **4**, 376–382 (1958).
43. H. V. Chamberlain, M. S. thesis in Chem. Eng., University of Washington, Seattle, Wash. (1957).
44. K. C. Channabasappa and H. R. Linden, *Ind. Eng. Chem.* **50**, 637 (1958).
45. E. D. Christiansen and D. H. Barker, *AIChEJ* **11**, 145–151 (1965).
46. J. C. Chu, J. Kalil, and W. A. Wetteroth, *Chem. Eng. Progr.* **49**, 141–149 (1953).
47. J. Ciborowski and A. Wlodarski, *Chem. Eng. Sci.* **17**, 23–32 (1962); see also S. Koncar-Djurdjeric, L. Capo, and D. Vukovic, *Genie Chimique* **86**, 110–115 (1961).
48. R. B. Cooper, M. S. thesis in Chem. Eng., The Ohio State University, Columbus, Ohio (1961).

REFERENCES

49. R. M. CORNFORTH, *Pet. Ref.* **36**, No. 9, 211 (1957).
50. S. CORRSIN and J. LUMLEY, *Appl. Sci. Res.* **6A**, 114–116 (1956).
51. C. B. COWAN, W. S. PETERSON, and G. L. OSBERG, *Pulp Paper Mag. Canad.* **58**, No. 13, 139 (1957); *Eng. J.* **41**, No. 5, 60 (1958).
52. P. V. DANCKWERTS, *Chem. Eng. Sci.* **2**, 1–13 (1953).
53. P. V. DANCKWERTS, J. W. JENKINS, and G. PLACE, *Chem. Eng. Sci.* **3**, 26–35 (1954).
54. J. F. DAVIDSON, *Trans. Inst. Chem. Engrs. (London)* **39**, 230–233 (1961).
55. J. F. DAVIDSON and D. HARRISON, *Fluidized Particles*, Cambridge University Press, Cambridge (1963).
56. J. F. DAVIDSON, R. C. PAUL. M. J. S. SMITH, and H. A. DUXBURY, *Trans. Inst. Chem. Engrs. (London)* **37**, 323–328 (1959).
57. G. DAVIES and D. B. ROBINSON, *Can. J. Chem. Eng.* **38**, 175–183 (1960); see also I. G. BLYAKHER and V. M. PAVLOV, *Khim. i Neft. Mash.* **7**, No. 6, 15–18 (1965); translation, *Int. Chem. Eng.* **6**, 47–50 (1966).
58. F. DEMARIA and J. E. LONGFIELD, *Chem. Eng. Progr. Symposium Series* No. 38, **58**, 16–27 (1962).
59. R. DIEKMAN and W. L. FORSYTHE, JR., *Ind. Eng. Chem.* **45**, 1174–1177 (1953).
60. K. DOICHER, *Khim. i Indust.* **35**, No. 4, 133–137 (1963); translation, *Int. Chem. Eng.* **4**, 192–197 (1964); see also E. V. DONAT, *Zhur. Prik. Khim.* **35**, 1516–1526 (1962).
61. E. V. DONAT, *Inzh. Fiz. Zh.* **5**, No. 6, 55 (1962); translation, *Int. Chem. Eng.* **2**, 550–553 (1962).
62. J. M. DOTSON, *AIChEJ* **5**, 169–174 (1959).
63. W. M. Dow and M. JAKOB, *Chem. Eng. Progr.* **47**, 637–648 (1951).
64. R. DURAND, *Proc. Minn. Hydraulic Conv.*, Part 1, pp. 89–103, University of Minnesota, Minneapolis, Minn. (1953).
65. R. EICHHORN, R. SHANNY, and U. NAVON, *Proj. Squid Tech. Rep. PR-107-P* (March 1964).
66. R. EICHHORN and S. SMALL, *J. Fluid Mech.* **20**, 513–527 (1964).
67. J. EICHHORN and R. R. WHITE, *Chem. Eng. Progr. Symposium Series* No. 4, **48**, 11–18 (1952).
68. A. EINSTEIN, *Ann. Phys.* **4**, 289 (1906); see also A. D. MAUDE, *J. Fluid Mech.* **7**, 230–236 (1960).
69. H. S. ELLIS, P. J. REDBERGER, and L. H. BOLT, *Ind. Eng. Chem.* **55**, No. 8, 18–26 (1963).
70. R. W. FAHIEN and J. M. SMITH, *AIChEJ* **1**, 28–37 (1955).
71. L. T. FAN, C. J. LEE, and R. C. BAILIE, *AIChEJ* **8**, 239–244 (1962).
72. L. T. FAN, J. A. SCHMITZ, and E. N. MILLER, *AIChEJ* **9**, 149–153 (1963).
73. L. T. FAN, and C. Y. WEN, *AIChEJ* **7**, 606–610 (1961).
74. L. T. FAN, Y. C. YANG, and C. Y. WEN, *AIChEJ* **5**, 407 (1959); **6**, 482 (1960).
75. L. FARBER and C. A. DEPAW, *Ind. Eng. Chem., Fund.* **2**, 130–135 (1963).
76. H. FAXÉN, *Arikv. Math., Astrn. Fysik*, **19a**, 13 (1925).
77. J. FEINMAN and T. D. DREXLER, *AIChEJ* **7**, 584–587 (1961).

78. J. R. FERRON, and C. C. WATSON, *Chem. Eng. Progr. Symposium Series* No. 38, **58**, 79–86 (1962).
79. B. A. FORSELL and R. KADEFORS, *Chem. Eng. Sci.* **20**, 168–169 (1965).
80. J. F. FRANTZ, *Chem. Eng. Progr.* **57**, No. 7, 35–42 (1961).
81. J. F. FRANTZ, paper presented at AIChE meeting Boston (Dec. 1964).
82. S. K. FRIEDLANDER, *AIChEJ* **3**, 381–385 (1957).
83. S. K. FRIEDLANDER and H. F. JOHNSTONE, *Ind. Eng. Chem.* **49**, 1151–1156 (1957).
84. H. L. FRISCH and R. SIMHA, *Rheology* (F. R. Eirich, editor), Vol. I, Chap. 14, Academic Press Inc., New York (1956).
85. J. FURUKAWA and T. OHMAE, *Ind. Eng. Chem.* **50**, 821–828 (1958); see also P. V. D. LEEDEN and G. J. BOVWHUIS, *Appl. Sci. Res.* **10A**, 78–80 (1961).
86. J. D. GABOR, *AIChEJ* **10**, 345–350 (1964).
87. J. D. GABOR and V. J. MECHAM, *Ind. Eng. Chem., Fund.* **3**, 60–65 (1964).
88. N. I. GEL'PERIN, V. G. AINSHTEIN, and I. D. GOIKHMAN, *Inzh. Fiz. Zh.* **7**, No. 7, 15–19 (1964); translation, *Int. Chem. Eng.* **5**, 55–57 (1965).
89. N. I. GEL'PERIN, V. G. AINSHTEIN, and N. A. ROMANOVA, *Khim. Prom.* No. 2, 21–41 (1964), No. 6, 16–22 (1965); translation, *Int. Chem. Eng.* **4**, 502–505 (1964), **6**, 67–74 (1966).
90. C. F. GERALD, *Chem. Eng. Progr.* **47**, 483–484 (1951).
91. B. GHOSH and G. L. OSBERG, *Can. J. Chem. Eng.* **37**, 205–207 (1959).
92. M. GILBERT, J. ALLPORT, and R. DUNLAP, *ARSJ* **32**, 1929–1930 (1962).
93. E. R. GILLILAND, W. K. LEWIS, and G. T. MCBRIDE, *Ind. Eng. Chem.* **41**, 1213–1226 (1949).
94. E. R. GILLILAND, W. K. LEWIS, and M. P. SWEENEY, *Chem. Eng. Progr.* **47**, 251–256 (1951).
95. E. R. GILLILAND and E. A. MASON, *Ind. Eng. Chem.* **41**, 1191–1196 (1949).
96. E. R. GILLILAND and E. A. MASON, *ibid.* **44**, 218–224 (1952).
97. E. R. GILLILAND, E. A. MASON, and R. C. OLIVER, *Ind. Eng. Chem.* **45**, 1177–1185 (1953).
98. I. GLASSMAN, *Jet Propulson* **27**, 542–543 (1957).
99. R. D. GLAUZ, *ARSJ* **32**, 773–775 (1962).
100. M. GOLDMAN, L. N. CANJAR, and R. B. BEEKMANN, *J. Appl. Chem.* **7**, 274–284 (1957).
101. D. GOLDSCHMIDT and P. LEGOFF, *Chem. Eng. Sci.* **18**, 805–806 (1963).
102. L. J. GORDON, Ph. D. thesis in Chem. Eng., University of Washington, Seattle, Wash. (1963).
103. E. GORIN and C. N. ZIELKE, *Ind. Eng. Chem.* **49**, 396–403 (1957).
104. K. GOTO and K. IINOYA, *Kagaku Kōgaku* **27**, 12 (1963); translation, *Chem. Eng. (Japan)* **1**, 7–12 (1963).
105. G. W. GOVIER and M. E. CHARLES, *Eng. J.* **44**, No. 8, 50–57 (1961).
106. F. Z. GREK and V. N. KISEL'NIKOV, *Izv. Vyss. Uch. Zav., Khim. i Khim. Tech.* **6**, No. 4, 659 (1963); translation, *Int. Chem. Eng.* **4**, 263–269 (1964).
107. V. D. GVOZDEV, A. A. SAL'NIKOV, A. G. FOMICHEV, V. A. TIKHONOV, and A. S. VASIL'EV, *Izv. Vyss. Uch. Zav., Khim. i Khim. Tech.* **6**, No. 2, 320–327 (1963); translation, *Int. Chem. Eng.* **3**, 562–566 (1963).

108. A. E. HANDLOS, R. W. KUNSTMAN, and D. O. SCHISSLER, *Ind. Eng. Chem.* **49**, 25–30 (1957).
109. T. J. HANRATTY and A. BANDUKWALA, *AIChEJ* **3**, 293–296 (1957).
110. T. J. HANRATTY, F. LATINEN, and R. H. WILHELM, *AIChEJ* **2**, 372–380 (1956).
111. J. HAPPEL, *AIChEJ* **4**, 197–201 (1958).
112. J. HAPPEL and H. BRENNER, *AIChEJ* **3**, 506–513 (1957).
113. D. HARRISON, J. F. DAVIDSON, and J. W. DE KOCK, *Trans. Inst. Chem. Engrs. (London)* **39**, 202–211 (1961).
114. D. HARRISON and L. S. LEUNG, *Trans. Inst. Chem. Engrs. (London)* **39**, 409–414 (1961).
115. D. HARRISON and L. S. LEUNG, *ibid.* **40**, 146–151 (1962).
116. D. HARRISON and L. S. LEUNG, in Third Congr. Europ. Fed. Chem. Engrs. (London), *Interaction Between Fluids and Particles*, pp. 127–134: Inst. Chem. Engrs. (London) (1962).
117. N. J. HASSETT, *Ind. Chemist* **34**, 116, 169, 489 (1958); **37**, 25 (1961); **40**, 25((1964); see also W. H. M. ROBINS, *Trans. Inst. Chem. Engrs. (London)* **42**, 158–163 (1964).
118. N. J. HASSETT, *Chem. Eng. Sci.* **19**, 987–989; **20**, 172 (1965); *Brit. Chem. Eng.* **6**, 777–780 (1961).
119. P. G. W. HAWKSLEY, *Some Aspects of Fluid Flow*, Arnold Press, New York (1950).
120. T. HAYAKAWA, W. GRAHAM, and G. L. OSBERG, *Can. J. Chem. Eng.* **42**, 99–103 (1964).
121. P. M. HEERTJES, *Can. J. Chem. Eng.* **40**, 105–109 (1962).
122. P. M. HEERTJES, H. G. J. DEBOOR, and A. H. DEHAAS VAN DORSSER, *Chem. Eng. Sci.* **2**, 97–107 (1953).
123. P. M. HEERTJES and S. W. MCKIBBINS, *Chem. Eng. Sci.* **5**, 161–167 (1956).
124. J. O. HINZE, *Turbulence*, p. 352, McGraw-Hill, New York (1959).
125. L. M. HOCKING, *J. Fluid Mech.* **20**, 129–139 (1964).
126. R. F. HOGLUND, *J. Am. Rocket Soc.* **32**, 662–671 (1962).
127. J. P. HOLMAN, T. W. MOORE, and V. M. WONG, *Ind. Eng. Chem., Fund.* **4**, 21–31 (1965).
128. S. HONARI, Ph. D. thesis in Chem. Eng., Northwestern University, Evanston, Ill. (1959).
129. G. A. HUGHMARK, *Ind. Eng. Chem.* **53**, 389–390 (1961).
130. R. L. HULL and A. E. VON ROSENBERG, *Ing. Eng. Chem.* **52**, 989–992 (1960).
131. J. A. HULTBERG, *Project Squid Tech. Rept. Ill-17-R* (Aug. 1964).
132. A. R. HUNTLEY, W. GLASS, and J. J. HEIGL, paper presented at Am. Chem. Soc. meeting, New York (Sept. 1960).
133. T. ISHII and G. L. OSBERG, *AIChEJ* **11**, 279–287 (1965).
134. R. JACKSON, *Trans. Inst. Chem. Engrs. (London)* **41**, 13–28 (1963).
135. J. K. JACOBS and J. D. MIRKUS, *Ind. Eng. Chem.* **50**, 24–26 (1958).
136. C. E. JAHNIG, paper presented at AIChE meeting, Chicago (Dec. 1957).
137. K. O. L. F. JAYAWEERA, B. J. MASON, and G. W. SLACK, *J. Fluid Mech.* **20**, 121–128 (1964).
138. H. F. JOHNSTONE, J. D. BATCHELOR, and C. Y. SHEN, *AIChEJ* **1**, 318–323 (1955).

139. H. F. JOHNSTONE, J. D. BATCHELOR, and H. N. STONE, *Ind. Eng. Chem.* **46**, 274–278 (1954).
140. H. F. JOHNSTONE and C. Y. SHEN, *AIChEJ* **1**, 349–354 (1955).
141. A. A. JONKE and S. LAWROSKI, paper presented at Am. Chem. Soc. meeting, Miami (April 1957).
142. R. JOTTRAND, *Chem. Eng. Sci.* **3**, 12–16 (1954).
143. H. KADA and T. J. HANRATTY, *AIChEJ* **6**, 625–630 (1960).
144. Y. KAMIYA, *Kagaku Kōgaku* **19**, 412 (1955).
145. Y. KATO, *Kagaku Kōgaku* **27**, 7 (1963); translation, *Chem. Eng., Japan* **1**, 3–7 (1963).
146. Y. KATO, *Repts. Fac. Eng.*, Yamanashi University, Kofu, Japan **7**, 111 (1956).
147. B. H. KAYE and R. P. BOARDMAN, in Third Congr. Europ. Fed. Chem. Engrs. (London), *Interaction Between Fluids and Particles*, pp. 17–21, Inst. Chem. Engrs. (London) (1962).
148. B. H. KAYE and R. DAVIES, in Third Congr. Europ. Fed. Chem. Engrs. (London), *Interaction Between Fluids and Particles*, pp. 22–25, Inst. Chem. Engrs. (London) (1962).
149. K. N. KETTENRING, E. L. MANDERFIELD, and J. M. SMITH, *Chem. Eng. Progr.* **46**, 139–145 (1950).
150. N. V. KHARCHENKO and K. E. MAKHORIN, *Inzh. Fiz. Zh.* **7**, No. 5, 11–17 (1964); translation, *Int. Chem. Eng.* **4**, 650–654 (1964).
151. D. S. KIM, PH. D. thesis in Chem. Eng., The Ohio State University, Columbus, Ohio (1965).
152. A. KIVNICK and N. A. HIXSON, *Chem. Eng. Progr.* **48**, 394–400 (1952).
153. J. KLASSEN and P. E. GISHLER, *Can. J. Chem. Eng.* **36**, 12–18 (1958).
154. J. R. KLIEGEL and G. R. NICKERSON, in *Detonation and Two-phase Flows*, (S. S. Penner and F. A. Williams, editors), p. 173: Academic Press, New York (1962).
155. N. B. KONDUKOV, A. N. KORNILAEV, I. M. SKACHKO, A. A. AKHROMENKOV, and A. S. KRUGLOV, *Inzh. Fiz. Zh.* **6**, No. 1, 13–18 (1963); translation, *Int. Chem. Eng.* **5**, 83–88 (1965).
156. H. KRAMERS, *Chem. Eng. Sci.* **1**, 35–37 (1951).
157. H. KRAMERS, M. D. WESTERMANN, J. H. DE GROOT, and F. A. A. DUPONT, in Third Congr. Europ. Fed. Chem. Engrs. (London), *Interaction Between Fluids and Particles*, pp. 114–119, Inst. Chem. Engrs. (London) (1962).
158. G. J. KYNCH, *Trans. Faraday Soc.* **48**, 166–176 (1952).
159. G. J. KYNCH, *J. Fluid Mech.* **5**, 193–208 (1959); see also S. M. ISAAKYAN and A. M. GASPARYAN, *Izv. Akad. Nauk Arm. SSR, Ser. Tekh. Nauk*, No. 3, 45–56 (1965); translation, *Int. Chem. Eng.* **6**, 74–81 (1966).
160. K. P. LANNEAU, *Trans. Inst. Chem. Engrs. (London)* **38**, 125–143 (1960).
161. L. LAPIDUS and J. C. ELGIN, *AIChEJ* **3**, 63–68 (1957).
162. C. Y. LEE, *J. Korean Inst. of Chem. Engrs.* **1**, 5–12 (1963); translation, *Int. Chem. Eng.* **4**, 204–211 (1964).
163. V. N. LEPILIN, N. B. RASHKOVSKAYA, and P. G. ROMANKOV, *Zh. Prik. Khim.* **33**, 2664–2671 (1960).

164. M. LEVA, *Fluidization*, McGraw-Hill, New York (1959).
165. M. LEVA, T. SHIRAI, and C. Y. WEN, *Génie chim.* **75** No. 2, 33–42 (1956).
166. M. LEVA, M. WEINTRAUB, and M. GRUMMER, *Chem. Eng. Progr.* **45**, 563–572 (1949).
167. N. M. LEVITZ, E. J. PETKUS, H. M. KATZ, and A. A. JONKE, *Chem. Eng. Progr.* **53**, 199–202 (1957).
168. W. K. LEWIS, E. R. GILLILAND, and W. L. BAUER, *Ind. Eng. Chem.* **41**, 1104–1117 (1949).
169. W. K. LEWIS, E. R. GILLILAND, and H. GIROUARD, *Chem. Eng. Progr. Symposium Series* No. 38 **58**, 87–97 (1962).
170. W. K. LEWIS, E. R. GILLILAND, and W. GLASS, *AIChEJ* **5**, 419–426 (1959).
171. W. K. LEWIS, E. R. GILLILAND, and P. M. LANG, *Chem. Eng. Progr. Symposium Series* No. 38 **58**, 65–78 (1962).
172. H. LITTMAN, *AIChEJ* **10**, 924–929 (1964).
173. V. C. LIU, *J. Meteorol.* **13**, 399–405 (1956).
174. J. P. LONGWELL and M. A. WEISS, *Ind. Eng. Chem.* **45**, 667–677 (1953).
175. L. A. MADONNA and R. F. LAMA, *Ind. Eng. Chem.* **52**, 169–172 (1960); and W. L. BRISSON *Brit. Chem. Eng.* **6**, 524–528 (1961).
176. M. A. MALEK and B. C. Y. LU, *Can. J. Chem. Eng.* **42**, 14–20 (1964); *Ind. Eng. Chem., Process Design Develop.* **4**, 123–128 (1965).
177. M. A. MALEK, L. A. MADONNA, and B. C. Y. LU, *Ind. Eng. Chem., Process Design Develop.* **2**, 30–34 (1963).
178. F. E. MARBLE, *AIAAJ* **1**, 2793–2801 (1963).
179. R. M. MARSHECK and A. GOMEZPLATA, *AIChEJ* **11**, 167–173 (1965).
180. J. MARTIN and B. ANDREIU, Fluidization Symposium, Association France de Fluidisation (June 11, 1956).
181. E. A. MASON, Sc. D. thesis, Massachusetts Institute of Technology, Cambridge, Mass. (1950).
182. L. MASSIMILLA, S. BRACALE, and A. CABELLA, *Ricerca Sci.* **27**, 1853 (1957).
183. L. MASSIMILLA and H. F. JOHNSTONE, *Chem. Eng. Sci.* **16**, 105–112 (1961).
184. L. MASSIMILLA, A. SOLIMANDO, and E. SQUILLACE, *Brit. Chem. Eng.* **6**, 232–239 (1961).
185. L. MASSIMILLA, G. VOLPICELLI, and F. ZENZ, *Ind. Eng. Chem., Fund.* **2**, 194–199 (1963).
186. L. MASSIMILLA and J. W. WESTWATER, *AIChEJ* **6**, 134–138 (1960).
187. G. L. MATHESON, W. A. HERBST, and P. H. HOLT, 2nd., *Ind. Eng. Chem.* **41**, 1099–1104 (1949).
188. J. F. MATHIS and C. C. WATSON, *AIChEJ* **2**, 518–524 (1956).
189. K. B. MATHUR and P. E. GISHLER, *AIChEJ* **1**, 157–164 (1955).
190. K. B. MATHUR and P. E. GISHLER, *J. Appl. Chem. (London)* **5**, 620 (1955).
191. W. MAY, *Chem. Eng. Progr.* **55**, No. 12, 49–56 (1959); *Dechama Monographien* **32**, 451–476, 261–282 (1959).
192. N. C. MEHTA, J. M. SMITH, and E. W. COMINGS, *Ind. Eng. Chem.* **49**, 986–992 (1957).
193. T. S. MERTES and H. B. RHODES, *Chem., Eng. Progr.* **51**, 429–432, 517–522 (1955).

194. A. S. MICHAELS and J. C. BOLGER, *Ind. Eng. Chem., Fund.* **1**, 24–33 (1962).
195. H. S. MICKLEY and D. F. FAIRBANKS, *AIChEJ* **1**, 374–384 (1955).
196. H. S. MICKLEY, D. F. FAIRBANKS, and R. D. HAWTHORN, *Chem. Eng. Progr. Symposium Series* No. 32 **57**, 51–60 (1961).
197. H. S. MICKLEY and C. A. TRILLING, *Ind. Eng. Chem.* **41**, 1135–1147 (1949).
198. R. G. MINET, J. HAPPEL, and W. KAPFER, paper presented at AIChE meeting, San Francisco (Dec. 1959).
199. B. V. MOLSTEDT and J. F. MOSER, JR., *Ind. Eng. Chem.* **50**, 21–23 (1958).
200. C. MORELAND, *Can. J. Chem. Engr.* **41**, 108–110 (1963).
201. R. D. MORSE and C. O. BALLOU, *Chem. Eng. Progr.* **47**, 199–204 (1951).
202. G. MURPHY, D. F. YOUNG, and R. J. BURIAN, *AEC Rpt. ISC-474* (April 1954).
203. J. D. MURRAY, paper presented at AIChE meeting Houston, Texas (Feb. 1965).
204. J. D. MURRAY, *J. Fluid Mech.* **21**, 465–493 (1965); **22**, 57–80 (1965).
205. M. NADER, *Ind. Eng. Chem.* **49**, 39–41 (1957).
206. H. NAKAMURA, and K. KURODA, *Keijo J. Med.* **8**, 256 (1937).
207. F. NAKASHIO and W. SAKAI, *Kagaku Kōgaku* **24**, 452 (1960).
208. D. M. NEWITT, J. F. RICHARDSON, and C. A. SHOOK, in Third. Congr. Europ. Fed. Chem. Engrs. (London), *Interaction Between Fluids and Particles*, pp. 87–100, Inst. Chem. Engrs. (London) (1962).
209. D. R. OLIVER, *Chem. Eng. Sci.* **15**, 230–242 (1961).
210. D. R. OLIVER and V. G. JENSON, *Can. J. Chem. Eng.* **42**, 191–195 (1964).
211. J. C. ORCUTT, J. F. DAVIDSON, and R. L. PIGFORD, *Chem. Eng. Progr. Symposium Series* No. 38 **58**, 1–15 (1962).
212. D. E. ORGEN, PH. D. thesis in Chem. Eng., University of Washington, Seattle, Wash. (1957).
213. C. W. OSEEN, *Neuere Methoden und Ergebnisse in der Hydrodynamik*, p. 132, Akademische Verlagsgesellschaft, Leipzig (1927).
214. K. ØSTERGAARD, in *Fluidization*, p. 58, Soc. Chem. Indust., London (1964).
215. K. ØSTERGAARD, *Chem. Eng. Sci.* **20**, 165–167 (1965).
216. R. H. OVERCASHIER, D. B. TODD, and R. B. OLNEY, *AIChEJ* **5**, 54–60 (1959).
217. K. W. PEARCE, in Third Congr. Europ. Fed. Chem. Engrs. (London), *Interaction Between Fluids and Particles*, pp. 30–39, Inst. Chem. Engrs. (London) (1962).
218. R. L. PESKIN, in *Proc. Heat Trans. Fluid Mech. Inst.*, p. 192, Stanford, Calif. (1960).
219. W. S. PETERSON, *Can. J. Chem. Eng.* **40**, 226–230 (1962).
220. R. L. PIGFORD and T. BARON, *Ind. Eng. Chem., Fund.* **4**, 81–87 (1965).
221. B. G. PRICE, L. LAPIDUS, and J. C. ELGIN, *AIChEJ* **5**, 93–97 (1959).
222. B. B. PRUDEN and N. EPSTEIN, *Chem. Eng. Sci.* **14**, 696–700 (1961).
223. D. L. PYLE and P. L. ROSE, *Chem. Eng. Sci.* **20**, 25–31 (1965).
224. D. L. PYLE and P. S. B. STEWART, *Chem. Eng. Sci.* **19**, 842–843 (1964).
225. J. A. QUINN, L. LAPIDUS, and J. C. ELGIN, *AIChEJ* **7**, 260–263 (1961).
226. I. M. RAZUMOV, *Khim. Tech. Topliv Masel* **7**, No. 4, 41 (1962); translation, *Int. Chem. Eng.* **2**, 539–543 (1962).
227. G. H. REMAN, *Chem. Ind.* 46–51 (Jan. 15, 1955).
228. H. REUTER, *Chem.-Ingr.-Tech.* **35**, 98–103 (1963).

REFERENCES

229. H. REUTER, *ibid.*, 219–228.
230. H. REUTER, paper presented at AIChE meeting, Houston, Texas (Feb. 1965); discussion in Third Congr. Europ. Fed. Chem. Engrs. (London), *Interaction Between Fluids and Particles*, pp. 165–170, Inst. Chem. Engrs. (London) (1962).
231. W. J. RICE and R. H. WILHELM, *AIChEJ* **4**, 423–429 (1958).
232. J. F. RICHARDSON and P. AYERS, *Trans. Inst. Chem. Engrs. (London)* **37**, 314–322 (1959).
233. J. F. RICHARDSON and M. MCLEMAN, *Trans. Inst. Chem. Engrs. (London)* **38**, **38**, 257–266 (1960).
234. J. F. RICHARDSON and W. N. ZAKI, *Trans. Inst. Chem. Engrs. (London)* **32**, 35–53 (1954).
235. J. B. ROMERO and L. N. JOHANSON, *Chem. Eng. Progr. Symposium Series* No. 38 **58**, 28–35 (1962).
236. P. N. ROWE, *Chem. Eng. Progr. Symposium Series* No. 38 **58**, 42–56 (1962).
237. P. N. ROWE, *Chem. Eng. Sci.* **19**, 75–77 (1964).
238. P. N. ROWE, *Chem. Eng. Progr.* **60**, No. 3, 75–80 (1964).
239. P. N. ROWE and B. A. PARTRIDGE, in Third Congr. Europ. Fed. Chem. Engrs. (London), *Interaction Between Fluids and Particles*, pp. 135–42, Inst. Chem. Engrs. (London) (1962); see also R. COLLINS, *Chem. Eng. Sci.* **20**, 851–853 (1965).
240. P. N. ROWE and B. A. PARTRIDGE, *Chem. Eng. Sci.* **18**, 511–524 (1963).
241. P. N. ROWE, B. A. PARTRIDGE, and E. LYALL, *Chem. Eng. Sci.* **19**, 973–985 (1964); **20**, 1151–1153 (1965).
242. E. RUCKENSTEIN, *Rev. Phys., Acad. Rep. Populaire Roumaine* **7**, 137 (1962).
243. E. RUCKENSTEIN, *Ind. Eng. Chem., Fund.* **3**, 260–268 (1964).
244. G. RUDINGER, *Phys. Fluids* **7**, 658–663 (1964).
245. G. RUDINGER, *Project Squid Tech. Rept. CAL-91-P* (Dec. 1964).
246. G. RUDINGER and A. CHANG, *Phys. Fluids* **7**, 1747–1754 (1964).
247. P. G. SAFFMAN, *J. Fluid Mech.* **13**, 120–128 (1962); see also J.T.C. LIU, *Phys. Fluids* **8**, 1939–1945 (1965).
248. K. SCHÜGERL, M. MERZ, and F. FETTING, *Chem. Eng. Sci.* **15**, 1–38 (1961); see also G. SCHIEMANN, K. SCHÜGERL, and F. FETTING, *Chem.-Ingr.-Tech.* **33**, 728–738 (1961); F. F. K. LIU and C. ORR, JR., *J. Chem. Eng. Data* **5**, 430–432 (1960); and W. W. SHUSTER and F. C. HAAS, *ibid.*, 525–530.
249. G. SEGRÉ and A. SILBERBERG, *J. Fluid Mech.* **14**, 115–157 (1962); see also P. G. SAFFMAN, *ibid.* **22**, 385–400 (1965); and R. C. JEFFERY and J. R. A. PEARSON, *ibid.*, 721–735.
250. P. T. SHANNON, R. D. DEHAAS, E. P. STROUPE, and E. M. TORY, *Ind. Eng. Chem., Fund.* **3**, 250–260 (1964).
251. P. T. SHANNON, E. STROUPE, and E. M. TORY, *Ind. Eng. Chem., Fund.* **2**, 203–211 (1963).
252. P. T. SHANNON and E. M. TORY, *Ind. Eng. Chem.* **57**, No. 2, 19–25 (1965).
253. W. W. SHUSTER and P. KISLIAK, *Chem. Eng. Progr.* **48**, 455–458 (1952).
254. H. C. SIMPSON and B. W. RODGER, *Chem. Eng. Sci.* **16**, 153–180 (1961); **17**, 951–953 (1962).
255. E. SINGER, D. B. TODD, and V. P. GUINN, *Ind. Eng. Chem.* **49**, 11–19 (1957).

256. J. W. SMITH and K. V. S. REDDY, *Can. J. Chem. Eng.* **42**, 206–210 (1964).
257. M. S. SMOLUCHOWSKI, *Bull. Intern. Acad. Sci., Cracovie* **1a**, 28 (1911); *Proc. Fifth Intern. Congr. Math.* **2**, 192 (1912).
258. P. SOLCHANI, *Kim. Prom.* No. **4**, 45 (1961); translation, *Int. Chem. Eng.* **2**, 258–260 (1962).
259. S. L. Soo, *Chem. Eng. Sci.* **5**, 57–67 (1956); as a general reference see S. L. Soo, *Fluid Dynamics of Multi-phase Systems*, Blaisdell, Waltham, Mass. (1967).
260. S. L. Soo, *AIChEJ* **7**, 384–391 (1961).
261. S. L. Soo, *Ind. Eng. Chem., Fund.* **1**, 33–37 (1962).
262. S. L. Soo, *ibid.* **3**, 75–80 (1964).
263. S. L. Soo, in *Proceedings of Symposium on Single- and Multi-Component Flow Processes* (C. F. Chen and R. L. Peskin, editors), Rutgers Eng. Pub. 45, 1, Rutgers University, New Brunswick, N.J. (1965).
264. S. L. Soo, *Project Squid Tech. Rept. Ill-18-P* (Sept. 1964).
265. S. L. Soo, H. K. IHRIG, JR., and A. F. ELKOUH, *J. Basic Eng.* **82D**, 609–621 (1960).
266. S. L. Soo and J. A. REGALBUTO, *Can. J. Chem. Eng.* **38**, 160–166 (1960).
267. S. L. Soo and C. L. TIEN, *J. Appl. Mech.* **27E**, 5–15 (1960).
268. S. L. Soo, C. L. TIEN, and V. KADAMBI, *Rev. Sci. Instr.* **30**, 821–824 (1959).
269. S. L. Soo, G. J. TREZEK, R. C. DIMICK, and G. F. HOHNSTREITER, *Ind. Eng. Chem., Fund.* **3**, 98–106 (1964).
270. K. E. SPELLS, *Trans. Inst. Chem. Engrs. (London)* **33**, 79–84 (1955).
271. W. T. SPROULL, *Nature.* **190**, 976 (1961).
272. H. H. STEINOUR, *Ind. Eng. Chem.* **36**, 618–624 (1944).
273. P. S. B. STEWART and J. F. DAVIDSON, *Chem. Eng. Sci.* **19**, 319–322 (1964).
274. M. STIMSON and G. B. JEFFERY, *Proc. Roy. Soc. (London)* **111A**, 110 (1926).
275. N. R. SUNKOORI and R. KAPARTHI, *Chem. Eng. Sci.* **12**, 166–174 (1960).
276. J. P. SUTHERLAND, *Chem. Eng. Sci.* **19**, 839–841 (1964).
277. J. P. SUTHERLAND, G. VASSILATOS, H. KUBOTA, and G. L. OSBERG, *AIChEJ* **9**, 437–441 (1963).
278. K. S. SUTHERLAND, *Trans. Inst. Chem. Engrs. (London)* **39**, 188–194 (1961).
279. K. S. SUTHERLAND, *ANL-6907*, Argonne National Laboratory, Argonne, Ill. (July 1964).
280. J. SZEKELY, in Third Congr. Europ. Fed. Chem. Engrs. (London), *Interaction Between Fluids and Particles*, pp. 197–202, Inst. Chem. Engrs. (London) (1962).
281. P. SZOLCSÁNYI, *Magyar Kémiai Folyoirat* **67**, No. 7, 320 (1961); translation, *Int. Chem. Eng.* **3**, 315–318 (1963).
282. S. R. TAILBY and M. A. T. COCQUEREL, *Trans. Inst. Chem. Engrs. (London)* **39**, 195–201, 237 (1961); see also D. R. MORRIS, K. E. GUBBINS, and S. B. WATKINS, *ibid.* **42**, 323–333 (1964).
283. E. TALMOR and R. F. BENENATI, *AIChEJ* **9**, 536–540 (1963).
284. A. I. TAMARIN, *Inzh. Fiz. Zh.* **6**, No. 7, 19–25 (1963); translation, *Int. Chem. Eng.* **4**, 50–54 (1964).
285. C. M. TCHEN, PH. D. thesis, Delft: Martinus Nijhoff, the Hague (1961).
286. I. TEOREANU, *Bull. Inst. Politech. Bucuresti* **24**, No. 2, (1962); translation, *Int. Chem. Eng.* **3**, 459–463 (1963).

287. I. TEOREANU, *Bul. Inst. Politech. Bucuresti* **24**, No. 1, 77 (1962); translation, *Int. Chem. Eng.* **3**, 239–250 (1963).
288. D. G. THOMAS, *AIChEJ* **7**, 423–437 (1961).
289. D. G. THOMAS, *ibid.* **8**, 373–378 (1962).
290. D. G. THOMAS, *ibid.* **10**, 303–308 (1964).
291. W. J. THOMAS, P. J. GREY, and S. B. WATKINS, *Brit. Chem. Eng.* **6**, 78–86, 176–81 (1961).
292. B. THORLEY, J. B. SAUNBY, K. B. MATHUR, and G. L. OSBERG, *Can. J. Chem. Eng.* **37**, 184–192 (1959).
293. C. L. TIEN, *J. Heat Trans.* **83C**, 183–188 (1961).
294. R. D. TOOMEY and H. F. JOHNSTONE, *Chem. Eng. Progr.* **48**, 220–225 (1952).
295. R. D. TOOMEY, and H. F. JOHNSTONE, *Chem. Eng. Progr. Symposium Series* No. 5, **49**, 51–63 (1953).
296. L. B. TOROBIN and W. H. GAUVIN, *Can. J. Chem. Eng.* **37**, 129–41, 167–176, 224–236 (1959); **38**, 142–153, 189–200 (1960); **39**, 113–120 (1961).
297. L. B. TOROBIN and W. H. GAUVIN, *AIChEJ* **7**, 406–410 (1961).
298. E. M. TORY, *Ind. Eng. Chem., Fund.* **4**, 106–107 (1965).
299. H. TRAWINSKI, *Chem.-Ingr.-Tech.* **25**, 229–238 (1953).
300. R. TURNER, *Fluidization*, p. 47, Soc. Chem. Indust, London. (1964).
301. I. YA. TYURAER, and A. L. TSAILINGOL'D, *Zh. Prik. Kim.* **33**, 1783 (1960).
302. I. A. VAKHRUSHEV and G. S. EROKHIN, *Khim. Prom.* No. 11, 30 (1962); translation, *Int. Chem. Eng.* **3**, 333–338 (1963).
303. V. VAND, *J. Phys. Colloid Chem.* **52**, 277 (1948).
304. J. J. VAN DEEMTER, *Chem. Eng. Sci.* **13**, 143–154, 190–191 (1961).
305. D. VAN ZOONEN, in Third Congr. Europ. Fed. Chem. Engrs. (London), *Interaction Between Fluids and Particles*, pp. 64–71, Inst. Chem. Engrs. (London) (1962).
306. W. VOLK, C. A. JOHNSON, and H. H. STOTLER, *Chem. Eng. Progr.* **58**, No. 3, 44–47 (1962).
307. P. F., WACE and S. J. BURNETT, *Trans. Inst. Chem. Engrs. (London)*, **39**, 168–174 (1961).
308. G. B. WALLIS, Third Congr. Europ. Fed. Chem. Engrs. (London), *Interaction Between Fluids and Particles*, pp. 9–16, Inst. Chem. Engrs. (London) (1962).
309. J. S. WALTON, R. L. OLSON, and O. LEVENSPIEL, *Ind. Eng. Chem.* **44**, 1474–1480 (1952).
310. W. E. WAMSLEY and L. N. JOHANSON, *Chem. Eng. Progr.* **50**, 347–355 (1954).
311. C. Y. WEN and H. P. SIMONS, *AIChEJ* **5**, 263–267 (1959).
312. L. WENDER and G. T. COOPER, *AIChEJ* **4**, 15–23 (1958).
313. E. WICKE and K. HEDDEN, *Chem.-Ingr.-Tech.* **24**, 82–91 (1952).
314. R. H. WILHELM, in *Proceedings of the Second Midwestern Conference on Fluid Dynamics*, p. 379, The Ohio State University Press, Columbus, Ohio (1952).
315. R. H. WILHELM and M. KWAUK, *Chem. Eng. Progr.* **44**, 201–218 (1948).
316. D. WOLF and W. RESNICK, *Ind. Eng. Chem., Fund.* **4**, 77–81 (1965).
317. S. YAGI and D. KUNII, *Chem. Eng. Sci.* **16**, 364–391 (1961); *Kagaku Kōgaku* **16**, 283 (1952); *see also* V. F. FROLOV and P. G. ROMANKOV, *Zhur. Prik. Chim.* **35**, 1526–1533 (1962).

318. G. YASAI and L. N. JOHANSON, *AIChEJ* **4**, 445–452 (1958).
319. N. YOSHIOKA, Y. HOTTA, S. TANAKA, S. NAITO, and S. TSUGAMI, *Kagaku Kōgaku* **19**, 616 (1955).
320. F. A. ZENZ, *Pet. Ref.* **36**, No. 4, 173–178; No. 5, 261–265; No. 6, 133–142; No. 7, 175–183; No. 8, 147–155; No. 9, 305–308; No. 10, 162–170; No. 11, 321–328 (1957).
321. F. A. ZENZ, D. CH. E. thesis, Polytechnic Institute of Brooklyn, New York (1961).
322. F. A. ZENZ, and D. F. OTHMER, *Fluidization and Fluid-Particle Systems*, Reinhold, New York (1960).
323. E. N. ZIEGLER and W. T. BRAZELTON, *Ind. Eng. Chem., Process Des. Dev.* **2**, 276–281 (1963).
324. E. N. ZIEGLER and W. T. BRAZELTON, *Ind. Eng. Chem., Fund.* **3**, 94–98 (1964).
325. E. N. ZIEGLER, R. W. FRISCHMUTH, JR., and W. T. BRAZELTON, *Ind. Eng. Chem., Process Des. Dev.* **4**, 239–240 (1965).
326. N. ZUBER, *Chem. Eng. Sci.* **19**, 897–917 (1964).
327. J. J. BARKER, *Ind. Eng. Chem.* **51**, No. 5, 33–39 (1965); see also N. I. GEL'PERIN, P. D. LEBEDEV, G. N. NAPALKOV, and V. G. AINSHTEIN, *Khim. Prom.* No. 6, 28–37 (1965); translation, *Int. Chem. Eng.* **6**, 4–15 (1966).
328. A. J. BOBKOWICZ and W. H. GAUVIN, *Can. J. Chem. Eng.* **43**, 87–91 (1965).
329. P. R. OWEN, *J. Fluid Mech.* **20**, 225–242 (1964).
330. C. A. SHOOK, and S. M. DANIEL, *Can. J. Chem. Eng.* **43**, 56–64 (1965).
331. E. M. TORY, and P. T. SHANNON, *Ind. Eng. Chem., Fund.* **4**, 194–204 (1965).

AUTHOR INDEX

AUTHOR INDEX

Abou-Sabe, A. H., 458, 460, 475, 534
Abramowitz, M., 24
Abramson, A. E., 458, 460, 530
Acrivos, A., 115, 116, 141, 142, 444, 448
Adams, G., 604, 615
Adams, H., 539, 540, 611
Adams, H. S., 563, 564, 613
Adamson, A. W., 601, 602, 611
Ainshtein, V. G., 652, 686, 692, 700
Akhromenkov, A. A., 675, 694
Aladyev, I. T., 526, 530
Alfrey, T., Jr., 380, 448
Allan, R. S., 604, 611
Allen, W. F., Jr., 501, 530
Allport, J., 646, 692
Al Sheikh, J., 458, 459, 460, 463, 530
Altman, D., 646, 689
Altman, M., 508, 530
Alves, G. E., 457, 458, 460, 475, 530
Amberkar, S., 583, 613
Ambrose, T. W., 458, 460, 530
Amick, E. H., Jr., 540, 544, 545, 547 570, 611
Amundson, N. R., 527, 537, 668, 689
Anderson, G. H., 458, 460, 481, 493, 521, 530
Anderson, J. D., 481, 490, 530
Anderson, L. B., 571, 612
Anderson, R. P., 520, 527, 530, 535
Anderson, T. B., 647, 648, 689
Anderson, T. T., 479, 530
Andersson, K. E. B., 633, 689
Andreas, J. M., 539, 611
Andreiu, B., 672, 695
Andrew, S. P. S., 601, 611
Angelino, H., 655, 689
Aoki, S., 482, 483, 486, 530
Appeldoorn, J. K., 450
Apuzzo, G., 587, 615
Aris, R., 49, 50, 349, 363, 686, 689
Arterburn, D. R., 444, 455
Askins, J. W., 680, 682, 689
Astarita, G., 587, 615
Atit, K. M., 527, 534
Auerbach, R., 370, 453
Ayers, P., 686, 697

Baerg, A., 686, 689
Bagley, E. B., 379, 411, 448
Bailey, J. F., 517, 518, 530
Bailey, W. S., 645, 646, 689
Bailie, R. C., 672, 685, 689, 691
Baird, M. H. I., 603, 611
Baker, J. L. L., 589, 617
Baker, O., 458, 459, 460, 475, 477, 530
Baker, W. O., 380, 451
Bakker, P. J., 657, 672, 673, 681, 683, 686, 689
Bakowski, S., 570, 616
Baldwin, L. V., 272, 322, 325, 355, 356
Ballou, C. O., 672, 696
Balzhiser, R. E., 508, 531
Banchero, J. T., 586, 587, 612
Bandukwala, A., 632, 693
Bankoff, S. G., 272, 355, 361, 479, 480, 482, 483, 487, 510, 524, 527, 530, 534, 536
Barker, D. H., 622, 690
Barker, J. J., 686, 700
Baroczy, C. J., 508, 531
Baron, T., 143, 161, 245, 355, 604, 613, 647, 689, 696
Bartholomew, R. N., 686, 689
Bashforth, E. Q., 494, 531
Bashforth, F., 539, 540, 611
Baskakov, A. P. 686, 689
Bass, J., 294, 355
Basset, A. B., 621, 689
Batchelor, G. K., 264, 265, 269, 273, 277, 281, 291, 293, 299, 300, 305, 307, 320, 326, 327, 334, 335, 355
Batchelor, J. D., 686, 693, 694
Bateman, H., 77, 82, 140, 142
Bates, R. L., 381, 448
Bauer, W. L., 652, 695
Baumgarten, P. K., 655, 670, 689
Beardmore, R. H., 593, 594, 617
Beaty, K. O., Jr., 440, 453
Becker, H. A., 686, 689
Becker, K. M., 526, 531
Beek, J., Jr., 325, 338, 339, 340, 341, 351, 355
Beek, W. J., 444, 448, 640, 689

Beekmann, R. B., 686, 692
Begell, W., 470, 507, 531
Bell, D. S., 526, 531
Bell, J. D., 410, 444, 449
Belner, R. J., 369, 451
Benard, H., 596, 611
Benenati, R. F., 676, 698
Benjamin, M. W., 494, 499, 500, 501, 518, 531
Brooke Benjamin, T., 461, 462, 531
Bennett, A. W., 526, 531
Bennett, C. A., 572, 574, 611
Bennett, J. A. R., 493, 520, 531
Benzing, R. J., 544, 611
Berbente, C., 601, 616
Bergelin, O. P., 458, 475, 476, 486, 531, 533
Bergen, J. T., 397, 448
Bergles, A. E., 521, 531
Berkowitz, N., 644, 649, 652, 690
Bernath, L., 526, 531
Bertoletti, S., 526, 531
Besford, H., 671, 690
Betchov, R., 233, 363
Bikerman, J. J., 539, 601, 602, 611
Bingham, E. C., 368, 369, 382, 448
Bird, R. B., 17, 25, 28, 35, 39, 40, 49, 50, 143, 157, 161, 385, 388, 390, 391, 399, 401, 402, 403, 406, 415, 416, 428, 431, 433, 434, 435, 436, 437, 444, 448, 451, 452, 454, 455
Birkhoff, G., 54, 58, 166, 169
Birks, A. M., 379, 448
Bizzell, G. C., 444, 448
Bizzell, G. D., 153, 161
Blasius, H., 120, 142, 151, 247, 254, 356
Block, M. J., 596, 611
Bloore, P. D., 671, 690
Blyakher, I. G., 652, 653, 691
Boardman, R. P., 629, 694
Bobkowicz, A. J., 652, 700
Boelter, L. M. K., 458, 460, 470, 500, 531, 535
Bogue, D. C., 410, 434, 437, 442, 448, 449
Bogusz, E., 589, 617
Bolger, J. C., 397, 452, 641, 696
Bollinger, R. E., 481, 490, 530

Bolt, L. H., 649, 652, 691
Bond, W. N., 552, 583, 584, 611
Borchers, E., 570, 589, 616
Botterill, J. S. M., 686, 690
Bottomley, W. T., 500, 501, 531
Boussinesq, J., 129, 142, 597, 611, 621, 690
Boussinesq, T. V., 239, 356
Bovwhuis, G. J., 653, 692
Bowling, K. McG., 676, 690
Bracale, S., 685, 695
Bradshaw, P., 232, 358
Bradshaw, R. D., 686, 690
Braun, I., 422, 448
Braun, W. H., 310, 361
Brazelton, W. T., 686, 700
Breach, D. S., 111, 115
Brenner, H., 111, 112, 113, 115, 622, 693
Bretherton, F. P., 595, 611
Brewer, L., 171, 222
Bridge, A. G., 631, 690
Bridge, T. E., 499, 531
Brink, J. C., 604, 614
Brinkman, H. C., 428, 633, 690
Brisson, W. L., 686, 695
Brodkey, R. S., 233, 252, 259, 289, 332, 335, 343, 349, 351, 359, 393, 395, 396, 408, 410, 411, 412, 413, 423, 425, 432, 435, 439, 441, 443, 448, 449, 452, 455, 457, 481, 489, 491, 534
Brodnyan, J. G., 370, 385, 396, 449
Brooks, L. H., 596, 617
Brookshire, A., 272, 356
Brötz, W., 676, 690
Brown, F. E., 542, 613
Brown, R. A. S., 464, 465, 482, 487, 531, 533
Brüchner, R., 596, 611
Bruinzeel, C., 639, 690
Brumm, E. A., 403, 455
Brundrett, G. W., 686, 690
Brunner, C. A., 602, 611
Bruss, H., 390, 449
Bryant, L. T., 527, 530
Buchanan, R. H., 686, 690
Buchdahl, R., 380, 449
Buckingham, E., 166, 169

Bueche, F., 396, 449
Buitelaar, A. A., 477, 478, 534
Bulkley, R., 389
Bupara, S. S., 601, 618
Burgers, J., 55, 58, 127, 142, 323, 356, 630, 690
Burian, R. J., 644, 649, 652, 696
Burnett, S. J., 659, 661, 699
Buthod, P., 523, 531

Cabella, A., 685, 695
Cacaso, A. G., 540, 569, 570, 579, 614
Cain, G. L., 686, 690
Cairns, E. J., 623, 636, 638, 639, 690
Cairns, R. C., 644, 690
Calderbank, P. H., 424, 449, 539, 602, 603, 604, 607, 611, 614
Calvert, S., 493, 531
Canjar, L. N., 686, 692
Cannon, M. R., 410, 444, 449
Cantow, M. J. R., 394, 451
Capo, L., 647, 653, 690
Carberry, J. J., 603, 615
Carley, E. L., 379, 380, 381, 411, 452
Carpenter, F. G., 475, 476, 531
Carrier, G. F., 645, 646, 690
Carter, C. O., 464, 531
Carter, J. C., 527, 530
Casson, N., 389, 444, 449
Caswell, B., 444, 449
Cauchy, A. L., 62
Cavers, S. D., 543, 613
Cengel, J. A., 479, 531
Cermak, J. E., 272, 356
Cess, R. D., 526, 531
Chamberlain, H. V., 652, 690
Chan, B. K. C., 588, 617
Chandrasekhar, S., 304, 356, 362, 596, 611
Chang, A., 645, 646, 697
Chang, C. C., 388, 449
Chang, Y. P., 524, 531
Channabasappa, K. C., 686, 690
Chao, B. T., 586, 587, 589, 603, 611, 617, 618
Charles, G. E., 604, 605, 611
Charles, M. E., 463, 478, 479, 494, 531, 536, 644, 649, 652, 692

Charronia, D. A., 456, 464, 489, 532
Charwatt, A. F., 289, 358
Charzat, C., 655, 689
Chase, R. C., 252, 356, 441, 449
Chen, J. C., 521, 525, 532
Chen, W. S., 272, 358
Cheney, A. G., 676, 690
Chenoweth, J. M., 475, 476, 477, 478, 532
Chien, S. F., 489, 532
Chisholm, D., 478, 532, 593, 614
Cholette, A., 349, 356
Chou, P. Y., 318, 356
Christensen, C. J., 381, 450
Christiansen, E. B., 142, 440, 444, 450, 452, 622, 690
Christiansen, R. M., 548, 550, 551, 552, 554, 611
Chu, J. C., 570, 604, 611, 686, 690
Chung, D. S., 685, 689
Church, J. M., 605, 616
Churchill, R. V., 64, 67, 82, 86, 103
Ciborowski, J., 647, 653, 690
Cicchitti, A., 526, 532
Clapp, R. M., 444, 449
Clare, H., 577, 611
Clark, J. A., 524, 534
Clauser, F. H., 249, 256, 356
Clay, D. R., 384, 449
Clemons, D. B., 289, 363
Cleveland, R. G., 459, 477, 501, 532
Cloutier, L., 349, 356
Clump, C. W., 571, 612
Cockbain, E. G., 604, 611
Cocquerel, M. A. T., 676, 678, 698
Cohen, L. S., 461, 537
Cohen, M. F., 272, 356
Coleman, B. D., 403, 404, 449
Coles, D., 233, 317, 356, 362
Collier, J. G., 481, 490, 520, 532, 535
Collins, M., 410, 449
Collins, R., 658, 659, 662, 697
Colwell, R. E., 380, 381, 411, 452, 455
Comings, E. W., 652, 695
Cooper, G. T., 686, 699
Cooper, R. B., 653, 690
Corder, W. C., 491, 532
Corino, E. R., 233, 259, 356, 439, 449

Cornforth, R. M., 686, 691
Corrsin, S., 274, 275, 277, 278, 289, 300, 304, 309, 317, 319, 322, 324, 325, 333, 334, 335, 336, 337, 338, 345, 346, 356, 357, 360, 624, 625, 691
Costisch, E. W., 606, 616
Coull, J., 589, 612
Couette, M., 88, 89
Cowan, C. B., 686, 691
Cox, B. G., 595, 611
Cox, J. T., Jr., 541, 613
Cox, R. T., 28, 40
Criminale, W. O., 233, 363
Crosby, E. J., 582, 605, 617
Crow, S. C., 232, 358
Cruver, J. E., 515, 537
Curl, R. L., 348, 357, 360, 606, 607, 608, 611, 615
Curtis, J. M., 646, 689
Curtiss, C. F., 25, 40, 390, 391, 451
Cutter, L. A., 345, 357

DaCruz, A. J. R., 493, 497, 501, 510, 512, 513, 514, 517, 534
DallaValle, J. M., 475, 476, 537, 539, 571, 615
Damerell, G. L., 606, 607, 614
Danckwerts, P. V., 289, 328, 329, 330, 331, 345, 348, 357, 364, 603, 611, 674, 682, 691
Daniel, S. M., 652, 700
Darwin, C., 71, 82
Datta, R. L., 544, 587, 589, 611
David, M. M., 521, 532
Davidov, B. I., 318, 357
Davidson, J. F., 487, 536, 591, 592, 603, 611, 617, 646, 647, 648, 654, 655, 658, 662, 663, 665, 668, 670, 674, 683, 686, 691, 693, 696, 698
Davidson, L., 540, 544, 545, 547, 570, 611
Davidson, W. F., 500, 502, 508, 532
Davies, C. N., 589, 612
Davies, G., 652, 653, 691
Davies, J. T., 539, 596, 602, 612
Davies, P. O. A. L., 261, 363
Davies, R., 643, 694
Davies, R. M., 487, 532, 591, 593, 612

Davies, S. J., 228, 357
Davis, E. J., 521, 532
DeBoor, H. G. J., 683, 684, 686, 693
De Groot, J. H., 639, 694
De Groot, S. R., 36, 40
Dehaas, R. D., 634, 635, 641, 643, 697
DeHaas van Dorsser, A. H., 683, 684, 686, 693
Deissler, R. G., 247, 248, 290, 298, 299, 323, 357, 360
DeKock, J. W., 646, 647, 658, 693
DeKraa, J. A., 608, 617
DeMaria, F., 674, 680, 691
Denbigh, K. G., 36, 40
Denn, M. M., 380, 455
Dennis, S. C. R., 142
Denny, D. A., 393, 395, 396, 449, 455
Depaw, C. A., 686, 691
Derbyshire, R. T. P., 464, 532
Dexter, F. D., 385, 449
Diehl, J. E., 475, 532
Diehl, Z. W., 310, 317, 358
Diekman, R., 653, 691
Diglio, A. J., 459, 475, 536
Dimick, R. C., 644, 645, 698
Dittus, F. W., 470, 500, 523, 532
Dodge, D. W., 435, 436, 449
Doherty, D. J., 426, 449
Doicher, K., 672, 691
Doll-Steinberg, A., 523, 532
Dombrowski, N., 555, 556, 557, 558, 561, 563, 565, 566, 567, 568, 578, 612
Domich, E. G., 563, 564, 613
Donat, E. V., 652, 672, 691
Doroshchuk, V. E., 526, 530
Dotson, J. M., 672, 691
Douglas, J. M., 349, 363
Dow, W. M., 686, 691
Drake, R. M., 153, 161
Dreier, W. M., Jr., 554, 615
Drexler, T. D., 686, 691
Dryden, H. L., 77, 82, 140, 142, 267, 357
Duffie, J. A., 551, 612
Dukler, A. E., 458, 459, 463, 464, 475, 477, 480, 481, 489, 501, 532, 535, 537
Dumas, R., 266, 279, 289, 310, 357, 363
Dumitrescu, D. T., 487, 532, 593, 612
Dunlap, R., 646, 692

Dunn, J. S. C., 464, 475, 533
Dunwoody, J., 142
Dupont, F. A. A., 639, 694
Durand, R., 644, 691
Dusinberre, G. M., 499, 532
Duty, R. L., 362
Duxbury, H. A., 655, 691

Eagleton, L. C., 349, 364
Eckert, E. R. G., 153, 161
Eddy, K. C., 479, 509, 534
Edgerton, H. E., 541, 542, 612, 613
Edwards, S. F., 363
Eggink, R., 444, 448
Eichhorn, J., 686, 691
Eichhorn, R., 622, 644, 645, 691
Einstein, A., 632, 691
Einstein, H. A., 253, 357
Eirich, F. R., 397, 449
Eisenklam, P., 539, 555, 558, 567, 578, 579, 612
Eisenschitz, R., 407, 449
Eissenberg, D. M., 434, 442, 449
Elbert, D. D., 362
Elgin, J. C., 631, 632, 635, 637, 690, 694, 696
Elkouh, A. F., 628, 645, 698
Elliott, D. E., 686, 690
Ellis, H. S., 649, 652, 691
Ellis, S. R. M., 461, 463, 489, 532
Elrod, W., 424, 451
Elzinga, E. R., 586, 587, 612
Emersleban, O., 112, 115
Emmons, H. W., 233, 318, 357
Engelberg-Forster, K., 524, 532
Engen, J. M., 457, 458, 461, 462, 489, 533
Engle, D. L., 563, 608, 616
Enskog, D., 31
Epstein, N., 486, 494, 533, 640, 643, 696
Ericksen, J. L., 375, 399, 403, 453
Erokhin, G. S., 676, 699
Euler, L., 59
Evans, H. D., 458, 462, 534, 544, 573, 577, 613, 615
Everett, H. J., 606, 616
Eyring, H. J., 381, 388, 389, 390, 392, 393, 449, 450, 451, 453

Fabula, A. G., 44, 449
Fage, A., 289, 357
Fahien, R. W., 639, 691
Fairbanks, D. F., 686, 696
Faletti, D. W., 512, 513, 514, 515, 532
Falkner, V. M., 135, 142
Fan, L. T., 652, 672, 685, 689, 691
Fanning, J. T., 254, 255
Farber, L., 686, 691
Farley, R. W., 583, 585, 603, 616
Farugui, A. A., 479, 531
Fauske, H. K., 234, 512, 513, 514, 515, 517, 526, 527, 532, 534, 538
Favre, A., 266, 279, 289, 357
Faxen, H., 112, 115, 629, 691
Feinman, J., 686, 691
Feldman, S., 461, 532
Ferron, J. R., 686, 692
Ferry, J. D., 381, 396, 450
Fetting, F., 653, 697
Fick, A., 39
Fincke, A., 381, 450
Findlay, J. A., 484, 487, 538
Firstenberg, H., 526, 527, 533
Fischer, E. K., 380, 450
Fisher, M. J., 267, 363
Fleishman, B. A., 325, 357
Flettner, A., 77
Flint, D. L., 327, 357
Fohrman, M. J., 456, 479, 480, 532
Foltz, H. L., 523, 524, 532
Fomichev, A. G., 652, 653, 692
Fordham, S., 540, 612
Forgrieve, J., 570, 604, 611
Forsell, B. A., 643, 692
Forster, H. K., 521, 524, 532
Forsyth, J. S., 543, 613
Forsythe, W. L., 653, 691
Foster, H. H., 555, 569, 578, 613
Foster, P. A., Jr., 456, 458, 460, 464, 520, 533
Foster, W. R., 408, 422, 424, 454
Fourier, J. B., 39
Foust, A. S., 508, 533, 536, 571, 612
Francis, G. C., 301, 342, 361
Franklin, N. L., 572, 574, 611
Frantz, J. F., 652, 686, 692
Fraser, J. B. P., 494, 531

Fraser, R. P., 555, 556, 565, 566, 567, 578, 579, 612
Frazier, A. W., 458, 460, 532
Fredrickson, A. G., 375, 381, 397, 399, 400, 402, 403, 415, 426, 444, 450, 455
Frenkiel, F. N., 281, 325, 357, 364
Freundlich, H., 375, 450
Fried, E., 520, 537
Fried, L., 470, 532
Friedlander, S. K., 603, 612, 624, 625, 626, 645, 692
Friedman, R., 520, 533
Friend, L., 213, 222
Friend, P. S., 444, 452
Frisch, H. L., 633, 692
Frischmuth, R. W., Jr., 686, 700
Fritsch, T. R., 583, 586, 613
Froessling, N., 603, 612
Fröhlich, H., 402, 450
Frolov, V. F., 678, 686, 699
Fromm, J. E., 84, 103
Fucks, W., 289, 357
Fujie, H., 484, 492, 493, 533
Fulford, G. D., 461, 538
Funnigan, J. W., 479, 531
Furukawa, J., 653, 692

Gabor, J. D., 686, 692
Gabrysh, A. F., 381, 450
Gaden, E. L., 539, 602, 616
Gaggioli, R. A., 154, 157, 161
Gal-Or, B., 608, 612
Gambill, W. R., 526, 533
Garner, F. H., 596, 612
Gaskins, F. H., 370, 417, 449, 450, 453
Gasparyan, A. M., 629, 694
Gauss, C. F., 18
Gauvin, W. H., 622, 624, 652, 699, 700
Gaviglio, J., 266, 279, 289, 357
Gavis, J., 417, 418, 450, 452
Gay, B., 461, 463, 489, 532
Gazley, C., Jr., 458, 475, 476, 486, 531, 533
Gee, R. E., 385, 450
Gegner, J. P., 350, 359
Gel'perin, N. I., 652, 686, 692, 700
Gemmel, A. R., 486, 494, 533
George, J. J., 671, 690

Gerald, C. F., 672, 692
Gerber, R., 304, 357
Gerrard, J. E., 415, 450
Ghosh, B., 686, 692
Gibbons, J. H., 589, 612
Gibson, C. H., 349, 350, 357
Gibson, M. M., 309, 357
Gilbert, M., 646, 692
Gilchrist, A. W. R., 461, 535
Gill, A. E., 233, 363
Gill, L. E., 481, 490, 533
Gill, W. N., 253, 357, 358
Gillespie, T., 604, 612
Gilliland, E. R., 583, 613, 652, 674, 680, 686, 692, 695
Ginn, R. E., 405, 419, 450
Girouard, H., 686, 695
Gishler, P. E., 686, 689, 694, 695
Glass, W., 674, 680, 685, 686, 693, 695
Glassman, I., 646, 692
Glasstone, S., 390, 393, 450
Glauz, R. D., 645, 646, 692
Goikhman, I. D., 652, 692
Goldman, M., 686, 692
Goldmann, K., 512, 526, 527, 533
Goldschmidt, D., 672, 692
Goldsmith, H. L., 594, 612
Gomezplata, A., 675, 676, 695
Gordon, L. J., 638, 639, 692
Gorin, E., 686, 692
Görtler, H., 135, 142
Goto, K., 644, 692
Gouse, S. W., Jr., 456, 533
Govier, G. W., 463, 464, 465, 470, 475, 478, 479, 487, 494, 531, 533, 536, 538, 644, 649, 652, 692
Graham, R. W., 520, 533, 534
Graham, W., 676, 693
Grant, H. L., 303, 309, 324, 358
Grassman, P., 166, 169, 459, 535, 549, 589, 590, 612
Graumann, R. E., 583, 613
Green, G., 18, 494, 533
Green, H., 369, 448
Green, H. L., 544, 555, 561, 563, 565, 583, 584, 585, 614
Green, S. J., 509, 537
Greenhalgh, R. E., 349, 358

Greif, R., 481, 532
Grek, F. Z., 672, 692
Gresham, W. A., Jr., 456, 458, 460, 464, 520, 533
Grey, P. J., 653, 699
Griffith, P., 98, 99, 100, 101, 237, 465, 482, 487, 489, 524, 525, 526, 533, 535, 538
Griffith, R. M., 583, 584, 585, 586, 603, 612
Groothuis, H., 606, 607, 612
Gross, B., 380, 450
Grossman, L. M., 289, 358
Grosso, R., 570, 604, 611
Grotz, L. C., 563, 608, 616
Grummer, M., 686, 695
Gubbins, K. E., 676, 678, 698
Guinn, V. P., 676, 678, 686, 697
Gunn, R., 584, 585, 612
Gunther, F. C., 524, 533
Gurnee, E. F., 380, 448
Gutfinger, C., 494, 533
Gutoff, E. B., 349, 358
Guttinger, C., 444, 450
Gvozdev, V. D., 652, 653, 692
Gwyn, J. E., 582, 617

Haas, F. C., 563, 564, 612, 653, 697
Haberman, W. L., 112, 115, 591, 612
Hadmard, M. J., 114, 115, 583, 604, 612
Haenlein, A., 554, 612
Hagen, G., 91
Hagerty, W. W., 557, 558, 613
Hahn, S. J., 393, 450, 453
Hall, C. D., 444, 453
Hama, F. R., 63, 82, 232, 233, 358
Hamielec, A. E., 583, 586, 613
Handlos, A. E., 544, 604, 613, 674, 680, 693
Hankel, R., 512, 533
Hanks, R. W., 440, 444, 450
Hanna, O. T., 153, 161
Hanratty, T. J., 249, 321, 327, 357, 358, 457, 458, 461, 462, 489, 490, 533, 535, 537, 632, 638, 645, 693, 694
Hanson, A. R., 563, 564, 613
Happel, J., 112, 115, 622, 633, 681, 686, 693, 696

Hardie, P. H., 500, 502, 508, 532
Hardy, B. W., 546, 617
Harkins, W. D., 542, 613
Harlow, F. H., 84, 103
Harmathy, T. Z., 593, 613
Harper, D. A., 602, 613
Harper, J. C., 381, 426, 450, 455
Harriott, P., 603, 604, 613
Harris, D. L., 362
Harris, J., 381, 434, 435, 437, 450
Harrison, D., 646, 647, 648, 654, 655, 658, 662, 665, 670, 686, 691, 693
Harrison, V. G. W., 397, 450
Hartree, D. R., 137, 142
Hartunian, R. A., 591, 613
Harvey, B. F., 508, 533
Haselden, G. G., 521, 530
Hassett, N. J., 643, 671, 693
Hasson, D., 555, 558, 579, 612
Haubenreich, P. N., 499, 501, 533
Hauser, E. A., 539, 541, 542, 611, 612, 613
Hawksley, P. G. W., 632, 693
Hawthorn, R. D., 686, 696
Hay, G. E., 10, 24
Hayakawa, T., 676, 693
Hayasi, N., 444, 448
Haydon, D. A., 596, 613
Hayes, J. W., 401, 405, 451
Hayes, W. D., 140, 142
Hayworth, C. B., 543, 554, 613
Heasley, J. H., 527, 537
Hedden, K., 646, 699
Hedström, B. O. A., 435, 438, 439, 451
Heertjes, P. M., 652, 657, 672, 673, 681, 683, 684, 686, 689, 693
Hegarty, J. C., 232, 358
Heideger, W. J., 596, 603, 617
Heidmann, M. F., 555, 569, 578, 613
Heigl, J. J., 674, 693
Heinz, W., 381, 389, 450, 451
Heisenburg, W., 273, 293, 325, 338, 358
Hendricks, R. C., 520, 533
Henwood, G. A., 676, 690
Herbst, W. A., 653, 695
Hernborg, G., 526, 531
Heroman, L. C., 501, 502, 508, 535
Herring, J. R., 363

Herschel, W. H., 389
Hershman, A., 457, 461, 463, 489, 533
Hess, K., 369, 453
Hewitt, G. F., 80, 87, 88, 109, 110, 464, 479, 481, 489, 490, 526, 532, 533, 535, 537, 538
Higbie, R., 603, 613
Hildebrand, A., 470, 500, 523, 532
Hill, B. A., 606, 613
Hinds, G. P., 680, 682, 689
Hinze, J. O., 239, 256, 265, 267, 272, 277, 279, 289, 293, 338, 350, 358, 539, 555, 563, 564, 567, 613, 621, 624, 626, 627, 693
Hirai, N., 392, 393, 449, 451
Hirschfelder, J. O., 25, 40, 390, 391, 451
Hitchon, J. W., 481, 490, 533
Hixson, A. N., 539, 540, 548, 550, 551, 552, 554, 611, 613, 686, 694
Hochschwender, Z., 561, 613
Hocking, L. M., 629, 693
Hodgson, G. W., 463, 478, 479, 487, 494, 531, 536
Hoelscher, H. E., 345, 364
Hoftijzer, P. J., 540, 617
Hoglund, B. M., 520, 535
Hoglund, R. F., 628, 693
Hohnstreiter, G. F., 644, 645, 698
Holcomb, E. G., 540, 569, 570, 579, 614
Holland, C. D., 516, 617
Holman, J. P., 686, 693
Holt, B. M., 541, 613
Holt, P. H., 653, 695
Honari, S., 683, 693
Hoogendoorn, C. J., 458, 460, 475, 478, 480, 534
Hooke, R., 366
Hooper, P. C., 555, 558, 561, 565, 568, 612
Hoopes, J. W., Jr., 470, 507, 508, 531, 534
Hopf, E., 303, 322, 358
Hopkins, I. L., 380, 451
Horton, T. J., 583, 586, 613
Hotta, Y., 643, 700
Hottel, H. C., 289, 361
Houghton, G., 588, 589, 603, 612, 614, 618

Houghton, W. T., 380, 418, 452
Howarth, L., 123, 142, 268, 286, 290, 297, 358
Howarth, W. J., 605, 613
Howells, I. D., 334, 355
Hsu, N. T., 603, 613
Hsu, Y. Y., 479, 520, 533, 534
Hu, S., 589, 613
Hubbard, P. G., 289, 358, 360
Hudson, J. L., 527, 534
Hughes, B. A., 461, 534
Hughes, R. P., 36, 40
Hughes, R. R., 338, 358, 458, 462, 534, 544, 583, 613
Hughmark, G. A., 468, 475, 476, 478, 480, 483, 487, 534, 649, 652, 693
Hull, R. L., 676, 693
Hultberg, J. A., 645, 646, 693
Humphrey, J. C., 569, 613
Humphreys, C. G. R., 500, 502, 508, 532
Huntington, R. L., 460, 464, 531, 536
Huntley, A. R., 674, 693
Hurd, R., 426, 449
Hutchison, H. P., 494, 531
Hyman, D., 539, 608, 613

Ibele, W., 489, 532
Ichiki, T., 482, 483, 486, 530
Ihrig, H. K., Jr., 628, 645, 698
Iinoya, K., 644, 692
Inoue, Eiichi, 319, 358
Isaakyan, S. M., 629, 694
Isbin, H. S., 456, 458, 459, 460, 464, 479, 491, 493, 497, 498, 501, 508, 509, 510, 512, 513, 514, 517, 526, 527, 534, 537
Ishii, T., 686, 693
Izard, J. A. W., 543, 613

Jackson, R., 647, 648, 667, 669, 689, 693
Jacmic, J. J., 540, 569, 570, 579, 614
Jacobs, J. K., 686, 693
Jacowitz, L. A., 457, 481, 489, 491, 534
Jahnig, C. E., 633, 693
Jakob, M., 686, 691, 524, 534
Janssen, F., 126, 142

Jayaweera, K. O. L. F., 628, 693
Jeffery, G. B., 112, 115, 629, 698
Jeffery, R. C., 622, 697
Jeffreys, H., 461, 462, 494, 534
Jenkins, J. W., 674, 682, 691
Jenson, V. G., 643, 696
Jervis, R. E., 603, 615
Jiji, L. M., 524, 534
Jobling, A., 380, 381, 401, 430, 431, 451
Johanson, L. N., 646, 647, 654, 655, 657, 670, 671, 672, 680, 685, 686, 697, 699, 700
Johns, W. R., 555, 556, 557, 558, 565, 612
Johnson, A. I., 583, 586, 603, 613, 617
Johnson, C. A., 685, 699
Johnson, E. F., 349, 360
Johnson, H. A., 458, 460, 475, 534
Johnson, J. F., 369, 370, 394, 451, 453
Johnson, M., 589, 617
Johnson, M. M., 440, 453
Johnson, R. L., 349, 358
Johnstone, H. F., 645, 653, 675, 680, 685, 686, 692, 693, 694, 695, 699
Jones, I., 553, 614
Jonke, A. A., 678, 686, 694, 695
Jonsson, V. K., 362
Joseph, J., 319, 358
Jottrand, R., 640, 694
Joubert, P. N., 256, 361
Joukowsky, N., 77
Juliusburger, F., 375, 450

Kada, H., 327, 357, 358, 645, 694
Kadambi, V., 628, 645, 698
Kadefors, R., 643, 692
Kalb, J. W., 403, 455
Kalil, J., 686, 690
Kalinske, A. A., 325, 358
Kaloni, P. N., 453
Kalvinskas, L. A., 519, 520, 526, 536
Kaminsky, A., 596, 613
Kamiya, Y., 676, 678, 694
Kampé De Feriet, J., 321, 358
Kanwal, R. P., 112, 115
Kaparthi, R., 686, 698
Kapfer, W., 681, 686, 696
Kaplun, S., 111, 115

Kapoor, N. N., 403, 455
Karim, S. M., 39, 40
Kato, Y., 686, 694
Katz, D. L., 33, 40, 62, 82, 137, 142, 148, 160, 161, 256, 259, 359, 686, 689
Katz, H. M., 686, 695
Kaye, B. H., 629, 694
Kearsey, H. A., 479, 526, 531, 533
Keeler, R. N., 344, 363
Keenan, J. H., 197, 222, 499, 534
Keeys, R. K. F., 526, 531
Kegel, P. K., 475, 476, 531
Keith, F. W., 539, 540, 552, 554, 613
Kelley, E. L., 370, 385, 396, 449
Lord Kelvin, 79
Kepner, R. H., 500, 531
Kepple, R. R., 456, 458, 464, 520, 534
Kettenring, K. N., 686, 694
Keulegan, G. H., 462, 490, 494, 534
Keyes, F. G., 499, 534
Kezios, S. P., 526, 534
Kharchenko, N. V., 686, 694
Kim, D. S., 684, 685, 694
Kim, H. T., 393, 395, 441, 451, 455
Kim, K. Y., 380, 381, 455
Kim, T. S., 526, 534
Kim, W. J., 345, 363
Kim, W. K., 392, 451
Kim, Y. G., 289, 363
Kintner, R. C., 539, 583, 585, 586, 587, 589, 595, 608, 613, 615, 617, 618
Kinzer, G. D., 584, 612
Kisel'nikov, V. N., 672, 692
Kisliak, P., 672, 697
Kistler, A. L., 272, 289, 317, 357, 358, 360
Kivnick, A., 686, 694
Klassen, J., 686, 689, 694
Klebanoff, P. S., 232, 233, 257, 269, 271, 281, 310, 311, 317, 358, 361, 364
Klee, A. J., 589, 613
Kliegel, J. R., 645, 646, 694
Kline, S. J., 130, 142
Klipstein, D. H., 459, 475, 536
Knox, J., 356
Knudsen, J. G., 33, 40, 62, 82, 137, 142, 148, 161, 256, 259, 359, 479, 531

AUTHOR INDEX

Knudsen, M., 36
Koestel, A., 461, 538
Kofoed-Hansen, O., 319, 359
Kolin, A., 289, 359
Kolmogoroff, A. N., 291, 359, 564, 565, 613
Koncar-Djurdjeric, S., 647, 653, 690
Kondukov, N. B., 675, 694
Korchinski, I. J. O., 604, 611
Kordyban, E. S., 482, 487, 533
Kornilaev, A. N., 675, 694
Kosterin, S. I., 457, 458, 460, 475, 476, 534
Kottler, J., 576, 614
Kovasznay, L. S. G., 232, 289, 293, 295, 359, 363
Kozlov, B. K., 464, 534
Kraichnan, R. H., 287, 290, 297, 300, 301, 303, 304, 322, 342, 359, 363
Kramers, H., 639, 653, 694
Krasiakova, L. Y., 458, 460, 489, 491, 534
Kreith, F., 524, 533
Krieger, I. M., 392, 424, 451
Kroepelin, H., 596, 614
Kronecker, L., 14
Kronig, R., 604, 614
Kruglov, A. S., 675, 694
Kubota, H., 686, 698
Kudirka, A. A., 524, 525, 535
Kuerti, G., 181, 222
Kunii, D., 678, 686, 699
Kunreuther, J., 680, 682, 689
Kunstman, R. W., 674, 680, 693
Kuroda, K., 643, 696
Kusui, M., 604, 617
Kutateladze, S. S., 526, 535
Kwauk, M., 621, 646, 699
Kyle, R. J., 456, 458, 460, 464, 520, 533
Kynch, G. J., 629, 635, 641, 694

Lacey, P. M. C., 86, 87, 88, 109, 145, 479, 481, 489, 490, 533, 535, 537
Ladenberg, R., 112, 113, 115
Lagerstrom, P., 111, 115
Lagrange, J. L., 61
Laird, A. D. K., 478, 532, 593, 614
Lama, R. F., 686, 695

Lamb, D. E., 289, 359, 465, 481, 490, 493, 530, 535
Lamb, H., 80, 81, 83, 104, 106, 111, 116, 461, 462, 535, 547, 553, 583, 614
Landau, L. D., 38, 40, 49, 50, 189, 220, 222
Lane, W. R., 539, 544, 555, 561, 563, 565, 583, 584, 585, 614
Lang, P. M., 653, 676, 695
Langhaar, H. L., 129, 142, 166, 169
Langlois, G. E., 565, 607, 617
Lanneau, K. P., 672, 694
Lapidus, L., 631, 632, 635, 637, 690, 694, 696
Laplace, P. A., 61, 105
Lapple, C. E., 181, 199, 222, 571, 614
La Rosa, P., 349, 363
Larson, H. C., 458, 459, 460, 479, 501, 508, 534
Latinen, G., 321, 358, 638, 693
Laufer, J., 256, 257, 268, 269, 270, 271, 279, 289, 307, 308, 309, 313, 314, 325, 359
Lawroski, S., 678, 686, 694
Laws, J. O., 584, 614
Lawther, K. R., 644, 690
Leaderman, H., 380, 451
Lebedev, P. D., 686, 700
Lee, C. J., 672, 691
Lee, C. Y., 672, 694
Lee, J., 252, 289, 336, 342, 343, 349, 351, 356, 363, 441, 449
Lee, K. S., 489, 538
Lee, S. M., 253, 358
Leeden, P. V. D., 653, 692
LeGoff, P., 672, 692
Leibnitz, Baron B. G., 17
Leibson, I., 540, 569, 570, 579, 614
Leidler, K. J., 390, 393, 450
Lemaire, L. H., 166, 169, 459, 535, 549, 589, 590, 612
Lemlich, R., 602, 611, 613, 614
Lemmon, H. E., 142
Lenard, P., 561, 614
Lendrat, E. G., 370, 449
Leonard, J. H., 588, 614
Leonard, R. A., 602, 614
Lepilin, V. N., 686, 694

Leslie, F. M., 444, 451
Leung, L. S., 654, 655, 658, 670, 693
Leva, M., 539, 614, 636, 650, 651, 652, 676, 679, 686, 695
Levenspiel, O., 343, 347, 348, 349, 359, 608, 618, 686, 699
Levich, V. G., 114, 116, 583, 588, 601, 614
Levitz, N. M., 686, 695
Levy, S., 485, 486, 490, 493, 510, 515, 522, 526, 535, 536, 538
Lewis, J. B., 553, 596, 614
Lewis, R. M., 304, 359
Lewis, W. K., 652, 680, 686, 692, 695
Li, H., 235, 357
Libby, P. A., 99, 103
Licht, W., 585, 586, 614
Lienard, P., 289, 359
Liepmann, H. W., 289, 307, 308, 309, 359, 360
Liepmann, K., 307, 308, 309, 359
Lifshitz, E. M., 38, 40, 49, 50, 189, 220, 222
Lightfoot, E. N., 17, 24, 35, 40, 49, 50, 153, 161, 444, 448
Lighthill, M. J., 71, 83, 461, 535
Lilleleht, L. V., 461, 463, 478, 479, 489, 490, 531, 535
Lin, C. C., 228, 360, 599, 614
Lin, C. S., 249, 360
Lin, S. H., 129, 142
Linde, H., 596, 614
Linden, H. R., 686, 690
Lindgren, E. R., 232, 233, 289, 360
Lindsey, E. E., 607, 616
Ling, S. C., 289, 300, 360, 362
Linning, D. L., 493, 497, 510, 535
Littman, H., 676, 695
Liu, F. F. K., 653, 697
Liu, J. T. C., 644, 697
Liu, V. C., 628, 695
Lobo, W. E., 213, 222
Lochiel, A. C., 588, 590, 603, 611, 614
Lockhart, R. W., 470, 473, 474, 475, 507, 508, 509, 510, 511, 535
Lodge, A. S., 376, 379, 405, 431, 451
Loitsiansky, L. G., 297, 360
Lombardi, C., 526, 527, 531, 533

Long, J. D., 232, 358
Longfield, J. E., 674, 680, 691
Longuet-Higgins, M. S., 461, 535
Longwell, J. P., 645, 695
Lottes, P. A., 520, 535
Lovegrove, P. C., 479, 533
Lowe, L. S., 491, 532
Lu, B. C. Y., 686, 695
Ludford, G. S. S., 99, 103
Lumley, J. L., 266, 304, 320, 322, 360, 443, 451, 624, 625, 691
Lundgren, T. S., 129, 142
Lyall, E., 659, 660, 667, 668, 669, 670, 676, 690, 697
Lyon, J. B., 385, 450
Lyon, R. N., 444, 451
Lyons, J. W., 380, 381, 455

MacKay, G. D. M., 604, 605, 614
MacPhail, A. C., 256, 360
Madden, A. J., 606, 607, 608, 614
Madonna, L. A., 686, 695
Maeda, S., 545, 546, 617
Magarvey, R. H., 563, 614
Magiros, P. G., 481, 535
Magnus, G., 76, 77
Mahoney, J. F., 545, 546, 614
Maier, C. G., 543, 614
Maker, F. L., 523, 535
Makhorin, K. E., 686, 694
Malek, M. A., 686, 695
Malkus, W. V. R., 319, 360
Manderfield, E. L., 686, 694
Manegold, E., 601, 602, 615
Manning, F. S., 289, 345, 349, 359, 360, 361, 363
Manning, R. E., 410, 444, 449
Mantzouranis, B. G., 458, 460, 481, 493, 521, 530
Manurung, F., 686, 690
Marble, F. E., 646, 695
Marchaterre, J. F., 520, 527, 530, 535
Margules, M., 421
Mark, H., 380, 451
Markovitz, H., 380, 401, 403, 405, 431, 449, 451
Markson, A. A., 500, 502, 508, 532
Maron, S. H., 369, 392, 424, 451

Maroudas, N. G., 596, 601, 615
Marr, G. R., 349, 360
Marsh, B. D., 603, 617
Marshall, W. R., Jr., 539, 549, 550, 551, 555, 567, 571, 574, 577, 579, 581, 582, 603, 612, 615, 617
Marsheck, R. M., 675, 676, 695
Martin, J., 672, 695
Martin, M. W., 475, 476, 477, 478, 532
Martinelli, R. C., 458, 460, 468, 470, 473, 474, 475, 498, 502, 503, 504, 505, 506, 507, 508, 509, 510, 511, 535
Mason, B. J., 628, 693
Mason, E. A., 674, 680, 682, 692, 695
Mason, S. G., 397, 451, 594, 604, 605, 611, 612, 614
Massimilla, L., 638, 675, 685, 695
Matheson, G. L., 653, 695
Mathis, J. F., 680, 686, 695
Mathur, K. B., 686, 695, 699
Maude, A. D., 632, 691
Maus, L., 571, 612
Maxwell, B., 411, 455
Maxwell, J. C., 376, 378, 451
May, W., 674, 675, 676, 677, 678, 680, 681, 682, 683, 686, 695
Maycock, R. L., 544, 613
McAdams, W. H., 501, 502, 508, 519, 520, 535
McBain, J. W., 596, 613, 614
McBride, G. T., 686, 692
McConnell, A. J., 398, 452
McCoy, B. J., 571, 608, 614
McDowell, R. V., 604, 614
McEachern, D. W., 435, 438, 452
McGinnis, P. H., 440, 453
McGoldrick, L. F., 461, 538
McKee, G. W., 356
McKelvey, J. M., 396, 415, 452
McKennell, R., 381, 426, 428, 452
McKibbins, S. W., 683, 684, 686, 693
McLeman, M., 645, 697
McManus, H. M., 457, 481, 489, 491, 535
McMillen, E. E., 435, 438, 452
McRoberts, T. S., 604, 611
Mecham, W. J., 686, 692
Mehta, N. C., 652, 695

Meissner, J., 369, 394, 452
Meksyn, D., 135, 140, 142
Merrill, E. W., 370, 433, 435, 436, 437, 440, 442, 452, 454
Merrington, A. C., 540, 564, 615
Mertes, T. S., 631, 635, 695
Merz, E. H., 411, 452
Merz, M., 653, 697
Mesler, P. B., 524, 536
Meter, D. M., 388, 433, 434, 435, 436, 452
Metzger, A. P., 408, 410, 411, 412, 413, 452
Metzner, A. B., 375, 379, 380, 381, 385, 402, 403, 405, 408, 411, 418, 419, 424, 433, 434, 436, 437, 442, 444, 448, 449, 450, 452, 454, 455
Mhatre, M. V., 587, 589, 615
Michaels, A. S., 397, 452, 641, 696
Mickelsen, W. R., 272, 322, 323, 355
Mickley, H. S., 10, 24, 61, 83, 86, 103, 686, 696
Middleman, S., 417, 418, 450, 452
Miesse, C. C., 539, 615
Milborn, H., 567, 613
Miles, J. W., 461, 462, 535
Miller, D. N., 603, 604, 615
Miller, E. N., 652, 691
Miller, J. G., 494, 499, 500, 501, 518, 531
Miller, R. S., 322, 338, 339, 340, 341, 348, 351, 355, 360, 539, 589, 602, 604, 606, 607, 615
Miller, W. S., 519, 520, 526, 536
Millionshtchikov, M., 290, 300, 360
Mills, J. W., 585, 617
Mills, R. R., 289, 360
Minet, R. G., 681, 686, 696
Mintzer, D., 293, 362
Mirkus, J. D., 686, 693
Mirpolsky, Z. L., 526, 530
Mitchell, J. A., 517, 537
Miura, Y., 608, 617
Mixon, F. O., 603, 615
Mock, W. C., Jr., 267, 357
Moen, R. H., 456, 458, 459, 460, 464, 491, 498, 501, 508, 509, 514, 534
Mohtadi, M. F., 596, 612

Moilliet, A., 303, 309, 324, 358
Moissis, R., 464, 482, 487, 489, 524, 535
Mologin, M. A., 458, 460, 536
Molstedt, B. V., 686, 696
Moody, F. J., 515, 538
Mooney, M., 407, 408, 413, 422, 452
Moore, D. W., 588, 615
Moore, F. D., 524, 536
Moore, F. K., 140, 142
Moore, T. W., 686, 693
Moo-Young, M. B., 424, 449
Morduchuw, M., 99, 103
Moreland, C., 633, 644, 649, 652, 690, 696
Morkovin, M. V., 232, 360
Morrin, E. H., 458, 460, 470, 535
Morris, D. R., 676, 678, 698
Morrison, S. R., 426, 455
Morse, R. D., 672, 696
Morton, R. K., 591, 612
Moser, J. F., Jr., 686, 696
Mosher, D. R., 456, 458, 459, 460, 464, 491, 498, 501, 508, 509, 514, 534
Moss, H., 554, 616
Moulton, R. W., 249, 360, 512, 513, 514, 515, 532, 537
Moy, J. E., 493, 497, 501, 510, 512, 513, 514, 517, 534
Mugele, R. A., 573, 577, 615, 647, 689
Mumford, A. R., 500, 502, 508, 532
Munro, W. D., 362
Murnaghan, F. D., 378, 452
Murnaghan, F. P., 77, 82, 140, 142
Murphy, G., 166, 169, 644, 649, 652, 696
Murray, J. D., 647, 658, 667, 669, 696
Murray, R. G., 523, 524, 532
Myers, J. E., 153, 161, 544, 604, 611, 614, 686, 690

Nader, M., 686, 696
Nagai, N., 555, 569, 617
Nagata, S., 608, 617
Naito, S., 603, 700
Nakajima, N., 392, 451
Nakamura, H., 643, 696
Nakashio, F., 676, 696
Naor, P., 349, 360

Napalkov, G. N., 686, 700
Napier, D. H., 544, 587, 589, 611
Narasimhamurty, G. S. R., 585, 586, 589, 614
Nassenstein, H., 596, 616
Navier, M., 41
Neal, L. G., 479, 536
Nedderman, R. M., 481, 489, 494, 531, 536
Nelson, D. B., 468, 498, 502, 503, 504, 505, 506, 507, 508, 509, 510, 535
Nesbit, J. R., 458, 460, 536
Neumann, E. P., 197, 222
Neumann, H. J., 596, 614
Navon, U., 644, 645, 691
Newitt, D. M., 544, 587, 589, 611, 644, 696
Newman, A. B., 604, 615
Newman, P. C., 544, 615
Newton, D. A., 552, 583, 584, 611
Newton, Sir I., 33, 37, 365
Nicholls, B., 464, 532
Nickerson, G. R., 645, 646, 694
Nicklin, D. J., 486, 487, 536
Nielson, E. N., 645, 646, 689
Nielson, J. N., 215, 222
Nielson, L. E., 604, 615
Nikuradse, J., 127, 142, 244, 254, 255, 360
Noll, W., 403, 404, 449, 452
Norris, R. H., 508, 530
Nott, H. D., 349, 358
Nukiyama, S., 577, 615
Nusselt, W., 494, 536
Nutt, C. W., 596, 612
Nye, J. O., 335, 349, 350, 351, 359, 364

Oberbeck, A., 111, 116
O'Brien, E. E., 301, 342, 361
O'Brien, V., 289, 360
Obukhov, A. M., 293, 295, 322, 333, 334, 360
Ogura, Y., 300, 361
Ohji, M., 299, 360
Ohmae, T., 653, 692
Ohnesorge, G., 551, 615
Oka, S., 410, 420, 426, 452
Olander, D. R., 601, 615

AUTHOR INDEX

Oldroyd, J. G., 397, 398, 399, 401, 402, 413, 426, 444, 452
Oliver, D. R., 633, 643, 696
Oliver, R. C., 674, 680, 692
Olney, R. B., 539, 579, 589, 602, 604, 615, 674, 678, 685, 696
Olson, R. L., 686, 699
Omer, M. M., 470, 478, 533
Orcutt, J. C., 674, 683, 686, 696
Orell, A., 596, 601, 615
Orgen, D. E., 653, 696
Orr, C., Jr., 539, 571, 615, 653, 697
Orr, W. M. F., 231
Osberg, G. L., 676, 686, 691, 692, 693, 698, 699
Oseen, C. W., 110, 111, 113, 116, 621, 696
Ostergaard, K., 696
Ostrach, S., 461, 538
Ostwald, W., 369, 370, 452, 453
Oswatitsch, K., 181, 222
Othmer, D. F., 570, 604, 611, 622, 633, 636, 644, 649, 650, 651, 652, 678, 679, 686, 700
Ousterhout, R. S., 444, 453
Overcashier, R. H., 674, 678, 685, 696
Owen, P. R., 652, 700

Pai, S.-I., 55, 58, 84, 89, 103, 106, 116, 140, 142, 148, 153, 161, 220, 222, 239, 245, 250, 259, 361
Pao, Y. H., 295, 334, 346, 363
Park, I. K., 379, 380, 381, 411, 452
Park, M. G., 444, 455
Park, R. W., 605, 617
Partridge, B. A., 72, 83, 658, 659, 660, 662, 663, 665, 666, 667, 668, 669, 670, 671, 676, 690, 697
Paslay, P. R., 388, 454
Pattie, B. D., 479, 534
Paul, H. I., 565, 616
Paul, R. C., 655, 691
Pavlov, V. M., 652, 653, 691
Pawlowski, J., 381, 453
Payne, L. E., 111, 116
Pearce, K. W., 643, 696
Pearson, J. R. A., 111, 116, 596, 615, 622, 697

Pell, W. H., 111, 116
Perry, A. E., 256, 361
Peskin, R. L., 628, 629, 696
Peter, S., 381, 453
Peterlongo, G., 526, 531
Petersen, A. W., 444, 452
Petersen, E. E., 344, 363, 444, 448
Peterson, W. S., 686, 691, 696
Petkus, E. J., 686, 695
Petrick, M., 479, 520, 535, 536
Philippoff, W., 320, 369, 388, 389, 415, 417, 431, 449, 450, 453
Phillips, O. M., 461, 536
Phinney, R., 457, 536
Pien, C. L., 325, 358
Pierce, D., 140, 142
Pierce, P. E., 392, 451
Pierson, W. J., 304, 361
Pigford, R. L., 647, 655, 670, 674, 683, 686, 689, 696
Pike, R. W., 493, 497, 511, 512, 514, 517, 536
Pings, C. J., 157, 161
Pipes, L. A., 10, 17, 19, 24, 86, 103
Piteer, K. S., 171, 222
Place, G., 674, 682, 691
Poiseuille, J. L., 90
Pollett, W. F. O., 401, 453
Polomik, E. E., 522, 526, 536
Popov, M., 569, 615
Popovich, A. T., 603, 615
Popper, G. F., 520, 535
Portalski, S., 462, 494, 536
Porter, R. S., 369, 370, 394, 451, 453
Potter, O. E., 603, 615
Powell, R. C., 388, 389, 453
Poynting, J. H., 401
Prandtl, L., 55, 58, 80, 83, 117, 142, 231, 240, 245, 247, 260, 361
Pratt, H. C. R., 553, 596, 614
Prausnitz, J. M., 289, 344, 361, 363, 623, 636, 638, 639
Pressburg, B. S., 468, 475, 476, 479, 534
Price, B. G., 631, 696
Priem, R. J., 569, 613
Prince, R. G. H., 588, 617
Proudman, I., 111, 116, 287, 290, 300, 325, 355, 361

Pruden, B. B., 640, 643, 696
Pulling, D. J., 479, 533
Putnam, G. L., 249, 360
Putnam, J. A., 470, 535
Pyle, D. L., 667, 696

Quandt, E. R., 489, 527, 536, 538
Quinn, J. A., 631, 696

Rabinowitsch, B., 407, 408, 449, 453, 455
Radcliffe, A., 577, 611
Radford, B. A., 464, 475, 533
Rafchiek, F. M., 520, 534
Ralph, J. L., 348, 360, 606, 607, 615
Ramanaiah, P., 388, 449
Rammler, E., 577, 616
Ranz, W. E., 501, 536, 539, 554, 571, 603, 615
Rashkovskaya, N. B., 696, 694
Ravese, T., 500, 502, 508, 532
Lord Rayleigh, 547, 558, 616
Razumov, I. M., 652, 696
Redberger, P. J., 649, 652, 691
Reddy, K. V. S., 686, 698
Reddy, L. B., 601, 615
Redfield, J. A., 588, 603, 618
Redish, K. A., 686, 690
Ree, T., 381, 392, 393, 449, 450, 451, 453
Reed, C. E., 10, 24, 61, 83, 86, 103
Reed, J. C., 385, 408, 433, 435, 436, 452
Regalbuto, J. A., 644, 698
Reichart, H. L., Jr., 458, 460, 536
Reichart, J., 247, 361
Reid, R. C., 459, 475, 536
Reid, W. H., 287, 290, 300, 325, 361, 362
Reiner, M., 369, 375, 378, 380, 385, 388, 389, 400, 401, 425, 453
Reman, G. H., 639, 686, 690, 696
Resnick, W., 349, 363, 608, 612, 678, 699
Reuter, H., 656, 658, 659, 662, 663, 669, 696, 697
Reynolds, A. B., 459, 475, 536
Reynolds, O., 228, 233, 234, 235, 236, 361, 371, 453

Rhodes, H. B., 631, 635, 695
Rice, A. W., 345, 361
Rice, W. J., 647, 697
Richardson, E. G., 540, 564, 615
Richardson, F. M., 440, 453
Richardson, J. F., 633, 644, 686, 696, 697
Richardson, L. F., 322, 361
Rideal, E. K., 604, 612
Riemann, B., 62
Rietema, K., 539, 605, 606, 607, 608, 615
Rivlin, R. S., 375, 378, 399, 400, 401, 403, 425, 453
Roberts, J. E., 380, 381, 401, 430, 431, 451, 453
Robins, W. H. M., 643, 693
Robinson, D. B., 652, 653, 691
Robinson, M. S., 289, 307, 360
Rodger, B. W., 646, 697
Rodger, W. A., 607, 616
Rodriguez, F., 563, 608, 616
Rodriguez, H. A., 479, 534
Rogers, J. D., 509, 536
Rohsenow, W. M., 519, 520, 521, 531, 536
Roller, P. S., 374, 453
Romankov, P. G., 678, 686, 694, 699
Romanova, N. A., 686, 692
Romero, J. B., 646, 647, 654, 672, 695, 697
Roos, F. W., 363
Roscoe, I. R., 370, 371, 453
Rose, P. L., 667, 696
Rosen, G., 304, 361
Rosenhead, L., 39, 40, 140, 142
Rosensweig, R. E., 289, 336, 345, 361
Rosin, P., 577, 616
Rosler, R. S., 272, 355, 361
Ross, D. K., 686, 690
Rotta, J., 233, 318, 361
Round, G. F., 644, 649, 652, 690
Rouse, H., 300, 361
Rouse, P. E., 396, 453
Routley, J. H., 555, 565, 566, 567, 578, 612
Rowe, P. N., 72, 83, 655, 656, 658, 659, 660, 662, 663, 664, 665, 666, 667, 668, 669, 670, 671, 675, 686, 690, 697

Rubanovich, M. N., 457, 458, 475, 476, 534
Rubin, E. E., 539, 602, 616
Ruckenstein, E., 583, 601, 603, 616, 639, 647, 697
Rudinger, G., 645, 646, 697
Rushton, J. H., 606, 607, 616
Russell, T. W. F., 463, 481, 487, 490, 494, 530, 536
Ryan, N. W., 440, 453
Rybczyński, D. P., 114, 116, 583, 616

Sack, R., 402, 450
Saffman, P. G., 326, 361, 589, 616, 622, 644, 697
Sage, B. H., 603, 613
Saidel, G. M., 345, 364
Sailor, R. H., 380, 418, 452
Saint-Guilhem, R., 166, 169
Sakai, W., 676, 696
Sakiadis, B. C., 153, 161, 417, 418, 419, 453
Sallans, H. R., 686, 689
Sal'nikov, A. A., 652, 653, 692
Sandborn, V. A., 256, 269, 310, 361
Sanghani, P. K., 233, 259, 356, 439, 440, 449, 453
Sargent, L. M., 232, 233, 358
Sasaki, S., 555, 569, 617
Satapathy, R., 586, 587, 616
Saunby, J. B., 686, 699
Savins, J. G., 380, 381, 408, 417, 419, 422, 424, 435, 444, 454, 455
Sawistowski, H., 596, 601, 615
Sawochka, S. G., 522, 526, 536
Sayre, R. M., 112, 115
Schechter, R. S., 583, 585, 603, 616
Scheppe, J. L., 508, 536
Scher, M., 253, 357
Schiemann, G., 653, 697
Schiller, L., 130, 142
Schissler, D. O., 674, 680, 693
Schlichting, H., 84, 103, 138, 140, 142, 153, 161, 220, 222, 228, 231, 232, 239, 255, 362
Schmitz, J. A., 652, 691
Schneider, F. N., 460, 536
Schowalter, W. R., 410, 444, 449, 454

Schroeder, R. R., 585, 618
Schubauer, G. B., 232, 233, 257, 258, 267, 271, 313, 315, 316, 317, 318, 357, 361
Schügerl, K., 653, 697
Schwartz, L. M., 289, 362
Schwarz, W. H., 349, 350, 357, 444, 449
Scott, D. S., 487, 536
Scriven, L. E., 583, 596, 597, 599, 616
Seader, J. D., 519, 521, 526, 536
Sears, W. R., 591, 613
Sedov, L. I., 166, 169
Seely, G. R., 388, 389, 455
Segré, G., 622, 697
Sendner, H., 319, 358
Serra, R. A., 645, 646, 689
Severo, N. C., 572, 575, 617
Shah, M. J., 444, 448
Shah, S. M., 570, 604, 611
Shames, I. H., 166, 169
Shannon, P. T., 633, 634, 635, 641, 643, 697, 700
Shanny, R., 644, 645, 691
Shapiro, A. H., 130, 142, 170, 174, 189, 190, 220, 222
Shaver, R. G., 433, 435, 436, 437, 440, 442, 454
Shea, J. F., 557, 558, 613
Shearer, C. J., 481, 489, 536
Shen, C. Y., 680, 686, 693, 694
Sher, N. C., 124, 479, 509, 534, 537
Shertzer, C. R., 381, 454
Sherwood, T. K., 10, 24, 61, 83, 86, 130, 596, 616
Shinnar, R., 349, 360, 565, 605, 606, 616
Shirai, T., 652, 695
Shook, C. A., 644, 652, 696, 700
Short, W. L., 464, 533
Shuster, W. W., 653, 672, 697
Sideman, S., 583, 616
Siemes, W., 539, 544, 570, 589, 591, 616
Sigwart, K., 596, 616
Silberberg, A., 622, 697
Silver, R. S., 517, 518, 537
Silvestri, M., 526, 532, 537
Simha, R., 633, 692

Simon, F. F., 479, 534
Simons, H. P., 644, 699
Simpson, H. C., 646, 697
Singer, E., 676, 678, 686, 697
Singh, S., 526, 527, 534
Sinnar, P., 44, 450
Sisko, A. W., 385, 386, 387, 392, 401, 451, 454
Skachko, I. M., 675, 694
Skan, S. W., 135, 142
Skaperdas, G. T., 213, 222
Skelland, A. H. P., 444, 455, 591, 604, 612, 616
Skinner, T., 289, 362
Skramstad, H. K., 232, 267, 357, 361
Slack, E. G., 289, 362
Slack, G. W., 628, 693
Slattery, J. C., 153, 154, 161, 401, 444, 448, 454, 455
Sleicher, C. A., Jr., 240, 362, 565, 616
Slibar, A., 388, 454
Small, S., 622, 691
Smith, C. R., 480, 537
Smith, J. M., 390, 454, 639, 652, 691, 694, 695
Smith, J. W., 686, 698
Smith, K. A., 596, 599, 616
Smith, L. R., 508, 531
Smith, M. J. S., 655, 691
Smith, R. V., 513, 518, 537
Smith, S. J., 554, 616
Smith, W., 586, 587, 616
Smoluchowski, M. S., 629, 698
Snyder, N. W., 524, 531
Solchani, P., 652, 698
Soldaini, G., 526, 532
Solimando, A., 686, 695
Sommerfeld, A., 231
Sonneman, G., 526, 537
Soo, S. L., 627, 628, 644, 645, 646, 698
Souders, M., Jr., 245, 355
Spalding, D. B., 253, 362
Sparrow, E. M., 129, 142, 362, 526, 531
Spells, K. E., 119, 190, 191, 570, 583, 616, 649, 652, 698
Spiegel, E. A., 319, 362
Spielberg, K., 232, 362
Spielman, L. A., 608, 618

Spiewak, I., 459, 475, 536
Spriggs, T. W., 403, 455
Sproull, W. T., 644, 698
Squillace, E., 686, 695
Squire, H. B., 232, 362, 557, 558, 616
Stalder, J. R., 289, 362
Stanton, T. E., 254
Staub, F. W., 508, 530
Stegun, I. A., 24
Steidler, F. E., 450
Stein, R. P., 512, 533
Steinour, H. H., 633, 698
Stergarrd, K., 686, 696
Sternling, C. V., 458, 462, 534, 596, 599, 616
Stewart, J. J., 672, 689
Stewart, P. S. B., 655, 686, 696
Stewart, R. W., 209, 286, 293, 299, 303, 306, 309, 324, 355, 358, 362, 461, 493, 534, 537
Stewart, W. E., 17, 24, 35, 40, 49, 50, 153, 161, 401, 406, 444, 448
Stimson, M., 629, 698
Stine, H. A., 289, 362
Stoddard, C. K., 374, 453
Stokes, G. G., 18, 41, 50, 84, 104, 108, 113, 116, 400, 621
Stone, H. N., 686, 694
Stotler, H. H., 685, 699
Streeter, V. L., 160, 161
Stroupe, E. P., 633, 634, 635, 641, 643, 697
Stuart, H. A., 380, 454
Stuart, M. D., 518, 537
Styrikovich, M. A., 526, 530
Sullivan, D. M., 607, 616
Sullivan, G. A., 464, 465, 531, 533
Sullivan, S. L., Jr., 546, 617
Sunkoori, N. R., 686, 698
Sutherland, J. P., 652, 686, 698
Sutherland, K. S., 655, 672, 676, 690, 698
Sutton, O. G., 589, 612
Swanson, B. S., 479, 536
Sweeney, M. P., 686, 692
Szekely, J., 686, 698
Szolcsányi, P., 686, 698
Szymanski, P., 91, 94, 103

AUTHOR INDEX

Tadaki, T., 545, 546, 617
Tailby, S. R., 676, 678, 698
Takahashi, T., 482, 483, 486, 530
Tallmadge, J. A., 494, 533
Talmor, E., 676, 698
Tamarin, A. I., 672, 698
Tan, H. S., 300, 362
Tanaka, S., 603, 700
Tanasawa, Y., 555, 569, 577, 615, 617
Tanenbaum, B. S., 293, 362
Tang, Y. S., 480, 537
Tanner, R. I., 113, 116, 401, 402, 405, 451, 454
Tate, R., 555, 579, 617
Tatsumi, T., 290, 362
Tausch, F., 397, 452
Taylor, B. W., 563, 614
Taylor, G. I., 200, 227, 228, 229, 230, 231, 242, 243, 262, 273, 320, 326, 362, 383, 384, 454, 487, 532, 563, 591, 593, 612, 617
Taylor, N. H., 489, 491, 537
Taylor, T. D., 115, 116
Taylor, T. H. M., 458, 460, 470, 535
Tchen, C. M., 621, 698
Tek, M. R., 508, 531
Teoreanu, I., 647, 672, 698, 699
Ter Haar, D., 36, 40
Terjesen, S. J., 585, 586, 617
Thomas, D. G., 369, 414, 433, 435, 436, 438, 454, 644, 648, 649, 652, 699
Thomas, W. J., 653, 699
Thomsen, E. G., 458, 460, 470, 535
Thomson, G. W., 213, 222
Thorley, B., 686, 699
Thornton, J. D., 493, 531
Thorsen, G., 585, 586, 617
Tidstrom, K. D., 232, 233, 358
Tien, C., 444, 454, 455
Tien, C. L., 627, 628, 644, 645, 698, 699
Tietjens, O. G., 24, 80, 83, 117, 142
Tikhonov, V. A., 652, 653, 692
Timan, H., 232, 362
Tippets, F. E., 526, 527, 537
Titt, E. W., 308, 358
Tobolsky, A. V., 380, 396, 451, 454
Todd, D. B., 674, 676, 678, 685, 686, 696, 697

Tollmien, W., 231, 232
Tomita, Y., 444, 454
Toms, B. A., 381, 414, 444, 454
Toomey, R. D., 653, 675, 680, 686, 699
Toor, H. L., 343, 344, 345, 361, 362, 364, 415, 454
Töpfer, C., 123, 142
Torobin, L. B., 622, 624, 645, 699
Tory, E. M., 250, 251, 253, 298, 331, 633, 634, 635, 641, 643, 697, 699, 700
Toupin, R., 63, 83
Towell, G. D., 348, 360, 606, 607, 615
Townsend, A. A., 289, 293, 306, 313, 317, 319, 320, 326, 327, 334, 355, 362
Trass, O., 603, 615, 617
Trawinski, H., 646, 699
Treloar, L. R. G., 380, 454
Treybal, R. E., 543, 554, 589, 613
Trezek, G. J., 644, 645, 698
Trice, V. G., Jr., 607, 616
Trilling, C. A., 686, 696
Tritton, D. J., 289, 362
Truesdell, C. A., 63, 83, 397, 399, 400, 404, 419, 454
Tsailingol'd, A. L., 676, 699
Tsugami, S., 603, 700
Tucker, M., 289, 362
Tucker, W. B., 539, 542, 611, 612
Tung, T. V., 456, 458, 464, 520, 534
Turian, R. M., 428, 448, 454
Turner, K. S., 644, 690
Turner, R., 686, 699
Tyler, E., 548, 551, 554, 617
Tyuraer, I. Ya., 676, 699

Uberoi, M. S., 289, 325, 363
Uno, S., 595, 617
Ursell, F., 461, 494, 537

Vakhrushev, I. A., 676, 699
Valentine, R. S., 596, 617
Vand, V., 632, 699
Van Deemter, J. J., 680, 682, 683, 686, 699
Van der Laan, E. Th., 639, 690
Vanderwater, R., 526, 527, 534
Van Dierendonck, L. L., 608, 617
Van Driest, E. R., 250, 357

AUTHOR INDEX 721

Van Dyke, M. D., 126, 140, 142
Van Krevelen, D. M., 540, 617
Van Rossum, J. J., 457, 458, 461, 463, 489, 537
Van Wazer, J. R., 380, 381, 455
Van Zoonen, D., 644, 645, 699
Vassilatos, G., 344, 364, 686, 698
Vasil'ev, A. S., 652, 653, 692
Veda, T., 464, 537
Vela, S., 403, 455
Vermeulen, T., 565, 607, 617
Vohr, J. H., 458, 460, 464, 465, 493, 537
Voigt, W., 378
Volk, W., 685, 699
Volpicelli, G., 638, 695
Volynski, M. S., 564, 617
Von Kármán, T., 243, 254, 260, 268, 286, 290, 297, 358
Von Mises, R., 99, 103
Von Rosenberg, A. E., 676, 693
Von Weizsäcker, C. F., 293, 363
Vukovic, D., 647, 653, 690

Wace, P. F., 659, 661, 699
Wakiya, S., 112, 113, 116
Walker, C. L., 480, 537
Wall, R., 604, 615
Wallick, G. C., 408, 422, 424, 444, 454, 455
Wallis, G. B., 482, 487, 527, 533, 537, 631, 634, 635, 636, 638, 641, 642, 643, 699
Walsh, T. J., 272, 322, 323, 324, 355
Walters, J. K., 591, 592, 617
Walton, J. S., 686, 699
Wamsley, W. E., 686, 699
Ward, D. M., 603, 617
Ward, F. H., 596, 617
Ward, H. C., 475, 476, 497, 511, 512, 514, 517, 536, 537
Warshay, M. W., 589, 617
Wasserman, M. L., 401, 444, 455
Watanabe, T., 604, 617
Watkin, F., 551, 554, 617
Watkins, S. B., 653, 676, 678, 698, 699
Watson, C. C., 680, 686, 692, 695
Watts, A., 676, 690

Weatherhead, R. J., 520, 535
Weber, C., 554, 558, 617
Wei, J. C., 596, 616
Weintraub, M., 686, 695
Weiss, M. A., 645, 695
Weissenberg, K., 378, 402, 407, 408, 419, 431, 455
Welch, J. E., 184
Wellek, R. M., 604, 616
Weltmann, R. N., 373, 385, 433, 435, 438, 455
Wen, C. Y., 644, 652, 686, 691, 695, 699
Wender, L., 686, 699
Wenzel, L. A., 545, 546, 571, 612, 614
Weske, J. R., 232, 363
Westermann, M. D., 639, 694
Westerterp, K. R., 608, 617
Westmoreland, J. C., 493, 537
Westover, R. F., 411, 455
Westwater, J. W., 519, 520, 537, 596, 601, 615, 675, 685, 695
Wetteroth, W. A., 686, 690
Whelan, P. F., 544, 615
Whitaker, S., 461, 538, 601, 617
White, C. M., 228, 357
White, E. T., 593, 594, 617
White, J. L., 380, 403, 417, 418, 452, 455, 465, 493, 535
White, P. D., 458, 460, 536, 537
White, R. R., 686, 691
Wicke, E., 646, 699
Wickey, R. O., 458, 459, 460, 501, 508, 534
Wicks, M., III, 458, 459, 463, 464, 475, 477, 480, 481, 489, 501, 532, 537
Wilhelm, R. H., 289, 321, 345, 358, 359, 360, 363, 638, 646, 647, 653, 693, 697, 699
Wilkes, J. O., 487, 536
Wilkinson, W. L., 380, 381, 455
Williams, B., 493, 531
Williams, G. C., 289, 361
Williams, G. M., 565, 607, 617
Williams, J. R., 686, 690
Williams, M. C., 49, 50, 399, 402, 431, 455
Williams, R., 655, 689

Williamson, R. V., 369, 455
Willmarth, W. W., 272, 289, 363
Wills, J. A. B., 267, 364
Willson, R. A. B., 289, 364
Winnilow, S., 586, 587, 618
Winning, M. D., 433, 435, 455
Winovich, W., 289, 362
Wissler, E. H., 527, 537
Wlodarski, A., 647, 653, 690
Wolf, D., 349, 363, 678, 699
Wong, V. M., 686, 693
Woo, T. M., 596, 614
Wood, R. K., 465, 533, 537
Woods, W. K., 501, 502, 508, 535
Woolley, R. H., 381, 450
Wooldridge, C. E., 272, 289, 363
Worrell, G. R., 349, 364
Wright, C. H., 479, 531
Wyld, H. W., Jr., 304, 363
Wylie, C. R., Jr., 10, 12, 24, 43, 50, 399, 455
Wyllie, G., 494, 537

Yabuta, S., 608, 617
Yagi, S., 478, 537, 678, 686, 699
Yamaguchi, I., 608, 617
Yang, Y. C., 686, 691
Yang, K. T., 153, 161
Yarnall, D. R., 518, 537
Yasai, G., 654, 655, 657, 670, 671, 680, 685, 700
Yau, J., 444, 455
Young, D. F., 644, 649, 652, 696
Yoshida, F., 608, 617
Yoshioka, N., 643, 700

Zaki, W. N., 633, 697
Zaloudek, F. R., 512, 514, 537
Zarattarelli, R., 526, 532
Zelen, M., 572, 575, 617
Zenz, F. A., 622, 633, 636, 638, 644, 649, 650, 651, 652, 678, 679, 686, 695, 700
Ziegler, E. N., 686, 700
Zielke, C. N., 686, 692
Zivi, S. M., 492, 493, 537
Zuber, N., 484, 487, 520, 521, 524, 532, 537, 538, 630, 631, 632, 635, 700
Zuiderweg, F. J., 606, 607, 612
Zupnik, T. F., 645, 646, 689
Zwietering, Th. N., 348, 363

SUBJECT INDEX

SUBJECT INDEX

Accelerative contribution in two-phase flow, 506
Adiabatic, evaporating, one-component, two-phase flow, 494
Adiabatic flow of a perfect gas, one-dimensional, steady, 178
Aerated flow, 463
Aerodynamic instability, 557
Age of a fluid at a point, 347
Aggregate fluidization, 670
Aggregate motion, 620
Agitator design, 606
Analysis methods, 162
Angular momentum, 154
Annular two-phase flow, 457, 463, 481, 489
  boiling, 525
  models, 511
Antithixotropic, 375
Apparent viscosity, 400
  true, 368, 400
Area fraction in two-phase flow, 468
Area plot in two-phase flow, 459
Atomization, 555
Atomizing systems, 565
Autocorrelation in turbulence, 266
Average, statistical, 233
  time, 233
  true, 233
Average velocity across a pipe, 145
Axial diffusion, 639

Balance, overall-mass, momentum, and energy, 56
Basic shear diagram, 365, 375
Bernoulli equation, 56, 155
Bias in distribution analysis, 582
Bingham ideal plastic, 368, 409, 425
Bingham plastic equation, 385
Bob in an infinite fluid, 423
Boiling in two-phase flow, 524
Boundary layer, 55, 117, 125
  axially symmetrical, 140
  chemical reaction, 140
  in compressible flow, 140
  equations, 117
  on a flat plate, 120
  instability, 231
  local drag coefficient, 124, 137
  mass and heat transfer, 140
  non-Newtonian flow, 444
  separation, 138
  similar solutions, 134
  with suction and injection, 140
  temperature, 211
  thermal, 140
  thickness, 86, 118, 120, 125
  three-dimensional, 140
  unsteady, 140
Boundary limit problem, 55
Bond number, 167
Boussinesq's theory, 239
Breakdown-reformation mechanism in rheology, 372
Breakup, of drops, 561
  effect of viscosity, 554
  of jets, 547
  of a liquid sheet, 555
Brinkman number, 164
Bubble, chains in solids–gas systems, 671
  coalescence in solids–gas systems, 670
  constant frequency, 541
  constant volume or static, 541
  dynamics, 653
  flow, 457, 463, 486
  formation, 539, 541
  formation at a submerged orifice, 654
  ideal model for solids–gas system, 661
  interactions, 604, 607
  in solids–gas systems, 670
  large, 591
  motions, 582
  phase, 680
  restricted in a tube, 593
  rise in solids–fluid systems, 655
  shape in solids–fluid systems, 655
Buckingham-Reiner equation, 410
Bulk deformation, 38
Bulk viscosity, 38
Burnout, 526

Capacitance number, 544
Capillary flow, 406
Capillary jet experiments, 417
Capillary shear diagram, 431
Capillary viscometers, 129

SUBJECT INDEX

Cartesian coordinates, 21
Cartesian tensor notation, 8
Casson model, 389
Cauchy-Green deformation tensor, 404
Cauchy-Riemann conditions, 62
Cellular patterns during interface
  motions, 596
Chamber volume effect in bubble
  formation, 544
Channel flow, entrance boundary layer,
  129
Characteristic funtional, 303
Chemical reaction and reactors, 343
Choked two-phase flow, 499
Choked velocity, 650
Choking, 184
Churn flow, 463
Circulation, 61, 73, 75, 76, 583
Closure problem of turbulence theory,
  288
Coalescence of drops and bubbles,
  604
Co-current annular two-phase flow
  upward, 493
Column matrix or vector, 15
Completeness of sets, 166
Complex flow problems, 380
Complex variables, 64
Compressible flow, 163, 170
  exact solution for a shock, 98
Concentric rotating cylinders, 95, 228
Conformal mapping, 64
Constant-area flow, 185
Constant-frequency bubble, 541
Constant-volume bubble, 541, 543
Constant wall-temperature, 215
Constitutive equations, 397
Continuity equation, 32, 35
  in turbulence, 235
Convected derivative, 398
Converging nozzles, 184
Converging-diverging nozzles, 185, 192
Coordinates, cartesian, 21
  curvilinear, 43
  cylindrical, 21
  spherical, 23
Correction for rotational viscometer
  data, 423

Correlation, lateral, 264
  longitudinal, 264
  triple velocity, 269
  in turbulence, 262
Correlations for pressure drop, 435
Couette flow, 88
Concentric cylinders, 95, 419
  simple, 89
Criteria for mixing, 330
Critical heat flux, 526
Critical pressure, 499
Critical Reynolds number, 439
Critical two-phase flow, 499, 512
Critical velocity, 179
Critical wavelength for stability, 553
Cross viscosity, 400
Cumulant-discard approximation for
  turbulence, 299, 300, 342
Cumulants, 273
Cumulative curve, 572
Curl, 11, 60, 77
Curvilinear coordinates, 43
Cylindrical coordinates, 21

Deep-layer model, Kelvin-Helmholtz,
  461
Deep liquid layers, 460
Deformation and flow, 365
Deformation, dilatational, 366
  extensional, 366
  shear, 366
Del, 11
Detachment of drops and bubbles, 541
Developing slug flow, 463, 487
Diffusion by kinetic and pressure effects,
  313
Diffusion of pressure energy, 315
Diffusion velocity, 26
Diffusivity, momentum, 37
  solids mixing, 697
Dilatancy, 371
Dilatant materials, 374
Dilational surface viscosity, 597
Dimensional analysis, 162, 165, 244
Dimensionless groups, 166
  wave formation, 462
Direct-interaction approximation, 342
Discharge coefficient, 181

SUBJECT INDEX   727

Dispersed flow, 457, 463
Dispersion, long time, 321
  of mass, radial, 638
  measurements, 645
  and molecular diffusion, 326
  small time, 320
Displacement thickness, 124
Dissipation, function, 50
  length, 305
  of turbulent energy, 287, 313
Distribution, analysis, 571
  bias in, 582
  empirical, 577
  function, 571
  normal, 573
Divergence, 11, 32
Drag, flat plate boundary layer, 124, 151
Drag coefficient, flat plate, 151
  on a sphere, 108
Drag reduction, 443
Drag on a sphere, 108
Drift line, 658
Drop(s), breakup, 561
    critical, 563
    and local isotropic turbulence, 564
  and bubble motions, analytical
      representations, 589
  coalescence, 604
  deformation, 563
  distributions, 551
  formation, 539, 541
    mechanism, 555
  on a horizontal circular tip, 540
  interactions, 604
  motions, 582
  and particle size distribution, 571
Dyads, 16
Dynamic similarity, 477

Eddy, diffusion coefficient, 321
  diffusivities of heat and mass, 239
  stress, 235
  viscosity, 239
Elastic characteristics, 375
Elasticity, linear, 366
Elasticoviscous solid, 378
Ellis model, 385
Empirical distributions, 577

Empirical two-phase correlations, 470, 478
Empirical models of rheology, 385, 401
Emulsion phase, 680
  for solids–gas mixture, 653
End effects, 381, 383, 410, 426
Energy, balance, 310
  overall, 170
  in two-phase flow, 495-497
  density in wave number space, 275
  equation, 34, 36
    temperature form, 35, 36, 49
  flux, 30
  loss term for laminar flow, 157
  production, 315
  spectrum tensor, 274
  tensor, 262, 275, 282
Engineering Bernoulli equation, 157
Enthalpy, 157, 171
Entrainment in two-phase flow, 481, 489, 490
Entrance effects in two-phase flow, 481
Entrance region, pressure drop, 131
Entropy, 171
Equation, change, 31
  of change, 25
    general integral, 143
  continuity, 32, 35
  energy, 34, 36
  Euler, 59
  for friction and heat transfer in a
      constant area duct, 212
  Kármán integral momentum, 150
  Laplace, 61
  momentum, 33
  motion, 33, 35
  of motion of a single accelerating
      particle, 621
  Navier-Stokes, 41–43, 53
  of statistical turbulence, 283
  of two-phase flow with interphase
      transfer, 495
Equilibrium range in turbulence, 309
Equilibrium or steady state in rheology, 395
Equimolar counter diffusion, 26
Error function, 86
Euler equation, 59

Euler number, 163
Eulerian concentration correlation, 330
Eulerian correlation function, 263, 281
Eulerian correlation tensor, 262, 281
Eulerian space-time correlation, 323
Eulerian time correlation, 266
Eulerian triple correlation function, 282
Exact solutions, 54, 84
Exchange coefficient, 239
Exponential decay law in turbulence, 335
Extension, linear, 367

False-body material, 370
Fanning friction factor, 253, 432
Fanno Line, 187, 198
Fick's first law, 39
Fick's second law of diffusion, 50
Film thickness in horizontal channels, 463
Films, thin liquid, 461
Final period of decay in turbulence, 297
First Law of thermodynamics, 170
First-order catalytic kinetic reaction, 683
First-order reaction, effect on spectra, 346
Flat plate(s), flow, 84
  flow between, 87, 89
Flettner rotor, 77
Flooding point, 494
Flow, adiabatic compressible, 178
  areas in two-phase flow, 549
  circular pipe, 90
  compressible, 170
  compressible exact, 98
  Couette, 88
  over a flat plate, 117
  Hagen-Poiseuille, 90
  hypersonic, 175
  incompressible, 174
  isentropic, compressible, 179
  in long ducts, compressible, 199
  between parallel plates, 87, 89
  parallel plates boundary layer, 129
  patterns in solids-gas systems, 657
  pipe boundary layer, 129
  Plane Poiseuille, 90
  rate and the wall-shear rate, relation between, 407
  between rotating concentric cylinders, 95
  slow motion, 54, 104
  Stokes', 54, 104
  subsonic, 174
  supersonic, 174
  transonic, 174
  units, heterogeneity of, in rheology, 392
  with flashing, 494
Fluctuations, in concentration, measuring, 289
  in pressure, measuring, 289
  in temperature, measuring, 289
  in velocity, measuring, 289
Fluidity, 367
Fluidization, 620
  aggregate, 670
  particulate, 619, 632, 636
  velocity, minimum, 619
Fluidized bed models, 679
Flux vectors, 26
Foam flow, 457
Foam fractionation, 602
Foams, 601
Fog model for two-phase flow, 498
Force balance, 28, 212
Forces, inertial, 163
  irrotational and rotational, 61
  pressure, 146
  viscous, 146, 163
Formation, of bubbles from orifice, 543
  of drops and bubbles, 539
  of drops from tips, 541
Fourier's law, 39
Fourier transforms, 273
Fourth-order cumulant discard, 300
Free energy of activation in rheology, 390
Free flow, 539
Frictional contribution in two-phase flow, 506
Friction factor, 91, 195, 500
  model for estimation of critical flow, 512
  model for two-phase flow, 498
  supersonic flow, 197
  two-phase flow, 465, 471
Friction factors in pipes, 253

SUBJECT INDEX 729

Friction velocity, 237
Froth flow, 463
Froude number, 164

Gas cloud in solids–gas systems, 667
Gas-dispersed flow, 463
Gas-piston flow, 463
Gas residence time distribution, 673, 681
Gauss' Theorem, 18
General equation of change, 31
General orthogonal coordinates, 46
General property balance, 30
Geometrically similar boundaries, 162
Goodness of mixing, 328
Grad, 11
Green's theorems, 18

Hagen-Poiseuille flow, 90
Heating effects, viscous, 415, 426
Heat transfer, 39
  coefficient in compressible flow, 211
  during forced-convection boiling, 520
  and friction in compressible flow, 211
  without friction in compressible flow, 206
  in solids–liquid systems, 686
  in two-phase flow, 519
Heisenberg's turbulence model, 293
Herschel-Bulkley rheological model, 389
Higher-order moment-discard approximation for turbulence, 297
Hindered settling, 640
Histogram, 571
Homogeneous turbulence, 261
Homogeneous two-phase flow, 477, 496, 498
  without mass transfer, 469
  models, 482, 510
Homogeneous two-phase mixture, 468
Hooke's law, 366
Horizontal two-phase annular flow, 457, 490, 525
Hot-film anemometers, 289
Hot-wire anemometers, 289
Hydraulic diameter, 471
Hydrodynamic instability, 599
Hypersonic flow, 175
Hysteresis loop in rheology, 373

Ideal flow, 53, 59
Ideal gas, 171
Ideal mixer, 346
Inertial-convective subrange in turbulence, 335
Inertial subrange in turbulence, 292
Initially together problem for turbulent mixing, 345
Inlet of a pipe, 129
Inspection analysis, 162
Instability of a liquid–gas interface, 601
Integral, angular momentum balance, 154
  approach for vertical solids–fluid systems, 630
  energy balance, 153
  equation of change, general, 143
  mass balance, 144
  methods of analysis, 56, 143
  momentum balance, 146
  momentum equation for solids–fluid systems, 630
Integrals, 16
  differentiation, 17
  line or curve, 16
  surface, 17
Intensity of turbulence, 269, 282
Intensity of segregation, 328, 332
  (age of fluid), 347
Interaction effects for drops and bubbles, 604
Interface dynamics, 596
Interface equation, 597
Interface hydrodynamics, 595
Interface oscillation, 595
Interfacial tension, evaluation of, 539
Interfacial turbulence, 596
Interior limit problem, 55
Intermittency in turbulence, 316
Internal energy, 34
Intuitive analysis for turbulence, 290
Intuitive physical model of turbulence, 293
Invariant to coordinate transformation, 397
Irreversible or real process, 171
Irrotational flow, 53, 77
Isentropic expansion, 172
Isentropic flow, 179, 182

730 SUBJECT INDEX

Isentropic retardation process, hypothetical, for compressible flow, 180
Isothermal flow, 212, 217
Isothermal expansion, 172
Isotropic turbulence, 261, 288
Isotropic viscous fluid, 400

Jaumann derivative used in rheology, 398
Jet, bubble, 548
  breakup, 547
  liquid, 547
  liquid in liquid, 548
  measurement of turbulence in, 309
  Rayleigh, 547
Joint probability distribution in turbulence, 281

Kelvin's theorem, 79
Kinematic viscosity, 37
Kinetic approach in rheology, 393
Kinetic energy, 29
  diffusion term in turbulence, 315
  in two-phase flow, 465
Kinetics in solids–liquid systems, 686
Kolmogoroff scale for turbulence, 292
Kovasznay local transfer model for turbulence, 293
Kronecker delta, 14

Lagrangian correlation, for turbulence, 266
  function, 281
Lagrangian and Eulerian systems, relations between, 322
Lagrangian time microscale, 325
Lagrangian models, 304
Laminar instability, 381, 383
Laminar single-particle motion, 621
Laminar turbulent transition, 227
Laminar velocity profiles in rheology, 440
Laplace equation, 61
Laplace transform, 86
Laplacian, 11
Layers, deep liquid, 460
Leading edge of boundary layer, 126
Line or curve integral, 16

Linear differential equation, 86
Linear momentum, 146
Linear viscoelasticity, 402
Liquid sprays, sampling, 571
Liquid–gas interface, 595, 597, 601
Liquid–liquid, pipe flow, 463
  systems, 494
Local drag coefficient, boundary layer similar solutions, 137
  flat plate boundary layer, 124
Local isotropic turbulence, 304, 309
  Reynolds number, 305
Lockhart and Martinelli correlation for two-phase flow, 470
Log-normal number distribution, 575
Log number distribution, 577

Mach cone, 175
Mach number, 163, 174
Magnus effect, 76
Margules equation for Newtonian viscosity, 421
Martinelli-Nelson correlation for two-phase flow, 502
Mass average velocity, 26
Mass flux, 30
Mass and heat transfer in free flow, 602
Mass transfer, 39
  between phases in two-phase flow, 494
  in solids–liquid systems, 686
Material derivative, 398
Materials, non-Newtonian, 367
Maximum flow rate in compressible flow, 181
Maxwell liquid, 376
Mean diameters, 573
Mean-squared displacement in turbulent dispersion, 320
Mean velocity profile, 239
Mechanical energy balance, 56, 157
Mechanical energy balance for two-phase flow, 465, 495, 517
Mechanism of isotropic turbulence, 290
Metastable two-phase flow conditions, 517
Metric coefficients, 19
Metric tensor, 19, 404
Microrheology, 396

SUBJECT INDEX 731

Microscale of turbulence, 273, 305
Minimum entropy production in
  two-phase flow, 492
Minimum fluidization velocity, 652
Minimum transport velocity, 648
Mist flow, 463
Mixer, nonideal, 347
Mixing, 327
  calculations, 350
  criteria, 328
  in an isotropic field, 333
  and molecular diffusion, 327
  results, experimental, 349
  scale-up, 337, 341
  between two or more streams, 343
Mixing-length approach for two-phase
  flow, 485
Mixing-length theory, 240
Mobility, 368
Modeling and models, 166
Modulus, bulk, 366
  shear, 367
Molecular-based theories of rheology,
  396
Momentum approach, for annular
  two-phase flow, 492
  for vertical solids–fluid systems, 630
Momentum, diffusivity, 164
  equation, 33, 35
  equation for two-phase flow, 467–495
  flux, 27, 37
  linear, 146
  thickness, 125
Motion, of air bubbles in water, 587
  of liquid–liquid drops, 585
  of single drops and bubbles, 582
Multiparticle flow, 644
Multiparticle systems, in homogeneous
  flow, 629
  in nonhomogeneous flow, 651
Multiphase flow, 226
  general, 36
Multizone model of a fluidized bed, 684

Nabla, 11
Navier-Stokes equation, 41–43, 53
Newtonian fluids, 366
Newton's law of viscosity, 37
Newton's second law of motion, 33

Nonhomogeneous horizontal two-phase
  flow, 469
Nonideal mixer, 347
Nonisentropic flow, 185
Non-Newtonian, 365
  fluid flow, 225, 405
  flow, heat transfer, 444
  slow motion, 444
Normal number distribution, 573
Normal shock, 189
Normal stress, 27, 38, 405
  effects, 375, 417, 430, 434
Notation, 4, 56, 353, 445, 528, 609, 687
  cartesian tensor, 8
Nucleate boiling, 524
Nukiyama-Tanasawa number
  distribution, 577
Number-distribution function, 572

Obukhov energy transfer model for
  turbulence, 293
One-dimensional spectrum, experimental,
  349
One-seventh power law for turbulence,
  247, 254
Operational diagrams for solids–liquid
  flow, 635
Operator D method, 96
Operators, 11
Order-of-magnitude estimation, 118
Orifice, horizontal, 540
Orr-Sommerfeld equation, 231
Orthogonal coordinate systems, 18
Oseen's analysis for slow motion, 110
Ostwald-de Waele model for rheology,
  385
Ostwald material or curve for rheology,
  369
Overall angular momentum balance,
  155
Overall approaches, 56
Overall energy balance, 154
Overall mass balance, 144
Overall mechanical energy balance, 157
Overall models for two-phase flow heat
  transfer, 522
Overall momentum balance, 146
Overall two-phase flow correlations,
  470, 478

Pao's continuously transferred model for turbulence, 296
Particle behavior in dilute systems, 621
Particle motion, 675
Particle-to-particle interactions, 628
Particle paths, 658
Particulate-to-aggregate fluidization transition, 646
Particulate fluidization, 619, 632, 636
Path lines, 63, 70
Peclet number, 164
  mass transfer, 165
Pendant drop, 539
Perfect gas, 171
Permutation symbol, 9
Perturbation technique, 304
Phenomenological theories of turbulence, 239
Pipe apparent viscosity, 407, 432
Pipe flow, entrance boundary layer, 129
  two-phase, 456
Pitot tube, 160
Plane Poiseuille flow, 90
Plastic materials, pressure drop, 438
Plastic viscosity, 368
Plate-and-cone, 381
  geometry, 419
  viscometer, 424
Plug flow, 381, 382, 457, 463
Pneumatic transport, 644
Point correlation function in turbulence, 268
Poisson's ratio, 367
Potential motion in solids–gas system, 662
Powell-Eyring equation for rheology, 392
Powell-Eyring model for rheology, 388
Power law for rheology, 385, 408, 423
Poynting effect in rheology, 401
Prandtl's mixing-length theory, 240
Prandtl number, 164
Pressure, dynamic, 88
  fluctuations in turbulence, 272
  gradient effect in integral momentum balance, 152
  hydrostatic, 88
  static, 27, 29
  tensor, 27
  total, 88
  wave, propagation of small, 173
Pressure drop, 434
  correlations, 465
  data, two-phase, 475
  in the entrance region, 131
  in plastic materials, 438
  reduction in two-phase flow, 478, 494
  in shear thinning materials, 436
  two-phase frictional, 467
  in viscoelastic materials, 437
Principal invariants of the shear-rate tensor, 399
Principle of material indifference, 397
Probability, 171
  distribution, 281
Production of turbulence, 257
Production of turbulent kinetic energy, 313
Propagation of a small pressure wave, 173
Properties and gravitation group, 168
Pseudoplastic, 369
Pseudo-shear rate, 407

Quality, 467
Quasinormal approximation for turbulence, 300

Radial dispersion in solids–liquid systems, 645
Rate of deformation, 37
Rate processes theory, 390
Ratio of specific heats, 163
Rayleigh jet, 547
Rayleigh line, 188, 208
Recovery factor, 211
Reiner-Philippoff model for rheology, 388
Reiner-Rivlin equation for rheology, 425
Relative dispersion between two particles, 322
Relative velocity, 479
Residence time distribution, 678
Response function, 302
Reverse transformation, 274
Reversible process, 171
Reynolds analogy, 212
Reynolds equations, 233, 235

Reynolds number, 163, 432
  in boundary layer flow, 119
  lower critical, 228
  modified, 433
  upper critical, 227
Reynolds stress, 235, 256
Rheogoniometry, 380
Rheological characteristics of
    materials, 365
Rheological equations of state, 384
  empirical, 385, 401
  semi-empirical, 385
  theoretical, 390
Rheological measurements, 380
Rheology, 365
Rheopectic, 375
Rheopexy, 375
Rigidity, 367
Rivlin and Ericksen constitutive
    equation for rheology, 403
Root normal volume distribution, 577
Rosin-Rammler volume-distribution
    function, 577
Rotation, 60
Rotational apparent viscosity, 422
Rotational flow, 419
Rotational instruments, 381
Rotational-type viscometers, 95, 381,
    423
Row matrix or vector, 15
Rules of averaging, 234

Saltation velocity, 649
Scalar quantity equation for turbulent
    mixing, 333
Scalar product, 8
Scalar transfer function for turbulent
    mixing, 333
Scalars, 7
Scale, in turbulence, 272
  Eulerian, 272
  factors, 19
  Lagrangian length, 273
  Lagrangian time, 272
  lateral integral, 272
  longitudinal integral, 272
  of segregation in turbulent mixing,
    328, 330

transverse Eulerian, 272
transverse Lagrangian, 273
Scales, Eulerian, 282
  Lagrangian, 282
Schmidt number, 165
Second law of thermodynamics, 171
Sedimentation, 632
Seely empirical formula in rheology, 388
Segregation, intensity of, 328, 332, 347
Self-mixing, 343, 347
Semi-annular flow, 463, 486
Separation, 232
  boundary layer, 138
  of variables method, 92
Sequence of plots method in two-phase
    flow, 459
Settling curve, 641
Settling, hindered, 640
Settling velocity, 644
Shear diagram, basic, 365, 375
Shear modulus, 367
Shear stress, 29, 37
Shear surface viscosity, 597
Shear-thickening materials, 371
Shear-thinning materials, 368, 373
Shear-thinning materials, pressure drop,
    436
Sheet, attenuating, 558
  breakup of liquid, 558
  viscous controlled, 558
Sheltering model in two-phase film
    flow, 462
Shock, 189
  exact solution, 98
Shock wave, 186
  thickness, 98
Similarity, 162
Simple fluid of Noll for rheology, 403
Single-particle motion, 621
Singular point in potential flow, 77
Sink in potential flow, 73
Sinuous motion of jets, 540
Sisko model for rheology, 385
Slip, 91
  ratio, 479
  velocity, 479, 486
  at the wall, 381, 413
Slow motion, 54

Slow motion flow, 104
  alternate shapes, 111
  mean free path effects, 114
  mobile interface effect, 114
  wall effects, 112
Slug flow, 457, 463, 486, 487
Slug-flow boiling, 525
Small-amplitude oscillatory motion, 431
Solid dynamics in solids–liquid systems, 674
Solids, 366
  distribution, 645
    measurement, 571
  flux plot, 635
  mixing, 676
  mixing coefficient, 677
  motion, 675
  transport systems, 643
Solids–fluid flow, 619
Solids–fluid systems, bubble formation, 654
  bubble shape, 655
  bubble rise, 655
Solids–gas bed characteristics at high velocities, 670
Solids–gas system, ideal bubble model, 661
  potential motion, 662
Source in potential flow, 73
Specific heats, 171
Spectra results in turbulence, 351
Spectrum, lateral one-dimensional, 277
  longitudinal one-dimensional, 277
  one-dimensional, 275, 282
  tensor, energy, 282
    integrated energy, 276, 282
  three-dimensional, 273
Spectrum function, three-dimensional energy, 276, 282
Spectrum in statistical turbulence, 273
Spherical coordinates, 23
Spontaneous emulsification, 596
Spontaneous motions at an interface, 595
Spray flow, 457, 463
Stability, 227
Stagnation flow, boundary layer similar solution, 137

Stagnation state, 178
Stagnation temperature, 211
Standard deviation, 552
Static bubble, 541
Statistical theory of turbulent flow, 260
Stokes', assumption, 400
  flow, 54, 104
  law, 104
  problem, 84
  theorem, 18
Strain, dilatational, 366
  extensional, 366
Stratified flow, 457, 486
Streak lines, 63, 72
Stream function, 61
  boundary layer, 120
  Oseen's analysis, 110
  slow motion, 107
Streamlines, 62, 68
  instantaneous, 69
Stress, eddy, 235
  normal, 27, 38, 405
  shear, 27, 367
  tangential, 27
Structural viscosity in rheology, 369
St. Venant body in rheology, 367
Sublayer and buffer zones in turbulence, 247
Submerged vertical slot for bubble formation, 540
Submicroscopic scale of turbulent mixing, 327
Subsonic flow, 174
Substantial derivative, 11, 398
Sudden contraction, 159
Sudden expansion, 158
Supersonic duct flow, 205
Supersonic flow, 174
Surface coefficients of viscosity, dilational and shear, 597
Surface force, 27, 165
Surface fraction mean, 577
Surface integral, 17
Surface tension contribution to interface motion, 598
Systems, isentropic, 158
  isothermal, 158

SUBJECT INDEX 735

Tangential stress, 27, 597
Taylor's vorticity-transport theory for turbulence, 242
Temperature heating effect in rheology, 381
Tensor, contraction, 14
  inner product, 14
  metric, 19, 404
  outer product, 14
  pressure, 27
  transposed, 13
  unit, 14
Terminal velocity, 109
Thermal diffusivity, 164
Thermodynamics, 170
Thickening-with-time materials, 375
Thin liquid films, 461
Thinning-with-time materials, 372
Thixotropic, 372, 375, 393
Time constant of mixing, 336
Time-dependent materials, 372
Time mean velocity, 145
Tollmien-Schlichting, instability, 232
  waves in the laminar-turbulent transition, 232
Total energy balance in two-phase flow, 497
Transfer function in statistical turbulence, 286
Transfer of energy by convection, 313
Transformations, 18
Transition between flows for solids–fluid systems, 646
Transition mechanism for laminar to turbulent flow, 228
Transmission coefficient, 390
Transonic flow, 174
Transport coefficients, 36
Transposed tensor, 13
Turbulence, 36, 261, 381, 383
  closure problem, 288
  description of, 261
  effect on a particle, 624
  interfacial, 596
  isotropic, 261, 288, 290
Turbulent core, 247
Turbulent diffusion, 319
Turbulent dispersion, 319

Turbulent energy dissipation, 287, 313
Turbulent flow, 225, 227, 463
Turbulent impulse, 234
Turbulent kinetic energy, 313
Turbulent microscale, 273, 305
Turbulent mixing, 328
Turbulent motion, 261
  mechanism, 260
Turbulent point correlation function, 268
Turbulent production, 257
Turbulent scale, 272
Turbulent shear flow, 262, 310
Turbulent single-particle motion, 624
Turbulent spectrum, 273
Turbulent spot, 228
  propagation, 233
Turbulent stress, 235
Turbulent velocity profile for non-Newtonian materials, 441
Two-component isothermal flow, 457
Two-phase flow, 226
  accelerative contribution, 507
  adiabatic, 494
  annular, 457, 463, 481, 489, 490, 511, 525
  area plots, 458
  choked, 499
  co-current annular, upward, 493
  correlations, overall, 470, 478
  critical, 499, 512
  dispersed, 457, 463
  entrainment, 481, 489, 490
  entrance effects, 481
  equations, 465, 495
  fog model, 498
  friction-factor models, 498
  with heat transfer, 519
  homogeneous, 468, 477, 482, 496, 498, 510
  mechanical energy balance, 465, 495, 517
  metastable flow conditions, 517
  minimum entropy production, 492
  mixing-length approach, 485
  momentum balance, 467, 495
  operational diagram in solids–liquid system, 635

patterns, 457
pipe, 456
pneumatic transport, 644
Two-phase model of a fluidized bed, 680
Two-phase pressure drop data, 475
Two-phase steam water flow, void fractions, 509

Universal equilibrium range in statistical turbulence, 291
Universal velocity distributions, 249
Upper-limit log normal distribution, 577

Variable-density homogeneous fluid model for two-phase flow, 482
Variable-density turbulent flow for two-phase flow, 485
Variance of turbulence, 269
Varicose motion for a jet, 540
Vector, product, 9
Vectors, 7
  column, 15
  derivative, 11
  row, 15
  unit, 7
Velocity, angular, 60
  choking, 650
  defect law, 243
  diffusion, 26
  distribution, 317
    Pohlhausen, 151
    Prandtl, 151
    in rough pipes, 253
    for turbulent flow, 245
  instantaneous, 233
  mass average, 26
  minimum fluidization, 652
  minimum transport, 648
  potential, 53, 60
  profiles for non-Newtonian flow, 440
  relative, 479
  saltation, 649
  slip, 479
  slow motion, 106
  of sound, 173
    for a perfect gas, 174
  terminal, 109

Vertical bubble flow, 486
Vertical liquid–liquid systems, 465
Vertical slug flow, 482
Vertical solids–fluid systems, integral approach, 630
  equations, 630
Vertical two-phase flow, 463, 482
Viscoelastic, 375
  fluid, 376
  materials, pressure drop, 437
Viscoelasticity, 405
  linear, 402
  nonlinear theories, 378
Viscometric flows, 406
Viscometry, 380
Viscosity, 91, 367
  apparent, 400
  bulk, 38
  cross, 400
  eddy, 239
  effective, 433
  function, 632
  pipe apparent, 407, 432
  plastic, 368
  rotational apparent, 422
  true apparent, 368
Viscous-convective subrange for turbulence, 334
Viscous dissipation, 164, 257, 315, 383
  in two-phase flow, 467
Viscous heating effects in rheology, 426
Viscous shear forces, 313
Viscous sublayer, 245
Viscous tensor, 39, 49
Void-fraction, correlations, 465
  in two-phase flow, 509
Voidage distributions in solids–liquid flow, 672
Voidage fluctuations, 672
Voigt body in rheology, 378
Volume scale, 330
Volume-surface or Sauter mean, 577
Volumetric extension coefficient, 399
Von Kármán-Howarth equation for turbulence, 286
Von Kármán's similarity hypothesis for turbulence, 243

Von Kármán theorem of integral momentum, 148
Vortex, in potential motion, 75
   filament, 78
   filaments, 80
   line, 78
   loops, 233
   motion, 77
   pair, 80
   ring, 78
   tube, 78
Vorticity, 60
   vector, 77

Wall, area, 245
   effects, 426
   friction, 193
   shear forces, 195
   shear stress, 433
   slip, 381, 413
Wavelength, 274
Wave number vector, 274
Wavy flow, 457
Weber number, 165
Wedge flow, boundary layer similar solution, 137
Weissenberg effect in rheology, 401
Weissenberg-Rabinowitsch-Mooney relation for rheology, 407
Well-stirred mixer, 346
Work done by viscous forces, 313

Zone of silence, 174

A CATALOG OF SELECTED
DOVER BOOKS
IN SCIENCE AND MATHEMATICS

A CATALOG OF SELECTED
DOVER BOOKS
IN SCIENCE AND MATHEMATICS

QUALITATIVE THEORY OF DIFFERENTIAL EQUATIONS, V.V. Nemytskii and V.V. Stepanov. Classic graduate-level text by two prominent Soviet mathematicians covers classical differential equations as well as topological dynamics and ergodic theory. Bibliographies. 523pp. 5⅜ × 8½. 65954-2 Pa. $10.95

MATRICES AND LINEAR ALGEBRA, Hans Schneider and George Phillip Barker. Basic textbook covers theory of matrices and its applications to systems of linear equations and related topics such as determinants, eigenvalues and differential equations. Numerous exercises. 432pp. 5⅜ × 8½. 66014-1 Pa. $10.95

QUANTUM THEORY, David Bohm. This advanced undergraduate-level text presents the quantum theory in terms of qualitative and imaginative concepts, followed by specific applications worked out in mathematical detail. Preface. Index. 655pp. 5⅜ × 8½. 65969-0 Pa. $13.95

ATOMIC PHYSICS (8th edition), Max Born. Nobel laureate's lucid treatment of kinetic theory of gases, elementary particles, nuclear atom, wave-corpuscles, atomic structure and spectral lines, much more. Over 40 appendices, bibliography. 495pp. 5⅜ × 8½. 65984-4 Pa. $12.95

ELECTRONIC STRUCTURE AND THE PROPERTIES OF SOLIDS: The Physics of the Chemical Bond, Walter A. Harrison. Innovative text offers basic understanding of the electronic structure of covalent and ionic solids, simple metals, transition metals and their compounds. Problems. 1980 edition. 582pp. 6⅛ × 9¼. 66021-4 Pa. $15.95

BOUNDARY VALUE PROBLEMS OF HEAT CONDUCTION, M. Necati Özisik. Systematic, comprehensive treatment of modern mathematical methods of solving problems in heat conduction and diffusion. Numerous examples and problems. Selected references. Appendices. 505pp. 5⅜ × 8½. 65990-9 Pa. $12.95

A SHORT HISTORY OF CHEMISTRY (3rd edition), J.R. Partington. Classic exposition explores origins of chemistry, alchemy, early medical chemistry, nature of atmosphere, theory of valency, laws and structure of atomic theory, much more. 428pp. 5⅜ × 8½. (Available in U.S. only) 65977-1 Pa. $10.95

A HISTORY OF ASTRONOMY, A. Pannekoek. Well-balanced, carefully reasoned study covers such topics as Ptolemaic theory, work of Copernicus, Kepler, Newton, Eddington's work on stars, much more. Illustrated. References. 521pp. 5⅜ × 8½. 65994-1 Pa. $12.95

PRINCIPLES OF METEOROLOGICAL ANALYSIS, Walter J. Saucier. Highly respected, abundantly illustrated classic reviews atmospheric variables, hydrostatics, static stability, various analyses (scalar, cross-section, isobaric, isentropic, more). For intermediate meteorology students. 454pp. 6⅛ × 9¼. 65979-8 Pa. $14.95

CATALOG OF DOVER BOOKS

RELATIVITY, THERMODYNAMICS AND COSMOLOGY, Richard C. Tolman. Landmark study extends thermodynamics to special, general relativity; also applications of relativistic mechanics, thermodynamics to cosmological models. 501pp. 5⅜ × 8½. 65383-8 Pa. $12.95

APPLIED ANALYSIS, Cornelius Lanczos. Classic work on analysis and design of finite processes for approximating solution of analytical problems. Algebraic equations, matrices, harmonic analysis, quadrature methods, much more. 559pp. 5⅜ × 8½. 65656-X Pa. $13.95

SPECIAL RELATIVITY FOR PHYSICISTS, G. Stephenson and C.W. Kilmister. Concise elegant account for nonspecialists. Lorentz transformation, optical and dynamical applications, more. Bibliography. 108pp. 5⅜ × 8½. 65519-9 Pa. $4.95

INTRODUCTION TO ANALYSIS, Maxwell Rosenlicht. Unusually clear, accessible coverage of set theory, real number system, metric spaces, continuous functions, Riemann integration, multiple integrals, more. Wide range of problems. Undergraduate level. Bibliography. 254pp. 5⅜ × 8½. 65038-3 Pa. $7.95

INTRODUCTION TO QUANTUM MECHANICS With Applications to Chemistry, Linus Pauling & E. Bright Wilson, Jr. Classic undergraduate text by Nobel Prize winner applies quantum mechanics to chemical and physical problems. Numerous tables and figures enhance the text. Chapter bibliographies. Appendices. Index. 468pp. 5⅜ × 8½. 64871-0 Pa. $11.95

ASYMPTOTIC EXPANSIONS OF INTEGRALS, Norman Bleistein & Richard A. Handelsman. Best introduction to important field with applications in a variety of scientific disciplines. New preface. Problems. Diagrams. Tables. Bibliography. Index. 448pp. 5⅜ × 8½. 65082-0 Pa. $12.95

MATHEMATICS APPLIED TO CONTINUUM MECHANICS, Lee A. Segel. Analyzes models of fluid flow and solid deformation. For upper-level math, science and engineering students. 608pp. 5⅜ × 8½. 65369-2 Pa. $13.95

ELEMENTS OF REAL ANALYSIS, David A. Sprecher. Classic text covers fundamental concepts, real number system, point sets, functions of a real variable, Fourier series, much more. Over 500 exercises. 352pp. 5⅜ × 8½. 65385-4 Pa. $10.95

PHYSICAL PRINCIPLES OF THE QUANTUM THEORY, Werner Heisenberg. Nobel Laureate discusses quantum theory, uncertainty, wave mechanics, work of Dirac, Schroedinger, Compton, Wilson, Einstein, etc. 184pp. 5⅜ × 8½. 60113-7 Pa. $5.95

INTRODUCTORY REAL ANALYSIS, A.N. Kolmogorov, S.V. Fomin. Translated by Richard A. Silverman. Self-contained, evenly paced introduction to real and functional analysis. Some 350 problems. 403pp. 5⅜ × 8½. 61226-0 Pa. $9.95

PROBLEMS AND SOLUTIONS IN QUANTUM CHEMISTRY AND PHYSICS, Charles S. Johnson, Jr. and Lee G. Pedersen. Unusually varied problems, detailed solutions in coverage of quantum mechanics, wave mechanics, angular momentum, molecular spectroscopy, scattering theory, more. 280 problems plus 139 supplementary exercises. 430pp. 6½ × 9¼. 65236-X Pa. $12.95

CATALOG OF DOVER BOOKS

ASYMPTOTIC METHODS IN ANALYSIS, N.G. de Bruijn. An inexpensive, comprehensive guide to asymptotic methods—the pioneering work that teaches by explaining worked examples in detail. Index. 224pp. 5⅜ × 8½. 64221-6 Pa. $6.95

OPTICAL RESONANCE AND TWO-LEVEL ATOMS, L. Allen and J.H. Eberly. Clear, comprehensive introduction to basic principles behind all quantum optical resonance phenomena. 53 illustrations. Preface. Index. 256pp. 5⅜ × 8½.
65533-4 Pa. $7.95

COMPLEX VARIABLES, Francis J. Flanigan. Unusual approach, delaying complex algebra till harmonic functions have been analyzed from real variable viewpoint. Includes problems with answers. 364pp. 5⅜ × 8½. 61388-7 Pa. $8.95

ATOMIC SPECTRA AND ATOMIC STRUCTURE, Gerhard Herzberg. One of best introductions; especially for specialist in other fields. Treatment is physical rather than mathematical. 80 illustrations. 257pp. 5⅜ × 8½. 60115-3 Pa. $6.95

APPLIED COMPLEX VARIABLES, John W. Dettman. Step-by-step coverage of fundamentals of analytic function theory—plus lucid exposition of five important applications: Potential Theory; Ordinary Differential Equations; Fourier Transforms; Laplace Transforms; Asymptotic Expansions. 66 figures. Exercises at chapter ends. 512pp. 5⅜ × 8½. 64670-X Pa. $11.95

ULTRASONIC ABSORPTION: An Introduction to the Theory of Sound Absorption and Dispersion in Gases, Liquids and Solids, A.B. Bhatia. Standard reference in the field provides a clear, systematically organized introductory review of fundamental concepts for advanced graduate students, research workers. Numerous diagrams. Bibliography. 440pp. 5⅜ × 8½. 64917-2 Pa. $11.95

UNBOUNDED LINEAR OPERATORS: Theory and Applications, Seymour Goldberg. Classic presents systematic treatment of the theory of unbounded linear operators in normed linear spaces with applications to differential equations. Bibliography. 199pp. 5⅜ × 8½. 64830-3 Pa. $7.95

LIGHT SCATTERING BY SMALL PARTICLES, H.C. van de Hulst. Comprehensive treatment including full range of useful approximation methods for researchers in chemistry, meteorology and astronomy. 44 illustrations. 470pp. 5⅜ × 8½. 64228-3 Pa. $11.95

CONFORMAL MAPPING ON RIEMANN SURFACES, Harvey Cohn. Lucid, insightful book presents ideal coverage of subject. 334 exercises make book perfect for self-study. 55 figures. 352pp. 5⅜ × 8¼. 64025-6 Pa. $9.95

OPTICKS, Sir Isaac Newton. Newton's own experiments with spectroscopy, colors, lenses, reflection, refraction, etc., in language the layman can follow. Foreword by Albert Einstein. 532pp. 5⅜ × 8½. 60205-2 Pa. $9.95

GENERALIZED INTEGRAL TRANSFORMATIONS, A.H. Zemanian. Graduate-level study of recent generalizations of the Laplace, Mellin, Hankel, K. Weierstrass, convolution and other simple transformations. Bibliography. 320pp. 5⅜ × 8½. 65375-7 Pa. $8.95

CATALOG OF DOVER BOOKS

THE ELECTROMAGNETIC FIELD, Albert Shadowitz. Comprehensive undergraduate text covers basics of electric and magnetic fields, builds up to electromagnetic theory. Also related topics, including relativity. Over 900 problems. 768pp. 5⅜ × 8¼. 65660-8 Pa. $18.95

FOURIER SERIES, Georgi P. Tolstov. Translated by Richard A. Silverman. A valuable addition to the literature on the subject, moving clearly from subject to subject and theorem to theorem. 107 problems, answers. 336pp. 5⅜ × 8½. 63317-9 Pa. $8.95

THEORY OF ELECTROMAGNETIC WAVE PROPAGATION, Charles Herach Papas. Graduate-level study discusses the Maxwell field equations, radiation from wire antennas, the Doppler effect and more. xiii + 244pp. 5⅜ × 8½. 65678-0 Pa. $6.95

DISTRIBUTION THEORY AND TRANSFORM ANALYSIS: An Introduction to Generalized Functions, with Applications, A.H. Zemanian. Provides basics of distribution theory, describes generalized Fourier and Laplace transformations. Numerous problems. 384pp. 5⅜ × 8½. 65479-6 Pa. $9.95

THE PHYSICS OF WAVES, William C. Elmore and Mark A. Heald. Unique overview of classical wave theory. Acoustics, optics, electromagnetic radiation, more. Ideal as classroom text or for self-study. Problems. 477pp. 5⅜ × 8½. 64926-1 Pa. $12.95

CALCULUS OF VARIATIONS WITH APPLICATIONS, George M. Ewing. Applications-oriented introduction to variational theory develops insight and promotes understanding of specialized books, research papers. Suitable for advanced undergraduate/graduate students as primary, supplementary text. 352pp. 5⅜ × 8½. 64856-7 Pa. $8.95

A TREATISE ON ELECTRICITY AND MAGNETISM, James Clerk Maxwell. Important foundation work of modern physics. Brings to final form Maxwell's theory of electromagnetism and rigorously derives his general equations of field theory. 1,084pp. 5⅜ × 8½. 60636-8, 60637-6 Pa., Two-vol. set $21.90

AN INTRODUCTION TO THE CALCULUS OF VARIATIONS, Charles Fox. Graduate-level text covers variations of an integral, isoperimetrical problems, least action, special relativity, approximations, more. References. 279pp. 5⅜ × 8½. 65499-0 Pa. $7.95

HYDRODYNAMIC AND HYDROMAGNETIC STABILITY, S. Chandrasekhar. Lucid examination of the Rayleigh-Benard problem; clear coverage of the theory of instabilities causing convection. 704pp. 5⅜ × 8¼. 64071-X Pa. $14.95

CALCULUS OF VARIATIONS, Robert Weinstock. Basic introduction covering isoperimetric problems, theory of elasticity, quantum mechanics, electrostatics, etc. Exercises throughout. 326pp. 5⅜ × 8½. 63069-2 Pa. $8.95

DYNAMICS OF FLUIDS IN POROUS MEDIA, Jacob Bear. For advanced students of ground water hydrology, soil mechanics and physics, drainage and irrigation engineering and more. 335 illustrations. Exercises, with answers. 784pp. 6⅛ × 9¼. 65675-6 Pa. $19.95

CATALOG OF DOVER BOOKS

NUMERICAL METHODS FOR SCIENTISTS AND ENGINEERS, Richard Hamming. Classic text stresses frequency approach in coverage of algorithms, polynomial approximation, Fourier approximation, exponential approximation, other topics. Revised and enlarged 2nd edition. 721pp. 5⅜ × 8½.
65241-6 Pa. $14.95

THEORETICAL SOLID STATE PHYSICS, Vol. I: Perfect Lattices in Equilibrium; Vol. II: Non-Equilibrium and Disorder, William Jones and Norman H. March. Monumental reference work covers fundamental theory of equilibrium properties of perfect crystalline solids, non-equilibrium properties, defects and disordered systems. Appendices. Problems. Preface. Diagrams. Index. Bibliography. Total of 1,301pp. 5⅜ × 8½. Two volumes.　　Vol. I 65015-4 Pa. $14.95
Vol. II 65016-2 Pa. $14.95

OPTIMIZATION THEORY WITH APPLICATIONS, Donald A. Pierre. Broad-spectrum approach to important topic. Classical theory of minima and maxima, calculus of variations, simplex technique and linear programming, more. Many problems, examples. 640pp. 5⅜ × 8½.　　65205-X Pa. $14.95

THE CONTINUUM: A Critical Examination of the Foundation of Analysis, Hermann Weyl. Classic of 20th-century foundational research deals with the conceptual problem posed by the continuum. 156pp. 5⅜ × 8½.　　67982-9 Pa. $5.95

ESSAYS ON THE THEORY OF NUMBERS, Richard Dedekind. Two classic essays by great German mathematician: on the theory of irrational numbers; and on transfinite numbers and properties of natural numbers. 115pp. 5⅜ × 8½.
21010-3 Pa. $4.95

THE FUNCTIONS OF MATHEMATICAL PHYSICS, Harry Hochstadt. Comprehensive treatment of orthogonal polynomials, hypergeometric functions, Hill's equation, much more. Bibliography. Index. 322pp. 5⅜ × 8½.　　65214-9 Pa. $9.95

NUMBER THEORY AND ITS HISTORY, Oystein Ore. Unusually clear, accessible introduction covers counting, properties of numbers, prime numbers, much more. Bibliography. 380pp. 5⅜ × 8½.　　65620-9 Pa. $9.95

THE VARIATIONAL PRINCIPLES OF MECHANICS, Cornelius Lanczos. Graduate level coverage of calculus of variations, equations of motion, relativistic mechanics, more. First inexpensive paperbound edition of classic treatise. Index. Bibliography. 418pp. 5⅜ × 8½.　　65067-7 Pa. $11.95

MATHEMATICAL TABLES AND FORMULAS, Robert D. Carmichael and Edwin R. Smith. Logarithms, sines, tangents, trig functions, powers, roots, reciprocals, exponential and hyperbolic functions, formulas and theorems. 269pp. 5⅜ × 8½.　　60111-0 Pa. $6.95

THEORETICAL PHYSICS, Georg Joos, with Ira M. Freeman. Classic overview covers essential math, mechanics, electromagnetic theory, thermodynamics, quantum mechanics, nuclear physics, other topics. First paperback edition. xxiii + 885pp. 5⅜ × 8½.　　65227-0 Pa. $19.95

CATALOG OF DOVER BOOKS

HANDBOOK OF MATHEMATICAL FUNCTIONS WITH FORMULAS, GRAPHS, AND MATHEMATICAL TABLES, edited by Milton Abramowitz and Irene A. Stegun. Vast compendium: 29 sets of tables, some to as high as 20 places. 1,046pp. 8 × 10½. 61272-4 Pa. $24.95

MATHEMATICAL METHODS IN PHYSICS AND ENGINEERING, John W. Dettman. Algebraically based approach to vectors, mapping, diffraction, other topics in applied math. Also generalized functions, analytic function theory, more. Exercises. 448pp. 5⅜ × 8¼. 65649-7 Pa. $9.95

A SURVEY OF NUMERICAL MATHEMATICS, David M. Young and Robert Todd Gregory. Broad self-contained coverage of computer-oriented numerical algorithms for solving various types of mathematical problems in linear algebra, ordinary and partial, differential equations, much more. Exercises. Total of 1,248pp. 5⅜ × 8½. Two volumes. Vol. I 65691-8 Pa. $14.95
Vol. II 65692-6 Pa. $14.95

TENSOR ANALYSIS FOR PHYSICISTS, J.A. Schouten. Concise exposition of the mathematical basis of tensor analysis, integrated with well-chosen physical examples of the theory. Exercises. Index. Bibliography. 289pp. 5⅜ × 8½. 65582-2 Pa. $8.95

INTRODUCTION TO NUMERICAL ANALYSIS (2nd Edition), F.B. Hildebrand. Classic, fundamental treatment covers computation, approximation, interpolation, numerical differentiation and integration, other topics. 150 new problems. 669pp. 5⅜ × 8½. 65363-3 Pa. $15.95

INVESTIGATIONS ON THE THEORY OF THE BROWNIAN MOVEMENT, Albert Einstein. Five papers (1905-8) investigating dynamics of Brownian motion and evolving elementary theory. Notes by R. Fürth. 122pp. 5⅜ × 8½. 60304-0 Pa. $4.95

CATASTROPHE THEORY FOR SCIENTISTS AND ENGINEERS, Robert Gilmore. Advanced-level treatment describes mathematics of theory grounded in the work of Poincaré, R. Thom, other mathematicians. Also important applications to problems in mathematics, physics, chemistry and engineering. 1981 edition. References. 28 tables. 397 black-and-white illustrations. xvii + 666pp. 6⅛ × 9¼. 67539-4 Pa. $16.95

AN INTRODUCTION TO STATISTICAL THERMODYNAMICS, Terrell L. Hill. Excellent basic text offers wide-ranging coverage of quantum statistical mechanics, systems of interacting molecules, quantum statistics, more. 523pp. 5⅜ × 8½. 65242-4 Pa. $12.95

ELEMENTARY DIFFERENTIAL EQUATIONS, William Ted Martin and Eric Reissner. Exceptionally clear, comprehensive introduction at undergraduate level. Nature and origin of differential equations, differential equations of first, second and higher orders. Picard's Theorem, much more. Problems with solutions. 331pp. 5⅜ × 8½. 65024-3 Pa. $8.95

STATISTICAL PHYSICS, Gregory H. Wannier. Classic text combines thermodynamics, statistical mechanics and kinetic theory in one unified presentation of thermal physics. Problems with solutions. Bibliography. 532pp. 5⅜ × 8½. 65401-X Pa. $12.95

CATALOG OF DOVER BOOKS

ORDINARY DIFFERENTIAL EQUATIONS, Morris Tenenbaum and Harry Pollard. Exhaustive survey of ordinary differential equations for undergraduates in mathematics, engineering, science. Thorough analysis of theorems. Diagrams. Bibliography. Index. 818pp. 5⅜ × 8½. 64940-7 Pa. $16.95

STATISTICAL MECHANICS: Principles and Applications, Terrell L. Hill. Standard text covers fundamentals of statistical mechanics, applications to fluctuation theory, imperfect gases, distribution functions, more. 448pp. 5⅜ × 8½. 65390-0 Pa. $11.95

ORDINARY DIFFERENTIAL EQUATIONS AND STABILITY THEORY: An Introduction, David A. Sánchez. Brief, modern treatment. Linear equation, stability theory for autonomous and nonautonomous systems, etc. 164pp. 5⅜ × 8¼. 63828-6 Pa. $5.95

THIRTY YEARS THAT SHOOK PHYSICS: The Story of Quantum Theory, George Gamow. Lucid, accessible introduction to influential theory of energy and matter. Careful explanations of Dirac's anti-particles, Bohr's model of the atom, much more. 12 plates. Numerous drawings. 240pp. 5⅜ × 8½. 24895-X Pa. $6.95

THEORY OF MATRICES, Sam Perlis. Outstanding text covering rank, non-singularity and inverses in connection with the development of canonical matrices under the relation of equivalence, and without the intervention of determinants. Includes exercises. 237pp. 5⅜ × 8½. 66810-X Pa. $7.95

GREAT EXPERIMENTS IN PHYSICS: Firsthand Accounts from Galileo to Einstein, edited by Morris H. Shamos. 25 crucial discoveries: Newton's laws of motion, Chadwick's study of the neutron, Hertz on electromagnetic waves, more. Original accounts clearly annotated. 370pp. 5⅜ × 8½. 25346-5 Pa. $10.95

INTRODUCTION TO PARTIAL DIFFERENTIAL EQUATIONS WITH APPLICATIONS, E.C. Zachmanoglou and Dale W. Thoe. Essentials of partial differential equations applied to common problems in engineering and the physical sciences. Problems and answers. 416pp. 5⅜ × 8½. 65251-3 Pa. $10.95

BURNHAM'S CELESTIAL HANDBOOK, Robert Burnham, Jr. Thorough guide to the stars beyond our solar system. Exhaustive treatment. Alphabetical by constellation: Andromeda to Cetus in Vol. 1; Chamaeleon to Orion in Vol. 2; and Pavo to Vulpecula in Vol. 3. Hundreds of illustrations. Index in Vol. 3. 2,000pp. 6⅛ × 9¼. 23567-X, 23568-8, 23673-0 Pa., Three-vol. set $41.85

CHEMICAL MAGIC, Leonard A. Ford. Second Edition, Revised by E. Winston Grundmeier. Over 100 unusual stunts demonstrating cold fire, dust explosions, much more. Text explains scientific principles and stresses safety precautions. 128pp. 5⅜ × 8½. 67628-5 Pa. $5.95

AMATEUR ASTRONOMER'S HANDBOOK, J.B. Sidgwick. Timeless, comprehensive coverage of telescopes, mirrors, lenses, mountings, telescope drives, micrometers, spectroscopes, more. 189 illustrations. 576pp. 5⅜ × 8¼. (Available in U.S. only) 24034-7 Pa. $9.95

CATALOG OF DOVER BOOKS

SPECIAL FUNCTIONS, N.N. Lebedev. Translated by Richard Silverman. Famous Russian work treating more important special functions, with applications to specific problems of physics and engineering. 38 figures. 308pp. 5⅜ × 8½.
60624-4 Pa. $8.95

OBSERVATIONAL ASTRONOMY FOR AMATEURS, J.B. Sidgwick. Mine of useful data for observation of sun, moon, planets, asteroids, aurorae, meteors, comets, variables, binaries, etc. 39 illustrations. 384pp. 5⅜ × 8¼. (Available in U.S. only)
24033-9 Pa. $8.95

INTEGRAL EQUATIONS, F.G. Tricomi. Authoritative, well-written treatment of extremely useful mathematical tool with wide applications. Volterra Equations, Fredholm Equations, much more. Advanced undergraduate to graduate level. Exercises. Bibliography. 238pp. 5⅜ × 8½.
64828-1 Pa. $7.95

POPULAR LECTURES ON MATHEMATICAL LOGIC, Hao Wang. Noted logician's lucid treatment of historical developments, set theory, model theory, recursion theory and constructivism, proof theory, more. 3 appendixes. Bibliography. 1981 edition. ix + 283pp. 5⅜ × 8½.
67632-3 Pa. $8.95

MODERN NONLINEAR EQUATIONS, Thomas L. Saaty. Emphasizes practical solution of problems; covers seven types of equations. ". . . a welcome contribution to the existing literature. . . ."—*Math Reviews.* 490pp. 5⅜ × 8½. 64232-1 Pa. $11.95

FUNDAMENTALS OF ASTRODYNAMICS, Roger Bate et al. Modern approach developed by U.S. Air Force Academy. Designed as a first course. Problems, exercises. Numerous illustrations. 455pp. 5⅜ × 8½.
60061-0 Pa. $9.95

INTRODUCTION TO LINEAR ALGEBRA AND DIFFERENTIAL EQUATIONS, John W. Dettman. Excellent text covers complex numbers, determinants, orthonormal bases, Laplace transforms, much more. Exercises with solutions. Undergraduate level. 416pp. 5⅜ × 8½.
65191-6 Pa. $10.95

INCOMPRESSIBLE AERODYNAMICS, edited by Bryan Thwaites. Covers theoretical and experimental treatment of the uniform flow of air and viscous fluids past two-dimensional aerofoils and three-dimensional wings; many other topics. 654pp. 5⅜ × 8½.
65465-6 Pa. $16.95

INTRODUCTION TO DIFFERENCE EQUATIONS, Samuel Goldberg. Exceptionally clear exposition of important discipline with applications to sociology, psychology, economics. Many illustrative examples; over 250 problems. 260pp. 5⅜ × 8½.
65084-7 Pa. $7.95

LAMINAR BOUNDARY LAYERS, edited by L. Rosenhead. Engineering classic covers steady boundary layers in two- and three-dimensional flow, unsteady boundary layers, stability, observational techniques, much more. 708pp. 5⅜ × 8½.
65646-2 Pa. $18.95

LECTURES ON CLASSICAL DIFFERENTIAL GEOMETRY, Second Edition, Dirk J. Struik. Excellent brief introduction covers curves, theory of surfaces, fundamental equations, geometry on a surface, conformal mapping, other topics. Problems. 240pp. 5⅜ × 8½.
65609-8 Pa. $8.95

ROTARY-WING AERODYNAMICS, W.Z. Stepniewski. Clear, concise text covers aerodynamic phenomena of the rotor and offers guidelines for helicopter performance evaluation. Originally prepared for NASA. 537 figures. 640pp. 6⅛ × 9¼.
64647-5 Pa. $15.95

DIFFERENTIAL GEOMETRY, Heinrich W. Guggenheimer. Local differential geometry as an application of advanced calculus and linear algebra. Curvature, transformation groups, surfaces, more. Exercises. 62 figures. 378pp. 5⅜ × 8½.
63433-7 Pa. $8.95

INTRODUCTION TO SPACE DYNAMICS, William Tyrrell Thomson. Comprehensive, classic introduction to space-flight engineering for advanced undergraduate and graduate students. Includes vector algebra, kinematics, transformation of coordinates. Bibliography. Index. 352pp. 5⅜ × 8½. 65113-4 Pa. $8.95

A SURVEY OF MINIMAL SURFACES, Robert Osserman. Up-to-date, in-depth discussion of the field for advanced students. Corrected and enlarged edition covers new developments. Includes numerous problems. 192pp. 5⅜ × 8½.
64998-9 Pa. $8.95

ANALYTICAL MECHANICS OF GEARS, Earle Buckingham. Indispensable reference for modern gear manufacture covers conjugate gear-tooth action, gear-tooth profiles of various gears, many other topics. 263 figures. 102 tables. 546pp. 5⅜ × 8½. 65712-4 Pa. $14.95

SET THEORY AND LOGIC, Robert R. Stoll. Lucid introduction to unified theory of mathematical concepts. Set theory and logic seen as tools for conceptual understanding of real number system. 496pp. 5⅜ × 8¼. 63829-4 Pa. $12.95

A HISTORY OF MECHANICS, René Dugas. Monumental study of mechanical principles from antiquity to quantum mechanics. Contributions of ancient Greeks, Galileo, Leonardo, Kepler, Lagrange, many others. 671pp. 5⅜ × 8½.
65632-2 Pa. $14.95

FAMOUS PROBLEMS OF GEOMETRY AND HOW TO SOLVE THEM, Benjamin Bold. Squaring the circle, trisecting the angle, duplicating the cube: learn their history, why they are impossible to solve, then solve them yourself. 128pp. 5⅜ × 8½. 24297-8 Pa. $4.95

MECHANICAL VIBRATIONS, J.P. Den Hartog. Classic textbook offers lucid explanations and illustrative models, applying theories of vibrations to a variety of practical industrial engineering problems. Numerous figures. 233 problems, solutions. Appendix. Index. Preface. 436pp. 5⅜ × 8½. 64785-4 Pa. $10.95

CURVATURE AND HOMOLOGY, Samuel I. Goldberg. Thorough treatment of specialized branch of differential geometry. Covers Riemannian manifolds, topology of differentiable manifolds, compact Lie groups, other topics. Exercises. 315pp. 5⅜ × 8½. 64314-X Pa. $9.95

HISTORY OF STRENGTH OF MATERIALS, Stephen P. Timoshenko. Excellent historical survey of the strength of materials with many references to the theories of elasticity and structure. 245 figures. 452pp. 5⅜ × 8½. 61187-6 Pa. $11.95

CATALOG OF DOVER BOOKS

GEOMETRY OF COMPLEX NUMBERS, Hans Schwerdtfeger. Illuminating, widely praised book on analytic geometry of circles, the Moebius transformation, and two-dimensional non-Euclidean geometries. 200pp. 5⅜ × 8¼.
63830-8 Pa. $8.95

MECHANICS, J.P. Den Hartog. A classic introductory text or refresher. Hundreds of applications and design problems illuminate fundamentals of trusses, loaded beams and cables, etc. 334 answered problems. 462pp. 5⅜ × 8½. 60754-2 Pa. $9.95

TOPOLOGY, John G. Hocking and Gail S. Young. Superb one-year course in classical topology. Topological spaces and functions, point-set topology, much more. Examples and problems. Bibliography. Index. 384pp. 5⅜ × 8¼.
65676-4 Pa. $9.95

STRENGTH OF MATERIALS, J.P. Den Hartog. Full, clear treatment of basic material (tension, torsion, bending, etc.) plus advanced material on engineering methods, applications. 350 answered problems. 323pp. 5⅜ × 8½. 60755-0 Pa. $8.95

ELEMENTARY CONCEPTS OF TOPOLOGY, Paul Alexandroff. Elegant, intuitive approach to topology from set-theoretic topology to Betti groups; how concepts of topology are useful in math and physics. 25 figures. 57pp. 5⅜ × 8½.
60747-X Pa. $3.50

ADVANCED STRENGTH OF MATERIALS, J.P. Den Hartog. Superbly written advanced text covers torsion, rotating disks, membrane stresses in shells, much more. Many problems and answers. 388pp. 5⅜ × 8½. 65407-9 Pa. $9.95

COMPUTABILITY AND UNSOLVABILITY, Martin Davis. Classic graduate-level introduction to theory of computability, usually referred to as theory of recurrent functions. New preface and appendix. 288pp. 5⅜ × 8½. 61471-9 Pa. $7.95

GENERAL CHEMISTRY, Linus Pauling. Revised 3rd edition of classic first-year text by Nobel laureate. Atomic and molecular structure, quantum mechanics, statistical mechanics, thermodynamics correlated with descriptive chemistry. Problems. 992pp. 5⅜ × 8½. 65622-5 Pa. $19.95

AN INTRODUCTION TO MATRICES, SETS AND GROUPS FOR SCIENCE STUDENTS, G. Stephenson. Concise, readable text introduces sets, groups, and most importantly, matrices to undergraduate students of physics, chemistry, and engineering. Problems. 164pp. 5⅜ × 8½. 65077-4 Pa. $6.95

THE HISTORICAL BACKGROUND OF CHEMISTRY, Henry M. Leicester. Evolution of ideas, not individual biography. Concentrates on formulation of a coherent set of chemical laws. 260pp. 5⅜ × 8½. 61053-5 Pa. $6.95

THE PHILOSOPHY OF MATHEMATICS: An Introductory Essay, Stephan Körner. Surveys the views of Plato, Aristotle, Leibniz & Kant concerning propositions and theories of applied and pure mathematics. Introduction. Two appendices. Index. 198pp. 5⅜ × 8½. 25048-2 Pa. $7.95

THE DEVELOPMENT OF MODERN CHEMISTRY, Aaron J. Ihde. Authoritative history of chemistry from ancient Greek theory to 20th-century innovation. Covers major chemists and their discoveries. 209 illustrations. 14 tables. Bibliographies. Indices. Appendices. 851pp. 5⅜ × 8½. 64235-6 Pa. $18.95

CATALOG OF DOVER BOOKS

DE RE METALLICA, Georgius Agricola. The famous Hoover translation of greatest treatise on technological chemistry, engineering, geology, mining of early modern times (1556). All 289 original woodcuts. 638pp. 6¾ × 11. 60006-8 Pa. $18.95

SOME THEORY OF SAMPLING, William Edwards Deming. Analysis of the problems, theory and design of sampling techniques for social scientists, industrial managers and others who find statistics increasingly important in their work. 61 tables. 90 figures. xvii + 602pp. 5⅜ × 8½. 64684-X Pa. $15.95

THE VARIOUS AND INGENIOUS MACHINES OF AGOSTINO RAMELLI: A Classic Sixteenth-Century Illustrated Treatise on Technology, Agostino Ramelli. One of the most widely known and copied works on machinery in the 16th century. 194 detailed plates of water pumps, grain mills, cranes, more. 608pp. 9 × 12. 28180-9 Pa. $24.95

LINEAR PROGRAMMING AND ECONOMIC ANALYSIS, Robert Dorfman, Paul A. Samuelson and Robert M. Solow. First comprehensive treatment of linear programming in standard economic analysis. Game theory, modern welfare economics, Leontief input-output, more. 525pp. 5⅜ × 8½. 65491-5 Pa. $14.95

ELEMENTARY DECISION THEORY, Herman Chernoff and Lincoln E. Moses. Clear introduction to statistics and statistical theory covers data processing, probability and random variables, testing hypotheses, much more. Exercises. 364pp. 5⅜ × 8½. 65218-1 Pa. $9.95

THE COMPLEAT STRATEGYST: Being a Primer on the Theory of Games of Strategy, J.D. Williams. Highly entertaining classic describes, with many illustrated examples, how to select best strategies in conflict situations. Prefaces. Appendices. 268pp. 5⅜ × 8½. 25101-2 Pa. $7.95

MATHEMATICAL METHODS OF OPERATIONS RESEARCH, Thomas L. Saaty. Classic graduate-level text covers historical background, classical methods of forming models, optimization, game theory, probability, queueing theory, much more. Exercises. Bibliography. 448pp. 5⅜ × 8¼. 65703-5 Pa. $12.95

CONSTRUCTIONS AND COMBINATORIAL PROBLEMS IN DESIGN OF EXPERIMENTS, Damaraju Raghavarao. In-depth reference work examines orthogonal Latin squares, incomplete block designs, tactical configuration, partial geometry, much more. Abundant explanations, examples. 416pp. 5⅜ × 8¼. 65685-3 Pa. $10.95

THE ABSOLUTE DIFFERENTIAL CALCULUS (CALCULUS OF TENSORS), Tullio Levi-Civita. Great 20th-century mathematician's classic work on material necessary for mathematical grasp of theory of relativity. 452pp. 5⅜ × 8½. 63401-9 Pa. $9.95

VECTOR AND TENSOR ANALYSIS WITH APPLICATIONS, A.I. Borisenko and I.E. Tarapov. Concise introduction. Worked-out problems, solutions, exercises. 257pp. 5⅜ × 8¼. 63833-2 Pa. $7.95

CATALOG OF DOVER BOOKS

THE FOUR-COLOR PROBLEM: Assaults and Conquest, Thomas L. Saaty and Paul G. Kainen. Engrossing, comprehensive account of the century-old combinatorial topological problem, its history and solution. Bibliographies. Index. 110 figures. 228pp. 5⅜ × 8½. 65092-8 Pa. $6.95

CATALYSIS IN CHEMISTRY AND ENZYMOLOGY, William P. Jencks. Exceptionally clear coverage of mechanisms for catalysis, forces in aqueous solution, carbonyl- and acyl-group reactions, practical kinetics, more. 864pp. 5⅜ × 8½. 65460-5 Pa. $19.95

PROBABILITY: An Introduction, Samuel Goldberg. Excellent basic text covers set theory, probability theory for finite sample spaces, binomial theorem, much more. 360 problems. Bibliographies. 322pp. 5⅜ × 8½. 65252-1 Pa. $8.95

LIGHTNING, Martin A. Uman. Revised, updated edition of classic work on the physics of lightning. Phenomena, terminology, measurement, photography, spectroscopy, thunder, more. Reviews recent research. Bibliography. Indices. 320pp. 5⅜ × 8¼. 64575-4 Pa. $8.95

PROBABILITY THEORY: A Concise Course, Y.A. Rozanov. Highly readable, self-contained introduction covers combination of events, dependent events, Bernoulli trials, etc. Translation by Richard Silverman. 148pp. 5⅜ × 8¼. 63544-9 Pa. $5.95

AN INTRODUCTION TO HAMILTONIAN OPTICS, H. A. Buchdahl. Detailed account of the Hamiltonian treatment of aberration theory in geometrical optics. Many classes of optical systems defined in terms of the symmetries they possess. Problems with detailed solutions. 1970 edition. xv + 360pp. 5⅜ × 8½. 67597-1 Pa. $10.95

STATISTICS MANUAL, Edwin L. Crow, et al. Comprehensive, practical collection of classical and modern methods prepared by U.S. Naval Ordnance Test Station. Stress on use. Basics of statistics assumed. 288pp. 5⅜ × 8½. 60599-X Pa. $6.95

DICTIONARY/OUTLINE OF BASIC STATISTICS, John E. Freund and Frank J. Williams. A clear concise dictionary of over 1,000 statistical terms and an outline of statistical formulas covering probability, nonparametric tests, much more. 208pp. 5⅜ × 8½. 66796-0 Pa. $6.95

STATISTICAL METHOD FROM THE VIEWPOINT OF QUALITY CONTROL, Walter A. Shewhart. Important text explains regulation of variables, uses of statistical control to achieve quality control in industry, agriculture, other areas. 192pp. 5⅜ × 8½. 65232-7 Pa. $7.95

THE INTERPRETATION OF GEOLOGICAL PHASE DIAGRAMS, Ernest G. Ehlers. Clear, concise text emphasizes diagrams of systems under fluid or containing pressure; also coverage of complex binary systems, hydrothermal melting, more. 288pp. 6½ × 9¼. 65389-7 Pa. $10.95

STATISTICAL ADJUSTMENT OF DATA, W. Edwards Deming. Introduction to basic concepts of statistics, curve fitting, least squares solution, conditions without parameter, conditions containing parameters. 26 exercises worked out. 271pp. 5⅜ × 8½. 64685-8 Pa. $8.95

CATALOG OF DOVER BOOKS

TENSOR CALCULUS, J.L. Synge and A. Schild. Widely used introductory text covers spaces and tensors, basic operations in Riemannian space, non-Riemannian spaces, etc. 324pp. 5⅜ × 8¼. 63612-7 Pa. $8.95

A CONCISE HISTORY OF MATHEMATICS, Dirk J. Struik. The best brief history of mathematics. Stresses origins and covers every major figure from ancient Near East to 19th century. 41 illustrations. 195pp. 5⅜ × 8½. 60255-9 Pa. $7.95

A SHORT ACCOUNT OF THE HISTORY OF MATHEMATICS, W.W. Rouse Ball. One of clearest, most authoritative surveys from the Egyptians and Phoenicians through 19th-century figures such as Grassman, Galois, Riemann. Fourth edition. 522pp. 5⅜ × 8½. 20630-0 Pa. $10.95

HISTORY OF MATHEMATICS, David E. Smith. Nontechnical survey from ancient Greece and Orient to late 19th century; evolution of arithmetic, geometry, trigonometry, calculating devices, algebra, the calculus. 362 illustrations. 1,355pp. 5⅜ × 8½. 20429-4, 20430-8 Pa., Two-vol. set $23.90

THE GEOMETRY OF RENÉ DESCARTES, René Descartes. The great work founded analytical geometry. Original French text, Descartes' own diagrams, together with definitive Smith-Latham translation. 244pp. 5⅜ × 8½.
60068-8 Pa. $7.95

THE ORIGINS OF THE INFINITESIMAL CALCULUS, Margaret E. Baron. Only fully detailed and documented account of crucial discipline: origins; development by Galileo, Kepler, Cavalieri; contributions of Newton, Leibniz, more. 304pp. 5⅜ × 8½. (Available in U.S. and Canada only) 65371-4 Pa. $9.95

THE HISTORY OF THE CALCULUS AND ITS CONCEPTUAL DEVELOPMENT, Carl B. Boyer. Origins in antiquity, medieval contributions, work of Newton, Leibniz, rigorous formulation. Treatment is verbal. 346pp. 5⅜ × 8½.
60509-4 Pa. $8.95

THE THIRTEEN BOOKS OF EUCLID'S ELEMENTS, translated with introduction and commentary by Sir Thomas L. Heath. Definitive edition. Textual and linguistic notes, mathematical analysis. 2,500 years of critical commentary. Not abridged. 1,414pp. 5⅜ × 8½. 60088-2, 60089-0, 60090-4 Pa., Three-vol. set $29.85

GAMES AND DECISIONS: Introduction and Critical Survey, R. Duncan Luce and Howard Raiffa. Superb nontechnical introduction to game theory, primarily applied to social sciences. Utility theory, zero-sum games, n-person games, decision-making, much more. Bibliography. 509pp. 5⅜ × 8½. 65943-7 Pa. $12.95

THE HISTORICAL ROOTS OF ELEMENTARY MATHEMATICS, Lucas N.H. Bunt, Phillip S. Jones, and Jack D. Bedient. Fundamental underpinnings of modern arithmetic, algebra, geometry and number systems derived from ancient civilizations. 320pp. 5⅜ × 8½. 25563-8 Pa. $8.95

CALCULUS REFRESHER FOR TECHNICAL PEOPLE, A. Albert Klaf. Covers important aspects of integral and differential calculus via 756 questions. 566 problems, most answered. 431pp. 5⅜ × 8½. 20370-0 Pa. $8.95

CATALOG OF DOVER BOOKS

CHALLENGING MATHEMATICAL PROBLEMS WITH ELEMENTARY SOLUTIONS, A.M. Yaglom and I.M. Yaglom. Over 170 challenging problems on probability theory, combinatorial analysis, points and lines, topology, convex polygons, many other topics. Solutions. Total of 445pp. 5⅜ × 8½. Two-vol. set.
Vol. I 65536-9 Pa. $7.95
Vol. II 65537-7 Pa. $6.95

FIFTY CHALLENGING PROBLEMS IN PROBABILITY WITH SOLUTIONS, Frederick Mosteller. Remarkable puzzlers, graded in difficulty, illustrate elementary and advanced aspects of probability. Detailed solutions. 88pp. 5⅜ × 8½.
65355-2 Pa. $4.95

EXPERIMENTS IN TOPOLOGY, Stephen Barr. Classic, lively explanation of one of the byways of mathematics. Klein bottles, Moebius strips, projective planes, map coloring, problem of the Koenigsberg bridges, much more, described with clarity and wit. 43 figures. 210pp. 5⅜ × 8½. 25933-1 Pa. $5.95

RELATIVITY IN ILLUSTRATIONS, Jacob T. Schwartz. Clear nontechnical treatment makes relativity more accessible than ever before. Over 60 drawings illustrate concepts more clearly than text alone. Only high school geometry needed. Bibliography. 128pp. 6⅛ × 9¼. 25965-X Pa. $6.95

AN INTRODUCTION TO ORDINARY DIFFERENTIAL EQUATIONS, Earl A. Coddington. A thorough and systematic first course in elementary differential equations for undergraduates in mathematics and science, with many exercises and problems (with answers). Index. 304pp. 5⅜ × 8½. 65942-9 Pa. $8.95

FOURIER SERIES AND ORTHOGONAL FUNCTIONS, Harry F. Davis. An incisive text combining theory and practical example to introduce Fourier series, orthogonal functions and applications of the Fourier method to boundary-value problems. 570 exercises. Answers and notes. 416pp. 5⅜ × 8½. 65973-9 Pa. $9.95

THE THEORY OF BRANCHING PROCESSES, Theodore E. Harris. First systematic, comprehensive treatment of branching (i.e. multiplicative) processes and their applications. Galton-Watson model, Markov branching processes, electron-photon cascade, many other topics. Rigorous proofs. Bibliography. 240pp. 5⅜ × 8½. 65952-6 Pa. $6.95

AN INTRODUCTION TO ALGEBRAIC STRUCTURES, Joseph Landin. Superb self-contained text covers "abstract algebra": sets and numbers, theory of groups, theory of rings, much more. Numerous well-chosen examples, exercises. 247pp. 5⅜ × 8½. 65940-2 Pa. $7.95

Prices subject to change without notice.
Available at your book dealer or write for free Mathematics and Science Catalog to Dept. GI, Dover Publications, Inc., 31 East 2nd St., Mineola, N.Y. 11501. Dover publishes more than 175 books each year on science, elementary and advanced mathematics, biology, music, art, literature, history, social sciences and other areas.